NEOTROPICAL BIRDS OF PREY

NEOTROPICAL BIRDS OF PREY

BIOLOGY AND ECOLOGY OF A FOREST RAPTOR COMMUNITY

Edited by **David F. Whitacre**

Foreword by J. Peter Jenny

Published in Association with The Peregrine Fund

Comstock Publishing Associates | *a division of*
Cornell University Press
Ithaca and London

First published 2012 by Cornell University Press

THE
PEREGRINE
FUND

Printed in the United States of America

Library of Congress Cataloging-in-Publication Data

Neotropical birds of prey : biology and ecology of a forest raptor community / edited by David F. Whitacre ; foreword by J. Peter Jenny.
    p. cm.
  "Published in association with the Peregrine Fund."
  Includes bibliographical references and index.
  ISBN 978-0-8014-4079-3 (cloth : alk. paper)
  1. Birds of prey—Guatemala—Parque Nacional Tikal. I. Whitacre, David F. (David Frederick) II. Peregrine Fund (U.S.)
  QL677.78.N46 2012
  598.9—dc23   2011046590

Cornell University Press strives to use environmentally responsible suppliers and materials to the fullest extent possible in the publishing of its books. Such materials include vegetable-based, low-VOC inks and acid-free papers that are recycled, totally chlorine-free, or partly composed of nonwood fibers. For further information, visit our website at www.cornellpress.cornell.edu.

Cloth printing 10 9 8 7 6 5 4 3 2 1

CONTENTS

Color plates follow page 78

METRIC TO ENGLISH EQUIVALENT VALUES

1 g = 0.04 oz	1 m = 3.28 ft
1 kg = 2.2 lb	1 km = 0.62 mi (3281 ft)
1 mm = 0.04 in	1 ha = 2.47 ac
1 cm = 0.4 in	°C = 5/9(°F − 32)

FOREWORD

J. Peter Jenny

The Peregrine Fund's interest in the Maya Forest had its origins in a chance conversation with Roger Tory Peterson in 1968. Roger had just returned from a birding trip to Tikal and, knowing of my passion for falcons, told me about a magnificent "orange-footed" falcon that he had just seen atop the ancient Maya ruins. He was surely referring to the Orange-breasted Falcon *(Falco deiroleucus)* and, in doing so, actually gave this falcon a more apt common name. Some years later, I had the opportunity to actually try and locate this enigmatic falcon known only from a handful of study skins and a few vague descriptions. With modest funding from my grandfather and at the encouragement of Tom Cade, I was able to mount two field trips to Guatemala, Ecuador, and Peru. I was assisted first by Mike Arnold in 1979 and then by Bill Sullivan in 1980, and together we were able to locate several breeding pairs in northern Guatemala and in eastern Ecuador and subsequently describe some of the first biology of this decidedly rare species. In 1984, I again had an opportunity to spend time in the field searching for the Orange-breasted Falcon, this time returning to eastern Ecuador accompanied by the late Bill Burnham.

Although this was Bill's first trip to the Neotropics, he quickly saw the need for studying the raptors of this region as well as the overwhelming challenges associated with working in this incredibly demanding habitat. Bill, himself a rare breed, was always one to take on the difficult challenge, almost preferring it. It was while Bill was visiting my home in Wyoming shortly after our return that we hatched the idea of what became the "Maya Project." We were discussing how difficult it was to observe raptors in a tropical forest, let alone study them. It was no wonder that so little was known about their biology or status. What was needed, Bill said, was to first develop a systematic method for locating raptors in a tropical forest environment. We both thought that whatever methods were developed, they should be low tech and fairly inexpensive so that they could be readily duplicated by others, particularly in developing countries where they were most needed.

We selected the semidry deciduous forest of northern Guatemala as a suitable location to develop these survey techniques because we had some familiarity with the region, and because the forest and associated fauna around the Maya ruins of Tikal appeared to be well protected while also providing adequate logistical access and support. Tikal National Park is a mystical place with stone pyramids built in AD 600 rising 50 m above the surrounding tropical forest. The ruins of this classic Maya city were first reported by early pilots who saw strange stone structures rising above the unbroken canopy of the northern Guatemalan forest.

We laid out a series of 14 rectangular transects, each 1.5 km long, emanating from the main road through the park, to systematically gain access to the forest. These were exciting times for the enthusiastic team composed of both graduate and undergraduate students, park guards, and local unemployed forest workers. The joy of working in the tropics is that information new to science is all around you; even the most basic information concerning the natural history of the raptor community remained, and in many cases still remains, largely undescribed. The tropics do not relinquish this knowledge without a fight, and accessing its secrets requires a significant investment in good old-fashioned fieldwork. Prior to our work in the Maya Forest, only a single nest for the entire forest-falcon genus *Micrastur* had ever been described. In the course of the Maya Project, Russell Thorstrom and his coworkers were able to not only describe the biology of two species of forest-falcons but also demonstrate that one species, the Barred Forest Falcon *(Micrastur ruficollis)*, at one pair per square kilometer, was, in fact, the most common raptor in the Maya Forest. This conclusion, however, required thousands of hours of fieldwork.

Fieldwork in the tropics is hot and sweaty work, and the well-armed vegetation, aggressive insects, and high temperatures all conspire to make life miserable. Working in these magnificent forests truly represents a love/hate relationship. Ultimately, more than 150 km of transects were laid out through the forest, and many thousands of hours of natural history observations were made.

Emergent trees were selected along each transect to provide visual access to raptors occurring within the forest canopy or soaring above. Many trees had to be climbed in order to select ones that provided the best view. We were very fortunate to have been assisted by "Estuardo" Aquiles E. Hernández and Feliciano Gutierrez, both former *chicleros*, adroit and highly skilled climbers who had harvested sap from the Chicozapote tree *(Manilkara zapota)* for the chewing gum industry. We quickly learned that no matter what the species of tree, any climbing technique that entailed hammering or penetrating the bark, such as the use of climbing spurs, would elicit a painful response from various insects.

In rigging a tree for multiple ascents, we shot a light messenger line over the upper limbs of the tree with a .22 caliber dummy launcher used for training hunting dog retrievers. The weight of the dummy led the trailing line down through the branches, where a heavier climbing rope could be attached and pulled back up into the canopy. We used great spools of caving rope specially designed for use with mechanical ascending devices called Jumars. The report from the .22 caliber dummy launcher was surprisingly loud, as was its recoil. Always the practical joker, Bill Heinrich handed me the launcher without warning me about its recoil, one result being a temporarily dislocated wrist. There was often a detachment of Guatemalan soldiers stationed in the park, and so as not to be confused with poachers or insurgents, we made a point of introducing ourselves and our launcher to the commander. Examining our .22 caliber launcher, the commander was obviously not impressed with our warning of its potent recoil as he assembled his company at attention before him. With the launcher held casually in one hand, he pulled the trigger. The launcher flew dramatically out of his hand, and grasping his wrist, he dismissed his men with as much military decorum befitting his rank as was humanly possible.

One of the great problems with work in tropical forests is the difficulty in observing all but the most conspicuous species. Raptors by their very nature are often extremely difficult to observe even under the best conditions. To increase detection rates, acoustical luring trials were conducted at several points along each transect. We used a simple portable tape recorder that broadcast a commercially available audiotape of distressed calls of various mammalian and avian prey. These trials were often exciting, eliciting responses from not only raptors but also mammalian predators such as the Ocelot *(Leopardus pardalis)* and Margay *(Leopardus wiedii)*.

Contained in the following chapters is the natural history of 20 resident raptor species found in the forests of northern Guatemala. No other attempt has been made to gather detailed natural history on an entire raptor community in a tropical forest habitat. Anyone who has had the opportunity to work with avian species in the field in North America will be well acquainted with the eminently useful *Life Histories* series by Arthur Cleveland Bent. If you want to find a particular species or its nest, or learn what it eats, your first clue is generally found within this series. I hope that the present work assembled by Dave Whitacre and his colleagues will represent a fundamental resource for those interested in tropical forest raptors and their ecology.

The Peregrine Fund employed and provided training for well over 150 local individuals during our nine-year tenure in Guatemala. We supported some 17 individuals in their successful completion of both secondary and advanced degree programs, and many have gone on to assume leadership roles in local government, NGOs, and administrative positions. It was indeed gratifying to return to Tikal last year and learn that my climbing partner Estuardo was now head of the park's Biological Monitoring Unit. Prior to working with The Peregrine Fund, many of these individuals had been involved in activities supported by their native forest, including lumber extraction, harvesting of chicle, gathering the ornamental palm *(Chamaedorea* spp.), and even searching for Maya ruins. These individuals are as much a part of the Maya Forest as are the other species with which they share it. They possessed a local knowledge of the forest that was, for us, of incalculable value. In our reports we say that we provided them with training and an opportunity for higher education, but in reality it was they who taught us.

ACKNOWLEDGMENTS

A study of this magnitude truly represents a team effort comprising contributions from hundreds of individuals and many institutions. The fieldwork that gave rise to this volume was the result of extensive cooperation with a number of international agencies and organizations. In Guatemala we worked with Comision Nacional del Medio Ambiente (CONAMA), Consejo Nacional de Areas Protegidas (CONAP), Instituto Nacional de Antropología (IDAEH), Fundacíon Interamerican de Investigacíon Tropical (FIIT), Fundación Mario Dary Rivera, the Universidad de San Carlos's Centro de Estudios Conservacionistas (CECON), and other groups and agencies and, above all, Tikal National Park. In Mexico we worked closely with Secretaria de Desarrollo Urbano y Ecología (SEDUE) and Pronatura México, and in Belize with the Ministry of Natural Resources and Department of Forestry, Belize Audubon Society, Programme for Belize, and the Belize Zoo.

Some of the many individuals who made this work possible, and whose friendship we value, include the following. Personnel of IDAEH include the directors Lic. Leopoldo Colóm Molina, Miguel Alvarez, Arq. Claudio Olivares, and Arq. Erick Cortez Serrano, and additional IDAEH personnel Lic. Rafael Torres Azurdia, Roxann Ortíz, and Elizabeth Toralla. Administrators of Tikal National Park who made the day-to-day operation of the project possible included Lic. Jose Rodolfo Morales Sanchez, Lic. Rolando Torres, and Sr. Rogel Chí Ochaeta. Personnel of Project Tikal who assisted us include Arq. Oscar Quintana and P.C. Edgar Raul Reyes Lee.

Personnel of San Carlos University's CECON who aided the project and whose friendship we value include Dr. Juan Carlos Godoy, Lic. Fernando Castro, Lic. Ismael Ponciano, Lic. Milton Cabrera, Otto Sandoval, Jorge Cardona, Rolando Escobar, Olga Valdez, Luís Villar, Claudio Méndez, Sergio Pérez, Herman Kihn, and the several CECON resource guards with whom we worked in the field. CONAMA personnel to whom we are grateful include Arq. Jorge Cabrera Hidalgo and Santiago Billy.

We are grateful to several directors of CONAP, including Arq. Andreas Lehnhoff, Arturo Duarte, and Ing. Milton Sarabia, and to other CONAP personnel, including Lic. Bayron Milian. We are very grateful for the assistance given to the project by Fundación Mario Dary individuals including Ing. Julio Obiols Gómez and Dr. Yuri Giovanni Melini.

For their continuing friendship, we are grateful to Lic. Jay Vannini, Juan Mario Dary, Licda. Maria José Gonzalez, Lic. Peter Rostroh, Licda. Hilda Rivera, Lorena Calvo, Ing. Hector Centeno, and Dr. Oscar Murga. At Secretaria de Planificación y Programación de la Presidencia (SEGEPLAN) we are grateful to Arq. Marco Antonio Palacios Méndez.

At Tikal National Park, we are forever grateful for the friendship of Don Mundo Solís and the late Pati de Solís. In the Petén, we are grateful for the friendship of Robert Heinzman, Conrad Reining, Carlos Soza, Scott Wilbur, Steve Gretzinger, Barbara Dugelby, John Beavers, Mario Jolón, Mike Lara, Jorge Roling, Mauro Salazar, Marie-Claire Paíz, Sofia Paredes, Roan Balas McNab, Robin Bjork, and others too numerous to name. We are deeply grateful for the support and friendship of Keith Kline of the U.S. Agency for International Development.

We value the friendship and goodwill of the many Tikal National Park guards who made us feel welcome during our nine-year presence in the park. We cherish the friendships we built with many family members of the extensive group of local men who made up a large part of our workforce, and with other local residents, especially of the village of Caoba, Petén.

We thank Ian Newton, Tom J. Cade, Richard O. Bierregaard, Richard T. Watson, Grainger Hunt, and James Enderson for reading through early drafts of this work and for providing valuable editorial suggestions. We also thank the many scientists who generously provided peer reviews of chapters both before and after submission of the text to Cornell University Press.

We thank Russell Thorstrom, Amy Siedenstrang, Susan Whaley, and Pat Burnham for their tireless editing, revisions, and formatting.

We thank Scott K. Robinson for his technical review of the final manuscript, and our publication editors, Peter Prescott and Heidi S. Lovette, for their patience and perseverance.

The following acknowledgments express the thanks of the respective chapter authors.

Chapter 2: For review of this chapter, we are grateful to Richard O. Bierregaard, Robin D. Bjork, Tom J. Cade, Scott K. Robinson, Nathaniel E. Seavy, and several anonymous reviewers.

Chapter 3: We are grateful to Lloyd Kiff for assistance in locating literature and interpreting published egg and clutch records. We thank Dr. Torsten Stjernberg for providing data on eggs in the University of Helsinki's Finnish Museum of Natural History, and Gene K. Hess for providing data and measurements of the eggs in the collection of the Delaware Museum of Natural History.

Chapter 4: We gratefully acknowledge Dr. Fred G. Thompson, curator of malacology at the Florida Museum of Natural History, for identifying snail remains and providing related natural history information, and to the many other malacologists who provided consultation on snail ecology and distribution. Lloyd Kiff assisted with compiling and reviewing relevant literature. Andy Mitchell helped explain Cuban geography and place names.

Chapter 5: We were blessed to have a great group, particularly of Guatemaltecos, working with us on this research. Those helping in nest observations included Apolinario de Jesus Mendoza, Cristobal Mateo, Israel Segura, Edi R. Martínez, and Tim Gerhardt. Normandy Bonilla, Tim Gerhardt, and Paula M. Harris assisted us in climbing. Ken Meyer provided insight throughout this research and made thoughtful comments on a draft of this chapter. We are grateful for the outstanding support of these friends.

Chapter 6: Portions of fieldwork were conducted by Tomas Dubón, Marcial Córdova, and Miguel Vásquez. For data and measurements of eggs at the Western Foundation of Vertebrate Zoology, we thank R. Corado, and for comments on the manuscript, we thank Richard Gerhardt, S. Olmstead, and an anonymous reviewer.

Chapter 7: We thank R. Gerhardt and Miguel A. Vásquez for sharing data gathered in 1991 and 1992. Portions of fieldwork were conducted by Tomás Dubón, Egidio Reyes, José L. Córdova, and Marcial Córdova. We are grateful to the curatorial staff of the following museums for sending data on egg clutches: Delaware Museum of Natural History, Western Foundation of Vertebrate Zoology, U.S. National Museum, The Carnegie Museum of Natural History, and American Museum of Natural History. Drafts of related manuscripts benefited from critique by Robin Bjork, Tom J. Cade, Lloyd Kiff, J. Parker, and anonymous reviewers. We thank Charlla Adams for assistance in preparing tables.

Chapter 8: I owe special thanks to Alenjandro M. Quixchán, Hector Girón, Eladio Martínez, Francisco Córdova, José Castillo, Amilcar Morales, and Ciriaco Marroquín for their assistance in the field. I thank William A. Burnham, J. Peter Jenny, David F. Whitacre, and Lloyd Kiff for their encouragement, assistance, and support in all phases of this study and for assistance in editing this chapter.

Chapter 9: I am deeply indebted to Walter E. Martínez, Nery "Pito" Oswaldo, and Francisco "Pancho" Oliva for their friendship and dedication to this research throughout the years. I thank Andres Rosales, Marcial Córdova, Juventino López, Julio A. Madrid, Ciriaco Marroquín, Amilcar Morales, O. René Ba, Eladio Martínez, Benjamin González, Angel Enamorado, José L. Córdova, Jeanette Waddell, and Laurin Jones for assisting with data collection. I also am indebted to Oscar A. Aguirre, Alejandro Manzaneros, Rodolfo Cruz, José D. Ramos, Estuardo Hernández, Gregorio López, Ricardo A. Madrid, Sixto H. Funes, Nehemias Castillo, Francisco Córdova, Theresa Panasci, Mark D. Schulze, Nathaniel E. Seavy, and Russell K. Thorstrom for providing additional information on Crane Hawk locations and behavior. In addition, I thank the directors and staff at Tikal National Park for their support and cooperation. I also thank Marc J. Bechard, Steven J. Novak, James F. Smith, and Ethan Ellsworth of the Department of Biology and Raptor Research Center, Boise State University, for their support, encouragement, and friendship. I also thank William A. Burnham and Lloyd Kiff of The Peregrine Fund. I especially thank David F. Whitacre for his encouragement and guidance in all aspects of this study. I thank Ernesto Iñigo for a review of the manuscript. Finally, I thank Kia Belice and Cheyton Sol Sutter for their support, patience, and love.

Chapter 10: We are grateful to Dr. Julian Lee for identification of several ophidian prey items, and to Santiago Choc Chocooj, Isabel Córdova, Nery Oswaldo, Antonio Aguirre, Feliciano Gutiérrez, Rigoberto Sandoval, and Gabriel Hernández for their capable assistance in the field.

Chapter 11: In 1991, fieldwork was conducted mainly by Richard P. Gerhardt, Paula M. Harris, and Miguel A. Vásquez. Fieldwork in 1993 and 1994 was supervised by David Whitacre; in 1993 it was conducted mainly by Angel M. Enamorado, assisted by Raul Ramírez, and in 1994 by Ricardo A. Madrid, assisted by Francisco Córdova. We are grateful to Scott K. Robinson for his helpful review of a chapter draft.

Chapter 12: I thank José M. Castillo, Nehemias Castillo, and Francisco O. Tovar for their diligence in collecting data, and also the multitude of Maya Project employees who assisted in preparing the study plots. I am grateful to Lloyd Kiff and the Western Foundation of Vertebrate Zoology for providing data on egg sets, to the various Guatemalan landowners who allowed us access to their land, and to Alejandro G. Di Giacomo for providing information published in Massoia (1988).

Chapter 13: We are grateful to Barbara Howard for providing records for Crested Eagles at the Oklahoma City Zoological Park, and to the staff of the vertebrate collections of the Field Museum of Natural History, especially Julian Kerbis and David Willard, for their assistance and hospitality. We are indebted to Ian Warkentin, Richard O. Bierregaard, David Ellis, and an anonymous reviewer for their critiques of an earlier draft.

Chapter 14: We are grateful for the able field assistance of Nehemías Castillo and for advice and assistance from Russell K. Thorstrom.

Chapter 15: We are grateful to several individuals who labored long and hard in this fieldwork: Roger R. Bótzoc, Marcos Castillo, José L. Córdova, Estuardo Hernández, Juventino López, Ricardo A. Madrid, Walter E. Martínez, and Antonio Ramos. Nathaniel E. Seavy assisted in extracting and summarizing data from field notebooks and carried out some initial data analyses.

Chapter 16: This study was conducted in cooperation with Guatemala's Instituto Nacional de Antropología e Historia, Centro de Estudios Conservationistas of Universidad de San Carlos, and the Consejo Nacional de Areas Protegidas. I thank Rogel Chí Ochaeta, administrator, and the staff of Tikal National Park. A special thanks to William A. Burnham, J. Peter Jenny, and David F. Whitacre of The Peregrine Fund for their assistance, support, and suggestions and to David F. Whitacre for his editorial perseverance. I thank José Ramos, Cristobal Mateos, Feliciano Ramírez, Oscar A. Aguirre, Eladio Ramírez, Carlos Mateo, Amilcar Gutierrez, and Alejandro Quixchán for their field assistance.

Chapter 17: This study was part of the Maya Project, a multiyear research effort conducted by The Peregrine Fund, in cooperation with the Instituto Nacional de Antropologia y Historia, Centro de Estudios Conservationistas, Guatemala, and Consejo Nacional de Areas Protegidas, Guatemala. We thank Rogel Chí Ochaeta, administrator, and the staff of Tikal National Park, Guatemala. A special thanks to William A. Burnham, J. Peter Jenny, and David F. Whitacre of The Peregrine Fund for their assistance, support, and suggestions. Special thanks to José Ramos, Cristobal Morales, Feliciano Gutierrez, José Castillo, Alejandro Quixchán, and Eladio Ramírez for their dedicated field assistance.

Chapter 18: We are greatly indebted to several individuals whose tireless fieldwork made this research possible: Mario Lima, Encarnación Paláez, Rolando Cruz, Manuel Chub, Benjamín González, Arturo Arévalo, José Osorio, Erwin Leonel, and Lázaro Carrillo. We are grateful also to Megan N. Parker's graduate committee members Nancy Clum and Jim Munger for their advice and support, to David F. Whitacre for his editorial efforts, and to William A. Burnham and The Peregrine Fund for making this work possible.

Chapter 19: We are grateful to Manuel Chub, Encarnación Peláez, Rolando Cruz, Mario Lima, and Sara Pedde for their roles in observing Bat Falcon nests, and to Pati Solís for her friendship.

Chapter 20: We gratefully acknowledge Robert B. Berry for major financial support for this research. Fieldwork was assisted by William Morales, Sherry Hudson, Nathaniel E. Seavy, Lenny Gentle, Erin B. Girdler, Benjamin González, Juventino López, and Mark Schulze. Flight surveys of Belize were made possible with help from LightHawk. Steve Herman, Hal Black, Mark Belk, Andrew Jenkins, and Lloyd Kiff provided comments on earlier drafts of the manuscript. Research in Guatemala and Belize was conducted under permit of the Consejo Nacional de Areas Protegidas and Belize Forestry Department, respectively. Special thanks are owed to Steve Herman, who pointed us toward the Orange-breasted Falcon and Tikal long ago, changing many lives in the process.

Chapter 21: Many Guatemaltecos assisted in data collection for this project. Chief among those was Normandy Bonilla. Others to whom we are indebted are Cristobal Mateo, Miguel A. Vásquez, Apolinario Mendoza, Israel Segura, Oscar Artola, Victor Quixchán, Luís Oliveros, Rony Hernández, Abel Maderos, Anibal Bonilla, and Eladio Martínez.

Chapter 22: This research was the result of the hard effort of numerous Guatemalan colleagues. These included Israel Segura, Miguel A. Vásquez, Cristobal Mateo, and Apolinario Mendoza. We are very grateful for their assistance and friendship.

ABOUT THE AUTHORS

Oscar A. Aguirre, native of San Benito, Petén, wore many hats as part of the Maya Project from 1991 to 1996. He played especially important roles in studies of the White Hawk and Orange-breasted Falcon. At the end of the Maya Project fieldwork, he continued as a field biologist for various nongovernmental agencies, and later, as a field technician, studying forest fragmentation and parrot conservation.

Aaron J. Baker earned degrees from The Evergreen State College (BSc) and Brigham Young University (MSc). As an undergraduate student he developed a keen interest in natural history and ornithology. Aaron first worked in the tropics in 1989 by volunteering in Belize on a vegetation survey project run by Manomet Bird Observatory and Programme for Belize. The following year he worked with The Peregrine Fund on the Maya Project in Guatemala where he eventually conducted his Master's thesis research on Orange-breasted Falcons. Aaron met his wife Serena Ayers at The Evergreen State College, and they worked together throughout Central America and in Uganda conducting and managing research projects on birds, primates, and fish. Aaron enjoys whitewater kayaking, surfing, mountain biking, and gardening. He now teaches environmental science, marine biology, and anatomy at Randolph High School in New Jersey where he lives happily with his wife and daughter, Violet.

Normandy Bonilla was a central figure in the Maya Project from 1989 to 1992, taking part mainly in field studies of the Mottled Owl and Black-and-white Owl. His knowledge, leadership skills, and sense of humor made him an invaluable colleague. Normandy also played a key role in our environmental education efforts, working with 600 local primary school children. With assistance from The Peregrine Fund, he completed a three-year course of study in Conservation and Management of Tropical Forests at the Petén branch of San Carlos University of Guatemala. He has worked several years for ProPetén, the local branch of Conservation International, as a specialist in management of community forestry concessions within the Maya Biosphere Reserve.

William A. Burnham's experience with birds of prey extended over 40 years, from arctic Greenland (since 1972) to the temperate mountains, plains, and forests of North America (since 1963) to the tropical forests of Latin America and Asia/Africa Pacific (since 1980). His 90+ publications, including one book (*A Fascination with Falcons*), reflect not only geographical regions of work but areas of interest from captive breeding and egg physiology to raptor ecology and species restoration. He received his PhD in wildlife biology from Colorado State University and an MSc in zoology from Brigham Young University. Bill managed his administrative responsibilities to allow for his continued involvement in fieldwork, especially in Greenland and the tropics where he co-founded the Maya Project with Peter Jenny. As time allowed, Bill was a practicing falconer, enjoyed fly fishing, big game and bird hunting, scuba diving, and he was an active rock climber to research raptor eyries. He developed and managed The Peregrine Fund's western program for Peregrine Falcon restoration beginning in 1974 when he joined the organization and was employed by Cornell University. He was elected to The Peregrine Fund Board of Directors in 1977 and assumed leadership of the organization in 1984 with the construction of its World Center for Birds of Prey, which he supervised and directed. He was elected a Founding Member of the Board of Directors in 1982 and President and CEO of The Peregrine Fund in 1986, a position he retained until his death in 2006.

José Luís "Chepe" Córdova is a native of El Caoba, just south of Tikal National Park. He participated in Maya Project fieldwork from 1992 to 1996. Chepe was a key field researcher on the Double-toothed Kite in 1995 and 1996, and played important roles in studying Ornate Hawk-eagles, Crane Hawks, and Barred Forest Falcons. He also played a major role in sampling the raptor community via visual/acoustical point counts. Using GPS and mapping skills gained in the Maya Project, he went on to work as a land surveyor in his native Petén.

Rodolfo Cruz, of Ixlú, Petén, was a vital part of the Maya Project from 1992 to 1996, studying the Ornate Hawk-eagle and leading the crew from 1993 to 1996. After the end of the Maya Project, he worked as a field technician for the Petén branch of Centro Agronómico Tropical de Investigación y Enseñanza, an agricultural and forestry research and management center based in Turrialba, Costa Rica.

Gregory S. Draheim conducted research on the White Hawk as part of the graduate program in Raptor Biology at Boise State University, receiving his BSc in 1995. A Colorado native with previous experience in raptor biology and falconry, Greg left a large part of his heart in Tikal, where he has many longtime friends. After receiving his BSc, Greg worked as a consulting field biologist and an observer on fishing boats in Alaska.

Angel M. "Pata" Enamorado, a native of Petén, worked as a resource guard for Tikal National Park for many years. From 1990 to 1995 he played major roles in the Laughing Falcon, White Hawk, and Great Black Hawk studies. He continued to work as a key member of the park staff after the completion of the Maya Project.

Craig J. Flatten worked on the Maya Project in 1989 and 1990, and was pivotal in getting the Ornate Hawk-eagle study off to an excellent start. He went on to become a wildlife biologist with the Alaska Department of Fish and Game, and studied Northern Goshawks and other wildlife in the temperate rain forests of southeast Alaska beginning in 1992. He is a nature and wildlife photographer, scuba diver, and seaplane pilot, who still travels to warmer climes to rekindle his appreciation of the beauty and diversity of tropical forests.

Sixto H. Funes participated in the Maya Project from 1990 to 1992, playing important roles in studies of the Ornate Hawk-eagle and Black Hawk-eagle. Sixto worked as a guard at Tikal National Park after the completion of the Maya Project.

Dawn M. Gerhardt, after completing her Master's degree at Boise State (studying breeding Ferruginous Hawks in the Idaho shrub-steppe), moved with her husband and co-worker, Rick, to Central Oregon to study owls and to raise a family. As a founder and co-owner of Sage Science Inc., a small ecological research firm, Dawn has home educated her four children, who have grown up believing it is only normal to know the names and behaviors of all of the birds and other animal life around them. Dawn treasures the time she spent in the Maya Forest, especially the hours of watching Swallow-tailed Kite nests from the tops of Tikal's temples, while around her Gray Fox pups crept from their dens in the temple aqueducts to get their first look at the world. Dawn also has fond memories of the Guatemaltecos who so ably assisted her and Rick in the owl and hawk studies at Tikal.

Richard P. Gerhardt's time in the forests of Tikal was life-changing, instilling in him a love of raptors as well as of the people and places of Central America. His experiences in Guatemala propelled him into a career of owl, hawk, and eagle research, which he continues as president of Sage Science Inc. Although his recent work has been mainly with the Strix owls, the migrating hawks, and the resident Golden Eagles of Oregon, he still jumps at every opportunity to take a trip to the American tropics, where he left a considerable portion of his heart. With his wife, Dawn, Rick continues to enjoy learning about the behavior of the predatory birds and raising another generation of raptor researchers.

Aquilas Estuardo Hernández was involved in the Maya Project from its inception in 1988 through 1996. He conducted more raptor point counts from Tikal's tree tops than any other individual, and is exceeded in this experience by few, if any, field workers worldwide. Before and after the Maya Project, Estuardo served as Chief Park Guard in Tikal National Park. He has taken part in numerous other ecological field studies in Petén and has served as assistant to the park biologist.

Gregorio López worked on studies of the Ornate Hawk-eagle, Black Hawk-eagle, Crested Eagle, Crane Hawk, and Gray-headed Kite as part of the Maya Project from 1992 to 1996. After that, he conducted bird surveys of the remote Sierra del Lacandón National Park for The Nature Conservancy and coordinated fieldwork for a study of Scarlet Macaws by Defensores de la Naturaleza.

Juventino "Tino" López participated in the Maya Project from 1991 to 1996, playing key roles in studies of the Ornate Hawk-eagle, Black Hawk-eagle, Crested Eagle, Crane Hawk, and Gray-headed Kite. He played a leadership role in the latter years of the project and continued to work as a biological field researcher in Petén.

Héctor D. Madrid participated in the Maya Project from 1990 to 1996. He played crucial roles in the Ornate Hawk-eagle study and later served as field logistics director and purchasing agent. With assistance from The Peregrine Fund, Héctor completed a three-year course of study in Conservation and Management of Tropical Forests at the Petén branch of San Carlos University of Guatemala. He became a regional forester for the Guatemalan National Forestry Institute, overseeing forest management operations within the large remaining forests of northern Petén.

Julio A. Madrid started with the Maya Project on the first day of fieldwork in 1988 and continued through 1996. He performed fieldwork on the Ornate Hawk-eagle and on songbirds, worked with local school children in our environmental education efforts, and served as the project field administrator and manager from 1993 to 1996. With assistance from The Peregrine Fund, Julio completed a three-year course of study in Conservation and Management of Tropical Forests at the Petén branch of San Carlos University of Guatemala. Since then he has been sought by local conservation groups and government agencies, serving as a resource guard for CECON and eventually as Regional Chief of Fauna for CONAP.

Ricardo A. Madrid worked on the Maya Project from 1992 to 1994, working on studies of the Great Black

Hawk and Ornate Hawk-eagle. After completion of the project, he worked as a tour guide in Tikal National Park.

Theresa Panasci received a Bachelor of Science degree in Environmental Science and Forest Biology from the State University of New York Environmental Science and Forestry College in 1991. She studied Roadside Hawks as part of the Maya Project in 1993 and 1994, receiving a Master of Science degree from Boise State University in 1995. She became a project coordinator for DST Innovis.

Margaret (Megan) N. Parker grew up in Montana, where she couldn't help but gain an appreciation for wildlife and wild places. She attended Middlebury College in Vermont for her BA and Boise State University for her MS in raptor biology, working on falcons in Tikal, National Park in Guatemala with The Peregrine Fund. She studied African wild dogs in the Okavango Delta of Botswana for her PhD work at the University of Montana. Megan has trained dogs since she was 10 years old and this led her to combine her passion for conservation biology with dog training to partner with other experts and co-found Working Dogs for Conservation. As director she develops projects, trains conservation detection dogs, and helps explore new avenues for noninvasive conservation applications with detection dogs, such as helping define wildlife corridors and identify areas of conflict, detect scats of endangered species, elusive live animals, and rare or invasive plants and animals. She has worked in Asia, Africa, the south Pacific, and North and South America on conservation projects.

Mark D. Schulze is Director of the H.J. Andrews Experimental Forest in Oregon. He has studied tropical and temperate forest ecology for 20 years. Recent emphases of his research are tropical tree population ecology as a foundation for sustainable forest management and response patterns of forest species to disturbance and climate variability.

Nathaniel E. Seavy worked for The Peregrine Fund's Maya Project from 1993 to 1998 and went on to receive his master's and PhD in zoology from the University of Florida. Nat has worked on research projects in North America, Central America, Africa, and Hawaii. These projects have included research on the breeding biology of raptors and owls, habitat associations of passerine birds, population ecology and monitoring of Pacific seabirds, and avian demography. Nat has published more than 35 peer-reviewed papers on topics including avian conservation and monitoring, restoration, and climate change. Today, Nat is based in California, where he is a Research Director for PRBO Conservation Science, a nonprofit organization that advances conservation through bird and ecosystem research. At PRBO, much of Nat's work has focused on population monitoring, the ecology and conservation of riparian ecosystems, and climate change. Most recently, Nat has led workshops in the Middle East on wildlife monitoring and statistical analysis for the US Forest Service International Program.

Jason Sutter's experience with birds of prey began with the Santa Cruz Predatory Bird Research Group, releasing Peregrine Falcons in northern California. He received his BA in anthropology from UCLA (1989) with an emphasis in Maya archaeology and his MSc in raptor biology from Boise State University (2000) based on his research of the Crane Hawk in Guatemala. He has coordinated projects involving raptors, migratory and sagebrush-obligate birds, sage-grouse, meso-predators, bighorn sheep, herpetofauna, small mammals, and other special status species throughout the western United States, Mexico, and Central America. His professional research interests include conservation biology, wildlife and landscape ecology, and conservation and management of shrub-steppe ecosystems. Jason is currently wildlife biologist for the Owyhee Field Office, Bureau of Land Management in southwestern Idaho.

Russell K. Thorstrom is currently Director of the Madagascar and West Indies Projects for The Peregrine Fund. Russell has been interested in birds and wildlife since childhood. He started working with Santa Cruz Predatory Research Group in 1983, the California Condor project in 1986, and The Peregrine Fund in 1988. Russell's Peregrine Fund career started as a member of the Maya Project in Guatemala where he described the first nests of Barred Forest Falcons. In 1993 while working for The Peregrine Fund in Madagascar, Russell rediscovered two species of raptors—one of which, the Madagascar Serpent Eagle, was thought to be extinct. His work in Madagascar fostered the establishment of the country's largest national park, which is intended to conserve habitat and biodiversity. In 2002, Russell began working in the Caribbean, where he conducted research on the ecology and distribution of the endangered Grenada Hookbilled Kite, and he continues studying and working on the conservation effort of the critically endangered Ridgway's Hawk. In 2007, he began assisting in research on the endangered White-collared Kite in Brazil. Russell has worked as a biologist in 13 countries. He received his MSc in biology from Boise State University and his BSc in Wildlife Biology from Washington State University. He has published and co-authored more than 60 articles in ornithological journals. Russell has a keen interest in nature and bird watching, cycling, running, hiking, and photography.

Miguel A. Vásquez played a key role in our studies of Swallow-tailed Kites, Plumbeous Kites, Hook-billed Kites, and Great Black Hawks. After working with the Maya Project for several years, he became lead field technician for the Wildlife Conservation Society in wildlife studies at Tikal and gained ample experience studying the endemic Ocellated Turkey. He also became senior park guard for Tikal National Park, where he served as a knowledgeable guide and interpreter of the antiquities, birds, and natural history of the park.

David F. Whitacre is a vertebrate ecologist who received his PhD in Zoology from the University of California at

Davis for research on the ecology of cave-nesting swifts in Mexico. In this work, he explored correlations among body size, nesting biology, and foraging strategies in the world's swift species. Other research has involved nesting montane songbirds, pesticide impacts on bird populations, predator-prey interactions between shorebirds and raptors, and the ecology of the Bat Falcon. Dave imprinted on snakes and the great diversity of broadleaf forests as a child in the Shawnee Hills of southern Illinois, where his father, Maurice A. Whitacre, a fish biologist, was his first biological mentor. Dave's main ecological interests center around the causation of organismic diversity and interspecific interactions in diverse communities, especially among predators. Introduced to the Mexican tropics at an early age, Dave learned Spanish largely during various trips there to bird and to explore caves. In addition to his dad, he counts among his most important biological mentors Steven G. Herman, Gary W. Page, and W. Grainger Hunt, along with his graduate school mentors Robert L. Rudd, George W. Salt, and Thomas W. Schoener. Dave directed The Peregrine Fund's Maya Project beginning in 1991. He currently teaches biology and statistics at the Treasure Valley Math and Science Center in Boise, Idaho.

NEOTROPICAL
BIRDS OF PREY

1   THE MAYA PROJECT

David F. Whitacre and William A. Burnham

In 1988, biologists of The Peregrine Fund set out to do something unprecedented—to study the ecology of the entire raptor community at a lowland forest site in the American tropics. Here we present our results, providing the first ecological portrait of virtually an entire tropical forest raptor community, results of a nine-year field study at a single locale. This effort, denoted the Maya Project, was centered on Tikal National Park in Guatemala's remote Petén Department, site of an ancient Maya city set within an immense area of lowland tropical forest. Over the nine years from 1988 through 1996, more than 100 field workers (up to 60 at a time) scoured the forests of Tikal to document the breeding biology, behavior, diet, space and habitat use, and ecological interactions of a Neotropical raptor community. We studied 20 of the 21 forest raptors common enough to study at Tikal. These species form the "core" species set of lowland tropical forest raptor assemblages throughout the New World tropical mainland (the Neotropics).[1] Here we present our results for the 20 species we studied, and summarize what we learned of their biology, ecology, and behavior.

WHY STUDY A NEOTROPICAL FOREST RAPTOR COMMUNITY?

There are many utilitarian reasons to study birds of prey, and those of tropical forests in particular. Predators, including raptors, play important ecological roles, sometimes affecting abundance patterns of their prey species, and can affect the structure of entire segments of biotic communities. They exert important selection pressures that shape characteristics of their prey from anatomy and coloration to habitat use, activity patterns, and social behavior. Predators are especially significant from the standpoint of conservation efforts. Often occurring at low densities and having large home ranges, they require large habitat areas. Thus, they can be a limiting factor in conservation efforts and can play an "umbrella" role—protecting sufficient habitat for predators can provide adequate habitat for many less space-demanding species. For us, the simple fascination of raptors was

additional justification for undertaking the work described here.

Prior to the Maya Project, only a few New World tropical forest raptors had been the subject of detailed studies of ecology or breeding biology (e.g., Beebe 1950; Rettig 1978; Delannoy and Cruz 1988; Thiollay 1991b; Alvarez-Cordero 1996). A number of studies had considered entire Neotropical raptor communities at some level. Most notably, several studies by Jean-Marc Thiollay and colleagues (e.g., Thiollay 1984, 1985, 1989b; Jullien and Thiollay 1996) considered patterns of abundance in different localities and habitats. These studies provided portraits of community composition, relative abundance patterns, and habitat affinities, and also tested methods for detecting and counting tropical forest raptors and made some initial density estimates (Thiollay 1989a). The only work approaching a detailed community-level ecological study was that of Scott Robinson (1994), who presented information on the diet, hunting behavior, and habitat use of raptors in Manu National Park in Amazonian Peru.

Why did we elect to study an entire assemblage of Neotropical forest raptors? The predator assemblages of Neotropical forests—some of the planet's most diverse— also are among the most poorly explored. In this biological terra incognita, the field naturalist may discover on an almost daily basis information new to science. Moreover, studying many species at a single locale provided insights not available from studying the same species at disparate sites, regarding community structure and function, comparative life history patterns, and comparative population densities and spatial use. To study the ecology of the entire raptor community at a single tropical forest locale was thus an irresistible prospect.

We also believed that such a project could yield benefits for conservation. Information we gathered on basic raptor ecology, habitat needs, population dynamics, and the effects of prevalent land uses might prove useful to land and wildlife managers and provide a basis for species conservation measures. We also felt that such a project would be a fertile training ground for local people who were or might become involved in conservation and management efforts in this globally significant forest area—a belief that proved well founded.

A PLETHORA OF PREDATORS

When we consider the epitome of terrestrial vertebrate predator communities, many of us think, no doubt, of the African plains, where the megafauna we associate with earlier geological epochs lives on today. There, a large mammal assemblage unparalleled elsewhere on Earth includes a diverse suite of mammalian carnivores. However, if one considers the total community of vertebrates that eat other vertebrates, among the planet's most diverse are those of Neotropical forests (Table 1.1). Vertebrate predators include not only mammals of the order Carnivora (cats, dogs, weasels, and the like) but also birds and reptiles, mainly snakes. Indeed four groups—the Carnivora, Falconiformes (raptors), Strigiformes (owls), and snakes (suborder Serpentes)—compose the bulk of vertebrate predators in Neotropical forests as they do in other low-latitude terrestrial communities.

For many groups of organisms, including birds, among the planet's most species-rich forests are those of the Amazonian headwaters, a crescent-shaped area along the eastern foothills of the Andes from southern Venezuela to northern Bolivia. For a vertebrate ecologist interested in tropical forest predators, to gain an understanding of the dynamics of predator communities in the upper Amazon is among the holiest of grails. This has not yet been attempted except, for the diurnal raptors, in the paper by Robinson (1994) mentioned earlier. Meanwhile, Harry Greene (1988) has written of the vertebrate predator community of La Selva, a lowland rain forest site in Costa Rica. Site of a renowned research station, La Selva is home to at least 56 species of snakes, up to 26 species of diurnal forest raptors, 6 owls, and 14 species of mammalian Carnivora (Table 1.1). As Greene points out, several of these feed mainly on invertebrates, and some on fruits. In addition, a number of species belonging to faunal groups not listed in Table 1.1 also prey partly on vertebrates at La Selva: these include at least two species of bats, a primate, an armadillo, a peccary, two species of lizards, three of frogs, several nonraptorial birds, and some large spiders, centipedes, and katydids (Greene 1988). A full tabulation of these other groups that prey on vertebrates would probably reveal an even steeper increase in predator species richness toward the equator than that suggested in Table 1.1.

In considering the ecology of tropical forest predator communities, it can be misleading to consider only one taxonomic group, because significant interactions such as food competition may occur even among distantly related groups. For example, a number of snakes feed mainly on lizards or other snakes, as do several raptors. Hence, raptors that specialize on snakes or lizards may overlap more in diet with these reptile-eating snakes than with other raptors.

Despite our interest in the overall vertebrate predator communities of Neotropical forests, it was beyond our means to extend our studies beyond raptors to other predator groups. Such studies to date have taken place elsewhere, including a study of the diet of several cats in Amazonian Peru (Emmons 1987) and that on snake diets at La Selva, Costa Rica (Greene 1988). Ultimately, it will be desirable to study the entire vertebrate predator community at a single Neotropical forest site—indeed, at many sites.[2]

RAPTORS OF THE MAYA FOREST

Including vultures, migrants, winter visitors, and open country and wetland species, as many as 50 raptor species occur in Petén, Guatemala, but the list of forest-dwelling raptors is much shorter (Table 1.2). Twenty-one raptor species made up the bulk of the non-vulture forest raptor community and were common enough to study at our project site, Tikal National Park (Table 1.2). We studied the nesting biology of all but one—the Guatemalan Screech Owl *(Megascops guatemalae).* Beyond these species, and disregarding the vultures, only a few other raptors can be confidently assumed to nest within Tikal National Park: the Short-tailed Hawk *(Buteo brachyurus)*, Black-and-white Hawk-eagle *(Spizaetus melanoleucus)*, and Central American Pygmy Owl *(Glaucidium griseiceps)*—all of which are uncommon there. Thus, the complete non-vulture forest raptor community at Tikal consists of 21 to 24 species that occur in ecologically significant numbers, 3 or 4 of them owls (Table 1.2).

If one includes a few forest species known from within 100 km of Tikal—the Harpy Eagle *(Harpia harpyja)*, Spectacled Owl *(Pulsatrix perspicillata)*, and Crested Owl *(Lophostrix cristata)*—the community expands to 27 species, 6 of them owls. Finally, if we include the Gray Hawk *(Buteo nitidus)* and Ridgway's Pygmy Owl *(Glaucidium ridgwayi)*—which in this region occur in human-modified, partly wooded habitats but rarely in mature forest—the forest raptor community expands to 29 species, 7 of them owls. Although the Red-throated Caracara *(Daptrius americanus)* is not known from Petén, it occurs in lowland forests nearby, expanding the regional lowland forest raptor list to 30 species (Table 1.2).

Of great interest, the set of raptor species we studied at Tikal is largely the same set that occurs in the hyper-diverse equatorial forests of the Amazon Basin: in essence, we studied the widely occurring set of "core" Neotropical lowland forest raptor species. A few additional falconiform and owl species are added closer to the equator, but the species we studied at Tikal comprise the bulk of the raptor assemblage throughout mainland Neotropical lowland forests (Table 1.2). While the ecology of our study species may vary geographically, describing their ecology at one locality is a first step toward a more global understanding (Plate 1.1).

Table 1.1 Species richness in temperate and tropical predator faunas[a]

Locality	Snakes	Owls[b]	Diurnal raptors[c]	Carnivora[d]	Total
Arctic sites					
Barrow, Alaska	0	2	2	3	7
Umiat, Alaska	0	1	7	7	15
Temperate chaparral					
Santiago, Chile	2	3	5	1	11
Doñana, Spain	5	3	9	5	22
Hastings, California	10	6	12	11	39
Temperate deciduous forest					
Shawnee Hills, S Illinois[e]	21[f]	4	9	13[g]	47
Great Smoky Mountain National Park	23[h]	4	8	13[i]	48
Temperate desert					
Big Bend, Texas[j]	32	10	26	13	81
Big Bend, Texas[k]	32	4–6	9–12[l]	16	61–66[l]
Tropical savanna					
Kruger, South Africa[m]	45	10	41	24	120
Kruger, South Africa[n]	45	9	27[l]	24	105[l]
Tropical moist forest					
Tikal, Guatemala	~43	4–6	20–21[l, o]	13	80–83[l]
Tikal, Guatemala	~43	7–8	29[p]	13	92–93
La Selva, Costa Rica	56	6	35[q]	14	112
La Selva, Costa Rica	56	6	24–26[l, r]	14	102[l]
Manaus, Brazil	61[s]	6[t]	21–27[u]	10–12[v]	98–106
Manu, Peru[w]	~65	8	35	16	~124
Manu, Peru[x]	~65	6–7	31[l]	16	118–119[l]

Source: Modified and expanded from Greene 1988.

[a] All numbers include total species richness for taxonomic group, not only species that eat vertebrates.

[b] *Tyto alba* omitted from consideration at all sites; not mainly a forest species.

[c] Falconiformes, excluding Old World and New World vultures.

[d] Members of the mammalian order Carnivora, not all of which are predominantly carnivorous; excludes other partly carnivorous mammals including opossums, armadillos, peccaries, and a few bats.

[e] Based on the editor's field experience, and on field guides.

[f] Shawnee Hills snakes: 9 eat mainly vertebrates and are terrestrial, 4 eat vertebrates and are aquatic, 2 eat invertebrates and vertebrates, and 6 eat invertebrates.

[g] Shawnee Hills Carnivora (original range; several now extirpated): 2 eat large vertebrates (> rabbit sized), 2 eat rabbit-sized vertebrates, 2 eat mouse-sized vertebrates, 1 eats mouse-sized invertebrates and fruits, 2 eat mouse-sized vertebrates and many invertebrates, 1 is an omnivore, 1 eats small, aquatic vertebrates, and 2 eat small aquatic vertebrates and invertebrates.

[h] Smoky Mountain snakes: 11 eat vertebrates and are terrestrial, 1 eats aquatic vertebrates, 2 eat invertebrates and vertebrates, 1 eats aquatic invertebrates, and 8 eat terrestrial invertebrates.

[i] Smoky Mountain Carnivora (original range; several now extirpated): 2 eat large (> rabbit-sized) vertebrates, 1 eats rabbit-sized vertebrates, 3 eat mouse-sized vertebrates, 1 eats small vertebrates and fruits, 2 eat small vertebrates and many invertebrates, 2 eat small aquatic vertebrates and invertebrates, 1 eats small aquatic vertebrates, and 1 is omnivorous.

[j] From Greene 1988, based on Wauer 1980.

[k] Based on analysis of same sources in Greene 1988, but omitting species rare or irregular in occurrence.

[l] Preferred estimate.

[m] All numbers from Greene 1988.

[n] All numbers from Greene 1988, except owl and diurnal raptor numbers based on Newman 1980, deleting species listed as rare or vagrant.

[o] Includes forest raptors only.

[p] Includes non-forest and wetland species occurring in northern Petén.

[q] Karr et al. (1990) state 35 species but this includes passage migrants, winter visitants, and non-forest species.

[r] Based on Karr et al. 1990 and Stiles and Levey 1994. Including only breeding-season residents and forest species, 24 species (without Gray Hawk and Harpy Eagle) or 26 species (including Gray Hawk and Harpy Eagle). Orange-breasted Falcon is considered absent. Red-tailed Hawk and Solitary Eagle omitted due to rarity, and Common Black Hawk omitted because not considered a forest species in our tabulations. Gray Hawk not a forest species at Tikal but can be elsewhere.

[s] Based on da Silva and Sites 1995.

[t] Including one rare species; based on Stotz and Bierregaard 1989 and Cohn-Haft et al. 1997.

[u] 21 including "rare" species; 27 including "rare" and "vagrant" species; based on Karr et al. 1990, Stotz and Bierregaard 1989, and Cohn-Haft et al. 1997; preferred total is 21, for 98–100 species total.

[v] From Malcolm 1990: *Speothos venaticus, Nasua nasua, Potos flavus, Eira barbara, Lutra longicaudis, Felis concolor, Panthera onca*; from Malcolm (pers. comm.): *Felis pardalis, Leopardus wiedii*, and probably *F. yagouaroundi*. Additional species that might be expected are *Galictis vittata* and *Procyon cancrivorus*. Several Carnivora genera occurring at the other Neotropical sites listed here do not range to the Manaus area (*Bassariscus, Bassaricyon, Mustela, Conepatus*: Emmons and Feer 1990).

[w] Based on Greene 1988.

[x] Data from Greene 1988 except that falconiform data are from Robinson 1994 and owl data are from Terborgh et al. 1990 and Karr et al. 1990.

Table 1.2 Raptors of the Maya Forest

No.	Common name	Scientific name	Status at Tikal
A. Raptor species studied as part of the Maya Project			
Accipitridae			
1	Gray-headed Kite	*Leptodon cayanensis*	Widespread forest resident
2	Hook-billed Kite	*Chondrohierax uncinatus*	Widespread forest resident
3	Swallow-tailed Kite	*Elanoides forficatus*	Widespread in forest; breeds at Tikal, migrating south in winter
4	Double-toothed Kite	*Harpagus bidentatus*	Widespread forest resident
5	Plumbeous Kite	*Ictinia plumbea*	Widespread in forest and modified habitats; breeds at Tikal, migrating south in winter
6	Bicolored Hawk	*Accipiter bicolor*	Widespread forest resident
7	Crane Hawk	*Geranospiza caerulescens*	Widespread forest resident
8	White Hawk	*Leucopternis albicollis*	Widespread forest resident
9	Great Black Hawk	*Buteogallus urubitinga*	Widespread forest resident
10	Roadside Hawk	*Rupornis magnirostris*	Widespread resident in forest and modified habitats
11	Crested Eagle	*Morphnus guianensis*	Widespread forest resident
12	Black Hawk-eagle	*Spizaetus tyrannus*	Widespread forest resident
13	Ornate Hawk-eagle	*Spizaetus ornatus*	Widespread forest resident
Falconidae			
14	Laughing Falcon	*Herpetotheres cachinnans*	Widespread resident in forest and more open habitats
15	Barred Forest Falcon	*Micrastur ruficollis*	Widespread forest resident
16	Collared Forest Falcon	*Micrastur semitorquatus*	Widespread forest resident
17	Bat Falcon	*Falco rufigularis*	Widespread resident, forest and modified forest
18	Orange-breasted Falcon	*Falco deiroleucus*	Rare resident, centered at tall cliffs amid forest
Strigidae			
19	Mexican Wood Owl	*Strix squamulata (Ciccaba virgata)*	Widespread forest resident
20	Black-and-white Owl	*Strix (Ciccaba) nigrolineata*	Widespread but local, thinly spread forest resident
B. Raptors occurring in Petén but not studied by us at Tikal			
Cathartidae			
21	American Black Vulture	*Coragyps atratus*	Widespread resident, mostly open habitats
22	Turkey Vulture	*Cathartes aura*	Widespread resident, forest and open habitats
23	Lesser Yellow-headed Vulture	*Cathartes burrovianus*	Wetlands of NW Petén
24	King Vulture	*Sarcoramphus papa*	Widespread forest resident
Pandionidae			
25	Osprey	*Pandion haliaetus*	Occasional; presumably a winter visitant; lakes
Accipitridae			
26	White-tailed Kite	*Elanus leucurus*	In pastures; uncommon near Tikal
27	Snail Kite	*Rostrhamus sociabilis*	Regular in marshes
28	Sharp-shinned Hawk	*Accipiter striatus*	Rare straggler/vagrant (Beavers 1992)
29	Cooper's Hawk	*Accipiter cooperii*	Hypothetical winter visitor (ibid.)
30	Common Black-Hawk	*Buteogallus anthracinus*	Straggler/vagrant (ibid.)
31	Black-collared Hawk	*Busarellus nigricollis*	Wetlands; rare near Tikal
32	Solitary Eagle	*Harpyhaliaetus solitarius*	If present, very rare (ibid.); in forest
33	Gray Hawk	*Buteo (Asturina) nitidus*	Widespread resident but not in mature forest at Tikal
34	Broad-winged Hawk	*Buteo platypterus*	Passes through in migration (ibid.); no sightings by us
35	Short-tailed Hawk	*Buteo brachyurus*	Widespread but uncommon in forest; presumably resident
36	White-tailed Hawk	*Buteo albicaudatus*	Common in marshes of NW Petén
37	Zone-tailed Hawk	*Buteo albonotatus*	Occasional during migration
38	Red-tailed Hawk	*Buteo jamaicensis*	Rare straggler at Tikal

Table 1.2—*cont.*

No.	Common name	Scientific name	Status at Tikal
39	Harpy Eagle	*Harpia harpyja*	Present in very low numbers elsewhere in Petén; historic presence at Tikal unknown; absent during recent decades; reintroduced since our fieldwork ended
40	Black-and-white Hawk-eagle	*Spizaetus melanoleucus*	Rare resident in forest at Tikal; 1 nest found nearby
Falconidae			
41	American Kestrel	*Falco sparverius*	Winter visitor in open country
42	Merlin	*Falco columbarius*	Hypothetical; presumably a rare, open-country visitor during migration or winter
43	Aplomado Falcon	*Falco femoralis*	Rare in open country near Tikal
44	Peregrine Falcon	*Falco peregrinus*	Winter visitor, open wetlands
Tytonidae			
45	Common Barn Owl	*Tyto alba*	Local resident, wooded and open areas
Strigidae			
46	Guatemalan Screech Owl	*Megascops guatemalae*	Widespread forest resident
47	Crested Owl	*Lophostrix cristata*	Forest resident; occurs farther south in Petén
48	Spectacled Owl	*Pulsatrix perspicillata*	A few sight records at Tikal (Beavers 1992); if resident, then rare and patchy; forest
49	Central American (Least) Pygmy Owl	*Glaucidium (minutissimum) griseiceps*	Rare at Tikal, in forest; common in wetter forest of NW Petén
50	Ferruginous Pygmy-owl	*Glaucidium brasilianum ridgwayi*	Common in modified habitats; rare in mature forest at Tikal

THE MAYA PROJECT

The Maya Project began in 1988, and fieldwork continued yearly through 1996. Two persons were responsible for the project's inception: Dr. William A. Burnham (Plate 1.2) and J. Peter Jenny of The Peregrine Fund. During their efforts to study the Orange-breasted Falcon *(Falco deiroleucus)* in Central and South America, Bill and Pete were impressed by the scant knowledge of most Neotropical raptors and the lack of proven methods for detecting, enumerating, and studying these birds in their forest home. Perceiving a need for research on these topics, they began the Maya Project, leading eventually to this book.

An initial trip was made in 1988, in which the Maya archaeological sites of Tikal (Plate 1.3), Calakmul (in southern Campeche, Mexico), and Caracol (Belize) were visited. The decision to focus at Tikal was made for several reasons, key among them being the protected status of the park and the year-round access provided by what was then the only paved, all-weather road in thousands of square miles—that linked the park with the airport at Flores, the political and commercial hub of Petén. The following year was the first one of research during a full breeding season at Tikal. Each year thereafter, through 1996, a team of researchers worked from February through July or August, studying nesting raptors at Tikal. From 1991 to 1996, we also employed a small crew of field workers during the nonbreeding season to achieve year-round coverage, especially in tracking radio-tagged birds.

A key ingredient in the project was the participation of seven graduate students from the raptor biology program of Boise State University who appear here as chapter authors. Other key participants were three students from Evergreen State College. These individuals and other participating field biologists from the United States and Mexico are listed in the acknowledgments.

A crucial ingredient was the involvement of a number of local people in field research, some of them for the entire nine-year duration of the project. Many appear here as chapter coauthors, and all are listed in the acknowledgments. Many did not have previous formal training in biology but became highly skilled biological research technicians. Several moved beyond the technician level to become full-fledged field biologists in their own right. They not only led crews in field research but also extracted and summarized data, wrote progress reports that contributed to the chapters appearing here, and in several cases presented results at international scientific meetings. The accomplishments reported here resulted in large part from the efforts of these dedicated individuals.

The Peregrine Fund assisted several local field technicians in completing high school degrees and supported four of them in a technical degree program in the conservation and management of tropical forests at the Petén branch of Guatemala's San Carlos University. Apart from such formal education, training within the Maya Project took place via an apprenticeship model. By working shoulder to shoulder during multiple years of fieldwork, experienced graduate students and professional biologists passed on to local participants the knowledge and skills needed to conduct field research on raptors, other birds, and vegetation.

Local participants, in turn, shared with U.S. project members their extensive knowledge of woodsmanship: how to extract water from a vine when lost in the forest, the names and uses of local plants, and the like. U.S. biologists learned a great deal from the local participants about land uses prevalent in the area: slash-and-burn farming, the chicle trade, the xate (pronounced shahtay) palm industry, and hardwood logging. In this way, the U.S. biologists gained a perspective on these economic activities from the practitioner's point of view— a perspective essential to any true understanding of the root causes of tropical deforestation and of the sorts of changes needed to achieve lasting conservation on the tropical forest frontier. Local project members also led U.S. biologists into two other endeavors: providing environmental education for 600 schoolchildren in local villages and introducing local farmers to nitrogen-fixing cover crops that could help them improve corn yields while reducing the amount of forest cleared for farming. These were rewarding aspects of the project that made a contribution toward conservation in the area.

The large amount of field research experience received by dozens of local people as part of the Maya Project is an important legacy of the project, as many of these individuals continue to employ this experience and training. Some of them now hold important positions with Guatemalan government agencies charged with managing the floral and faunal resources of Petén, while several others work for various non-governmental organizations, conducting ecological research in the forests of the Maya Biosphere Reserve.

Our research included work with migrant and resident songbirds and allies, examining abundance patterns in a variety of natural forest types and the effects of selective logging and shifting agriculture on the forest bird community (Madrid et al. 1995; Whitacre et al. 1993, 1995a). We also compared bat assemblages in extensive forest with those of forest fragments in the farming landscape (Schulze et al. 2000b). Studies of the woody vegetation of Tikal (Schulze and Whitacre 1999) supported our studies of raptors and other birds. The overall project is further described in Burnham et al. 1994, Whitacre 1998, and Whitacre and Burnham 2001.

FIELD METHODS

During the nine years of the project, we developed a suite of methods for studying raptors in continuous tropical forest. The most challenging task was finding nests. In general, a team of two to six people focused on a given species, searching for and observing the birds until their behavior revealed a nest. For the Forest Falcons (*Micrastur* spp.) this entailed stealthy observation on foot through the understory beginning well before dawn. For most other species, which were often active above the canopy, we would repeatedly ascend trees in areas where activity had been noted. Observing birds flying above or perched in the canopy, we would take compass bearings and move progressively closer, often watching from several different treetops over a period of days before the birds' behavior revealed a nest. On occasion, once a general vicinity had been determined, we were able to search through the forest on foot, scanning likely trees until a nest was found. As we searched on foot, vocalizations of adults or juveniles sometimes revealed a nest locality. In a few cases, radio tags placed on adults we had trapped led us to a nest. Finally, nests were also brought to our attention by park guards and other local people knowing of our interests.

We made heavy use of radiotelemetry. This allowed determination of home range areas and habitat use and was helpful in determining the duration of post-fledging dependency and in maintaining contact with adults from year to year, in order to find subsequent nesting efforts or to verify non-nesting. Transmitters were in the 216 kHz range, and mostly made by Holohil of Carp, Ontario. They were nearly always 3% or less of the birds' body weight. Most transmitters were mounted as backpacks, using Teflon ribbon joined via cotton thread over the breast bone, in the hope that the thread would eventually deteriorate and break, freeing the bird of the transmitter. In several cases transmitters eventually fell off the birds, while in other cases we do not know the transmitter's eventual fate. Some transmitters were mounted on central tail feathers, and a few juveniles were fitted temporarily with tarsal-mounted transmitters, which were replaced by tail-mount or backpack arrangements once the bird reached near-adult size. Receivers were TRX-1000, made by Wildlife Materials of Carbondale, Illinois. We used hand-held, three-element, folding directional yagi antennas from the same supplier.

We conducted most telemetry on foot, with the objective of making visual contact with the bird or pinning down its location to one or two trees. We tied locations to known geographic points and our base map by taking a GPS (global positioning system) point (usually 20 or more, averaged) in the vicinity, or making a pace and compass map to a known point. When conducting such "direct pursuit" telemetry, it was often necessary to climb a Maya temple or tree in order to pick up a bird's signal. On occasion, we resorted to triangulation to estimate locations. Through experience, we learned to discern and discard spurious bearings resulting from signal bounce. At times when we could not contact birds for a triangulation fix or could not find them on foot, they may sometimes have been at the peripheries of their home ranges, especially species with large home ranges, perhaps leading to conservative estimates of home range area. At other times, failure to obtain a radio signal was due to the bird being low in the forest or behind a hill, and such instances probably had little effect on home range estimates. When visual contacts were made, a data sheet was filled out, describing the habitat, perch characteristics, and the bird's behavior. In some cases, we made concerted efforts to observe hunting behavior of radio-tagged individuals.

Most trapping of adult raptors was conducted at nests, using the bal-chatri (Plate 1.4; Berger and Mueller 1959)—a wire cage covered with monofilament nooses that snared the raptors' toes when they attacked a live bait animal placed within. These traps were placed on nests (sometimes temporarily removing the egg or chick for safekeeping) or on limbs or the ground nearby. For hole-nesting species, we mostly used a net, hinged via stapling to the tree, which was pulled closed over the nest hole using monofilament lines after an adult entered. Other methods occasionally used were noose carpets (a variant of the *bal-chatri*, placed over a nest or perch), mist nets, and Swedish Goshawk trap (Thorstrom 1996).

Nest observations were made either from the ground or from a platform in a nearby tree, often concealed via camouflage netting or foliage. Platforms were made by lashing or nailing stout limbs in place, and usually employed a small piece of plywood as a floor, allowing use of a tripod and spotting scope (Plate 1.5). Observation platforms were built at sufficient distance that the birds' behavior was unaffected.

We conducted training sessions to give observers experience identifying prey and estimating its size. We estimated prey length by comparison with the size of the raptor's head, beak, and feet. To estimate mass, we practiced with tracings of scores of lizards that we captured, measured, and weighed. Prey mass was generally estimated using the following categories: less than 10 g, 11–20 g, 21–50 g, 51–100 g, 101–200 g, 201–400 g, 401–600 g, 601–1000 g, and more than 1000 g, in addition to point estimates of mass. We paid careful attention to the degree of reliability of prey species identifications; where there was any doubt as to species identity, we simply listed prey as snakes, lizards, vertebrates, and so on. Most species visually identified as prey at nests were also represented among prey remains collected at nests and identified in the hand, often by comparison with museum specimens or with the aid of specialists.

We concentrated our study in the central portion of Tikal National Park, where our field camp was located. We found nests in part by chance—having noted activity of a species in a certain area, we would search there for a nest. As the study progressed, we tried to find all the nests of a given species in a specific portion of the park, such that our study nests would be those of neighboring pairs, helping us estimate nesting density and degree of territoriality. In addition, we created a 20 km² study plot in continuous, mature forest, within which we made an effort to find nests and delineate territories. On this plot, we took advantage of survey lines cut by archaeologists years earlier. This yielded a 5 × 4 km study plot with a north-south access trail every 50 m in some areas and every 100 m in others. In 1994, we devoted a two-person team to count as many occupied raptor nests as possible within this 20 km² plot. This met with limited success, as a two-person team was not sufficient for this immense task.

ANALYTICAL METHODS

Because of the difficulty of detecting birds that did not defend a territory or engage in nesting behavior, we did not attempt to estimate the total population density of any species. Rather, we estimated the number of territorial pairs per unit area. We based these estimates on the spacing among neighboring pairs, indicated by active nests. Home range size estimates based on radiotelemetry did not lead directly to density estimates except where we obtained home range data for many neighboring individuals. Telemetry, however, aided in distinguishing neighboring pairs, especially when they were not nesting, and allowed estimation of home range sizes, which helped validate pair density estimates based on inter-nest distances.

Since our study nests were not generally in predetermined study plots that had been exhaustively searched for nests, we used plotless methods to estimate densities of territorial pairs, the basic datum being the average distance between neighboring nests. To determine this, we used the minimum spanning tree method (Gower and Ross 1969; Selas 1997); each of a group of n nests is joined by a straight line to its nearest neighbor, yielding n–1 inter-nest distances and a single mean inter-nest distance. We used this mean distance to create three complementary estimates of nest density.

The first, and preferred method, we refer to as the "polygon" method. Using CAMRIS GIS software (Ecological Consulting, Portland, Oregon), we created a convex polygon around the largest cluster of nests believed to be nearest neighbors. We extended the polygon beyond the outermost nests a distance equal to half the mean inter-nest distance. The outer edges of these circular "nest buffers" were joined to form a minimum convex polygon. The area of this polygon was then determined and divided by the number of known territories within, to yield the mean exclusive area per pair—the reciprocal of pair density.

The second method we use is the Maximum Packed Nest Density (MPND) method, as it sets an upper limit on the density possible under a given mean inter-nest distance. This method assumes that nests are spaced as in a close-packed crystal lattice, with all nests at the apices of equilateral triangles. This is unlikely to be strictly true, but it is a reasonable model for cases in which nests are overdispersed, that is, spaced more regularly than expected by chance—clearly the case for most raptors at Tikal. The area per nesting pair is given by $\varpi r2$ (where r is half the mean inter-nest distance), adjusted by multiplying by 1.158 to account for the interstices among the abutting circles thus generated.

The third method, referred to as the "square" method, is that used by Newton (1979). In this case, the mean inter-nest distance is simply squared to give the estimated space per pair. This method assumes nests are regularly spaced at the intersections of a grid of squares, but not as closely spaced as under the MPND method. All of these methods give accurate results only for densities within the group of nests yielding the mean inter-nest distance. If nests are not similarly spaced over the entire landscape, density estimates can be extrapolated to larger areas only with caution or not at all.

In most raptor species, females are larger than males, a pattern referred to as reverse sexual size dimorphism because it runs contrary to the more general pattern in birds and mammals, in which males are the larger sex. Many researchers have attempted to explain this phenomenon, with little consensus. For each species, we present values based on various measures of body size. We used the Dimorphism Index (DI) developed by Storer (1966), which, for any given measurement, is:

$$DI = (\text{female mean} - \text{male mean})/$$
$$(\text{female mean} + \text{male mean})(1/2)$$

We multiply the above by 100, to express the DI as a percentage. This index expresses the difference between the male and female measurements as a percentage of the mean value for the two sexes. This widely used index does not bear a linear relation to the simple ratio of female-to-male measures. As females become increasingly larger than males, the DI value increases, but at a diminishing rate. Hence, this index bears a downward-concave relationship to the percentage by which females exceed males in size. From the data presented, one may also calculate the percentage by which females exceed males in size. As an indication of overall size dimorphism, we present for each species the measures used by Snyder and Wiley (1976)—the DI value[3] for (cube root) body mass, and the mean of DI values for wing chord (wing length from wrist to tip), culmen (exposed beak, from front of cere to tip), and (cube root) body mass.

ABOUT THE BOOK

To provide all of our principal results in one place, we summarize here all of our published and unpublished results. Nearly half the material included is published here for the first time. In the species account chapters, our main goal is to present our results from Tikal. In

Table 1.3 Body size data for Maya Forest raptor species and temperate North American raptor species

Location of species	Common name	Scientific name	Mean female mass, g (n)	Mean male mass, g (n)	Percentage by which female mass exceeds male mass (%)
+	American Kestrel	*Falco sparverius*	119.0 (67)	109.0 (50)	9.2
+	Sharp-shinned Hawk	*Accipiter striatus*	179.0 (92)	102.0 (98)	75.5
•	Bat Falcon	*Falco rufigularis*	206 (9)	139 (14)	58.5
•	Double-toothed Kite	*Harpagus bidentatus*	208 (3)	182 (4)	14.3
+	Merlin	*Falco columbarius*	213 (15)	158 (14)	35.3
•	Barred Forest Falcon	*Micrastur ruficollis*	238 (17)	168 (13)	41.7
•	Plumbeous Kite	*Ictinia plumbea*	257 (7)	243 (16)	5.8
•	Hook-billed Kite	*Chondrohierax uncinatus*	280 (10)	263 (11)	6.4
•	Roadside Hawk	*Rupornis magnirostris*	286 (4)	244 (4)	17.2
+	Mississippi Kite	*Ictinia mississippiensis*	314 (6)	248 (14)	26.8
•	Swallow-tailed Kite	*Elanoides forficatus*	401 (6)	382 (5)	5.0
•	Bicolored Hawk	*Accipiter bicolor*	457 (11)	234 (11)	95.3
• +	Short-tailed Hawk	*Buteo brachyurus*	467 (3)	417 (3)	12.0
•	Crane Hawk	*Geranospiza caerulescens*	495 (3)	358 (3)	38.3
+	Broad-winged Hawk	*Buteo platypterus*	490 (13)	420 (14)	16.7
•	Gray-headed Kite	*Leptodon cayanensis*	521 (5)	461 (11)	13.0
+	Northern Harrier	*Circus cyaneus*	531 (97)	350 (90)	51.6
+	Cooper's Hawk	*Accipiter cooperii*	561 (143)	380 (34)	47.6
•	Orange-breasted Falcon	*Falco deiroleucus*	605 (4)	339 (2)	78.5
+	Swallow-tailed Kite	*Elanoides forficatus*	612 (5)[a]	542 (4)[a]	12.9
• +	Gray Hawk	*Buteo (Asturina) nitidus*	637 (4)	416 (5)	53.1
•	Laughing Falcon	*Herpetotheres cachinnans*	675 (8)	601 (13)	12.3
+	Red-shouldered Hawk	*Buteo lineatus*	701 (24)	550 (25)	27.5
•	White Hawk	*Leucopternis albicollis*	710 (3)	646 (4)	10.0
+	Prairie Falcon	*Falco mexicanus*	801 (34)	496 (10)	61.6
•	Collared Forest Falcon	*Micrastur semitorquatus*	869 (6)	587 (5)	48.0

Table 1.3—*cont.*

Location of species	Common name	Scientific name	Mean female mass, g (n)	Mean male mass, g (n)	Percentage by which female mass exceeds male mass (%)
+	Peregrine Falcon	*Falco peregrinus*	952 (19)	611 (12)	55.8
+	Harris's Hawk	*Parabuteo unicinctus*	1064 (11)	737 (22)	44.4
+	Swainson's Hawk	*Buteo swainsoni*	1069 (7)	908 (5)	17.7
+	Northern Goshawk	*Accipiter gentilis*	1095 (114)	860 (62)	27.3
•	Great Black Hawk	*Buteogallus urubitinga*	1111 (6)	1036 (5)	7.2
•	Black Hawk-eagle	*Spizaetus tyrannus*	1115 (5)	911 (3)	22.4
•	Black-and-white Hawk-eagle	*Spizaetus melanoleucus*	1191 (1)	780 (1)	ca. 53
+	Red-tailed Hawk	*Buteo jamaicensis*	1224 (100)	1028 (108)	19.1
+	Ferruginous Hawk	*Buteo regalis*	1231 (4)	1059 (15)	16.2
+	Rough-legged Hawk	*Buteo lagopus*	1278 (17)	1027 (11)	24.4
•	Ornate Hawk-eagle	*Spizaetus ornatus*	1452 (11)	1028 (4)	41.2
+	Gyrfalcon	*Falco rusticolus*	1470 (10)	1112.7 (4)	32.1
•	Crested Eagle	*Morphnus guianensis*	1849 (2)	1275 (1)	45
+	Golden Eagle	*Aquila chrysaetos*	4133 (3)	3400 (1)	21.5
+	Bald Eagle	*Haliaeetus leucocephalus*	5244 (37)	4123 (35)	27.2
•	Harpy Eagle	*Harpia harpyja*	8300	4400	ca. 89
Owls					
•	Central American Pygmy Owl	*Glaucidium griseiceps*	~56	~51	~10
+	Flammulated Owl	*Otus flammeolus*	57.2 (9)	53.9 (56)	6.1
+	Northern Pygmy Owl	*Glaucidium gnoma*	73.0 (10)	61.9 (42)	17.9
•	Ferruginous Pygmy Owl	*Glaucidium brasilianum ridgwayi*	80.4 (22)	68.6 (73)	17.2
+	Northern Saw-whet Owl	*Aegolius acadicus*	90.8 (18)	74.9 (27)	21.2
+	Whiskered Screech Owl	*Otus trichopsis*	92.2 (8)	84.5 (23)	9.1
•	Guatemalan Screech Owl	*Megascops guatemalae*	112.6 (5)[b]	105 (1)[b]	?
+	Boreal Owl	*Aegolius funereus*	140 (4)	102 (5)	37.3
+	Burrowing Owl	*Speotyto cunicularia*	151 (15)	159 (31)	−5.3
+	Eastern Screech Owl	*Otus a. asio (naevius)*	184 (36)	160 (38)	15.3
+	Long-eared Owl	*Asio otus*	279 (28)	245 (38)	13.9
•	Mexican Wood Owl	*Strix squamulata*	335 (11)	240 (7)	39.6
+	Northern Hawk Owl	*Surnia ulula*	345 (14)	299 (16)	15.2
+	Short-eared Owl	*Asio flammeus*	378 (27)	315 (20)	20.1
•	Black-and-white Owl	*Strix nigrolineata*	487 (4)	418 (3)	16.5
+	American Barn Owl	*Tyto furcata*	490 (21)	442 (16)	10.8
•	Crested Owl	*Lophostrix cristatus*	620 (1)	468 (2)	32.5
+	Spotted Owl	*Strix occidentalis*	637 (10)	582 (10)	9.4
+	Barred Owl	*Strix varia*	801 (24)	632 (20)	26.7
•	Spectacled Owl	*Pulsatrix perspicillata*	834 (5)	729 (4)	14.4
+	Great Gray Owl	*Strix nebulosa*	1298 (6)	935 (7)	38.7
+	Great Horned Owl	*Bubo virginianus*	1509 (94)	1142 (94)	32.1
+	Snowy Owl	*Bubo scandiaca*	1963 (30)	1642 (27)	19.5

Notes: • = Maya Forest raptor; + = Temperate North American raptor
[a] Mass data from Meyer 1995b, birds in premigratory condition.
[b] Mean index value based on wing chord and culmen only; no mass data available.

addition, we conducted a thorough literature review and present enough review material so that our chapters might serve as thorough synopses of the current knowledge on these species. While attempting to give uniform coverage to each topic for each species, we also allowed chapter authors to highlight certain aspects. Thus, for some chapters, additional headings are used. Chapters vary in length because we gathered more data on some species than others.

To facilitate comparisons with species more familiar to many readers, Table 1.3 gives some basic size data for the 29 Maya Forest raptor species and for most temperate-zone North American raptors. Using this table, the reader can select a series of touchstone species to aid in visualizing the raptors treated here. Unless otherwise noted, species names of all raptors are consistent with the Global Raptor Information Network (www.globalraptors. org). Common names of non-raptors, animals, and plants are taken from the Encyclopedia of Life (www.eol.org).

So that this book might serve as a scholarly source, we present some statistical results, including probability values—"P values." P values state the probability that a pattern existed beyond that likely due to chance. By convention, a P value of 0.05 or smaller indicates a statistically significant pattern; at this value, there is only a 1 in 20 chance that the pattern observed was due to chance.

Because much of the material presented here has not been published in scientific journals, it was important to us that this book serve as an original, scholarly source. Thus we secured a similar degree of peer review to that achieved in the normal journal publication process. We sent all chapters to one or more reviewers prior to submitting them to Cornell University Press. In turn, the Press sent each chapter to at least two academic peer reviewers. We are grateful to these reviewers, and any shortcomings of the book are the sole responsibility of the chapter authors and the editor.

Chapter 1 Notes

1. "Neotropics" refers to the biogeographic region extending from the Tropic of Cancer in Mexico, the Caribbean islands, south to Tierra del Fuego. Only the lowlands of this region have tropical climates.

2. The most detailed study of a tropical forest food web—that of Luquillo, Puerto Rico—is of a species-poor island ecosystem with limited predator diversity, including only three diurnal raptors and one owl (Reagan and Waide 1996); thus, it sheds limited light on trophic dynamics within the much more species-rich forests of the Neotropical mainland.

3. In calculating the DI based on body mass, it is customary to use the cube root of body mass. Because volume and body mass scale as the cube of linear dimensions, basing the body mass DI on the cube root of body mass makes it comparable to DI values based on wing or tarsus length or other linear dimensions.

2 THE MAYA FOREST

David F. Whitacre and Mark D. Schulze

To best understand the natural history of the raptors we studied, it is helpful to know something of the ecological community of which they form a part. Here we describe our study area and its present-day biota and physical environment. We also present information on the climatic and vegetational history of Neotropical lowland forests and on the evolutionary history of raptors, to help reveal how long our study species have likely occurred together. We focus on the "Maya Forest" or "Selva Maya" region—the basal, moister portions of the Yucatán Peninsula—within the context of the entire peninsula, Central America, and the American tropics as a whole.

PALEOHISTORICAL CONTEXT

It is interesting to know how long our study species and their forest home have occupied our study site at Tikal, but it is far more important to ask how long these species have coexisted, whether at Tikal or elsewhere. The key question is whether these species have occurred together over long time spans, affording opportunities for evolutionary accommodations to coexistence. Today these raptors occur together over a very large area of the humid Neotropical lowlands (see Plate 1.1). How long has this been the case? The short answer is that tropical forests have been present in the Americas since prior to the origin of raptors. Moreover, most Neotropical raptor lineages appear to have long been present in the American tropics. Thus, it seems likely that the raptors we studied have coexisted in the American tropics for several million years, and possibly throughout the entire evolutionary histories of the genera to which they belong.

Tropical forests have existed in the Neotropics throughout the past 65 million years (Behrensmeyer et al. 1992), whereas modern bird families arose by 35 million years ago (mya) and modern genera, by 23 mya (Feduccia 1996). Hence, tropical forests have been available to Neotropical raptors for as long as raptors have existed. The antiquity of living raptor species is less certain. Recently it was believed that much of the speciation producing the stunning diversity of the Neotropical

region took place during the late Pleistocene, the last million years or so. However, current evidence indicates that many Neotropical species originated earlier, more than 2 mya.

The earliest falconid fossils date from 23–36 mya, and a fossil from Nebraska 23 mya is attributed to the genus *Falco* (Olson 1985; del Hoyo et al. 1994), though fossils of present-day *Falco* species are from about 2 mya (del Hoyo et al. 1994). The earliest fossils of the Accipitridae, or true hawks and eagles, also date from 36 mya (Olson 1985). By 5–10 mya, a number of accipitrid fossils have been referred to living genera (Brodkorb 1964). Although the earliest-known owls date from 55 mya, the time of appearance of modern genera of strigid (typical) owls is uncertain (Olson 1985). The earliest recognizable tytonid (Barn Owl) is from France, about 20 mya, and represents an extinct genus. Three species of *Tyto* (modern Barn Owls) have been named from deposits 5–15 million years old in France and Italy.

Based on this information, we conclude that the genera we studied at Tikal likely differentiated by at least 10–15 mya. Many present-day raptor species are known from Pleistocene deposits of various ages (Brodkorb 1964), and we assume that our study species had speciated by early Pleistocene, at least 2 mya. In any event, most raptor diversity at Tikal is at the generic level and thus likely represents 10–15 million years since divergence of these lineages. Most raptor genera at Tikal are endemic to the Neotropics and probably evolved there. Thus it seems likely that the raptors we studied have coexisted in the American tropics for several million years and, in many cases, throughout their entire evolutionary histories.

Systematics and Biogeography of Diurnal Raptors

Falconidae. Seven of the 10 falconid genera and two of the family's three major clades are restricted to the Neotropics, with only *Falco* and the falconets *Microhierax* and *Polihierax* occurring in the Old World (Becker 1987; Kemp and Crowe 1990; Griffiths 1994a, 1994b).

The Neotropics are likely the family's center of origin and certainly the area where most higher-level diversity is concentrated, largely in forest habitats (Griffiths 1994a, 1994b).

The contribution of falconids to raptor faunas thus varies greatly in different parts of the globe and may affect evolutionary and ecological patterns seen in different tropical raptor communities. The falconid genera *Micrastur*, *Herpetotheres*, *Daptrius*, and *Ictyber*, all New World forest dwellers, present a unique set of traits, and if these birds have ecological counterparts among falconiforms in other tropical realms, those will inevitably be of accipitrid rather than falconid stock. Among other things, this means that they will be stick nesters rather than cavity nesters.

Accipitridae. Phylogenetic relationships within the Accipitridae remain uncertain. Sibley and Ahlquist (1990) review attempts to classify this large and varied group. Brown and Amadon (1968) describe a number of traditionally recognized, informal groupings within the family, and two cladistic analyses have been based largely on skeletal characters (Kemp and Crowe 1990; Holdaway 1994). Ongoing studies based on molecular genetics are revising our understanding of phylogenetic relationships among Neotropical accipitrids and demonstrate that some genera, as currently defined, contain species groups that are not monophyletic (e.g., Riesing et al. 2003; Helbig et al. 2005; Lerner and Mindell 2005; Raposo do Amaral et al. 2006), which will require redefinition of these genera.

Most generic diversity within the Accipitridae occurs in the tropics, which are populated also by genera widespread in the north temperate zones (Kemp and Crowe 1990). There is evidence of notable adaptive radiations in South America and Africa, while other areas resemble these two regions in proportion to their distance from them, and reflect the relative vagility of different taxa (Holdaway 1994). Many genera are endemic to tropical regions, and few to temperate zone regions.

Geologic and Biotic History of Central America

Central America as we know it has been recognizable for only the past few million years. Once united as portions of the supercontinent Pangea, North and South America began to rift apart 140 mya, and by 80 mya, South America had separated from Africa as well, beginning a long isolation that produced a strikingly unique South American biota (Coates 1997). While the landmass comprising most of Mexico has been above sea level for at least 140 million years, portions of Central America are mostly younger. By 65 mya, Mexico was present as a peninsula protruding from North America, and the beginnings of southern Central America came into existence as an arc of volcanic islands. By 12 mya, this volcanic chain collided with South America, and the deep ocean floor that had preserved South America's isolation for

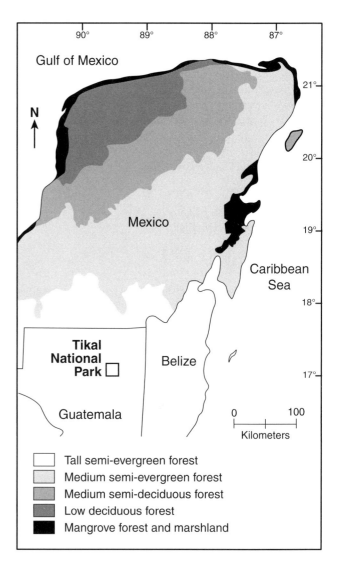

2.1 Vegetation of the Yucatán Peninsula.

so long lifted to form a floor only 1000 m deep (Coates 1997). Over the next few million years an archipelago formed here. By 6–8 mya, the ocean floor was only about 150 m deep, and faunal interchange between North and South America became evident, as raccoons spread southward and ground sloths northward, apparently by hopping islands. By 2.4–3.0 mya, the Central American isthmus was complete, allowing faunal interchange over a continuous landmass. The ensuing interchange of the vastly different faunas of North and South America—the "Great American Interchange"—ranks among the most striking biogeographic events known (Stehli and Webb 1985).

Physiography and Geology of the Maya Forest Region

The Yucatán Peninsula is a gently rolling limestone platform 500 km long and 300 km wide, protruding only modestly above the sea from which it continues to emerge (Fig. 2.1). The base of the peninsula has risen from the sea more than the northern tip, commonly achieving elevations of 300–350 m in and near Tikal. Lee (1980) concludes that the Yucatán Peninsula began emerging from the sea during the Miocene (5–23 mya), from south to north such that deposits in the north are progressively younger. While the northernmost tip of the peninsula did not emerge from the sea until the last million years or so, the basal portion around Tikal has been above sea level for perhaps some 10 million years (Lee 1980).

FORESTS OF THE SELVA MAYA FROM ICE AGE TO PRESENT

Throughout the past 2 million years, global climate has fluctuated dramatically along with the wax and wane of great ice sheets in the Northern Hemisphere. No great ice coverage was experienced in Central America, except atop the tallest mountains such as Guatemala's Cuchamatanes massif and the Talamanca range in southern Costa Rica. While a Pleistocene vegetation map for Central America is not yet possible, much has been learned from lake-bottom pollen cores, phytoliths (silica deposits laid down by some plants within their cells), and radiocarbon dating of associated organic matter. The best data are available for the past 19,000–33,000 years, but some inferences may be made back to 120,000 years, and a tentative scenario painted back to the closing of the sea portal 2.4 mya (Colinvaux 1997).

Pollen cores from two lakes a few kilometers south of Tikal suggest that forest was absent here 10,000 years ago, with a more open, arid-adapted vegetation apparently present (Leyden 1984). For a brief period, perhaps 8,000–10,000 years ago, forests here were largely of pine, oak, juniper, and temperate hardwoods, suggesting a cool interval before tropical forests similar to today's were most recently established. Hence, tropical forests were apparently reduced or absent in some portions of the isthmus at times during the ice age. However, abundant evidence from the last glacial maximum (19,000 years ago) to the present suggests that much of Central America was more or less continuously covered with forest, though of varying composition and distribution, throughout this period (Colinvaux 1997).

Lake-bottom cores indicate that the Panamanian lowlands remained primarily tropical forest from sea to sea throughout the past 120,000 years, embracing the Wisconsin glacial period and part of the previous interglacial (Colinvaux 1997). During cooler periods, plant species today restricted to montane forest migrated downward approximately 800 m, mixing with lowland plants. A cooling of 5–6 °C in Panama, and perhaps as much as 8 °C in our study area, seems well established. Such cooling was probably the dominant glacial influence on climate and vegetation of Central America prior to the advent of the Holocene warm interval 10,000 years ago. Locally, however, low rainfall at times is believed to have been an important influence, as in our study area.

In the vicinity of Tikal, during glacial advances of the past 2 million years, tropical forest may have been replaced by savannas, with the most recent round of tropical forest expansion beginning a mere 10,000 years ago. Reclamation of the Petén by tropical forest likely represented forest expansion from moist enclaves along the foot of neighboring mountain ranges of Chiapas, Alta Verapaz, and Quiché. Thus, while the raptor community we studied may not have co-occurred most recently at the precise site of Tikal until approximately 10,000 years ago, there is reason to believe these birds have co-occurred nearby in Central America throughout the past 2 million years or more, and for much longer in South America.

The present-day forest stands of our study site are likely younger than the 10,000 years just given, because of deforestation by the Mayan during their cultural apogee. Arising in this area more than 3000 years before present (BP), the Maya civilization flourished for nearly two millennia, then collapsed mysteriously during the tenth century AD (Curtis et al. 1996). Population densities during the Maya apogee are believed to have been 200–300 individuals/km² (Rice and Rice 1990)—much higher than today—and the population of the Yucatán Peninsula may have numbered several million (Curtis et al. 1998). During the early Holocene (approximately 9000–6800 years BP), evidence indicates moist conditions and semi-deciduous lowland forest in the area. By 5780 years BP, decline of lowland forest taxa in pollen cores may indicate early human disturbance, and substantial deforestation by the Mayan by 2800 years BP appears generally accepted among archaeologists, with widespread deforestation during the Classic Period (AD 250–850; Curtis et al. 1996). In many areas a layer of "Maya clay" as thick as 7 m was deposited in lakes over the 2600-year period leading up to European contact; this clay is believed to reflect soil erosion resulting from widespread Maya farming (Curtis et al. 1996).

Although the degree of deforestation engendered by Maya agriculture is not well known, we consider it reasonable to assume relatively great deforestation and forest modification. We assume also that forest refugia probably remained in areas not suitable for farming, for example, on rugged karst uplands and perhaps in the extensive wetlands of northwestern Petén. Unoccupied forest "buffers" between competing Maya polities and "forest gardens" within occupied landscapes may also have served as refugia for forest species (Hayashida 2005). Multiple lines of evidence indicate forest recovery beginning 1000–1100 years ago, soon after the Maya collapse. Evidence also indicates that different population centers had different dates of prominence and collapse, suggesting that dates of deforestation and recovery may have varied across the region (ibid.). Postclassic Maya populations continued to occupy the region until European conquest in 1597, but never regained the numbers existing prior to the collapse. Pollen evidence indicates that agriculture occupied much less land during the Postclassic Period, with longer fallows in swidden plots allowing establishment of secondary forests (ibid.). Hence we assume that the forests of Tikal and much of northern Petén were but lightly influenced by people from the time of the Maya collapse in AD 900 until the past couple of centuries, when inroads were made in search of Honduras Mahogany *(Swietenia macrophylla)* and chicle harvested from the Chicozapote *(Manilkara zapota)* tree (Schwartz 1990). Even these latter activities had little effect on forest cover until 1965 or later.

Are the forests of our study area, representing 1000 years of development since presumably massive alteration by the Maya, mature, "primary" vegetation, or are they successional in nature? Hatshorn (1980) concluded that 1000 years since abandonment of Maya crop fields is ample time for succession to have produced forests of climax status. Based on our own vegetation studies (Schulze and Whitacre 1999), we agree. In most of Tikal's forest, tree species that dominate the canopy are also abundant in the understory and appear to be regenerating successfully, suggesting stable species composition rather than continuing successional change. The legacy of Maya occupation and forest clearing must still be evident in regional patterns of species composition and even landscape forms—some species having achieved greater densities in response to ancient anthropogenic disturbance, and soil and topographic conditions having been modified by Maya works. However, current forest composition and species distribution patterns in Tikal likely reflect more recent disturbance processes, such as individual treefalls and larger-scale storm blowdowns, acting in ways both stochastic and predictable on a landscape of varying physical conditions.

CLIMATE OF NORTHERN PETÉN

Located at latitude 17° North, Tikal is well south of the Tropic of Cancer. Holdridge (1967) considered those portions of Central America north of 12–13° to be in the subtropical realm, corresponding to the "outer tropics" of geographers. While the biota is distinctly tropical, the Selva Maya experiences a seasonal climate, especially in terms of rainfall. The mean daily temperature at Tikal changes little through the year, ranging from the mid-70s to 80 °F (23–27 °C: Fig. 2.2). Average monthly maxima range from a low of 82 °F (28 °C) in December to 94 °F (nearly 35 °C) in May, with the hottest days of the year commonly falling from March–June, during the dry season and early rainy season. Temperatures drop at night, often into the 50s–60s °F (11–20 °C). The daily range of temperatures is greatest during the dry season, from February through May, with distinctly less variation during the rainy season (Fig. 2.2).

Rainfall in the Selva Maya is far more seasonal than temperature (Fig. 2.3). A dry season holds sway from roughly January through April or May. This is generally a severe dry season of reliably rainless, often windy days. It is not rare for a month or more to pass without measurable rain, though thick morning fog is frequent, sometimes depositing dew. Leaf litter becomes crunchy underfoot, and understory vegetation often takes on a wilted appearance. Onset of the rainy season is variable; on average, the rains begin gradually in April and May, building to an initial peak in June. Thereafter, rains generally decrease during July and August, with a dramatic peak in September, usually the rainiest month. Rainfall drops off but is still high in October, and declines steadily in November and December. By January or February the dry season is again underway.

Rainfall maps for the Yucatán Peninsula vary greatly; Figure 2.4 presents the one we consider most credible. Rainfall decreases markedly from south to north across the peninsula. In southern Belize, on the windward slopes of the Maya Mountains, rainfall reportedly may average 4 m annually, while in the highlands of Alta Verapaz along Petén's southern border, up to 5 m/yr is reported. Much of southern Petén appears to receive approximately 2 m annually. Moving northward, rainfall

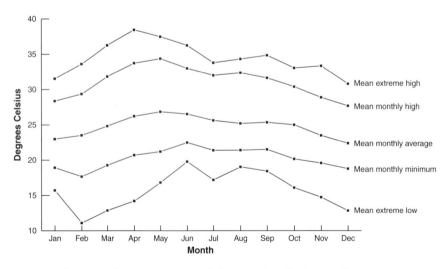

2.2 Annual pattern of temperatures in Tikal National Park, Guatemala.

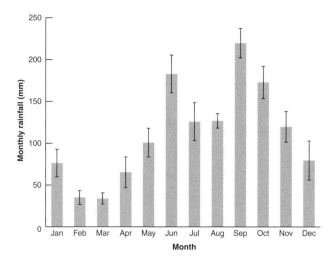

2.3 Mean monthly rainfall throughout the year in Tikal National Park.

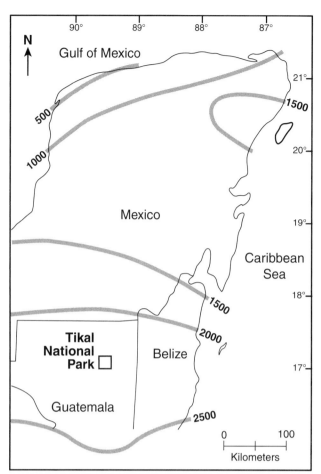

2.4 A rainfall map of the Yucatán Peninsula.

decreases to a mean of about 1480 mm at Flores (23 years of data) and 1360 mm at Tikal (9 years of data) and thence to about 1300 mm across much of southern Campeche and Quintana Roo, and to 450 mm in the extreme northwest of the peninsula, near Mérida, Yucatán. A lens of relatively high rainfall extends northward along the Caribbean coast of Quintana Roo, producing a marked gradient of decreasing moisture from east to west across the northern portions of the peninsula (López Ornat et al. 1989; see Fig. 2.4).

Mean yearly and monthly rainfall totals are deceptive, however, as rainfall in the Maya Forest varies strongly in quantity and timing from year to year. Reliability of total rainfall is substantially greater during the latter half of the rainy season, from August to November, than during the first half, from May or June to July (Table 2.1). Rainfall is quite unpredictable during the dry season and transitional periods, and variability between years is also high. Based on nine years of data (1960–62 and 1989–94), mean annual rainfall at Tikal was 1360 mm (52 inches). However, the annual total ranged from 1019 mm (40 inches) in 1992 to 1731 mm (68 inches) in 1993, with a coefficient of variation of 18%.

Climatic Variability: A Long-Term View

A detailed climatic reconstruction for the Yucatán Peninsula for the past 3500 years showed not only pronounced rainfall variation on a multi-decadal scale, but also a pronounced dry period from 1785 to 930 years BP, with several extreme droughts occurring during this and other time spans (Curtis et al. 1996). The same study showed correlations between rainfall variations and transitional periods in Maya culture, and the researchers favored the interpretation that drought caused the collapse of Classic Maya civilization 1000 years ago (Curtis et al. 1996).

During our nine years of fieldwork at Tikal and periodic visits since, we witnessed some effects of the vari-

able rainfall regime. During a drought in the mid-1990s, many aguadas (water holes) that are normally reliable throughout the year dried up. Since virtually all households rely on such natural or man-made rainwater receptacles, this drought resulted in occupants vacating entire villages for a time. During dry years, agricultural fire often escapes and passes through primary forest. Ground fires burned beneath the canopy in two areas of Tikal National Park in the dry season of 1993, providing a tangible example of the interaction between climate variability and forest disturbance. An extreme example was seen in 1998, when drought conditions credited to El Niño led to large forest fires in Mexico and Central America, with resultant smoke affecting air traffic as far north as Texas. Drought frequently leads to partial failure of the corn crop on which many families depend. Finally, two massive insect outbreaks we witnessed in Tikal's forests during the 1995 rainy season may have been causally related to a preceding seven-month drought period. Larvae of the dynastine beetle *(Enema endemion)* removed most leaf litter from large areas of forest, and an unidentified larva, probably a lepidopteran, caused extensive defoliation of *Pouteria reticulata*, one of the most common trees at Tikal (D. Whitacre and M. D. Schulze, unpubl. data).

Table 2.1 Variability of monthly rainfall at Tikal[a]

	Monthly mean (mm)	Standard deviation (mm)	Sample size (no.)	Coefficient of variation (%)
January	76.3	52.2	10	68.5
February	35.2	27.7	11	78.5
March	34.0	22.1	11	65.0
April	65.4	63.0	12	96.2
May	100.9	59.6	12	59.1
June	183.0	81.5	13	44.5
July	126.0	81.7	13	64.9
August	127.0	28.7	11	22.6
September	219.9	60.8	12	27.7
October	172.9	67.2	12	38.9
November	119.9	60.8	11	50.7
December	79.6	77.7	11	97.7

[a] Based on data from 1959–63 and 1988–1995 (n = 10 to 13 for different months).

Biological Seasonality of the Selva Maya

The longest day of the year at Tikal is about 12.95 hours from official dawn to dusk, and the shortest day, about 11.08 hours. Hence the longest day is about 16.9% longer than the shortest.

Arthropods. The abundance of many insect groups at Tikal fluctuates strongly with the march of the seasons. From September 1994 to June 1997, we sampled insect abundance weekly by shining an ultraviolet light against a suspended white sheet at our field camp in the forest near the park center. We operated the light for 90 minutes at a standardized time after sunset and, using a set protocol, counted the insects of different taxonomic orders and size categories that were attracted. From a distinct population low from November through March, insect numbers increased from April to a maximum in July, about a month after the initial rainfall peak, and decreased rapidly thereafter (Fig. 2.5). Other observations confirmed a dramatic peak in abundance of some insect groups early in the rainy season. At this time, overhead lights near park facilities often attracted many insects, including large beetles and moths, and the first rains invariably provoked mating flights of winged termites and ants.

The seasonal pattern of insect abundance depicted in Figure 2.5 is typical of Central American lowland tropical forests. Virtually all studies of arthropod populations in Neotropical forests with distinct wet and dry seasons have shown that overall insect abundance or biomass is greater during at least some portion of the wet season than during the dry season (Janzen 1973b, 1973c; Wolda 1978, 1982; Gradwohl and Greenberg 1982; Smythe 1982; Tanaka and Tanaka 1982; Levings and Windsor 1985; Pearson and Derr 1986; Boinski and Fowler 1989; Poulin et al. 1992). In most studies, arthropod abundance reaches a strong annual peak soon after the onset of rains, dropping during the latter half of the rainy season to lower, often dry season levels (Gradwohl and Greenberg 1982; Smythe 1982; Tanaka and Tanaka 1982; Levings and Windsor 1985; Pearson and Derr 1986;

Boinski and Fowler 1989; Poulin et al. 1992), and in some cases, mid to late rainy season populations appear to be the lowest of the year (Boinski and Fowler 1989). The multiple by which arthropod abundance or biomass during the wet season exceeds that during the dry season usually ranges between 2 and 8, but multiples as high as 2–12 (numbers) and 20–50 (biomass) were noted in a western Mexican dry forest (Lister and Aguayo 1992). A multiple of 2–4 seems most common (see summary in Lister and Aguayo 1992).

Different sampling methods at a given site often yield differences in seasonality estimates, because they selectively sample insects with divergent behavior, and because different insects occupy different microenvironments (Boinski and Fowler 1989). Insect taxa often diverge in seasonal demographic patterns within orders, families, and genera (Wolda 1978, 1982; Gradwohl and Greenberg 1982; Smythe 1982; Tanaka and Tanaka 1982; Levings and Windsor 1985; Boinski and Fowler 1989).

Seasonality of insect populations may be driven in large part by the appearance of new leaves. A flush of new leaves at the onset of the rainy season takes place at Tikal, as in other seasonal tropical forests (e.g., Wolda 1978, 1982).[1] New foliage—which is often relatively unprotected (chemically and structurally) and high in usable nutrients—often provides food for herbivorous insects (Feeny 1970; Rockwood 1974; Coley 1983; Coley and Barone 1996), and in many cases, seasonal peaks of arthropod populations seem clearly tied to episodes of new leafing (Janzen and Schoener 1968; Fogden 1972; Wolda 1978). Insect population peaks may be associated with each major peak of leafing (Boinski and Fowler 1989), often after a 2–3-week lag (Dunham 1978; Tanaka and Tanaka 1982). While the advent of rains after an intense dry season provides new vegetation as insect food, torrential rains may be partly responsible for the common decline in insect abundance during mid to late rainy season (Boinski and Fowler 1989).

Some insect groups, however, do not find the dry season inhospitable. Owing to abundant flowering of

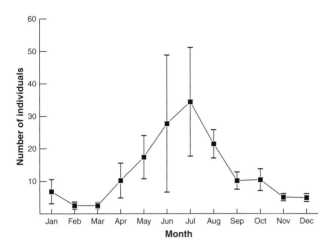

2.5 Seasonal patterns of insect abundance in Tikal National Park, as sampled via ultraviolet light.

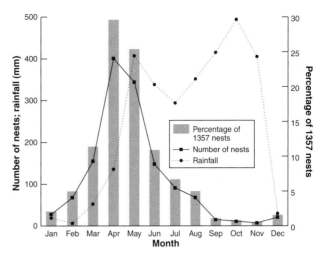

2.6 Seasonality of bird nesting in relation to rainfall at a Central American site (adapted from Skutch 1950).

woody plants during the dry season, Janzen (1973b, 1973c) believed that solitary bees peaked in abundance at that time. Moreover, with fruit and seed production greatest during the dry season, seed predators, including certain true bugs and beetles, reached peak abundance then (ibid.). Arthropods occupying dead, hanging leaves had relatively consistent abundance throughout the year (Boinski and Fowler 1989). Among 13 major groups of leaf litter arthropods, 9 increased in abundance during the rainy season, 2 increased during the dry season, and 2 showed irregular patterns not directly linked to season (Levings and Windsor 1985).[2]

For insectivorous vertebrates at Tikal, the net result of the patterns just described is a peak in prey abundance early in the rainy season, including many of the large insects that commonly serve as prey of certain raptors. This seasonal peak in arthropod abundance may affect the timing of raptor reproduction directly in the case of raptors that feed on arthropods, or indirectly by affecting the seasonal patterns of growth, reproduction, and activity of the many insectivorous vertebrates that are in turn consumed by raptors: lizards, frogs, snakes, many birds and bats, and perhaps rodents.

Birds. Skutch (1950) reviewed the nesting seasons of Central American birds. Adapted from his paper, Figure 2.6 shows that most small land birds in this region nest at the onset of the rainy season. Egg laying, peaking in April and May, slightly anticipates the initial peak of the rainy season, such that hatching and chick rearing take place during this initial rainfall peak. Our observations at Tikal agreed; most small birds here nested mainly at the outset of the rainy season. This resulted in a pulse of availability of nestling and fledgling birds in early to mid rainy season for those raptors taking such prey.

Reptiles and amphibians. In the Maya Forest, essentially all the anurans (frogs and toads) breed early in the rainy season, when surface water becomes common after months of drought (Appendix 2.1). At this time, the din of frog calls around water bodies can be deafening. Duration of egg and tadpole stages in anurans is brief, and species breeding early in the rainy season gives rise to froglets in a matter of weeks. For most snake and lizard species in our study area, hatching or birth also takes place mainly during the rainy season (Appendix 2.1). Since gestation and incubation in reptiles often require 2–3 months or longer (Zug 1993), to achieve such timing, egg laying occurs early in the rainy season or late in the preceding dry season (Appendix 2.1).[3]

Due to seasonal reproduction and growth of young, reptile and amphibian populations typically show an annual cycle of population size, age structure, and body size. Three Anolis species studied in Panama (Sexton et al. 1971) appear representative of many members of this genus (e.g., Andrews and Rand 1982; Andrews 1983; Campbell et al. 1989). Since most reproduction occurred during the rainy season, early in this season populations were composed mainly of large adults and tiny hatchlings. Through the course of the wet season, large individuals were progressively lost from the population, and hatchlings grew in size and number; by late rainy season, fewer hatchlings appeared and many young lizards were of medium size (Plate 2.1). During the dry season there was little or no addition of new hatchlings, and young lizards continued to grow (Plate 2.2) (Sexton et al. 1971).[4]

With regard to production of new raptor prey, timing of breeding in reptiles and amphibians has different implications than that in mammals and birds. Since birds, bats, and most rodents are nearly adult size when they become independent, the breeding season produces a pulse of new raptor prey. In contrast, most neonatal reptiles and amphibians are tiny and may require months or years to reach adult size. As newborns, these species are generally too small to serve as prey for most raptors.

Hence, the breeding season in amphibians and reptiles is not likely to produce an immediate pulse of raptor prey.

Activity patterns, however, probably make many reptiles and amphibians more available to raptors during the rainy season. These creatures are often so cryptic while motionless that activity must greatly enhance their detectability by raptors. Frogs and toads are clearly most active during the rainy season, and tropical lizards and snakes may also reach peak activity then (Duellman 1965; D. Whitacre, pers. observ.). In fact, it is possible that seasonal differences in activity levels of reptiles and amphibians, as well as annual rhythms of age structure and body size, may influence their availability to raptors more strongly than do seasonal changes in abundance resulting from reproductive pulses.

One study of seasonal patterns of snake activity should be especially applicable to Tikal, as it was conducted in nearby Belize. In this study, snake detections were rare during the dry season and far more frequent during the wet season, peaking in September, the wettest month. The authors concluded that most snakes in this area became inactive during the February–May dry season, possibly due to direct effects of dryness on snake activity and to very low levels of frog activity (important snake prey) during the dry season (Henderson and Hoevers 1977).

In addition to foraging, important causes of movement in snakes are mate searching by males and movements to egg-laying or birthing sites by females (e.g., Madsen 1984; Gregory et al. 1987). Males of some species are more active and make greater movements during mate searching than at other times (Madsen 1984; Gibbons and Semlitsch 1987). In one tropical Australian snake, high encounter rates with males during the mating season apparently resulted from their mate-searching activities (Brown et al. 2002). Gravid females of many snake species are relatively sedentary (Gibbons and Semlitsch 1987), but they may make large movements when moving to or searching for an egg-laying or birthing site (Madsen 1984; Gibbons and Semlitsch 1987). In a number of snakes, females have been found to aggregate in communal egg-laying sites: such females are relatively active in reaching these sites, but rather inactive once there (Gibbons and Semlitsch 1987; Gregory et al. 1987).

Diurnal or nocturnal activity patterns are often shown by snakes, depending partly on temperature. Seasonal shifts in activity time have been documented for many species, while others are consistently nocturnal or diurnal (Gibbons and Semlitsch 1987). It is common knowledge that activity of tropical forest frogs and toads is greatest during and immediately after a rain, especially the first downpours of the rainy season (D. Whitacre, pers. observ.). Activity of snakes and lizards may also be greater during periods of high ambient humidity or during or after a rain. In a Malaya Pitviper (*Calloselasma rhodostoma*), frequency and length of movements were strongly and positively correlated with ambient humidity, but not with rainfall or temperature (Daltry et al. 1998). In a species of Anolis, drought led to reduced ac-

tivity, and rainfall or artificial sprinkling restored activity to pre-drought levels (Stamps 1976). In another Anolis species, activity levels during the rainy season were 2–10 times higher and feeding rates 5–10 times higher than during the dry season (Lister and Aguayo 1992).

While for many frogs and toads standing water is important in providing breeding opportunities, for reptiles that feed on arthropods (mainly lizards but also some snakes), breeding during the rainy season may be advantageous because many insect populations reach annual peaks then. In turn, since many snakes prey largely on anurans or lizards, reproductive seasonality of the latter may affect the timing of snake feeding and reproduction.

Mammals. The limited information available suggests that small mammals may also tend to reproduce late in the dry and early in the rainy season at Tikal. In Desmarest's Spiny Pocket Mouse (*Heteromys desmarestianus*), most reproduction at Tikal took place during the dry season, with juveniles frequently encountered at the onset of the rainy season (Jolón 1997). In Gaumer's Spiny Pocket Mouse (*H. gaumeri*), reproduction began at the end of the dry season and continued during the rainy season (ibid.). In the Big-eared Climbing Rat (Plate 2.3: *Ototylomys phyllotis*), some reproduction was observed throughout the year, with markedly less during the dry season. All of these rodents were preyed on by raptors at Tikal. In western Mexico, population densities of the Painted Spiny Pocket Mouse (*Liomys pictus*) increased 35-fold during the first two months of the rainy season (Ceballos 1995).

Bats, which also were important raptor prey at Tikal, often have reproductive timing keyed to seasonal rainfall patterns elsewhere in the tropics (Heideman 1995; Porter and Wilkinson 2001), and likely also at Tikal. Insectivorous bats in Costa Rica generally produce a single litter yearly, timed such that young are weaned during the period of maximum prey availability, usually early in the rainy season: a few species raise up to three young in succession during the rainy season (Janzen and Wilson 1983). Frugivorous and nectarivorous bats in the Neotropics often have two litters yearly, weaning the first young at the onset of rains and a second later in the rainy season (ibid.). In the montane cloud forests of Monteverde, Costa Rica, two annual peaks of fruiting occur: one in the mid to late dry season (March–May) and the second late in the wet season (September–October; Dinerstein 1986). Two species of fruit bats here showed two seasonal peaks in reproduction, often involving the same females and corresponding closely with these two peaks in fruit production (ibid.).

Migrant birds. Another conspicuous seasonal phenomenon in the Yucatán Peninsula is the influx each year of millions of songbirds that breed in the United States and Canada and winter in this region. Species that winter abundantly at Tikal include several warblers and a few species each of vireos, flycatchers, thrushes, tanagers, mimids, and others. Densities of many of these species are high, and during a winter bird walk, easily half the individual birds seen may be wintering migrants. Several species hold winter territories, with individuals often returning to the same or a nearby territory in subsequent years (D. Whitacre, unpubl. data). Kentucky Warblers *(Oporornis formosus)* wintering in mature Upland forest at Tikal averaged one bird per hectare (Madrid et al. 1995); hence Tikal's 576 km^2 may provide winter quarters for more than 50,000 Kentucky Warblers. For at least some migrant species, most wintering individuals are present at Tikal by the end of October, and the vast majority leave the area by the time resident songbirds begin to nest in the late dry and early rainy season. Many migrants, however, are still present early in the breeding cycle of some bird-eating raptors at Tikal, including the Bat Falcon *(Falco rufigularis)* and Bicolored Hawk *(Accipiter bicolor)*.

Summary

In sum, many faunal groups in the Maya Forest show dramatic seasonal peaks in reproduction. Collectively, many of these reproductive peaks occur from just prior to the onset of rains until some point well into the rainy season. In addition, heightened activity of some faunal groups during the rainy season may make them more available to raptors, regardless of seasonal changes in population size. There appears to be no other time of year showing a comparable spike in reproduction across many types of organisms. Thus, the rainy season—especially the early part—is the only time of year boasting a simultaneous burst of reproduction (and activity, in many cases) by many kinds of organisms. One major departure from this pattern is that millions of migrant songbirds leave the area shortly before the rainy season. However, this exodus of migrants may be compensated, in part, by production of nestlings in the resident avifauna.

VEGETATION OF TIKAL AND THE MAYA FOREST

Except in some extensive areas of wetlands and natural savanna, and where human activities have cleared or modified the forest cover, tropical forest has covered most of the Yucatán Peninsula during modern times. Over the length of the peninsula, the forest varies greatly in structure and species composition, in concert with a rainfall gradient from wettest in the southeast to quite dry in the northwest corner (see Figs. 2.1, 2.4). Along this gradient, forest decreases in height from a mean of approximately 20 m on well-drained sites at Tikal to less than 10 m in the extreme northwest near Mérida, Yucatán, and the degree of deciduousness increases in parallel (see Fig. 2.1). Some tree species that are rare in Tikal are important components of the wetter forests to the west or drier forests to the north.

The forests of northern Petén have been called many things, including rain forest (Leopold 1950; Paynter 1955), quasi-rain forest (Lundell 1937), and tall semi-evergreen tropical forest (Pennington and Sarukhan 1968; López Ornat et al. 1989). Holdridge (1967) classified the area from Lake Petén Itzá southward as belonging to the subtropical moist forest life zone, with the subtropical dry forest life zone extending northward, including Tikal (Plates 2.4 and 2.5). Bullock et al. (1995) regarded these as tropical dry forests, which they define as forests in tropical regions that experience several months of severe drought; within this category they include forests exhibiting a great range of canopy heights and deciduousness. While the forests of Tikal are clearly too dry and seasonal to be classed as rain forests, use of the term "dry forest" is unfortunate, as is grouping these forests along with the much drier, largely deciduous forests of the Pacific coasts of Costa Rica and Mexico. We prefer Hatshorn's (1980) interpretation, which regards the forests of northern Petén as lowland humid forests, corresponding to Holdridge's tropical or subtropical moist forest life zones. On balance, the descriptors "medium to tall," "humid" or "subhumid," "semi-evergreen," "seasonal," and "tropical" apply well to forests on well-drained sites at Tikal and throughout northern Petén.

Mature forest at Tikal varies strongly with topographic position and associated soil and drainage characteristics. Topography in Tikal National Park is mostly gently rolling, 160–400 m above sea level (Fig. 2.7, Plate 2.6). Except for a large karst area in the northwest corner of the park, where slopes often exceed 25°, hills at Tikal are gentle, often standing only 30 m above adjacent basins. In contrast, vegetation may vary drastically over short distances. Tall-canopied, open-understoried, palm-rich forest can be found within a few hundred meters of low, nearly impenetrable, xerophytic Scrub Swamp Forest. Tikal's forests were described by Schulze and Whitacre (1999); here we give a thumbnail sketch based on that account. Throughout the book, names of forest "types" follow Schulze and Whitacre (1999), and hence are capitalized.

Soils at Tikal are limestone-derived clays, but they vary in texture and depth from the upper regions of slopes, referred to hereafter as "Upland" sites, to low-lying depressions known locally as bajos. On Upland sites, soils are shallow, rocky, well-drained, high in pH and organic matter, and have only modest clay content, while in adjacent bajos, soils are often more than 1.4 m deep, free of rocks, very rich in clay, and low in pH and organic matter. Drainage is good on Upland sites, whereas bajos may be inundated for weeks at a time during the rainy season, with some areas subject to strong currents of flowing water. The high clay content and visible soil cracking suggest that edaphic drought affects these low-lying sites during the dry season.

Mature forest at Tikal varies strongly along this modest topographic gradient, both in structure and in tree species composition (Figs. 2.8, 2.9). As one descends from a hilltop, mean canopy height remains at 20–21 m

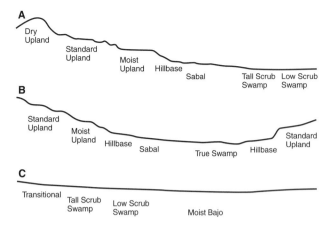

2.7 An idealized scheme of the topographic forest type gradient in Tikal National Park.

through four types of Upland Forest, decreases to 18 m in Hillbase and Sabal Forest (both occurring near the foot of hills), to 15 m in Transitional Forest, and thence to 13 and 10 m in Tall and Low Scrub Swamp Forest, respectively (Fig. 2.8). This series of forest types completes the topographic sequence in most areas; the Mesic Bajo Forest represented in Figures 2.8 and 2.9 is quite local in extent. Canopy opening and light penetration vary in mirror image to canopy height (Fig. 2.8). Tall, Upland Forests have relatively closed canopy and open understory, while the Low Scrub Swamp and Bajo Forest types have more open canopy and extremely dense understory; in many cases it is impossible to move through such forest off-trail without liberal use of a machete.

The environment for plants along the topographic gradient can be summarized as follows. Seasonal moisture stress is greatest on the rocky hilltops and in low-lying bajos, with an alternating inundation and drought cycle in the latter subjecting plants to edaphic extremes. Soil fertility is highest in Upland areas and at the bases of slopes. Dense understory shade in Upland and Hillbase Forests of the middle of the topographic gradient limits establishment opportunities for light-demanding seedlings to treefall gaps, whereas ample light penetrates Scrub Swamp and hilltop forest canopies. Soil instability, high vine densities, and low canopies result in more frequent treefall gaps in forests of low-lying areas and greater light penetration into gaps of a given size than in Upland forests.

Many forest attributes vary along this topographic-forest type continuum. In Dry Upland Forest, the understory is dominated by a single shrub species, *Piper psilorachis*, while in the more Mesic Upland Forest types and in Hillbase and Sabal Forests, this shrub decreases greatly in prominence while the Rootspine Palm *(Cryosophila stauracantha)* becomes a conspicuous element of the understory, along with juveniles of the Bay Palmetto *(Sabal mauritiiformis)* and many tree saplings. While grasses and sedges are rare beneath

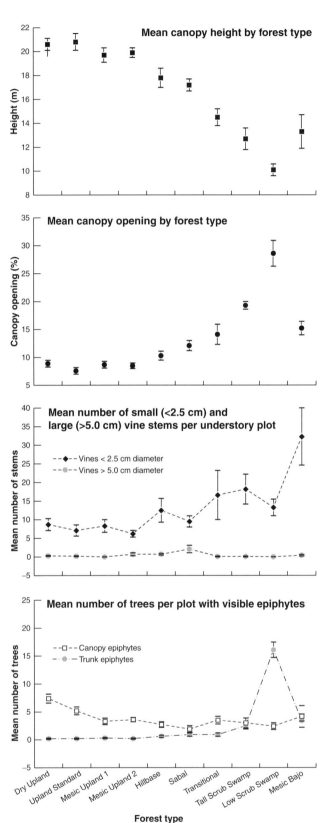

2.8 Structural characteristics of 10 forest types occurring along the topographic gradient in Tikal National Park.

the taller canopies, in Low Scrub Swamp Forest these often dominate the ground cover. One factor that appeared important to raptors at Tikal was the presence of canopy-emergent trees, which were often used as nest trees. While emergent trees can be found nearly anywhere in the forest, they are less frequent in the tall Upland Forest types, where the canopy's upper surface is relatively even, and are most common in Hillbase and Sabal Forests and Scrub Swamp Forest. Epiphytes are particularly abundant in Low Scrub Swamp, where trunks and limbs are often spangled with tank bromeliads (Fig. 2.8). Large vines (> 5 cm in diameter) are most frequent in Sabal Forest (Fig. 2.8), while small vines (< 2.5 cm) are often exceedingly abundant in the low-lying forest types (Hillbase through Low Scrub Swamp Forest; Fig. 2.8), contributing greatly to the dense understory of these forest types.

Virtually all tree species at Tikal showed pronounced differences in abundance along the gradient from well-drained to low-lying sites; Figure 2.9 depicts a few examples. We observed 185 tree species at Tikal and estimate that at least 200 occur there (Schulze and Whitacre 1999). Tree species diversity did not decrease from Upland Forest to seasonally flooded areas, despite the harsher environmental extremes of the latter. Ten-acre (0.041 ha) sample plots generally had 50–60 tree species in all forest types by the time 15 such plots were sampled. However, each portion of the topographic gradient was characterized by a suite of species that achieved their greatest abundance under those edaphic and micro-environmental conditions and collectively dominated the vegetation, numerically and structurally. Plates 2.4 and 2.5 depict a range of forest types at Tikal.

The bird community differs markedly along this topographic-forest type continuum. Many birds that are common in tall Upland Forest largely shun the Low Scrub Swamp Forest, while a few are Scrub Swamp specialists, sometimes using young second growth as well (Whitacre et al. 1995b). Species composition of the butterfly community also differs along this Upland to Bajo gradient (Austin et al. 1996). We suspect that with sufficient study, most groups of organisms would show differences in species occurrence along this gradient. Hence topography leads to an important degree of beta diversity, or turnover in species composition of localized areas within the forest.

While this description has emphasized the predictability of forest composition and structure, it is often impressive how much structural variation may be noted even over a span of 50 paces. Forests are in fact mosaics of patches with differing disturbance histories. Canopy height varies continually, dipping to ground level in recent treefall gaps, and density and size of vines vary greatly, largely reflecting the local history of treefalls. Where dense vine tangles exist, at any height from ground level to the upper canopy, these often produce a dense mass of foliage, beneath which may be dense tangles of leafless vine stems and dark, secluded nooks devoid of foliage. Different disturbance histo-

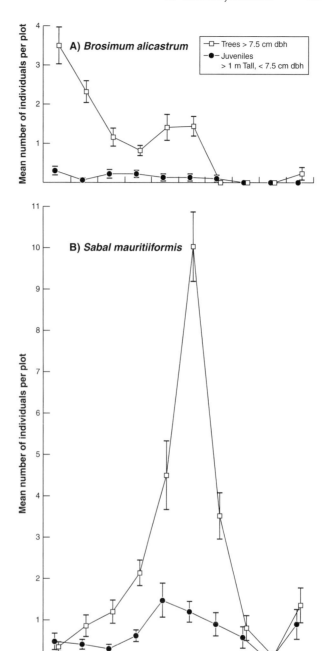

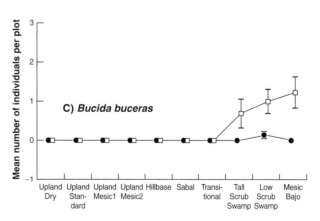

2.9 Examples of patterns of tree species abundance along the topographic/forest type gradient in Tikal National Park.

ries can also lead to variation in species composition from one site to the next within the same topographic situation or forest type. For example, Upland Forest stands where the canopy is dominated by large tree species with light-demanding seedlings but long life spans—sometimes termed long-lived pioneers or late-successional species (typified by Honduras Mahogany [*Swietenia macrophylla*], Spanish Cedar [*Cedrela odorata*], and Shaving Brush Tree [*Pseudobombax ellipticum*])—contrast with the more typical Upland Forest dominated by shade-tolerant species, and likely reflect a large-scale disturbance (e.g., blowdown) at some point in the distant past.

Spending much time in the treetops, we had the opportunity to study the canopy architecture, and a couple of points bear emphasis. First, canopy architecture varies greatly, from a relatively dense, level upper surface (though never truly continuous or level when viewed from above), to an extremely varied upper surface, with tall emergents, treefall gaps extending to the ground, and every height variant between. While some canopy trees are densely cloaked with thick evergreen leaves and are essentially visually opaque, other trees have small leaves and open architecture within which one can visually scan a great many surfaces. The result is a tremendous variety of vegetation volumes, surfaces, and arrangements, and one can envision the potential for partitioning among different raptor species of those portions of the vegetation surface toward which they devote most attention while foraging. We also noted that when looking steeply downward, even from the top of a canopy emergent, it is often possible to see portions of the forest floor. This suggests that a raptor perched in a treetop in continuous forest might be scanning any portion of the forest, from the airspace above to the ground below—an unexpected result.

FAUNA OF THE MAYA FOREST

A large part of our study revolved around who ate whom, and how they did so. Hence it is worth characterizing the fauna that comprises the prey base for Tikal's raptors, as well as potential food competitors and predators on raptors, their eggs, and young.

Birds

Since birds of the Maya Forest region are well known, we do not list them here. Beavers (1992) lists 404 bird species for the Petén, with at least 352 reported from Tikal. Of the Petén total, Beavers describes 250 species as residents, 6 as summer residents, 88 as winter residents or transients, and 60 as visitors. Applying these same ratios to the total for Tikal, about 300 bird species are found with some regularity in the park, though the vast majority of individuals are contributed by a modest subset of these species. Though distinctly tropical in composition, the bird community shows low diversity of some families that become much more diverse nearer the equator, for example, the antbirds and ovenbirds.

Mammals

The mammal fauna of Tikal is summarized in Appendix 2.2. Five species of opossums (all nocturnal and largely arboreal) occur here, compared to one in temperate North America. Among the Xenarthra, two anteaters range north to Tikal, but sloths do not. The Northern Tamandua *(Tamandua mexicana)*, an anteater, is not especially common at Tikal, and the diminutive Silky Anteater *(Cyclopes didactylus)* is uncommon or at least rarely seen. The Nine-banded Armadillo *(Dasypus novemcinctus)* is also present but rarely observed.

At least 39 bat species are known from Tikal (T. McCarthy, pers. comm.), and additional species probably occur there. About 69 bat species are expected in Guatemala's Caribbean lowlands (T. McCarthy, pers. comm.). These numbers reflect the generally greater bat diversity of tropical than temperate zone communities. Three bats at Tikal prey largely on small vertebrates, and hence may overlap in diet with some raptors: the Fringe-lipped Bat *(Trachops cirrhosus,* 28–45 g), feeding largely on frogs and lizards with some insects and small mammals; the Woolly False Vampire Bat *(Chrotopterus auritus,* 61–94 g), taking small mammals, birds, lizards, and large insects (prey up to 70 g but more often 10–35 g); and the False Vampire Bat *(Vampyrum spectrum).* With a weight of 125–190 g and wingspan up to 90 cm, the False Vampire Bat is the largest bat in the New World and feeds on birds, bats, and rodents. Birds taken are mostly 20–150 g, including cuckoos, anis, parakeets, trogons, and orioles (Emmons and Feer 1990).

A striking difference between the Maya Forest and forests closer to the equator is the low diversity of primates. Only two—the Black-handed Spider Monkey *(Ateles geoffroyi)* and the Guatemalan Howler Monkey *(Alouatta pigra)*—occur here, and adults of both are too large for most raptors to prey on. The only canid at Tikal is a diminutive subspecies of Gray Fox *(Urocyon cinereoargenteus)* endemic to the Yucatán Peninsula and abundant at Tikal. The area boasts four procyonids: the Cacomistle *(Bassariscus sumichrasti)*, which appears uncommon and local; the Raccoon *(Procyon lotor)*, not uncommon but mostly found near water; the White-nosed Coati *(Nasua narica)*, which is abundant at Tikal; and the Kinkajou *(Potos flavus)*, also common. Owing to their nocturnal habits, all but the diurnal coati are seldom seen.

Among the mustelids, most often seen is the diurnal and partly arboreal Tayra *(Eira barbara)*, a pugnacious animal about the size and shape of a Fisher *(Martes pennanti)*. We observed Tayras depredating nests of raptors and other birds, and we believe this mustelid to be a major nest predator. Also present are the Striped Hog-nosed Skunk *(Conepatus semistriatus)* and Long-tailed Weasel *(Mustela frenata)*, though neither is common. Tikal is home to five of the six widespread forest cats of Neotropical lowlands. From smallest to largest, these are the Margay (*Leopardus wiedii:* Plate 2.7), Jaguarundi *(Felis [Herpailurus] yagouarundi)*, Ocelot *(Felis pardalis)*, Puma *(Felis concolor)*, and Jaguar *(Panthera onca)*. All except the Jaguarundi seem reasonably common in the park. The area is home to six ungulates: Baird's Tapir *(Tapirus bairdii)*, Collared Peccary *(Tayassu tajacu)*, White-lipped Peccary *(T. pecari)*, White-tailed Deer *(Odocoileus virginianus)*, Red Brocket Deer *(Mazama americana)*, and the Yucatán Brown Brocket Deer *(Mazama pandora)*.

Two tree squirrels are common at Tikal. Deppe's Squirrel *(Sciurus deppei)*, averaging about 200 g, is half the size of the Yucatán Squirrel *(S. yucatanensis)* (mean = 420 g). The two are readily distinguished by pelage coloration and size, and we were often able to identify squirrels to species when they were brought to nests by raptors. Diversity of small rodents is reasonably high. Two Spiny Pocket Mice (Heteromyidae) occur here and are among the most common forest-floor rodents. Among the several murid rodent genera occurring here, only the Harvest Mouse *(Reithrodontomys gracilis)*, the Hispid Cotton Rat *(Sigmodon hispidus)*, and the rice rats *(Oryzomys* spp.) are shared with temperate North America (along with the human commensals *Rattus* and *Mus)*. A number of highly arboreal rodents occur at Tikal, including the Naked-tailed Climbing Rat *(Tylomys nudicaudus)*, Big-eared Climbing Rat *(Ototylomys phyllotis)*, Vesper Rat *(Nyctomys sumichrasti)*, and Yucatán Vesper Rat *(Otonyctomys hatti)*.

Three large rodents are present, all belonging to Caviomorpha, a suborder with South American affinities: the nocturnal and arboreal, prehensile-tailed Mexican Hairy Porcupine *(Coendou mexicanus)*, the Paca or Tepescuintle *(Agouti paca)*, and the Central American Agouti *(Dasyprocta punctata)*. While the nocturnal Paca, prized as a delicacy, is heavily hunted by people, the smaller, diurnal agouti is rarely hunted. Both live as territorial pairs on the forest floor, though are most often seen alone. The Brazilian Rabbit *(Sylvilagus brasiliensis)*, a typical cottontail, is uncommon and local at Tikal.

Two studies have characterized the small mammal community in various habitats in and near Tikal, and have evaluated vertical levels within the forest used by different small mammal species. These results are summarized in an endnote.[5] As in many ecosystems, most rodents at Tikal are nocturnal, conspicuous exceptions being the two squirrels, both abundant and diurnal; the agouti, which was rarely taken by raptors; and the Hispid Cotton Rat, which is partly active during the day. From the standpoint of potential raptor prey, probably the most notable difference between Tikal's mammal fauna and that of the north temperate and boreal zones is the absence of microtine rodents (voles and lemmings). While microtines often comprise a large fraction of the diets of many raptors in north temperate and boreal zones, they are practically absent in the New World south of Oaxaca, Mexico, with only one species restricted to the highlands of Guatemala (Hall and Kelson 1959). In addition to hav-

ing astonishing fecundity and often producing huge population irruptions, many microtines are active by day as well as night—both factors increasing the importance of this group in the diets of many northern raptors. In contrast, most rodents at Tikal have low fecundity and are mainly nocturnal; only the Hispid Cotton Rat resembles the microtines in ecology and behavior. Most of Tikal's rodents typically have litters of 1–4, with mean litter size often around 2 (G. Roling, pers. comm.). In Desmarest's Spiny Pocket Mouse at Tikal, females may give birth to five litters per year, averaging three pups per litter (G. Roling, pers. comm.). The Hispid Cotton Rat, however, has litters of up to 15 (commonly 5–7) and is prone to population irruptions (G. Roling, pers. comm.; Nowak and Paradiso 1983), in these features resembling voles and lemmings.

The great importance of microtines to raptors throughout the Holarctic realm, coupled with the practical absence of equally prolific, diurnal rodents at Tikal, suggests that diurnal raptors at Tikal have relatively low access to rodents compared to raptors studied in North America and Europe. In inter-regional comparisons, this biogeographic pattern may influence the relative importance of rodents in raptor diets. This, together with the much higher diversity of bats in the tropics compared to temperate and boreal regions, may give rise to regional dietary differences among raptor communities. Mammalogists, who are largely mouse trappers in the north temperate zone, often become bat netters in the tropics (Janzen and Wilson 1983); one might expect some raptors to do the same.

Reptiles and Amphibians

We estimate that 21 amphibian and 67 reptile species may serve as raptor prey at Tikal (see Appendix 2.1).[6] Some are uncommon, however, and several are tiny or fossorial, probably rarely finding their way into raptor diets. Among reptile and amphibian species, four axes of variation might largely determine their availability to different raptors: activity time, height within the forest, body size, and foraging strategy (i.e., movement pattern).

Activity time. The diurnal habit is significantly more common among the lizards (20 of 24 species) than among the snakes (18 of 43 species) and frogs at Tikal (6 of 20 species; Table 2.2).[7] The only nocturnal lizards are two tiny geckos (too small to serve often as raptor prey) and two medium-small geckos. Among the 20 anurans (frogs and toads), 14 are nocturnal and 6 are active day and night. All nine arboreal frogs are mainly nocturnal, as are the two true toads (terrestrial), the fossorial Mexican Burrowing Toad *(Rhinophrynus dorsalis)* and the two fossorial, litter-inhabiting Narrow-mouthed Toads (Microhylidae). The only day-active frogs (also active at night) are two leptodactylids and two ranids associated with water bodies, and two terrestrial *Eleutherodactylus* species that are uncommon forest-floor dwellers. Most snakes at Tikal (25 of 43 species) are believed to be mainly nocturnal, 7 nocturnal-diurnal, and 11 mainly diurnal.

Nocturnal raptors might thus have at their disposal about 14 species of night-active anurans, most of them tree frogs (Table 2.2). For diurnal raptors, access to active anurans appears more limited. Except along water bodies, any diurnal raptor taking anurans at Tikal must largely find them hidden in daytime retreats such as bromeliads. However, when frogs are aggregated at breeding sites or moving about during or after a rain, many species might be more accessible than usual to both nocturnal and diurnal raptors, and we observed cases in which raptors took unusual numbers of frogs on rainy days.

Diurnal raptors at Tikal should have access to 6 frog species, 20 lizard species, and 18 snake species active during the day. Nocturnal raptors should have access to 20 frog species, 4 lizard species, and about 32 snake species active at night.[8] In sum, more anuran species should be available to nocturnal than to diurnal raptors, unless the latter often find nocturnal frogs in their daytime retreats. In contrast, lizards should be much more available to diurnal than nocturnal raptors, unless nocturnal raptors often find sleeping lizards. A diversity of snake species is available to both diurnal and nocturnal raptors. However, snakes may be more dangerous to catch at night if distinguishing venomous from nonvenomous species is important to raptors.

Activity level within the forest. Among the 20 diurnal lizards, 11 are mainly arboreal, 2 terrestrial-arboreal, and 7 mainly terrestrial. Most snakes were deemed nocturnal and terrestrial (Table 2.2). Of the 18 day-active snakes, 11 are mainly terrestrial, 2 terrestrial-arboreal, and 4 mainly arboreal. Thus, among Tikal's diurnal reptiles, the arboreal habit is more common among lizards than among snakes. Both groups, however, include a diversity of arboreal and terrestrial species, and should thus be available to raptors hunting at different levels in the forest.

Most frogs, in contrast, are both nocturnal and arboreal, and the few diurnal frogs are terrestrial and partly associated with water bodies. Diurnal raptors, then, may take mainly the few terrestrial frogs, except for those raptors that find tree frogs in their daytime hiding places. Nocturnal raptors may be able to take active frogs from more forest strata than do most diurnal raptors.

Body size. Most anurans and lizards at Tikal are small and presumably most attractive to small raptors. A few medium-sized lizards also occur, but no truly large ones. Snakes at Tikal range from tiny to very large and thus may be taken by raptors of all sizes. The great diversity of snake sizes and habits would seem to open the door to some partitioning of snake prey among raptor species.

Foraging strategy and movement pattern. While most anurans are stationary much of the time, lizards and snakes range in foraging style from "sit-and-wait" predators to active-search foragers that move a great deal while foraging. These distinct movement patterns may render different lizard and snake species differentially vulnerable to raptors in general, and possibly to different raptor species. Some snakes may remain in one spot for several days before moving a few meters to a new ambush spot (Greene and Santana 1983; Daltry et al. 1998). This foraging strategy presumably makes such species less detectable by raptors than are snakes that search actively for prey. At Tikal, sit-and-wait predators include the pitvipers, while active searchers include species of *Leptophis, Oxybelis, Drymobius,* and many other genera, both nocturnal and diurnal.

While most diurnal snakes at Tikal might be deemed active searchers, among these there is a span of behaviors that might affect their vulnerability to raptors. Four diurnal, arboreal snakes (two species of *Leptophis* and two of *Oxybelis*) have obvious adaptations for concealment, in terms of form, color, and locomotory behavior. The latter features very slow motion and movements that mimic vegetation swaying in a breeze (D. Whitacre, pers. observ.). These especially cryptic species might be difficult for many raptors to detect and may be most vulnerable to raptors most specialized on ophidian prey.

Likewise, lizards at Tikal include sit-and-wait predators such as species of *Corytophanes* and *Anolis,* and active searchers such as *Ameiva* species, which are relentlessly mobile, terrestrial foragers. Some other lizards also move around a good deal, though not as conspicuously as *Ameiva*; these include some *Sceloporus* species and skinks (Scincidae). Among the diurnal, arboreal lizards are some marvelously cryptic forms that spend much time motionless, presumably difficult for raptors to detect. The two *Corytophanes* species are most exemplary of this category (Davis 1953; Andrews 1979), but *Laemanctus* may also qualify (Lee 1996), as well as several *Anolis* species.

PEOPLE AND THE FOREST

Among the fascinating aspects of the Maya Forest are the social and cultural traditions of its human occupants, described by Schwartz (1990). Here we briefly describe some of the ways in which local people—Peteneros—use the forest.

Most of Guatemala's population—including much of the large indigenous population, with more than 20, mostly Mayan, languages—is concentrated in the volcanic highlands that form a portion of the mountainous backbone of Central America. Petén, a massive lowland area, has long been Guatemala's hinterland—a sparsely populated forest frontier barely connected economically or politically to the rest of the nation. Until a gravel road was completed in about 1970, the northern Petén was not connected by road to the rest of the country. Prior to that time, several small villages in the north-

Table 2.2 A summary of activity height and time for reptiles and amphibians occurring at Tikal

	Mainly diurnal	Diurnal and nocturnal	Mainly nocturnal	Total
Frogs and Toads				
Arboreal	—	—	9	9
Terrestrial-aquatic	—	4	—	4
Terrestrial	—	2	2	4
Fossorial-litter	—	—	3	3
Total	0	6	14	20
Lizards				
Arboreal	11	—	1	12
Arboreal-terrestrial	2	2	—	4
Terrestrial	7	—	1	8
Total	20	2	2	24
Snakes				
Arboreal	4	—	4	8
Terrestrial-arboreal	2	1	3	6
Terrestrial (including fossorial)	5	6	17	28
Terrestrial-aquatic	—	—	1	1
Total	11	7	25	43
Grand totals	31	15	41	87[a]

[a] Plus 1 salamander, arboreal and terrestrial, hiding in bromeliads and beneath surface debris, active on surface probably mainly during rainy season.

ern Petén were serviced by scheduled flights of DC-3s that used grassy runways carved from the forest. In 1941 there were five automobiles and 11,000 people in the entire Petén, an area of 36,000 km²—twice the size of New Jersey (Schwartz 1990). This low population density has recently changed, but vast forests and a tiny human population have characterized the historical development of cultural, social, and economic conditions in the Petén.

Two industries were long the mainstays of the small human population of the Petén—selective logging, mainly of mahogany, and extraction of chicle. Chicle, the latex of the Chicozapote tree *(Manilkara zapota)*, was once the primary elastic ingredient of chewing gum, and has other uses as well. Chicle export often brings several million dollars annually into the Guatemalan economy (Heinzman and Reining 1990; Reining et al. 1992; Dugelby 1995, 1998). Using climbing spikes and a flip-rope, the chiclero ascends the tree, and using a machete, chops a herringbone pattern of cuts down to the tree's cambium (Plate 2.8). A collecting bag is affixed at the bottom to collect the flowing latex. Organized into camps of a couple-dozen men, with costs often fronted by camp organizers, a few thousand chicleros fan out over the northern Petén and adjacent portions of Belize and Mexico each rainy season for the chicle harvest. The chicle industry dominated the economy of the Petén from the 1890s until the 1970s and helped give rise to a distinctive culture and regional pride still evident among Peteneros today (Schwartz 1990; Heinzman and Reining 1990; Reining et al. 1992; Dugelby 1995, 1998).

In recent years demand for chicle has been low, and the contribution of this forest-reliant industry to the economy has declined. Among the surest signs of rapid social change now underway in the Petén is this: while once it was unthinkable to cut down a Chicozapote tree, this terrifically dense wood is now widely used as firewood (most rural families cook over wood fires) and is rapidly being removed from the increasingly deforested agricultural frontier. Cutting this wood and selling it along the highway is one of the few ways in which country folk—mostly subsistence farmers barely participating in the wage economy—can generate cash for necessities such as clothing and medicine.

More recently, two additional non-timber forest product industries have emerged. The larger is the trade in leaves of xate (pronounced *shah-tay*) palms, including at least three species of small understory palms of the genus *Chaemadorea* (Plate 2.9). Exported to the United States and Europe, these leaves are used in floral arrangements. Important at least since the 1960s, this trade amounts to hundreds of millions of palm leaves yearly and generates some $3.5 million in annual foreign trade for Guatemala, providing partial employment for some 4000–8000 people yearly (Heinzman and Reining 1990; Reining et al. 1992). In comparison to chicle harvest, the xate trade demands less skill and equipment. Partly for this reason, more people, including young boys and occasionally women, participate as xateros, or xate collectors. Typically equipped with a short knife, a machete, and water bottle, a xatero walks the forest, covering many miles daily, pausing only long enough to collect one or more leaves from each eligible plant. In the more inaccessible regions, xateros are hired and financed by organizers, or contratistas, much as in the chicle industry. In a good day a xatero might make \$10–\$12 (Heinzman and Reining 1990; Reining et al. 1992). Another important forest product industry is collection of allspice. Known locally as Pimienta Gorda, the tree *Pimenta dioica* yields its small fruits to forest collectors, who generally collect them by lopping off smaller limbs bearing fruit.

An important amount of hunting also takes place in the Maya Forest, especially in the more traditional, remote forest villages. The hunted species most important in the village of Uaxactún include seven mammals ranging in size from the Central American Agouti to the White-tailed Deer, and three large birds—the Great Curassow *(Crax rubra)*, Crested Guan *(Penelope purpurascens)*, and Ocellated Turkey *(Meleagris ocellata)*—with at least 34,000 kg of these species taken by Uaxactún's hunters from August 1992 through November 1994 (Roling 1995). This may be regarded as subsistence hunting, although meat is frequently sold within local areas, mostly within a village. In addition, some meat is sold to restaurants and other buyers outside the forest, especially in the town of Flores, the commercial hub of the northern Petén.

The three plant product industries just described, along with subsistence hunting, in large part form the sustenance of villagers in the northern Petén who still exemplify the traditional forest society that has developed here over the past century or more. The visitor today can witness this forest society intact in certain villages, notably Uaxactún (just north of Tikal) and Carmelita, farther west. In these villages, residents engage to only a limited extent in slash-and-burn farming and have not engaged in cattle ranching at the time of writing. Many horses and mules are in evidence, as these provide transport for the chicle and xate industries. While in the forest where grass is scarce, mule drivers climb Ramón, also known as Breadnut *(Brosimum alicastrum)*, trees and lop off nutritious foliage, relished by these beasts of burden.

Farming and cattle ranching have been most responsible for removing forest cover in this region, while road building associated with commercial hardwood logging has opened access to hitherto inaccessible forest, facilitating immigration and deforestation. Other important economic activities are the oil and tourism industries. The entire Maya Biosphere Reserve has been partitioned into petroleum exploration concessions. Time will tell what impact this industry will have on the forest. With its unique blend of wild nature and Maya ruins, the northern Petén has great tourism potential, and many observers believe tourism can provide more economic benefit to Guatemala than any other use of this unique region.

Today, the northern Petén is the scene of rapid and startling changes in human population and in the landscape. With the completion of the road linking the Petén to the rest of the country in about 1970, it became government policy to encourage immigration to the region, and the population of the department has soared rapidly, to an estimated half million or more as of 2005. Population growth of the department is 10% annually, mainly through immigration from other parts of the nation, and the population may reach one million during the first half of the twenty-first century. Many newcomers rely heavily on slash-and-burn farming of corn, beans, and other crops, and some aspire to build a small cattle herd, perceived as one of the few avenues to prosperity for country folk with limited resources. Thus, population growth fuels continued pressure on the region's forests.

Although conservation challenges posed by rapid population growth and other factors continue, the area has witnessed profound conservation advances, as described by Primack et al. (1998). In 1990, Guatemalan law decreed the Maya Biosphere Reserve, which encompasses the entire northern extreme of the Department of Petén. This reserve totals 16,000 km^2—14% of Guatemala's national territory.[9] Many local and international non-governmental organizations and foreign aid agencies have struggled long and hard to assist Guatemala's Consejo Nacional de Areas Protegidas (CONAP) in achieving effective protection of this immense reserve. At stake is the largest contiguous area of lowland tropical forest remaining in Mesoamerica. Spanning the boundaries of Guatemala, Mexico, and Belize, the forest is protected not only in Guatemala's Maya Biosphere Reserve, but also in the large, contiguous Calakmul Biosphere Reserve in Mexico, and in the Rio Bravo Resource Management and Conservation Area in adjacent Belize. In addition, other nearby forest reserves in Belize, well-managed *ejido* (communal) lands in Mexico, and nearby reserves such as Montes Azules Biosphere Reserve in Chiapas, Mexico, form a constellation of protected areas of global importance (ibid.). While this area is not stunningly rich in endemic species and has a biota characteristic of broad areas of Mesoamerican Caribbean lowlands, it has unique conservation value by virtue of the enormous areas of pristine and near-pristine forest and wetland habitats contained therein. Although the Harpy Eagle *(Harpia harpyja)* may well have disappeared from some areas and the Scarlet Macaw *(Ara macao)* population declined, the full complement of large carnivores is intact, and no human-caused regional extirpations are known to us since the passing of the megafauna 11,000 years ago. Here a unique opportunity exists to conserve large areas of intact Central American lowland forest. We hope this book may aid in this task.

Some characteristics of reptiles and amphibians known or believed to occur in or near Tikal National Park, exclusive of crocodilians and turtles[a]

Species	Snout-vent length (mm)	Habitat	Habits, notes	Breeding seasonality[b]
Salamanders				
Plethodontidae—lungless salamanders				
Bolitoglossa mexicana	55–70	Humid lowland and premontane forests	Arboreal and terrestrial; hides in bromeliads and beneath surface debris; surface activity probably restricted to rainy season	No data
Frogs and toads[c]				
Rhinophrynidae—Mexican burrowing toad				
Rhinophrynus dorsalis	60–65	Savanna, tropical dry and moist forest, usually associated with water hole	Below ground most of year; active on surface mainly at onset of rainy season	Breeds Jun–Sep in Petén, especially with first heavy rains of the rainy season (Foster and McDiarmid 1983)
Leptodactylidae				
Eleutherodactylus leprus	27–29	Humid lowland forest	Forest floor, little known, uncommon; presumably diurnal/nocturnal	Nothing known; presumably terrestrial eggs and without free-living tadpole stage
Eleutherodactylus rhodopis	30–40	Humid lowland forest	Forest floor, little known, uncommon, diurnal/nocturnal	Terrestrial eggs, without free-living tadpole stage; appears to breed much of year
Leptodactylus labialis	35	Varied habitats including forest, especially with water	Common, widespread, diurnal and nocturnal	Breeds throughout summer rainy season; aquatic tadpoles
Leptodactylus melanotus	36–41	Varied forest and disturbed habitats with temporary or permanent water	Common, terrestrial, largely nocturnal but calling during day as well; often seen on roads at night after a rain	Breeding mainly during summer rainy season, but males may call throughout year; aquatic tadpoles
Bufonidae—Toads				
Bufo marinus	150–200	Open habitats, secondary forest; human commensal	Terrestrial, nocturnal, often numerous on roads after rain	May breed throughout year after heavy rains, but peak of breeding at onset of the rainy season; metamorphosis occurs during periods of high humidity and abundant prey (Zug 1983).
Bufo valliceps	73–84	Varied habitats, especially open, non-forested ones; disturbed and primary forest	Nocturnal; mainly terrestrial, but may climb to 5 m	May breed throughout year, but especially at onset of rainy season
Hylidae—Tree frogs				
Agalychnis callidryas	45–65	Lowland tropical forest	Common; arboreal, nocturnal; climb with slow, hand-over-hand movements; hide in epiphytes and palm fronds by day and during dry season	Breeding begins with onset of summer rains; extended breeding season during rainy season
Hyla ebraccata	25–30	Forest and forest edge; common	Arboreal, nocturnal	Breeds during summer rainy season (Jun–Sep)
Hyla loquax	40	Lowland forest, dry to wet, primary to secondary; savanna	Common; mainly arboreal, nocturnal; in dry season, has been found in epiphytes high in trees	Breeds during summer rainy season (May–Aug), beginning only after several rains

Species	Snout-vent length (mm)	Habitat	Habits, notes	Breeding seasonality[b]
Hyla microcephala	23–25	Abundant in disturbed habitats; uncommon to absent in deep primary forest	Nocturnal, arboreal	Breeds during rainy season (late May–Dec) and also sometimes after heavy rains at other times of year
Hyla picta	20–21	Tall forest, forest edge, second growth, savanna; common	Nocturnal and arboreal; in dry season, has been found in bromeliads and banana plants	Breeds early in rainy season (late May–Jul)
Phrynohyas venulosa	75–79	Varied habitats; common, widespread	Nocturnal, arboreal and terrestrial; spends dry season hiding in bromeliads, leaf axils, tree crevices, beneath bark of standing trees; skin rich in poison glands	Breeds after heavy rains during rainy season (late May–Aug)
Scinax staufferi	25–27	Open, second-growth, and forest-edge habitats; avoids deep forest	Abundant; presumably arboreal and nocturnal; during dry season may be found in bromeliads, beneath bark of standing trees, or under logs or rocks	Breeds during summer rainy season, late May–Aug, possibly Sep
Smilisca baudinii	55–65	Nearly all habitats	"Perhaps the most abundant and ubiquitous amphibian in the Yucatán Peninsula" (Lee 1996); nocturnal, terrestrial and arboreal; by day and during dry season, hide in bromeliads, leaf axils of aroids, beneath tree bark or in other crevices	Breeds during most of rainy season
Triprion petasatus	55–70	Savannas, tropical dry and moist forest	Nocturnal, terrestrial and arboreal; active above ground only during rainy season; in dry season aestivates in dried mud or in crevices, sealing the opening with casque-shaped head	Breeds after heavy rains, late May–Aug
Microhylidae—Narrow-mouthed toads and sheep frogs				
Gastrophryne elegans	24–27	Humid lowland forests	Nocturnal, terrestrial, uncommon; inhabits leaf litter on forest floor	Breeds during rainy season, Jun–Aug
Hypopachus variolosus	33–38	Forested and open situations; most common in open situations	Terrestrial, fossorial, nocturnal; seen at night on roads after heavy rains; else usually burrowing or beneath leaf litter; aestivate, buried, during dry season	Breeds early in rainy season, late May–early Jun; also heard calling during heavy rains in Aug
Ranidae—True frogs				
Rana berlandieri	65–80	Virtually all freshwater habitats	Nocturnal and diurnal, terrestrial; during dry season, seen mainly at water bodies; during rainy season often seen far from water	Breeds during summer rains; tadpoles have been found Feb–Sep
Rana vaillanti	80–100	Humid lowland forest, in association with lakes, woodland pools, or streams	Common, terrestrial; may be found at night on forest floor; active year-round	Protracted breeding season during summer rains

Lizards

Eublepharidae—Banded and leopard geckos				
Coleonyx elegans	108	Forests and savannas	Terrestrial, mainly nocturnal; by day hidden beneath logs or other surface debris; often seen on road or forest floor at night during rainy season	Long egg-laying period during rainy season (females containing eggs Feb–Aug); usual clutch 2; females may produce more than 1 clutch yearly
Gekkonidae—Geckos				
Sphaerodactylus glaucus	29–30	Varied habitats including a range of forest types	Diurnal and nocturnal; beneath surface debris, under loose bark on standing trees and fallen logs; often seen climbing on thatched huts	Eggs laid during dry or rainy season; in Chiapas eggs laid Dec–Feb, hatching in 80–85 days; hatching mainly May–Aug in Petén (rainy season); clutch 1 or 2
Sphaerodactylus millepunctatus	24–31	Mainly forest; not often in houses	Presumably mainly nocturnal; found under surface debris and beneath loose bark on trees	Most egg-laying records from dry season, with hatching early in the rainy season; clutch 1 or 2
Thecadactylus rapicauda	90–100	Tropical forest	Mainly nocturnal, arboreal, hiding by day among tree buttresses, beneath loose bark, and in crevices of outcrops, Maya ruins	Clutch 1 or 2, laid during dry season (hatching during rainy season?)
Corytophanidae—"Basiliscine iguanids"				
Basiliscus vittatus	120–140	Varied habitats, especially near water	Often abundant; diurnal; mainly terrestrial but also climbs a good deal; sleeps perched conspicuously on vegetation; runs bipedally, sometimes across water	Lays eggs Feb–Aug (mainly May–Aug); many records of hatchlings in Jun and Jul (early rainy season) or through Sep, sometimes Oct and Nov; females lay multiple clutches, mean clutch 4 eggs; eggs hatch after 50–55 days
Corytophanes cristatus	110	Primary and secondary forests	Diurnal, mainly arboreal, often in lower 5 m of forest; a sit-and-wait predator usually clinging stationary to vertical trunk; relies heavily on crypsis and takes very large prey, remaining stationary most of the time (antipredator adaptations)	Lays 5–8 eggs per clutch, which hatch after approx. 5 months; eggs apparently laid mainly Jun–November (rainy season); one Panama captive laid in Feb
Corytophanes hernandezii	100	Lowland tropical forests, including drier forest than that favored by *C. cristatus*; more common at Tikal than *C. cristatus*	Diurnal, largely arboreal, often perched less than 1 m above ground; habits much like those of *C. cristatus*	Lays eggs May–Sep, hatching after 67–70 days; juveniles collected Mar–May in Petén; females probably lay several clutches (3–7 eggs/clutch) over an extended laying season
Laemanctus longipes	125–130	Lowland tropical forest, dense second growth	Arboreal, diurnal; spends most time in tree canopies, but also occurs in shrubs or on ground	Nests during summer rainy season; lays 3–6 eggs/clutch, Jun–Sep; hatching Jul–Sep, after approx. 45 days
Iguanidae				
Sceloporus lundelli	85–95	Forest, forest edge	Uncommon; diurnal, arboreal, ascending to 20 m or more; alert and wary	Reproduction unknown, probably a live-bearer
Sceloporus serrifer	80–90	Forest, forest edge	Diurnal, arboreal, often found on large trees, Maya ruins; moderately common; alert, very wary	Believed to be a live-bearer; neonates found around mid-May, just before onset of the rainy season

APPENDIX 2.1—*cont.*

Species	Snout-vent length (mm)	Habitat	Habits, notes	Breeding seasonality[b]
Sceloporus (variabilis) teapensis	52–56	Forest edge, savanna, clearings, Maya ruins; avoids deep forest interior	Diurnal, mainly terrestrial; a sit-and-wait predator found on logs, rocks, or low on tree trunks	Females lay probably multiple clutches of 3–5 eggs/clutch over a prolonged breeding season; hatchlings seen in Petén Jul–Sep, also Dec
Polychrotidae—Anoles				
Anolis biporcatus	85–90	Humid lowland forests	Diurnal, arboreal, largely in canopy, but occasionally seen on lower limbs or ground	Usual clutch 1; eggs laid throughout most of year; hatchlings found in nearly every month
Anolis lemurinus (bourgeaei)	67–70	Lowland forests; common	Diurnal, arboreal/scansorial; perches in shade low on tree trunks, often found on forest floor	Females lay multiple, 1-egg clutches, at least Apr–Jul (apparently mainly during the rainy season)
Anolis (pentaprion) beckeri	50–65	Humid forest, forest edge	Diurnal, arboreal; uncommon or rarely detected; frequent tree trunks where they are extremely cryptic; trees with many epiphytes seemingly preferred	Eggs laid and hatch during summer rainy season; females lay multiple, 1-egg clutches; in Chiapas, hatchlings reported early Jul–early Aug; eggs laid late dry season/early rainy season, hatching Jun–Aug; others laid Sep, Oct, hatching Oct, Nov
Anolis rodriguezii	40–45	Varied forest types, especially in openings and at forest edge	Common; diurnal, arboreal; found on vegetation up to several meters, also on ground, rocks, ruins	Females lay multiple, 1-egg clutches throughout year, but less frequently during dry season
Anolis sericeus	41–46	Open areas, savannas, forest edge, disturbed forest	Diurnal, arboreal; occasionally found on ground	Females lay multiple, 1-egg clutches during the summer rainy season
Anolis tropidonotus	50–55	Common in shaded areas of primary and secondary forest	Diurnal, mainly on forest floor, also lower parts of tree trunks	Lay multiple 1-egg clutches, mainly during summer rainy season; oviposition observed several times in Jul
Anolis uniformis	37	Mature, moist forest; deep forest interior, in heavily shaded areas	Diurnal, terrestrial; mainly on forest floor but occasionally lower parts of trunks	Some breeding throughout year, but dramatic peak in egg-laying early in rainy season, declining prior to dry season; females lay multiple 1-egg clutches
Scincidae—Skinks				
Eumeces schwartzei	110–115	Varied forest types	Diurnal, terrestrial, mainly on forest floor; alert and fast-moving, wary	Presumably lays eggs
Mabuya unimarginata (= brachypoda)	60–75	Savannas, varied forest types	Diurnal, terrestrial and arboreal; usually near escape shelter (rocks, logs, loose bark, crevices, etc.)	Live-bearer, broods of 4–9 born May–Aug (Jun–Jul)

	Size (mm)	Habitat	Ecology	Reproduction
Sphenomorphus cherriei	50–55	Lowland forests	Common; strictly diurnal, terrestrial; inhabitant leaf litter; wary and elusive; most active mid-morning, scarce during mid-day heat; minor midafternoon activity peak; not often seen during dry season but often seen during rainy season; population turnover rapid, with no marked individuals recaptured after longer than 7 months; mean distance between recaptures 13.4 m (Fitch 1983)	Clutch size 1–3; protracted breeding season during spring and summer rains, peaking Jun–Jul; in Belize, hatchlings noted in Jul
Teiidae				
Ameiva festiva	120	Humid, mostly primary, lowland forests; occupies deeper forest and shadier positions than does *A. undulata*; in dry areas seen most often near water	Diurnal, terrestrial; active searcher on forest floor; alert, wary, difficult to approach; most active in morning, rarely seen during hot mid-day hours; active again in late afternoon	Clutch size approx. 2; hatchlings have been seen in Feb–Mar; laying reported May–Jul; breeding season poorly known, and may extend over much of the year as it does in Costa Rica (Echternacht 1983)
Ameiva undulata	120	Open situations, second growth, forest edges; replaced in deep shade of tall, humid forest by *A. festiva*	Common, diurnal, terrestrial, active searcher on ground; greatest activity 0900 to noon on sunny days; remains inactive on cool or overcast days (Scott and Limerick 1983)	Breeding season apparently extended; females lay eggs Jul–Sep; average clutch approx. 5 eggs. In Costa Rica, reproduction "acyclic" but varies seasonally with rainfall (Echternacht 1983).
Anguidae				
Celestus rozellae	80–90	Humid forests; uncommon	Diurnal, arboreal; lives high in trees	Live-bearer; brood size 3–5, born May–Jul
Snakes				
Boidae				
Boa constrictor	1500–2000	Varied habitats from savanna to primary forest	Terrestrial and arboreal, mostly nocturnal but partly diurnal; diet lizards, birds, mammals	Live-bearing, broods 12–50; young born during rainy season (May–Aug in Petén)
Colubridae				
Adelphicos quadrivirgatus	280–300	Humid forests	Secretive, terrestrial/fossorial; found beneath ground litter; diet mainly earthworms	Eggs laid early in rainy season (May, Jun)
Coniophanes bipunctatus	500–550	Moist lowland forest; swampy areas	Uncommon; terrestrial; mainly nocturnal, but also diurnal; diet: frogs	One clutch of eggs found in Jul
Coniophanes imperialis	250–300	Forest, savanna, agricultural areas	Most common member of this genus in Yucatán Peninsula; terrestrial, mainly nocturnal but also found abroad during day; found beneath surface debris; diet: insects, frogs, toads, and lizards	Lays eggs in spring (May); clutch 3–10, hatching in approx. 40 days; juveniles reported Feb and Mar; female contained eggs mid-May

APPENDIX 2.1—cont.

Species	Snout-vent length (mm)	Habitat	Habits, notes	Breeding seasonality[b]
Dryadophis melanolomus	980–1010	Varied forest and savanna habitats	Common; terrestrial; diurnal; alert, fast moving; diet mostly lizards, also anurans, snakes, reptile eggs, small mammals	2–5 eggs, probably laid late in rainy season (Aug–Nov)
Drymarchon corais	to nearly 3 m; mostly 1.5 m	Varied forests and savannas; often found near water	Terrestrial; diurnal; alert, fast moving; diet hugely varied: all classes of vertebrates are eaten	Clutch 4–11; laid over extended period in rainy season, May–Jul; hatchlings seen Jul–Oct
Drymobius margaritiferus	500–600	Forested and open habitats, especially near water	Common; terrestrial; diurnal; alert, fast moving; diet: mainly frogs and toads	Clutch 2–7; eggs laid throughout the rainy season, Apr–Aug; incubation 64–68 days; hatchlings have been found Jun–Oct
Elaphe flavirufa	up to 1.3 m	Varied forest types	Moderately common; nocturnal; terrestrial and arboreal; diet: small mammals, birds	Lays eggs
Ficimia publia	230–280	Virtually all habitats	Moderately common, secretive; terrestrial, nocturnal; diet largely spiders	Lays eggs
Imantodes cenchoa	600–700	Primary and secondary forest	Nocturnal and crepuscular, arboreal; by day hides in epiphytes, tree cavities; diet lizards, largely *Anolis*	Females contained well-formed eggs (1–3) in late Apr and mid-Jun; hatchlings found Jun–Aug
Lampropeltis triangulum	1.0–1.5 m	Forests, agricultural habitats	Terrestrial; nocturnal and diurnal; moderately common; diet: lizards, snakes, frogs, mammals	Clutch 5–12; incubation period 35–50 days; hatching Jul–Aug
Leptodeira frenata	450	Humid lowland forest	Terrestrial and arboreal; nocturnal; found in bromeliads, under bark and logs; diet frogs, toads, lizards	Eggs apparently laid Feb–Apr; clutch 3–7
Leptodeira (septentrionalis) polysticta	800	Humid lowland forest	Common; nocturnal, mainly arboreal, found in bromeliads and palm axils; forages to at least 10 m above ground; diet: frogs and frog eggs; often common around frog choruses	Clutch 5–13, laid during late dry season, Feb–May; incubation 70–85 days; hatchlings found Jul and Aug
Leptophis ahaetulla	1200–1450	Primary and secondary forest, forest edges	Moderately common; diurnal, arboreal, climbing to considerable heights; alert, active searcher; diet mainly frogs, also lizards, snakes, birds, bird eggs	Clutch 1–5, laid early in rainy season
Leptophis mexicanus	880	Forest edge, second growth	Common; diurnal, arboreal, active, fast moving; diet mainly frogs but also lizards, salamanders, tadpoles, bird eggs, small snakes	Clutch 2–11; eggs laid during rainy season, Mar–Oct; juveniles seen in Aug
Masticophis mentovarius	1900	Open habitats such as savannas, forest edge	Diurnal, terrestrial; alert and fast moving; diet: lizards, mammals, nestling birds	Eggs laid Mar–Apr; clutch 7–30
Ninia diademata	250–270	Humid forests	Secretive, nocturnal, terrestrial; under leaf litter and surface debris; diet: snails and slugs	Lays 2–4 eggs during rainy season
Ninia sebae	220–270	Forest, savanna, agricultural areas	Secretive; terrestrial; mainly nocturnal but also active by day; beneath surface debris; diet of earthworms, snails, slugs, leeches	1–4 eggs/clutch; eggs laid during rainy season, Mar–Sep; juveniles found May–Sep; breeding season probably longer in wetter areas, from Jan–Sep

Species	Size (mm)	Habitat	Ecology and diet	Reproduction
Oxybelis aeneus	700	Forest, forest edge	Common; diurnal, arboreal; diet: mainly lizards, occasionally insects, frogs, birds, small mammals	Clutch 3–5, laid Mar–Aug; hatching during rainy season; juveniles seen from late rainy season to early dry season, suggesting eggs laid throughout much of rainy season; may be quiescent during dry season (Scott 1983)
Oxybelis fulgidus	up to 1.5 m	Second growth, forest edge	Moderately common; diurnal, arboreal; diet: lizards, birds, mammals	Eggs laid Apr–Jun, and probably earlier; juveniles seen Apr and May; a captive laid 10 eggs in May, which hatched in Aug, after 105 days
Oxyrhopus petola	600	Humid lowland forests	Uncommon; terrestrial, mainly nocturnal; diet: mostly lizards, snakes	Clutch 5–10, hatching in Mar or Apr, after approx. 90 days
Pseustes poecilonotus	750–1000	Humid lowland forests, savannas	Terrestrial and arboreal, diurnal; diet: birds, bird eggs	Lays eggs
Rhadinaea decorata	265–280	Humid lowland forests including disturbed areas, second growth	Rare; diurnal, terrestrial, inhabiting forest floor; beneath surface debris; diet: frogs, salamanders, and their eggs	Lays eggs
Scaphiodontophis annulatus	250–300	Humid lowland forest	Secretive; diurnal, terrestrial; found in leaf litter, possibly fossorial; diet: lizards, mainly skinks	Mating observed 18 and 21 Jun; clutch 1–4 in Petén, reportedly 6–10 in Chiapas; eggs laid and hatch during rainy season
Senticollis triaspis	1000	Varied forest types; partially cleared areas	Uncommon; terrestrial, nocturnal; diet: mainly small mammals, also birds, lizards	Breeding season apparently protracted, possibly throughout the year; clutch 3–7
Sibon dimidiata	500	Humid lowland forests	Uncommon; arboreal, nocturnal; found in bromeliads, base of palm frond, on low vegetation at night; beneath loose bark on fallen trees; diet mainly tree snails	1 female laid 2 eggs
Sibon nebulata	500	Varied lowland forest types	Moderately common; nocturnal, terrestrial and arboreal; beneath surface litter; diet: snails and slugs, sometimes concentrating in a tree where snails are abundant	Hatchlings seen Jul and Aug; juveniles collected at Tikal mid-Mar to mid-Apr; 3–9 eggs laid, usually May to Aug, but breeding season possible longer in some cases: e.g., 6 eggs found in Huehuetenango, Guatemala, on 25 Jul hatched 4–7 Oct
Sibon sartorii	450	Lowland forests, second growth	Common; nocturnal, crepuscular, terrestrial (nearly always found on ground); movements quick and nervous; diet: snails and slugs	3–5 eggs laid late in the dry season or early in rainy season
Spilotes pullatus	2000	Forests, second growth, savannas	Moderately common; diurnal, terrestrial and arboreal, from low vegetation to high in canopy; alert and fast moving; diet: birds, bird eggs, and small mammals	7–10 eggs laid late in dry season or early in rainy season, Apr–Jul; hatchlings found Jun–Nov

APPENDIX 2.1—*cont.*

Species	Snout-vent length (mm)	Habitat	Habits, notes	Breeding seasonality[b]
Tantilla cuniculator	155	Varied tropical forest types	Secretive, fossorial, terrestrial, nocturnal; beneath surface litter; diet: centipedes, beetle larvae	Presumably lays eggs
Tantilla moesta	490	Varied tropical forest types	Uncommon; secretive, terrestrial, fossorial, nocturnal; diet presumably invertebrates	Presumably lays eggs
Tantillita canula	145	Varied tropical forest types	Uncommon; secretive, terrestrial, fossorial, probably nocturnal; beneath surface litter; diet presumably invertebrates	Presumably lays eggs
Tantillita lintoni	125–135	Humid lowland forests, second growth	Seemingly rare; secretive, terrestrial, nocturnal; diet: presumably invertebrates	Presumably lays eggs
Thamnophis proximus	350–400	Near water; moderately common	Terrestrial, active day or night; diet: tadpoles, frogs, toads, fish	Live-bearer; 1 specimen gave birth 3 Jul; broods of half-dozen young born in Jun–Jul
Tretanorhinus nigroluteus	350–400	Aquatic; near water in humid lowland forests	Nocturnal and crepuscular; terrestrial/semiaquatic; diet: fish, tadpoles, frogs	6–9 eggs, laid over a seemingly protracted breeding season; females with unlaid eggs collected Jan and Jul; a newborn collected in Apr; a related species in Cuba lays approx. 8 eggs, which hatch in 35 days
Urotheca elapoides	300–350	Lowland forests, uncommon	Crepuscular, diurnal, nocturnal, terrestrial; a coral snake mimic; diet: salamanders, frogs, amphibian eggs	Lays 4–8 eggs, apparently during an extended period coinciding with the rainy season; one from Huehuetenango laid 5 eggs on 8 Aug, which hatched after 108 days, which may be longer than usual
Xenodon rabdocephalus	650–700	Moist lowland forests	Uncommon; terrestrial, diurnal and nocturnal; diet: mainly toads, fewer frogs	9–10 eggs laid in rainy season; juveniles seen Jun–Nov
Elapidae—Coral snakes				
Micrurus diastema	550–650	Varied habitats, especially tall forest	Common; secretive, terrestrial, active day or night; diet: mainly snakes, occasionally lizards, especially skinks	Probably lays eggs during rainy season, roughly Apr–Aug; probably hatch in approx. 2 months; juveniles seen Jun and Jul; an Apr female from Alta Verapaz was gravid with 4 eggs
Viperidae—Pitvipers				
Atropoides nummifer	400–600	Primary moist lowland forest	Mainly terrestrial, nocturnal; diet: smalls mammals; juveniles probably frogs, lizards, and invertebrates	Give birth to 13–36 young during summer rainy season, Aug–Nov; newborns seen in Belize Jul–Sep
Botriechis schlegelii	400–500	Moist lowland forests	Rare at Tikal (2 records); arboreal, mainly nocturnal; diet: lizards, frogs, mice, bats, possibly birds	Give birth to 12–20 young during rainy season; 166-day gestation reported for a captive female from Honduras

Species	Size	Habitat	Behavior/Diet	Reproduction
Bothrops asper	1.0–1.7 m	Varied forest types, second growth, milpas	Common; mainly terrestrial, nocturnal, not very active during dry periods; diet: small mammals, birds, some lizards; juveniles eat frogs, lizards, snakes, centipedes	Live-bearers; newborns appear during middle-late rainy season; brood size 5–86; breeding probably throughout rainy season, perhaps some of early dry season too; gestation 212 days in Honduras female
Crotalus durissus	1.3–1.6 m	Mainly savannas and other open areas; sometimes in forest	Terrestrial, diurnal, crepuscular, or nocturnal; especially active on drizzly nights and when humidity is high; rare but present at Tikal; diet mainly rodents; occasional birds and lizards	Give birth to 15–47 (21) young, during mid to late rainy season (Jul–Sep)
Porthidium nasutus	255–315	Humid forest	Terrestrial, nocturnal; beneath surface debris; found coiled on forest floor by day; diet: mostly lizards, frogs; also mice, snakes, invertebrates	Live-bearing; brood 14–18, born during rainy season
Species excluded from list:				
Ctenosaura similis	275–350	Very localized in distribution; absent or essentially absent from Tikal	—	—
Typhlops microstomus	270	This tiny fossorial snake is probably rarely encountered by raptors	—	—
Stenorrhina freminvillei	680–715	Thorn forests, savannas; moderately common	Terrestrial, fossorial, nocturnal; beneath surface litter; diet: scorpions and spiders; no records for Tikal region but likely present in open habitats in the region	—

[a] A few other species may also occur at Tikal, but lack records in the vicinity.

[b] Unless otherwise indicated, all information reported here comes from Campbell 1998 and/or Lee 2000.

[c] All frogs and toads breed in temporary or permanent water bodies unless otherwise stated.

APPENDIX 2.2
Some characteristics of mammals known or believed to occur in Tikal National Park[a]

Common name	Scientific name	Weight[b]	Strata of activity[c]	Activity time[c]
Marsupials: Opossums—Didelphidae				
Central American Woolly Opossum	*Caluromys derbianus*	245–370 g	Arboreal	Nocturnal
Common Opossum	*Didelphis marsupialis*[d]	565–1610 g	Terrestrial-arboreal	Nocturnal
Virginia Opossum	*Didelphis virginiana*[d]	500–2300 g	Terrestrial-arboreal	Nocturnal
Common Gray Four-eyed Opossum	*Philander opossum*	200–660 g	Terrestrial-arboreal	Nocturnal
Mexican Mouse Opossum	*Marmosa mexicana*	40–100 g	Arboreal	Nocturnal
Xenarthra: Anteaters, sloths, and armadillos				
Northern Tamandua	*Tamandua mexicana*	3.6–8.4 kg	Terrestrial-Arboreal	Diurnal-Nocturnal
Silky Anteater	*Cyclopes didactylus*	155–275 g	Arboreal	Nocturnal
Nine-banded Armadillo	*Dasypus novemcinctus*	2.7–6.3 kg	Terrestrial	Diurnal-Nocturnal
Bats: Chiroptera—At least 38 bat species are known from Tikal (McCarthy 1992); additional species are known from northern Petén and likely occur at Tikal.				
Primates				
Black-handed Spider Monkey	*Ateles geoffroyi*	6.6–9.0 kg	Arboreal	Diurnal
Guatemalan Howler Monkey	*Alouatta pigra*	5–8 kg	Arboreal	Diurnal
Carnivora				
Dog family: Canidae				
Gray Fox	*Urocyon cinereoargenteus*	~2–3 kg	Terrestrial	Diurnal-Nocturnal
Raccoon family: Procyonidae				
Cacomistle	*Bassariscus sumichrasti*	900 g	Terrestrial-Arboreal	Nocturnal
Raccoon	*Procyon lotor*	2.7–6.4 kg	Terrestrial-Arboreal	Nocturnal
White-nosed Coati	*Nasua narica*	4.5 kg	Terrestrial-Arboreal	Diurnal-Nocturnal
Kinkajou	*Potos flavus*	2.0–3.2 kg	Arboreal	Nocturnal
Weasel family: Mustelidae[e]				
Tayra	*Eira barbara*	2.7–7.0 kg	Terrestrial-Arboreal	Diurnal
Striped Hog-nosed Skunk	*Conepatus semistriatus*	1.4–3.4 kg	Terrestrial	Nocturnal
Long-tailed Weasel	*Mustela frenata*	85–340 g	Terrestrial	Diurnal-Nocturnal
Cat family: Felidae				
Margay	*Leopardus wiedii*	3–9 kg	Arboreal-Terrestrial	Mainly nocturnal
Jaguarundi	*Felis yagouarundi*	4.5–9.0 kg	Mainly terrestrial	Nocturnal
Ocelot	*Felis pardalis*	8–12 kg	Terrestrial	Nocturnal and Diurnal
Puma	*Felis concolor*	29–120 kg	Terrestrial	Nocturnal and Diurnal
Jaguar	*Panthera onca*	31–158 kg	Terrestrial	Nocturnal and Diurnal
Ungulates				
Baird's Tapir	*Tapirus bairdii*	150–300 kg	Terrestrial	Mostly nocturnal
Collared Peccary	*Tayassu tajacu*	17–30 kg	Terrestrial	Diurnal
White-lipped Peccary	*Tayassu pecari*	25–40 kg	Terrestrial	Mostly diurnal
Red Brocket Deer	*Mazama americana*	24–48 kg	Terrestrial	Diurnal and nocturnal
Yucatán Brown Brocket Deer	*Mazama pandora*	?	Terrestrial	Diurnal and nocturnal
White-tailed Deer	*Odocoileus virginianus*	30–50 kg	Terrestrial	Diurnal and nocturnal

Rodents

Squirrels: Sciuridae				
Deppe's Squirrel	*Sciurus deppei*	190–220 g	Arboreal	Diurnal
Yucatán Squirrel	*Sciurus yucatanensis*	420 g	Arboreal	Diurnal
Spiny Pocket Mice: Heteromyidae				
Gaumer's Spiny Pocket Mouse	*Heteromys gaumeri*	55–65 g	Terrestrial	Nocturnal
Desmarest's Spiny Pocket Mouse	*Heteromys desmarestianus*	60–110 g, usually 60–85	Terrestrial	Nocturnal
Murid rodents: Muridae				
Big-eared Climbing Rat	*Ototylomys phyllotis*	70–100 g	Arboreal	Nocturnal
Naked-tailed Climbing Rat	*Tylomys nudicaudus*	200–260 g	Arboreal	Nocturnal
Harvest Mouse	*Reithrodontomys gracilis*	8–29 g	Terrestrial-Arboreal	Nocturnal
Hispid Cotton Rat	*Sigmodon hispidus*	50–65 g	Terrestrial	Diurnal and nocturnal
Pygmy Rice Rat	*Oligoryzomys fulvescens*	9–40 g	Terrestrial	Nocturnal
Black-eared Rice Rat	*Oryzomys melanotis*	~40–120 g	Terrestrial	Nocturnal
Coues' Rice Rat	*Oryzomys couesi*	~40–120 g	Terrestrial	Nocturnal
Vesper Rat	*Nyctomys sumichrasti*	22–46 g	Arboreal	Nocturnal
Yucatán Vesper Rat	*Otonyctomys hatti*	24–43 g	Arboreal	Nocturnal
Roof or Black Rat	*Rattus rattus*	100–240 g	Terrestrial-Arboreal	Nocturnal
Norway Rat	*Rattus norvegicus*	200–505 g	Terrestrial-Arboreal	Nocturnal
House Mouse	*Mus musculus*	10–21 g	Terrestrial	Mainly nocturnal
Cavy-like rodents: Suborder Caviomorpha				
Mexican Hairy Porcupine	*Couendou mexicanus*	1.4–2.6 kg	Arboreal	Nocturnal
Paca (Tepescuintle)	*Agouti paca*	5–13 kg	Terrestrial	Nocturnal
Central American Agouti (Cereque)	*Dasyprocta punctata*	3.2–4.2 kg	Terrestrial	Diurnal and nocturnal
Rabbits: Lagomorpha				
Brazilian Rabbit	*Sylvilagus brasiliensis*	450–1200 g	Terrestrial	Nocturnal

[a] Throughout the book, mammal names conform to Emmons and Feer 1990.
[b] Weights are mostly from Emmons and Feer 1990.
[c] Information on stratum and time of activity are from Emmons and Feer 1990; Nowak and Paradiso 1983; and Jolón 1996, 1997.
[d] We did not distinguish between the opossums *Didelphis marsupialis* and *D. virginiana*, and we refer to all of these as *D. marsupialis* throughout the book.
[e] The Grisón (*Galictis vittata*) has been sighted in northern Petén but not yet at Tikal.

37

Chapter 2 Notes

1. At least some seasonal tropical forests also show a flush of leaf production at the end of the wet season (Boinski and Fowler 1989).

2. Levings and Windsor (1985) also found that nearly all leaf litter arthropod groups found certain (the same) years to be more or less favorable than average—as shown also in a Peruvian study (Pearson and Derr 1986). On average, all groups were favored by some rain during the dry season, and by a short wet season: a short rainy season is probably beneficial to arthropods in this detritus-based food chain because decomposition reduces leaf litter rapidly during the rainy season (ibid.).

3. The rainy season reproduction typical of most reptiles in our study area accords with many earlier generalizations regarding a rainy season peak in reptile reproduction in highly seasonal tropical environments (Fitch 1970, 1982; Saint Girons 1982; citations in James and Shine 1985). Some recent reviews of breeding seasons in tropical reptiles have emphasized the diversity of patterns, with some species breeding mainly in the wet season, others the dry season, and others virtually year-round, especially in areas without a strong dry season (James and Shine 1985; Vitt 1992; Zug et al. 2001). Where tropical snake assemblages have been studied in detail, a wide range of reproductive patterns have been found (ibid.). This diversity is thought to be linked to phylogeny (James and Shine 1985), and to differences in diet and hence temporal patterns of prey availability (ibid.; Vitt 1983). Still, several studies have found a predominance of reptile breeding during tropical rainy seasons. Fitch (1982) summarized reproductive seasonality of 30 species of *Anolis* from Central America and Mexico. While some species in areas without a strong dry season were known or believed to breed year-round, 14 species from areas with pronounced dry seasons showed a great preponderance of rainy season reproduction (ibid.). Regarding tropical snakes, Seigel and Ford (1987) cautioned that studies based on dissection of museum specimens taken over large areas and in different years cannot accurately assess reproductive seasonality; hence, prior statements that reproduction in snakes often takes place year-round in tropical environments must be treated as suspect. Of 20 well-studied snake species, all but one showed a discrete reproductive season, each of these during the rainy season (ibid.).

4. In addition to body size, fat content of adults often varies through the year in *Anolis* species, as in many reptiles, with fat bodies being smallest during egg or embryo formation. For three species, fat bodies of adults were maximal at the end of the dry season (just prior to the onset of breeding) and dropped rapidly during breeding, with reserves building again as reproduction tapered off and the next dry season approached (Sexton et al. 1971). Thus, caloric content of reptiles, and their energetic value to raptors, may vary through the year.

5. Roling (1992) characterized the small mammal community in various habitats in and near Tikal. In active slash-and-burn crop fields (corn, beans, tomatoes), the Hispid Cotton Rat *(Sigmodon hispidus)* made up 80% of captures, with two species of rice rats and a harvest mouse making up 14, 4, and 2% of captures. The Hispid Cotton Rat lives in shrubby and grassy areas, where, volelike, it creates runways through the grass. In five-year-old successional vegetation, Gaumer's Spiny Pocket Mouse *(Heteromys gaumeri)* made up more than half of trap captures, the remaining portion shared among Desmarest's Spiny Pocket Mouse *(H. desmarestianus)*, the Hispid Cotton Rat, three species of rice rats *(Oryzomys* and *Oligoryzomys)*, the Harvest Mouse *(Reithrodontomys gracilis)*, and the Big-eared Climbing Rat *(Ototylomys phyllotis)*. The latter was commonly captured 10–15 m up in trees (G. Roling, pers. comm.), whereas the other species were found mostly on the ground, with occasional captures up to 3 m above ground in some species. Desmarest's Spiny Pocket Mouse was largely restricted to forest while Gaumer's Spiny Pocket Mouse occupied a broad range of habitats from mature forest to young second growth and slash-and-burn crop fields. In 15-year-old second growth, Roling (1992) found Black-eared Rice Rat *(Oryzomys melanotis)* comprising just over half the captures and the two spiny pocket mice next in abundance, with Desmarest's contributing 22% of captures and Gaumer's 15%. The remaining captures were of the Big-eared Climbing Rat, the Harvest Mouse, and the Black Rat *(Rattus rattus)*.

In mature forest, Roling (1992) trapped mammals on the forest floor as well as a few meters above ground. On the forest floor the two *Heteromys* species together made up 88% of captures, with the Big-eared Climbing Rat and the Black-eared Rice Rat making up 6 and 5% of captures, and Hispid Climbing Rat 1%. A few meters above ground, the Mexican Mouse Opossum *(Marmosa mexicana)* made up 63% of captures, the Big-eared Climbing Rat 30%, and the highly arboreal Vesper Rat *(Nyctomys sumichrasti)* and Naked-tailed Climbing Rat *(Tylomys nudicaudis)* 4% each. Similar patterns of vertical occurrence were found by Jolón (pers. comm.); among six species of rodents and three of opossums, he encountered five species at ground level, seven at 1–3 m, and four 10–15 m up in trees. Desmarest's Spiny Pocket Mouse is one of the most abundant small mammals at Tikal, with population densities in forest ranging from 10 to 49 individuals/ha (G. Roling, pers. comm.).

6. Because turtles and crocodilians were not noted as raptor prey at Tikal, we omitted them from Appendix 2.1.

7. MRPP chi-square test, $\chi^2 = 20.36$, $P < 0.0001$.

8. In this accounting, six frog species and seven snake species are counted as active both night and day.

9. The reserve totals 21,000 km^2 including a buffer zone.

3 GRAY-HEADED KITE

Russell K. Thorstrom, David F. Whitacre, Juventino López, and Gregorio López

The Gray-headed Kite *(Leptodon cayanensis)*, a strikingly handsome raptor, remained largely an enigma to us, despite our persistent efforts to find and study nests. We most often observed these secretive kites early in the breeding season as they soared high above the forest, sometimes giving a conspicuous aerial display accompanied by loud calling. At other times, Gray-headed Kites were cryptic and easily overlooked as they moved about below the forest canopy. We found four nests and added to the scant information available on the species's nesting biology and behavior. Still, this remains one of the least known of widespread Neotropical raptors.

GEOGRAPHIC DISTRIBUTION AND SYSTEMATICS

In earlier literature, the Gray-headed Kite is referred to as the "Cayenne Kite" and bears the Latin name *Odontriorchis palliatus*. The genus *Leptodon* is considered closely related to the Old World cuckoo-falcons or bazas (genus *Aviceda*: Brown and Amadon 1968), and Friedmann (1950) found little difference between the two other than the proportionally longer primaries and more rounded tail of *Leptodon*. A second species of *Leptodon* occurs in northeast Brazil; this form, *L. forbesi*, was until recently considered by many a variant of *L. cayanensis* (del Hoyo et al. 1994). Recent discoveries, however, leave little doubt that *L. forbesi* is a distinct species, endemic to an extremely small area in eastern Brazil (Ferguson-Lees and Christie 2001).

The Gray-headed Kite occurs from Tamaulipas and Oaxaca, Mexico, south throughout the lowland tropics to Ecuador west of the Andes and, east of the Andes, to eastern Bolivia, northern Argentina, and Paraguay (Brown and Amadon 1968). Two subspecies are generally accepted: *L. c. cayanensis* occurs from the northern extent of the range south to central Brazil, while south of there the purportedly larger race *L. c. monachus* has been described (del Hoyo et al. 1994). This is a bird of lowlands and foothills, occurring up to 750 m in Costa Rica (Stiles et al. 1989), 1000 m in Colombia (Hilty and Brown 1986), 1500 m in Honduras (Monroe 1968), and 2000 m in Panama (Wetmore 1965).

MORPHOLOGY

Gray-headed Kites have relatively small heads, with notably short, stout, heavily scaled tarsi and toes, and a tomial notch on the upper mandible—a trait shared with few other raptors beyond the true falcons (Plate 3.1). Appendix 1 presents body mass data that triple the sample size given by Dunning (1993). With females averaging 521 g and males 461 g (Appendix 1), these kites are similar in mass to a Broad-winged Hawk *(Buteo platypterus)* or a female Northern Harrier *(Circus cyaneus;* Snyder and Wiley 1976). While the sexes are alike in plumage, females are about 13% heavier than males, for a body mass Dimorphism Index (DI) value of 4.1. The mean DI value for folded wing length, culmen, and mass is 2.24 (Appendix 1), making this species less dimorphic than nearly all North American falconiforms, but fairly typical for a kite (Snyder and Wiley 1976).

Two members of a pair trapped at Tikal had blue-gray iris, cere, and tarsi, and gray-brown pupils. The bird, believed to be a female, had white underwing linings while the presumptive male had black underwing linings.

PREVIOUS RESEARCH

The Gray-headed Kite has never been the subject of a detailed field study, and nearly all existing information appears in field guides and brief accounts by museum collectors. Land (1970) regarded the Gray-headed Kite as rare in Guatemala and reported it only from the northern extent of the Pacific coastal plain, although he expected it to be found elsewhere in the country. Apparently this kite remained unknown from the Caribbean slope of Guatemala until quite recently (Beavers et al. 1991). Beavers (1992) knew of no breeding records from the Tikal area and regarded this kite as seen mainly in open areas with scattered trees, especially along roadsides south of the park.

RESEARCH IN THE MAYA FOREST

At Tikal we studied four nesting attempts in three territories: Chikintikal 1991, Airstrip 1991, and Bajada de la Pina 1993 and 1996. All four nests failed before many data could be gathered. In 1991, two adults were radio-tagged at the Chikintikal nest, and a few periods of detailed nest observation were conducted before the nest failed. The radio-tagged female was soon found dead, but the male provided two weeks of movement data before it cut the surgical tubing and cable, freeing itself of the transmitter. From 1992 to 1995 we devoted some effort each year to nest search; we had little luck, but we did find a nest in 1993. In 1996 we dedicated two persons full-time in an effort to study this species. We (J. López and G. López) began searching for nests in April and found areas of activity we believed to represent four or five pairs: Bajada de la Pina, Chikintikal, Garita de Cobros, Airstrip, and Arroyo Negro. We located only one nest, that of the Bajada de la Pina pair, on 25 April, in a Kapok *(Ceiba pentandra)*. On 7 May we began dawn-to-dusk observations three days weekly at this nest. After the first week of observation this nesting failed, apparently owing to interference by Black-handed Spider Monkeys *(Ateles geoffroyi)*. Additional results are given by Thorstrom (1997).

DIET AND HUNTING BEHAVIOR

Gray-headed Kites take a varied diet of small prey, particularly insects, and especially adults and brood of social hymenoptera, including large wasps. Also taken are cicadas, termites, beetles, orthopterans, mollusks, frogs, lizards, snakes, birds, and bird eggs (Haverschmidt 1962; Brown and Amadon 1989; Ferrari 1990; del Hoyo et al. 1994). Thiollay (1985) characterized this kite as a slow-hunting raptor of the canopy that eats mainly inactive and mimetic large insects, or tree frogs and lizards; he also believed some snails may be taken. Brown and Amadon (1968) cited Salvin and Godman (1897) to the effect that mollusks are taken, the birds' plumage growing soiled by contact with the ground.

An individual collected in Venezuela had fed on adult and larval wasps, ingesting a good deal of the paper-like nest as well (Friedmann and Smith 1955). In stomachs Haverschmidt (1962) found insects, including great numbers of termites (Isoptera), as well as bees and wasps (Hymenoptera: Apidae, Vespidae), beetles (Coleoptera: Curculionidae), orthopterans (Locustidae), and caterpillars (Lepidoptera), in addition to remains of a frog and the nearly intact shell of a bird egg. One kite collected in the Yucatán Peninsula had a large lizard in its stomach (Paynter 1955), and Slud (1964) watched one eat a small snake. In Panama, one bird's stomach contained 55 pupae and 38 adults of a wasp, *Odynerus pachyodynerus*, and a few bits of an ant, *Azteca* species (Wetmore 1965). Two specimens from Surinam had each fed on grasshoppers, and one of them had taken a large caterpillar (Voous 1969). One individual collected in El Salvador had eaten insects, and another had fed on large wasp larvae (Dickey and van Rossem 1938). In Peru, Robinson (1994) observed a pair of these kites carrying food to a presumed nest, including seven lizards, two frogs, one snake, one unidentified vertebrate, five katydids, and two caterpillars. The contrast between Robinson's prey list and the insect-dominated stomach contents just cited may indicate that adult kites tend to eat many invertebrates, while delivering mainly larger prey to the nest.

We have only one prey record to report from Tikal, from forest 6 km south of the park. In April 1999, a Gray-headed Kite was seen perched, clutching in its talons a globular wasp nest approximately 5 cm in diameter (M. Córdova, pers. comm.). The two kites we captured in 1991 gave off a strong odor reminiscent to one of us (R. K. T.) of that of certain colubrid snakes. Whether this odor was a characteristic of the birds themselves or stemmed from prey they had captured was unknown.

Hunting Behavior

While some authors have commented that Gray-headed Kites often perch visibly on a prominent limb, at least in early morning, several have commented that these kites often perch high within the canopy and are easily overlooked. These kites are often described as sluggish; remaining perched for long periods before flap-gliding slowly to another tree, and as moving deliberately and inconspicuously as they forage within thick vegetation (Slud 1964; Stiles et al. 1989). Some authors have also described Gray-headed Kites as unsuspicious and easily approachable (Dickey and van Rossem 1938). Haverschmidt (1962) described a Gray-headed Kite darting suddenly from its perch to pursue an insect, "sideslipping with great velocity and impetus." Robinson (1994) described these kites as making short (1–10 m) flights in the upper canopy and making short dives from branch to branch.

At Tikal, we once observed a Gray-headed Kite fly into a large Honduras Mahogany *(Swietenia macrophylla)* and grasp at some type of small prey. While following a radio-tagged adult male in 1991, we did not observe any attacks on prey. The bird flew from perch to perch in and below the canopy, slowly turning its head while perched, to scan its surroundings from canopy to ground level. After several minutes on a perch the kite typically flew to another perch and began scanning again. In Brazil, a pair of these kites was observed to forage in association with a group of Buffy-headed Marmosets *(Callithrix flaviceps:* Ferrari 1990). The kites captured mainly cicadas startled into flight by the primates.

HABITAT USE

Gray-headed Kites are generally regarded as uncommon and local in occurrence, being found mainly in heavily forested humid lowlands, often near marshes and streams, and less often in open woodland or arid situations in tropical and subtropical zones (Dickey and van Rossem 1938; Blake 1958; AOU 1983; Hilty and Brown 1986; Thurber et al. 1987). Ridgely and Gwynne (1989) described the habitat as forest, second-growth woodland, and borders in more humid lowlands and foothills. In Costa Rica, Slud (1964) found Gray-headed Kites in both humid and drier forests and described the species as being found mainly along forest edges, in patches of woodland, and in trees in the semi-open near woodland. Wetmore (1965) regarded these kites as most frequent in areas of unbroken forest, though occurring also in gallery forest along streams or marshes in savanna regions. Similar comments regarding use of savannas, especially near marshes, were made by Olrog (1985) and Peterson and Chalif (1973). While some authors have regarded this as strictly a marsh-associated species, Paynter (1955) emphasized that, of five specimens he collected in the Yucatán Peninsula, only one was collected within several miles of water; the remaining four were collected in "rain forest" (i.e., tropical, probably semi-deciduous forest).

Most Gray-headed Kite nests at Tikal were in Bajo or Transitional Forest, but it is premature to generalize regarding the forest types most often used for nesting. These kites seemed to occur broadly throughout Tikal's mature forests. We saw no indication of an association with marshes or other wetlands as suggested by many published comments. However, it remains unknown whether areas of forest that are seasonally inundated held special importance to these kites.

Gray-headed Kites occurred in roughly similar abundance at three other sites near Tikal. In a series of point counts conducted each of two years in similar forest at Zotz (just west of Tikal), Dos Lagunas (60 km north of Tikal), and in Mexico's Calakmul Biosphere Reserve (100 km north of Tikal), we detected similar numbers of Gray-headed Kites at each site; on average we detected them in 3.5 of 20 point counts at Zotz, 3 of 20 counts at Dos Lagunas, and 5.5 of 20 counts at Calakmul (Jones and Sutter 1992).

BREEDING BIOLOGY AND BEHAVIOR

Nests and Nest Sites

We documented four nests in two or three territories at Tikal. All four were delicate structures of dried twigs placed loosely together near the top of a tall tree. At least two of the nests were so loosely constructed that one could see the egg through the bottom of the nest. Nests were 22–26 m up in trees that were 25–28 m tall and 55–88 cm in diameter at breast height. The few other known nest descriptions agree (Brown and Amadon 1968; data on cards accompanying five clutches from Trinidad).

Nest 1: Chikintikal Nest. Investigating a soft, trogon-like call, on 20 April 1991, we located a pair of Gray-headed Kites building a nest high in a tall Palo de Danto, also known as Bitter Angelim *(Vatairea lundellii)*, in an area of Transitional Forest with emergent mahogany trees (Table 3.1). Supported by three limbs 10–12 cm thick (Table 3.1), the nest was loosely constructed of approximately 100 dried twigs and was tucked up just beneath the dense upper canopy of the tree. The nest was 50 m from an active Bicolored Hawk *(Accipiter bicolor)* nest (Table 3.1).

Nest 2: Airstrip Nest. A second nest was found on 4 July 1991 after we observed an adult repeatedly visiting the same area the previous day (Table 3.1). One nestling was found dead in the nest and another at the base of the nest tree. The nest was in an area of Bajo Forest, 150 m from Tikal's airstrip; dimensions appear in Table 3.1. The downy white nestlings had olive-yellow iris and yellow cere and legs; we judged them to be approximately 2 weeks old, as the mantle and flight feathers were just beginning to emerge. (See Plate 3.2 and 3.3 for examples from Brazil.) The nestling on the ground had been partly eaten, apparently by a mammal. This nest and nest 1 (simultaneously active) were 4.48 km apart, as determined by GPS, and were thought to be in neighboring territories.

Nest 3: Bajada de la Pina Nest, Site A. A third nest was found on 20 April 1993, 15 m from a lightly traveled gravel road through the park (Fig. 3.1). Like nest 1, this one was also high in a tall Palo de Danto in an area of Transitional Forest. Built of twigs averaging 38 cm long and 5 mm thick, the nest was 6 m from the main trunk, in a crotch of five slender limbs (Table 3.1). We checked the nest daily until the birds began incubating on 5 May. With binoculars one egg could be seen through the nest bottom. In late May a severe thunderstorm destroyed the nest (Table 3.1).

Nest 4: Bajada de la Pina Nest, Site B. On 25 April 1996, we found this nest, with incubation underway, high in a tall Kapok 150 m south of the 1993 nest just described; it seems clear this was an alternate nest site within the same territory as nest 3. One adult appeared to be banded, but we could not confirm this. If it was banded, it must have been the male from the 1991 Chikintikal nest, since the banded 1991 female had died. The nest failed by 9 May, apparently destroyed by spider monkeys (see Causes of Nest Failure and Mortality, in this chapter).

Table 3.1 Characteristics of Gray-headed Kite nests and nest sites in Tikal National Park

Nest number	Nest height (m)[a]	Nest tree dbh (cm)	Nest tree species	Nest dimensions (cm)	External nest depth (cm)	Diameter of support limbs (cm)	Nest habitat	Nest description	Notes
1 Chikintikal	24.4 (27)	65.3	Palo de Danto, also called Bitter Angelim (*Vatairea lundellii*)	40 × 33	8	10, 10, 12	Transitional forest with emergent mahoganies	Loosely built of approx. 100 twigs; could see egg through nest bottom	Tucked just beneath dense upper canopy of tree; 50 m from active Bicolored Hawk nest
2 Airstrip nest	22 (25)	88	Chicozapote (*Manilkara zapota*)	43 × 35	7	—	Bajo forest	Similar to nest 1, but slightly more compact and securely anchored	On discovery, one dead nestling in nest and one at base of nest tree
3 Bajada de la Pina, A	26 (28)	54.5	Palo de Danto	—	—	1.5, 2.5, 2.5, 4, 5	Transitional forest	Flimsily built of twigs approx. 38 cm x 5 mm; could see egg through nest bottom	Nest 6 m from main bole; nest destroyed in late May by thunderstorm
4 Bajada de la Pina, B	High in a tall tree	—	Kapok (*Ceiba pentandra*)	—	—	—	Transitional forest	Similar to others	150 m S of Bajada de la Pina nest A; one adult was probably banded
Mean	24.1 (26.7)	69.3		41.5 × 34	7.5	—			

Note: dbh = diameter at breast height.

[a] Nest tree height in parentheses.

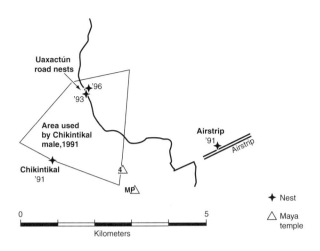

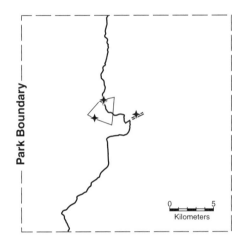

3.1 Gray-headed Kite nests located in Tikal National Park, Guatemala, and telemetry points and a tentative home range for the 1991 adult male of the Chikintikal pair.

Egg and Clutch Size

Two nests at Tikal contained 1-egg clutches (Plate 3.4), and one contained two nestlings. Published records of clutch size for this species are partly in error. In standard references, we find mention of only two clutches—a clutch of two and a clutch of three (Brown and Amadon 1968; del Hoyo 1994). Brown and Amadon (1968) give dimensions of "three eggs said to be this species," and Wetmore (1965) gives the same measurements of "3 eggs from Trinidad," attributing the data to Schönwetter (1961); both authors apparently are referring to the same three eggs of Schönwetter's. Brown and Amadon (1968) and del Hoyo et al. (1994) state or imply that these eggs comprised a clutch of three, which is unjustified; Schönwetter indicates nothing about clutch size. Schönwetter (1961) attributes the three eggs to Kreuger. These three eggs, of the Museum Oologicum R. Kreuger collection, now belong to the Finnish Museum of Natural History at the University of Helsinki. In the early 1950s, Dr. Kreuger purchased the West Indian egg collection compiled by Mr. G. D. Smooker, and these three eggs are from that collection (T. Stjernberg, in litt.). Copies

of the original data cards were kindly provided us by Dr. T. Stjernberg. Smooker's hand-written notes leave no doubt—these were all 1-egg clutches, two collected in 1926, and one in 1932 (Table 3.2). Hence, all published references to a 3-egg clutch appear to be in error.

Brown and Amadon (1968) also state "Another set of two from Trinidad are unmarked" and this is no doubt the same 2-egg clutch alluded to by del Hoyo et al. (1994). We have not been able to find the current location or original collection account of this alleged 2-egg clutch from Trinidad. Interestingly, both of the above authorities failed to mention two single-egg clutches from Trinidad that reside in the collection of the Western Foundation for Vertebrate Zoology, although Wolfe (1964) described them. We suspect that the "unmarked clutch of two" from Trinidad may in fact refer to these two 1-egg clutches. Thus it is not clear whether a 2-egg clutch has ever been collected. The only other museum clutch of which we are aware is in the Delaware Museum of Natural History. This 1-egg clutch was collected near Guanoco, in Sucre, Venezuela, on 26 April 1936 (Table 3.2).

Based solely on original museum data, we have data on six 1-egg clutches. At Tikal we found two 1-egg clutches and a brood of two, which we assume originated from a clutch of two. These data yield a mean clutch of 1.1 ± 0.3 eggs (n = 9). Adding to these the questionable 2-egg clutch mentioned by Brown and Amadon (1968) and del Hoyo et al. (1994) yields a mean clutch of 1.2 ± 0.4 eggs (n = 10). Hence, the best current estimate of mean clutch size is 1.1–1.2 eggs. In any event, the modal clutch is clearly one egg.

Based on available data (Table 3.2), mean egg dimensions are 54.4 × 43.2 mm (n = 6: Table 3.2), which yields an estimated fresh egg mass of 55.7 ± 6.0 g (n = 6: Table 3.2). The modal 1-egg clutch is thus equivalent to 10.7% of the 521 g mean female body mass, and the 1.15-egg mean clutch amounts to 64.1 g, or 12.3% of female body mass.

Nesting Phenology

Nest 1 was being built on 20 April, and an egg was laid in early May. Assuming incubation and nestling periods each require slightly more than a month, fledging would probably have occurred during the latter part of July. Nest 2, discovered on 3 July, held a half-grown chick. Hence, the approximate laying date would have been in late May, and fledging would have occurred in late July had the nestling survived. At nest 4, located on 25 April, incubation was underway; the nest failed by 9 May.

These dates indicate a high degree of nesting synchrony at Tikal. In addition, intense display activity we observed yearly during late winter and early spring also suggests that nesting is highly seasonal. To summarize, these four nests suggest that laying often took place during late April and May (during the latter part of the dry season), with hatching occurring about a month later (around the onset of the rainy season) and fledging in roughly late July, a month or so into the rainy season.

Table 3.2 Dimensions and descriptions of Gray-headed Kite eggs

Dimensions (mm)	Clutch size	Collection locality	Date	Data source
55.8 × 46.1 (Gene K. Hess)	1	Guanoco, Sucre, Venezuela	26 Apr 1936	Delaware Museum of Natural History
53.6 × 43.6	1	Arima, Trinidad	2 Apr 1933	Smooker collection; Western
54.2 × 40.4 (Wolfe 1964)	1		16 Apr 1933	Foundation for Vertebrate Zoology; Wolfe 1964
3 eggs: 53.4–54.3 × 42.2–45.0 (Schönwetter 1961)	1 1 1	Spring Hill Estate, Trinidad	23 April 1926 21 May 1926 9 April 1932	Smooker collection; Museum Oologicum R. Kreuger, Finnish Museum Natural History; Dr. T. Stjernberg
54.8 × 42.1	1	Tikal, Guatemala	21 May 1991	Thorstrom 1997[a]
Mean: 54.4 × 43.2 (n = 6)	Mean: 1 (n = 7)			

Dimensions (mm)	Specimen number	Estimated mass (after Hoyt 1979)	Egg description
55.8 × 46.1 (Gene K. Hess)	DMNH 14978	64.9 g	—
53.6 × 43.6	WFVZ 16314	55.7 g	"Grayish white, lightly marked w/thin peppering and scrawls of yellowish brown" (Wolfe 1964)
54.2 × 40.4 (Wolfe 1964)	WFVZ 16315	48.4 g	
3 eggs: 53.4–54.3 × 42.2–45.0 (Schönwetter 1961)	Kreuger 9571, 9601, 9602	52.0 g 60.1 g	"Grayish white, lightly marked with a thin peppering of dots and scrawls of reddish brown" (Brown and Amadon 1968)
54.8 × 42.1	WFVZ 164697	53.1 g (actual mass 49.0 g— incubated > 2 weeks)	Large end of egg a solid, faded purplish brown extending one-third its length; other two-thirds dirty white, spotted throughout with small rusty brown spots (Thorstrom 1997)
Mean: 54.4 × 43.2 (n = 6)		Mean: 55.7 ± 6.0 g (n = 6)	

[a] We collected this egg after the female died and male abandoned nest.

The scanty data available from elsewhere suggest nesting phenology similar to what we observed. In Belize, two females on 18 and 20 April had an enlarged ovary and an egg in the oviduct, respectively (Barlow et al. 1969), and five clutches from Trinidad were collected on 2 April 1933, 9 April 1932, 16 April 1933, 23 April 1926, and 21 May 1926 (museum data). In El Salvador, a recently fledged juvenile was collected in mid-August (Dickey and van Rossem 1938).

Length of Incubation, Nestling, and Post-fledging Dependency Periods

Nothing is known of the duration of the incubation, nestling, or dependency intervals. Based on body size, one might guess that the incubation period is probably around 32–35 days and that the nestling period may be roughly 35–40 days, but these are only crude guesses.

VOCALIZATIONS

We heard at least three calls given by these kites. Perhaps most often heard was a loud, hollow-sounding, tro-gon-like *kek kek kek kek* that was given in the presence of a conspecific when both were perched in a treetop. This call, with up to 40 syllables voiced at a time, has been likened to that of the Lineated Woodpecker (*Dryocopus lineatus*: Howell and Webb 1995), but is perhaps more reminiscent of a Slaty-tailed Trogon (*Trogon massena*: Slud 1964). A second call was a feline *myowr* given when flying low over the canopy or during incubation exchanges. The third call was a ringing *caw* or *cow* given repeatedly during display flights high above the canopy. This call was reminiscent of the *Spizaetus* hawk-eagles.

BEHAVIOR OF ADULTS

Pre-laying and Laying Period

Wetmore (1965) noted that these kites were most vocal toward the end of the dry season (i.e., early in the breeding season), when they began to voice their loud *kek kek kek kek* call, often from soaring flight high above the forest. We made similar observations at Tikal; these kites were rarely observed except when engaged in such vocal display flights high overhead (see Territorial Behavior and Displays, in this chapter). One such calling bird that Wetmore shot proved to be a female, and he assumed that both sexes gave this call.

Nest-building Behavior

When we discovered the Chikintikal nest on 20 April 1991, it was under construction. Using the beak, one or more of the pair brought small sticks to the nest, seemingly gathering them within 50–100 m. Later, when incubation was underway, at least once, the bird arriving at the nest to take over incubation brought a small twig in its beak, adding it to the nest.

Incubation Period

Nest 1 (Chikintikal nest) was discovered on 20 April 1991, while still under construction. At our next visit, on 1 May, one individual incubated throughout our 4.8-hour observation period, which began at first light; no prey deliveries were made. On 5 May, during 2.5 hours of morning observations, we observed an incubation exchange between the two adults, but again no prey deliveries. The entering bird (sex unknown) uttered a soft, catlike *meow* and perched 5 m from the nest. The incubating bird stood, stretched, made a weak chatter call, and flew off below the canopy, and the incoming bird assumed incubation.

On 9 May we observed for 13.25 hours, from dawn (0515) to dusk (1830). One kite was on the nest incubating at first light. At noon another kite voiced a feeble catlike *meow* to the east. Six minutes later a Gray-headed Kite entered the nest area, perched for 8 minutes, then flew to the nest. The incubating bird stood, stretched, and flew east, disappearing from view. The second bird settled on the nest and began incubating 2 minutes later, continuing until nightfall. Of the 795-minute observation period, the first bird spent 56% (445 minutes) of the observation time on the nest and the second bird, 44% (350 minutes). Thus, during 20 hours of observation of nest 1 on three days, we saw no prey deliveries to the incubating bird, and we witnessed three incubation exchanges among adults. It appeared that both pair members played roughly equal roles in incubation, and that each foraged to meet its own needs.

Results at the Bajada de la Pina nest in 1996 were somewhat different. Here we observed from dawn to dusk on 7, 8, and 9 May, for 13.8 hours per day, totaling 41.5 hours. On 7 May one bird (which we presumed to be the female) incubated all day with only brief respites and remained on the nest during the night. The other bird delivered prey at 0822, effecting a transfer within 20 m of the nest, then left the area after 5 minutes, remaining absent all day. The presumed female spent 14 minutes receiving and consuming the prey item out of view near the nest, then resumed incubation. Just after noon, the presumed female left the nest vicinity for 31 minutes, possibly to hunt. The female was on the nest or within a few meters of it for 96.3% of the day, and incubated 90.1% of the day.

On 8 May one bird incubated throughout the day from first light. Again the presumptive male brought a prey item at 0837, effecting a transfer out of sight but probably within 50 m of the nest, and perched in the nest tree until the female resumed incubation after a 28-minute period in which she presumably ate the prey item out of sight nearby. The presumed male left the area when the female returned to the nest, and remained absent the rest of the day. The female was on the nest 95.9% of the day, incubating constantly except for several interludes of up to 50 seconds spent shifting position, moving the egg, and manipulating nest material. Again this bird remained on the nest until dark and presumably during the night.

On 9 May, the pattern of adult roles differed from that on the previous two days. Again the presumed female incubated from first light, but at 0711 the presumed male flew in, voiced a loud catlike call from a perch a half meter below the nest, and the female quickly stood and flew off. The male then incubated all day until the first bird returned, taking up incubation again at 1749 and continuing until after dark. We witnessed no prey exchanges on this day. At 0943 the male left the nest for 3 minutes to attack and drive away two spider monkeys that had moved into the nest tree, and at 1754, shortly after the female returned to incubation duty, the presumed male drove a Crested Guan *(Penelope purpurascens)* from the nest tree, called for a bit near the nest, then flew off.

In summary, during two days of observation at this nest, Gray-headed Kites showed the most typical falconiform pattern, with one bird, presumably the male, delivering prey once daily to the presumed female, who performed all incubation, including at night. On the third day, one adult performed 17.0% of incubation (during the first 2.18 hours after dark and the 58 minutes before dark), and the other performed 83.0% of incubation, during a single 10.63-hour midday stint. No prey transfers took place. By 13 May, the nest had failed (see Causes of Nest Failure and Mortality, in this chapter).

Nothing is known of nestling development or behavior, nor of behavior or ecology during the post-fledging period. We did observe several immature Gray-headed Kites at Tikal, including one that was following an adult from perch to perch late in the breeding season.

SPACING MECHANISMS AND POPULATION DENSITY

Territorial Behavior and Displays

It seems likely that the primary mechanism by which these kites maintain spacing between nests is by conspicuous, noisy, aerial display. Beginning in about February (i.e., early in the breeding season), these kites were often seen giving a unique courtship or territorial display high above the forest. These distinctive displays consisted of rapid, shallow wing beats, wherein the wings were held above the horizontal plane, after a soar or glide—a behavior we have termed the "butterfly display." A displaying kite broadcasts a ringing, hollow-sounding call, consisting of a repeated note sounding like *caw*, the series beginning slowly and accelerating rapidly and lasting about three seconds. Another vocalization frequently heard is a two-note call that begins at a low frequency and shifts up the scale to a higher-frequency second note. Occasionally these kites can be seen soaring high over the canopy, gliding off to a distant locale.

Interspecific Interactions and Nest Defense

At the 1996 nest, the adult male attacked two spider monkeys as they moved within the nest tree. He struck one monkey, causing it to fall. On another occasion this male attacked a Crested Guan in the nest tree, causing it to flee; we presumed this to be an instance of nest defense rather than a prey capture attempt. At the same nest, an American Black Vulture *(Coragyps atratus)* flew very low over the incubating adult as if attempting to flush it from the nest. As the vulture approached, the kite stood, spreading its wings over the nest.

Constancy of Territory Occupancy, Use of Alternate Nest Sites

Though we obtained few concrete data on this, we had the impression that territories were occupied each year, in about the same localities. Two nests found in 1993 and 1996 were approximately 150 m apart, no doubt being alternate sites within the same territory.

Home Range Estimates

An adult believed to be a male was radio-tagged at the Chikintikal nest in 1991. During a two-week period we located this bird 12 times, during which it used at least 6.2 km^2 (Minimum Convex Polygon [MCP]). Clearly these data are inadequate to suggest a home range size, but presumably the actual home range size must be larger than this.

Inter-nest Spacing and Density of Territorial Pairs

Although we gathered only limited data on nesting density and spatial use, our impression was that these birds ranged over quite large areas. We found four nests, but only the Airstrip and Chikintikal nests were active in the same year, being 4.48 km apart. The two Bajada de la Pina nests, 150 m apart and active in two years, may conceivably represent a third territory. If so, then mean inter-nest distance among these three territories was 2.9 km. However, two lines of evidence suggest that the two Bajada de la Pina nest sites were alternate sites within the same territory as the Chikintikal nest. First, the radio-tagged Chikintikal male in 1991 ranged close to the 1993 and 1996 Bajada de la Pina nest sites. Second, one adult at the 1996 Bajada de la Pina nest was almost certainly banded, in which case it was the same male that nested at Chikintikal in 1991. If the Chikintikal and Bajada de la Pina nest sites were alternate sites within one territory, they were relatively far apart—2.02 km.

Our best estimates as to density are probably those based on the 4.5 km distance between the Airstrip and Chikintikal nests, both active in 1991. Using this datum, the square method yields an estimated one pair per 20 km^2, and the Maximum Packed Nest Density (MPND) estimate is one pair per 18.2 km^2. Use of the 2.9 km inter-nest distance given earlier yields an estimated one pair per 8.4 km^2 (square method) and one pair per 7.65 km^2 (MPND method), quite a different result from the one just given. As a final and maximally conservative approach, a circle with an area of 72 km^2 embraced an area within which we believed there to be 3–5 pairs of Gray-headed Kites, for an estimate of one pair per 14.4–24.0 km^2. All in all, we believe the most defensible estimate of density is roughly one pair per 20 km^2, though this is best regarded an educated guess.

Causes of Nest Failure and Mortality

Stormy weather and predation on adults and young contributed to nest failure, and disturbance by spider monkeys probably also played a role. Failure of the 1993 Bajada de la Pina nesting resulted from a windstorm that toppled the nest. On 28 May 1991 at the Chikintikal nest, one of the adult birds, thought to be the female, was found dead on the ground below the nest. The bird had been plucked, and its breast and wing tissue eaten. Evidently the incubating female had been captured on the nest, perhaps by another raptor. Several days later, when it became apparent that the male had abandoned the nest, we collected the egg, depositing it at the Western Foundation of Vertebrate Zoology (specimen 164,697).

At the 1996 Bajada de la Pina nest, at the outset of the second week of observation (during the incubation period), we found the nest abandoned. Searching below, we found feathers, eggshell fragments, and 14 of the living limbs that had shaded the nest. Based on prior experience with spider monkeys breaking off limbs that shaded raptor nests and throwing a young hawk-eagle from the nest (see Chapter 14), we speculate that spider monkeys caused this nest failure. Thereafter, the kites were observed in the vicinity but did not re-nest here.

CONSERVATION

Little is known of this species's population status, but its fate is likely tied to rates of deforestation. Ramos (1986) believed populations in Mexico to be declining, probably owing to habitat destruction and shooting, and in El Salvador, where nearly all mature forest is gone, the species is believed to have largely disappeared (Thurber et al. 1987). In western Ecuador, Ortiz-Crespo (1986) described the species as very rare, due to deforestation. In southern Brazil, Albuquerque (1986) reported that gallery forest, the preferred habitat of these kites in that area, was declining because of conversion to soybean farming. In point counts comparing Tikal's intact forests with the adjacent farming landscape, we had more detections in the forest than in the farming landscape, but detections were too few overall to have much confidence in this result (Whitacre et al. 1992a, 1992b).

Some evidence suggests that Gray-headed Kites may show some tolerance of deforestation. In Argentina, Olrog (1985) speculated that these kites might not be adversely impacted by forest clearing, since they occupy savanna woodlands in part. Likewise, at least as recently as 1978, these kites still occurred in the isolated, 87-ha patch of primary forest constituting Rio Palenque reserve in lowland Ecuador, whereas several other forest raptors were no longer observed there (Leck 1979). Jullien and Thiollay (1996) classified this as a species that, while widespread in primary rain forest, also occurred nearly as often in a wide array of forest types such as drier coastal, mangrove, palm swamp, and fragmented forests. Species included in this grouping were considered to use forest edges, gaps, upper canopy, and river banks more readily than did several "obligate mature forest" raptors, and to tolerate a higher degree of deforestation. Finally, Thiollay (1991a) regarded this as one of several essentially primary forest species in Colombia that may survive forest disturbance and fragmentation better than obligate forest species. Still, he concluded that these kites probably would not be able to persist in heavily deforested areas; we share this opinion.

HIGHLIGHTS

Other than brief notes on data cards in egg collections, the nesting observations reported here are the first for the Gray-headed Kite. This species is not rare at Tikal as suggested by Beavers (1992), and we believe it is often overlooked when present. The key to effective detection is to recognize the ringing call that is frequently given from flight high overhead; this call was often our first notice of a kite nearby. Searches for this kite also could profitably focus on the early (pre-laying) portion of the breeding cycle, when aerial display appears to be most frequent.

At the two nests we observed, a range of incubation patterns was noted. On some days, one bird incubated all day, receiving a single prey item from its mate during the morning. On other days there were no prey deliveries, and both birds incubated for long spells, each adult presumably foraging for itself while off the nest. In future nest observations, it will be valuable to learn more of the degree to which adults meet their own food demands during the nesting cycle, as opposed to the male feeding the female during incubation and early in the nestling phase—the most frequent pattern among raptors.

Pondering this kite's apparent predilection for feeding on nests of large wasps, Dickey and van Rossem (1938) wondered how these hawks could pursue such a diet without being stung to death. This question seems as relevant today as it was in 1938. The strong odor given off by the kites we handled at Tikal is intriguing. Though wildly speculative, one is tempted to wonder whether this could indicate a chemical defense against the stinging wasps on which these kites apparently often prey. Another mystery is the following. Foster (1971) described puzzling behavior in which an individual hung upside down from a perch for 15 minutes, then casually flipped back upright. This did not appear to be sunning or anting behavior, or juvenile clumsiness. Whoever first manages to observe in detail the foraging of this kite will no doubt be treated to some fascinating results.

For further information on this species in other portions of its range, refer to Cabanne 2005; Carvalho Filho et al. 2005; and Di Giacomo 2005.

4 HOOK-BILLED KITE

David F. Whitacre and Miguel A. Vásquez

Secured to perches in the top of an emergent tree, we sway in a light breeze as our three-hour point count reaches the halfway point. "Mira, otro picoganchudo . . . un macho, no?," says Miguel, pointing toward the center of our westward viewshed. "Ah, sí, es el tercero, entonces," replies Dave—a third individual, each in a different plumage—this one a male. The Hook-billed Kite *(Chondrohierax uncinatus)* rises on the freshening breeze and tilts, exposing its grayish underparts and unique wing profile. Another burst of flaps and the kite soars, wobbling, its wedge-shaped tail twitching as it circles into the wind, rising quickly. Another moment and it passes directly above, giving a perfect underside view. The paddle-shaped wings emerge from an unusual, pinched-in base and extend outward like broad planks, culminating in a rounded series of deeply emarginated primaries, producing a notably "fingered" wing tip. The outer primaries are boldly barred black and translucent white, and the head appears smallish for the bird's size. The kite soars with its wings pushed distinctly forward, further emphasizing the narrow wing base and broad, deeply indented outer wing.

Completing a few more circuits, the kite eases into a descending glide path back toward us, and drops into the open upper canopy of a Chicozapote *(Manilkara zapota)*, where we are able to see the fine whitish barring of the gray breast, its orange feet, and excessively hooked bill. The black pupil set amid a piercingly white iris gives the bird a demented expression, as suggested by Snyder and Snyder (1991). We can also see the lime-green loral region and bright orange crescent of bare skin above and in front of the eye—features whose significance is unknown. The kite leans forward, spreads its wings, and drops steeply to a lower perch in the same tree. Later in the census we catch a final glimpse of the kite as it lofts steeply upward to another perch within the same tree.

Although the Hook-billed Kite ranges north to Texas and has been the subject of a detailed study of its plumage and bill polymorphisms and its method of prey handling, this strange raptor retains, for us, a distinct air of mystery. The few facts known of this bird do little more than deepen the secrets hidden within its strange anatomy, hypervariable plumage, and specialized feeding habits—as this kite is a nearly absolute devotee of escargot! The Hook-billed Kite is of special interest in showing the greatest degree of variation known in any raptor, both in plumage coloration and in bill size. Large-billed individuals were once allocated to a separate species, *C. megarhynchus*, but it is now believed that large and small-billed forms belong to a single species. Bill size is unrelated to age, sex, and, for the most part, geographic origin, and in some regions is bimodal, with both small-billed and large-billed forms present, apparently without intermediates (Smith and Temple 1982a). This kite is also of urgent interest from a conservation perspective, in possessing two distinctive populations that are in danger of extinction. Here we review what is known of the Hook-billed Kite and present new data on adult roles during nesting, on diet and prey delivery rates, and on other aspects of nesting biology at Tikal.

GEOGRAPHIC DISTRIBUTION AND SYSTEMATICS

Brown and Amadon (1968) regarded this genus as allied with *Leptodon* and *Aviceda*. Such an affinity was supported at a gross level in a cladistic study based on aspects of behavior, osteology, anatomy, plumage, pigmentation, karyology, and egg characters (Kemp and Crowe 1990).

Two subspecies of *Chondrohierax uncinatus* are currently recognized—one mainland form and one island form. While mainland populations have at times been divided into three subspecies, all mainland birds are now considered to belong to the nominate subspecies, which ranges from southern Sinaloa, south Texas, and Tamaulipas, Mexico, southward, west of the Andes to western Ecuador, and east of the Andes to central Bolivia, southern Brazil, Paraguay, and northern Argentina (del Hoyo et al. 1994; AOU 1998). The island subspecies, the Grenada Hook-billed Kite *(C. u. mirus)* occurs on the island of Grenada in the Lesser Antilles. The Cuban Hook-billed Kite (formerly subspecies *C. u. wilsonii*) is now accorded species status *(C. wilsonii)* and is found in eastern Cuba.

The Cuban Kite has at times been considered merely rare (Collar 1986), but it is now generally conceded to be critically endangered (BirdLife International 2000), and the Grenada form has long been considered critically endangered, with possibly fewer than 30 individuals extant (del Hoyo et al. 1994).

Hook-billed Kites were first recorded in the United States—in southernmost Texas—in 1964, when a nesting pair was discovered along the Rio Grande in Santa Ana National Wildlife Refuge, about 100 km upstream from the river's mouth (Fleetwood and Hamilton 1967). A second U.S. record and a successful nesting were noted in 1975–76 in the same area (Delnicki 1978). Since that time, Hook-billed Kites have been sighted frequently, in both summer and winter, but apparently not every year, at several localities in Texas's lower Rio Grande Valley, and a number of nestings have been recorded (Brush 2005).

In Mexico the usual vertical range is sea level to 900 m (tropical and lower subtropical zones), and rarely to 2500 or even 2900 m (Friedmann et al. 1950; Schaldach 1963; Blake 1977; Binford 1989; Howell and Webb 1995). In South America this kite also occurs mainly in the tropical and subtropical zones, and rarely in temperate forest up to 2500 m (Venezuela), 2700 m (Ecuador), and 3100 m (Colombia; Fjeldsa and Krabbe 1990). In northern Argentina, a pair was found nesting in wet mountain forest at 1600 m (Olrog 1985).

MORPHOLOGY

The Hook-billed Kite has, arguably, the most unusual flight profile of any raptor in the Americas, with the broad outer wing and many deep wing-tip slots among the notable features. Wing-tip slots are generally believed to minimize drag due to wing-tip vortex (Norberg 1985) and to prevent loss of lift from the wing tip (Graham 1930), facilitating a wing that is broad from front to back over much of its total length—allowing a relatively short wing without sacrificing lift. Short wings, in turn, are thought to be advantageous for birds living in the cluttered airspace of forest interior. In addition, wing-tip slots, along with a broad, deeply cambered wing, are believed to maximize the ability to fly at high angles of attack and at relatively slow speeds, maximizing climbing ability and maneuverability (Norberg 1985). We speculate that the foraging style of these kites, which seems to involve frequent changes of height over short distances within and below the canopy, may place a premium not only on maneuverability but also on the ability to climb steeply in flight. The oddest feature of this kite's wing shape, however, is the distinctly "pinched-in" wing base. We have seen no potential adaptive role attributed to this feature, and we speculate that perhaps this mechanism allows a rapid spilling of air, permitting the kite to drop steeply while retaining aerodynamic control.

As mentioned also by Howell and Webb (1995), these kites often fly with the wings pressed slightly forward—as portrayed in Plate 4.1, or even more so. Our impressions also agree with these authors' description of the wing beats as often loose and floppy, and with Stiles and Skutch's (1989) description of this kite's few quick flaps followed by "a wobbly glide."

Males average 263 g and females 280 g in body mass (see Table 1.2), for a Dimorphism Index (DI) of 2.07. A larger data set on linear measurements (Appendix 2) yields DI values of 1.3 for flattened wing length, 5.9 for tail length, 0.95 for culmen length, –3.8 for tarsus length, and –0.64 for middle toe without claw. The mean of DI values for wing length, culmen length, and body mass is 1.42—among the lowest degrees of reversed sexual size dimorphism shown by any North American raptor (Snyder and Wiley 1976) (Plates 4.2 and 4.3).

Smith and Temple (1982a) found that bill size was independent of sex and age class. Mean bill size differed among geographical regions, but within each mainland region, variation in bill size was extreme, with the coefficient of variation of bill area ranging from 15 to 42%—compared to 10% in two other widespread raptors—the Red-tailed Hawk *(Buteo jamaicensis)* and American Kestrel *(Falco sparverius)*. The regions with greatest bill size variation also showed bimodal distributions of bill size, having large- and small-billed individuals, but few or none with intermediate bill size. Everywhere that large-billed birds were found, they occurred along with small-billed individuals. Wing length (as an indicator of body size) varied among regions, but it was much less variable than bill size and revealed no discernible geographic pattern (Smith and Temple 1982a). On the islands of Cuba and Grenada, bill size variation was 30% less than on the mainland, a reversal of the frequently observed phenomenon of "ecological release" and expansion of feeding niches on islands (Crowell 1962; MacArthur et al. 1972; Wright 1980). No geographic variation in plumage coloration was evident, except that both island populations differed significantly from mainland populations, and no dark-phase birds are known from either island.

Hook-billed Kites have notably short tarsi. The only diurnal raptor at Tikal (of the 22 most common species) with proportionally shorter tarsi than the Hook-billed Kite is the Swallow-tailed Kite *(Elanoides forficatus)* (Appendix 2). Snyder and Snyder (1991) also note that snails are often grasped in the foot and manipulated and eaten in a parrot-like fashion, "much as a small child might attack an ice-cream cone"—an unusual mode of prey handling among raptors.

PREVIOUS RESEARCH

The only thorough study of the species is Smith and Temple's (1982a) analysis of the feeding mechanism and geographic patterns of plumage and bill size variation. A thorough summary of existing knowledge is given by Smith (1988), with briefer summaries by Brown and Amadon (1968), Johnsgard (1990), del Hoyo et al. (1994), and Ferguson-Lees and Christie (2001).

RESEARCH IN THE MAYA FOREST

In 1992 we studied four nesting attempts of Hook-billed Kites at Tikal. Two nests were observed from a similarly tall perch atop a Maya temple approximately 50 m away. One of these (nest 2) was observed every three days from dawn to dusk on 15 days from 17 June to 10 August, for a total of 168.57 hours. Observations here averaged 11.24 ± 1.5 hours per day (*n* = 15). We paid special attention to identifying prey brought in and estimating its size and mass, and to documenting behavioral roles of pair members.

Nest 0 was 15 m up in a Gumbo-limbo tree, also known as West Indian Birch *(Bursera simaruba)*, 16.5 m tall and 55.5 cm in diameter at breast height (dbh). This nest, discovered when under construction on 22 May, was in disturbed, tall Upland Forest in the Mundo Perdido area of Tikal's ruins. It was abandoned before being used. We do not consider it further.

Nest 1, in a Jesmó tree *(Acacia* sp.) 50 m from the large Mundo Perdido temple, was discovered on 13 May, when the female was incubating a single egg. The nest was 24 m up in a 25 m tree, 43.6 cm dbh, amid disturbed Upland Forest of this major temple complex. On 25 May we found the nest abandoned (during incubation), the egg(s) having been destroyed by a predator, possibly Brown Jays *(Cyanocorax morio)*. It was not possible to climb to the nest, which was precariously situated in a fork of a thin limb.

Nest 2, discovered on 15 June with incubation underway, was 37 m up in a 45 m Ramón, also known as Breadnut *(Brosimum alicastrum)*, 119 cm dbh. This nest was also amid disturbed Upland Forest at Mundo Perdido, and we believe it was a re-nesting attempt by the same pair of kites responsible for nest 1. One chick, first observed on 13 July, we believed to have hatched on 11 July, and the second chick hatched by 14 July. Hence, the incubation period was at least 30 days for one egg. One chick here disappeared from the nest, presumably fledged, on August 4 or 5, and the other fledged between August 11 and 16. Assuming the first-hatched chick fledged, these chicks fledged at about 24–25 days and 28–33 days.

Nest 3, containing two nestlings, was discovered on 1 July by a park guard and shown to us that evening. It was 17 m up in a 19 m Honduras Mahogany *(Swietenia macrophylla)* in an area of Bajo Forest near Tikal's airstrip. The nest was very flimsily constructed of dried twigs and was situated in a fork of a limb. We again visited the nest early on the afternoon of 2 July, finding one dead nestling on the ground below the nest; we estimated its age as approximately 20 days, as black feathers were beginning to emerge on the tips of the wings and on the back. The dead chick may have fallen because of wind, as the nest was on a very thin branch and was moving around a good deal in the wind during our visit. One chick, which we believe was the older of the two, remained in the nest, and we climbed to examine it. We estimated the surviving chick to be approximately

25 days old and estimated rough hatch dates of 7 June and 12 June for the two chicks. There were many snail shells in the nest, and on the ground below, 12 m from the nest, we found many more large snail shells.

On 2 July, we were at the nest from 1330 to 1632. At 1550 the female arrived at the nest with food for the chick, then perched 30 m from the tree, calling. At 1615 the female made what appeared to be an unsuccessful prey capture attempt near the nest, then perched again 15 m from the nest and continued to call. As the female called, the male flew to the nest tree at 1625 with a small snail, but he did not deliver it to the nest. At 1632 the female flew to where the male was perched, and continued to call.

On 6 July we discovered the juvenile 9 m from the nest, at a height of 6 m. On 7 July the chick was 8 m from the nest tree at a height of 7 m, vocalizing; all day we attempted, without luck, to trap an adult to fit with a radio transmitter. On 8 July we briefly caught the adult female but she escaped. The nestling remained in the same tree, constantly begging for food. The male almost never came to the nest but perched instead in a Yoruba Indigo *(Lonchocarpus* sp.) tree nearby. The female played the role of delivering prey to the chick, mainly snails. We trapped until mid-afternoon, then built a tree platform girded with nooses, in the center of which we placed the chick as an attractant. On 9 July we placed the fledgling (which could not yet fly well and hence was easily captured by hand) in the trap, but we failed to capture an adult. We weighed, measured, banded, and released the chick in early afternoon. On 14 July we attempted to place a trap near nest 2 in the Ramón at Mundo Perdido, but gave up as the effort was too dangerous owing to the thin limb supporting the nest.

DIET AND HUNTING BEHAVIOR

Smith (1988) lists all Hook-billed Kite prey known through 1982. Adding the observations of Thorstrom et al. (2001) brings that account up to date. In every case, the diet has been entirely snails, or nearly so. In most cases land or tree snails are the main prey, although in a few cases aquatic snails are said to have been taken (Van Tyne 1935; Stiles and Skutch 1989). Non-snail prey listed in the past, but without supportive data or details, includes toads (Meyer de Schauensee 1964); frogs, salamanders, insects, and caterpillars (Brown and Amadon 1968); reptiles, insects, and birds (Meyer de Schauensee and Phelps 1978); and frogs and birds (Fjeldsa and Krabbe 1990). However, the only actual data we could find indicating non-snail prey was a small batrachian (frog or toad) in one kite's stomach (Dickey and van Rossem 1938). We have the impression that inclusion of non-snail prey items in this raptor's diet has been repeated from author to author, and in some cases may not rest on actual observations. Overall, the prevalence of non-snail prey has perhaps been exaggerated.

As elsewhere, Hook-billed Kites at Tikal fed predominantly on arboreal snails. At nest 2, we observed 63 identified prey items brought to the nest and 3 unidentified items. Prey items were 58 snails (92%) and 5 small lizards (8%). Snail shells collected below a nest at Tikal were *Orthalicus princeps* (23), *Helicina rostrata* (6), and *Drymaeus* species (5) (identifications by Dr. Fred G. Thompson, Florida Museum of Natural History). According to Dr. Thompson, *O. princeps* is an arboreal snail broadly distributed in lowland tropical forest from Tampico, Veracruz, south through Panama and possibly South America. *Helicina rostrata* is a smaller, arboreal snail of lowland mesic forest, occurring in Petén and Alta Verapaz, Guatemala, and in Belize. The *Drymaeus* species was unknown, not previously reported for the area.

At nest 2, the adult kites brought in prey from 0635 to 1755. Prey delivery rates, adjusted for observation effort per hourly period, are shown in Figure 4.1. Of all the raptors we studied at Tikal, these kites showed the least tendency toward a dip in prey delivery rates during the midday to early afternoon hours. In contrast to other raptor species, these kites had highest delivery rates from mid-morning to early evening, with lower rates in early morning and late evening. Presumably this may relate to patterns of activity or detectability of their mollusk prey, or both. Alternatively, the graph in Figure 4.1 may not relate to capture rates—perhaps adults fed themselves mainly during early morning and late evening and fed nestlings proportionally more often at midday.

Details of feeding behavior are given by Smith and Temple (1982a, 1982b), who observed kites extract more than 60 snails from their shells. Hook-billed Kites extract snails in a manner very different from several other snail-eating birds (Snyder and Snyder 1969; Snyder and Kale 1983). Bill size in the kites was closely matched to sizes of tree snails common in the same region (Smith and Temple 1982a). In Tamaulipas, Mexico, where only small-billed kites were observed, only one, small, tree snail species was available to them. Likewise, in

Grenada, only small-billed kites were found, and they fed mainly on two snail species, of similar size. In contrast, in western Mexico, where large- and small-billed kites occurred together, they fed on snails of two distinct size classes.

Few observations of hunting behavior have been reported. In Grenada, a female was seen feeding on small tree snails. She walked along a limb in search and, finding one, jumped to a small limb, hung upside down, and plucked the snail from a limb with her bill, then flew to a larger limb to extract the prey from its shell (Smith and Temple 1982b). In south Texas, Brush (1999:32) made a similar observation. A female that landed in an open stand of mesquite "immediately began turning her head scanning for snails. At least five times in five minutes she sallied to a nearby branch or to the ground, picked off a snail, and then perched on a fallen log to consume it."

HABITAT USE AND ABUNDANCE

While earlier accounts often portrayed these kites as occurring mainly in marshy areas or swampy forest (e.g., Griscom 1932; Land 1970), more recent accounts show clearly that these kites frequently occur in a variety of forest types—often quite arid—in the absence of surface water. Habitats used include tropical lowland evergreen and tropical deciduous forest, gallery forest, and montane evergreen forest, ranging from sea level up to 2800 m, and, for the Cuban form, tropical lowland evergreen forest (AOU 1998; Ferguson-Lees and Christie 2001). Smith and Temple (1982b) found the Grenada Hook-billed Kite mainly in thorny, xeric woodlands less than 5 m in height, although they historically have occurred in other habitats as well. Thorstrom and McQueen (2008) also found them nesting in montane rain forest above 400 m. Paynter (1955) emphasized that while this species is found "chiefly in marshy areas," it was found in abundance by Traylor (1941) at Chichén Itzá, an area of low, dry deciduous forest devoid of marshes, the only surface water consisting of a few cenotes (vertical-walled pit-caves). In Tamaulipas, Smith (1982) studied several nests in low semiarid Tamaulipan thorn forest. In Costa Rica, Stiles (1985) regarded the Hook-billed Kite as occupying the interior of intact tropical evergreen forest. In Colima, Mexico, Schaldach (1963) described this kite as being "normally seen only in or over heavy Tropical Deciduous Forest." In Oaxaca, Binford (1989) regarded this kite as an uncommon resident on the Pacific slope, in palm forest, swamp forest, mangrove swamp, and near water in tropical deciduous forest, although he does give one record (in May) from sparse arid pine-oak forest at 2000 m elevation. In Oaxaca, Rowley (1984) found a nest at 240 m elevation, in a dense thicket amid tall tropical deciduous vegetation in an area that was "very definitely arid," the nearest water being a small creek about 800 m away.

Many authors have described these kites as being distinctly patchy and local in occurrence (Griscom 1932;

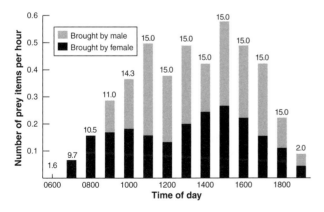

4.1 Time of day adults delivered prey items to a Hook-billed Kite nest at Tikal, Guatemala, 1994. Number at top of bar = total observation time, in hours.

Ridgely and Gwynne 1989; Stiles and Skutch 1989). In Guatemala, Land (1970) considered these kites rare in the Pacific and probably the Caribbean lowlands, and uncommon and very local in distribution in the Petén. In Colombia, this kite is described as "thinly spread in humid forest, swampy areas, second growth, and gallery woodland; also occasionally thorn scrub" (Hilty and Brown 1986). In Panama, Ridgely and Gwynne (1989) described this kite as "uncommon in humid forest and second-growth woodland in lowlands" and as being quite local in occurrence, and not yet recorded from large areas such as the Darién, where it is probably present. They indicate that this kite is especially fond of swampy areas in Panama and is not often recorded at higher elevations as it is in many regions. In Costa Rica, this kite has been called "widespread but generally uncommon to rare and local," mainly in lowlands and foothills (Stiles and Skutch 1989).

A few authors, especially in northern South America, have regarded these kites as somewhat more abundant than suggested above. In Surinam, Hook-billed Kites have been described as "rather common in wet, thick woodland and also in coffee plantations" (Haverschmidt and Mees 1994). In Venezuela, these kites were considered "not uncommon" in lowland seasonal forest (Friedmann and Smith 1950).

Our comparative point count results at Zotz (adjacent to Tikal, to the west), Dos Lagunas (60 km north of Tikal), and Calakmul (100 km north of Tikal) were instructive. In two years of point counts at each site (20 counts: 2 at each of 10 points), out of a possible 20 detections in 1991 and repeated in 1992, detections of Hook-billed Kites were as follows: 2 at Zotz; 4 and 6 at Dos Lagunas; and 14 and 8 at Calakmul (Jones and Sutter 1992). Regardless of any possible difference in detectability between sites, Hook-billed Kites appeared to be substantially more common at Calakmul than at the other two sites, or at Tikal. The abundance of Hook-billed Kites at Calakmul may be understandable in terms of habitat. Forest at our Calakmul sampling area was low statured with dense understory, similar to seasonal swamp forest at Tikal. We hypothesize that this forest type may be especially favorable for this kite.

MIGRATION AND OTHER MOVEMENTS

One of the most exciting things we have to report here is the recent discovery that at least some individuals of some Hook-billed Kite populations migrate. Although we detected no evidence of Hook-billed Kites migrating at Tikal, researchers elsewhere have recently shown beyond a doubt that regular seasonal movements occur in some areas. For three years, Lee Jones (2002, pers. comm.) has detected significant autumn migration of Hook-billed Kites southward along the coast of Belize. In flocks as large as 120, as many as 338 birds have been counted in a day, and Jones estimated that perhaps 5000 Hook-billed Kites passed over this site in autumn 2001, during October and November. Along the central coast of Veracruz, Mexico, Hook-billed Kite migration has been studied annually since 1991 by Ernesto Ruelas and associates (Ruelas et al. 2002). From 1991 to 2001, a mean of 60 kites has been recorded during spring migration (March–May), and 120 during the autumn (August–November).

At least two or three additional observations of flocking by Hook-billed Kites have also been made, although it is not clear whether these were associated with migration. In Venezuela, Paulson (1983) observed a flock of 25 Hook-billed Kites on 2 September, soaring up from semi-wooded country in an afternoon thermal and disappearing to the northeast; this appeared to be migration or at least a substantial movement. Similarly, in French Guiana, Thiollay (pers. comm.) once saw a flock of 11 Hook-billed Kites in apparent migration over the extensive primary forest. On the Pacific coast of Chiapas, Mexico, Tom Dietch (pers. comm.) was told by plantation owner Walter Peters of flocks of Hook-billed Kites feeding in mangroves he owns near Puerto Madero, Chiapas.

In both of the clear-cut cases of migration just cited (along the coasts of Veracruz and Belize), the geographic origin of migrating birds is unknown. It has not been reported to date that this kite completely vacates any portion of its breeding range, and birds breeding in south Texas have often been noted wintering there. Along with Jones's (2002) observations of a large southward autumn migration in Belize, it is interesting to note that Traylor (1941) found these kites "surprisingly common" at Chichén Itzá, in the northwestern corner of the Yucatán Peninsula, where he collected five individuals between 10 October and 2 November 1939. Given the collection dates, these specimens could have been migrants, and conceivably their apparent abundance could even have been due to migratory movement.

Of course, even if these kites routinely make local to large-scale migrations, there is no reason why these need to be along a north-south axis. Since the entire range for this species occurs within the tropics and subtropics, it is perhaps more likely that movements would be between areas where tree snail populations differ in phenology, so that areas of maximal snail density or availability are occupied throughout the year—and this may be independent of latitude. Hence it may be that drier regions are occupied only during the wet season, and that wetter regions are occupied either year-round or during the dry season. One wonders whether local movements between habitats may also take place, perhaps with seasonally flooded and drought-stressed Bajo Forests occupied at one time of year, and nearby, better-drained Upland Forests occupied during another time of year. Finally, it may be worth considering the possibility that elevational migration occurs. Hook-billed Kites have been observed at strikingly high elevations in a number of cases, and we wonder whether the kites are present year-round at upper elevations. In Colima, Schaldach (1963) collected a specimen at about 2450 m in January, and he observed one at 2900 m in December, in heavy fir forest on a north slope. Ferguson-Lees and Christie (2001) indicated that some altitudinal movement occurs in the lower temperate zone of the Andes, but they gave no details. At any rate, it is best to bear in mind that snail availability to the kites may not be the result simply of snail population size. In Tamaulipas, Smith (1982) found Hook-billed Kites foraging very rapidly on a dense population of torpid, aestivating snails (Snyder and Snyder 1991). Conceivably the degree of snail aggregation, positions of snails in the environment, and snail activity levels may all affect snail availability to Hook-billed Kites.

In contrast to migratory movements per se, it has also been suggested that Hook-billed Kites may often aggregate at rich prey resources (Paulson 1983), and that the species may lead a somewhat nomadic existence, "moving around as snail populations vary" (Brush 1999:29). In south Texas, Brush (1999) reported intriguing patterns of occurrence over multiday periods. In this area, the kites appeared to forage in certain areas for a few days, during which time they followed a similar routine, often passing the same point at the same time each day. Then, the observer would often go several weeks without seeing a kite. Brush (1999:31) surmised that the kites "may deplete local populations of tree snails before moving to another area," which seems to us a reasonable conjecture. Similarly, Stiles and Skutch (1989) stated that a pair of Hook-billed Kites may frequent an area for several months, breed, and then apparently move elsewhere; this observation, if generally true, could apply either to a situation of regular migration, to seasonal changes of habitat, or to a somewhat itinerant or wandering mode of habitation. Paulson (1983), in reporting a sighting of two-dozen Hook-billed Kites soaring directionally across the countryside together, emphasized that their snail prey may well occur patchily in time and space, and that these kites may prove to be more gregarious than is generally appreciated. Gregariousness is a frequent correlate of patchy prey resources, probably due to one or more mechanisms through which flocking aids animals in more reliably finding patchy, unpredictable food resources (Pulliam and Millikan 1982).

Hook-billed Kite observations were by far most frequent during June and July in our study area, but kites were also seen in November, December, and January. Thus it does not appear that Hook-billed Kites vacate Tikal during the winter.

BREEDING BIOLOGY AND BEHAVIOR

Nests and Nest Sites

The most detailed published descriptions of nesting are those of Haverschmidt (1964a, 1965, 1968; Haverschmidt and Mees 1994) and Smith (1988). Adding to these the nesting data provided by Fleetwood and Hamilton (1967), Orians and Paulson (1969), Delnicki (1978), Kiff (1981), Smith (1982), Rowley (1984), Snyder and Snyder (1991),

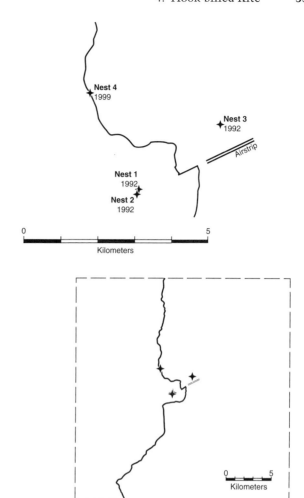

4.2 Hook-billed Kite nests located in Tikal National Park.

Thorstrom et al. (2001), and Thorstrom and McQueen (2008) completes a survey of published nesting information. Nests are typically rather small and shallow, perhaps 30 cm in diameter, and somewhat flimsily constructed of small twigs, such that eggs or chicks are often visible through the nest bottom (Smith 1982; Haverschmidt and Mees 1994; D. Whitacre, pers. observ.).

All four Tikal nests were stick structures precariously situated on thin branches near the tops of trees. Nests were an average 23 ± 10 m (range = 15–37 m, *n* = 4) up in trees that averaged 26.5 ± 13 m tall (range = 16.5–45 m, *n* = 4). Three nest trees measured 56, 44, and 119 cm dbh. Three of four nests were within 1–2 m of the top of the nest tree, and the fourth was in the upper fifth of the tree. Nests described earlier, in lower forest types, were much closer to the ground (Smith 1988), and nest height no doubt reflects the predominant height of trees in the area. Nest locations at Tikal are shown in Figure 4.2.

Egg and Clutch Size

The egg description given by Schönwetter (1961) is apparently in error, almost certainly referring instead to eggs of the Gray-headed Kite (*Leptodon cayanensis*: Haver-schmidt 1964a; Kiff 1981). Kiff (1981) reviewed existing information on Hook-billed Kite eggs and gave measurements of seven eggs in collections. These eggs averaged 45.0 ± 1.9 by 36.5 ± 0.9 mm (n = 7), for an estimated fresh mass of 32.8 ± 2.8 g (n = 7) per egg. Two eggs weighed by Haverschmidt (1968) were 32 and 35 g. For the seven museum clutches, estimated clutch masses were 59.2, 68.0, 67.8, and 34.7 g (three 2-egg and one 1-egg clutch).

Table 4.1 lists Hook-billed Kite nests described in published accounts of which we are aware. In addition, nestings have been discovered in Santa Ana National Wildlife Refuge and other nearby locations in the lower Rio Grande Valley of Texas a number of times since the first known Texas nesting in 1964 (Webster 1978), but generally few details are given. In Costa Rica, a male was observed incubating on 30 May and 5 and 12 June, and by 19 June the eggs had hatched; clutch size was not determined (Orians and Paulson 1969). In Surinam, Haver-schmidt (1964a, 1965) observed two active nests. His results are summarized in Haverschmidt and Mees 1994, as follows. Both nests were in Erythrina (*Erythrina sp.*) trees in coffee plantations. The first was under construction on 24 October and 4 November 1964, and contained one egg on 24 November and two eggs on 27 November. The second nest, in a similar situation, held two newly hatched nestlings on 20 April 1963. For the Grenada form, Bond (1936, 1984) states that a nest was found in March high up in an inaccessible tree 800 meters from the coast, about 30 meters above sea level, but no details are given. Two other nests in Grenada were described by Thorstrom et al. (2001).

To summarize Table 4.1, all known clutches are of 2 eggs (SD = 0, n = 7), and we conclude that the modal and mean clutch sizes are reliably of 2 eggs. A non-overlapping set of nests with chicks held a mean of 1.78 ± 0.67 chicks (n = 9; range = 1–3); one brood of three chicks is known, as are five broods of two and three broods of one, which may have been noted after egg or chick loss in some cases. The 32.8 g mean egg mass amounts to 12.8% of adult female body mass, and the 2-egg mean and modal clutch amounts to 25.6% of female body mass.

Nesting Phenology

Apart from one Surinam nest that was under construction 24 October and had a 2-egg clutch completed between 24 and 27 November (Haverschmidt and Mees 1994), all other known nest records (all from Surinam north) indicate nesting relatively late in the north temperate spring (Table 4.1). Both eggs and nestlings have been noted from early May through the end of June, with the latest hatch date of 5–15 July indicated at one Grenada nest.

Although we visited four historic nest sites every few days beginning in mid-February 1992, we saw no clear-cut nesting behavior until mid-March. Checking these areas on six days from 15 February to 15 March, we observed no activity around nests, but we did observe pairs and singles soaring on a few occasions. We observed copulation on 17 March and 24 April, and nest building on 15 and 24 April. We first located a nest on 13 May, with incubation of a single egg underway. Another nest under construction on 22 May was abandoned before laying

Table 4.1 Known clutch and brood sizes of Hook-billed Kites

Locality	Number	Nest contents	Date	Source
Texas	2	Eggs early May	Newly hatched 6 Jun	Delnicki 1978
Texas	3	Downy young	3 May 1964	Fleetwood and Hamilton 1967
Texas	—	Male nest building	29 Apr	
same nest as above	2	Half-grown young	Later in summer	Webster 1978
Nuevo Leon, Mexico	—	Unknown	1 May 1975	Montiel de la Garza and Contreras-Balderas 1990
Nuevo Leon, Mexico	—	Unknown	28 Jul 1977	Montiel de la Garza and Contreras-Balderas 1990
Nuevo Leon, Mexico	2	Pigeon-sized young	Jun 1978	Smith 1988
Tamaulipas, Mexico	2	Eggs	23 Jun 1979	Smith 1982
Tamaulipas, Mexico	2	Eggs	2 May 1910	Kiff 1981
Tamaulipas, Mexico	2	Eggs	14 May 1908	Kiff 1981
Tamaulipas, Mexico	1	Large nestling	18 Jun 1979	Smith 1982
Tamaulipas, Mexico	1	Large downy chick	21 Jun 1979	Smith 1982
Tamaulipas, Mexico	2	Old nestlings	21 Jun 1978	Smith 1982
Tamaulipas, Mexico	—	Empty	23 Jun 1979	Smith 1982
Tamaulipas, Mexico	—	Empty	2 Jun 1980	Smith 1982
Oaxaca, Mexico	2	Eggs	28 May 1966	Rowley 1984
Tikal, Guatemala	2	Eggs	Nest 2: 15 Jun	This study
Tikal, Guatemala	2	2-week-old nestlings	Nest 3: 1 Jul	This study
Costa Rica	?	Egg(s)	Incubating 30 May, 5–12 Jun	Orians and Paulson 1969
Surinam	2	Eggs	24–27 Nov 1964	Haverschmidt and Mees 1994
Surinam	2	Downy young	20 Apr 1963	Haverschmidt and Mees 1994
Grenada (*C. u. mirus*)	?	?	March	Bond 1936, 1984
Grenada (*C. u. mirus*)	1	2- to 3-week-old nesting	2 Aug 2000	Thorstrom et al. 2001

took place. A nest found on 15 June with incubation underway probably contained eggs during the previous week (hatching took place 27–30 days later); these two chicks fledged between 4 and 16 August. Finally, a nest discovered on 1 July held two well-grown chicks, and we estimated a hatching date during the second week of June, hence a laying date during the second week of May; one chick here was nominally fledged on 6 July and more thoroughly fledged by 9 July.

To summarize, nesting activity at Tikal began in March, with nest building from mid-March through at least late May. Approximate laying dates were the second week of May at one nest, prior to mid-May at another, and the second week of June at a third. Hatching dates were approximately 7–12 June at one nest and 11–14 July at another, while fledging dates were 6–9 July and 4–16 August at two nests. As noted also by other workers (Rowley 1984; Smith 1988), these kites were relatively late nesters. Hook-billed Kites at Tikal fledged young in August, when Swallow-tailed and Plumbeous Kites *(Ictinea plumbea)* were beginning to migrate southward. With laying dates from early or mid-May to mid-June, Hook-billed Kites laid eggs at the onset of the rainy season or just before, and hatched young early in the rainy season. This timing, among the latest of any raptor at Tikal, may be related to the species's largely snail-based diet.

Length of Incubation, Nestling, and Post-fledging Dependency Periods

We measured no exact incubation intervals. At one nest, one chick disappeared, presumably fledged, at approximately 24–25 days after hatching, and the other chick fledged at 28–33 days. Although little is known about the duration of post-fledging dependency in Hook-billed Kites, a few observations indicate that fledglings may remain in the company of adults for some time. In South Texas a "family group" of four remained together all winter (Smith 1988), and in Mexico an immature was observed foraging in close company with an adult pair (Smith and Temple 1982a, 1982b).

VOCALIZATIONS

Dickey and van Rossem (1938) described one of this kite's vocalizations as a "very musical whistle which very much resembles three notes of an oriole's song"; they recount that they searched for days for this unfamiliar oriole before discovering the song's author. While these notes were voiced by kites perched in treetops, the kites gave a "harsh chattering and screaming" when harassing Common Black Hawks *(Buteogallus anthracinus)* that were presumably near the kite pair's nest site. This latter vocalization is no doubt the same one that has been described by other authors as a "defense call . . . very similar to the call of the Common Flicker *(Colaptes auratus)* but louder" (Smith 1982), and as "possibly an

alarm note . . . a loud, rattling call, descending in pitch" (Fleetwood and Hamilton 1967). Conceivably, this latter call is also the same described by Willis and Eisenmann (1979) as "a rapid chuckling *wi-i-i-i-i-i-i-i-uh*" and by Howell and Webb (1995) as "a rapid, slightly clipped, clucking chatter, *weh keh-eh-eh-eh-eh-eh-eh-eh-eh-eh* or *w-kehehehehehehehehe."*

BEHAVIOR OF ADULTS

Pre-laying and Laying Period

At Tikal, both males and females took part in nest building, as noted by previous observers (Orians and Paulson 1969; Haverschmidt and Mees 1994). In mid-February 1992, we (M. A. V.) began checking regularly for activity in association with several nests that had been used in previous years, including two in the Mundo Perdido section of the ruins, a third at the Plaza Oeste, and a fourth area at Group F where a pair was seen delivering prey the year before. Checking these areas on six days from 15 February to 15 March, we observed no activity around nests but did observe pairs and single individuals soaring on a few occasions. On 16 March we first noted a pair in the Mundo Perdido area, soaring overhead, and the following day we observed a pair copulate in this area. On 26 March there was activity at one of these nests, and on 27 March we observed three individuals soaring nearby. On 10 April we noted a male in the area. On 15 April a pair was observed breaking off limbs and perching in the area, and on 24 April we also noted activity in the area, beginning at 0845. The birds copulated 20 m from one of the old nests and delivered limbs to a Gumbo-limbo tree to build a nest. We noted the male delivering dead limbs and the female vocalizing very softly; they perched 50 cm apart. On 1 May we again noted a pair in the area, and from atop the Temple of the Inscriptions we observed five pairs at one time over the forest, one of which perched 75 m from the temple.

Incubation and Nestling Periods

Total nest attendance (incubation of eggs plus brooding of young) remained high and quite constant throughout the latter three weeks of incubation and the first week of hatchling life (Fig. 4.3). After nestlings reached the age of one week, the percentage of time adults spent on or near the nest rapidly declined, to about 30% at chick age two weeks and 5% or less at chick age four weeks (Fig. 4.3). Overall, based on an observation time of 11 days with good data, of the total nest attendance time, males were there $39.8 \pm 10.5\%$ $(n = 11)$ of the observation time, and females $60.2 \pm 10.5\%$ $(n = 11)$.

With regard to the duration of stints spent on the nest incubating or brooding, we present two versions. In the first, we used all intervals on the nest, including those that were truncated by the beginning or end of observations (i.e., true duration unknown). This approach may

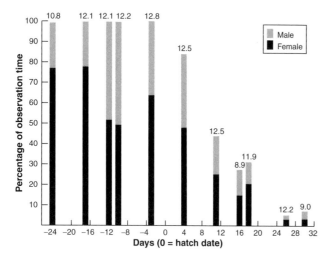

4.3 Total nest attendance by male and female Hook-billed Kites during the incubation and nestling periods in Tikal National Park. Number at top of bar = total observation time, in hours.

yield a duration estimate shorter than the true duration. In the second approach, we use only stints observed in their entirety. However, this approach is also likely to be biased, since longer intervals are more likely to be truncated, hence omitted. Because nest attendance decreased strongly after chicks reached the age of one week, we divided the data set into two parts, the first including all observations during incubation and up until the chick reached one week of age, and the second consisting of all observations thereafter. For the first period, referred to hereafter as the "incubation" period, male stints on the nest averaged 45.5 ± 64.5 minutes (range = 3–348 minutes, *n* = 42, including truncated sessions), and female stints averaged 71.2 ± 102.3 minutes (range = 1–445 minutes, *n* = 41, including truncated sessions), and male attendance totaled 31.95 hours, versus 48.78 hours for females). If truncated sessions are deleted, male stints averaged 30.6 ± 35.7 minutes (range = 3–148 minutes, *n* = 37) and females averaged 38.4 ± 53.2 minutes (range = 1–205 minutes, *n* = 32), with males totaling 18.98 hours of nest attendance, versus 20.57 hours for females. Hence, female stints on the nest during the incubation phase averaged 25–55% longer than male stints.

During the nestling period, male stints (including truncated sessions) averaged 20.6 ± 29.8 minutes (range = 3–148 minutes, *n* = 30) and female stints averaged 28.7 ± 43.0 minutes (range = 1–184 minutes, *n* = 28). If truncated sessions are deleted, male stints averaged 19.0 ± 29.0 minutes (range = 3–148 minutes, *n* = 29), and female stints averaged 23.2 ± 38.9 minutes (range = 1–184 minutes, *n* = 25). Thus, during the nestling phase, female stints on the nest were 22–39% longer than male stints.

Addition of Twigs and Greenery to Nest

Like most of the accipitrids we studied at Tikal, Hook-billed Kites brought sprigs of green foliage and, more rarely, dry sticks to the nest with some frequency. At

nest 2, we observed for 71.9 hours during incubation (from just after laying until the first chick hatched) and 96.65 hours during the nestling stage (from the first chick's hatching until both chicks fledged). During incubation, we saw nine sprigs of foliage brought in, or one sprig per 8.0 hours. In contrast, during the nestling period, we saw only three sprigs brought in, or one sprig per 32.2 hours of observation (Fig. 4.4). Hence this pair of kites showed a stronger tendency to bring foliage to the nest during incubation than during the nestling period. Both sexes played a similar role in this; the female brought in seven sprigs and the male four. Most sprigs were brought in during the afternoon, when at least one was noted each hour from 1200 until 1725, and the rate per hour of observation was notably higher between 1300 and 1500 than at any other time (Fig. 4.5). Only one sprig was noted in the morning.

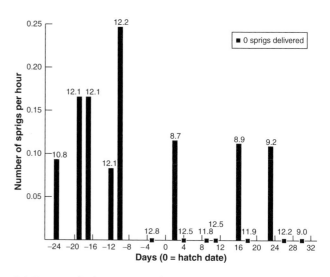

4.4 Rate at which sprigs were brought to Hook-billed Kite nest 2 during the incubation and nestling periods in Tikal National Park. Number at top of bar = total observation time, in hours.

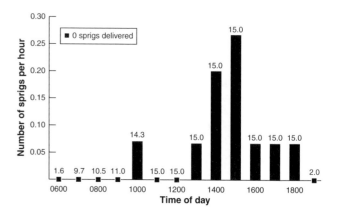

4.5 Hourly rate sprigs were brought to Hook-billed Kite nest 2 in Tikal National Park. Number at top of bar = total observation time, in hours.

Roles of Adults as Providers

It has been previously noted that both male and female Hook-billed Kites play a role in nest building (Haverschmidt and Mees 1994), incubation (Orians and Paulson 1969; Haverschmidt and Mees 1994), foraging for the young, and feeding the young at the nest (Snyder and Snyder 1991), and at least one adult male has proved to possess a well-defined brood patch (Rowley 1984). At the one nest we observed in detail, the male and female played quite similar roles in provisioning the chicks. Of 65 prey items brought to the nest, the male brought in 35 (54%) and the female 30 (46%). Thorstrom et al. (2001) found similar results for the Grenada Hook-billed Kite; of 156 prey items, the male delivered 47% and the female 53%.

Prey Delivery Rates

In Tamaulipas, Mexico, Smith (1982) and Snyder and Snyder (1991) observed Hook-billed Kites nesting in a situation in which a single arboreal snail species was present in extraordinary numbers, as clusters of motionless, dormant snails. They described the snails as being conspicuous, and kites delivered them to nests as often as one every 6–7 minutes, or roughly 8–10 snails per hour. A pair of Grenada Hook-billed Kites delivered prey to a nest at the rate of 4.7 snails per hour (Thorstrom et al. 2001).

During our observations, kites at Tikal delivered prey at much lower rates (Fig. 4.6). In 71.9 hours of observation during incubation, only six prey items were delivered to the adult on the nest (0.083 item/hr). This very low rate resulted from the fact that during the incubation phase, adults mainly hunted for themselves and rarely brought prey to one another at the nest. During the nestling period, 60 items were delivered during 96.65 hours of observation, for a rate of 0.62 prey item/hr. The highest delivery rate we observed was 0.81 item/hr during one 8.67-hour session. The prey delivery rate at this nest was an order of magnitude lower than at the two nests mentioned earlier.

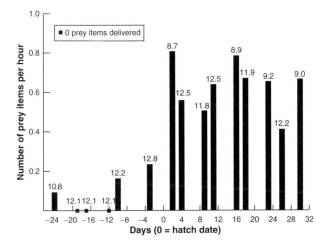

4.6 Hourly rate of prey items delivered to a Hook-billed Kite nest during the incubation and nestling period in Tikal National Park. Number at top of bar = total observation time, in hours.

Prey Transfers, Preparation, and Caching

The manner in which Hook-billed Kites remove snails from their shells is described in detail by Smith and Temple (1982a), and is very different from methods used by other snail-eating birds (Snyder and Snyder 1969; Snyder and Kale 1983). We know of no evidence of prey caching by this kite.

BEHAVIOR AND DEVELOPMENT OF NESTLINGS

We did not take detailed notes on chick behavior or development. One barely fledged chick weighed 160 g and had a flattened wing length of 191 mm, bill length from the cere of 20.9 mm, tarsus that was 63 mm long, hallux that was 50.8 mm long, middle toe with claw that was 26.2 mm long, tail 80 mm long, total length of 237 mm, and wing span of 62 mm. The eyes were grayish green, the bill black, and the cere and feet yellow, with black claws. Dorsal plumage was dark black, and the back, wings, and tail appeared fully feathered (Plate 4.4).

SPACING MECHANISMS AND POPULATION DENSITY

Territorial Behavior and Displays

We observed Hook-billed Kites in soaring flight fairly often at Tikal. Most authors agree that these kites soar occasionally to regularly, although several authors comment that they do not often soar for long periods or high above the canopy. In South Texas, Brush (1999) observed soaring mostly during mid-morning (08:30–10:30), which agrees with our observations. Our interpretation of these soaring flights is the same as for other frequently soaring raptor species at Tikal. We suggest that these flights function largely in intraspecific spacing, and possibly also in pair-bond maintenance and mate attraction.

In Grenada, Smith and Temple (1982b) observed a pair soaring over woodland, engaged in apparent courtship flight. They circled closely together, diving at each other repeatedly and calling frequently. In Texas, Brush (1999) twice saw possible courtship displays by male kites, once in the company of females; the kites made several very deep, slow wing beats in the midst of other flight. Although Ferguson-Lees and Christie (2001) speculate that perhaps these kites rely more on vocal cues than on soaring as an intraspecific spacing mechanism, this seems unlikely, as their mellow calls do not carry far (Snyder and Snyder 1991), and they are observed soaring with some frequency.

Nest Defense and Interactions with Other Species

At Tikal, these kites had a great many interactions with other birds at and near the nest. At nest 2, during 15 days

(180 hours) of observation, we observed 11 cases of strife, mainly with Brown Jays and Keel-billed Toucans *(Rhamphastos sulfuratus)*, both known nest robbers. Six cases involved groups of Brown Jays, and two cases involved at least one toucan. In one case, the adult kites interacted aggressively with a group of jays for a half hour, and in another case, a group of toucans harassed the chicks until the female kite arrived and drove them away. The male kite once attacked a vulture.

In Panama, Van Tyne (1950) once observed a group of five Keel-billed Toucans closely following an adult male Hook-billed Kite as it flew from tree to tree. Van Tyne speculated that the toucans followed the kite in order to preempt snails found by the kite. Such a possibility has not, to our knowledge, been verified, and such a verification (or falsification) would be a worthwhile endeavor for Neotropical birders. If these kites are known to other birds as "easy marks" from which to preempt or rob prey (especially an abundant prey type generally difficult for them to find on their own), this would offer a possible explanation of the polymorphic and phenomenally variable plumage coloration of these kites. Parading under varied disguises might facilitate the kites in foraging with minimal interference by would-be prey robbers. This, however, is nothing more than wild speculation, and only further observations can reveal whether these kites suffer at all from such a hypothetical parasitism by other birds.

Inter-nest Spacing and Density of Territorial Pairs

The pattern of nest dispersion in Hook-billed Kites is not yet clear. In Tamaulipas, Mexico, Smith (1982) found four active nests and two empty nests with adults nearby. Three were in an area 2 km across, and two others were within 5 km of these. This situation was described as a loose nesting colony (Snyder and Snyder 1991). Torpid, aestivating tree snails were very abundant here and nesting dispersion in this case may represent a response to unusually high prey density. Still, if these kites often nest in areas of high snail density, close spacing of nests may also be common. Although this situation may well qualify as a loose colony, the spacing between nests was far greater than what we observed in Swallow-tailed Kites (see Chapter 5).

At Tikal, two Hook-billed Kite nests simultaneously active in 1992 were only 75 m apart. A third simultaneous nest was 2790 m from these two, and a 1999 nest was 2825 m from the location of the same two (see Fig. 4.2). Using these two larger distances yields a mean of 2810 ± 25 m among "nesting areas," which conceivably could indicate substantial spacing among nesting areas. However, the fact remains that two simultaneous nests were only 75 m apart.

All of the remaining eight nests mentioned in the literature appear to represent cases in which only a single nest was discovered. Hence we tentatively conclude that these kites often nest as lone pairs and likely at some substantial distance from one another, but in some cases

they nest relatively close together and perhaps in a somewhat colonial fashion. Whether intraspecific territoriality occurs is not known; much remains to be learned of nesting dispersion and the degree of territoriality in this species.

DEMOGRAPHY

Causes of Nesting Failure and Mortality

The four nests at Tikal had the following fates. One was abandoned prior to use. In another, the eggs were destroyed by a predator. A third nest, possibly a re-nesting of the pair just mentioned, fledged at least one young—probably two. A fourth nest had a three-week-old nestling dead on the ground at the time of discovery, and a live one in the nest. We believed it likely the nestling fell to its death because of wind, which moved the nest a good deal during our observations. The surviving nestling fledged a week later.

CONSERVATION

Unlike all other raptors discussed in this book, the Hook-billed Kite possesses one distinct population that is threatened with extinction. Classified as a subspecies, the Grenada Hook-billed Kite has a very small total population and is generally regarded as in grave peril of extinction (Bond 1979; Smith and Temple 1982b), having at times even been thought extinct (King 1978). Blockstein (1988) estimated the kite's total population as 15–30 birds, occurring almost exclusively in xeric woodlands in the south and west of the island. More recently, Thorstrom et al. (2001) observed at least 15 individuals and estimated the total population as 40 or more individuals. While nearly all earlier kite observations had been in scrubby, xeric woodland, the latter authors observed the kites up to 400 m and in rain forest habitat. If use of rain forest and higher elevations is common, this implies that more habitat is available to the kites than has been estimated in the past. Further surveys are clearly needed in order to more accurately estimate this kite's population size and to verify the range of habitats used by the kites.

Because of their specialized diet, Hook-billed Kites arguably possess an unusually great degree of intrinsic vulnerability to the impacts of human activities. If introduction of a predatory snail, disease, changes in atmospheric chemistry, or global warming had large impacts on snail populations or community composition, it would not be surprising to see Hook-billed Kites greatly affected.

As for all forest-dwelling Neotropical raptors, deforestation is certainly among the foremost threats to the Hook-billed Kite. Hook-billed Kites are often described as being uncommon to rare, and distinctly patchy in distribution. While driving around 29,000 km of South

America, surveying raptors, Ellis et al. (1990) detected very few of these distinctive, easily identified raptors. For mainland populations of Hook-billed Kites, explicit data on abundance or changes in abundance are largely lacking. Several regional accounts, however, present general impressions regarding population status and the adverse impacts of deforestation. For northern Argentina, where the species is said to prefer tall forest with underbrush, Olrog (1985) described this kite as being not uncommon 20 years prior, but scarce at that time, apparently owing to extensive clearing. He stated that no recent record was available from Misiones province, where it had been found common in the late 1950s. A poignant example was provided when, in the span of a single year, a stretch of Tamaulipan thorn forest that had housed a colony of kites studied by Smith (1982) in northeast Mexico was bulldozed into cropland (Snyder and Snyder 1991). In Colombia's Cauca Valley, the Hook-billed Kite is regarded as a forest raptor that has probably become regionally extinct (Alvarez-López and Kattan 1995), as forests are "now virtually non-existent," and the only records in the prior 15 years were from small forest remnants on the valley slopes. The Hook-billed Kite was last detected on the island of Trinidad more than a century ago (ffrench 1973); its disappearance there presumably is due to habitat loss.

HIGHLIGHTS

The main highlight of our own research was the further quantitative demonstration of the fact that male and female Hook-billed Kites perform largely the same roles in nesting and divide the work quite evenly. Both adults feed themselves during incubation by hunting while off the nest, unlike the majority of raptors, in which the male supplies the female during incubation. This discovery, in combination with the type of prey used by these kites, further strengthens a hypothesis we have developed elsewhere (Seavy et al. 1998; Schulze et al. 2000a) concerning correlations between characteristics of prey resources and the form of role division among nesting raptor pair members.

Another highlight is the fact that these kites are relatively late nesters. It seems likely that this has something to do with these kites being snail specialists. However, it remains unknown how the seasonal patterns of snail populations and body size, and patterns of snail activity and torpor, may affect seasonal prey abundance and availability for Hook-billed Kites. Other questions related to a snail diet are these: What is the biomass of arboreal and terrestrial snails in tropical forests at Tikal and elsewhere? What are the production dynamics and seasonality of these snails? To what extent are these snails eaten by other predators, including certain snakes, non-raptorial birds, and perhaps mammals? Are Hook-billed Kites in a dietary adaptive zone that is unusually free of overlap with potential competitors? How do all of these factors vary closer to the equator where another snail-specialist raptor occurs in at least some forest areas (the Slender-billed Kite [*Rostrhamus hamatus*])?

Recent studies by others highlight the discovery of migration of at least some individuals in some populations. This discovery, and a few observations of flocking and of possible flock feeding in mangroves, poses interesting questions regarding movement patterns among habitats, local areas, and larger regions, especially in relation to the seasonal dynamics of snail availability.

Other questions ripe for testing are our twin hypotheses, (1) that these kites do an unusually large amount of upward and downward flying while foraging, and (2) that their broad wings and extensive wing-tip slots confer an unusually great ability to climb steeply in flight. The latter question should be easily testable using captive birds in a tall flight chamber, while the former would require field observation, probably of radio-tagged birds.

Other highlights come largely the research of Smith and Temple (1982a), and may be posed in the form of questions for future research. Is there assortative mating of kites of similar bill sizes? Is there more than one species of Hook-billed Kite—a large-billed and a small-billed form? If not, is this then a case of polymorphism resulting from disruptive selection for large- and small-billed extremes, as suggested by Smith and Temple (1982a), and if so, what are the genetics of bill size determination? Finally, while this may be intractable to research, it would be gratifying to have greater insight into why plumage coloration is so variable in this species—as well as the adaptive significance, if any, of the oddly colored iris and loral region in these strange raptors.

For further information on this species in other portions of its range, refer to Brush 2005, Di Giacomo 2000 and 2005, and Krügel 2003.

5 SWALLOW-TAILED KITE

Richard P. Gerhardt, Dawn M. Gerhardt, and Miguel A. Vásquez

The scene is spectacular—a low, forested hilltop that rises above a sea of forest in the southern Yucatán Peninsula. On this hilltop were built the temples of the Maya city of Tikal. Since the waning of this civilization a thousand years ago, the varied flora and fauna of the tropical forest have largely reclaimed the ancient ruins. Today, visitors to this site are inspired by both the shadows of the human past and the brilliance of the natural world of the present.

Between early February and late August, Swallow-tailed Kites *(Elanoides forficatus)* soar and career above Tikal's forests, and human observers treated to the sight are awed by the aerobatics of these graceful raptors. Nesting semicolonially in the tops of the tallest trees, these kites can be seen from vantage points among the ruins as they forage, mate, and rear their young. These handsome black-and-white birds with their long forked tails are creatures of the sky and light, but they hide a dark secret. The first young to hatch, if healthy, invariably kills its younger nestmate. This behavior, known as obligate siblicide, appears to be genetically fixed in this population: although two eggs are generally laid, one young is all that ever fledges from a successful nest. By the end of August, adults and young alike migrate; they spend the remainder of the year in South America, but exactly where is a mystery.

Our research at Tikal, which focused on the diet at the nest and on breeding ecology, was the first in-depth study of the southern subspecies of this beautiful raptor. The results, which include interesting similarities and differences between the temperate and tropical subspecies, are highlighted in the following account.

GEOGRAPHIC DISTRIBUTION AND SYSTEMATICS

Two subspecies of Swallow-tailed Kites are recognized. *Elanoides f. forficatus* breeds in coastal plains and riparian areas of North America from east Texas to South Carolina, with the majority in Florida. This present range is but a small fraction of the species's former U.S.

range. Prior to 1880, Swallow-tailed Kites nested up the Mississippi River valley as far north as Minnesota. The species declined precipitously between 1880 and 1940, retracting from much of its historic range. There has been no evidence of population recovery in the subspecies's present range, nor of reoccupation of vacant portions of its former range (Meyer 1995b). This northern subspecies migrates to South America.

The southern subspecies, *E. f. yetapa*, is found from southern Mexico to northern Argentina, Paraguay, and southeast Brazil. Our study site at Tikal is near the northern edge of this subspecies's range. Swallow-tailed Kites breeding in Central America are absent September through January, migrating somewhere to the south. These kites are resident at least in humid parts of Ecuador, Colombia, Venezuela, Guyana, Surinam, and northwestern Brazil, and their status (migrant or resident) is poorly known farther south. As in Meyer (1995b), there is a gap of some 1500 km in the species's breeding range, between east Texas *(E. f. forficatus)* and Chiapas, Mexico *(E. f. yetapa)*, though some range maps show a continuous distribution.

MORPHOLOGY AND DESCRIPTION

The Swallow-tailed Kite is strikingly patterned in black and white, and it is easily recognized by its long, pointed wings and deeply forked tail (Plate 5.1). Each tail feather, or rectrix, increases in length from the short inner pair to the long outer pair, and throughout Latin America the Gavilán Tijereta (scissor-tailed hawk) is recognized. The back, wing, and tail feathers are black, while the head, neck, and undersides are white. An iridescent sheen (purple in *E. f. forficatus*, green in *E. f. yetapa*) can be seen on the black scapulars and upper secondary coverts. The bill is black, the cere blue-gray, and the tarsi a somewhat paler blue-gray. The iris is dark brown. There are no known external characteristics that allow one to distinguish the sexes (Plate 5.2). Young of the year can be distinguished from adults by their shorter tails.

Swallow-tailed Kites are medium-sized raptors, with small feet and short tarsi. The total length of E. f. forficatus is 50–64 cm, and mean body mass about 465 g (Cely and Sorrow 1990; Meyer 1995b). The southern subspecies, which we studied, is believed to be smaller, but the differences are slight, and the validity of the subspecific separation has been questioned (Robertson 1988). The few published body mass data for E. f. yetapa are summarized in Appendix 1, along with masses from Tikal. Our best estimate of body mass is that of Haverschmidt and Mees (1994): five males averaged 382 g and six females 401 g. The largest set of linear measurements—that of Friedmann (1950) for 26 males and 14 females—is given in Appendix 2. The sexes do not differ markedly in size. Males in this data set were slightly larger than females in folded wing length, tail, and tarsal length (Dimorphism Indexes (DI) of –1.8, –4.6, and –0.3, respectively), whereas females were slightly larger than males in culmen length and middle toe length (DI of 3.0 and 0.4, respectively: Appendix 2). Females average 5% heavier than males, for a (cube root) DI of 1.6 (Appendix 1). It is questionable whether these data demonstrate any sexual size dimorphism at all.

PREVIOUS RESEARCH

Because its range includes part of North America, the Swallow-tailed Kite is probably the best studied of all the raptors that breed at Tikal. Studies of the species's ecology, including breeding biology, home range size, food habits, and roosting behavior, have been conducted in South Carolina (Cely and Sorrow 1990) and Florida (Meyer 1995a, 1995b; Meyer and Collopy 1995). The Swallow-tailed Kite is treated to its own species account in the Birds of North America project (Meyer 1995b); that account addresses all aspects of what is known of the species, but is based mainly on the northern subspecies, E. f. forficatus.

Prior to our research at Tikal, no detailed studies had been published on the southern subspecies, E. f. yetapa. Reports on diet (Haverschmidt 1962; Skutch 1965; Voous 1969; Lemke 1979), roosting behavior (Haverschmidt 1977), movements (Fjeldsa and Krabbe 1990), and distribution are largely anecdotal. The little information available is further clouded by questions of subspecies and migratory status. The present account examines this subspecies as we observed it at Tikal, making comparisons with E. f. forficatus where appropriate.

RESEARCH IN THE MAYA FOREST

Our choice of study area was partly dictated by logistic considerations—Swallow-tailed Kite nests were easy to find and observe from the temple tops of Tikal's central ruins. During the years of our study, the hilltop where the main temples are found was also an area where these birds concentrated their nests, and surrounding areas of forest were not similarly occupied.

We conducted observations, then, through binoculars and spotting scopes while sitting on temple tops. We did intensive observations at nine nests—four in 1990 and five in 1991—with the focus originally on food brought to nests. Reported here are prey data for eight nests during the incubation period (representing 58 nest-days or approximately 754 hours of observation) and for five nests during the nestling period (representing 88 nest-days or approximately 1144 hours of observation).

In addition, we monitored all nests found, with the goals of determining clutch size, duration of incubation, hatch dates, and other aspects of nesting phenology. Although studies of this species in the United States have generally avoided climbing to nests for fear of abandonment, we realized early on that the kites we studied tolerated frequent climbs to their nests. On numerous occasions, incubating individuals did not leave the nest until we peeked over the rim, and returned as soon as we began our descent. When we expected hatching to occur, and subsequently to measure and weigh chicks, we climbed to some nests every three days. For the nests placed so precariously that we could not actually reach them, we were able to view their contents with a small mirror affixed to a long pole. The initial climb often involved free climbing and rope throwing or the use of spurs. Thereafter, we left a thin line in the tree, with which we raised a stronger line, used in turn to raise a climbing rope. We then climbed with ascenders and rappelled back to the ground.

DIET AND HUNTING BEHAVIOR

Foraging adults delivered food to incubating mates at a rate of 1.0 ± 1.3 items/day (range = 0–5 items/day; $n = 17$ observation days). During our observations in the incubation phase, 47 identifiable food items were delivered to nests. Most were vertebrates, including 22 lizards, one snake, and 15 nestling birds. Only six insects were observed during this period, along with four deliveries of unidentified fruit (R. Gerhardt et al., unpubl. data).

During the nestling period, food was delivered to nests at a rate of 15.9 ± 10.8 items/day (range = 1–45 items/day; $n = 91$ observation days). This consisted primarily of insects, nestling birds, and lizards ($n = 1496$ food items at five nests in 1990 and 1991; Gerhardt et al. 1991). Four unidentified fruits were delivered, as well as four hylid frogs. By frequency, insects comprised 68% of the nestling diet, while birds and lizards contributed 21% and 11% of observed prey, respectively. On a biomass basis, however, vertebrates—particularly nestling birds—comprised a larger portion of the diet than did insects. Indeed, avian biomass was greater than that of all other food types combined (R. Gerhardt et al., unpubl. data). All birds delivered to nests appeared to be nestlings too young to fly.

Insects in the nestling diet belonged mainly to six orders (R. Gerhardt et al., unpubl. data). Kite nests differed greatly from one another in frequency of delivery of these orders. Beetles (Coleoptera), especially scarabs, ranked first or second in frequency among insect orders at all nests. Even so, bees and wasps (Hymenoptera) were easily the most numerous insect order at one nest and were second to beetles at another. Orthopterans (katydids and grasshoppers) formed only a small part of the diet at most nests but were the most numerous insects delivered to one 1990 nest; this was a re-nesting begun much later in the season than the other nests. Butterflies (Lepidoptera) were rarely delivered, but the adults at the late nest brought in a number of caterpillars on a single day. Cicadas (Homoptera: Cicadidae) were delivered to all nests infrequently. Dragonflies (Anisoptera) were observed as prey at all nests but one, though only in small numbers (R. Gerhardt et al., unpubl. data).

Little information exists concerning the diet of these kites outside the breeding season (Meyer 1995b). It is believed that adults are largely insectivorous throughout the year, capturing vertebrate prey mainly for delivery to nestlings. Our data suggest a seasonal pattern in the abundance or availability of the three main prey taxa. Lizards were delivered to nests in much greater numbers in April and May, and infrequently thereafter. Insects appeared to become more abundant at Tikal as the nesting season progressed (see Chapter 2); the one nest at which insects were not the most frequent prey item was the earliest nest, at which the young fledged in early June. Nestling birds appeared to be available throughout the Swallow-tailed Kite nesting period and are likely far less abundant at other times of year (see Chapter 2), unless the kites migrate into regions where nesting is underway. While only a few fruits were seen in kite diets at Tikal, observations of fruit eating elsewhere (Buskirk and Lechner 1978; Lemke 1979) suggest intriguingly that this behavior may not be rare, perhaps especially during times when animal prey is scarce. Further dietary data from various times and places during the nonbreeding season may hold interesting surprises.

Swallow-tailed Kites capture prey with their feet while flying (Meyer 1995b). They catch flying insects while soaring high in the open sky or while coursing low over open bodies of water, the forest canopy, or other types of vegetation. In addition, these kites often fly low over the canopy, snatching insects and lizards from treetop foliage and reaching out to seize nestling birds without pausing to perch. We observed adults carrying entire birds' nests in their feet as they soared above the canopy, extracting the young one by one for delivery to their own offspring. Swallow-tailed Kites appeared to congregate at hatches or swarms of flying insects and sometimes delivered identical insects to the nest every minute or two for most of an hour. These kites often gathered in groups of eight or more while hunting and were frequently joined by numbers of Plumbeous Kites *(Ictinia plumbea).*

HABITAT USE

Swallow-tailed Kites generally inhabit areas below 500 m (Meyer 1995b), but they can also be found in cloud forests above 1600 m in Mexico and Central America (Skutch 1965; Brown and Amadon 1968). In the southeastern United States, an important habitat feature seems to be an association of tall trees (for nesting and roosting) with open areas used in foraging. At Tikal, such a juxtaposition of open and forest habitat did not seem important to our study pairs, which routinely hunted over the sea of mature forest. Our study area at Tikal was notable for its lack of water during the breeding season—many breeding populations of these kites are associated with swamps, lakes, rivers, or marshes (Meyer 1995b).

Radio-tagged kites in Florida and South Carolina made daily foraging trips of up to 22 and 24 km, respectively, from the nest (Meyer and Collopy 1995). As our research was limited to observations at nests, we cannot address the distance flown or habitats used by foraging kites at Tikal. We believe, however, that our study area represented high-quality nesting habitat. We suspect that the density of kites nesting around Tikal's central ruins at the time of our study was higher than that found in other areas of the park or of the surrounding, largely forested region, although we have no quantitative data on this question. Habitat use and nesting dispersion of tropical populations of Swallow-tailed Kites remain fertile areas for research.

MIGRATION

Swallow-tailed Kites arrive in Tikal beginning in early February each year (M. A. Vásquez, pers. observ.). After breeding, kites seem to congregate in the vicinity of Tikal's ruins and leave by early September. This migration timing is rather synchronous with that of *E. f. forficatus.* The latter arrive in Florida some two weeks later each spring and vacate Florida some two weeks earlier in late summer than do the Tikal birds. Whether individuals of the two populations migrate together remains to be discovered. Likewise, the whereabouts of Central American kites during the nonbreeding season is unknown.

It is only in the past decade that migration of the northern subspecies has received some research attention, as satellite transmitters have been made small enough to use on this species. Previously, only a single data point existed regarding the wintering area—a kite banded as a nestling in Florida was shot six months later in southeastern Brazil (Mager 1967). Migration across the high Andes has been documented (subspecies unknown) in Ecuador and Peru (Fjeldsa and Krabbe 1990). That some kites from the United States migrate along the east coast of Mexico is apparent from observations in coastal Texas (Shackleford and Simons 2000; Smith et al. 2001) and Veracruz, Mexico (Goodrich et al. 1993). Evidence suggests that adults leave Florida before young birds do, and that the young undertake their first migration largely without the company of experienced birds (Meyer and Collopy 1995).

BREEDING BIOLOGY AND BEHAVIOR

Nests and Nest Sites

Nests in Tikal were situated precariously in slender limbs in the tops of the tallest lone or emergent trees. Twenty-two nests found in 1990 and 1991 were in five tree species. Twelve (55%) were in Ramón or Breadnut *(Brosimum alicastrum)*, four in Honduras Mahogany *(Swietenia macrophylla)*, three in *Ficus* species, two in Spanish Cedar *(Cedrela odorata)*, and one in Chechén Blanco or White Poisonwood *(Sebastiana longicuspis)*. Mean diameter of nest trees was 94.3 ± 46.5 cm (n = 22); mean tree and nest heights were 32.6 ± 4.8 m and 30.3 ± 4.8 m, respectively. Nests were made of dead branches and vines, with the nest cup formed of *Usnea* lichens and Spanish Moss *(Tillandsia usneoides*; Gerhardt et al. 1997). Both adults took part in nest construction.

Egg and Clutch Size

We found 18 nests during incubation. Modal clutch size was two (15 nests), with three nests containing a single egg each. Thus, mean clutch size was 1.83 eggs (Gerhardt et al. 1997), but two of the 1-egg nests failed early and these clutches may not have been complete. Three-egg clutches have been found in nests of *E. f. forficatus*, and the mean clutch size of the North American population (Cely and Sorrow 1990) is greater than what we observed in the Tikal population (Gerhardt et al. 1991). This is in keeping with a general trend among birds—clutch size tends to decrease with proximity to the equator, both within species and among species of similar taxa, size, and ecology (Moreau 1944; Lack 1966; Ricklefs 1969b). We weighed nine eggs at five nests. On average, eggs weighed 36.8 ± 4.7 g and measured 49.3 ± 2.9 mm by 37.6 ± 1.7 mm. Second eggs (33.6 ± 3.4 g, n = 5) were significantly smaller than first eggs (40.7 ± 2.1 g, n = 4, Mann-Whitney U = 20, P = 0.02; Gerhardt et al. 1997). First and second eggs represented 9% and 7%, respectively, of female body mass; the modal 2-egg clutch was approximately 16% of female mass. Using Hoyt's (1979) formula, fresh mass of first eggs is estimated at 42.0 g, and that of second eggs at 33.9 g. Hence, fresh mass of the modal 2-egg clutch is estimated to be 18.9% of mean female body mass (401 g). With respect to these characteristics (single egg per clutch mass as percentage of female mass), however, Swallow-tailed Kites are not among the most productive of the raptors breeding in Tikal.

Nesting Phenology

Pairs began courtship and nest building shortly after their arrival in February. The earliest egg laying we documented was 12 March (in 1990) and 18 March (in 1991), and the latest was 11 May (in 1990) and 9 May (in 1991). The latter dates were anomalous and probably represented re-nestings; most egg laying was completed between mid-March and mid-April. The peak of hatching occurred during the last week of April and first week of May in both years.

Length of Incubation, Nestling, and Dependency Intervals

Eggs were laid 3–4 days apart, and incubation began after the first egg was laid; hatching was asynchronous, at 3–5-day intervals. Mean incubation period was 31.5 ± 0.9 days (n = 6 eggs for which both laying and hatching dates were verified by climbing the nest tree). Two eggs of the northern subspecies were believed to have hatched 28 days after laying (Snyder 1974). Using standard artificial incubation temperatures and procedures, however, J. Coulson (pers. comm.) found that two eggs from Florida nests required slightly over 30 days of incubation before hatching.

We determined exact dates of hatching and fledging for six Swallow-tailed Kites: four in 1990 and two in 1991 (Gerhardt et al. 1991). Mean duration of the nestling period for these six individuals was 52.3 ± 5.2 days, significantly longer than the 35–42 days in *E. f. forficatus* (Sutton 1955; Wright et al. 1970; Snyder 1974; Meyer and Collopy 1995). Young were presumably independent of their parents by the onset of migration, some two months after fledging, as is the case with kites of the northern subspecies (Meyer 1995b).

VOCALIZATIONS

Swallow-tailed Kites were often quite vocal, particularly during nest building, in nest defense, and whenever numerous kites were gathered. We did not analyze calls, but vocalizations were apparently similar to those described in Florida (Snyder 1974). The *klee-klee-klee* call was most common among larger groups of kites, whether foraging, mobbing intruders, or entering or leaving a roost. Between mates at a nest, a two-syllable, rising whistle seemed to convey several meanings. A single-syllable whistle, often repeated monotonously, served as a begging call. This call was used both by incubating or brooding adults and by young.

BEHAVIOR OF ADULTS

Both members of a pair participated in nest building, incubation (including at night), and hunting. It is uncertain whether males ever brooded young. We were unable to sex birds, but one member of each pair (likely the female) spent more time engaged in incubating (Plate 5.3), while the other (presumably the male) did more of the foraging during this time. Adults delivered food to incubating mates at a rate of only 1.0 ± 1.3 items/day (range = 0–5 items/day). Thus, both birds foraged primarily for themselves while their mate relieved them at the nest. When one bird relieved another without bringing food, it frequently brought nest material such as a twig or a clump of Beard Lichen (*Usnea* sp.) or Spanish Moss. An adult—presumably the female—brooded young throughout much of the day for the first week and through the night for the first three weeks, and shaded young from the midday sun during the first 4–5 weeks of nestling life (R. Gerhardt, unpubl. data). These behavioral patterns are similar to those observed in *E. f. forficatus* (Sutton 1955; Snyder 1974).

BEHAVIOR AND DEVELOPMENT OF NESTLINGS

At nine nests at Tikal (five in 1990, four in 1991) both eggs hatched, with hatching intervals of 3–5 days between eggs (Gerhardt et al. 1997) (Plate 5.4). At eight of these nests, the second chick died between the ages of three and five days. In the ninth nest, both chicks were taken by predators when the second chick was four days old. We were able to observe siblicide (one chick killing the other) at eight nests and conducted all-day observations at seven of these. In all cases, the first-hatched chick was seen attacking its younger nestmate. Attacks began shortly after the second chick hatched, and continued intermittently until its death. Most frequently, the larger young pummeled the smaller with its bill, directing blows toward the head and neck. Sometimes the older young would grab its sibling with its bill, pulling with an up-and-sideways motion. There was no battle; the undersized, second nestling never fought back or displayed any effective defense (Gerhardt et al. 1997).

Aggressive attacks came in bouts lasting from a few blows to more than 100 blows delivered over a period of nearly 30 minutes. The behavior occurred both when the chicks were alone and when a parent was present. Adults did not intervene. Attacks occurred both when food was present and when food was absent, and continued in spite of prey deliveries. The presence of food did not appear to stimulate this aggression, but begging by the younger chick did appear to stimulate attacks by the older. A prone and silent smaller young was less likely to be attacked than one that was upright and calling. Nonetheless, smaller young continued raising themselves and vocalizing as long as strength allowed (Gerhardt et al. 1997).

The blows inflicted did not lacerate the skin but may have caused internal bleeding. The continued attacks weakened the recipients to the point when they could no longer raise themselves to receive food. They received little or no food to begin with, since the aggression of their older nestmates often placed these smaller, younger chicks in a subordinate position when a parent arrived with prey. If second chicks did receive food, it did not result in weight gain. We do not believe that any second chicks died as a result of falling from a nest, but suspect that starvation, possibly combined with internal injuries, caused their deaths (see Gerhardt et al. 1997 for details).

If some resource limits (or limited) optimal brood size to one, why do Swallow-tailed Kites lay two eggs? We can only speculate that, as in several other obligately siblicidal species (e.g., many eagles, boobies, pelicans, and penguins: Dorward 1962; Warham 1975; Mock 1985; Cash and Evans 1986; Anderson 1990), a small percentage of successful fledglings in this kite population will come from second-laid eggs, which thus act as insurance (Dorward 1962) against loss of an entire reproduction attempt should the first egg fail to hatch or result in a weak or deformed young. The cost to Swallow-tailed Kites of laying a second egg is presumably small, since the mass of the second egg is only about 7% of female body mass and since little or no time and effort are spent feeding second young (Gerhardt et al. 1997).

SPACING MECHANISMS AND POPULATION DENSITY

Inter-nest Spacing and Density of Territorial Pairs

As in other parts of the species's range (Meyer 1995b), Swallow-tailed Kite nests at Tikal exhibited a clumped distribution. Numerous pairs nested in loose "neighborhoods" (Newton 1979) and much similar habitat was devoid of any nests. In 1990, we observed 12 nests in an area 1 km², and nests were situated as little as 35 m apart (R. Gerhardt et al., unpubl. data).

In Florida (Meyer and Collopy 1995) and South Carolina (Cely 1973), one or two additional nonbreeding adults were associated with most nests. These generally attempted, unsuccessfully, to assist in nest building or food delivery. In Tikal, we never observed such supernumerary birds or "helping" behavior; the number of individuals observed at any one time during nesting could be accounted for by the pairs known to be nesting.

Nest Defense, Interspecific Interactions

Swallow-tailed Kites defended only the immediate nest area from others of their species. Such defense was, for the most part, limited to vocalizing when another bird flew over. In rare instances, a member of the nesting pair, already in flight, would dive, screaming, toward the intruder. Since most or all kites in the area were themselves busy nesting nearby, intraspecific agonistic behavior was minimal. Away from nests, these kites foraged in loose aggregations and roosted side by side in groups.

Swallow-tailed Kites defended a larger area from monkeys and from other large raptors. Defense included screaming, diving on, and occasionally making contact with such intruders. When a large hawk flew by, it was generally escorted from the entire hilltop by a group of Swallow-tailed Kites, each of which continued the chase far beyond its own nest and the area it normally defended against others of its species. Raptors attacked in this way included Ornate Hawk-eagles (*Spizaetus ornatus*), Black

Hawk-eagles *(S. tyrannus)*, Roadside Hawks *(Rupornis magnirostris)*, and Crane Hawks *(Geranospiza caerulescens)*. We did not observe similar defense against Hook-billed Kites *(Chondrohierax uncinatus)*, and in two cases, a nest of this species succeeded within 30 m of a Swallow-tailed Kite nest. Plumbeous Kites were treated much like conspecifics—they were chased only from the immediate nest area, and sometimes they joined in chasing other intruders from the larger area. We observed one instance in which a pair of Plumbeous Kites and a pair of Swallow-tailed Kites initiated nests in the same tree; the former remained and the latter were eventually driven off to seek another tree in which to build.

We observed a troop of Black-handed Spider Monkeys *(Ateles geoffroyi)* that were repelled by a pair of Swallow-tailed Kites. The monkeys were moving through the trees, climbing higher and nearer to the kite nest, though there was no indication that their movements were directed at reaching the nest. When the monkeys were still about 50 m away, the pair of kites began screaming and diving at the primates. One of the kites hit a young monkey, which lost its grip, falling some 20 m before catching some lower branches. The other monkeys descended to where the fallen one was, and when they moved off, it was in another direction.

Smaller birds, whose nestlings represented potential prey for kites, were frequently seen attacking and screaming at kites flying near the forest canopy. Prominent among these were Social Flycatchers *(Myiozetetes similis)*, Sulphur-bellied Flycatchers *(Myiodynastes luteiventris)*, and Brown-crested Flycatchers *(Myiarchus tyrannulus)*. One pair of Social Flycatchers had a nest quite near one of our focal kite nests. Although the flycatchers were diligent in chasing the kites from the vicinity of their nest for several weeks, it was ultimately with apparent ease and impunity that the adult kite grabbed the entire nest and delivered the nestlings, one by one, to its own chick.

Constancy of Territory Occupancy, Use of Alternate Nest Sites

Swallow-tailed Kites appeared to build a new nest each year, and nests maintained little, if any, integrity from one year to the next. Several trees held nests in consecutive years, and in two cases, nests were situated in the exact same location in both years of our study. Whether the same individual kites were involved from one year to the next, we cannot say. The situation is apparently similar in South Carolina (Cely and Sorrow 1990) and Florida (Meyer and Collopy 1995)—kites may build new nests near old ones, may refurbish old nests before constructing a new one, or may refurbish and use an old nest, either from the previous year or from prior years. We doubt that individuals retain any "ownership" of a particular nest from one year to the next. Within a breeding season, we saw a pair of kites re-nest (their first nest failed shortly after hatching); the second nest was approximately 200 m from the first, and two other

pairs were actively nesting in the intervening space. This highlights the fact that "territory" is not as meaningful a concept for this species as for most raptors.

Home Range Estimates

We did not radio-tag these kites, nor could we estimate the area used by individuals during the breeding season. Little such information exists for this species elsewhere. In South Carolina, five breeding adults had home ranges (Minimum Convex Polygon (MCP) method) of 232.6 ± 116.2 km^2 (range = 89.2–360.5 km^2; Cely and Sorrow 1990). In Florida, a breeding female had a home range estimated at 10.6 km^2, and two breeding males used areas of 30.8 and 117.1 km^2 (Meyer and Collopy 1995). The male using the latter area made long, daily foraging trips, and the resulting home range estimate was quite linear in shape.

DEMOGRAPHY

Frequency of Nesting, Percentage of Pairs Nesting

We can say very little, with any certainty, about either the frequency with which individuals nested or the percentage of pairs nesting. At most nests in Florida (Meyer and Collopy 1995) and South Carolina (Cely 1973), one or more nonbreeding adults were present in addition to the breeding pair. The situation in Tikal differed markedly. During the breeding season, most if not all of the Swallow-tailed Kites present in our study area were believed to be members of a nesting pair. We did not see "extra" individuals, or birds in numbers greater than those expected based on the number of known nests. Nonetheless, we have no idea whether other portions of the park or nearby areas supported nonbreeding individuals, nor in what numbers such "floaters" may have existed. Such information is needed for an accurate understanding of the demography of this subspecies.

Productivity and Nest Success

One fledgling was produced at each of nine successful nests in 1990 and 1991; 13 of 22 nesting attempts resulted in failure. Thus, reproductive success was 0.41 young fledged per nesting attempt and 1.00 young per successful nest (Gerhardt et al. 1991). These rates are clearly low, and one wonders whether they are high enough to maintain a viable population. However, Plumbeous Kites in Tikal had similar productivity, of 0.37 fledging per nest attempt and 1.00 young per successful nest (Seavy et al. 1998). No information exists for Swallow-tailed Kites at Tikal or elsewhere regarding lifespan, survivorship, or lifetime reproductive success. In South Carolina, productivity was 1.14 young per nest with eggs and 1.57 young per successful nest (*n* = 29 nests over 8 years; Cely and Sorrow 1990), and in Florida productivity was 0.90 young per nest with eggs and 1.40 young per successful nest (*n* = 49 nests in 2 years; Meyer and Collopy 1995).

Causes of Nest Failure and Mortality

Not surprisingly, given the vulnerable position of nests on slender treetop limbs, wind was responsible for the failure of four nests during incubation and the loss of one egg from a fifth nest. Avian predators, including Brown Jays *(Cyanocorax morio)*, Keel-billed Toucans *(Ramphastos sulfuratus)*, and possibly Bat Falcons *(Falco rufigularis)* were implicated in the failure of six nests, four of which held young. One nest containing two eggs was destroyed by spider monkeys. Another failure was likely due to a heavy parasite load (botflies and unidentified others) in the single nestling. Six nest failures occurred during incubation (mainly due to wind), and seven nests failed during the nestling stage, with predation the leading cause.

Parasites

Botflies, probably *Philornis* species (Diptera, Muscidae), were found infesting the skin of many Swallow-tailed Kite chicks. We attributed the death of only one chick to botflies, but we believe that botfly-related mortality may not be rare and can cause the failure of entire nesting attempts. In 1999, the only chick known to fledge from our study site (see below) was handled one week after hatching. Already it had 25 botfly larvae, which were concentrated near the eyes, head, and wing pits. We believe these would have killed this individual had we not removed them; in a study of the Puerto Rican Sharp-shinned Hawk *(Accipiter striatus venator)*, all nestlings with six or more *Philornis* botflies perished (Delannoy and Cruz 1991). A variety of trematodes, nematodes, acanthocephalans, mites, chewing lice, and mosquitoes have been found on dead Swallow-tailed Kite nestlings in Florida (Meyer 1995b).

CONSERVATION

On a range-wide scale, there seems little doubt that deforestation is the main threat to Swallow-tailed Kite populations in the tropics. Although these kites forage over a variety of cover types, they are basically forest-associated birds, and outright deforestation can safely be assumed inimical to their populations. The extent of direct persecution by humans—that is, shooting as a food source, out of curiosity, or for sport—is undocumented, but we cannot assume it to be minimal. In their annual migrations, these birds may cross much of the Western Hemisphere, perhaps predictably in certain places, and that some shooting occurs is a given. Whether other, more subtle factors are presently exerting a negative effect on populations of this kite is unknown. We suspect, however, that the timing and spatial pattern of these kites' migration movements may be attuned to subtle, seasonal patterns in plant phenology and prey populations. We are nervous, therefore, about the potential impact on these birds of weather and climate changes resulting from global warming.

With regard to the local breeding population we studied, a decline in the number of kites breeding in the heart of Tikal National Park appears to have taken place. Whereas in 1990 and 1991 we studied 12 pairs nesting around Tikal's central ruins, in 1999 only six nests were documented in this area (R. P. Gerhardt and M. A. Vásquez, pers. observ.) and only one was known to have succeeded. In 2001 only a single pair of these kites was noted in the vicinity of Tikal's ruins (M. A. Vásquez, pers. observ.), and no nest was found. We hope this recent trend represents merely a short-term change in choice of nesting locale. We believe, however, that some region-wide monitoring of this species's population status is warranted. Since these birds are long-distance migrants, factors operating in disparate geographic locations may potentially affect their populations.

HIGHLIGHTS

Our extensive nest observations enabled us to compare diet and breeding habits with those reported in similar studies, most notably of Swallow-tailed Kites in Florida (Meyer and Collopy 1995) and of Plumbeous Kites at Tikal (Seavy et al. 1997b; see Chapter 7). Swallow-tailed Kites in Guatemala delivered significantly more insects and fewer frogs to nestlings (Meyer 1995b) than did kites in Florida (Sutton 1955; Wright et al. 1970; Snyder 1974; Meyer and Collopy 1995). Frogs comprised between 20 and 45% of nestling diets in the four Florida studies but less than 1% at Tikal. Plumbeous Kites and Swallow-tailed Kites in Tikal delivered similar prey to their young, but nestling birds were far less important in the diets of Plumbeous Kites, and insects of greater importance. Both kites took largely the same insects, but in much different proportions. Most insects delivered by Swallow-tailed Kites were beetles or hymenopterans, with insects of other orders observed much less frequently. Plumbeous Kites took mainly cicadas, beetles, and dragonflies, with lesser numbers of orthopterans, lepidopterans, and hymenopterans (Seavy et al. 1997b; see Chapter 7).

By far the most novel outcome of our research was our discovery of obligate siblicide in this population. This reproductive strategy is not seen in the northern subspecies, where two and even three young often fledge from successful nests. We suspect that either presently or in the evolutionary past, a brood size of one is or was optimal within the more tropical population of kites we studied. It is noteworthy that Plumbeous Kites, the sympatric raptor most similar to Swallow-tailed Kites in size, migratory strategy, feeding behavior, and prey at Tikal (Seavy et al. 1998), also raise only a single young (Vasquez et al. 1992). Plumbeous Kites, however, lay only one egg (Vasquez et al. 1992; Seavy et al. 1998), whereas Swallow-tailed Kites lay two eggs but rear only one young.

Swallow-tailed Kites share certain traits with other obligately siblicidal falconiforms: a tropical distribution, a maximum clutch size of two (Simmons 1988), a longer hatching interval, and a greater size difference between eggs within a clutch (and hence between hatchlings: Gerhardt et al. 1997) than in species that are facultatively siblicidal (Edwards and Collopy 1983). But these kites are anomalous in several respects: other obligately siblicidal raptors tend to be large, long-lived, solitary species (Gargett 1970, 1993; Simmons 1988). Our findings demonstrate that the obligate siblicidal reproductive strategy is not confined among falconiforms to the large eagles, and that evolution of this phenomenon can occur at the subspecific level (Gerhardt et al. 1997) since no similar siblicide has been observed in *E. f. forficatus*.

Finally, it is interesting to contemplate why the northern population has supernumerary, unmated adult "helpers" at the nest, whereas the Tikal population does not. Does this result from a larger clutch and absence of siblicide in the north, together resulting in greater population productivity? Perhaps coupled with limited habitat, such enhanced productivity results in a paucity of breeding opportunities for northern birds compared to those amid the massive forested areas of our Guatemalan study area. However, kite researchers in Florida believe there is a good deal of suitable, but unoccupied habitat there, in which case the above hypothesis would not apply. This is but one of many questions deserving further research on the lifestyles of these elegant raptors.

For further information on this species in other portions of its range, refer to Azevedo et al. 2000, Azevedo and Di-Bernardo 2005, Coulson 2006, and Gerhardt et al. 2004.

6 DOUBLE-TOOTHED KITE

Mark D. Schulze, José L. Córdova, Nathaniel E. Seavy, and David F. Whitacre

From atop the weathered limestone of Tikal's Temple IV, our eyes roam over an expanse of undulating green forest that extends to every horizon. The early morning rush of parrots and forest pigeons commuting above the canopy has waned, and the dawn chorus of Bright-rumped Attilas *(Attila spadiceus)* and forest falcons gives way to the first, tenuous keenings of cicadas and the monotonous croaking of Keel-billed Toucans *(Ramphastos sulfuratus)*. Thick morning fog breaks into tendrils that fade into nothingness as the first stir of a breeze brings rich odors of the warming forest to us on our exotic perch. The tropical sun, though still low, drives us to the shady side of the temple's roof comb. At length, a Double-toothed Kite *(Harpagus bidentatus)* rises from the forest nearby, soars up to 200 m above the canopy, and begins a series of short stoops broken periodically by bursts of flapping to regain altitude. Another kite circles upward near the first. Before long, scanning with binoculars, we can see five of these small kites engaged in similar soaring and stooping display. Finally, the kite nearest us tucks its wings and plummets in one breath-taking stoop until swallowed up by the dense, green foliage.

Such display activity seems to be confined largely to the breeding season. During the rest of the year, one may spend weeks without sighting this small, unobtrusive forest raptor—for when not engaged in soaring display flights, these kites remain in the shaded forest interior, where they are inconspicuous and easily overlooked. Double-toothed Kites hunt within and beneath the forest canopy, gracefully navigating a sea of branches as they make short flights from perches to pluck prey from the air or, more often, from vegetation. An opportunistic hunter, this kite captures large insects and small vertebrates, with an emphasis on cicadas and small arboreal lizards.

GEOGRAPHIC DISTRIBUTION AND SYSTEMATICS

The Double-toothed Kite is distributed widely in moist tropical forests from southeastern Mexico to eastern

Bolivia and southeast Brazil, and in Trinidad (del Hoyo et al. 1994). Two subspecies are recognized: the nominate race *H. b. bidentatus* east of the Andes, and *H. b. fasciatus* from southern Mexico to Colombia and Ecuador west of the Andes (Brown and Amadon 1968). These kites are found from tropical lowlands up to 1500 or even 2100 m (Hilty and Brown 1986; Stiles et al. 1989; Howell and Webb 1995). The only other member of the genus, the Rufous-thighed Kite *(H. diodon)*, is restricted to South America.

Morphological similarities to the genus *Accipiter* as well as to other kites and even falcons has led to debate about the phylogenetic position of the Double-toothed Kite (Miller 1937; Amadon 1961). For some time it has been regarded as one of the "milvine kites," a traditional but informal grouping (Brown and Amadon 1968). However, the milvine kite group is probably polyphyletic, based on osteology (Holdaway 1994), syringeal morphology (Griffiths 1994a, 1994b), and other evidence (Kemp and Crowe 1990). Holdaway (1994) portrays *Harpagus* as forming a clade with *Ictinia*, as sister group to the *Buteo-Accipiter* radiation. Further research is needed before one can be confident of *Harpagus*'s phylogenetic position within the Accipitridae.

MORPHOLOGY

The Double-toothed Kite resembles an *Accipiter* in form, with a long tail—appearing distinctly narrow in soaring flight—and medium-length, round-tipped wings. The oval-shaped wings, however, are proportionately longer than in accipiters (see Plates 2, 8, and 13 of Howell and Webb 1995). In soaring flight, these kites are distinctive because of the combination of a crosslike silhouette with slightly downward-bowed, oblong wings and a long, narrow tail held tightly closed, with fluffy white undertail coverts visible from above and below.

At 175–230 g, the Double-toothed Kite is the second smallest diurnal raptor at Tikal. For *H. b. fasciatus*, the race occurring at Tikal, the best available body mass estimates are 182 g (male) and 208 g (female; Appendix 1), giving a mass dimorphism value of 4.5. For the South American race *H. b. bidentatus*, males average 167.5 g and females 201.5 g (Appendix 1), for a body mass Dimorphism Index (DI) of 6.2.

Linear measurements for the northern subspecies *H. b. fasciatus* are given in Appendix 2. Based on wing chord, this kite's DI value of approximately 8–10 is similar to that of the Merlin (*Falco columbarius*; 10.0), less than that of the Cooper's Hawk (*Accipiter cooperi*; 12.7) and Sharp-shinned Hawk (*A. striatus*; 17.7), greater than that of all North American *Buteo* species, and far greater than that of the Swallow-tailed Kite (*Elanoides forficatus*; 1.8) and Mississippi Kite (*Ictinia mississippiensis*; 2.2: Snyder and Wiley 1976). The mean of dimorphism values for body mass, folded wing length, and culmen length equals 4.6 for *H. bidentatus fasciatus* and 5.2 for *H. b. bidentatus* (data from Appendixes 1, 2).

The ratio of wing to tail length is sometimes considered reflective of the degree of adaptation to open versus closed habitats, with woodland raptors having proportionally short wings and long tails compared to open-country raptors (Brown 1976a). The wing-to-tail ratio for the Double-toothed Kite is 1.43, similar to that of the American Kestrel (*Falco sparverius*; 1.42), less than that of most North American buteos (1.55–1.71), and greater than that of the three North American *Accipiter* species (1.16–1.32; data from Palmer 1988). This magnitude of wing-to-tail ratio appears to reflect adaptation of this kite to its largely sub-canopy lifestyle, but not to the degree typical in *Accipiter* species. Flight morphology in the Double-toothed Kite may represent a compromise between features favorable for hunting below the canopy and those advantageous in its extensive soaring displays.

In addition to their substantial size dimorphism, these kites are strikingly dichromatic for a kite, with females generally more rufous below and males more grayish. Among the world's roughly 33 species of "kites" (admittedly a "wastebasket" grouping of often distantly related taxa), only two other species, both Neotropical, show pronounced sexual dichromatism—the Hook-billed Kite (*Chondrohierax uncinatus*) and Snail Kite (*Rostrhamus sociabilis*)—with the African Cuckoo-Hawk (*Aviceda cuculoides*) showing a more subtle dichromatism. Additional species of *Harpagus*, *Rostrhamus*, and *Aviceda* fail to show differences in coloration between the sexes. These genera might be interesting material for a study of factors correlated with sexual dichromatism, since some species show it and others do not.

Another distinctive feature of *Harpagus* species is the presence of two notches or tomial teeth on the upper mandible, giving rise to the specific epithets of both members of the genus—*Harpagus bidentatus* and *H. diodon* (Plate 6.1). Though this feature may superficially suggest a close relationship with the falcons, such an affinity apparently is not real, and in fact such teeth occur in some other kites, for example, the Old World genus *Aviceda* and New World *Leptodon* (Amadon 1961).

PREVIOUS RESEARCH

Prior to our work, most knowledge of the natural history of this species was based on opportunistic observations rather than deliberate study. Six nests are mentioned in the literature (Laughlin 1952; Skutch 1965; Wetmore 1965; Brown and Amadon 1968; ffrench 1976), with brief observations made at two of them (Laughlin 1952; Skutch 1965). Double-toothed Kites are best known through a series of notes describing these kites foraging in association with monkey troops (see below).

RESEARCH IN THE MAYA FOREST

During four breeding seasons, we studied 11 nesting attempts in six territories at Tikal—one each in 1992 and 1993, five in 1995, and four in 1996 (Table 6.1; Schulze et al. 2000a). We obtained data on prey delivery rates and diet at six nests and observed adult behavior at five nests during the incubation period (one nest in 1993, two each in 1995 and 1996) and four during the nestling period (two each in 1995 and 1996). We observed nests for 295 hours during incubation and 521.9 hours during the nestling phase. Using a 30X spotting scope at 10–30 m, we observed nests from treetop platforms, emergent Maya temples, and in one case the ground. Double-toothed Kites were sufficiently dimorphic in size and coloration that we distinguished the sexes with confidence. By comparison with the kites' bill, we estimated prey length in millimeters and rounded the resulting statistics to the nearest 0.5 cm. We measured nests after young had fledged and avoided climbing to active nests, except that we climbed two 1996 nests to place radios on juveniles and adults. By radio tracking on foot, we documented dispersal of two fledglings and spatial use of an adult female. In our characterization of nest sites, the sample size is 10, as one nest was occupied in two breeding seasons.

To analyze time-of-day effects, we divided observations into four blocks: early morning (0500–0859), late morning (0900–1159), early afternoon (1200–1459), and late afternoon (1500–1859). For tests involving a portion of the nesting cycle, we grouped observations by week from the date of egg laying or hatching. Values reported are means ± SD.

DIET AND HUNTING BEHAVIOR

Table 6.2 is based on 517 prey items observed (463 identified) during the nestling phase at the four best-studied nests. In total, however, we recorded 622 prey items (550 identified) during observations at seven nests from courtship through the nestling phase, and during nest searching and radio tracking. Here we discuss results based on Table 6.2, except as noted.

Insects and lizards together made up 99% of identified prey. More than 15 taxa were recorded, and variation in prey composition between nests and years was

Table 6.1 Breeding phenology of Double-toothed Kites at Tikal

Territory number, name, and year	Dates of incubation	Duration of incubation (days)	Dates of nestling period	Duration of nestling period (days)	Number of young fledged	Reason for failure
1) Barens 1992	—	—	8–10 Jun to 2–3 Jul	—	0	Predation of nestlings
1) Barens 1996	9–10 Apr to 21–22 May	42–44	21–22 May to 18 Jun	28	1	One chick suffered predation approx. 1 week before fledgling; second chick fledged, found dead within 1 week
2) Bajada la Pina 1993	14 May to 30 Jun	—	—	—	0	Eggs did not hatch
2) Bajada la Pina 1995	6 May to 19 Jun	—	—	—	0	Eggs did not hatch
3) Parcela Pajaritos 1995	—	—	7–8 Jun to 14 Jul	> 36–37	0	One chick taken from nest by unidentified raptor; second chick fell from nest 14 Jul
3) Temple IV 1996[a]	—	—	? to 18–19 June	—	2	—
4) Temple III 1995	21–22 May to 2–3 Jul	42–44	2–3 Jul to 29–31 Jul	27–30	1	—
4) Temple III 1996	? to 17–18 Jun	—	18 Jun to 19 Jul	31	1	—
5) Aguada Seca 1995	—	—	—	—	0	Nest abandoned prior to egg laying, no re-nesting observed in territory
5) Temple IV 1996[b]	—	—	—	—	0	Nest abandoned prior to egg laying, no re-nesting observed in territory
6) Parcela Norte 1995	—	—	—	—	0	Nest abandoned prior to egg laying, no re-nesting observed in territory

Note: — = data unavailable or inapplicable (no egg laid, discovered during incubation, did not fledge, exact hatch and fledge dates unknown, etc.).
[a] Believed to be same territory as Parcela de Pajaritos 1995, but not certain.
[b] Believed to be same territory as Aguada Seca 1995, but not certain.

Table 6.2 Prey items delivered to four Double-toothed Kite nests (nestling phase only) at three territories in Tikal National Park

Prey items	Temple III 1995	Pajaritos 1995	Temple III 1996	Barens 1996	Total
Homoptera (mainly cicadas)	28 (44.4)	149 (70.6)	24 (23.8)	26 (29.5)	227 (49.0)
Coleoptera	0 (0.0)	2 (1.0)	2 (2.0)	0 (0.0)	4 (0.9)
Orthoptera	4 (6.4)	4 (1.9)	2 (2.0)	4 (4.6)	14 (3.0)
Lepidoptera	0 (0.0)	0 (0.0)	0 (0.0)	1 (1.1)	1 (0.2)
Hymenoptera	0 (0.0)	1 (0.5)	0 (0.0)	0 (0.0)	1 (0.2)
Unidentified insects	2 (3.2)	15 (7.1)	3 (3.0)	7 (8.0)	27 (5.8)
Total number of insects	34 (54.0)	171 (81.0)	31 (31.0)	38 (43.2)	274 (59.2)
Lizards	27 (42.9)	39 (18.5)	69 (68.3)	50 (56.8)	185 (40.0)
Snakes	0 (0.0)	1 (0.5)	0 (0.0)	0 (0.0)	1 (0.2)
Bats	1 (1.6)	0 (0.0)	0 (0.0)	0 (0.0)	1 (0.2)
Birds	1 (1.6)	0 (0.0)	0 (0.0)	—	1 (0.2)
Rats (Rodentia)	0 (0.0)	0 (0.0)	1 (1.0)	0 (0.0)	1 (0.2)
Total number of vertebrates	29 (46.0)	40 (19.0)	70 (69.3)	50 (56.8)	189 (40.8)
Unidentified prey	20	1	27	6	54
Total number of identified prey	63	211	101	88	463
Total number of prey	83	212	128	94	517

Note: Numbers given are number of prey items, with percentages (based on the number of identified prey items) given in parentheses.

considerable. During the nestling phase, insects comprised 59.2% of identified prey, lizards 40.0%, with birds, bats, rodents, and snakes together comprising 0.8% (Table 6.2). Proportions in the overall diet sample were nearly identical, with insects making up 60.5% of identified prey, lizards 38.1%, and birds, bats, rodents, and snakes together 1.4% (n = 550 identified items). We suspect that most unidentified items were insects.

Most lizards were arboreal *Anolis* species, with geckos (Gekkonidae), four *Corytophanes* species (Iguanidae), and a skink (Scincidae) also recorded. Based on Table 6.2, cicadas (Homoptera: Cicadidae) made up 92% of insect prey, grasshoppers and katydids (Orthoptera) 5.7%, beetles (Coleoptera) 1.6%, and caterpillars (Lepidoptera) and wasps (Hymenoptera) 0.4% each. In the overall diet, a few cockroaches (Blattidae) were also included. If unidentified insects differed in composition from the results just given, the percentage for the contribution of cicadas might be slightly less than that just stated, but could not have been lower than 83% of all insects.

In terms of biomass, lizards and other vertebrates in the diet assumed greater importance. Insect prey averaged 4 ± 1 cm in length (n = 333, range = 1.5–9.5 cm), with an estimated mass of 1–3 g. Most lizards were small, with 77% of the lizards (total number = 277) estimated as having a mass less than 10 g, while lizards estimated at 10–15 g accounted for 19% of lizard prey. However, lizards with an estimated mass up to 30 g were recorded, and those with a mass up to 15 g comprised 5% of the total number of lizards (n = 211). Hence, a conservative estimate is that vertebrate prey items were on average five times the mass of insect prey items. Based on this estimate, vertebrate prey accounted for at least 75% of total prey biomass. Lizards were the only vertebrate group well represented in the diet. The comment by Todd and Carriker (1922) that this kite "feeds entirely upon small birds," which has plagued the literature since, appears to be without basis, apart from the fact

that a captive individual kept by Mrs. Carriker for several months was fed on the carcasses of birds skinned by these collectors!

During the incubation phase, we observed 48 prey items delivered to nests and identified 31 of these. Lizards, at 61% of identified prey, were nearly twice as frequent as insects (35.5%) and provided roughly 10 times the biomass. Most of these items were delivered by males to incubating females. Of eight items captured by females during brief absences from the nest, five were insects. It is perhaps not surprising that males brought in mainly larger prey (lizards) during this period. Males provided most food for incubating females and appeared to hunt at some distance from the nests. This implies significant travel costs, likely making it more efficient to deliver a few large prey items rather than many small ones.

During the nestling period, diets varied considerably between nests, even within the same season (Table 6.2). In 1995, at the Pajaritos nest, insects made up 81.0% of identified prey, whereas at the Temple III nest, insects were only marginally more frequent (54.0%) than lizards (42.9%). The Pajaritos pair took many more cicadas than did the Temple III pair, despite a high degree of temporal overlap between these two nesting efforts. The difference in diet between these nests probably stemmed from the much larger role the female played as a hunter at the Pajaritos than at the Temple III nest. The Pajaritos female spent nearly twice as much time away from the nest, apparently hunting, as did the Temple III female, who mostly remained perched near the nest tree. The Pajaritos male was rarely observed near the nest, and the female played a major role in provisioning the young. The combined duties of hunting and nest vigilance may have led this female to rely more heavily on prey available in the immediate nest vicinity (i.e., cicadas) than was the case at the Temple III nest, where hunting was performed mainly by the male. At the two 1996 nests, insects formed fewer than half the identified prey items (31 and 43%), while lizards accounted for most (68 and 57%) of the identified prey and at least 85% of prey biomass.

In 1995, diet composition varied with time of day. Cicadas in the diet were common in early morning (0600–0900), fewer at midday, and again more common in late afternoon. Conversely, lizards were most frequent in the diet during midday. Conceivably this pattern may relate somehow to the relative availability—a function of abundance, detectability, and ease of capture—of these two prey types as temperatures change through the day. Captures of other insects and non-lizard vertebrates showed no apparent correlation with time of day.

Hunting Behavior

Despite their aerial display flights, we never saw Double-toothed Kites hunt above the forest canopy. These kites hunted exclusively within the forest, mainly in the upper to lower canopy, though occasionally descending to ground level in pursuit of prey. All observed hunts were launched from a perch, and captures were made with the talons. We observed 51 hunts: 45% were level or nearly level, powered flights to intercept flying quarry or culminating in contact with vegetation; 49% were hard glides or swoops at steep angles (25–45°) downward from a perch to snatch prey from the air or vegetation with a quick turn; and 6% were fluttering "parachute" drops straight down to the surface of a lower tree crown. Some parachuting flights were direct attacks on prey, while others preceded active scanning from sub-canopy perches. Attempts at quarry perched on vegetation were twice as common (63%) as those at quarry in flight (31%); the remaining hunts were directed at quarry on the ground.

The rate of hunting success was 73% overall, and it did not differ among hunting methods. This figure, however, may be artificially high, as we could record only those hunts occurring within view of our observation platforms, and most of these were attempts on insect prey. Hunts on vertebrates likely have lower success rates (Temeles 1985). We did, however, witness four captures of lizards, using both straight, powered flight to the vegetation surface and hard swoops culminating in a crash into a tree crown. We also frequently observed individuals on extended, erratic flights through tree crowns—apparently hunting attempts directed at quarry unseen by us.

Reports of hunting behavior elsewhere are consistent with our conclusion that this species is mainly a perch hunter, and add three additional hunting methods: (1) pursuing lizards along branches by hopping with outstretched wings (Laughlin 1952; Wetmore 1965); (2) taking bats on the wing at dusk at a presumed roost site (Baker et al. 1999); and (3) attending troops of White-throated Capuchin (*Cebus capucinus*), Common and Red-backed Squirrel Monkeys (*Saimiri sciureus* and *S. oerstedi*), and Spix's Moustached and Saddle-backed Tamarins (*Saguinus mystax* and *S. fuscicollis*),

capturing prey flushed by troop movements (Greenlaw 1967; Moynihan 1976; Fontaine 1980; Terborgh 1983; Boinski and Timm 1985; Boinski and Scott 1988; Egler 1991; Heymann 1992).

Kites typically have been noted to follow troops at a distance of 5–40 m, spending most of the time on perches at or below the level of the monkeys, and making occasional short flights after flying insects or bats flushed by primates (Fontaine 1980; Boinski and Timm 1985; Boinski and Scott 1988; Heymann 1992). Primate species regularly followed by kites are small monkeys that travel extensively and include a fair number of invertebrates in their diet—traits that make them effective at flushing quarry for the kites (Fontaine 1980). In contrast, howler (*Alouatta* spp.) and spider monkeys (*Ateles* spp.), which are inactive for long periods of the day, eat few arthropods, and often follow regular routes through the canopy, were rarely followed by Double-toothed Kites (Fontaine 1980), presumably because they served poorly as prey flushers.

Association time, recorded as the percentage of scan samples or of total observation time in which a kite was present, varied from 2 to 19% among studies and monkey species (Fontaine 1980; Boinski and Timm 1985; Boinski and Scott 1988; Heymann 1992). Variations in association time have been attributed to differences in troop size or in the vigorousness of foraging activity among monkey species (Heymann 1992). Boinski and Scott (1988) also noted a seasonal component to Double-toothed Kite monkey-following behavior in Costa Rica, with twice the frequency of *Harpagus* sightings around squirrel monkeys in the middle of the wet season (a period of low insect abundance) compared with more bountiful times of year. Most prey taken during primate-following bouts were arthropods—principally orthopterans and cicadas, but also including beetles (Coleoptera), dragonflies (Odonata), scorpions (Scorpionida), cockroaches (Blattidae), mantises (Mantidae), and walking sticks (Phasmatidae). Tent-making bats (Phyllostomidae) and a lizard were also taken in this fashion.

In Tikal we did not observe Double-toothed Kites hunting in association with monkeys, a result probably attributable to two factors. First, Black-handed Spider Monkeys (*Ateles geoffroyi*) and Guatemalan Howler Monkeys (*Alouatta pigra*) are the only nonhuman primates at Tikal, and these are not good candidates as prey flushers for the kites (see above). Highly active, partially insectivorous monkeys—the types seemingly preferred by the kites—are absent. Second, our observations were made mainly during the late dry and early wet season when insect prey is abundant (see Chapter 2) and kites are nesting. We suspect that if Double-toothed Kites display monkey-following behavior in our study region, it is not particularly common. There is some suggestion that these kites follow army ants or bird flocks at times, but there are no published details to our knowledge (Brown and Amadon 1968; Ferguson-Lees and Christie 2001).

Space Use in Hunting

We did not determine the maximum distance adult kites traveled from the nest on hunting forays. However, at the Pajaritos nest, at least 19% of the prey items (total number = 212) were captured within 100 m of the nest. The remaining items were captured out of view—either by a nearby adult shielded from view by vegetation, or by an adult farther than 100 m from the nest. Because a number of these prey were captured by the female during very brief absences from the nest, we estimate that at least 30–40% of the prey items delivered to this nest were captured within a 100 m radius. This case, however, seemed unusual: the other females we observed did not play as much of a hunting role. Males were visible near nests only 25 min/day on average, and likely foraged much more widely than females. Virtually all (99.5%) of the prey items known to be captured in the nest vicinity were insects, suggesting that adults had to travel farther to catch lizards and other vertebrates than they did to capture insects.

HABITAT USE

As nesting habitat, Double-toothed Kites displayed a strong preference for tall, dense-canopied, open-understoried Upland Forest. Ten of 11 nests were surrounded by Upland Forest, while the eleventh was situated in Sabal Forest, a relatively tall forest type occurring just below Upland Forest on the topographic gradient (see Chapter 2). No nests were found in Scrub Swamp and other lowland forest types despite the fact that these types covered about 20% of the area we intensively surveyed for nests.

In this respect, Double-toothed Kites differed from a number of raptors in Tikal, which preferentially nested in Transitional and Lowland Forest, and in emergent trees, which are common in those forest types, owing to the low or irregular canopy there. Emergent trees are thought to provide a measure of safety from climbing nest predators. As Double-toothed Kites nested in Upland Forest and in trees that were integral with the canopy, proximity to favorable foraging habitat may have been of greater importance in nest site selection than was availability of emergent trees. That female kites were often observed hunting near the nest supports this notion. For species nesting within the forest canopy, as did these kites (and not in emergent trees), Upland Forest may provide safer nest sites than does low Scrub Swamp Forest, where vine densities are considerably higher (Schulze and Whitacre 1999), resulting in more access for climbing predators.

Telemetry data from an adult female suggested that affinity for tall, closed-canopy forest was not limited to nest placement; we detected this female mainly in such forest (12 of 15 sightings) and rarely in the shorter, more open-canopied forest types (1 of 15 sightings, the 2 remaining detections being in Transitional Forest types). Nest observations revealed that other females also hunted largely in Upland Forest, mainly within 300 m of the nest. Adult males no doubt traveled more extensively in search of prey, and hence may have used additional forest types more than did females. Still, we observed soaring Double-toothed Kites descend into the forest canopy on dozens of occasions and never witnessed a bird descending into a Lowland Forest stand—another indication that these kites used mainly Upland Forest at Tikal. Other data also support this apparent habitat preference: point counts at Tikal found these kites significantly associated with structural variables indicative of tall, closed-canopy, Upland Forest (Whitacre et al. 1990).

Most published literature agrees that this kite is mainly a bird of mature forest, though also occurring in forest edge or older second growth (Hilty and Brown 1986; Stiles et al. 1989; Howell and Webb 1995). Some accounts, however, describe a somewhat broader habitat amplitude, including pastures and wooded savanna (Meyer de Schauensee and Phelps 1978; Ferguson-Lees and Christie 2001). Slud's (1960, 1964) impression that this hawk rarely enters far into mature forest is misleading—likely an artifact of the difficulty of observing them within the forest. Published observations of foraging Double-toothed Kites are mainly of individuals in the forest interior. Likewise, at two forest edge nests, adults were seen foraging only within the forest (Laughlin 1952; Skutch 1965; Greenlaw 1967; Heymann 1992). While these kites do use forest edge and other somewhat open habitats at times (D. Whitacre, pers. observ.), we saw no indication that our study pairs sought out such habitat—indeed, it was not available to most of them (Fig. 6.1). Although we detected Double-toothed Kites during point counts in the farming landscape, we do not know what habitat elements they used within this habitat mosaic.

BREEDING BIOLOGY AND BEHAVIOR

Nests and Nest Sites

Nests were shallow cups composed of small sticks about 1 cm in diameter and 20–40 cm long (Plate 6.2). Four nests averaged 32 ± 6 cm in diameter and 10 ± 1 cm in external depth. Nests were typically placed in a fork of two or more major limbs (9–25 cm in diameter), 0–12 m from the bole. Four of 10 nests were built on supporting bromeliads. All nests were placed in the upper third of the tree and ranged in height from 20 to 27 m. We observed no obvious selection of nest trees on the basis of species or morphology, aside from the exclusive use of non-emergent, canopy trees. Of 10 nests, only 3 were placed in the same tree species, Kapok (Ceiba pentandra).

Aspects of nest placement suggested that nest trees may have been selected in a manner that reduced predation risk. All nests were in relatively vine-free trees, without connections to neighboring trees, and eight

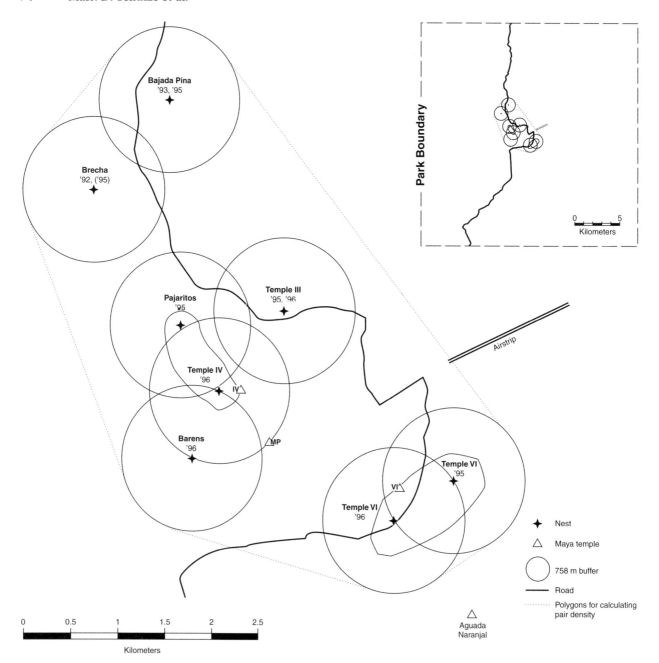

6.1 Locations of Double-toothed Kite nests we found in Tikal National Park, Guatemala. Area shown is the central portion of the park.

were in trees bordering a canopy gap larger than 100 m², also reducing connectivity with the surrounding canopy. Two other nests in Central America were also placed in mature forest trees on the edges of clearings (Laughlin 1952; Skutch 1965).

Four nest trees were dry season deciduous, but only one was lacking leaves during any part of the nesting epoch—the first few days of incubation. Although all nests were at least partly shaded by upper limbs of the nest tree, there was considerable variation in the amount of direct sunlight reaching nests, owing to variation in the thickness of vegetation overhead and in nest location relative to canopy gaps. In addition to camouflage, the partial protection from sun and rain afforded by nest

location may be critical in allowing females to leave the nest periodically in search of prey—apparently a characteristic feature of the species's breeding ecology.

Egg and Clutch Size

At Tikal we documented five 2-egg and two 1-egg clutches (consecutive years in the same nest). For an eighth nest, the female's behavior indicated the presence of more than one egg; we assume this was a 2-egg clutch. The only other report of a 1-egg clutch is by Laughlin (1952) in Panama; however, that clutch may not have been complete. Skutch (1965) observed a nest in Costa Rica with two chicks. Although Euler (1900) gave the clutch size in Brazil as three or four eggs, the only actual record of a clutch larger than two eggs that has come to our attention is a 3-egg clutch from Trinidad attributed to this species (see below). Thus while clutch size apparently ranges from one to three eggs, the modal clutch at Tikal was two and the mean clutch 1.75 eggs ($n = 8$).

We did not weigh or measure any eggs. Only one clutch identified as *H. bidentatus* exists in U.S. museums (Kiff 1979a). Measurements for this 3-egg clutch (Western Foundation of Vertebrate Society specimen 16323, taken in Trinidad in 1933) are as follows: 39.13 × 32.32 mm, 36.54 x 30.98 mm, and 42.13 x 31.56 mm, for a mean of 39.3 ± 2.8 x 31.6 ± 0.7 mm ($n = 3$). Using Hoyt's (1979) formula, these measurements give an estimated fresh egg mass of 21.5 g. Depending on which estimate of female body mass is used, a single egg thus amounts to 10.3–10.7% of female body mass, the mean 1.75 egg clutch at Tikal amounts to 18.1–18.7%, and the modal 2-egg clutch to 20.7–21.3%.

Nesting Phenology

Nesting was relatively synchronous among pairs and was concentrated between early April and late July, during the late dry season and early wet season. All study pairs began nest building in April or May—late in the dry season—with 8 April the earliest confirmed date of nest building and 25 May the latest observed date of nest building prior to incubation. Though at one nest, incubation began on 9–10 April, at four other nests, incubation was begun in May and continued to June or early July (see Table 6.1). The earliest hatching date was 6–7 June and the latest 1–2 July. Fledging occurred from mid-June to late July (see Table 6.1). Chicks hatched early in the wet season, coinciding with a pronounced increase in abundance of large-bodied adult insects, including the orders Homoptera, Coleoptera, and Orthoptera (see Chapter 2), and adults fed the nestlings increasingly on insects as the nestling period progressed. We speculate that such timing of nesting serves to maximize the hunting efficiency of adults at a time when energy demands are highest.

Other records from the northern Neotropics also indicate nesting of this species during the Northern Hemisphere spring as at Tikal. Slud (1980) witnessed nest building at the end of January in Costa Rica; Olivares (1962, cited by Hilty and Brown 1986) collected a male in breeding condition on 4 February in Colombia; Wetmore (1965) found a pair at a partly constructed nest in Panama on 26 March; Brown and Amadon (1968) described a pair building a nest on 1 April in Oaxaca; Hilty and Brown (1986) reported nest building in early April in Costa Rica; Laughlin (1952), in Panama, observed incubation beginning on 3 July; and Skutch (1965) found a nest in Costa Rica that held two well-feathered nestlings on 1 May.

Length of Incubation and Nestling Periods

At two nests, incubation lasted 42–44 days (see Table 6.1). The nestling period ranged from 27–37 days (31.0 ± 3.9 days; $n = 4$), though three out of four nestling periods were 27–31 days (mean = 29.2 ± 1.6; $n = 3$).

VOCALIZATIONS

The principal vocalization of adult Double-toothed Kites was a thin, high-pitched, two-note *cheee-weet* generally given in a series. This vocalization reminded us more of a songbird than a raptor, and Skutch (1965) likened it to the call of a flycatcher. This was used as a contact call between pair members, as a female begging call prior to prey exchanges, and during nest defense and aggressive encounters. In the latter case the call was typically louder, sharper, and with greater accent on the first note, resembling *cheeeeee-it*.

We also noted three other calls. When arriving near the nest with prey, males gave a repeated chirping call, described in the same context by Brown and Amadon (1968). Females, prior to arriving at the nest with food, voiced a single-note *cheeep*, seemingly an abbreviation of the common two-note call. The begging call of chicks was similar to the typical adult call but was higher pitched, raspier, and less clearly two noted. Other authors have reported a high-pitched, single-note whistle, a lisping *tsiip-tsiip*, and a rapid *tsup-sup-sup-sup* (Brown and Amadon 1968; Howell and Webb 1995).

BEHAVIOR OF ADULTS

Pre-laying and Laying Period

Females did most of the nest building, although males added nest material on occasion. Typically, there was a flurry of nest building in the early morning, prior to the breakup of the chronic morning fog and onset of display flights. Often no further activity was observed around a nest until late afternoon, when activity sometimes resumed. During courtship and nest building, males regularly brought prey (always lizards) to females in the nest vicinity, during early to mid morning. The male would fly into the nest area with prey in his talons, typically landing in a tree 10–20 m from the nest, repeatedly voicing a single-note chirping call. The female would fly to the perched male, usually voicing a two-note *cheee-weet*, and receive the prey. On one occasion the Temple III male copulated with the female while she ate the lizard he had brought her.

Double-toothed Kites frequently soared in pairs high above the forest during mid to late morning, often maintaining a distance of less than 20 m between the two. We observed no contact between soaring birds. The aerial display of this kite has been likened to that of an accipiter (Howell and Webb 1995); males made repeated short stoops, interrupted periodically by rapid flapping to regain altitude. In addition, we frequently observed a soaring individual or pair stoop from a height of more than 300 m down into the forest canopy. In two such cases we confirmed that the point of entry into the canopy was within 100 m of a nest under construction. The peak of aerial display activity occurred during April and May, before and during nest construction. During this period, several displaying pairs could often be observed from a single vantage point. After this time, we continued to see soaring individuals, but stooping activity was much reduced. To date we have not observed Double-toothed Kites soaring over the canopy outside of the breeding season. Aside from these aerial displays, which likely functioned in both courtship and territorial advertisement, no obvious courtship displays were witnessed.

Incubation Period

At all five nests observed during incubation, only females were seen to incubate or spend time on the nest. Overall, females were on the nest 85.3% of the time during this phase of nesting (n = 38 observation periods)—either incubating or standing in the nest (standing included rolling and shading the eggs, preening, and stretching). Females were off the nest but visible and within 40 m 10.7% of the time (n = 38). Only rarely was a female out of view (4.1% of observation time), and in many of these cases we lost contact as she dropped into the canopy near the nest tree; hence even then she may often have been near the nest. We saw no change in female activity as incubation progressed (Schulze et al. 2000a).

Incubation bouts at five nests averaged 42 ± 29 minutes in duration (n = 38 observation periods) and were significantly longer in early morning and late afternoon than at midday (Schulze et al. 2000a). The percentage of time females spent incubating was also greatest in early morning (87%) and late afternoon (77%), and least during midday (59–66%). Standing bouts, averaging 21 minutes (n = 38 observation periods), were half as long as incubation bouts, and birds stood marginally more often at midday, when shading eggs became a dominant activity, than in the morning and late afternoon (Schulze et al. 2000a). At one nest, the female stood or shaded the nest 90% of the time during the heat of the day (Schulze et al. 2000a).

On average, females incubated five times as much as they stood, shading the nest and eggs, although this ratio varied between nests from 12:1 to 1.1:1. Variation in this ratio was best explained by the degree of canopy cover above nests, with more shaded nests receiving relatively more incubation, and exposed nests receiving more shading by females. Females rarely moved out of sight of nests during the cool early morning and late evening, or the hot midday hours—the times when temperatures are most likely to reach dangerous extremes for unprotected eggs. Unlike several other accipitrid hawks we studied at Tikal, these kites were never observed bringing sprigs of green foliage to the nest.

Males were rarely seen near nests during the incubation phase, except during prey deliveries, which were like those described for the nest-building period. Receiving prey 10–20 m from the nest, females usually ate it before returning to the nest, but several times they

brought prey, especially lizards, to within a few meters of the nest before eating it. Males supplied most of the food for incubating females. Males delivered 2.3 ± 0.7 prey items per day during incubation (n = 5 nests), but this varied among nests from a high of 1 item per 4.4 hours of observation to slightly more than 1 item per day (9 hours). Of 48 prey items recorded at nests during incubation, 40 were delivered by males and 8 by females, apparently captured by them during incubation breaks. We regularly observed females leaving the nest on what appeared to be foraging excursions and at times watched them launch prey capture attempts from the nest.

Nestling Period

After the chicks hatched, females spent progressively less time on the nest (median for week 1 = 71.2% of time on the nest, for week 4 = 15.7%) and more time hunting. Overall during the nestling phase, females were on the nest on average only 34.5% of the time (n = 40 observation days). Of time spent on the nest, only 24.2% was spent in brooding position, with the remainder spent standing (including shading or feeding chicks, preening, or stretching). Throughout the nestling phase, females brooded chicks significantly more during early morning and late afternoon than during the midday hours (Schulze et al. 2000a), and individual brooding bouts were longer on average during these times of day than at midday. Once the chicks reached the age of one week, brooding by females was largely restricted to early morning (0500–0800), late afternoon (1600–dusk), and at night. Exceptions to this pattern occurred mainly on rainy days, when females brooded nestlings.

Female activity varied considerably among nests. At one nest (Pajaritos) the female spent substantially more time away (mean = 61%, n = 10 observation days) than did females at all other nests (means = 22, 32, and 40%; n = 8, 13, and 9 observation days, respectively). At all nests except this one, males provided most of the prey during the nestling phase. Whereas at other nests females were observed hunting only sporadically and captured an estimated 13–35% of the prey fed to nestlings, at the Pajaritos nest the female consistently hunted in the nest vicinity and captured at least 70% of the prey delivered to the nest. In nearly all cases, males delivered prey to females, who then fed the nestlings; on only eight occasions (1.4% of deliveries) did males feed chicks directly. As during incubation, prey transfers usually took place at least 10 m from the nest tree. Prior to all observed prey exchanges, the female emitted a two-note call; we sometimes heard males calling as well.

Throughout the nestling phase, adults fed insect prey to the nestlings in small pieces torn off with the bill as the adult pinned the prey down with its talons. Skutch (1965) reported that the adult he observed always fed entire lizards to chicks; he speculated that Double-toothed Kites lacked the strength to dismember vertebrate prey. We also observed that most lizards were fed to chicks in one piece, following removal of the head. However, several observations of adults feeding larger lizards, birds, bats, rats, and a snake to young bit by bit showed that these kites were capable of tearing apart vertebrate prey. We were not able to observe in detail the manner in which larger prey were killed, and hence cannot verify whether the toothed mandible is used to kill prey, as it is in falcons (Cade 1982).

During the nestling phase, a mean of one prey item was delivered per hour of observation. At the two 1995 nests, delivery rates followed a bimodal pattern, with most deliveries in mid to late morning (0800–1100, mean = 1.6/hr, n = 40 observation days), and a second, lower peak in late afternoon (1500–1700, mean = 1.4/hr; Fig. 6.2). In 1996, delivery rates peaked from mid-morning through early afternoon, decreasing in late afternoon. Combining both years, delivery rates during the late morning were higher than at any other time, and significantly higher than during late afternoon (Schulze et al. 2000a; Fig. 6.2).

The mean number of prey delivered per day varied among nests and was correlated with the number of chicks in the nest and with the percentage of vertebrates in the diet. The mean prey delivery rate for four nests during the nestling phase was 13.6 items/day. The Pajaritos pair, with two chicks, averaged nearly twice this number (24.3 ± 5.8 items/day, range = 22–31, n = 10 observation days), a significantly higher rate than recorded at any other nest (Schulze et al. 2000a). However, when rates were corrected for the number of chicks in the nest, nest means ranged from 7.8 to 12.1 feedings/chick/day, and only one nest had a significantly lower delivery rate per chick than that recorded for the Pajaritos nest. Per-chick daily prey delivery rates across all four nests showed a non-significant increase from weeks 1 to 4 of the nestling period (7.4 deliveries/chick in week 1, to 11.1 deliveries/chick in week 4).

Nests differed in both the number of vertebrate prey delivered per day (range = 2.4–5.4 items/chick/day), and the proportion of prey contributed by vertebrates (Schulze et al. 2000a). At the Pajaritos nest, where the female did most of the provisioning, chicks received the fewest vertebrates per day. Across nests, as the nestling period progressed, insects made up more of the diet at the nest, and vertebrates less. Further details of adult behavior are given by Schulze et al. (2000a).

BEHAVIOR AND DEVELOPMENT OF NESTLINGS

Nestling Period

Upon hatching, chicks were covered in white down, with dark eyes and beak, and yellowish cere and legs. By day 10, chicks had wing feathers in pin but no visible body or tail feathers. By 12 days, eyes had begun to lighten to a dark yellow-orange. By day 16, chicks were estimated at half the size of attending females, wing feathers were one-fourth to one-third grown, and tail feathers were emerging from feather tubes, as well as some contour feathers on the back, forming a V pattern. After 23 days, nestlings were typically full sized or nearly so, fully plumed on the body except the head and part of breast, and with flight feathers nearly or fully grown. Less developed birds at this age had flight feathers about half grown and were sparsely feathered above and below. At fledging, chicks appeared fully feathered.

Chicks initially were quiet, spending much of the day prostrate. Within a week of hatching they became more active, periodically standing and stretching, as well as more vocal, issuing soft begging calls when an adult arrived at the nest. At one of the two-nestling nests, we observed a very uneven distribution of food items, with the larger of the two chicks consuming 38% more prey than the smaller, including 28 of 39 lizards. This chick was not observed attacking the smaller one, but it would aggressively pursue prey morsels, often obtaining pieces that initially appeared destined for the second chick. At other two-chick nests we observed much less dominance of one nestling over the other.

Within 5–7 days prior to fledging, chicks began to leave the nest to clamber on supporting limbs. They often flapped while perched on the nest, and several days prior to fledging, they began making short hop-flights from branch to branch. Occasionally, after a long hiatus in prey deliveries, chicks would begin vocalizing, but in general, their begging calls appeared to begin when they sighted or heard a returning adult, usually the female.

Post-fledging Dependency Period

Of two juveniles outfitted with radios, one was found dead within a week of fledging, in the immediate nest vicinity. The other fledgling remained within 250 m of the nest for 35 days, during which time we observed the adults feeding it. By 42 days after fledging, the juvenile had moved more than 2 km from the nest, and over the next 10 days we repeatedly detected it roughly 4–8 km northeast of the nest. The last detection, 52 days after fledging, was 6–8 km northeast of the nest. After this

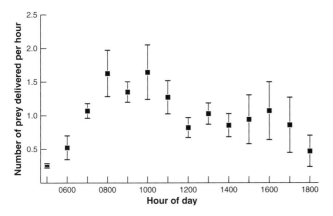

6.2 Daily pattern of prey deliveries to four Double-toothed Kite nests at Tikal (two nests in 1995 and two in 1996). Shown are means ± 1 SE.

time, repeated telemetry efforts suggested the bird had moved at least 10 km from the nest, more than six times the mean inter-nest distance.

SPACING MECHANISMS AND POPULATION DENSITY

Territorial Behavior and Displays

We observed only one or two aggressive encounters between conspecifics at nests. On one occasion an immature Double-toothed Kite landed within 3 m of a nest and was rapidly driven off by the returning female. In a second event at the same nest, a nestling was killed and carried off by a raptor, possibly a conspecific. Soaring individuals frequently appeared to follow or, on a few occasions, to pursue conspecifics soaring in the vicinity. These encounters resembled ritualized displays more than outright aggression, and were similar to those of Plumbeous Kites (*Ictinia plumbea*) at Tikal (see Chapter 7).

The habitual late-morning soaring flights of these kites served, we believe, in territorial advertisement, and possibly also in courtship and pair bond maintenance. It seems likely that these aerial maneuvers were the primary mechanism for maintaining territories and inter-nest spacing. In addition, these kites sometimes made spectacular, 300 m stoops into the forest canopy, and in two cases these were found to terminate quite near a nest under construction. It would be interesting to know whether such stoops serve as a conspicuous and accurate signal for maintaining inter-nest spacing.

Nest Defense and Interspecific Interactions

Double-toothed Kites aggressively defended their nests from Brown Jays (*Cyanocorax morio*), Keel-billed Toucans, Collared Aracaris (*Pteroglossus torquatus*), Pale-billed Woodpeckers (*Campephilus guatemalensis*), Roadside Hawks (*Rupornis magnirostris*), spider monkeys, a Montezuma's Oropendola (*Psarocolius montezuma*), and a Crane Hawk (*Geranospiza caerulescens*). They swooped on intruders, occasionally making contact, and called aggressively, tail-chasing fleeing interlopers. All of these species, except perhaps the woodpecker and oropendola, are potential predators of the kites' eggs or young. At one nest, where eggs did not hatch, the female was repeatedly harassed by Brown Jays and even driven from the nest in what may have been an attempt to prey on the eggs. This interference with the female's shading duties may have been lethal to the eggs, which failed to hatch; of our 11 study nests, this nest was most exposed to the sun. At several nests we observed repeated harassment of the female by Keel-billed Toucans, and we consider it likely that these toucans occasionally eat Double-toothed Kite eggs as Laughlin (1952) observed for Chestnut-mandibled Toucans (*Ramphastos swainsonii*) in Panama. We often observed subdued defensive behavior by incubating or brooding females when vultures or

raptors soared near the nest; typically the female mantled over the nest, erecting her plumage and sometimes vocalizing. We never observed Double-toothed Kites pursuing soaring raptors of other species.

Constancy of Territory Occupancy, Use of Alternate Nest Sites

Territory occupancy appeared to be relatively stable. Of five territories active in 1995, courtship activity was observed in four during 1996, and active nests were located in three of these. One territory was known to be active in 1993, 1995, and 1996 (not checked in 1994), and another in at least 1992 and 1996. Within territories, in two cases later nestings were in the same site as earlier nestings, while in two cases the second nest was 758 and 805 m from the first. Hence subsequent nestings within a territory either were in the same site as earlier nestings (*n* = 2), or were on average 780 ± 35 m (*n* = 2) from an earlier nesting.

Home Range Estimates

We radio-tracked one adult female regularly from 21 July to 18 September 1996, beginning one week after a fledgling left this nest. In 15 independent sightings, this female used an area of 2 km², ranging from 150 m to 2 km from the nest. She was found in a 75-degree-wide wedge to the north of the nest on all but one occasion, when she was located 500 m to the southeast.

Inter-nest Spacing and Density of Territorial Pairs

Results given here represent a slight refinement over those given by Schulze et al. (2000a). Spacing of nests at Tikal is shown in Figure 6.1. To estimate the density of pairs, we used the Maximum Packed Nest Density (MPND), square, and polygon methods described in Chapter 1. Nests that were simultaneously active in territories believed contiguous averaged 1.52 ± 0.65 km apart (*n* = 7, range = 0.76–2.49 km). We also calculated mean inter-nest distance regardless of year, using midpoints between alternate nest sites within a given territory. This version yielded a mean inter-nest distance of 1.58 ± 0.67 km (*n* = 5, range = 1.04–2.09 km). The shorter inter-nest distance yielded an MPND estimate of 2.09 km²/pair, a square estimate of 2.3 km²/pair, and a polygon estimate of 3.05 km²/pair, while the larger inter-nest distance yielded corresponding estimates of 2.27, 2.50, and 3.14 km²/pair, very similar values. All of our best estimates fell within the range of 2.1 to 3.14 km²/pair, with 3.14 km²/pair being the most conservative and hence appropriate estimate, equivalent to 30 pairs/100 km² of similar habitat. In 1994, we attempted to determine densities of several raptor species on a 20 km² study plot of mostly Upland Forest. Results suggested that 10 or 11 pairs of Double-toothed Kites occupied this plot, for a density of about one pair/2 km², according well with the estimates just given.

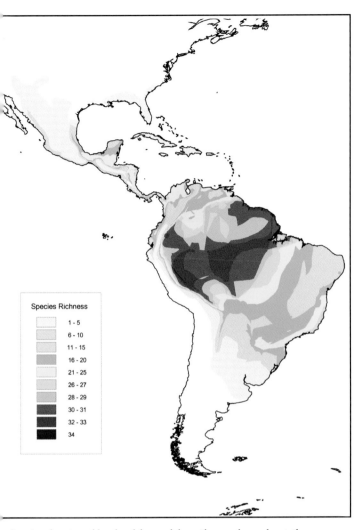

Species density of lowland forest falconiforms throughout the [Ne]otropics. Pale blue through pale orange indicate virtually the same set [of] species throughout. Courtesy Robin Bjork

1.3 View of Temple II with Temple III and Temple IV in the background, Tikal National Park, Guatemala. Courtesy Craig Flatten

1.4 Hector Madrid removes a *bal-chatri* after trapping for Ornate Hawk-eagles, Tikal National Park, Guatemala. Courtesy David Whitacre

[1.?] Megan Parker measures an adult Laughing Falcon with Bill Burnham [hol]ding the bird and Dawn Gerhardt observing. Courtesy Rick Gerhardt

1.5 Julio Madrid observes a White Hawk nest from a canopy blind, Tikal National Park, Guatemala. Courtesy David Whitacre

2.1 Anolis lizard (*Norops* sp.) in Tikal National Park, Guatemala. Courtesy Craig Flatten

2.2 Helmeted Iguana (*Corytophanes* sp.) in Tikal National Park, Guatemala. Courtesy Craig Flatten

2.3 Big-eared Climbing Rat (*Ototylomys phyllotis*), a native rodent in Tikal National Park, Guatemala. Courtesy David Whitacre

2.4 Maya forest surrounding Temple IV and Temple III, Tikal National Park, Guatemala. Courtesy Craig Flatten

2.5 Maya forest looking east from Tikal National Park, Guatemala. Courtesy Russell Thorstrom

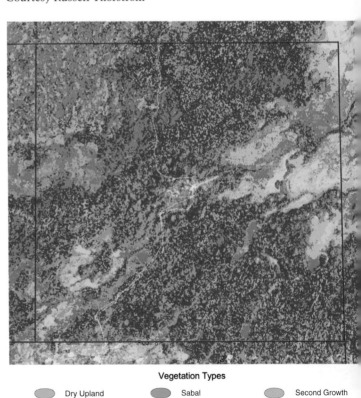

Vegetation Types

Dry Upland
Upland Standard (dry)
Upland Standard (wet)
Hillbase
Sabal
Transitional
Scrub Swamp
True Swamp
Second Growth
Farming Landscape
Urban/Bare

0 1 2 Kilometers

2.6 Topography and forest types of Tikal National Park. The Park is 36 km on each side. Courtesy Robin Bjork

2.7 Margay *(Leopardus wiedii)*, one of five native cats, in the canopy of the Maya Forest in Tikal National Park, Guatemala. Courtesy Russell Thorstrom

Chicozapote *(Manilkara zapota)* tree showing scars from machetes of ·vesting the milky chicle latex used as a component for chewing gum. ·urtesy David Whitacre

2.9 Xate *(Chamaedorea* sp.), an understory palm collected in Tikal National Park for floral arrangements in florist shops in North America and Europe. Courtesy David Whitacre

3.1 Gray-headed Kite adult in hand in Tikal National Park, Guatemala. Courtesy Russell Thorstrom

3.3 Gray-headed Kite nestling in Brazil. Courtesy Eduardo Pio Mendes Carvalho Filho

3.2 Gray-headed Kite adult standing over two nestlings in Brazil. Courtesy Eduardo Pio Mendes de Carvalho Filho

3.4 Gray-headed Kite egg from nest in Tikal National Park, Guatemala. Courtesy Russell Thorstrom

1 Hook-billed Kite soaring. Courtesy Russell Thorstrom

4.3 Dark morph Hook-billed Kite. Courtesy Ryan Phillips, Belize Raptor Research Institute

2 Hook-billed Kite in Grenada. Courtesy Russell Thorstrom

4.4 Hook-billed Kite nestling in Grenada. Courtesy Russell Thorstrom

5.1 Swallow-tailed Kite soaring over Tikal National Park, Guatemala. Courtesy Russell Thorstrom

5.3 Swallow-tailed Kite on nest at Tikal National Park, Guatemala. Courtesy Rick Gerhardt

5.4 Swallow-tailed Kite nestling in Tikal National Park, Guatemala. Courtesy Rick Gerhardt

5.2 Swallow-tailed Kite prey transfer at Tikal National Park, Guatemala. Courtesy Greg Lavaty

6.1 Double-toothed Kite incubating in Tikal National Park, Guatemala. Courtesy Russell Thorstrom

6.2 Double-toothed Kite adult feeding two nestlings in Tikal National Park, Guatemala. Courtesy Richard R. Jackson

7.1 Plumbeous Kite in flight. Courtesy Ryan Phillips, Belize Raptor Research Institute

7.2 Plumbeous Kite pair perched together. Courtesy Angel Muela

7.3 Plumbeous Kite adult at Tikal National Park, Guatemala. Courtesy Russell Thorstrom

7.4 Plumbeous Kite on nest in Tikal National Park, Guatemala. Courtesy Russell Thorstrom

7.5 Plumbeous Kite juvenile in Tikal National Park, Guatemala. Courtesy Rick Gerhardt

8.1 Adult Bicolored Hawk. Courtesy Jose Luis Dibos

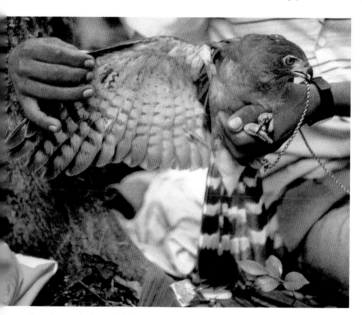

Bicolored Hawk female (dark morph) in hand in Tikal National Park, atemala. Courtesy Russell Thorstrom

8.3 Adult Bicolored Hawk. Courtesy Ryan Phillips, Belize Raptor Research Institute

8.4 Bicolored Hawk nest and eggs in Tikal National Park, Guatemala. Courtesy David Whitacre

9.1 Adult Crane Hawk in hand in Tikal National Park, Guatemala. Courtesy Craig Flatten

9.3 Nestling Crane Hawk in hand in Tikal National Park, Guatemala. Courtesy Jason Sutter

9.4 Crane Hawk nest and eggs in Tikal National Park, Guatemala. Courtesy Jason Sutter

9.2 Crane Hawk adult perched in forest in Tikal National Park, Guatemala. Courtesy Russell Thorstrom

9.5 Crane Hawk nestlings in Tikal National Park, Guatemala. Courtesy Jason Sutter

White Hawk soaring over Maya forest, Tikal National Park, Guate-

10.3 White Hawk subadult in Tikal National Park, Guatemala. Courtesy David Whitacre

10.4 White Hawk adult delivering prey to young at nest, Petén, Guatemala. Courtesy Rick Gerhardt

2 White Hawk perched in forest in Tikal National Park, Guatemala. rtesy Robert Berry

11.1 Great Black Hawk adult. Courtesy Ryan Phillips, Belize Raptor Research Institute

11.2 Great Black Hawk juvenile perched in forest in Tikal National Park, Guatemala. Courtesy Russell Thorstrom

11.3 Great Black Hawk juvenile. Courtesy Rick Gerhardt

11.4 Great Black Hawk nestling in Tikal National Park, Guatemala. Courtesy Rick Gerhardt

11.5 Great Black Hawk adult on nest in Tikal National Park, Guatema Courtesy Rick Gerhardt

12.1 Roadside Hawk adult perched in tree in Tikal National Park, Guatemala. Courtesy Russell Thorstrom

12.2 Roadside Hawk adult in hand in Tikal National Park, Guatemala. Courtesy Russell Thorstrom

3 Roadside Hawk nestlings in Tikal National k, Guatemala. Courtesy Theresa Panasci

13.1 Crested Eagle adult in Tikal National Park, Guatemala. Courtesy Richard R. Jackson

13.3 Crested Eagle adult guarding a nestling in Petén, Guatemala. Courtesy Richard R. Jackson

13.2 Crested Eagle adult perched near nest in Petén, Guatemala. Courtesy Richard R. Jackson

13.4 Crested Eagle juvenile perched in Tikal National Park, Guatemala. Courtesy Russell Thorstrom

1 Black Hawk-eagle in flight. Courtesy W. S. Clark

14.3 Black Hawk-eagle nestling in hand in Tikal National Park, Guatemala. Courtesy Russell Thorstrom

2 Black Hawk-eagle in Tikal National Park, Guatemala. Courtesy vid Whitacre

14.4 Black Hawk-eagle adult held by Julio Madrid in Tikal National Park, Guatemala. Courtesy Craig Flatten

15.1 Ornate Hawk-eagle subadult in hand in Tikal National Park, Guatemala. Courtesy Russell Thorstrom

15.2 Ornate Hawk-eagle in flight over Tikal National Park, Guatemala. Courtesy Russell Thorstrom

15.3 Ornate Hawk-eagle young standing in nest in Tikal National Park, Guatemala. Courtesy Craig Flatten

15.4 Ornate Hawk-eagle juvenile perched in Tikal National Park, Guatemala. Courtesy Craig Flatten

15.5 Remains of a young Ornate Hawk-eagle with a leg-mounted radio t in Tikal National Park, Guatemala. We observed a subadult Ornate Hawk-eagle eating the carcass of this bird in a tree, and later collected t remains on the ground directly beneath the perch. Although the death o the young Ornate was not observed, it's very likely that it was both kille and eaten by the subadult. Courtesy Craig Flatten

16.1 Adult male Barred Forest Falcon perched below nest in Tikal National Park, Guatemala. Courtesy Russell Thorstrom

16.5 Barred Forest Falcon eggs, complete clutch, in Tikal National Park, Guatemala. Courtesy Russell Thorstrom

16.2 Male Barred Forest Falcon in Tikal National Park, Guatemala. Courtesy Craig Flatten

16.6 Barred Forest Falcon nest inside tree cavity of Spanish Cedar (Cedrela odorata) tree in Tikal National Park, Guatemala. Courtesy Russell Thorstrom

16.3 Adult male Barred Forest Falcon perched near an army ant swarm in Tikal National Park, Guatemala. Courtesy Russell Thorstrom

16.7 Barred Forest Falcon female guarding young inside nest in Tikal National Park, Guatemala. Courtesy Russell Thorstrom

16.4 Juvenile young of the Barred Forest Falcon in Tikal National Park, Guatemala. Courtesy Craig Flatten

16.8 Three Barred Forest Falcon nestlings in Tikal National Park, Guatemala. Courtesy Russell Thorstrom

17.1 Adult male Collared Forest Falcon held by Sixto Funes in Tikal National Park, Guatemala. Courtesy Craig Flatten

17.3 Juvenile female Collared Forest Falcon perched in forest in Tikal National Park, Guatemala. Courtesy Russell Thorstrom

17.2 Adult female Collared Forest Falcon perched in tree near nest in Tikal National Park, Guatemala. Courtesy Russell Thorstrom

17.4 Adult female Collared Forest Falcon at nest entrance, tree cavity, in Spanish Cedar *(Cedrela odorata)* tree in Tikal National Park, Guatemala Courtesy Russell Thorstrom

18.1 Adult Laughing Falcon perched in Tikal National Park, Guatemala. Courtesy Craig Flatten

18.2 Adult Laughing Falcon in Tikal National Park, Guatemala. Courtesy Craig Flatten

18.3 Nestling Laughing Falcon in nest in Tikal National Park, Guatemala. Courtesy Craig Flatten

19.1 Bat Falcon adult perched on Temple IV in Tikal National Park, Guatemala. Courtesy Craig Flatten

19.2 Typical Bat Falcon habitat. Courtesy David Whitacre

20.1 Adult Orange-breasted Falcon perched in an area west of Tikal National Park, Guatemala. Courtesy Russell Thorstrom

20.3 Orange-breasted Falcon adult guarding eggs. Courtesy Angel Muela

20.2 Orange-breasted Falcon nest site on karst sink cliff in Belize. Courtesy Robert Berry

21.1 Adult Mexican Wood Owl. Courtesy Ryan Phillips, Belize Raptor Research Institute

21.3 Mexican Wood Owl nest in allspice *(Pimienta dioica)* tree cavity in Tikal National Park, Guatemala. Courtesy Rick Gerhardt

21.4 Mexican Wood Owl nestlings in Tikal National Park, Guatemala. Courtesy Rick Gerhardt

21.2 Mexican Wood Owl adult in hand in Tikal National Park, Guatemala. Courtesy Craig Flatten

22.1 Black-and-white Owl in Tikal National Park, Guatemala. Courtesy Craig Flatten

22.3 Black-and-white Owl nest and egg in arboreal epiphyte in Tikal National Park, Guatemala. Courtesy Rick Gerhardt

22.2 Black-and-white Owl perched near nest in Tikal National Park, Guatemala. Courtesy Craig Flatten

22.4 Black-and-white Owl nestling in Tikal National Park, Guatemala. Courtesy Craig Flatten

23.1 Anolis lizard (*Norops* sp.), one of the many species of lizards in Tikal National Park, Guatemala. Courtesy Rick Gerhardt

23.2 Casque-headed Iguanid (*Laemanctus* sp.), one of the larger species of lizards in Tikal National Park, Guatemala. Courtesy Rick Gerhardt

23.3 Close up of Gray Hawk, one of the uncommon hawks in Tikal National Park, Guatemala. Courtesy Craig Flatten

23.4 Female Black-handed Spider Monkey (*Ateles geoffroyi*) with young clinging to its back in the Maya Forest in Tikal National Park, Guatemala. Courtesy Russell Thorstrom

23.5 Kapok tree *(Ceiba pentandra)* in the Maya Forest south of Tikal National Park, Guatemala. Courtesy David Whitacre

23.6 Successional stages after subsistence (milpa) farming, like that south of Tikal National Park. Courtesy David Whitacre

23.7 Guatemalan Howler Monkey (*Alouatta pigra*) assuming a typical resting pose in the canopy of the Maya Forest in Tikal National Park, Guatemala. Courtesy Russell Thorstrom

Assuming these kites do not nest in the Low Scrub Swamp Forest covering roughly 12% of Tikal National Park, the remaining 510 km² of the park may support as many as 160 pairs of Double-toothed Kites—or fewer if the kites avoid Transitional as well as Scrub Swamp Forest. This estimated density is probably the fourth highest of all forest raptors at Tikal, ranking behind that of the Mexican Wood Owl (*Strix squamulata*; 3.5–4.4 pairs/km²; see Chapter 21), Barred Forest Falcon (*Micrastur ruficollis*; one pair/km²; see Chapter 16), and probably the Guatemalan Screech Owl (*Megascops guatemalae*), which may well be the most abundant raptor at Tikal.

The high density we documented for the Double-toothed Kite is interesting, inasmuch as several authors have commented on this kite's rarity in Central America north of Panama (Paynter 1955; Skutch 1965; Monroe 1968; Land 1970). As we have found this small kite to be quite inconspicuous except during display flights, we suspect it has often been overlooked, and that characterizations of it as a rare species may be far off the mark. Several other authors have accorded this kite an abundance more closely agreeing with our observations, regarding it as uncommon to fairly common (Wetmore 1965; Ridgely and Gwynne 1976; Hilty and Brown 1986; Stiles et al. 1989).

Our density estimate for Double-toothed Kites at Tikal is much higher than the 1 pair/14 km² estimated by Thiollay (1989a, 1989b) for a site in French Guiana. It would be interesting to know whether densities of this species are consistently lower near the equator than at our outer-tropical study site. Certainly the diversity of species that overlap dietarily with this kite must be far higher near the equator than at Tikal. These kites overlap in diet not only with other raptors but also with many non-raptorial birds and with certain mammals and reptiles.

DEMOGRAPHY

Frequency of Nesting, Percentage of Pairs Nesting

We have few data on the proportion of pairs nesting in a given year. Of 11 pair-years in which breeding behavior was initially noted, eight pairs laid eggs and three did not. Hence a tentative estimate is that about 70–75% of territorial pairs nested in a given year. The fact that kite pairs on a given territory sometimes nested in consecutive years suggests that dependency of juveniles was not sufficiently protracted to prohibit this, and we suspect that pairs routinely nest in consecutive years.

Productivity and Nest Success

Young were fledged in 4 (36.4%) of 11 territorial pair-years and four (50%) of eight nestings. Of 14 eggs laid in eight nests, 10 (71%) hatched, 5 (10%) of these producing a fledgling; hence 36% of eggs resulted in fledglings. Productivity was 0.45 fledgling per territorial pair-year, 0.63 fledgling per nesting attempt, and 1.25 fledglings

per successful nest. Three nests were abandoned during construction (two in the same territory in consecutive years), with no evidence of a later nesting effort. Of eight nests receiving eggs, two failed during incubation and two with nestlings. In addition, at one 1996 nest, one chick was killed and carried off by an unidentified raptor 7–10 days prior to fledging age, and the other was discovered dead, with a head wound, seven days after fledging. One nest fledged two chicks; the other three fledglings were singletons. The productivity reported here, 0.63 fledgling per nesting attempt, compares to 0.37 fledgling per nesting attempt in the Plumbeous Kite at Tikal, and 0.41 in the Swallow-tailed Kite—both of which rear a single chick.

Sources of Nest Failure and Mortality

In the two cases just described in which nestings failed during incubation, the eggs failed to hatch; females incubated them beyond the normal incubation interval and eventually abandoned the effort. In one of these cases harassment by Brown Jays may have led to overheating and embryo death. In the two nests that failed during the nestling phase, in one case the chick fell to its death, and in the other case, the chick was taken by an unknown predator. In sum, we observed nest failure or chick mortality twice due to failure to hatch, once due to falling, and three times due to predation on nestlings or young fledglings.

CONSERVATION

The main threat to Double-toothed Kite populations is deforestation. At Tikal, these kites showed a strong tendency to nest in tall, dense-canopied Upland Forests—and not in the lower, more open-canopied forest types. Moreover, a radio-tagged female was located mainly in tall Upland Forest. Finally, point counts at Tikal also found these kites significantly associated with closed-canopy Upland Forest. In sum, while these kites are sometimes observed in forest edge or other somewhat open situations, all evidence in our study pointed to an association with tall, mature forest.

Our point counts in the farming landscape also detected these kites, but we do not know what habitat elements they used there. If these kites nest in slash-and-burn farming landscapes, we suspect they do so mainly in mature forest remnants. Their modest spatial requirements presumably should facilitate their use of the small forest fragments often persisting in such farming landscapes.

HIGHLIGHTS

The breeding biology and hunting behavior of the Double-toothed Kite differed markedly from those of Plumbeous and Swallow-tailed Kites at Tikal—despite substantial dietary overlap among the three. Incubation in Double-toothed Kites (42–44 days at two nests) was remarkably long for a raptor of this size (Newton 1979), and confirmation at additional nests is desirable. While incubation was at least 10 days longer than for Plumbeous (32–33 days; see Chapter 7) and Swallow-tailed Kites (31.5 days; see Chapter 5) at Tikal, duration of the nestling period (averaging 29–31 days) was shorter than for Plumbeous (38.5 days) and Swallow-tailed Kites (52.3 days) at Tikal. Hence the interval from laying to fledging of Double-toothed Kites (71–76 days) was similar to that of Plumbeous Kites (70–72 days), while the equivalent figure for Swallow-tailed Kites at Tikal was substantially greater—approximately 83–84 days.

The fact that one radio-tagged kite dispersed at least 10 km from its natal site (six times the mean distance between nests) 6–8 weeks after fledging, and the fact that territories were sometimes the scene of nesting in consecutive years, suggest that these kites do not have a protracted dependency period as do several larger raptors at Tikal. We conclude that young at Tikal may typically reach independence within a few months after fledging. However, the frequent observation of juvenile kites in the company of one or two adults attending monkey troops (Fontaine 1980; Boinski and Scott 1988), suggests that juvenile kites and parents may at times associate for a longer period than suggested by our radio-tagged fledgling. Further data on this question are needed.

Inter-nest distances for Double-toothed Kites (1.5 km) were substantially greater than for Plumbeous Kites (500 m; see Chapter 7) and Swallow-tailed Kites (as little as 35 m at times; see Chapter 5). Moreover, Double-toothed Kite nests were evenly distributed within areas of tall Upland Forest. In contrast, Swallow-tailed Kites nested in loose colonies, often on hilltops or in other situations creating high availability of emergent trees, and large areas of forest had few if any nests. Plumbeous Kite nests, although evenly spaced within local areas, appeared to occur in loose neighborhoods, and their spacing seemed to be affected by the occurrence of clearings with which they were often associated. The more homogeneous spacing of Double-toothed Kite nests may be facilitated by these kites' use of nest trees within the forest canopy, rather than emergent trees in association with clearings or other landscape features—intrinsically uncommon sites. In addition, the prey base of Double-toothed Kites may be more homogeneously distributed and perhaps more defensible than that of these other kites, whose diets included some prey types that occur sporadically in time and space. Despite the closer spacing of Plumbeous and Swallow-tailed Kite nests within local areas, we suspect that in Tikal as a whole, Double-toothed Kite nesting density was substantially higher than that of these other kites.

During incubation and nestling phases, the role of males was limited to prey acquisition. Rarely did males feed nestlings, and we never witnessed a male incubating, brooding, or otherwise engaged in the care of eggs or nestlings. However, neither were males the sole providers. During the incubation period, females regularly left the nest vicinity for several minutes to half an hour, pos-

sibly to hunt, and after chicks were a week old, females hunted frequently. Hence, these kites did not conform to the most common pattern among raptors, in which males relieve females periodically from incubation duty (Newton 1979). They also departed from the model exemplified by some other raptors that feed largely on insects and small vertebrates—for example, Plumbeous and Swallow-tailed Kites and some small insectivorous falcons—in which pair members share incubation duties relatively equally and meet their own food demands throughout the nesting cycle (Seavy et al. 1998). Double-toothed Kites conformed more nearly to the pattern typical in accipiters, in which males provide most food for the female and young nestlings and commonly play no role in incubation and feeding young (Newton 1979). In 12 hours of observation at a Double-toothed Kite nest in Costa Rica, Skutch (1965) observed both adults brooding and feeding nestlings. This suggests that the full range of variation in parental role division may be greater than what we observed.

Double-toothed Kites took larger prey relative to their body size than did Plumbeous and Swallow-tailed Kites. This may make it more energetically efficient for one *Harpagus* adult to provision the nest than is the case for these other two kite species. The regularity with which Double-toothed Kite females obtained insect prey in the nest vicinity appeared to allow females to combine nest care with limited hunting activity—a compromise that would not be possible for a raptor preying on large vertebrates or other prey types requiring extended foraging trips. Males delivered mostly lizards rather than insects to the nest, especially early in the nesting cycle, suggesting that it was energetically more efficient to provision the nest with lizards than with insects. We did not observe males hunting near the nests, and they probably often brought prey from some distance, which would emphasize the energetic advantage of delivering few large items rather than many small ones. At all nests but one, males brought most prey, and lizards made up most of the prey biomass. At the remaining nest, the adult female provided most of the food for the nestlings, feeding them mainly cicadas she captured nearby. Females tending nestlings no doubt are under different time constraints than are males acting as providers, and likely remain nearer the nest, quickly catching whatever they can. They may forage more like "time-minimizers" than like "energy-maximizers" (Schoener 1974).

Diets of Double-toothed, Plumbeous, and Swallow-tailed Kites at Tikal overlapped substantially, with all three species taking many large insects and small lizards, but they also differed in important ways. The food the Double-toothed Kites fed their nestlings comprised 59% insects, 40% lizards, and 1% other vertebrates, and the food the Plumbeous Kites fed their nestlings, 93% insects, 5% lizards, and 2% other vertebrates (Seavy et al. 1997b). The proportion of cicadas in the Double-toothed Kite's diet is noteworthy. Numerically, cicadas made up 49% of the diet (Table 6.2) and 92% of all identified insects, compared to 33% of the overall diet and 37% of identified insects taken by Plumbeous Kites at Tikal (Seavy et al. 1997b; see Chapter 7). While Plumbeous and Mississippi Kites *(Ictinia mississippiensis)* have sometimes been regarded as cicada specialists (R. Gerhardt, pers. comm.), the Double-toothed Kite, according to this sample, is even more so. Still, in terms of biomass, lizards figured much more prominently in the diet than did cicadas (see below). Unlike Double-toothed Kites, Plumbeous Kites took nearly as many beetles as they did cicadas, as well as a fair number of dragonflies (see Chapter 7).

The diet of nestling Swallow-tailed Kites at Tikal resembled that of Double-toothed Kites in terms of percentage of insects (68%) and lizards (11%), but Swallow-tailed Kites also took many nestling birds (21%) and a few fruits and frogs. Swallow-tailed Kites took a wider variety of insects than did Double-toothed Kites—especially beetles, wasps, and locusts (Orthoptera)—with dragonflies, butterflies, cicadas, and others figuring less prominently (Gerhardt et al. 1991; see Chapter 5). Overlap of Double-toothed Kite diets is perhaps more significant with the Barred Forest Falcon than with Plumbeous or Swallow-tailed Kites, as this forest falcon, like the Double-toothed Kite, hunts below the canopy, with diet biomass at Tikal comprising 37% lizards, including many small *Anolis* of sizes like those taken by Double-toothed Kites (see Chapter 16).

Hunting behavior of these three kites differed considerably as well. Double-toothed Kites hunted mainly from perches within the forest, taking prey from the canopy, sub-canopy, and occasionally the ground, while the other two kites captured prey from the upper canopy surface and the air above. Part of the observed difference in diet is attributable to differences in hunting methods, with, for example, no dragonflies observed among Double-toothed Kite prey, as these raptors did not forage in open airspace. Likewise, the much greater proportion of beetles in the diet of Plumbeous than Double-toothed Kites may be related to these beetles' relative availability, while in flight, to the aerially hunting Plumbeous Kite. Double-toothed Kites take most prey from a substrate, and beetles may be more difficult and costly to find when at rest on a substrate than while in flight. It is conceivable that vertical stratification of hunting activity also provides a degree of resource partitioning among these kite species—to the extent that insect and lizard prey available below the canopy and those in the upper canopy and air above represent different prey populations—a topic worthy of study.

These three kites also differed in their timing of nesting. Double-toothed Kites laid eggs about six weeks later than Plumbeous and Swallow-tailed Kites, and most nestlings of the latter two species fledged by the time Double-toothed Kite eggs hatched, or soon thereafter. Whether this difference in timing is related to seasonal abundance patterns of the differing prey of these kite species, and whether such a difference holds true elsewhere in their large zone of sympatry, would be interesting to know. While this staggering of reproductive timing may result from chance or other factors, it could also result from natural selection to minimize simultaneous reliance on prey resources used in common, such as cicadas and *Anolis* lizards.

This kite's degree of sexual size dimorphism (7.5–10.0%) is relatively large compared to that of other raptors taking a diet of insects and small reptiles (Newton 1979), and is unusually large for a kite (Snyder and Wiley 1976). Such ecological comparisons between this and other raptors will take on more meaning when this enigmatic raptor's evolutionary relationships are better understood.

Raptors in this size range (200–300 g) at Tikal had mean clutch sizes ranging from one to three eggs. Thus, with a modal clutch of two eggs, these kites had a clutch size only slightly below average. The fact that pairs failed to lay eggs in 3 of 11 territorial pair-years was not unusual among our study species. At 36% of eggs, fledging success was low, resulting twice from failure of eggs to hatch, once from a chick's falling, and three times from predation on nestlings or fledglings. In all probability, this raptor's small body size heightens its vulnerability, not only to nest predation but probably also to occasional predation on adults. The observed productivity, 0.455 fledgling per territorial pair-year, implies that merely to replace themselves (to fledging age), a pair of adults must achieve at least this degree of success in each of five years. Taking into account mortality between fledging and reaching adulthood, adults must produce yet more fledglings. One can conclude only that these raptors have relatively low adult mortality, or that the productivity we observed is not typical. The fact that post-fledging dependency does not prohibit annual nesting by a pair no doubt facilitates the success of this species despite relatively low productivity.

Among the accipitrids we studied at Tikal, this kite was the only one that did not commonly bring green foliage to the nest. Newton (1979) generalized that small accipiters and kites do not, as a rule, bring foliage to nests, and data given by Wimberger (1984) largely agree. We offer the following hypothesis to explain this pattern of small species failing to bring in foliage: whatever the advantage of bringing greenery to the nest, in small raptors that are vulnerable to many kinds of nest predators, the risk of attracting attention to the nest by carrying green twigs outweighs the advantages of such behavior. Larger raptors are relatively more secure from nest predators, hence may respond more strongly to whatever advantages are provided by green foliage in the nest. Still, in future nest observations of Double-toothed Kites, it would be worth verifying whether green foliage is ever brought in.

Finally, the rarity of visits to the nest by the 1995 Pajaritos male, and the large role played by this female in provisioning young, makes one wonder whether this male had reproductive interests elsewhere. Conceivably, the fact that females are able to help provision young by catching cicadas near the nest may facilitate occasional polygyny in this species, although this is purely speculative.

For further information on this species in other portions of its range, refer to Boinski et al. 2003, Greeney and Gelis 2008, Greeney et al. 2004, and Jullien and Thiollay 2001.

7 PLUMBEOUS KITE

Nathaniel E. Seavy, Mark D. Schulze, David F. Whitacre, and Miguel A. Vásquez

From atop a small Maya temple, we peer through a spotting scope into the slender, arching leaves of bromeliads that spangle the leafless upper limbs of a tall Spanish Cedar Tree *(Cedrela odorata)*. In the crown of this tree, a twig nest is built amid the varied shades of green and maroon of the fine bromeliad leaves. From our position, we can see only the tail and wing tips, the ashy gray face, dark beak, and wine-colored eye of an incubating Plumbeous Kite *(Ictinia plumbea)* that is settled deep in the nest. The kite cocks an eye skyward, and we follow its gaze to see the pointed wings of its mate cutting gracefully across the sky. As this bird circles above, the morning sun reflects from the rusty orange patches in its slate-colored wings and from its narrow white tail bands, field marks that distinguish the Plumbeous Kite from its closest relative, the Mississippi Kite *(Ictinia mississippiensis)*. In the distance, a half-dozen more of these kites are soaring and dipping as they capture flying insects above the canopy. While their silhouettes are falcon-like—with long, pointed wings and a long tail—their deep, slow wing beats and buoyant flight reveal their identity (Plate 7.1).

Commonly seen foraging in groups during mid-morning, these kites make quick turns, seemingly effortless climbing pursuits, and even spectacular, falcon-like stoops as they pluck insect after insect from the air. A successful kite grasps an insect in its talons and extends its leg forward to meet its bent head. Nimbly the kite consumes the prey on the wing, never interrupting the graceful circles it scribes overhead. Reliance on flying insect prey has undoubtedly played a major role in the evolution of the flight attributes of these kites, easily observed in their finely tuned aerial performance and effortless grace. Not as easily observed, but equally elegant, the breeding biology and social system of this species reflect this influence as well.

GEOGRAPHIC DISTRIBUTION AND SYSTEMATICS

The monotypic Plumbeous Kite inhabits humid tropical lowlands from eastern Mexico south to Bolivia, Argentina, and Paraguay. Kites breeding in Central America are migratory, moving to South America during the Northern Hemisphere autumn and winter (Eisenmann 1963). At the southern limit of its breeding range, in Argentina, this kite is also migratory, arriving in mid-September and departing in the Southern Hemisphere autumn (Brown and Amadon 1968). Also migratory, the Mississippi Kite shares the South American wintering range, but it moves northward to breed in the southern United States. The only members of the genus *Ictinia*, Plumbeous and Mississippi Kites comprise a superspecies and are even considered by some to be conspecific (Sutton 1944; AOU 1983; Parker 1988; del Hoyo et al. 1994). Thus, we freely compare their ecology and behavior.

The anatomical similarities between the Plumbeous-Mississippi Kite *(Ictinia)*, Swallow-tailed Kite *(Elanoides forficatus)*, and Black-shouldered Kites *(Elanus* spp.) illustrate some of the difficulties hampering our understanding of evolutionary lineages within the Accipitridae. Similarities among these genera result both from primitive character states resulting from a common ancestry and from derived states resulting from convergent evolution. As long ago as 1891, Shufeldt (cited by Holdaway 1994) noted skeletal similarities between *Ictinia* and *Buteo*, but Jollie (1977) placed *Ictinia* with the "elanine" kites, while Brown and Amadon (1968) regarded it as between the elanine and "milvine" kites (Holdaway 1994). A cladistic analysis based on skeletal characters confirmed Shufeldt's (1891) opinion, and placed *Ictinia* and *Harpagus* squarely in the *Accipiter-Buteo* radiation, as sister group to a clade containing *Accipiter, Leucopternis, Buteo,* and others (Holdaway 1994). Hence, rather than being a primitive, basal accipitrid as *Elanus* and *Elanoides* apparently are (being independent sister groups to virtually the entire remainder of the family), *Ictinia* and *Harpagus* appear, on current evidence, to be members of a highly derived and diverse radiation within the accipitrids. As with other accipitrid genera, however, future research may greatly revise our current conception of generic relationships.

MORPHOLOGY

Plumbeous Kites have a slender build, with long, pointed wings and a long tail (Plate 7.2). Published data on mass and linear dimensions are given in Appendixes 1 and 2. In Surinam, males averaged 243 g in mass ($n = 16$) and females 257 g ($n = 7$: Appendix 1), with females 5.8% heavier than males, for a mass Dimorphism Index (DI) value of 1.9. Bierregaard (1978) found that females were slightly larger than males in total length (DI = 3.7) and wing length (DI = 1.3), and in three measures of mandible size (DI = 2.0–6.5), indicating a slightly larger body overall. Still, these kites are among the least size dimorphic of North American raptors (Snyder and Wiley 1976).

Males, however, had slightly longer tails than did females (DI = −2.7), longer tarsi (DI = −2.1), and larger feet (three measurements with DI of −1.9 to −5.3). This pattern, with males having proportionally larger prey capture apparatus (tarsi and feet) than females—in opposition to the prevailing pattern of larger overall female size—is commonly seen among diurnal raptors, as indicated by values in Appendix 2. Compared with other raptors, the beak and feet of these kites are small, apparently well suited to capturing and handling insects and occasional small vertebrates (Plate 7.3).

PREVIOUS RESEARCH

While the closely related Mississippi Kite has received a good deal of study (Evans 1981; Glinski and Ohmart 1983; Parker 1988), the Plumbeous Kite, until now, has received little attention. The first description of Plumbeous Kite breeding biology was published by Alexander Skutch in 1947. His account of a nest in Ecuador gives information on parental care, diet, and hunting behavior. Although his observations were limited to 22 hours at a single nest, many of his conclusions held true in our more extensive study. Additional information is mainly anecdotal, with notes by Haverschmidt (1962) regarding diet and aerial foraging by groups of kites, and by Ferrari (1990) on Plumbeous Kites catching cicadas flushed by primate troops. Among the most informative of the many regional accounts is that of Haverschmidt and Mees (1994).

RESEARCH IN THE MAYA FOREST

We studied six Plumbeous Kite nests at Tikal—two each in 1991, 1992, and 1994—and documented nesting histories totaling 20 territory-years from 1991 to 1996 (Table 7.1). Most nest observations were dawn to dusk, but some were 2–6 hours in duration. In 1994, we studied hunting behavior concurrently with nest observations: at each of two nests we kept one adult under observation as much as possible while it hunted. Unless stated, values given are means ± SD. Further results are given by Seavy et al. (1997a, 1997b, 1998).

DIET AND HUNTING BEHAVIOR

Plumbeous Kites at Tikal relied heavily on insect prey, although they were not exclusively insectivorous as some authors have implied (Skutch 1947). Of 655 identified prey items delivered to six nests, 93% were insects, 5% lizards, 0.9% snakes, 0.5% birds, 0.3% bats, and 0.2% frogs (Table 7.2). At this level of taxonomic distinction, variation between nests was relatively low.

About 7% of the prey delivered to nests were small vertebrates, mainly lizards. Lizards in the diet included both the arboreal genus *Anolis*, which is common throughout the forest at Tikal, and the terrestrial and rock-loving genus *Sceloporus*, which is uncommon in the forest but abundant in recently felled sections of forest and on limestone Maya temples. *Anolis* lizards in the diet represent captures from the forest canopy. Although we never observed the capture of a spiny lizard (genus *Sceloporus*), we witnessed several attempts in which the kites came in contact with the temples or the ground, and we assume these lizards were taken in this fashion. We were unable to identify the few snakes and frogs that were delivered to nests. All three of the birds taken were Ridgway's Rough-winged Swallows *(Stelgidopteryx serripennis)*, which nested abundantly in the ruins. One pair of kites was observed capturing one of these swallows via a cooperative hunt (Seavy et al. 1997b). Only two bats were observed as prey, although bats were abundant in the ruins and the forest. All vertebrates in the diet were small, estimated to weigh less than 15 g. Though vertebrates consistently made up a small portion of the diet, even some large insects are more than Plumbeous Kites are equipped to handle. We once watched a kite arrive at the nest with a live Rhinoceros Beetle (Scarabaeidae) about 8 cm in length. After struggling with the beetle for over a minute, the bird gave up, dropping the still very much alive beetle to the forest floor.

In terms of identified insect prey ($n = 400$), 37% were Homoptera (nearly all cicadas: Cicadidae), 29% beetles (Coleoptera), 15.8% dragonflies (Odonata), 8.8% Orthoptera (mainly katydids), 5% Lepidoptera, and 4.5% Hymenoptera (Table 7.2). In total, 455 prey were identified to class (vertebrates) or order (insects), and another 210 were identified only as insects; the latter may have included additional orders and may have systematically included smaller insects (Table 7.2). Overall diet percentages depend on whether unidentified insects are omitted from calculations (i.e., are assumed similar in identity to identified insects) or retained in calculations (i.e., are assumed to possibly differ from identified insects). In either case, cicadas were most important (22–33% of prey), followed by beetles (18–26%) and dragonflies (10–14%), with orthopterans, lizards, lepidopterans, and hymenopterans being less important (Table 7.2). The relative importance of these orders varied among years, with cicadas and beetles predominating in 1991 and 1994 and being practically absent from the diet in 1992 (Table 7.2). In 1992, when cicadas and beetles in the diet were few, dragonflies, lizards, and orthopterans assumed greater importance.

Table 7.1 Nest site descriptions, reproductive histories, and data collected for Plumbeous Kite nests studied in Tikal National Park

Territory name and year	Nest tree		Nest height (m)	Nest situation	Reproductive success	Observation time (hr), data recorded
	Species	Height (m)				
Mundo Perdido						
1992	?	?	?	?	Fledged 1	88.65, nestling phase
1994	Cedrela odorata	39.0	33.0	Fork, 10–15-cm limb, 4 m from bole, on bromeliad	Fledged 1	19.28, incubation phase; 72.62, nestling phase; diet and adult hunting studied
1995	C. odorata	same	same	same	Hatched; fate unknown	24.32, incubation phase
1996	C. odorata	same	same	same	Fledged 1 on 1 Jun (± 3 days)	Occupancy, productivity only
Temple I						
1991	?	?	?	?	Laid 1 egg; chick killed at 5 days old	Productivity only
1994	Bernoulia flammea	25.5	22.5	Against bole on supporting limb, on bromeliad	Fledged 1	22.7, incubation phase; 98.35, nestling phase; diet and adult hunting studied
1995	C. odorata	36.0	27.0	Fork in major limb, 3 m from bole, 93 m from 1994 nest	Failed during incubation	24.32, incubation phase
1996	C. odorata	same	same ?	same ?	Failed during incubation; no re-nesting observed	Occupancy, productivity only
Grupo G						
1995	Bernoulia flammea	37.5	36	On 20-cm limb, at intersection of three 5-cm limbs, 4 m from bole	Nested, success unknown; no pair known here in prior years; may be alternate site of Champon pair, though a non-nesting pair was in evidence at the latter this year	27.78, incubation phase
1996	—	—	—	—	No activity seen despite 3 mornings observed in late May; either failed early or pair nested distant from prior year's site	Occupancy only
Champon						
1993	Ceiba pentandra	23.3	19.5	6 m from trunk in fork of 20 cm limb	Failed one day after hatching	Productivity only
1994	Cedrela odorata	17.3	12.8	Against 15 cm bole on base of 10 cm limb; 96 m from 1993 nest	Laid 1 egg; failed during incubation	Few hours, incubation period; data not used
1995	—	—	—	—	Pair copulated, defensive, but did not nest; may conceivably have nested at Grupo G ?	—
1996	Ceiba pentandra	?	35 m	10–15 m from bole on epiphyte on 20 cm limb; 105 m S of 1994 nest	Fledged 1 on 6 Jun	Productivity only
Bajada la Pina						
1994	Cedrela odorata	27.5	22.5	Far out in slender secondary limb, 8 m from bole	Failed during windstorm during incubation phase	Productivity only
1995	Aspidosperma megalocarpon	31.2	27.1	5 m from bole on fork of two limbs (7 and 3.5 cm diameter)	Hatched 1; chick fell from nest	23.13, during incubation phase

Year	Nest tree species			Nest site description	Nesting outcome	Notes
1996	*Ceiba pentandra*	?	26	10–12 m from bole in small crotch without epiphytes; 150 m NW of 1995 nest	Failed prior to nestling stage; no re-nesting seen	Productivity only
Airstrip						
1991	?	?	?	?	Laid 1 egg; nestling fell during windstorm; died	Diet studied during nestling period
1992	?	?	?	?	Fledged 1	95.08 during nestling phase
1994	*Ceiba pentandra*	22.5	16.5	8 m from bole on 30 cm limb at insertion of two 2–5-cm limbs	Failed during incubation	Productivity only
1995	*Swietenia macrophylla*	39.7	27.5	10 m from bole at trifurcation of 3 limbs, 12, 2, and 3 cm diameter	Nested: no data taken	None
1996	—	—	—	—	No activity seen in vicinity of prior years' nests	Occupancy only
Bodega						
1991	?	?	?	?	Laid 1 egg; fledged 1 nestling	Dietary data collected
Brecha Anabela						
1992	?	?	?	?	Laid 1 egg; failed during incubation	Observed during nestling phase; productivity only
San Antonio						
1992	?	?	?	?	Laid 1 egg; failed during incubation	Productivity only

Table 7.2 Diet at six Plumbeous Kite nests in four territories (a, b, c, and d) in Tikal National Park[a]

Prey category	Year						Total number of prey items	Partly identified prey (%)[b]	Fully identified prey (%)[c]
	1994a	1994b	1992b	1992c	1991c	1991d			
Homoptera	41	57	1	7	29	12	147	22.4	33.0
Coleoptera	37	53	0	0	6	20	116	17.7	26.1
Odonata	2	2	19	33	3	4	63	9.6	14.2
Orthoptera	2	3	8	16	4	2	35	5.3	7.9
Lepidoptera	2	0	8	8	2	1	21	3.2	4.7
Hymenoptera	6	0	1	2	8	1	18	2.8	4.0
Unidentified insects	70	44	27	31	21	17	210	32.1	—
Total number of insects	160 (95.8)	159 (98.2)	64 (78.0)	97 (87.4)	73 (98.7)	57 (96.6)	610	93.1	89.9
Lizards	2 (1.2)	2 (1.2)	14 (4.9)	12 (10.8)	1 (1.4)	2 (3.4)	33	5.0	7.4
Snakes	0	0	4	2	0	0	6	0.9	1.3
Frogs	0	1	0	0	0	0	1	0.15	0.2
Birds	3	0	0	0	0	0	3	0.5	0.7
Bats	2	0	0	0	0	0	2	0.3	0.4
Total number of vertebrates	7 (4.2)	3 (1.9)	18 (22.0)	14 (12.6)	1 (1.4)	2 (3.4)	45	6.9	10.0
Unidentified prey	2	8	5	19	10	3	47	—	—
Total number of identified prey	167	162	82	111	74	59	655	100	100
Total number of prey	169	170	87	130	84	62	702	655	445

[a] Numbers given are numbers of prey items, with percentages (based on the number of insects identified only to class and identified to order) given in parentheses.
[b] Percentage of 610 items, including 210 unidentified insects.
[c] Percentage of 445 fully identified items; 210 unidentified insects are omitted.

We often saw adult kites feeding on the wing, catching insects too small to be identified—or at times even seen—through binoculars. Such small insects, however, were rarely if ever delivered to nests, presumably because the time and energy costs of repeatedly delivering very small prey items would exceed the benefits. We consider it especially likely that adult kites fed at times on winged termites and ants, which sometimes swarmed briefly in huge numbers, especially just after the first few rains of the rainy season. These winged reproductives are rich in fat and hence a desirable prey item (Wiegert and Coleman 1970; Baroni-Urbani et al. 1978), often heavily used by many bird species, including birds that do not normally feed on insects (Thiollay 1970). Haverschmidt (1962) described an observation of four Plumbeous Kites hawking termites during a nuptial swarm.

In 1994, we recorded 403 hunting attempts and their outcomes. Plumbeous Kites used two foraging techniques—aerial and perch hunting. Aerial attacks were launched at flying prey from extensive prospecting flights in which the kites soared anywhere from canopy level to several hundred meters above. During aerial hunts insects were captured via (1) quick grabs—when birds did not perceptibly alter their flight path, merely reaching out to grab nearby insects; (2) short flapping climbs to seize an insect above; and (3) diving pursuits—when kites attacked an insect below them by dropping into a steep dive or shallow swoop, often lofting upward at the end to seize the insect. These falcon-like dives were up to 100 m but more often 10–40 m in length. When hunting from perches, the birds would often sit still for long periods then suddenly bolt from the perch, making either a stoop or a level flapping flight to capture prey. While attacks launched from flight were invariably directed at insects in flight, perch hunts were sometimes directed at insects or lizards on the ground or in the forest canopy.

Overall, 246 of 403 attacks, or 61%, were successful. Aerial attacks comprised 69% of all hunts and were more successful (65% success rate; 179 of 276) than were hunts from perches (53% success rate; 67 of 127). Of the 127 hunts initiated from perches, 87% were launched at flying insects and 13% were directed at quarry in the forest canopy or on the ground. Perch hunts at flying insects were less successful (55% success rate; 60 of 110) than were aerial hunts at the same quarry (65% success rate; 179 of 276), but were similar in success rate to perch hunts at quarry in the canopy or on the ground (41% successful; 7 of 17: Seavy et al. 1997a).

Mid-morning was by far the most active foraging period, followed by a long afternoon lull and a resurgence of hunting activity in late afternoon and evening (Fig. 7.1). Most early morning and late evening hunts were launched from perches, while most hunts during the mid-morning peak hunting hours were launched from soaring flight. These daily patterns are probably the combined result of temporal patterns of aerial insect activity and atmospheric conditions providing lift for economical flight by the kites.

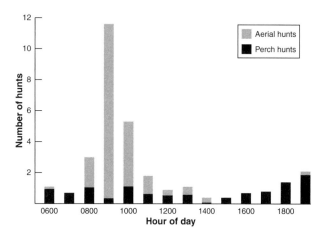

7.1 Daily pattern of Plumbeous Kite hunts launched from flight and from perches at two nests at Tikal, Guatemala (nests 1994a and 1994b, combined data).

An interesting component of Plumbeous Kites' foraging behavior was their tendency to forage in flocks in mid-morning. We often observed flocks of 5–14 individuals soaring over Tikal's forests between the hours of 0800 and 0930. Swallow-tailed Kites numbering from one to five also were usually present in these foraging flocks, and at times Bat Falcons (*Falco rufigularis*) also foraged with these groups. During these flock-foraging bouts, Plumbeous kites foraged intently 50–200 m above the forest canopy, repeatedly seizing quarry with their feet and then feeding on the wing. The insects captured were generally too small for us to see with binoculars. As aptly described by Skutch (1947), "One watches the big bird soaring about overhead, striking out with one foot and then the other and since its prey is mostly invisible at so great a height, one is at first puzzled to account for this apparently purposeless behavior. The bird appears to be boxing with phantoms!" In such flocking incidents, the kites likely cue in on the foraging motions of distant individuals—often termed "network foraging" or "local enhancement" (Thorpe 1956; Pulliam and Millikan 1982)—a sort of information parasitism that does not require intentional sharing of information among individuals. Network foraging may aid Plumbeous Kites and other flock participants in finding rich but ephemeral food resources such as insect nuptial swarms, but this remains to be proved. Being short-lived, spatially erratic, and composed of many thousands of insects, nuptial swarms cannot be economically defended by an individual kite against conspecifics or other would-be foragers.

HABITAT USE

Plumbeous Kites are clearly forest- and woodland-dwelling birds, yet they seem to have an affinity for open areas and appear able to tolerate a range of forest disturbances. Although these kites are often observed foraging over expanses of primary forest, all nests that we studied in Tikal were associated with some type of forest opening. These disturbances ranged from the large main plaza of Tikal's central ruins to smaller clearings associated with park buildings or simply dirt roads through the forest. The few nests we observed amid large tracts of forest have often been associated with natural clearings such as small ponds or marshes. The degree to which Plumbeous Kites nest throughout large areas of undisturbed forest remains poorly known. In other parts of Central America, nests have been observed in mangrove forest and pine ridges (Dickey and van Rossem 1938; Brown and Amadon 1968).

Nests at Tikal were found not only in the park but also in the farming landscape nearby. Unlike many other forest-dwelling raptors at Tikal, which utilized large forest fragments when nesting in the farming landscape, Plumbeous Kites often nested in lone remnant trees (living or dead) surrounded by fields. Moreover, we detected Plumbeous Kites more often in point counts in the farming landscape than in Tikal's primary forests (Whitacre et al. 1992a, 1992b). Several factors probably facilitate use of the slash-and-burn mosaic by this species. Nests were generally built in trees whose crowns emerged above surrounding vegetation, a condition often met by remnant trees in farming areas. A shaded nest site does not seem necessary, and many of the nests we observed were exposed to direct sun for much of the day. Much of the perch hunting performed by these kites took place from emergent trees, with flights occurring out into open areas and prey captures taking place near, or at times even on, the ground. This hunting style lends itself to the largely open nature of the farming landscape. The presence of spiny lizards (*Sceloporus* spp.) in the diets of the birds indicates that they are capable of exploiting terrestrial as well as arboreal and aerial prey. Finally, though we have not studied these kites in detail in the farming landscape, the number of kite nests we observed there suggests that adequate prey were available. Even if prey are less abundant in the farming landscape than in the forest, the high mobility of these kites may allow them to forage at sites distant from the nest, for example, over nearby forest remnants and over marshy areas where dragonflies are abundant. However, we do not know how Plumbeous Kite productivity in the farming landscape compares to values within extensive forest.

MIGRATION

The Plumbeous Kite is one of but a few raptors that nest in the northern Neotropics and migrate southward for the nonbreeding season. The only other known migrant among the breeding raptors at Tikal is the Swallow-tailed Kite, which shares an insect-dominated diet and highly aerial foraging habits. Plumbeous Kites arrived at Tikal in late February to early March and departed in August, as verified by general observations and a year-round series of point counts (Vásquez et al. 1992; D. F. Whitacre and A. E. Hernández, unpubl. data).

BREEDING BIOLOGY AND BEHAVIOR

Nests and Nest Sites

Plumbeous Kites built their nests 13–36 m above ground (mean = 26 ± 7 m, *n* = 12), in the upper portions of trees averaging 30 ± 8 m tall (range = 17–40 m). Nests were generally situated in forks created by two or more branches and were most often on large branches or against the main trunk (*n* = 8), but sometimes they were placed on more slender limbs toward the outer crown (*n* = 4). Nest trees used at Tikal were dry season deciduous (Table 7.1); as a result, during the end of the dry season, when incubation began, these nests were conspicuously visible (Plate 7.4). Nest trees often leafed out during the nestling phase, rendering nests much less conspicuous. Though all the nests we observed in the park were in live trees, we observed several nests in dead snags in the farming landscape. Three of 12 nests in the park were built on top of epiphyte masses, mainly bromeliads (see Table 7.1). Vine tangles, commonly used as nest substrate or concealment by some raptors at Tikal, were never used at the Plumbeous Kite nests we observed.

Nest trees were generally emergent above surrounding forest or isolated within a clearing. Such choice of nesting trees with minimal crown contact with the forest canopy probably reduces the threat of nest predation by climbing predators including snakes, primates, and other mammals. Nests, small and distinctly cup shaped, were built of dry sticks and lined with green leafy twigs, which were replenished by adults throughout the nesting season.

Egg and Clutch Size

The normal clutch at Tikal was one egg (*n* = 8 clutches and 9 other nests with single nestlings). Plumbeous Kites demonstrate a latitudinal gradient in clutch size similar to that known for many birds, with smallest clutches occurring near the equator and larger clutches at higher latitudes (Seavy et al. 1998). Throughout the bulk of the species's range, only 1-egg clutches are known. However, three 2-egg clutches were collected at the southern extreme of the species's range on the Tropic of Capricorn in Paraguay, and 2-egg clutches are also known from the northern extreme of the range; Belcher and Smooker (1934) documented a 2-egg clutch in Trinidad, and Sutton et al. (1950) described a nest with two chicks in Tamaulipas, Mexico. All other museum records (35 clutches) and primary literature accounts (e.g., Dickey and van Rossem 1938; Thurber et al. 1987) mention evidence of only 1-egg clutches. The very similar or conspecific Mississippi Kite has a normal clutch size of two (Parker 1988), further evidence of latitudinal variation. Schönwetter (1961) gave mean egg dimensions as 42.2

mm by 34.3 mm (n = 11). Using Hoyt's (1979) formula, this yields an estimated fresh egg mass of 27.2 g. Hence the modal 1-egg clutch amounts to 10.5% of mean female body mass.

Nesting Phenology

Plumbeous Kites at Tikal were highly seasonal and synchronous in breeding. Kites began nesting nearly immediately after their arrival in late February to early March. Ten clutches were completed between 18 March and 5 April, during the mid to late dry season. Most young hatched during the last week of April (late dry season), with extreme hatch dates being 20 April and 3 May. Young fledged from the last days of May to 11 June, coinciding with the onset of the rainy season and the increased insect abundance that accompanies it (see Chapter 2). Plumbeous Kites left Tikal on their southward migration in August, two months after the peak fledging period.

Nesting of this species also appears highly seasonal elsewhere and similarly timed with respect to the onset of rains. These kites are present year-round in Surinam, breeding during March–April, while laying dates in nearby Trinidad are given as February–April (Brown and Amadon 1968)—in both cases similar to the timing at Tikal. The nestling observed by Skutch (1947) in Ecuador fledged in mid-September. At the southern limit of their range, in Misiones, Argentina, these kites breed during October and November, the Southern Hemisphere spring (Brown and Amadon 1968).

Length of Incubation, Nestling, and Post-fledging Dependency Periods

Two incubation periods at Tikal were 32 and 33 days, similar to the 29–32-day period reported for the Mississippi Kite (Parker 1988). Tikal young fledged on average 38.5 days after hatching (n = 4, range = 36–40 days). For at least a week after leaving the nest, young remained in the nest tree, fed by adults. The dependency period, if any, of the juveniles once they have left the nest tree is unknown but presumably brief.

VOCALIZATIONS

Adult Plumbeous Kites commonly gave three types of vocalizations: (1) Two-note whistle: This is probably the most common vocalization, and it was used in several contexts. This call is a whistled *CHU-Huiiiiiiii*, with the first note brief and sharp, and the second longer in duration and dropping in pitch. This call is often exchanged between pair members during incubation switches, brooding switches, and prey deliveries, and at times is heard when both birds are near the nest but not engaged in any obvious activity. This two-note whistle is also given by the birds as they drive potential avian predators from the nest vicinity. It may serve as an alarm call, as

it was sometimes given by perched kites when Ornate Hawk-eagles *(Spizaetus ornatus)*, or other large raptors flew over the nest area. (2) Three-note whistle: This call is very similar to, and possibly a variation of, the two-note whistle. Softer than the two-note whistle, this call is a high-pitched *tee-tee-teeee*, with nearly equal accent on all three syllables, but the last being longer and dropping slightly in pitch. This call was mainly given between pair members, as described for the two-note whistle. At times it was heard in a defensive or alarm context, though not as often as the two-note whistle. (3) Single-note whistle: This is a very soft call, at times barely audible to an observer 15–30 m distant. It is a high-pitched but soft *teeee*, descending slightly in pitch and often given repeatedly. This vocalization was heard only in the context of incubation switches, brooding switches, and prey deliveries, and at times was given by lone, incubating birds. This call was very similar to the vocalizations given by older chicks while being fed by adults.

BEHAVIOR OF ADULTS

Pre-laying and Laying Period

We observed no conspicuous courtship displays, and we speculate that adults may often arrive on the breeding grounds already paired. It is likely, however, that some type of display occurs at times, perhaps employing the same swooping flights and ritualized postures used by the Mississippi Kite (Parker 1988). Both sexes participated in nest building. Copulation took place in the nest tree or nearby emergent trees.

Incubation Period

Pair members at six nests shared the task of daytime incubation in a 60:40 ratio, with females presumably providing the greater portion (Seavy et al. 1998). Plumbeous Kites were vigilant during incubation, with at least one adult remaining at the nest or within sight during most of the day. During this time, defense of the nesting area was confined to the nest tree and most often occurred when potential avian predators approached within a few meters of the nest. Prey deliveries during incubation were rare, and each pair member met its own food demands by foraging while away from the nest. This no doubt was facilitated by the relatively even sharing of incubation duties (Seavy et al. 1998), allowing each pair member sufficient foraging time.

Nestling Period

Immediately after the egg hatched, one pair member, presumably the female, assumed the majority of brooding and feeding duties. Prey were delivered at the nest to the brooding bird, which then would feed the nestling. The brooding adult rarely consumed much of the delivered prey, and both adults continued throughout the nesting cycle to meet their own energy needs via their own hunting efforts. As the chick grew, the chores of brooding it, standing guard at the nest, and feeding the chick bits of prey were shared increasingly equally by the male and female. After the chick was about two weeks old, both pair members delivered and fed most prey directly to the chick, that is, without any transfer between adults.

Prey delivery rates varied throughout the day and mirrored the daily pattern of hunting activity, with highest delivery rates during mid-morning, at times reaching 7 prey items/hr. Throughout the rest of the day, delivery rates ranged from 0.5–3.0 items/hr. Overall, delivery rates averaged 1.75 items/hr throughout the nestling period (n = 49 observation sessions). Although delivery rates did not change significantly with the age of nestlings, the highest rates (up to 11 items/hr) were noted betweeen nestling ages 16 and 32 days and occurred in sessions when it appeared that both adults were bringing prey with roughly equal frequency (Seavy et al. 1998).

BEHAVIOR AND DEVELOPMENT OF NESTLINGS

Upon hatching, nestlings were essentially immobile, raising their heads only to receive food. They were covered in white down, with dark eyes, yellow cere, and black beak. Adults delivered prey to the nest as described by Skutch (1947), carrying them in either the beak or the talons while in flight and transferring them to the beak before landing in the nest tree and walking to the nest. The adult would hold the prey in its talons and, using its beak, tear off and feed small pieces to the nestling.

At about the age of one week, while still covered in white down, nestlings became increasingly active, standing for short periods, lifting their heads above the nest rim during feeding, and stretching their wings. At about 10 days, development of the flight feathers first became noticeable, with the dark tips of the primary feathers beginning to poke through the downy coat. By the fourth week, the juvenal plumage was well feathered out, though patches of down remained, especially around the head. Nestlings were increasingly active during the fourth and fifth weeks, standing on the edge of the nest, flapping, and at times eating unassisted. The young left the nest while flight feathers were still growing, and flew awkwardly from branch to branch during the next week, until leaving the nest area (Plate 7.5).

Nestlings were generally silent, though toward the end of the nestling period they occasionally voiced a softly whistled *teeeeee*, often repeating this numerous times. Chicks voiced this call as adults arrived with prey, or as nestlings perched alone in the nest tree, awaiting the adults' return.

SPACING MECHANISMS AND POPULATION DENSITY

Territorial Behavior and Displays

Plumbeous Kites at Tikal did not demonstrate the tendency toward colonial nesting that is pronounced in Mississippi Kites, especially in the U.S. Great Plains where the latter species often nests in aggregations within shelterbelts (with nests as close as 13 m apart), leaving similar sites unoccupied and sometimes relocating en masse to other shelterbelts (Parker 1988). While Plumbeous Kites at Tikal may have shown a tendency to occur in loose "neighborhood" aggregations, we observed an average spacing of 500 ± 52 m between neighboring nests, with surprisingly little variation (range = 453–570 m, n = 4). Swallow-tailed Kites clearly grouped their nests into loose colonies, leaving extensive areas of forest with few if any nests. In the area occupied by two of our Plumbeous Kite nests, as many as 12 pairs of Swallow-tailed Kites nested simultaneously, sometimes as little as 35 m apart (R. Gerhardt, pers. comm.).

We observed behavioral interactions indicative of intraspecific territoriality. In a few cases, kites displayed overtly aggressive behavior, chasing away conspecifics that approached within 100 m of a nest. More commonly, these kites exhibited a more subdued form of agonistic behavior when conspecifics approached a pair's nest. Often one or both of the nesting birds would fly up to meet the approaching kite(s) and then follow behind at a constant distance of 10–15 m, flying in formation for up to 1 km, until out of sight of the nest tree. The birds seemingly made no attempt to overtake the interlopers, and appeared to escort rather than pursue them.

Along with the substantial and regular spacing we observed between nests, this aggressive behavior—ranging from muted to overt—suggests that territoriality influenced nesting dispersion and density at Tikal. In contrast, intraspecific territorial defense is said not to occur in the Mississippi Kite (Parker 1988). Mississippi Kites nesting in wooded areas of the southeastern United States and expanses of oak savanna and mesquite in the southwestern states may show a somewhat reduced tendency toward coloniality, and nest farther apart than in the Great Plains (Parker 1988), but published literature presents no clear evidence of regular, substantial spacing of nests like what we observed at Tikal.

We hypothesize that the degree of tolerance among conspecifics near the nest is facultative in the Mississippi-Plumbeous Kite superspecies and variable in expression depending on conditions, which may sometimes favor colonial nesting and sometimes favor nesting farther apart. According to conventional ecological wisdom, expression of territorial behavior may be largely mediated by the degree to which prey resources (or the space within which they occur) are economically defensible (Gill and Wolf 1975). Defensibility, in turn, hinges on both resource abundance and dispersion and on the intensity of intruder pressure (Gill and Wolf

1975). Cicadas, beetles, and lizards, which together contributed at least 45% of the diet of Plumbeous Kites at Tikal, are probably spread fairly evenly in both space and time and thus may be defensible via a territorial strategy. Ephemeral swarms of dragonflies, termites, or other flying insects are probably not defensible, but made up only 10% of identified prey items at the nest at Tikal.

In the Great Plains, the clumped nature of the Mississippi Kite's shelterbelt nesting habitat amid large potential foraging areas probably results in high conspecific intruder pressure near nests. This, and the fact that kites there commute between clumped nests and foraging habitat some distance away, may have led to a loss of the tendency to defend all-purpose territories surrounding nests. Moreover, kites there may conceivably even benefit from gregarious nesting by reducing nest predation via group vigilance or by gleaning information from one another regarding the location of rich, ephemeral prey sources (Thorpe 1956; Ward and Zahavi 1973; Pulliam and Millikan 1982). In contrast, both nest sites and foraging habitat are more homogeneously distributed at Tikal, and this may give free rein to any tendencies toward development of wide spacing between nests and defense of nest and foraging areas. Still, the evidence presented here does not indicate that kites at Tikal defended mutually exclusive, all-purpose territories, and the frequent aggregation of up to a dozen or more kites in feeding flocks demonstrates a high degree of intraspecific tolerance, if not gregariousness, during foraging. Admittedly, such social foraging may occur mainly or only when kites converge on ephemeral prey patches that cannot be successfully defended by any individual kite.

Nest Defense and Interspecific Interactions

With young in the nest, adults aggressively defended a 100–200 m radius against certain bird species. Roadside Hawks *(Rupornis magnirostris)*, Bat Falcons, Swallow-tailed Kites, Keel-billed Toucans *(Ramphastos sulfuratus)*, and Brown Jays *(Cyanocorax morio)* were all commonly chased from the nest area. In contrast, some large birds, notably parrots, were commonly tolerated in the nest tree. The kites were frequently harassed by flycatchers and passerines including Sulphur-bellied Flycatchers *(Myiodynastes luteiventris)*, Social Flycatchers *(Myiozetetes similis)*, and Brown Jays. Mobbing was sometimes so intense that kites attempting to reach the nest to deliver prey were temporarily turned back. We saw no evidence of these kites depredating bird nests, unlike Swallow-tailed Kites at Tikal, which commonly did so.

Constancy of Territory Occupancy, Use of Alternate Nest Sites

Nesting territories of Plumbeous Kites remained largely stable from year to year. When prior-year nesting areas were checked, they rarely lacked a current nesting attempt. Nestings in successive years occurred in the same

nests, nest trees, or in nearby trees within 150 m of the previous nest.

Inter-nest Spacing and Density of Territorial Pairs

We did not estimate home range size for any Plumbeous Kites, and our information on densities of pairs over large areas of habitat is sketchy. Though four nests near the central ruins were quite evenly spaced only 500 ± 52 m apart, we consider it unlikely that this density holds true over larger areas. This spacing implies a maximum density of 4.4 pairs/km² (Maximum Packed Nest Density method), 4.0 pairs/km² (square method), or 3.36 pairs/km² (polygon method). Densities of this magnitude are probably attained only in local situations where adequate nest trees (open-topped emergent or isolated trees) and favored nest environs (forest clearings) are abundant. In large, contiguous areas of unbroken forest, breeding densities are probably lower.

In our 20 km² study plot amid extensive mature forest, we found evidence that probably three pairs of these kites were present, for a density of roughly one pair per 6.7 km², equivalent to 14.9 pairs/100 km² of forest. Estimated pair density in this extensive forest study plot was 20–30 times lower than that cited above for an area with many clearings—the latter being a situation of probably maximal density. Hence we tentatively conclude that availability of clearings, be they natural or anthropogenic in nature, may limit the density of nesting pairs of Plumbeous Kites in large expanses of mature forest.

To conclude, while all density estimates for this species are quite speculative, the average density in Tikal's forests likely falls between 15 and 115 pairs/100 km², with the best estimate probably being 15 pairs/100 km², or an estimated 75 pairs in Tikal's roughly 510 km² of non-Bajo Forest. We believe it likely that breeding density at Tikal is limited not by available foraging habitat, but by the interaction between territorial behavior and the availability and spacing of suitable nest sites. In largely deforested landscapes, however, we consider it likely that the amount of forest habitat may limit reproduction and population size.

DEMOGRAPHY

Age of First Breeding, Attainment of Adult Plumage

Howell and Webb (1995) reported that the first basic plumage, attained on the wintering grounds, resembles adult plumage but retains juvenile flight feathers and most underwing coverts. These authors further reported that adult plumage is attained with the second prebasic molt on the wintering grounds. Apparently nothing is known of the age of first breeding in the Plumbeous Kite. In the Mississippi Kite, however, some breeding by yearlings is reported, and yearlings appeared to be paired with an adult (two years or older) at 35 (17%) of 209 nests studied (Parker 1988). Also in the Mississippi Kite, yearlings have been reported as helpers at the nest, contributing to incubation, brooding, and nest defense (Parker and Ports 1982). We saw no indication of helpers at the nest in Plumbeous Kites.

Frequency of Nesting, Percentage of Pairs Nesting

In six territories with known occupancy status over 2–4-year periods, totaling 20 territory-years, birds nested in 18 territory-years, were present but non-nesting in 1 territory-year, and absent in another. In both the latter cases they may have nested nearby. These data suggest that at least 90% of territorial pairs nested in a given year. As most territories were the site of nesting each year, it appeared that many adults routinely nested in consecutive years, although this was not established via marked birds.

Productivity and Nest Success

Between 1991 and 1996 we documented the fate of 19 nests in which eggs were laid. At 8 nests, eggs failed during incubation, while at each of 11 nests, a single chick hatched. Of the latter, four were lost and seven fledged for a productivity of 0.37 fledgling per nesting attempt and 0.33–0.35 fledgling per territory-year. This is a substantially lower productivity rate than that shown by Mississippi Kites in the Great Plains, Arizona, and Illinois (0.60, 0.61, and 0.63 fledgling per nesting attempt, respectively: Evans 1981; Glinski and Ohmart 1983; Parker 1988).

Sources of Nesting Failure and Mortality

Details as to the cause of failure during incubation were rarely available. However, as nests were high in trees and often quite exposed to the elements, we believe that wind may often have led to a loss of eggs. Of four nestlings lost, two fell during windstorms and two apparently were taken by predators.

CONSERVATION

Plumbeous Kites often nested in the slash-and-burn farming landscape near Tikal and were detected more often in point counts in the farming landscape than within the primary forest. They seemed to nest preferentially in or near clearings, both natural and man-made, and are capable of hunting over a wide range of vegetation types. Thus, Plumbeous Kites should prove fairly resilient to the effects of limited deforestation by humans. Still, these kites typically occur in fairly well-wooded environs, and extensive deforestation may be expected to diminish their populations. In some cases, Plumbeous Kites have shown serious local population declines. In El Salvador, Thurber et al. (1987) found that this species had declined drastically since the late 1920s, and Komar (1998) considered this kite in danger of extirpation in El Salvador. Thurber et al. (1987) believed that suitable habitat still exists in El Salvador, and that the causes of this decline merit investigation. This is one Central American raptor that often nests in mangrove forests, and in Costa Rica and El Salvador, mangroves are reportedly the principal nesting habitat. Thus one may expect that the catastrophic ongoing loss of mangroves in many parts of Central and South America (Valiela et al. 2001) may have proportionally more impact on these kites than on many other raptors.

HIGHLIGHTS

Perhaps the most notable aspect of our findings was the relatively even sharing of parental duties between pair members, with both sexes foraging for themselves throughout the nesting cycle. This is in marked contrast with most birds of prey, in which males are generally the sole providers for their mates and nestlings until the latter are half grown (Newton 1986). A similar pattern, however, is seen in the large Old World vultures of the genus *Gyps* (Houston 1976; Newton 1979), in all three U.S.-breeding cathartid vultures (Jackson 1988a, 1988b; Snyder 1988), and in some other largely insectivorous raptors—for example, the Lesser Kestrel *(Falco naumanni)* and Western Red-footed Falcon *(Falco vespertinus:* Cade 1982).

We hypothesize that such even sharing of parental roles, along with self-provisioning, results from conditions that make it energetically more efficient for each pair member to conduct its own foraging than for the male to provision the female at the nest. This would certainly be the case for raptor species relying on small prey such as insects, but it would also apply to species reliant on prey that is sparsely and patchily distributed, ephemeral, and unpredictable in time and space, so that large blocks of time are required to find and exploit it. These adjectives apply to carrion, the main food of the vultures just mentioned, and to nuptial swarms and other aggregations of ants, termites, dragonflies, and other insects.

In most respects, the breeding biology and ecology of Plumbeous Kites at Tikal are quite similar to those of the Mississippi Kite. The main differences known at this point between these two species are the smaller clutch size and greater manifestation of regular nest spacing and territoriality in the Plumbeous Kite than in the Mississippi Kite. The apparent difference in nest spacing and degree of territoriality between these two species requires verification. In future research, it would also be interesting to examine whether nonparental helpers at the nest occur in the Plumbeous as in the Mississippi Kite, as well as further documenting the ecological and demographic correlates of such behavior in either species.

For further information on this species in other portions of its range, refer to Cabanne 2005, Carvalho and Bohórquez 2007, de la Peña 2004, Di Giacomo 2005, and Lourdes-Ribeiro et al. 2004.

8 BICOLORED HAWK

Russell K. Thorstrom

Rarely seen by the casual birder, the Bicolored Hawk *(Accipiter bicolor)* is one of the more enigmatic raptors at Tikal. Nearly all diurnal raptors in our study area can be detected either through their habit of soaring above the forest in probable territorial display or through conspicuous vocalizations in the case of the two *Micrastur* forest falcons and the Laughing Falcon *(Herpetotheres cachinnans)*. The Bicolored Hawk is practically unique in indulging in neither soaring nor loud, conspicuous vocalizations. Once recognized, the call aids in locating nests, but it does not carry far through the forest.

This striking bird, with bright rufous pantaloons contrasting with a usually gray breast, is about the size of the closely related Cooper's Hawk *(Accipiter cooperii)* and has typical accipiter proportions—rounded wings and a long, rounded tail (Plate 8.1). A bold and dashing bird predator, the Bicolored Hawk is one of the few raptors of the Maya Forest that preys nearly exclusively on birds. Territories at Tikal were remarkably evenly spaced and appeared to fill the tall Upland Forest. No doubt often more abundant than suggested by casual sightings, these dedicated bird hunters have likely provided a good deal of the selection pressure leading to flocking behavior among small and medium-sized Neotropical forest birds.

GEOGRAPHIC DISTRIBUTION AND SYSTEMATICS

The Bicolored Hawk, Cooper's Hawk, and Gundlach's Hawk *(A. gundlachi)* of Cuba comprise a superspecies (Amadon 1964; Brown and Amadon 1968; Bierregaard 1994), collectively ranging from southern Canada to southern South America. The Bicolored Hawk is widely distributed throughout tropical and subtropical forests from southern Tamaulipas, Mexico, to northern Argentina, Bolivia, and Paraguay. Four subspecies are recognized, with two southern forms, *A. b. guttifer* and *A. b. chilensis*, sometimes regarded as distinct species (Brown and Amadon 1968; AOU 1983; Bierregaard 1994). Most raptor taxonomists regard *A. b. guttifer* as a subspecies of

A. bicolor or *A. chilensis*. At Tikal we studied the nominate subspecies, *Accipiter bicolor bicolor*, which ranges from southern Mexico to Amazonia and eastern Bolivia (Brown and Amadon 1968). Widespread in tropical lowlands, the Bicolored Hawk ranges well into subtropical or montane elevations, reaching 1830 m in Ecuador, 2290 m in Bolivia, 2500 m in Venezuela, and 2745 m in Peru (Wattel 1973).

MORPHOLOGY

Combining body weight data from Tikal and elsewhere, males of *A. b. bicolor* averaged 234.1 ± 24.8 g (*n* = 11) and females 457.0 ± 57.2 g (*n* = 11; Appendix 1). Based on linear dimensions (Appendix 2), Bicolored Hawks had the following Dimorphism Index (DI) values: 14.5 for wing chord, 8.8 for tail length, 19.3 for culmen, and 10.3 for tarsus. Body mass (cube root) dimorphism is 22.2, with females averaging 95% heavier than males. This exceeds the degree of mass dimorphism of all North American raptors, including the Sharp-shinned Hawk *(Accipiter striatus)*, for which the body mass DI is 18.6 (Snyder and Wiley 1976). For the Bicolored Hawk, the mean DI for wing chord, culmen, and mass was 18.7, exceeded among North American raptors only by the Sharp-shinned Hawk (20.4: Snyder and Wiley 1976).

Although belonging to the same superspecies, the Bicolored Hawk appears more dimorphic than the Cooper's Hawk in terms of wing chord (DI = 14.5 vs. 12.7), culmen (19.3 vs. 18.7), and especially body mass (22.2 vs. 12.9). Although female Cooper's Hawks in Oregon (474.1 g, *n* = 18: Henny et al. 1985) were nearly identical in size to (< 1% larger than) female Bicolored Hawks, male Cooper's Hawks averaged 277.3 g (*n* = 31), or 10% heavier than male Bicolored Hawks at Tikal, resulting in a lesser degree of dimorphism in the Cooper's Hawk. In comparison with other members of the genus, Wattel (1973) characterized the Bicolored Hawk as having a long and heavy tarsus and middle toe, a rather large hind claw and bill, a medium-long wing tip, and a long tail.

Plumage and Eye Color

Plumage of adult Bicolored Hawks at Tikal resembled that described by Bierregaard (1994). Immatures were dusky brown above, blacker on the crown, usually showing a partial buff or whitish collar on the hind neck. Below they were usually buffy white, but occasionally ferrous, with thighs usually darker or mottled rufous (Plates 8.2, 8.3). With their nuchal collar, juveniles somewhat resembled adult Collared Forest Falcons (*Micrastur semitorquatus*) or juvenile Barred Forest Falcons (*M. ruficollis*). In adults, the cere and legs were yellow, and the iris varied from cadmium yellow to orange or orange-red (Chubb 1910; Van Tyne 1935; Wetmore 1957). Nestlings and juveniles at Tikal had tan to light brown iris.

PREVIOUS RESEARCH

Although the Bicolored Hawk is widely distributed throughout the Neotropics, little has been published on its biology or behavior, no doubt in part due to its secretive nature. Mader (1981) reported on two nests in Venezuela, and more recently we published on the eggs, breeding biology, and nest characteristics of this species (Thorstrom and Kiff 1999; Thorstrom and Quixchan 2000). Beyond these accounts, the Bicolored Hawk has received only brief treatments in field guides and regional works.

RESEARCH IN THE MAYA FOREST

From 1991 to 1994 we studied 17 Bicolored Hawk nestings in seven territories—a total of 21 territory-years (Table 8.1). Using 7–10X binoculars, we observed nests at 35–50 m, from the ground or a nearby tree platform. Incubation period is defined as the interval from clutch completion to hatching of the last egg. Eggs were measured to the nearest 0.1 mm with dial calipers, and egg and body mass determined using 100 and 500 g Pesola spring scales to the nearest 1 and 5 g, respectively. To band and place radio transmitters on adults, we trapped them using a *bal-chatri* or noose carpet placed on the nest during incubation. Several young were fitted with backpack-mounted transmitters weighing 4.5 g, and adults were fitted with tail- or backpack-mounted transmitters weighing 4.5–6.0 g (< 3% of body weight in each case). Six birds were color banded, allowing individual recognition. Measures given are means ± 1 SD unless otherwise noted.

DIET AND HUNTING BEHAVIOR

Published diet records for the Bicolored Hawk are few. In Venezuela, two birds have been reported as prey—a Squirrel Cuckoo (*Piaya cayana*: Mader 1981) and a Tropical Mockingbird (*Mimus gilvus*: Friedmann and Smith 1955). Brown and Amadon (1968) characterized the diet as birds, especially doves, but gave no data other than

the mockingbird just mentioned, and Stiles et al. (1989) reported the diet as mostly thrush- to pigeon-sized birds and some lizards, again with no particulars. The most complete published observations of Bicolored Hawk diet aside from those presented here, are those of Robinson (1994) from southeast Peru, where 13 birds and one squirrel were observed captured or carried as prey.

From 1991 to 1994, we observed 315 Bicolored Hawk prey items at Tikal—mainly those brought to nests—and identified 219 of these at least to class. Birds comprised 95.4% (209/219) of identified prey, mammals (bats and rat-sized rodents) 3.2% (7/219), and lizards 1.4% (2/219; Table 8.2). The 97 unidentified prey items were often plucked and partly eaten before delivery to the nest, rendering their identification impossible.

Adult males during the breeding season brought in birds ranging in size from euphonias and greenlets (8–16 g) to Collared Aracari (*Pteroglossus torquatus*, 226 g), while adult females brought in (and presumably captured) birds ranging from trogons (50–140 g) and puffbirds (96 g), to Spotted Wood-Quail (*Odontophorus guttatus*, 300 g; Table 8.2). Emerald Toucanets (*Aulacorhynchus prasinus*, 155 g) and woodcreepers (mostly about 40 g) were among the birds most often identified among male prey items, while trogons (50–140 g) and bats (probably 50 g or less) were among the prey types most often delivered to young by females (Table 8.2).

Mean size of prey species taken by males (unweighted by number of prey individuals) was 84 ± 63 g (range = 8–226 g, $n = 22$) and for females was 116 ± 100 g (range = 15–300 g, $n = 6$), with female prey species averaging 38% larger than male prey species. Taking male and female prey together, prey species averaged 90.3 ± 74.0 g ($n = 26$: Table 8.2). Individual prey items of males averaged 89.1 ± 60.3 g ($n = 38$), whereas prey of females averaged 94.1 ± 76.3 g ($n = 11$), only a 5.8% difference between male and female prey mass. Median mass of male prey species was 76.5 g ($n = 22$) and for male prey items was 76 g ($n = 38$). For females, median prey mass was 96 g for both prey species ($n = 6$) and individual prey items ($n = 11$). Thus, median prey mass for female Bicolored Hawks (96 g) was 26% greater than median prey mass of males (76.5 g). This is perhaps the best measure of relative prey size for the two sexes. None of these apparent male-female differences in prey size were statistically different, however (mass of prey species [i.e., not frequency weighted]: Mann-Whitney $U = 52.0$, 1 df, $P = 0.43$, $n = 22$ for males, 6 for females; mass of individual prey items [i.e., frequency weighted]: Mann-Whitney $U = 191.0$, 1 df, $P = 0.665$, $n = 38$ for males, 11 for females). Additional data, especially on prey of females, are needed before a fair test can be made of whether male and female Bicolored Hawks take differently sized prey.

Hunting Behavior

Two radio-tagged adult males followed during their hunting excursions perched from the sub-canopy to the upper canopy. During visual contacts, the mean perch height for one of these males was 7.5 ± 4 m (range = 1–20 m, $n = 39$).

Table 8.1 Nesting histories of seven territorial pairs of Bicolored Hawks during 21 pair-years[a]

Territory	1991	1992	1993	1994	Total
Barens	(3) 1	(1) 1	*	(2) 0	2
Bejucal	(3) 2	(2) 2	(3) 1	(3) 3	8
Chikintikal	(3) 0	(3) 2	(3) 2	*	4
Caoba Felipe	nd	(?) 2[b]	(?) 0[c]	*	2
Bajada	nd	(3) 3	*	*	3
San Antonio: nest 1	nd	nd	(2) 2	(2) 0	2
nest 2[d]	—	—	—	(1) 1	1
Brecha 20	nd	nd	nd	(?) 2[a]	2
Total (eggs) fledglings:					
Using only nests found before hatching	(9) 3	(9) 8	(8) 5	(8) 4	(34) 20
Including nests found after hatching[e]	(9) 3	(11.5) 10	(8) 5	(10.5) 6	(39) 24

Notes: * = pair was on territory but nesting did not occur; nd = we have no data (territory not yet known to us).
[a] Numbers given are number of eggs laid and young fledged.
[b] Two nestlings present when nest discovered.
[c] Failed just before or just after laying; we assume the latter.
[d] In 1994 the San Antonio pair had failed at nest 1 then switched to nest 2 and re-nested.
[e] Assuming these two nests contained the mean clutch size of 2.5 eggs.

In 38 cases, the position of the male's perch relative to the forest canopy was noted; twice (5%) it was on perches protruding above the canopy, twice (5%) on perches just barely emergent from the canopy, 15 times (40%) within the canopy, 10 times (26%) at the lower canopy boundary, and 9 times (24%) in the understory or sub-canopy space. All observed perches of this male were in live trees on horizontal limbs, and perch limbs averaged 15.0 ± 3.6 cm in diameter ($n = 35$, range = 1.0–45.3 cm). The hunting male would remain on a given perch for several minutes, apparently perch hunting, then move on, quickly arriving at a new perch, often 100–400 m distant.

During radio-tracking sessions, we observed captures of a Blue-crowned Motmot *(Momotus momota)*, a Wood Thrush *(Catharus mustelina)*, and several woodcreepers. We observed captures by both radio-tagged males, resulting from surprise attacks launched from a concealed, higher perch. The hawks attacked in a direct flight at a downward angle, slamming into the quarry with a burst of speed, seizing and carrying it to a perch. On 23 May 1991, during a mid-morning radio-tracking session, the adult male of the Bejucal pair was perched 10 m above ground in forest at the edge of one of the cleared plazas around Tikal's Maya ruins. After 8 minutes he flew, turned 180 degrees while in flight, and pursued a small passerine into the forest. In 1992, we observed a capture by the Caoba Felipe male. A trogon flew below the perched male, who left his perch and descended quickly, seizing the bird and carrying it off into the forest. On 22 April 1991, during a pre-dawn raptor census in similar forest some 60 km northwest of Tikal, colleagues observed an adult Bicolored Hawk fly in and seize a squirrel that had begun calling a short time before (A. Baker and J. Sutter, pers. comm.). Whether the hawk was attracted by the squirrel's calling, or perhaps by the tail jerking that accompanied it, one can only guess. Elsewhere, most information on hunting behavior comes from Amazonian Peru, where Robinson (1994) and colleagues observed several dozen attacks, mostly on birds.

HABITAT USE

Bicolored Hawks are said to occupy a variety of forest, forest-edge, and woodland habitats. Wattel (1973) describes this species as inhabiting "forest edges, stands of trees in savannas, scrub, open deciduous woodland, dense gallery forest, second-growth, patches of native cultivation, plantations, woodland villages, and tropical rainforest." Bierregaard (1994) likewise regarded these raptors as unspecialized in terms of habitat use, occupying rain forest in the Guianas but elsewhere being more common in drier, open or thinned forest, occurring also in palm savanna with gallery forest, and in scattered forest patches and second growth.

At Tikal, we studied Bicolored Hawks in the essentially unbroken mature forest of the park, where these birds ranged from Bajo or Swamp Forest to Hillbase and Upland Forest types for nesting and foraging. Radio-tagged individuals were found in a variety of these forest types during the course of a day. Most observations of successful hunts took place in Upland Forest, during radio-tracking sessions. During 31 sightings of a radio-tagged male, he was perched 23 times (74%) in Upland Forest, 7 times (23%) in Bajo Forest, and once (3%) in Transitional Forest. Of 19 nest trees documented at Tikal, 8 (42%) were situated in Bajo Forest, 8 (42%) in Transitional Forest, and 3 (16%) in Upland Forest. From these numbers, it is apparent that these hawks nested most often in Bajo and Transitional Forest, whereas this radio-tagged individual hunted largely in the taller Upland Forest, with its relatively closed canopy and open understory.

BREEDING BIOLOGY AND BEHAVIOR

Nests and Nest Sites

From 1991 to 1994, we observed 17 nesting attempts, using 15 nest structures in seven territories; one other

Table 8.2 Prey items delivered by male and female Bicolored Hawks to nests at Tikal (and believed captured by them) during the breeding season, 1991–94

Prey type	Prey mass (g)[a]	Male (number of items)	Female (number of items)	Total number of items
Birds				
Ocellated Turkey poult (*Agriocharis ocellata*)	~100	1	—	1
Spotted Wood-Quail (*Odontophorus guttatus*)	300	—	1	1
Pigeon (*Columba* sp.)	176[b]	2	—	2
Ground-dove (*Columbina* sp.)	40[b]	1	—	1
Dove (*Leptotila* sp.)	155[b]	3	—	3
Aztec Parakeet (*Aratinga nana*)	76.9	1	—	1
Parrot (*Pionopsitta/Pionus*)	149–212 (180.5)[c]	1	—	1
Trogon spp.	51–141 (96)[b,c]	1	3	4
Blue-crowned Motmot (*Momotus momota*)	133	1	—	1
White-necked Puffbird (*Bucco macrorhynchos*)	96	1	1	2
Emerald Toucanet (*Aulacorhynchus prasinus*)	155	5	—	5
Collared Aracari (*Pteroglossus torquatus*)	226	1	—	1
Woodpecker (*Melanerpes* sp.)	83[b]	1	—	1
Ruddy Woodcreeper (*Dendrocincla homochroa*)	41.1	1	—	1
Strong-billed Woodcreeper (*Xiphocolaptes promeropirhynchus*)	136	—	1	1
Woodcreeper spp. (several species)	13.6–64.2 (41.0)[b]	6	—	6
Dot-winged Antwren (*Microrhopias quixensis*)	7.9	1	—	1
Flycatcher (Tyrannidae)	15[b]	—	1	1
Wood Thrush (*Catharus mustelina*)	47.4	2	—	2
Red-legged Honeycreeper (*Cyanerpes cyaneus*)	14.0	1	—	1
Olive-backed Euphonia (*Euphonia gouldi*)	14.0	1	—	1
Gray-headed Tanager (*Eucometis penicillata*)	27.0	2	—	2
Ant-tanager (*Habia* spp.)	36[b]	1	—	1
Total number of identified birds	—	34	7	41
Total number of unidentified birds	—	131	37	168
Total number of birds	—	165	44	209
Mammals				
Bat (Chiroptera)	50[b,d]	—	4	4
Rats and mice (Rodentia)	76[b,e]	3	—	3
Reptiles				
Ameiva sp.	25	1	—	1
Unidentified lizard	—	1	1	2
Unidentified to class	—	67	31	97
Total		237	79	315
Mean[f] mass of prey species ± SD (sample size)		83.9 ± 62.7 g (n = 22)	115.5 ± 99.6 g (n = 6)	
Median[f] mass of prey species (sample size)		76.5 (n = 22)	96 g (n = 6)	
Mean[g] mass of individual prey items ± SD (sample size)		89.1 ± 60.3 g (n = 38)	94.1 ± 76.3 g (n = 11)	
Median[g] mass or individual prey items		76 g (n = 38)	96 (n = 11)	

[a] Bird weights were taken from Dunning 1993, and other weights from our own data or other sources as indicated. Where Dunning (1990) gives separate male and female masses, we use the mean of the two.

[b] Mean of species most likely taken.

[c] Midpoint of range (used in prey mass calculations).

[d] Mean mass of *Artibeus jamaicensis*.

[e] Midpoint of *Heteromys* mass range given in Emmons and Feer 1990.

[f] Each species entered only once into calculations.

[g] Weighted by number of individuals of each prey type taken by hawks.

nest was built but did not receive eggs. Nests were built of dead branches and twigs collected from nearby trees (Plate 8.4). Several contained dried leaves in the nest cup—possibly brought in while green—but others had none. Fifteen nests were supported by 1–4 single or forked branches and two others were supported by hanging vines that had captured fallen bark, which helped support nest material. Excluding the largest (71 cm) limb, support limbs ranged from 4 to 15 cm in diameter, averaging 9.5 ± 5 cm (n = 11 nests). Nests averaged 3.5 ± 3.7 m laterally from the principal trunk (range = 0–9 m, n = 11); none was against the trunk on a lateral branch, but one was in a fork of the main trunk.

Nests averaged 21.7 ± 2.3 m up (range = 17.2–25 m, n = 15) in good-sized trees. All nest trees were living and averaged 25 ± 5 m in height (n = 15) and 75 ± 17 cm in diameter at breast height (range = 58–124 cm, n = 17). Nests were moderate-sized stick nests averaging a half meter across (51 ± 16 cm, range = 30–97 cm, n = 15), and with a very shallow cup (3.5 ± 2.0 cm, range = 2–9 cm, n = 15). Nests averaged 27 ± 12 cm in external depth (range = 12–30 cm, n = 15) and were constructed in 16 trees of six species. These were 6 (38%) Ramón (Brosimun alicastrum), 5 (31%) Honduras Mahogany (Swietenia macrophylla), 2 (13%) Manchiche (also known as Black Cabbage-bark, Lonchocarpus castilloi), 1 (6%) Yaxníc (also known as Fiddlewood, Vitex gaumeri), 1 (6%) Silión (Pouteria amygdalina), and 1 (6%) Chicozapote (Manilkara zapota).

Bicolored Hawks often nested in the same swales of Scrub Swamp, Transitional, or other intermediate forest types frequently used by Crane Hawks (Geranospiza caerulescens), Roadside Hawks (Rupornis magnirostris), and other raptors—and at times, in the same exact nests or nest sites. In 1994, a pair of Bicolored Hawks used a nest that had been occupied by Crane Hawks the previous year. In 1991, one Bicolored Hawk nest was 40 m north of a Gray-headed Kite (Leptodon cayanensis) nest and 50 m northwest of a Crane Hawk nest. We observed the Bicolored Hawks behave territorially toward the Crane Hawks but not toward the Gray-headed Kites. Bicolored Hawk nest trees in areas of low-canopied Bajo Forest tended to be large, isolated emergents, whereas nest trees in tall Upland Forest were canopy members.

Egg and Clutch Size

Our work at Tikal enabled us to dispel confusion surrounding some published egg descriptions for this species (Thorstrom and Kiff 1999). The modal clutch size at Tikal was three (n = 8), followed by five 2-egg and two 1-egg clutches, one of these a replacement clutch. Hence, 14 first clutches ranged from one to three eggs and averaged 2.5 eggs. Mean dimensions of 14 eggs from Tikal were 47.1 ± 1.1 mm by 36.5 ± 1.1 mm, and mean egg mass at various stages of incubation was 33.5 ± 3.5 g (28.0–38.0 g, n = 14). Using Hoyt's (1979) formula, mean fresh mass for these eggs is estimated at 34.3 g, or 7.3% of the 470 g mean adult female body mass at

Tikal. Hence the most common clutch, of three eggs, amounts to 21.9%, and the mean clutch of 2.5 eggs to 18.2%, of mean female mass. All eggs at Tikal were unmarked and off-white, with pale greenish blue on the inner surface.

Nesting Phenology

Bicolored Hawks began building nests in February, early in the dry season. Our earliest observation of nest building occurred on 5 February 1992, and copulation was first observed on 28 February 1992, about 5 weeks before egg laying. The nest-building and courtship period spanned some 90 days (5 February–8 May) and peaked in mid-March. Laying of first clutches spanned 36 days, from 2 April to 8 May, and peaked around 11 April ± 10 days at 15 nests. The earliest recorded laying dates were 4 April 1991, 2 April 1992, 6 April 1993, and 2 April 1994. Nearly all hatching occurred in May, the exception being a re-nesting after a failed earlier attempt, in which one young hatched on 26 June. For 10 eggs with well-documented hatch dates, hatching occurred between 9 and 23 May, with peak hatching on 16 May ± 5 days. Taken as a group, nestlings were present during a period of nearly seven weeks, from the first week of June to late July, that is, the early portion of the rainy season. To summarize, egg laying occurred during April and May (late dry season), incubation and nestling periods in May and June (earliest rainy season), and fledging from late June to July (a month into the rainy season), with the post-fledging dependency period extending to September.

Duration of the breeding cycle at Tikal was about six months (February–late July) from nest building to dispersal of young (see also Mader 1981). This is slightly longer than the breeding cycle of the Cooper's Hawk and other temperate zone accipiters (Fig. 8.1; Brown and Amadon 1968; Newton 1978, 1986; Reynolds and Wight 1978; Millsap 1981; Rosenfield and Bielefeldt 1993), and is comparable to the relatively long breeding cycle of the Puerto Rican Sharp-shinned Hawk (Accipiter striatus venator), in which the nest-building and courtship period was estimated as 82 days, incubation as 32 days, the nestling period as 30 days, and post-fledging dependency as 40 days (Delannoy and Cruz 1988). Duration of the period between occupation of nest sites and initiation of laying accounted for much of the difference in length of the breeding cycle between these tropical and temperate zone studies (Fig. 8.1).

Length of Incubation, Nestling, and Post-fledging Dependency Periods

The interval from laying to hatching of the penultimate egg was 33–35 days for five clutches at Tikal. Young made their first flight beyond the nest tree at the ages of 30–36 days. Three males fledged at an earlier age (30–32 days) than did two females (34–36 days). Two males reached independence at the ages of 95–100 days, nine weeks after fledging.

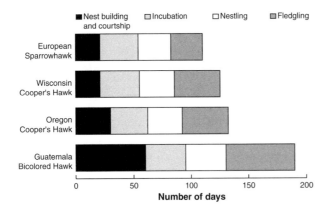

8.1 Duration of Bicolored Hawk (*Accipiter bicolor*) breeding cycle in Guatemala, two populations of Cooper's Hawks (*A. cooperii*) in North America, and a population of European Sparrowhawks (*A. nisus*) in Scotland.

VOCALIZATIONS

Males and females voiced *kek-kek-kek* as a contact call, softer when given by the male. A louder, more rapid version of the same call was used in nest defense. Females also voiced a slurred, hoarse growl—*keer*—when soliciting food. Young voiced a high-pitched *keeyaaa* when soliciting food, as also noted by Stiles et al. (1989). Our impression was that Bicolored Hawk vocalizations, in general, were not as loud as those voiced by Cooper's Hawks (R. Thorstrom, pers. observ.).

BEHAVIOR OF ADULTS

Courtship and Nest-building Period

Categories of behavior that we deemed to play a role in courtship included vocalizations, nest building, supplemental feeding, copulation, and nest defense. Temporal occurrence of these behavioral patterns is given under Nesting Phenology, in this chapter. Both adults took part in nest construction, collecting dry sticks from trees within 50 m of the nest site. During 100 hours of observation at three 1991 nests, we observed deliveries of 81 dry sticks, males bringing 41 and females 40; 70% were delivered between 0600 and 0800. Of 47 copulations, 49% were not associated with any special activity we could detect, whereas 38% took place after nest-building activity (delivering sticks to the nest), and 13% following courtship feedings. On one occasion a pair was observed to copulate three times between 0530 and 0630. Females generally remained within 50 m of the nest site throughout much of the day.

Although occasional soaring has been reported for the Bicolored Hawk by some authors (Thiollay 1989a, 1989b; Fjeldsa and Krabbe 1990; Bierregaard 1994), we did not see it at Tikal, where these birds moved primarily below and through the canopy. We observed no aerial displays or soaring above the forest during courtship activities or at any other point of the breeding cycle. In this regard, this species appears to differ from other accipiters studied to date, which commonly make conspicuous display flights near the nest vicinity (Brown 1976a; Newton 1979; Rosenfield and Bielefeldt 1993; Bildstein and Meyer 2000).

Incubation Period

By mid-April, females remained near nest sites and spent most of their time in "pre-laying lethargy" (Newton 1979). Females laid eggs on alternate days until clutches were complete (*n* = 3 clutches). We suspect that full incubation began when the penultimate egg was laid. Females performed virtually all incubation, and males conducted nearly all hunting during the pre-laying and incubation periods. Occasionally, males attempted to incubate after delivering prey to the female off the nest, but their incubation stints ranged only from 30 seconds to 5 minutes, at which time the female returned to the nest and the male moved off.

Hatching was asynchronous within clutches, spanning 1–3 days in two 3-egg clutches. In one case the first egg hatched on 18 May, the second on 19 May, and the third on 20 May; in the other case, two eggs hatched on 21 May and the third was pipping at 1300 on 22 May. The 33–35-day incubation period we recorded for the Bicolored Hawk is similar to the 30–36-day (typically 34–36-day) interval reported for the Cooper's Hawk (Reynolds and Wight 1978; Rosenfield and Bielefeldt 1993). Of the four cases of nest failure we documented, in only one instance did we observe Bicolored Hawks re-nest; this pair built a new nest from 5 to 23 May and laid again on the latter date, three weeks after losing the first clutch.

Nestling Period

As in other accipiters, the male provided all or nearly all prey for the female during the pre-laying and incubation phase, and for the female and nestlings while the latter were young. Females began to hunt and bring prey when nestlings were two weeks old. Only in the latter part of the post-fledging dependency period (time between fledging and dispersal of young) did males decrease their prey delivery rate (and presumably their hunting effort), as females brought in an increasing proportion of prey, and as the total number of deliveries to young probably decreased. Of 248 prey deliveries observed from courtship through the post-fledging period at several nests from 1991 to 1993, the male and female delivered 177 (71%) and 71 (29%) prey items, respectively. Of 45 items delivered during the post-fledging period, males delivered 27 (60%) and females 18 (40%).

BEHAVIOR AND DEVELOPMENT OF NESTLINGS

Nestling Period

At one nest three young hatched on consecutive days, weighing 25–28 g. The following is based on two males.

Young hatched with light pinkish natal down, a black beak, whitish silver nails, and a yellow cere. After one week the natal down had changed to a creamy white, the cere was yellowish orange, nails grayish white, and the egg tooth began disappearing; body weight approached 120 g, total length (beak to tail tip) was 140 mm, and wing length 90 mm. By two weeks, the young weighed 200–220 g, primaries were emerging from their sheaths, body length was nearly 220 mm, and wing length was 170 mm. At three weeks young weighed approximately 240 g and were 230 mm in total length, with wings 225 mm and tails 40 mm long. By four weeks, young were active in the nests, moving and calling constantly; tails were approaching 80 mm, body length was 305 mm, and wing length 253 mm. During their fifth week—the week prior to fledging—young were very active, moving in and out of the nest as "branchers." At times we observed them in limbs 3–4 m above and beyond the nest, occasionally fluttering, hopping, and flying back to the nest.

Post-fledging Dependency Period

After fledging at the ages of 30–36 days, young initially remained in the nest vicinity and returned to the nest frequently. Adults continued delivering prey to the nest during the first few weeks of the fledging period. During the first few days, adult females often fed the fledglings bit by bit at the nest, but as young quickly developed and moved farther from the nest, females often simply left prey with them and departed. The fledglings constantly begged for food in the nest area, and when an adult entered the vicinity, it was mobbed by the hungry young. During the first several days, adults would call to alert the fledglings of their arrival, but later they just flew in silently and transferred the prey to one of the young. As the fledgings became more capable of handling and transporting prey, they carried food items away from the nest area, seemingly to avoid having it stolen by a sibling.

Dispersal of Fledglings

We radio-tagged three young in order to study their dispersal movements. At the Bejucal nest in 1991 we radio-tagged two young males prior to fledging and made nest observations totaling 103.5 hours during the post-fledging period. The young fledged (first flight away from the nest tree) on 18 June at the ages of 30 and 32 days and, during the first week, ranged up to 20 m from the nest. Adults delivered prey to the nest and the young ate there. By 4 July the fledglings began intercepting the adults and transporting prey to a perch to eat, and on 24 July one fledgling was observed chasing a woodpecker. On 25 July, six weeks after fledging, the young ranged up to 50 m from the nest and, on 12 August, were 300 m from the nest. On 18 August, the fledglings became more difficult to find via radiotelemetry and were located 400 m northeast of the nest. Between 22 and 25 August, at the ages of 95–100 days, and nine weeks after fledging, these two young males dispersed beyond radio contact;

we believe they reached independence at this point. One young female, after fledging at 36 days, apparently dispersed six weeks later, when she moved out of range of the radio equipment.

SPACING MECHANISMS AND POPULATION DENSITY

Territorial Behavior and Displays

Although we observed no territorial disputes among conspecifics, we suspect that the regular spacing of nest sites resulted from territorial behavior. As these birds did not soar at Tikal and had no loud, far-carrying vocalizations, one wonders what mechanisms they used to maintain spacing among neighbors. The fact that nest locations were largely consistent from year to year would appear to facilitate neighbors accumulating knowledge about the locations of nearby pairs. Also, in one case, a bird first captured and banded at one nest later turned up as a breeder in a completely different territory, suggesting that the birds at least sometimes knew the locations of neighboring nests.

Nest Defense

Female Bicolored Hawks were quite vocal around the nest site. They defended a radius of about 40 m around the nest tree, within which they frequently chased Black-handed Spider Monkeys (Ateles geoffroyi), Keel-billed Toucans (Ramphastos sulfuratus), Brown Jays (Cyanocorax morio), and researchers climbing to nests. Both males and females voiced a loud kek-kek-kek while defending the nest area. Most nest defense was by females, as males were often absent from the vicinity. If the male was present, he too would sometimes vocalize and make passes at intruders near the nest. During the nestling period at the Bejucal nest site in 1993, a group of spider monkeys was observed traveling toward the Bicolored Hawk nest. As they approached within 30 m of the nest, both the male and the female hawk took flight, calling loudly and making quick passes at the monkeys. On one aggressive pass, the female bound momentarily to the upper back of a spider monkey, causing the monkey to screech and jump.

Constancy of Territory Occupancy, Use of Alternate Nest Sites

Territories were occupied with great constancy; when checking occupancy of a previously known territory, we never found one vacant (Table 8.1). Nearly all pairs built a new nest each year, and the average distance between alternate nests in a given territory in successive years was 98 ± 43 m (range = 25–200 m, $n = 14$), a similar pattern to that seen in the Cooper's Hawk (Rosenfield and Bielefeldt 1993). However, one nest site was used by the same banded pair for three consecutive years (1991–1993).

Home Range Estimates

Two radio-tagged adult females showed no movement away from their nests during the incubation and nestling periods; they guarded the nest vicinity, chasing away any large bird or mammal that passed within 40 m of the nest. Unfortunately, the transmitters on these two females failed soon after the young fledged. For the adult male radio-tagged in 1991, we made breeding season home range estimates of 5.0 km^2 using the 85% Harmonic Mean (HM) method and 5.9 km^2 using the 100% Minimum Convex Polygon (MCP) method (n = 68 locations). The male radio-tagged in 1992 yielded breeding season home range estimates of 4.0 km^2 (85% HM) and 3.8 km^2 (100% MCP; n = 50 locations). These estimates (4–6 km^2) fall within the lower end of the range observed in Cooper's Hawks in temperate North America (4.0–18.0 km^2: Rosenfield and Bielefeldt 1993). This accipiter also fits well the relationship between home range area and female body weight given by Newton (1979). Neighbors at Tikal appeared to overlap home ranges to some extent, as one of the radio-tagged males passed close to two neighboring nests during his hunting forays. Male Bicolored Hawks were extremely active hunters, traveling a fair distance from nest sites to capture birds and restlessly moving from perch to perch through the canopy in search of prey.

Inter-nest Spacing and Density of Territorial Pairs

Of the seven territories we studied, six were contiguous (Fig. 8.2). Our best estimate of the mean distance between neighboring nests was obtained by using these six contiguous territories, and average distance between nest sites within each. This method yielded a mean inter-nest spacing of 2.14 km ± 0.34 km (n = 5), showing a high degree of constancy in nest spacing (CV = 16%). This spacing yields a Maximum Packed Nest Density (MPND) estimate of 4.17 km^2/pair, a square estimate of 4.58 km^2/pair, and a polygon estimate of 5.33 km^2/pair (see Chapter 1). The polygon estimate is likely conservative, as the resulting polygon embraced areas used to some extent by a pair whose nesting area we never found. These estimates (4.2–5.3 km^2/pair) agree well with our home range estimates for two radio-tagged males (3.8–5.9 km^2; see earlier), and imply a density of 19–24 pairs/100 km^2 of appropriate habitat.

At a site in French Guiana, Thiollay (1989a) estimated a mean density of about 1 pair/21 km^2, one-fourth to one-fifth the density we estimated for Tikal. Thiollay based his estimate on both soaring birds and strip-transects through the understory. He indicated that this species soared only occasionally, and his detection rates of soaring birds were very low (Thiollay 1989b). Thus he may have underestimated the true density of pairs, but to what extent is impossible to say. At one territorial pair per 4.2–5.3 km^2, Bicolored Hawks at Tikal had a density as high as, or slightly higher than, the normal density

range reported for active Cooper's Hawk nests in various parts of temperate North America (Rosenfield and Bielefeldt 1993: highest densities = 1 pair/3.3, 6.4, 6.7, or 8.0 km^2 at various sites).

DEMOGRAPHY

Frequency of Nesting, Percentage of Pairs Nesting

Unlike several raptors at Tikal in which prolonged juvenile dependency prohibits annual nesting by pairs, Bicolored Hawk pairs routinely nested in consecutive years (Table 8.1), including banded birds in many cases. Nesting efforts took place (eggs were laid) in 16 of 21 territory-years (with one re-nesting after a failure, 17 nestings total); hence nesting occurred in 76% of territorial pair-years (Table 8.3).

Surplus Breeders, Breeding Dispersal, and Evidence of Polyandry

Three adult females and one adult male were radio-tagged during the 1991 breeding season. One adult female (call her "A") was banded and radio-tagged at the Bejucal nest in early April 1991—during the courtship and nest reconstruction period. She was present for a week and was the only adult female we observed here, engaged in courtship with the resident male. She disappeared two days after banding, apparently leaving this territory. Although she subsequently proved to remain alive, her radio signal was not detected again during this breeding season. Within a week the male at this site was observed with a different, unbanded female (B), and they later fledged two young. The following year, the missing female (A) turned up as a breeder at a previously unknown, neighboring nest site 2 km north of the Bejucal nest. At this new site, named "Bajada," she fledged three young in 1992. She may well have moved here and bred in 1991, though this is unproved. Another instance of disappearance occurred at the Bejucal site when the male banded and radio-tagged here in 1991 was not present in 1992; he had been replaced by a new, unbanded male.

During the 1993 breeding season two adult males frequented the Bejucal nest area, both copulating with the banded, resident female. One of these males (the presumptive "main" male) was banded in 1992 and was present through 1993. The "extra," unbanded male, noted only in 1993, was never observed to feed the female, but he did mount her in what appeared to be completed copulations. At times both males were present near the nest at the same time, and we noted no aggressive interactions between them. By late in the incubation period, the unbanded male was no longer seen.

Productivity and Nest Success

We studied seven pairs of Bicolored Hawks for a total of 21 territorial pair-years (Tables 8.1, 8.3), during which 17

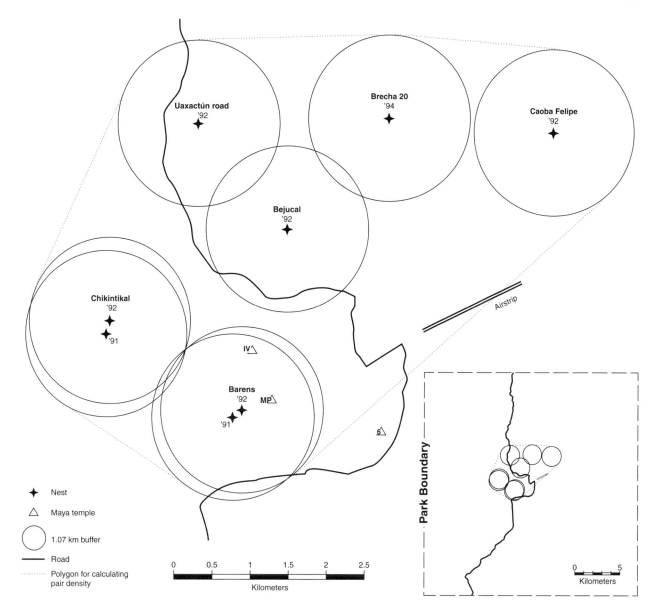

8.2 Locations of Bicolored Hawk nests within Tikal National Park, Guatemala, with 1.07-km buffers around nests and 31.97-km² polygon constructed to aid in estimate of nesting density.

nesting attempts were made (i.e., eggs were laid). Including one of these that was a re-nesting after initial failure, clutch size was documented for 15 nesting efforts, totaling 34 eggs (mean = 2.4 eggs per clutch but 2.5 for first clutches).

Based on 15 nesting attempts located prior to hatching, 23 (68%) of 34 eggs hatched, and 20 young (87%) fledged, or 1.33 per nesting attempt (Tables 8.1, 8.3). Including two nests located after hatching (and assuming each held the mean 2.5 egg clutch), 17 nesting attempts held 39 eggs, of which 27 (69%) hatched, with 24 young (89%) fledging, for a productivity of 1.4 fledglings per nesting attempt (Tables 8.1, 8.3). The 10 fully documented nests that were successful contained 27 eggs, of which 20 (74%) hatched, and all the resulting young fledged. In total, 24 young fledged from 13 successful breeding attempts, for a productivity of 1.8 young

fledged per successful attempt. This includes two nests with two chicks each, discovered after hatching. Overall productivity was 1.14 (24/21) young per territorial pair-year (Table 8.3). Nest success (percentage of nests fledging one or more young) for the four years was 76% (13/17). One pair raised young in all four years of study, producing eight (33%) of the fledged young (Table 8.1). This includes two nests with two chicks each, discovered after hatching. This pair used the same nest for three successive years (1991–1993) and in 1994 moved to another nest site 35 m distant.

Sources of Nest Failure and Mortality

In 17 nesting attempts, three failures occurred during the incubation period and one during the nestling phase. In

Table 8.3 Overall reproductive success of Bicolored Hawks studied at Tikal National Park, 1991–94

Year	Territorial pair-years (number)	Nesting attempts (number)[a]	Eggs laid (number)[b]	Clutch size	Eggs hatched (number)
1991	3	3	9 (3)	3.0	4 (44%)
1992	5	5	9 (4)	2.25	8 (89%)
1993	6	4	8 (4)[f]	2.7	5 (63%)
1994	7	5	8 (4)	2.0	6 (75%)
Total[d]		15[d]	34 (15)[d]	2.43[g]	23 (68%)[d]
Total[e]	21	17[e]	39 (17)[e,h]	—	27 (69%)[h]

Year	Young fledged (number)[c]	Fledglings/ nesting attempt	Fledglings/ territorial pair	Nest success[d]
1991	3 (75%)	1.0	1.0	2/3 (66%)
1992	8 (10)[e] (100%)	2.0	2.0	5/5 (100%)
1993	5 (100%)	1.25	0.83	3/4 (75%)
1994	4 (6)[e] (67%)	1.2	0.86	3/5 (60%)
Total[d]	20 (87%)[d]	1.33[d]	—	11/15 (73%)
Total[e]	24 (89%)[e]	1.4[e]	1.14[e]	13/17 (76%)[e]

[a] Nests with 1 or more eggs.
[b] Number of nests is given in parentheses.
[c] Two nests containing two nestlings each when discovered.
[d] Using only nests discovered before hatching.
[e] Including two nests that contained two nestlings each upon discovery, and assigning them 2.5 eggs each (the mean observed clutch size).
[f] 8 eggs laid in three nests; fourth nest had clutch size unknown, failed soon thereafter.
[g] 34 eggs in 14 nests with known clutch = 2.43 mean; 15th nest had clutch size unknown.
[h] Clutch size for nests discovered with two chicks in each assumed to be 2.5 eggs each (i.e., the mean observed clutch size).

two of these nests, eggs remained present but failed to hatch, while predation, possibly by a mammal, caused the failure of one nest with eggs and one with nestlings. One adult death was observed in August 1992 when a female banded and radio-tagged as an adult in 1991 was found floating dead, along with a rat, in a water tank amid the Maya ruins. Apparently she entered this deep tank in pursuit of prey or in order to drink, and could not escape. The following year a new female was resident on the territory that had been occupied by the drowned female. We suspect that adult mortality is relatively low, and one adult female banded in 1991 was still on her territory in 1994. No immature-plumaged birds were detected on nesting territories, further evidence of relatively low adult turnover.

Parasites and Pests

Occasionally adult and nestling Bicolored Hawks were bothered by sweat bees and blowflies. Some nestlings were observed with ectoparasitic fly larvae attached to blood feathers in the primary and secondary tracts, but we observed no deaths attributable to these parasites. This is in marked contrast with Delannoy and Cruz's (1988) findings that botfly parasitism caused 69% of nest failures during the nestling stage in the Puerto Rican Sharp-shinned Hawk.

CONSERVATION

According to some authors, these hawks are not greatly affected by forest clearing. Olrog's (1985) comment to this effect was applied to montane populations of *A. b. guttifer* nesting up to 2700 m in northern Argentina. Fjeldsa and Krabbe (1990) characterized the species as occupying forest, forest edge, and parkland, with *A. chilensis* occurring largely in mosaics of open land and deciduous *Nothofagus* forest and being little influenced by deforestation. It is questionable whether these comments, which refer to south temperate and largely mid- to high-elevation populations, adequately describe the species's compatibility with forest clearing elsewhere in its broad range. Different authors have opposing impressions. In Costa Rica, Slud (1964) described this species as "seldom seen inside virgin forest," being more of a woodland and forest edge species, but Slud made similar comments about many raptors that are patently forest species at Tikal. In contrast, Howell and Webb (1995) regard this as "a rarely encountered species of deep forest." At Tikal, our experience accords closely with that of Howell and Webb; this is a secretive bird of the mature forest interior. In Central America and Mexico, though these birds are sometimes seen in second-growth and forest-edge habitats, we know of no evidence that they thrive in truly savanna-like or largely deforested areas.

Many authors have described this hawk as uncommon to rare (e.g., Ridgely and Gwynne 1989; Stiles et al. 1989; Bierregaard 1994). However, Wetmore (1965) speculated that these birds may be more common than indicated by the infrequent sightings, and we believe he is correct. Our results at Tikal suggested that these birds filled suitable habitat with adjacent territories. On the other hand, we believe it likely that this species's ability to persist in the face of deforestation has been somewhat overstated. On balance, this bird is probably just as likely to suffer population declines due to forest loss as any other forest-reliant raptor of equivalent body size and spatial needs.

HIGHLIGHTS

Of the hawks that hunt within the forest interior at Tikal, the Bicolored Hawk, with a 96% avian diet in our sample, was the preeminent bird-eating specialist. Bicolored Hawks took a higher percentage of avian prey than what was reported in any of nine Cooper's Hawk diet studies cited by Rosenfield and Bielefeldt (1993). Only two other species at Tikal demonstrated an equivalent reliance on birds, the Bat Falcon (*Falco rufigularis*), which takes much smaller birds—captured mainly above the forest canopy or when crossing openings—and the Orange-breasted Falcon (*Falco deiroleucus*), which also takes prey mainly in flight above the forest. The two forest falcons at Tikal relied substantially less on avian prey, with the large Collared Forest Falcon (*Micrastur semitorquatus*) taking many mammals, and the small Barred Forest Falcon (*M. ruficollis*) taking many small lizards in addition to birds. The Ornate Hawk-eagle (*Spizaetus ornatus*), while a significant bird predator, had a diet of only 56% birds (70% biomass) and took much larger birds, on average, than did Bicolored Hawks.

At the level of prey species, Bicolored Hawks overlapped most with the Orange-breasted Falcon (Pianka's [1974] overlap measure = 0.58), and next with the Barred Forest Falcon (overlap = 0.30), Great Black Hawk (*Butegallus urubitinga*), and Collared Forest Falcon (for both, overlap = 0.07). That this sub-canopy hunter should overlap most with the highly aerial Orange-breasted Falcon we found somewhat surprising, but the falcons took a number of typically forest-interior birds such as trogons—presumably while the latter crossed openings or were vulnerable in treetops—giving rise to this degree of overlap. Mean prey size of Bicolored Hawks (90 g) and Orange-breasted Falcons (85–90 g; see Chapter 20) was essentially identical. In contrast, birds taken by Collared Forest Falcons and Barred Forest Falcons were mostly larger (median weight 373 g) and smaller (62 g), respectively, than those taken by Bicolored Hawks.

Whatever the evolutionary causality of reversed sexual size dimorphism in raptors, this bird specialist, among the most dimorphic raptors known, accords with the well-known correlation between strong dimorphism and a diet largely of birds (Newton 1979). With a mean clutch size of 2.5, these accipiters had one of the largest clutches among Tikal's raptor community, and with individual pairs often breeding every year—and occasionally re-nesting after losing a first clutch early in the season—this species must show one of the higher rates of productivity and presumably population turnover among this raptor community. Still, the clutch size and productivity were low by temperate zone standards, and presumably adult turnover is as well.

In comparison to temperate zone accipiters, among our most notable findings were the longer duration of the breeding season and the lack of soaring and flight displays in Bicolored Hawks at Tikal. The lack of soaring flights and of loud vocalizations made these hawks very difficult to enumerate, except by searching out active nests, a laborious proposition. Both the Bicolored Hawk and the Puerto Rican Sharp-shinned Hawk have shown a breeding cycle 180 days or more in length—substantially longer than that of their temperate zone counterparts (Fig. 8.1; Delannoy and Cruz 1988). In both species, the long duration resulted from a lengthy period between the initiation of nesting-related behavior and egg laying. The shorter duration of the pre-laying period in temperate zone accipiters may relate to their being migrants or partial migrants, whereas Bicolored Hawks at Tikal and Sharp-shinned Hawks in Puerto Rico remain on their territories year-round. In contrast, it seems plausible that migrant accipiters may arrive on their nesting grounds only shortly before the optimal nesting date.

Bicolored Hawks also may have a slightly longer period of post-fledging dependency than does the temperate zone Cooper's Hawk. Two of our young males reached independence nine weeks after fledging, and Mader (1981) observed a fledgling begging for food near its natal nest 8.7 weeks after fledging. In comparison, the longest reported durations of dependency in the Cooper's Hawk appear to be 7.0–7.5 weeks after fledging (Rosenfield and Bielefeldt 1993) and up to 8 weeks (R. Rosenfield, pers. comm.). Whether this apparent slight difference is significant awaits further study.

The mixed-species avian–foraging flocks so characteristic of Neotropical forests have often been interpreted as arising largely because of raptor predation, with flocking behavior offering a survival advantage to individual flock participants (Buskirk et al. 1972; Buskirk 1976; Munn and Terborgh 1979). While several raptors no doubt play an important role in applying this predation pressure, we consider it likely that the Bicolored Hawk is among the species most directly responsible for shaping the social systems and antipredator behavior of small and medium-sized birds of the Neotropical forest interior.

For further information on this species in other portions of its range, refer to Di Giacomo 2005 and Marini et al. 2007.

9 CRANE HAWK

Jason Sutter

Deep wing beats, interspersed with short sails on slightly arched wings, carry a dark form lethargically above the forest's humid penumbra. A pair of white tail bands, a white crescent across the base of the outer primary feathers, and long, slender, reddish legs identify this as a Crane Hawk *(Geranospiza caerulescens)*—a little-known denizen of the forests and swampy woodlands of tropical America. Rarely soaring or flying high above the canopy except during courtship or territorial displays, Crane Hawks rely less on speed to secure prey than on a methodical reconnaissance of their large, diverse home ranges. Facilitated by several unique anatomical features—long legs, an unusually flexible intertarsal joint, smooth, scutellate tarsi, and miniature outer toes—these hawks forage largely by reaching within nooks and crannies for hidden prey. This raptor's foraging ecology is somewhat paradoxical. Crane Hawks are anatomically and behaviorally specialized for taking a functionally specialized set of prey—mainly nocturnal species in their daytime retreats. However, the prey taken are taxonomically diverse, resulting in a broad dietary niche. Often said to be found in forested habitats near water, at Tikal these hawks nested throughout much of the forest, where surface water is absent most of the year. They did, however, nest selectively in a specific forest type—Sabal and Hillbase Forest occupying a narrow portion of the predominant topographic gradient. Inconspicuous flight behavior and dark plumage made the Crane Hawk a challenging, elusive research subject.

MORPHOLOGY AND PLUMAGE

Crane Hawks are lightly built, medium-sized raptors about the size of a Common Barn Owl *(Tyto alba)* or Broad-winged Hawk *(Buteo platypterus)*. Slender and rather small headed in appearance, they have relatively long and broad, rounded wings, a long tail, and noticeably long, thin legs (Plates 9.1 and 9.2). Their appearance and perceived behavior and diet are reflected in their Latin American vernacular names—Gavilán Zancón ("long-legged hawk"), Gavilán Zancudo ("wading hawk"), and Gavilán Ranero ("frog-eating hawk": Smithe 1966; Alvarez del Toro 1980; Stiles et al. 1989; Bierregaard 1994; Howell and Webb 1995).

Adult Crane Hawks at Tikal were blackish slate, with whitish vermiculation on the lower belly, thigh feathers, and underwing and undertail coverts. Juveniles were paler than adults, with the underparts washed with white. The forehead, supercilium, auriculars, chin, and upper throat were also streaked with white, and undertail coverts conspicuously buff colored, barred with gray. Legs were yellowish orange in juveniles and dark reddish orange in adults (Plate 9.3).

Like the Old World Harrier-hawks or Gymnogenes *(Polyboroides* spp.), Crane Hawks possess unusually long legs and a uniquely mobile intertarsal joint that flexes both forward and backward and has a wide range of lateral mobility. These seem to be adaptations permitting an unusual foraging style that relies heavily on extracting prey from holes, crevices, and other hideaways. For five Crane Hawks at Tikal, backward flexion of the tarsal joint averaged 34 ± 8° (range = 20–40°, n = 5), lateral swing averaged 16 ± 6° (range = 13–25°, n = 5), and medial swing was 15° for all birds, demonstrating the extreme flexibility of the species's "double-jointed" legs. Most raptors have much more limited tarsal mobility, without backward flexion.

Several other features facilitate the Crane Hawk's habit of reaching its legs into cavities in search of prey. The tarsal scales are fused into a few large plates, presenting an almost smooth surface (Ridgway 1873; Friedmann 1950; Sutton 1954; Wetmore 1965), which may help prevent catching on the sides of nooks and crannies. Crane Hawks also have an unusually small outer toe (Friedmann 1950; Wetmore 1965), which may enable them to extract prey from confined spaces where a larger clenched foot could not pass. Presumably the smallish head is also related to these foraging habits.

Male Black Crane Hawks (*G. caerulescens nigra*: see below) averaged 358 g (n = 3) and females 495 g (n = 3), or 38% more, for a mass Dimorphism Index (DI) of 10.8 (Appendix 1). The largest set of linear measurements is that of Friedmann (1950), given in Appendix 2 along with data from Tikal. Based on Friedmann's data, DI values are as follows: 7.6 for wing chord, 2.3 for tail length, 6.3 for culmen length, 8.2 for tarsus length, and 11.2 for middle toe length. The mean of dimorphism values for wing, culmen, and body mass is 8.2 (Appendix 2). Thus, the Crane Hawk is more size dimorphic than most North American *Buteo* species but less so than most falcons and accipiters (Snyder and Wiley 1976).

GEOGRAPHIC DISTRIBUTION AND SYSTEMATICS

Crane Hawks occur in wooded tropical lowlands from southern Sonora and Tamaulipas, Mexico, to northern Argentina (Brown and Amadon 1968). Often judged uncommon, these raptors are in some cases reasonably common, as at Tikal. Most often noted in lowlands, Crane Hawks have been recorded from sea level to 250 m in Oaxaca (Binford 1989), 500 m in Costa Rica and Colombia (Hilty and Brown 1986; Stiles et al. 1989), 700 m and higher in southern Sonora (Sutton 1954), 1350 m in Guatemala (Land 1970), and to 1500 m in Honduras and Mexico (Monroe 1968; Howell and Webb 1995). At Tikal, these hawks are resident year-round; we know of no evidence of migration here or elsewhere.

Based on shared anatomical peculiarities, some researchers (e.g., Friedmann 1950) have believed the Crane Hawk to be closely related to *Polyboroides* species—the African and Madagascar Harrier-hawks. In contrast, Amadon (1982) believed these resemblances to result from convergent evolution and thought *Geranospiza* to be a sub-buteonine, allied to New World *Leucopternis* and *Buteogallus*, and not far from *Buteo*. Amadon's view is supported by recent molecular genetics research. Using the nuclear RAG-1 gene and other genetic markers, Griffiths et al. (2002, pers. comm.) found strong support for a phylogeny in which the Old World *Polyboroides* belongs in a clade along with Old World *Gypaetus* snake eagles—with *Geranospiza* belonging to a distant "buteonine" clade containing *Buteo*, *Accipiter*, *Circus*, and other genera. Lerner and Mindell (2005), using the nuclear c-*myc* gene and other evidence, also place *Geranospiza* in a clade containing *Buteo*, *Accipiter*, *Circus*, and allies. Thus, it now seems clear that New World *Geranospiza* and Old World *Polyboroides*—both showing a hyper-mobile tarsal joint, other anatomical specializations related to foraging, and similar search-and-probe foraging behavior—represent a notable case of convergent evolution.

Six subspecies of the Crane Hawk are recognized, differing dramatically in body size. These are described by Brown and Amadon (1968) and are divisible into three well-defined groups, perhaps approaching species status (Blake 1977; Bierregaard 1994; Ferguson-Lees and Christie 2001). The form occurring at Tikal—*G. c. nigra*—was previously regarded a distinct species and given the common name Black Crane Hawk, which we use here. This blackish slate subspecies is the darkest and one of the largest, and ranges from Tamaulipas, Mexico, to the

Panama Canal Zone. Because the subspecies differ so greatly in size, we restrict ourselves mainly to treatment of this subspecies.

PREVIOUS RESEARCH

Brief accounts of some aspects of Crane Hawk biology have been published in regional works and field guides, mainly based on subspecies other than the Black Crane Hawk, the one occurring at Tikal. Attracting most attention have been the species's unusual foraging behavior (Sutton 1954; Jehl 1968; Olmos 1990a) and associated morphology (Burton 1978). The little that is known of nesting is based on four nests observed briefly in Sonora, Mexico (Sutton 1954), five nests in Surinam (Haverschmidt 1964b), and seven in Venezuela (Mader 1981)—all of subspecies other than the one we studied. Several authors have commented on the diet, foraging, or habitat of the Black Crane Hawk: Lawrence (1876), Carriker (1910), Dickey and van Rossem (1938), Wetmore (1943, 1965), Sutton (1951), Felten and Steinbacher (1955), Storer (1961), Slud (1964), and Weyer (1984). Prior to our work, however, the Crane Hawk had never been the subject of a detailed field study.

RESEARCH IN THE MAYA FOREST

In 1993 we found three Crane Hawk nests and collected limited data on breeding biology. In 1994 and 1995, we studied 11 breeding pairs in eight territories, and monitored five additional territories. We studied breeding biology at seven nests in 1994 and four in 1995.

We observed nests using 15–60X spotting scopes from 35–60 m distant in blinds built in trees or on the ground. We estimated prey length by comparison with the hawks' bill and head size, and categorized prey by estimated size class: from 0 to 20 g, more than 20 to 50 g, more than 50 to 100 g, and more than 100 g. We documented hunting behavior, habitat use, and home range size by use of radio tags on two adults, supplemented by opportunistic observations on other individuals. We radio-tagged one fledgling to aid in measuring the duration of post-fledging dependency. To place radio transmitters, we captured adults by using a *bal-chatri* (Berger and Mueller 1959) in the nest tree. Both adults and juvenile were fitted with 18-month transmitters (216 kHz; Holohil Systems, Carp, Ontario), in a backpack arrangement, using Teflon ribbon sewn with cotton thread. The entire assembly weighed 11.0 g and was less than 3% of adult and less than 4% of fledgling body mass. We conducted radio tracking on foot, using hand-held three-element yagi antennas. We often achieved visual contact, but we sometimes used triangulation to estimate locations. We radio-tracked adults 1–8 times per month, taking one or two locations per 4–8-hour tracking session. Spatial use data were analyzed using ArcView GIS 3.0a, and the Animal Movement Analysis ArcView Extension (Hooge and

Eichenlaub 1997) and RANGES V software (Kenward and Hodder 1996).

To compare Crane Hawk diets with those of other raptors at Tikal, we used Levins's (1968) niche breadth metric, B, and Pianka's (1974) niche overlap index, O_{jk}. Levins's B ranges in value from 1 (minimal niche breadth; all prey of a single category) to n, the total number of prey categories. A value of n would indicate maximal niche breadth, the situation in which each prey category is equally represented in the diet. Pianka's overlap measure ranges from 0 (no resources used in common) to 1.0 (complete overlap). During foraging observations we recorded details of behavior, substrate, and foraging height, using 5 m intervals. Vegetation at foraging sites was characterized using the classification of Schulze and Whitacre (1999). To characterize forest in nest vicinities, we sampled, at each nest tree, a 0.04 ha plot (40.5 m by 10.0 m) laid out in each of the cardinal directions (four plots per nest), originating in a nonoverlapping fashion 1 or 5 m from the tree base. Sampling methods followed those of Schulze and Whitacre (1999). Further details regarding these and other methods are given in Sutter 1999 and Sutter et al. 2001.

DIET AND HUNTING BEHAVIOR

Diet

Crane Hawks are reported to feed on a broad assortment of small to medium-sized animals. Based on stomach contents and direct observations, the species's diet is known to include large insects—Orthoptera: Tettigoniidae (katydids), Gryllidae (crickets), Locustidae (short-horned grasshoppers), Blattidae (roaches)—arachnids (Mygalidae), lizards (Iguanidae and Teiidae), frogs (Hylidae and Ranidae), rodents (Muridae: *Tylomys*), nestling birds (Psittacidae: *Forpus*), bats (Vespertilionidae: *Eptesicus melanopterus*), mollusks, and fruit. Of 21 published sources we found that report Crane Hawk stomach contents or observed prey items, nine pertain to the Crane Hawk subspecies we studied at Tikal. None of these report snakes, bats, or birds in the diet; five report insects, four report lizards, three mention frogs, and two found rodents (Lawrence 1876; Dickey and van Rossem 1938; Wetmore 1943; Sutton 1951, 1954; Felten and Steinbacher 1955; Storer 1961; Smithe 1966; Jehl 1968).

From 1993 to 1995, we observed 227 Crane Hawk prey items and identified 181. Rodents (Muridae and Heteromyidae) made up the largest portion of the diet, with terrestrial, cursorial, and arboreal rodents of at least eight species taken (Table 9.1). Rice Rats (*Oryzomys* spp.), Spiny Pocket Mice (*Heteromys* spp.), Vesper Rats (*Nyctomys sumichrasti*), and Hatt's Vesper Rats (*Otonyctomys hatti*) were the rodents identified. Rodents were delivered to three nests in 1994 and two in 1995, and were observed as prey at five other (non-focal study) territories.

Lizards and frogs were common in the diet, though contributing far less biomass than rodents. Lizards in the diet averaged 14 ± 5.5 cm in estimated length (range = 8–25 cm, $n = 28$), and at least four species were taken

(Table 9.1). Frogs were the only amphibian prey identified; most were small and some had toe pads—possibly tree frogs. Frogs were sporadically abundant in Crane Hawk diets when frog activity increased following rains, especially at the onset of the rainy season in late May and June. For example, on 21 June 1994, six frogs were brought to two nests after heavy morning rains. Bats, birds, snakes, and other mammals made up the remainder of prey items, with bats and birds both twice as frequent as snakes. Snakes averaged 75 ± 28 cm in estimated total length (range = 40–100 cm, $n = 5$). All snakes were relatively slender and green to brown; we believed most were arboreal species. On one occasion, we watched a female Crane Hawk deliver a juvenile Striped Hog-nosed Skunk (*Conepatus semistriatus*) to her young.

Of 224 prey items we assigned to weight classes, most (52%) were estimated to weigh less than 20 g, and 40% weighed more than 50 g (Sutter et al. 2001). Rodents, the most frequent prey type, comprised most (77%) of the biomass consumed (Table 9.1). In contrast, lizards—the second most frequent prey type—made up only a small fraction (2.8%) of prey biomass. The diet in 1994 and 1995 did not differ markedly in the proportion of major prey types ($P = 0.097$, $\chi^2 = 10.7$, df = 6). Both males and females captured prey of all types and from all weight classes, and did not differ detectably in terms of prey types ($P = 0.847$, $\chi^2 = 3.4$, df = 7) or weight classes of prey taken ($P = 0.789$, $\chi^2 = 3.2$, df = 6).

The diet was largely consistent among nests. Among the four best-studied nests ($n = 173$ identified prey items), we observed rodents, lizards, and frogs at all four, birds at three, and bats and snakes at two. Based on identified prey, the mean contribution of rodents to the diet was $48.9 \pm 11.2\%$ (range = 37.3–60.7%, $n = 4$ nests), mean percentage of lizards was $17.8 \pm 6.7\%$ (range = 8.3–22.9%, $n = 4$), and mean percentage of frogs was $16.7 \pm 3.9\%$ (range = 12.0–21.4%, $n = 4$). Unidentified prey comprised $25.3 \pm 22.6\%$ of observed prey (range = 10.7–58.6%, $n = 4$). These four nests did not differ significantly in the proportion of prey provided by the three most frequent prey types—rodents, lizards, and frogs ($P = 0.672$, $\chi^2 = 4.0$, df = 6).

Nocturnal animals (bats, most rodents, some frogs, a skunk, and a snake) made up at least 48% of Crane Hawk prey items at Tikal (Table 9.1), even though all prey were apparently captured during daylight. The diet of Crane Hawks at Tikal resembled that previously reported, with the notable exception of invertebrates, which were conspicuously absent from the diet at the nests we observed. It is possible that adults ate invertebrates away from nests but did not bring them to nestlings owing to their small size and hence low energetic reward.

Foraging Behavior

No doubt the most frequent mode of hunting among buteonine hawks is "perch hunting"—scanning the surroundings from a perch. This mode of hunting has only occasionally been reported for Crane Hawks (Sutton 1954). We did see a Crane Hawk catch a Big-eared Climb-

Table 9.1 Breeding season diet at nests of the Crane Hawk (*Geranospiza caerulescens nigra*) in Tikal National Park, 1994–95

Prey type	Frequency		Biomass	
	Number	Percentage (%)	Grams	Percentage (%)
Mammalia	100	55.2	6465	83.6
Rodents (Rodentia)[a]	86	47.5	5980	77.3
Bats (Chiroptera)[b]	12	6.6	268	3.5
Carnivores (Carnivora)[c]	1	0.6	175	2.3
Unidentified mammal	1	0.6	43	0.5
Reptilia	41	22.7	478	6.1
Lizards (Sauria)[d]	36	19.9	220	2.8
Snakes (Ophidia)[e]	5	2.8	258	3.3
Amphibia[f]	29	16.0	485	6.3
Aves[g]	11	6.1	313	4.0
Total number of identified prey	181	100.0	8415	100.0
Number of unidentified prey	46	—	485[h]	—

Note: Percentages are based on total number of identified prey items.

[a] 19 *Oryzomys* spp.; 18 *Oryzomys, Nyctomys,* or *Otonyctomys* spp.; 14 *Heteromys desmarestianus*; 12 *Nyctomys sumichrasti*; 9 unidentified rodents; 4 *Ototylomys phyllotis*; 4 *Tylomys nudicaudus*; 2 *Heteromys* spp.; 2 *Otonyctomys hatti*; 1 *Heteromys gaumeri*; 1 *Sigmodon hispidus*. At least 76 of 86 rodents were nocturnal species.

[b] 10 unidentified bats; 1 *Artibeus jamaicensis*; 1 *Dermanura* spp. (all nocturnal).

[c] *Conepatus semistriatus* (nocturnal).

[d] 26 unidentified lizards; 7 *Anolis* spp.; 1 *Ameiva undulata*; 1 *Basiliscus vittatus*; 1 *Mabuya unimargata* (all diurnal).

[e] 4 unidentified colubrids (activity time unknown); 1 *Leptodeira septentrionalis* (nocturnal).

[f] 23 unidentified anurans; 3 Hylidae, Leptodactylidae or Microhylidae (nocturnal); 3 *Rana berlandieri* (diurnal).

[g] 3 unidentified passerines; 2 unidentified birds; 2 Psittaciformes; 1 Columbiformes; 1 Troglodytidae; 1 *Contopus cinereus*; 1 *Empidonax* spp. (all or most diurnal).

[h] Based on 43 unidentified prey items.

ing Rat *(Ototylomys phyllotis)* in this fashion, but perch hunting comprised only 16% of the hunting incidents we observed, and we believe most prey were obtained by other methods. Crane Hawks have occasionally been reported to hunt by coursing over open areas, harrier-like (Sutton 1954; Olmos 1990), sometimes behind fire lines in burning pastures. Such fires were regarded by Dickey and van Rossem (1938) as "a sure attraction" for these hawks. We did not observe this behavior in our forested study area, nor in the farming landscape nearby, where we occasionally witnessed field burning.

The most frequently reported Crane Hawk foraging tactic might be termed the "search-and-probe" technique, in which individuals use their long, flexible legs, and less often their rather slender head, to search within potential prey refugia in trees or on the ground (Sutton 1954; Friedmann and Smith 1955; Haverschmidt 1962; Slud 1964; Wetmore 1965; Jehl 1968; Mader 1981). Crane Hawks have often been observed clinging acrobatically to tree trunks and probing their head or legs into holes while using their tail and wings to balance and prop themselves (Sutton 1954; Friedmann and Smith 1955; Haverschmidt 1962). Jehl (1968) observed a Crane Hawk repeatedly extracting large katydids (Tettigoniidae) from behind loose bark of a tree trunk. The hawk clung to the trunk in a variety of positions, including upside down, and removed prey with its beak and feet, at times bending its legs backward to probe deeply into crevices. On the ground, Crane Hawks have been observed capturing lizards by reaching their legs into crevices in rock piles (Sutton 1954), and searching on foot along stream banks (Wetmore 1965).

At Tikal, we observed 19 hunting attempts by at least seven individuals, involving several of the hunting techniques reported for this species (Sutter et al. 2001). In 15 of these cases, Crane Hawks actively searched for or pursued prey, in 3 cases they scanned quietly from a perch, and in the remaining case, a Crane Hawk attacked an *Empidonax* flycatcher caught in a mist net. In 3 of the 15 cases, Crane Hawks pursued prey on foot along the top of large tree limbs, while in 12 cases, they probed with their head or legs, searching for prey within holes in snags, tree trunks, and limbs; within bromeliads, other epiphytes, and behind bark; and in palm leaf axils, leaf litter on the forest floor, and in a shallow puddle. The many nocturnal prey we observed in the diet were most likely captured in this search-and-probe fashion.

During daylight, rodents shelter in dens on the ground or in tree cavities (Emmons 1990), while bats roost in tree holes, among foliage, behind bark, and beneath hanging palm fronds (Foster and Timm 1976; Fleming 1988; Altringham 1996). Several species of tree frogs, all nocturnal, are known to breed or hide by day in bromeliads or palm leaf axils (Meyer and Foster 1996), and the Northern Cat-eyed Snake *(Leptodeira septentrionalis)* identified among prey (Table 9.1) is a nocturnal, arboreal, frog predator that sometimes hides in bromeliads during the day (Lillywhite and Henderson 1993; Greene 1997). Prey hidden in such sites would presumably often be undetectable visually even at close quarters. Thus, we hypothesize that Crane Hawks focus their visual search at least partly on locating potential hiding places of prey rather than the individual prey items themselves.

Some diurnal prey may also have been captured using the search-and-probe method, especially birds in nests. Although a few adult birds were taken, these are probably too elusive for Crane Hawks to often capture, except perhaps from a nest. Among the birds taken were several nestlings, including two hole-nesting parrots, and Mader (1981) observed a Crane Hawk using one foot to hang beneath palm fronds while grabbing at a hanging bird nest with the other. We suspect that Crane Hawks frequently rob bird nests. We captured a Crane Hawk in a mist net after it attacked a trapped flycatcher (*Empidonax* spp.), and Ouellet (1991) made similar observations in Venezuela. Crane Hawks at Tikal were often harassed by Social (*Myiozetetes similis*) and Sulphurbellied Flycatchers (*Myiodynastes luteiventris*), and harassment by flycatchers and oropendolas (*Psarocolius* spp.) has been reported previously (Slud 1964; Wetmore 1965).

Wetmore (1965) observed Crane Hawks clambering around awkwardly, investigating the undersides of limbs, and other researchers have observed them energetically chasing prey through the limbs on foot (Lawrence 1876; Friedmann and Smith 1950; Slud 1964). On various occasions we observed Crane Hawks chasing what we believed to be arboreal lizards on tree trunks and branches, and such lizards (*Anolis* spp.) were frequent in Crane Hawk diets at Tikal (Table 9.1).

HABITAT USE

Crane Hawks are generally found in lowland forests or woodlands, often in association with wetland habitats (Lawrence 1876; Friedmann and Smith 1955; Haverschmidt 1964b). These hawks have often been observed foraging near rivers, marshy streams (Dickey and van Rossem 1938; Wetmore 1943, 1965; Sutton 1951, 1954), lakes, ponds, and temporary pools (Carriker 1910; Dickey and van Rossem 1938), as well as in mangrove lagoons, gallery forest, swampy forest, and estuaries (Carriker 1910; Dickey and van Rossem 1938; Felten and Steinbacher 1955; Weyer 1984). Nevertheless, Crane Hawks seem somewhat flexible in habitat requirements, as evidenced by their use of more open areas including semiarid, scattered woodlands (Sutton 1954; Jehl 1968), seasonally inundated palm savannas of the Venezuelan llanos and Brazilian pantanal (Mader 1981; Olmos 1990a; Ouellet 1991), and boggy pastures and meadows (Dickey and van Rossem 1938). Crane Hawk nests have been found in semi-deciduous woodlands along arroyos in Sonora, Mexico (Sutton 1954), and in gallery forest and coffee plantations in Surinam (Haverschmidt 1964b). It remains unclear to what extent this species relies on the proximity of water. Sutton (1954) emphasized that while Crane Hawks in Sonora nested in riparian corridors, surface water was rare there during the nesting season. Although few perennial water sources exist in our forested study area, the intermittent arroyos and ephemeral pools of seasonally inundated bajos may be important to Crane

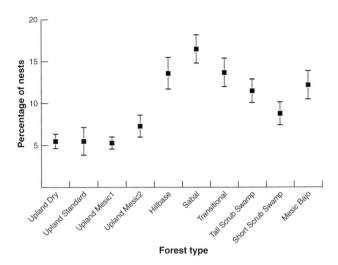

9.1 Habitat in Crane Hawk nest vicinities: percentage of 12 nests with mean values of habitat variables falling within 95% confidence intervals typical of each forest type. Confidence interval values were taken from Schulze and Whitacre 1999.

Hawks. We found 13 nest sites in three forest types in Tikal. Nearly all were in Sabal or Hillbase Forests (Fig. 9.1), suggesting strong positive selectivity by the hawks of these forest types, which occur along a restricted portion of the topographic gradient (Schulze and Whitacre 1999; see Chapter 2). These two forest types have a notably irregular canopy with many large emergents such as Mahogany (*Swietenia* spp.) and Quebracho (*Aspidosperma* spp.) trees, interspersed with canopy palms (*Sabal morrisiana*)—and a relatively dense understory rich in palms (*Cryosophila stauracantha*: Schulze and Whitacre 1999; see Chapter 2).

We observed Crane Hawks hunting in many forest types at Tikal, including Dry Upland Forest on hilltops, the ubiquitous Upland Standard Forest, and Hillbase, Sabal, and Transitional Forests. Because passage through Scrub Swamp vegetation was difficult, we dedicated less effort to monitoring this forest type. Hence, while we did not observe Crane Hawks hunting in Scrub Swamp Forests, this may not indicate that the hawks avoided this habitat.

BREEDING BIOLOGY AND BEHAVIOR

Nests and Nest Sites

Crane Hawk nests averaged 24.2 ± 4.5 m in height (*n* = 14), and were in trees averaging 26.8 ± 3.5 m in height (*n* = 14), usually in the upper 3 m of the crown. Eight of 14 nests were located in crotches along the main trunk, and trunk diameter at this point ranged from 7.7 to 32.0 cm (mean = 14.1 ± 10.4 cm, *n* = 5). The remaining six nests were built on large limbs 60 cm to 12 m from the trunk (mean = 7.5 ± 4.3 m, *n* = 6).

Nests were platform-like and round (*n* = 7) or oval (*n* = 3) in shape, averaging 51 ± 8 cm long, 47 ± 12 cm wide, and 19 ± 7 cm in external depth, with a shallow, leaf-lined cup (mean = 3.7 ± 3.1 cm deep, n = 10). Nests were constructed of twigs averaging 30 cm long and 5 mm in diameter (n = 10 nests). All nests were situated in dense vine tangles, which usually formed masses around several smaller limbs that supported both the vine tangle and the nest. These vine masses provided nests with excellent concealment from above and below. Of 14 nests, 6 were completely and 8 partially covered with vegetation from above. Three of the 14 were undetectable from the ground, 9 were partially visible, with effort, and the remaining 2 were easily visible from the ground.

All nests were built in large, living trees, which averaged 82.0 ± 45.0 cm in diameter at breast height (n = 14). Nest trees were tall emergents, and the crowns of most (86%, n = 14) were judged to be disconnected from the surrounding forest canopy. Nest trees tended to have many vines and at least some epiphytes (aeroids, orchids, bromeliads). All nest trees supported 4 to more than 100 small vines (< 5 cm in diameter). Most nest trees had some (mean = 5.9 ± 5.0, n = 10) large-diameter vines (> 5.0–7.5 cm) growing on them, and more than half had at least one (mean = 3.3 ± 3.8, n = 8) very large-diameter vine (> 7.5 cm). Most nest trees had some epiphytes (> 20 cm diameter aggregation) on their trunks (79%, n = 14) and in their crowns (86%, n = 14).

Crane Hawks showed substantial overlap in nest site selection with some other raptors—especially with the Bicolored Hawk *(Accipiter bicolor)*. Crane Hawks and Bicolored Hawks frequently nested near one another, often in a single swale supporting Sabal Forest (J. Sutter, pers. obser.)—evidence that they shared a similar choice of nesting habitat. Moreover, they sometimes used the same precise nests. The "Barens" nest used in 1991 and 1992 by Bicolored Hawks was used by nesting Crane Hawks in 1995, at which time Bicolored Hawks almost certainly nested quite nearby, as they were frequently in the area and attacked Ornate Hawk-eagles *(Spizaetus ornatus)* several times (J. Sutter, pers. obser.). Additional places where Crane Hawks and Bicolored Hawks frequented (and presumably nested) in the same vicinities were Brecha 80 (in 1994), Chikintikal, and Bajada de la Pina (J. Sutter, pers. obser.). Use of the same exact nest site was also observed between Crane Hawks and Great Black Hawks *(Buteogallus urubitinga)*. The Brecha 20 nest used by Crane Hawks in 1993 was used by nesting Great Black Hawks in 1994, at which time Crane Hawks attempted to nest 100 m away. Roadside Hawks *(Rupornis magnirostris)* also were frequently found nesting in the same swales of Sabal Forest along with Crane Hawks and Bicolored Hawks.

Egg and Clutch Size

Eggs were oblong and mostly white, with varying amounts of sepia spotting (Plate 9.4). Only 1- and 2-egg clutches have been reported to date. At Tikal we documented four 2-egg and two 1-egg clutches, for a mean clutch of 1.7 (n = 6). The same mean clutch size (1.7 ± 0.49, n = 18) was obtained by pooling all published data we could find on clutches or broods up to two weeks old, for all subspecies (data from Hewitt 1937; Sutton 1954; Haverschmidt 1964b; Wetmore 1965; and Mader 1981). Mean egg dimensions at Tikal were 49.6 ± 1.2 mm by 38.0 ± 0.8 mm (n = 6), and mean egg mass was 35 ± 3 g (n = 6). Using Hoyt's (1979) formula, estimated fresh egg mass was 39.2 g (n = 6), or 7.9% of mean adult female mass. Hence the mean clutch of 1.7 amounts to 13.5% of adult female mass.

Nesting Phenology

Crane Hawk breeding activity began in March and early April, when we observed courtship behavior including aerial displays, vocalizations, copulation, and nest building. Eggs were laid from late March to mid-April and hatched from late April to mid-May, with chicks fledging in mid to late June. Egg laying coincided with the driest time of the year, and chicks fledged early in the rainy season. Although the egg-laying period has been described as "prolonged" based on four Mexican nests with laying dates from about 1 April to 1 June (Brown and Amadon 1968), the laying period at Tikal was no more protracted than in other raptors there.

Length of Incubation, Nestling, and Post-fledging Dependency Periods

The incubation period was 39 days or longer at one nest, and the nestling period averaged 36.8 ± 4.3 days (n = 4). Crane Hawks had slightly longer incubation and nestling periods than do similarly sized temperate zone raptors such as the Cooper's Hawk *(Accipiter cooperii)* and Broad-winged Hawk *(Buteo platypterus*: Palmer 1988).

Nestlings first ventured out of the nest at the ages of 26–36 days, and fledged at 32–44 days, 2–11 days after leaving the nest for the first time (n = 4). Fledglings remained within 500 m of the nest for 11–21 days (n = 4), and three fledglings dispersed beyond this distance 46–52 days after hatching (though likely were not yet independent; see below). At one site, the fledgling was still being provisioned near the nest at 62 days. This was a case in which adults were obliged to continue visiting the nest to feed this fledgling's botfly-infested sibling, which remained in the nest an unusually long time.

We ascertained the post-fledging dependency period of one radio-tagged male. This bird fledged at 38 days and was periodically relocated in the company of an adult until reaching independence at about the age of 166 days—some four months after fledging. After this, the fledgling dispersed southward and frequented an area 8.8 km south of its natal nest; here it was eaten by some predator by mid-January, about 230 days after hatching.

While our sample is small, it appears that juveniles often abandon the nest vicinity within a week or two of fledging, but may be fed by adults elsewhere for several months. Still, there is no evidence that the dependency period is long enough to prevent adults from nesting in consecutive years.

VOCALIZATIONS

We recognized three distinct Crane Hawk vocalizations, and named these the "conspecific recognition call," the "begging" or "hunger call," and the "defensive" or "agonistic call." A fourth vocalization, termed the "proximity call," was occasionally given; this appeared to be a variation or composite of two or more of the other calls.

The conspecific recognition call can be described as a siren-like *WHEEe-oo* repeated a few to many times in succession, sometimes growing in intensity to a final crescendo. This call was not loud and probably carried only several hundred meters. Early in the breeding season, we usually heard this call during early to mid morning. Like the early-dawn vocalizations of Collared Forest Falcons *(Micrastur semitorquatus)*, Laughing Falcons *(Herpetotheres cachinnans)*, Blue-crowned Motmots *(Momotus momota)*, and Great Curassows *(Crax rubra)*, this Crane Hawk vocalization makes use of low-frequency notes, which are effective for long-range communication in forests, because they experience less attenuation in obstructed habitats than do higher-frequency calls (Wiley and Richards 1982). The normally still air at dawn is ideal for singing (i.e., less ambient noise and signal degradation; Henwood and Fabrick 1979), which may be a reason why all these species concentrate their vocalizations in the early hours of the day.

Crane Hawks seemed to use the conspecific recognition call for mate attraction and coordination of activities between pair members. During the courtship and nest-building period, this call seemed to attract the mate's attention and to precipitate nest building and copulation. During the nesting period, both males and females gave this call when they arrived near the nest with prey. Most often it was the male who gave this call, apparently to alert the nest-attending female of his presence. On a few occasions we heard this call at dusk.

The begging or hunger vocalization can be described as a quick, single-note call, descending in tone but rising in amplitude: *p-weeOO*. This call at times resembled a cat's *meow*, and was strikingly similar to the Gray Hawk's *(Buteo nitidus)* defensive vocalization, and to the call of the Rufous Piha *(Lipaugus unirufus)*. This call was uttered singly or repeated about every 10 seconds for short periods or up to an hour and a half and was used almost exclusively by the female while attending the nest and waiting for the male to bring prey. Females gave this call anywhere in the nest area, but usually from the nest or the prey exchange tree. In addition, this call sometimes was given by the female as she mantled over prey received from the male. On a few occasions we heard males use this call while attending the nest and awaiting the female's return. This was the first call uttered by nestlings, in which it also appeared to function in food

begging; they often voiced this call, especially during lapses between feedings.

The defensive or agonistic vocalization can be described as a sharp, nasal, drawn-out scream: *MEEeahh*. This call is very similar to that of the Roadside Hawk, though with less of a whistled quality and less drawn out. It was usually uttered when Crane Hawks appeared to feel threatened. It was given most often by Crane Hawks at the nest when other raptors or potential predators (e.g., toucans, monkeys, or humans) were nearby, and less often by Crane Hawks when approached far from the nest. In the nest area, this call was given when intruders were as far away as 100 m, or was withheld until intruders were only meters from the nest.

The call we termed the "proximity call" usually was given when pair members were touching or very close to one another, for example, during nest building, nest attendance switches, prey exchanges, and copulation. This call was variable and seemed to combine the conspecific recognition and begging calls, along with a fair amount of chatter, apparently produced by beak clacking. Although this call was recognizable because it consistently contained some element of the Crane Hawk's other calls, it rarely was exactly the same from one occasion to another, owing to the hawk's apparent agitation and the erratic nature of the associated chatter.

BEHAVIOR OF ADULTS

Pre-laying and Laying Period

The onset of breeding was signaled by courtship displays (see later), copulation, and nest building. Copulation sometimes followed aerial displays and usually occurred high (20–30 m) in emergent trees near the future nest site; 7 of 12 copulations were within 50 m of an eventual nest site. Four of 12 copulations were accompanied by the proximity vocalization. Copulation often occurred during nest-building activities. Crane Hawk mating behavior at Tikal, including pre-copulation aerial displays and the use of a variety of vocalizations, resembled that previously described (Brown and Amadon 1968). Although courtship feeding was reported by Ouellet (1991), we did not witness it.

All nest building was observed in the morning. Both sexes took part in building and were quite vocal during construction. While adults arranged sticks in the nest, we often heard them softly voicing the proximity call. Bouts of nest building usually ended with the pair flying off together through the forest.

Incubation Period

We quantified incubation behavior of four pairs in 1994 and two in 1995. During the incubation phase, adults either incubated or stood in the nest during the majority of dawn-to-dusk observation time (mean = 89.7% of time, median = 94.9%, n = 17 observation sessions totaling 216.6 hours). Both sexes participated in incubation and nest attendance, which concurs with previous reports (Haverschmidt 1964b). Females provided 87% of incubation and nest attendance, and males 13%. When no hawk was on the nest, an adult, usually the female, was generally nearby (mean = 90.5% of the time within 50 m of the nest, median 99.2%, n = 17). During most of the nest observation time (83.3%, n = 17), males were away from nests, presumably hunting, while females remained vigilant at nests.

During 17 observation sessions, male Crane Hawks usually relieved females on the nest once a day (mean daily nest attendance switches = 1.2 ± 0.75, range = 0–3, median = 1). Males usually relieved females during the morning, generally when they delivered prey to the female (81%, n = 21); time of relief averaged 0943 ± 192 minutes (range = 0550–1642, median = 0836, n = 21). Males normally spelled females for an hour at a time (mean = 73 ± 83 minutes, range = 19–397 minutes, median = 60 minutes, n = 21) while females perched or ate in nearby trees, generally within 50 m of the nest.

Males provided 95% of prey deliveries during incubation (n = 22), normally bringing prey to females once a day (mean = 1.3 ± 0.77 prey deliveries per day-long observation session, range = 0–3, median = 1, n = 17). Most prey deliveries occurred in the morning (mean = 0935 CST ± 205 minutes, range = 0550–1642, median = 0837, n = 22). Except for one occasion when the female returned with prey she had apparently captured herself (her only meal of the day), afternoon prey deliveries were either the second or third of the day (80%, n = 5).

Males arriving with prey near the nest most often gave the conspecific recognition call (77%, n = 22), though in 14% of cases some other vocalization was used (begging, agonistic, or proximity calls), and in 9% of cases (n = 22) they were silent. The female would immediately fly toward the male, who consistently used the same one or two trees for prey exchanges. Upon reaching the male, the female would take the prey, talon to talon, with or without some form of vocalization (begging call 41%, agonistic call 18%, proximity vocalization 5%, the rest silent, n = 22).

During the incubation phase, Crane Hawks periodically added to the nest small twigs with or without green leaves (mean = 0.6 ± 0.6 twig/day-long observation session, range = 0–2, median = 1, n = 17). Females brought 9 out of 10 twigs to the nest, males the remainder. Twigs were usually brought in the morning (n = 10). Half had green leaves attached, and half did not (n = 10). Of 10 deliveries of nesting material, 4 occurred as part of a switch in nest attendance.

Nestling Period

During the nestling period, we quantified adult behavior at three nests in 1994 and two in 1995. In the first week after hatching, an adult, usually the female, was in the nest or nest tree at all times (98:2 female-to-male ratio). We observed males near the nest more often during this period than at any other time of the nestling period. Still, they were away from the nest most of the time and provided all prey consumed by females and nestlings.

By the second week, females began spending less time in the nest and nest tree and more time perched 25–50 m distant. At this time, males began delivering prey to the nests and feeding chicks. Females, however, were usually nearby and would hastily usurp feeding duties from the male. Female presence on and near the nest declined steadily from soon after hatching until fledging, as males began feeding nestlings more often and females began hunting and bringing in prey. During the fourth and fifth weeks, males regularly dropped off prey at the nest, leaving the nestlings to feed themselves.

In 30 dawn-to-dusk observations during the nestling period, we recorded 120 prey deliveries. Prey deliveries ranged from 2 to 8/day-long observation session (mean = 4.0 ± 1.3, n = 30 sessions). With the exception of the fourth week, prey delivery rates were fairly constant from week to week. During the fourth week, deliveries were more frequent, as a result of many frogs being delivered on rainy days. Prey deliveries occurred during every daylight hour, but they were most frequent from 1000 to 1100 and least frequent from 1400 to 1500—usually the hottest time of day. Nest maintenance during the nestling period involved the removal of prey remnants (bones and viscera, n = 10), and occasional addition of nesting material (mean = 0.1 ± 0.35 twig/day-long observation session, n = 30).

Adult behavior and vocalizations during prey exchanges were similar to those during incubation. The arrival of the male with prey was usually accompanied by vocalizations (64%, n = 120), more than a fourth of these cases involving two or more call types (29%, n = 77). The begging call was most often used (46%, n = 77 prey deliveries), the conspecific recognition and agonistic calls less often (39 and 34%), and the proximity vocalization occasionally (14%). The actual moment of prey exchange between adults was nearly always accompanied by vocalizations (89%, n = 35).

Post-fledging Dependency Period

After young fledged, adults provisioned them in and near the nest tree for two to three weeks. After this time we were unable to locate most fledglings, and at four nests, fledglings were known to have dispersed from the nest vicinity by this time. However, we did observe one radiotagged fledgling accompanying and being provisioned by an adult (usually the male) over a four-month period after fledging. During the first two weeks after fledging, this young male was found several times, on average 295 ± 130 m (n = 6) from the nest, always changing perches frequently, and at times following the adult male, who was actively hunting. In one case the fledgling received prey from the adult female. After dispersing from the natal area 14 days after fledging, this young bird remained elusive until 11 weeks after fledging, when we found him 1 km away; he remained in this area for 10 weeks.

During this period we observed the juvenile five times, always actively changing perches and usually begging (four of five times). An adult (usually the male) was nearby during four out of five observations, and three times we observed the juvenile following close behind the male as he hunted.

At the age of 151 days (113 days after fledging), we saw the juvenile probing his head and feet into the leaves of a small *Sabal* palm while following the adult, who was also using the same manner of searching. At 156 days (118 days after fledging), the fledgling was seen with prey it apparently had captured on its own, and appeared to have reached independence. It was not seen in the company of an adult again and rapidly dispersed to another area, 8.8 km from its natal nest. Here it was eaten at 215–230 days (177–192 days after fledging), probably by some other raptor, in a bromeliad high in a tree.

A radio-tagged female that successfully fledged one chick (Chikintikal 1994) was never observed provisioning or accompanying the juvenile hawk during the subsequent 15 months of radio tracking. We do not know if this fledgling survived beyond the age of 62 days, but if it did, the female played little or no role in providing for it during the post-fledging period. The adult male here was not radio-tagged.

BEHAVIOR AND DEVELOPMENT OF NESTLINGS

We tracked the development of six nestlings from hatching to the time of fledging or nest failure. At hatching, whitish-buff natal down covered most of the nestlings' bodies. Nestlings hatched with black cere and talons, which are retained in adults. The beak was black, with a white egg tooth that disappeared within two weeks (Plate 9.5). In the first week after hatching, nestlings were active only during feeding and were brooded by females most of the time. By the end of the first week, nestlings were quite active and moved to the edge of the nest to defecate, and they often voiced their loud, high-pitched begging calls.

By two weeks after hatching, prejuvenal molt was underway, and wing and tail feathers emerging. By three weeks, contour feathers were abundant on the back, flanks, and upper legs, and feathers were emerging on the nape and crown. By the end of the second week or early in the third, nestlings began standing, preening, and flapping their wings. These behaviors were considerably more frequent during the fourth week, and older nestlings began footing the nests and nearby vines as if seizing prey.

By five weeks, natal down had mostly disappeared and contour feathers nearly covered the body. Wing and tail feathers were more than half grown and the buff-colored upper- and under-tail coverts of the juvenal plumage were strikingly evident. Also in the fifth week, nestlings flapped vigorously, gaining lift while holding onto the nest, and all nestlings footed and seized nest material

with their talons. Although one nestling ventured out of the nest 1 m during the fourth week, the remaining nestlings first departed the nest in their fifth and sixth weeks.

By six weeks after hatching, few natal down feathers remained and young were near fledging. By this age, nestlings probed with their legs and made grabbing motions behind bark and among epiphytes in the nest trees, and one nestling used its legs and head to search holes in a rotten branch. Depending on nest tree architecture, nestlings moved about the crown either via 5–10 m flights or by walking and hopping from limb to limb.

Leg color in six adult males at Tikal was dark reddish orange, and in five females a lighter orange. Chicks at Tikal hatched with pink legs that gradually changed to yellowish orange within two weeks and to pale orange by fledging time. Likewise, Dickey and van Rossem (1938) state that the legs are yellowish orange in immature birds and orange in adults.

Eye color in Tikal adults ranged from scarlet-orange to crimson. Possibly this is related to age, as in the Cooper's Hawk (Rosenfield and Bielefeldt 1997). Dickey and van Rossem (1938) state that the iris is orange in immature birds and red in adults. Sutton (1951) commented that an immature Black Crane Hawk he collected had red eyes "but less intensely so than those of the fully adult specimen," and Storer (1961) noted that an immature he collected in December, that is, during its first winter, had yellow-gold eyes. Six chicks at Tikal hatched with dark brown eyes which lightened to a yellowish orange hue within three weeks and progressively grew darker orange from the pupil outward during the nestling period; presumably these continued to become redder after fledging.

SPACING MECHANISMS AND POPULATION DENSITY

Territorial Behavior and Displays

Pair members sometimes soared together in tight circles, flapping vigorously in short bursts to gain altitude and then abruptly dropping. This resulted in an undulating, aerial display, which we observed on five occasions. Similar, "inverted V" flight displays have been interpreted previously as courtship or territorial displays (Friedmann and Smith 1955; Brown and Amadon 1968; Mader 1981). Except for the display flights just mentioned, we never saw Crane Hawks indulge in prolonged soaring and rarely saw them fly more than 5–10 m above the canopy. Crane Hawks typically moved about below the canopy with short flights when actively searching for prey. Longer flights above the canopy were characterized by flap-gliding and were strongly directional and slow. Therefore, we were confident that the soaring aerial displays we observed were directly related to courtship or territorial behavior. We observed interactions between nesting Crane Hawks and other adult Crane Hawks on two occasions. In both cases the nesting hawks intercepted conspecifics 70–200 m from the nest and had a long (> 5 minute) exchange of agonistic vocalizations. We witnessed no chasing or stooping during these interactions. It appeared that the nest vicinity, rather than the entire home range, was vocally defended against conspecifics.

Unlike most diurnal raptors at Tikal, Crane Hawks did not routinely soar above their home ranges. The aerial displays performed early in the breeding season, while seemingly courtship displays, may have assisted pairs in spacing themselves across the landscape. However, frequent perching in treetops and use of the siren-like conspecific recognition call were likely the main spacing mechanisms used by these birds at Tikal.

Nest Defense and Interspecific Interactions

Crane Hawks did not often defend their territories aggressively. Nevertheless, some raptors and other potential nest predators were occasionally met with vocal harassment or were attacked when within 50 m of the nest. During incubation and nestling periods, we recorded Crane Hawk behavioral responses to raptors and other potential nest predators within 500 m of the nest. We noted most responses during the nestling period (66%, n = 90). In most cases, Crane Hawks gave no apparent reaction to raptors and other potential nest predators (54%, n = 90), which were usually some distance from the nest (mean = 85 ± 75 m, range = 2–400 m, n = 49). Sometimes Crane Hawks would assume an alert posture (19%, n = 90), staring intently at the intruder or standing ready in the nest, raising the hackles at the back of the head. This response was usually given when the interloper was fairly near the nest tree (mean = 70 ± 80 m, range = 5–300 m, n = 17).

Most aggressive responses occurred near the nest (mean = 30 ± 25 m, range = 1–100 m, n = 24). Aggressive responses (24 total) ranged from swooping near and chasing predators away (25%), and attacks with contact (17%), to prolonged vocal harassment (58%). The most aggressive nest defense was directed at Black-handed Spider Monkeys (*Ateles geoffroyi*), Ornate Hawk-eagles, White Hawks (*Leucopternis albicollis*), Bicolored Hawks, Keel-billed Toucans (*Ramphastos sulfuratus*), and Brown Jays (*Cyanocorax morio*). Nevertheless, many of these species were ignored when more than 50 m from the nest. During our visits to nests, Crane Hawks usually tolerated human observers in the vicinity. However, on several occasions, the hawks protested loudly when we were 10–100 m away. When we climbed to nests, one or both adults would perch nearby, vocalizing frequently. In several cases, Crane Hawks made daring passes within 1 m of the climber.

Constancy of Territory Occupancy, Use of Alternate Nest Sites

Crane Hawk territories were consistent in location from year to year, and once located, no territory was found unoccupied in a subsequent year. For 10 territories we know the occupancy status for 3–4 consecutive years; each was occupied by a pair during each year it was checked. In some cases, our repeated observations revealed no nesting effort, though a pair was present. In such cases, we believed the pair either failed very early in a nesting attempt or, more likely, did not lay eggs. In six territories, we documented nesting attempts in two or more years. No pair ever used the same nest or nest tree a second time. Alternate nest sites used within a given territory averaged 340 ± 388 m apart (range = 38–1042 m, n = 6) and were usually within the same stand of Sabal Forest in a swale.

Home Range Estimates

We radio-tagged an adult male and female during the 1994 breeding season and took locational data during 9 and 16 months, respectively. We radio-tagged the Arroyo Negro male on 9 June 1994 during the fifth week of the nestling period and relocated him every 10 days on average (SD = 11 days, n = 27) from 10 June 1994 to 22 February 1995, recording 27 relocations—7 by triangulation and 20 by visual contact. Only this male's first relocation occurred before the nesting failed on 12 June; hence the resulting home range estimate was for a non-breeding-season male without dependent offspring. On 22 February 1995, we found the male's radio on the ground with the Teflon harness intact. Searching a 70 m radius, we found no evidence of the male. That a female in this territory mated with a different male later in the year suggests that this male perished when the radio became detached. The estimated home range for this male was 755 ha using the 100% Minimum Convex Polygon (MCP) method and 392 ha using the 85% Harmonic Mean (HM) method (Table 9.2). The sample size appeared adequate, as the estimate no longer grew with additional data points. We favor the 755 ha MCP estimate.

We captured the Chikintikal female on 13 April 1994 during the second week of incubation and relocated her with radio-tracking every 8 days (mean = 8.2 ± 8.0 days, n = 55) from 13 April 1994 to 6 July 1995. With 71 relocations on 56 days (all with visual contact), this period spanned most of two breeding seasons and the intervening nonbreeding season. On eight days we recorded multiple relocations for the female (mean = 2.8 ± 1.2 relocations/day, n = 8). During the incubation phase, the female spent most of her time in or near the nest, and not until the third week of the nestling period did she leave the nest vicinity regularly for extended periods.

Eleven locations were obtained during the 1994 breeding season, indicating that the female used at least 140 ha (MCP). During the first nonbreeding season (August 1994–February 1995), we located the female 24 times. Her estimated home range during this interval was 506 ha (100% MCP method) or 270 ha (85% HM method; Table 9.2). Due to small sample size, these data probably underestimate actual space use. As the next breeding season approached (January–February 1995), the female began using areas farther west, and in early March we observed her carrying twigs in the area where we later found her new nest. Localizations beginning with this incident were considered to constitute her 1995 breeding home range. Based on 18 locations, this female's estimated 1995 breeding home range was 347–452 ha (100% MCP) or 226–284 ha (85% HM; Table 9.2). We favor the 452 ha estimate, though it too is likely an underestimate. The nesting failed in late May, during incubation, but we continued relocating the female until mid-July. The estimated home range after nest failure (18 locations) was 343 ha (MCP) or 177 ha (85% HM; Table 9.2); we favor the 343 ha estimate, again likely an underestimate.

The cumulative home range estimated for this female during the entire 15-month period (71 relocations) ranged from 1038–1045 ha (100% MCP) to 542–575 ha (85% HM; Table 9.2). Despite a shift of nest sites during this 15-month period, the female used much the same area throughout. Hence, we believe that the 15-month home range estimate does not seriously overstate her annual home range. It appears safe to base density estimates on the 7.5 km² used by an adult male during an eight-month period, with perhaps up to 10 km² also used at times.

Inter-nest Spacing and Density of Territorial Pairs

We determined the distance between three neighboring nest sites each in 1993 and 1994, and four in 1995 (Table 9.3, Fig. 9.2). Inter-nest distance was quite comparable between years, averaging 2607 ± 446 m (n = 3 years). Resulting estimates of space per pair (mean of three years) were 6.80 km² for the "square" method, 6.18 km² for the Maximum Packed Nest Density (MPND) method, and 6.25 km² for the polygon method (Table 9.3). These estimates are in good agreement with the home range estimates just given. Hence, we estimate an average of 6.25 km²/territorial pair of Crane Hawks, for a density of 16 pairs/100 km², or roughly 82 pairs in Tikal National Park (based on approximately 510 km² of non–Scrub Swamp Forest).

In Venezuela, Mader (1981) found three pairs of Crane Hawks in an area of approximately 7 km² of mixed palm savanna and forest, or about one pair per 2.33 km². Thiollay (1989a) made no density estimate for this species in French Guiana, where he regarded it a forest-edge or woodland species rarely if ever encountered in primary forest (Thiollay 1984, 1985; Jullien and Thiollay 1996). In southeastern Peru, Robinson (1994) also found this species uncommon, sighted no more than five times during several years of research.

DEMOGRAPHY

Age of First Breeding, Attainment of Adult Plumage

Dickey and van Rossem (1938) stated that adult plumage is attained in the second fall, along with the red iris and orange leg color of adults, which replace the orange iris and yellowish orange leg color of immature birds. Presumably these birds may begin nesting at the age of two or three years, but data are lacking.

Frequency of Nesting, Percentage of Pairs Nesting

We documented 12 nesting efforts in 14 territorial pair-years monitored. In the 2 remaining cases we found nests and observed nest building, aerial displays, and copulation, but we were unable to confirm that eggs were laid. We believe these two nesting efforts were abandoned before egg laying, and subsequent observations gave no indication that these pairs later nested. Hence pairs made nesting attempts (i.e., laid eggs) in 85.7% of territorial pair-years. Moreover, pairs at least sometimes nested in consecutive years. At both the Curva Peligrosa and Chikintikal nests, one or more young fledged in 1994 and the pair again nested in 1995; at Chikintikal, the same banded adult female nested in both years. In neither case was it known whether the 1994 fledglings remained alive in 1995, but our clear impression was that pairs

Table 9.2 Home range estimates (in hectares) for breeding and nonbreeding adult Crane Hawks at Tikal, 1994 and 1995[a]

Individual	Dates radio-tracked	Period	100% Minimum Convex Polygon (ha)		85% Harmonic Mean (ha)	
			A	R	A	R
Arroyo Negro adult male	10 Jun 1994–22 Feb 1995	Nonbreeding	755 (27)	755 (27)	392	392
Chikintikal adult female	13 Apr 1994–6 Jul 1995	Overall	1045 (71)	1038 (56)	575	542
Note: too few to serve as a formal home range estimate	13 Apr–13 Jul 1994	Breeding	139 (11)	139 (10)	35	0.8
—	10 Aug 1994–23 Feb 1995	Nonbreeding	506 (24)	506 (24)	270	270
—	7 Mar–26 May 1995	Breeding	452 (18)	347 (16)	226	284
—	23 Jun–6 Jul 1995	Nonbreeding	343 (18)	Too few for home range estimate	177	Too few for home range estimate

[a] Estimates are based on all points (A), or on all visual relocations and triangulations, and on one randomly chosen relocation from each day with multiple relocations (R). Number of location points per analysis is given in parentheses and is the same in each category for 85% Harmonic Mean.

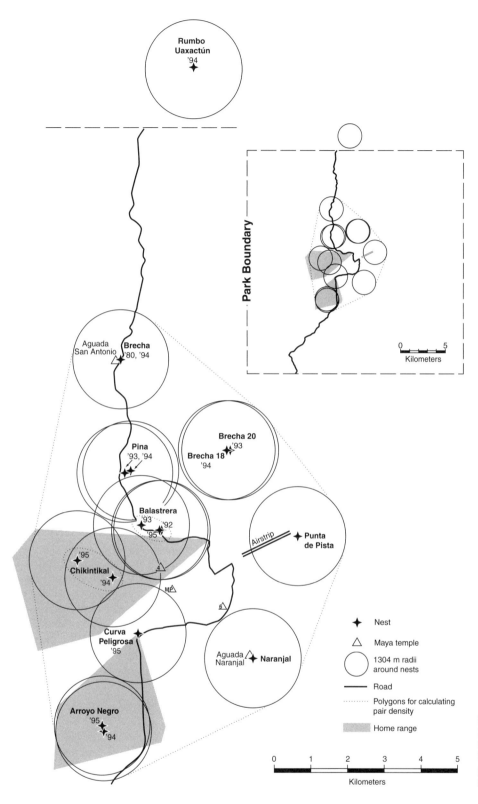

Nest

△ Maya temple

◯ 1304 m radii around nests

── Road

⋯⋯ Polygons for calculating pair density

▨ Home range

9.2 Nest site location of Crane Hawks in Tikal National Park, Guatemala, showing 1304-m buffers around nests, and polygons used in estimating nesting density.

routinely nested in consecutive years, probably including years when the prior year's progeny remained alive.

Productivity and Nest Success

For the 12 nesting efforts monitored, results were as follows. At 6 nests where we verified clutch size, 10 eggs were laid, for a mean of 1.67 eggs per clutch. Eight (80%) of 10 eggs hatched, and 50% (4/8) of hatchlings fledged. At 4 of the 12 nests, we did not verify clutch size but did verify brood size: one nestling at each of two nests, and two nestlings at each of two other nests. Taking these 10 nests together, 43% (6/14) of the nestlings fledged. At one other nest we observed incubation but never verified

Table 9.3 Annual and cumulative breeding density estimates for Crane Hawks in Tikal National Park, 1993–95

Year	Sample size (inter-nest distances)	Mean distance (m) between neighboring nests (mean ± 1 SD)	Method		
			Polygon[a]	Square[a]	MPND[a]
1993	3	2429 ± 846	5.47[b]	5.90	5.37
1994	3	2840 ± 100	6.88[b]	8.06	7.34
1995	4	2564 ± 160	6.63[b]	6.57	5.98
Overall	10	2607 ± 446	6.25	6.80	6.18

[a] See Chapter 1 for explanations of polygon, square, and MPND methods; values are number of square kilometers per territorial pair.

[b] 1993: 3 pairs in 16.4 km²; 1994: 6 pairs in 41.3 km²; 1995: 4 pairs in 26.5 km².

clutch size before the nest failed. Finally, we found one nesting attempt by locating two recently fledged young begging near a nest (see Table 9.4).

Eight nestlings fledged in 12 nesting attempts (though one of these attempts was discovered after fledging), giving a mean productivity of 0.67 fledging per nesting attempt and 0.57 fledgling per territorial pair-year (Table 9.5). If we assume that nests with unknown clutch size contained the mean clutch of 1.7 eggs (or 2 eggs, in cases of two observed chicks), then 21.1 eggs produced 16 nestlings (75.8% hatching success) and 8 fledglings (50% of nestlings fledged). Overall, 5 (41.7%) of 12 nests produced one or more fledglings.

Sources of Nest Failure and Mortality

Of 12 nesting attempts, 7 failed completely: 3 during the egg stage and 4 during the nestling stage (Table 9.4). Three nests were completely successful, fledging two young each (Table 9.4). The remaining two nests were partially successful, each fledging one of two chicks (Table 9.4). We treat the seven complete and two partial nest failures separately. For the seven complete nest failures, predation and wind were the most frequent causes. For three of these nests that failed during the egg stage, in one case the cause was unknown, in another case wind was suspected, and in the third case wind tipped the nest, after which adults ceased incubating and the egg was later eaten by a Brown Jay. For the four nests that failed during the nestling stage, two cases were definitely due to predation and a third was very likely as well; the fourth case was almost certainly due to wind (Table 9.4). The two cases of partial failure (one of two chicks lost) involved apparent starvation in one case, and heavy botfly infestation followed by starvation or predation in the other case. Details of the cases are as follows.

At the 1993 Brecha 20 nest, two nestlings were killed and eaten by a Black-and-white Hawk-eagle *(Spizaetus melanoleucus)* while the female Crane Hawk perched 25 m away, voicing the agonistic call. At the 1994 Arroyo Negro nest, two nestlings were killed during the fifth week after hatching; both had been partly devoured in the nest, and a trail of feathers led down and away from the nest tree. We had observed a Tayra *(Eira barbara)* approaching within 10 m of this nest tree on two previous occasions. These mustelids have been seen climbing trees and preying on nestlings of other birds, including raptors, in Tikal (J. Madrid, D. Whitacre, pers. comm.).

The 1995 Chikintikal nest was built on a mass of vines that swayed in the wind. After a particularly strong storm, the nest had begun to fall and the adults had all but abandoned incubation and nest attendance duties. Within several days, we observed a Brown Jay enter the nest and fly away with the egg while the adults were absent. We believe wind played a role in some of the other nest failures as well. Because Crane Hawks usually nested in vine tangles high in tall emergent trees, storms and strong winds led to nests being shaken and dislodged. On one occasion shortly after several strong windstorms, park personnel found a nestling that had fallen from its nest and broken a wing. Increased vulnerability to wind may be a price paid by raptors that nest high in isolated or emergent trees.

Nestling mortalities that were not associated with total nest failures were apparently caused by starvation and ectoparasite infestation (Table 9.4). At the 1994 Brecha 80 nest, the smaller, younger nestling disappeared during the second week after hatching, and we found no evidence of it within 50 m of the nest tree. When we last climbed to the nest prior to the nestling's disappearance, the nestling had been very small, inactive, and too weak to hold up its head; we did not observe any parasites on this chick, but it appeared to be starving.

Parasites and Pests

At the 1994 Chikintikal nest, the younger nestling was heavily infested with botfly larvae (possibly *Philornis*) on its head and body. The nestling's health was severely affected as evidenced by weight loss and the poor condition of its flight feathers. Several botfly larvae in the chick's intertarsal joints prevented it from standing and moving effectively in the nest. By the sixth week after hatching, all of the nestling's tail feathers were missing, and most of its remiges broken. When we returned to check on it during its thirteenth week, the nestling was gone; it almost certainly died before fledging.

Table 9.4 Reproductive success and probable causes of failure in Crane Hawk nesting attempts in Tikal National Park, 1993–95

Year	Nest site	Laid eggs	Clutch size	Number of nestlings	Number of fledglings	Cause of failure	Level of certainty about cause of failure[a]	Date of failure	Nesting stage at failure
1993	Bajada de la Pina	y	1	0	0	[b]	0	18 May	3 weeks[c]
	Balastrera	y	2	2	2	—	—	—	—
	Brecha 20	y	[2][d]	2	0	Predation[e]	3	14 Jun	2–3 weeks[f]
1994	Arroyo Negro	y	2	2	0	Predation[g]	3	12 Jun	5 weeks[f]
	Bajada de la Pina	n	—	—	—	Wind	2	13 Mar	1 week[c]
	Brecha 20	y	[1.7][d]	0	0	Wind	1	28 Apr	3 weeks[c]
	Brecha 80	y	2	2	1	Malnutrition[h]	2	2 Jun	2 weeks[f]
	Chikintikal	y	2	2	1	Ectoparasites[h]	3	8 Jul	7 weeks[f]
	Curva Peligrosa	y	[2][d]	[2][d]	2[i]	—	—	—	—
	Rumbo Uaxactún	y	[1.7][d]	1	0	Wind	2	23 May	3–4 weeks[f]
1995	Arroyo Negro	y	[1.7][d]	1	0	Predation	2	29 Apr	1–2 weeks[f]
	Balastrera	n	—	—	0	Disturbance[j]	1	19 Mar	1 week[c]
	Chikintikal	y	1	0	0	Wind followed by predation[k]	3	14 Jun	4 weeks[c]
	Curva Peligrosa	y	[2][d]	2	2	—	—	—	—

[a] Level of certainty: 0 = zero; 1 = fair indication; 2 = strong indication; 3 = certain.
[b] Unknown.
[c] After nest found.
[d] Brackets indicate values assigned as most likely for some analyses.
[e] Black-and-white Hawk-eagle (*Spizaetus melanoleucus*).
[f] After hatching.
[g] Possibly Tayra (*Eira barbara*).
[h] Cases in parentheses refer to cases of partial nest failure.
[i] Found fledglings begging near nest.
[j] Abandoned after monkeys climbed within 0.5 m of nest, or in nest.
[k] Nest was falling because of wind and adults perched near nest but didn't incubate; subsequently, a Brown Jay took the egg.

Table 9.5 Reproductive performance of Crane Hawks in Tikal National Park, 1993–95

Year	Territories monitored (number)	Nesting attempts[a] (number)	Nesting incidence[b] (%)	Nestlings (and per nesting attempt) (number)	Fledglings (and per nesting attempt) (number)	Proportion of eggs hatched[c]	Proportion of eggs hatched[d]	Proportion of eggs producing a fledgling[c]	Proportion of eggs producing a fledgling[d]
1993	3	3	100.0	4 (1.33)	2 (0.67)	2/3 (0.67%)	4/5 (0.80%)	2/3 (0.67%)	2/5 (0.40%)
1994	7	6	85.7	9[e] (1.5)	4[e] (0.67)	6/6 (1.00%)	9/11.4 (0.79%)	2/6 (0.33%)	4/11.4 (0.35%)
1995	4	3	75.0	3 (1.00)	2 (0.67)	0/1 (0%)	3/4.7 (0.64%)	0/1 (0%)	2/4.7 (0.43%)
Overall[f]	**14**	**12**	**85.7**	**16 (1.33)**	**8 (0.67)**	**8/10 (0.8%)**	**16/21.1 (0.76%)**	**4/10 (0.40%)**	**8/21.1 (0.38%)**

[a] Cases with one or more eggs laid.
[b] Percentage of territorial pair-years in which one or more eggs was laid.
[c] Using only fully known clutch and brood sizes.
[d] Includes estimated clutch and brood sizes in some cases; for unknown clutches, assumes mean of 1.7 eggs; for unknown clutches in cases with two nestlings, assumes clutch of 2; total eggs estimated at 21.1.
[e] Includes brood of two fledglings discovered after fledging.
[f] Boldface numbers are considered the most reliable estimates.

CONSERVATION

It is safe to say that deforestation is the main threat to Crane Hawk populations. This species has been described as declining in Mexico (Ramos 1986) and threatened with extirpation in El Salvador (Komar 1998), due to deforestation in both cases. These birds do at times occur in gallery forest, coffee plantations, scattered woodlands, and other partially open habitats, and thus are likely more resilient to deforestation than are species tied exclusively to mature forest interior. As this is one raptor often reported to use mangroves, the rapid loss of mangroves underway in Latin America (Jackson and D'Croz 1997) may have a proportionally large impact on this species.

Although Crane Hawks are often found in the vicinity of water, they have otherwise been described as tolerant and adaptable in habitat use (Brown and Amadon 1968; Bierregaard 1994). Nevertheless, during our several years of fieldwork in and near Tikal National Park, we made few observations of Crane Hawks in the modified environments outside the park, and our impression was that they were uncommon there. Detection rates in our point counts comparing modified and pristine habitats were too low to allow comparisons.

The tendency shown by Crane Hawks at Tikal to nest mainly in Sabal and similar forest types has implications for conservation efforts. These most-used nesting habitats are restricted to a narrow segment of the topographic gradient near hill bases, where level areas with relatively deep soil occur. These areas are also preferred by farmers and are often among the first to be cleared for cultivation. As a result, most forest remnants in the farming landscape are on hilltops and steep slopes and are mainly of Upland Forest types little used by Crane Hawks for nesting. This is but one example of the importance of preserving representative segments of the entire complement of regional and local habitat types, if we are to maximize the conservation value of human-dominated landscapes.

Although Crane Hawks usually nested in trees whose crowns were isolated from surrounding forest canopy, the more severely isolated trees left standing in the farming landscape may not meet nest site criteria. Besides being unprotected from the elements, solitary trees may not provide adequate concealment to minimize nest predation. In addition, raptors nesting in isolated trees in farming landscapes are conspicuous targets for human persecution (see Chapters 12, 18).

HIGHLIGHTS

Crane Hawks exhibited foraging habits that were uncommon among raptors at Tikal. Their frequent use of the search-and-probe foraging technique and the predominance of nocturnal animals in their diets suggest that Crane Hawks encountered many prey through a process of inspecting potential refugia. As pointed out by Jehl (1968), such behavior may lend access to resources largely unavailable to other raptors. Anatomical specializations, along with specialized search behavior, enable Crane Hawks to exploit a specific set of structural features in the environment. The prey types available in these sites belong to a great many taxa, from arthropods to mammals, the features they have in common being their nocturnality and their tendency to hide during the day in cavities, behind bark, within epiphytes, or in similar sites. Hence, the Crane Hawk is a predator quite specialized both anatomically and behaviorally at exploiting a functionally specific but taxonomically diverse array of prey types.

The larger Black Hawk-eagle *(Spizaetus tyrannus)* and Crested Eagle *(Morphnus guianensis)* likewise preyed heavily on nocturnal prey species that were captured throughout the day, as did the Great Black Hawk to a lesser extent. Unlike the Crane Hawk, these other raptors do not show conspicuous morphological adaptations that facilitate capture of nocturnal prey in their daytime retreats. Conceivably, these other raptors are restricted to capture of prey in more accessible sites than many of those exploited by Crane Hawks. Still, these other species provide an interesting example of dietary convergence through behavior in the absence of obvious morphological convergence—providing a reminder that behavioral plasticity may often precede, and facilitate, morphological adaptation.

That older nestlings were sometimes seen to insert their legs or heads into holes in limbs suggests that the search-and-probe foraging tactic is based partly on an innate behavioral tendency. However, a radio-tagged fledgling accompanied foraging adults for several months, indicating the potential for fledglings also to learn by observing adults.

Although Crane Hawks took a wide variety of prey types, their diet at Tikal was heavily dominated by rodents, lizards, and frogs, resulting in a dietary niche-breadth value (3.33) that was only moderately high (Sutter et al. 2001). In comparison, Collared Forest Falcons, Roadside Hawks, and Great Black Hawks had dietary niche-breadth indices ranging from 4.04 to 4.73 (Sutter et al. 2001). Crane Hawks overlapped in diet far more with the Roadside Hawk (Pianka's O_{jk} = 0.76) and the Great Black Hawk (O_{jk} = 0.51) than with any other raptor species at Tikal (Sutter et al. 2001).

Finally, Crane Hawks were interesting in their pattern of post-fledging behavior. Juveniles left the nest area within a few weeks, but then at least one radio-tagged juvenile followed adults around for several months as the latter foraged, and received prey from them. This resembles the pattern typical of many small songbirds—though on a much more protracted time line—and is in marked contrast to that shown by several other raptors with extended dependency at Tikal, in which juveniles waited in a more-or-less central location for adults to bring them prey. While details are lacking, it appears that the Laughing Falcon and possibly the Double-toothed Kite *(Harpagus bidentatus)* may show behavior similar to the Crane Hawk, with juveniles following foraging parents during a dependency period of substantial duration (see Chapter 23).

These contrasting behavior patterns of fledgling raptors have not, to our knowledge, received attention. We speculate that this dichotomy of central prey delivery versus fledglings following adults may in part be in response to prey size. Raptors taking large prey apparently find it profitable to deliver these some distance to a centrally located juvenile, whereas delivery of small prey perhaps is worthwhile only if the recipient juvenile is nearby. This may apply in the case of Crane Hawks, which are known to prey often on arthropods. Although adults at Tikal did not bring arthropods to nests, it is not known whether they fed arthropods to dependent fledglings away from the nest. Another in-fluencing factor may be prey dispersion and resultant patterns of adult foraging movements: the wider the wandering required in foraging, the greater may be the advantage of juveniles following adults during their dependency period. This may be especially important in the Laughing Falcon, in which adults used large home ranges relative to their body size, particularly during the post-fledging period. Finally, in raptors that rely heavily on stealth and surprise in hunting—for example, the Ornate Hawk-eagle—a juvenile following an adult would likely erode the latter's hunting success. In comparison, species such as the Crane Hawk that take mainly prey types that are not particularly wary may be uniquely able to use a strategy in which dependent juveniles follow adults.

10 WHITE HAWK

*Gregory S. Draheim, David F. Whitacre,
Angel M. Enamorado, Oscar A. Aguirre,
and Aquiles E. Hernández*

From a high, forested ridge, many a sweat-soaked tropical field biologist has been stirred by the sight of a pair of White Hawks *(Leucopternis albicollis)* soaring below. Seen from above, the snow-white plumage, neatly offset with a black tail band and black-tipped outer primaries—as these birds soar languidly against the varied shades and textures of the forest canopy below—is a delight to behold. Owing to both their snow-white plumage and their penchant for soaring in pairs above the forest, these are among the easiest of Neotropical forest raptors to observe—at least while in flight (Plate 10.1). Not uncommonly one may view three soaring pairs at once from a single vantage point, and Monroe (1968) tells of counting more than 35 individuals soaring over the forest during a plane ride over a mountain in Honduras. Finding nests, alas, is not so simple and requires much patience. And most hunting is silent perch hunting, often below the forest canopy, making these hawks inconspicuous much of the time.

Referred to in earlier literature as the "Snow Hawk" (Griscom 1932), "White-collared Hawk" (Hellmayr and Conover 1949), and "White Snake Hawk" (Lowery and Dalquest 1951), the race of White Hawk occurring at Tikal, *L. a. ghiesbreghti*, is one of the world's few virtually snow-white raptors. In published literature this handsome raptor is often associated with mature moist forest and is said to take many snakes and lizards. Both patterns were strongly supported by our results. In addition, these hawks preyed more heavily on small mammals than suggested by some earlier authors. We found some evidence that these stocky buteonines hunt preferentially at treefall gaps amid mature forest, as suggested by previous authors. We also found evidence suggesting White Hawks may respond differently to habitat modification in disparate geographic regions. These handsome raptors showed a surprisingly low reproductive rate for their body size, resulting from a small clutch, a long nesting cycle, and a very long post-fledging dependency period.

GEOGRAPHIC DISTRIBUTION
AND SYSTEMATICS

The genus *Leucopternis* occurs throughout the humid forested Neotropics (Brown and Amadon 1968). Com-

prising 9 or 10 species, this is among the half-dozen most species-rich genera of falconiforms. *Leucopternis* has long been regarded a "sub-buteonine" genus—one of several primarily New World genera allied with *Buteo*. One or more species currently allocated to this genus have at times been placed in *Buteogallus* (Amadon 1982), but an osteological study (Holdaway 1994) placed these genera some distance apart, and considered *Leucopternis* to be allied with *Buteo*, *Parabuteo*, and *Accipiter*, and more removed, but not drastically, from the Old World genera *Melierax* (Chanting Goshawk), and *Kaupifalco* (Lizard Buzzard). The same study regarded *Buteogallus* as more closely allied with the sea eagles, "milvine kites," and harriers (*Circus*).

Another analysis (Kemp and Crowe 1990) placed *Leucopternis* among a group of buzzard-like clades, but not particularly close to *Buteo*, *Buteogallus*, or *Accipiter*. Rather, it placed *Leucopternis* in a clade with the African genera *Melierax* and *Kaupifalco* (see above) and *Micronisus* (Gabar Goshawk)—sister group to a clade containing the New World Crane Hawk *(Geranospiza)* and *Asturina (Buteo) nitidus*, the Gray Hawk. This arrangement, in suggesting a close affinity of *Leucopternis*, *Asturina*, and *Kaupifalco*—despite an ocean separating the former two from the latter—mirrors Ridgway's (1914) conclusion nearly a century ago—with which Amadon (1982) concurred. All of this is subject to revision, however, based on molecular studies. These studies suggest that *Leucopternis* belongs in a clade containing *Buteo*, *Parabuteo*, *Buteogallus*, the sea eagles, *Heterospizias*, *Accipiter*, *Circus*, and others (Braun and Holznagel 2002; Braun and Kimball 2002).

Del Hoyo et al. (1994) recognized 10 species of *Leucopternis*, with 7 of these grouped into three superspecies. The White Hawk is currently regarded as comprising four subspecies, of which that studied at Tikal, *L. a. ghiesbreghti*, is the northernmost, occurring from southeast Mexico to Guatemala and Belize. Collectively, the four subspecies occupy much of the moist forested tropics and subtropics from Mexico to the Amazon Basin. Two other forms, which at times have been regarded as subspecies of *L. albicollis*, have been given species status and considered to form a superspecies with *L. albicollis*: these are the Gray-backed Hawk *(L. occidentalis)*, which replaces *L. albicollis* in western Ecuador and adjacent northwestern Peru, and the Mantled Hawk *(L. polionota)* of the east coast of South America.

Known locally as Gavilán Blanco, the White Hawk is similar in form to, though somewhat larger than, the Red-shouldered Hawk *(Buteo lineatus)*. In contrast to the warm reds and browns common in the genus *Buteo*, plumage in *Leucopternis* hawks is almost entirely white and gray, with black markings on the wing tips and tail varying among species (Plates 10.2 and 10.3). Immatures differ little in plumage from adults, which is unusual among accipitrid genera. Of all the species and subspecies of *Leucopternis*, the one we studied at Tikal—*L. a. ghiesbreghti*—is the most completely white; only the primaries are black tipped, and the subterminal tail band is reduced to a series of large black spots.

This is mainly a bird of humid lowlands and foothills, occurring up to 1400 m in Colombia (Hilty and Brown 1986), Panama (Blake 1958), and Costa Rica (Stiles et al. 1989), and 1500 m in Guatemala (Land 1970). In Honduras, Monroe (1968) found this species recorded up to 1800 m, but rare above 1000 m. Likewise, in Ecuador, it is described as regular up to 1100 m, with one record at 1500 m (Ridgely and Greenfield 2001).

MORPHOLOGY

While these birds are mildly size dimorphic, we found it impractical to sex them by size in the field or in hand. We relied on breeding behavior to distinguish the sexes. Several breeders were color-banded or radio-tagged, allowing distinction of pair members and confirmation of our behavior-based sexing. We trapped and measured three adult females and three adult males at Tikal. Based on our own and published data (Appendixes 1, 2), females were slightly heavier than males (9.9%), and values of the Dimorphism Index (DI) were as follows: 3.1 for (cube root) body mass, 5.9 for wing chord, and 4.5 for culmen, for a mean DI of 4.5 among the three. While results might differ slightly in a larger sample, it is apparent that this species shows very little size dimorphism. In Surinam, White Hawks *(L. a. albicollis)* were heavier and seemingly more size dimorphic than indicated here, with a body mass dimorphism value of 9.5 (Haverschmidt and Mees 1994; Appendix 1).

In many past accounts, this hawk has been said to have yellow eyes (Friedmann 1950; Brown and Amadon 1968; Blake 1977), yet all 23 birds observed closely at Tikal had brown eyes. White Hawks observed in southeast Mexico also had brown eyes (E. Iñigo, pers. comm.), and Howell and Webb (1995) described the species as dark eyed. It appears safe to state that individuals of this subspecies are brown eyed, the same as the other races, which are also brown or gray eyed (Ferguson-Lees and Christie 2001).

PREVIOUS RESEARCH

Prior to our work, no detailed study had been made on this species, and the entire genus is among the most poorly known genera of accipitrid raptors. Concurrent with this study, the Gray-backed Hawk—sometimes regarded as a subspecies of *L. albicollis*—was studied in western Ecuador by Hernán Vargas (1995). We draw on Vargas's study for comparative material.

RESEARCH IN THE MAYA FOREST

From 1991 to 1995, we studied 11 nestings in six White Hawk territories in Tikal National Park and monitored these six territories for 2–4 years each (Appendixes 10.1, 10.2). In 1994 we collected limited data on three additional territories: La Paila, Brecha 50, and Brecha 64 (Appendixes 10.1, 10.2). Also in 1994 we endeavored to determine how many White Hawk territories occupied a 20 km² study plot in primary forest. Assisted by walkie-talkies, we triangulated soaring birds simultaneously, from treetop lookouts scattered throughout the plot, and subjected these results to traditional spot-mapping techniques. Research in 1991–92 was conducted by one of us (G. S. D.), and fieldwork from 1993 to 1995 was conducted mainly by another (A.M. E.), supervised by a third (D. F. W.). Most fieldwork was conducted from early February through July or August, though telemetry efforts were continued year long.

We searched for active territories mostly in February and March. We found nests mainly by observing from treetops during the morning hours of peak soaring and vocal activity (0800–1000), taking compass bearings on areas where birds were seen, and moving successively closer until the birds' behavior revealed a nest. Two nests were found by following radio-tagged adults. Using binoculars and a 20X spotting scope, we observed White Hawks at the nest from blinds built in trees 30–40 m away and camouflaged with vegetation or burlap. In 1991 and 1992 nest observations were generally conducted from 0630 to 1200 and 1230 to 1800 (i.e., dawn to dusk with a 30-minute hiatus at noon), and in 1993 and 1994 were usually dawn to dusk without breaks.

We climbed to several nests to confirm nest contents. Eggs were weighed to 0.1 g with a 100 g Pesola scale and measured to 0.1 mm using Vernier calipers. We banded all nestlings, and the adults that we trapped, with unique combinations of colored, numbered metal leg bands. We trapped adults at or near nests, using a *bal-chatri* (Berger and Mueller 1959). Four nestlings (slightly before or after fledging), one yearling, and six adults were fitted with 18–20 g radio transmitters via a backpack arrangement, with straps of Teflon ribbon (Dunstan 1972). In most cases telemetry was used to gain visual contact with a bird, whose location was then tied to known reference points via GPS or a pace and compass map. Triangulation was occasionally used. Usually, a single location was obtained per bird per day, but in a few cases two or three locations were obtained, usually several hours apart. During visual contacts we took detailed data on perch characteristics and on vegetation within an 11.3 m radius.

For one dependent juvenile (the Brecha Sur yearling), we studied habitat use in some detail. Based on 72 perch locations during a 2.5-month period, we determined the 85% Minimum Convex Polygon (MCP) home range and established a 950 m transect across the home range's long axis. Using a stratified random procedure, we selected 92 points within the home range, for which we compared habitat traits and treefall gap occurrence to the 72 locations where we found this bird perched. We used a chi-square goodness-of-fit test to compare characteristics of perch sites and random points.

Our use of the term Bajo Forest corresponds mainly to the Hillbase, Sabal, and Transitional Forest types of Schulze and Whitacre (1999). Unless stated, means are presented ± 1 SD.

DIET AND HUNTING BEHAVIOR

Most published accounts portray the White Hawk as feeding mainly on snakes and lizards, but some accounts also mention occasional mammals, frogs, insects and other arthropods, and, rarely, birds in the diet (Plate 10.4). Slud (1964) and Hilty and Brown (1986) described the diet as reptilian, mainly of snakes, and Slud attributed to this hawk "a marked preference for coral snakes, which it swallows whole." Brown and Amadon (1968) stated that "tree snakes and lizards are the preferred food" and ffrench (1973) commented in a similar vein. Stiles et al. (1989) described the diet as snakes, lizards, frogs, small mammals, large insects, and occasionally birds, whereas Wetmore (1965) described the diet as small mammals such as mice, rats, and small opossums, and lizards, snakes, frogs, and large Orthoptera. Haverschmidt and Mees (1994) described the diet as snakes, frogs and lizards, and arthropods: Coleoptera (Passalidae), Orthoptera (Locustidae), lepidopteran larvae, and centipedes (Chilopoda).

Griscom (1932) reported on several White Hawks collected in Alta Verapaz, Guatemala, "all being filled to capacity with either small snakes or lizards." Voous (1969) gave dietary records of a snake (probably a coral snake, *Micrurus*), a caecilian (Amphibia: Caecilidae), and a fish (*Erythrinus erythrinus*). In southern Veracruz, the stomach of one individual contained several lizards (*Anolis* and *Ameiva*) and another held a small colubrid Speckled Racer snake (*Drymobius margaritiferus*) (Lowery and Dalquest 1951). Wolfe (1962) mentioned a single stomach "packed with small lizards" apparently of several species, as well as a 10 cm snake and a large beetle. Van Tyne (1950) described the diet as principally snakes and lizards, including basilisks (*Basiliscus*) up to 38 cm long; he also reported stomach contents of one male as a few large Orthoptera

and a small snake, Degenhardt's Scorpion-eating Snake (*Stenorrhina degenhardtii*). Skutch (1971) described this raptor as feeding largely on snakes, though he also witnessed the capture of a fledgling aracari (*Pteroglossus* sp.). Trail (1987) witnessed one unsuccessful attack at a Cock-of-the-Rock (*Rupicola rupicola*) lek. Lamm's (1974) observation of a White Hawk at Tikal on a freshly killed Great Tinamou (*Tinamus major*) is, we believe, unusual—such large prey probably are rarely taken. In addition to the taxa mentioned here, del Hoyo et al. (1994) listed amphisbaenids, and Brown and Amadon (1968) listed fiddler crabs.

While published accounts represent many localities, no single account mentions more than a handful of actual prey records. Hence the data we present here are the first quantitative dietary data of any magnitude for this species. From 1991 to 1994 we observed 260 prey items, nearly all brought to nests (Table 10.1). In most cases we observed adults deliver items to nests, while in several cases we found prey items when we climbed to nests. Of 210 items identified to class, reptiles made up 68.1%, mammals 18.6%, birds 5.7%, frogs 4.3%, and insects 3.3% (Table 10.1). Of the reptile portion, two-thirds were snakes and one-third lizards. On a biomass basis, the contribution of insects would be far less than 3%, but the proportional composition of the more important prey types would probably differ little from that just given. Dietary composition was quite consistent across nests. Based on 165 items at the five nests that contributed most prey records, reptiles comprised 50–72% of prey items (mean = 64.9 ± 8.6%, n = 5), and mammals were second in importance in each case, comprising 16–23% of prey (mean = 20.0 ± 2.7%, n = 5). The contribution of birds, insects, and amphibians was more variable, ranging from 0 to 14.8%, 12.5%, and 12.1%, respectively.

Most species identifications were achieved through inspection of prey items or remains in the hand. Mammalian prey included rat-sized rodents, the small arboreal Deppe's Squirrel (*Sciurus deppei*), two fruit bats (*Artibeus* spp.), and one unidentified bat, while birds included Keel-billed Toucan (*Ramphastos sulfuratus*), a juvenile Mexican Wood Owl (*Strix squamulata*), an adult Guatemalan Screech Owl (*Megascops guatemalae*), and a White-breasted Wood Wren (*Henicorhina leucosticta*). Snakes taken included coral snakes (*Micrurus*) and mimics (most likely *Lampropeltis triangulum*), two Indigo Snakes (*Drymarchon corais*), several individuals of *Coniophanes* species including at least one Faded Black-striped Snake (*C. schmidti*), one Blotched Hooknose Snake (*Ficimia publia*), a slender tree snake (possibly *Imantodes* sp.), one Dark Tropical Racer (*Dryadophis melanolomus*), one Neotropical Whip Snake (*Masticophis mentovarius*), and other colubrid snakes. Among lizards in the diet, two Schwartze's Skink (*Eumeces schwartzei*) and one Eastern Casquehead Iguana (*Laemanctus longipes*) were identified in the hand. Lizards identified visually but not in the hand were believed to include Hernandez's Helmeted Iguana (*Corytophanes*

Table 10.1 Diet of White Hawks at six nests in five territories at Tikal

Nest	Reptiles		Mammals		Birds	
	Number	Percentage	Number	Percentage	Number	Percentage
Baren 1991	18	66.7	5	18.5	4	14.8
Trans 9, 1992	49	72.1	11	16.2	2	2.9
Naranjal 1992	9	69.2	3	23.1	1	7.7
Brecha 0 1993	12	50.0	5	20.8	3	12.5
Other 1992–93	27	—	7	—	1	—
Brecha Sur 1994	22	66.7	7	21.2	0	0
Brecha 0 1994	6	66.7	1	11.1	1	11.1
Overall	143	68.1	39	18.6	12	5.7

Nest	Amphibians		Insects		Total number of identified	Total number of unidentified
	Number	Percentage	Number	Percentage		
Baren 1991	0	0	0	0	27	—
Trans 9, 1992	2	2.9	4	5.9	68	—
Naranjal 1992	0	0	0	0	13	—
Brecha 0 1993	1	4.2	3	12.5	24	—
Other 1992–93	1	—	0	—	36	22
Brecha Sur 1994	4	12.1	0	0	33	22
Brecha 0 1994	1	11.1	0	0	9	4
Overall	9	4.3	7	3.3	210	50

hernandezii), several *Anolis* species, *Ameiva*, and probably Brown Basilisk *(Basiliscus vittatus)*.

At the 1994 Brecha Sur nest, the following detailed prey tabulation was made. Of 55 prey items observed, 22 were unidentified, 14 were snakes, 8 lizards, 6 rodents, 1 squirrel, and 4 frogs. Among the 14 snakes, 4 were noted as coral snakes (possibly including mimics), 1 as a Fer-de-lance *(Bothrops asper)*, 1 was identified in hand as a Black-striped Snake *(Coniophanes* probably *imperialis)*, and another as a Green Vine Snake *(Oxybelis* probably *fulgidus)*. Three others were also noted as "vine snakes," a common name that applies to *Leptophis*, *Oxybelis*, *Imantodes*, and possibly other genera. Four other snakes were slender colubrids. Also in 1994, 13 prey items were observed at the Brecha 0 nest. Among the nine identified were five snakes, a lizard, a frog, a Deppe's Squirrel, and a Guatemalan Screech Owl, the latter two identified via prey remains. The five snakes included a large coral snake or mimic and three that were green.

Both terrestrial and arboreal reptiles were commonly taken. Terrestrial forms included coral snakes and mimics, indigo snakes, fer-de-lance, *Coniophanes*, *Ficimia*, *Masticophis*, *Dryadophis*, and the lizard *Ameiva*. Principally arboreal reptiles in the diet included numerous "vine snakes," one of which was confirmed as *Oxybelis*, as well as the lizards *Corytophanes*, Eastern Casquehead Iguanas, and *Anolis*. We observed a few successful attacks on lizards in trees, as described later. By and large, our quantitative data agree with the impression one obtains from published, largely anecdotal accounts of the White Hawk's diet. Perhaps the greatest surprise was that mammals comprised approximately 20% of the diet,

consistently at all of our study nests—a greater incidence of mammals than suggested by published accounts. Our data also showed a greater prevalence of birds in the diet (5.7%) than one might expect based on the literature. Prey items were often headless when brought to the nest, the adult presumably having eaten the head prior to delivering the prey.

Hunting Behavior

Though we had little luck observing hunting behavior of White Hawks, our observations generally accord with the view expressed in many accounts, that this hawk is largely a patient perch hunter. We particularly agree with the comments of Stiles et al. (1989) that these hawks hunt within the canopy and at edges and gaps, waiting quietly for prey to expose itself on the ground, tree trunk, or branch. We witnessed two captures by adult females who spotted quarry close to their nest trees. In each case, the female left the nest tree to swoop down on a lizard in a neighboring tree, then fed it to the young nestling. Radio-tagged birds were often hunting as we followed them, but only once did we witness in this fashion a successful hunt. On 23 February 1994, while radio-tracking the young male that had fledged from the Transect 9 nest in 1992—now nearly two years old—we observed the following successful attack. From a perch in the sub-canopy space, the juvenile suddenly appeared to see something and launched, landing on an epiphyte on a dead log among some vines. Here he remained standing, wings extended from his sides, for 90 seconds, then flew back to the earlier perch with a lizard in his right foot.

In addition to these few successful attacks, we made a number of observations of apparent foraging behavior. Radio-tagged adults often made frequent perch changes within the canopy and sub-canopy space, and appeared to be still-hunting for a period from each perch, then moving on to hunt a new area. In one case we watched a White Hawk flying from tree to tree very low beneath the forest canopy, apparently hunting, and carefully inspecting a dead tree trunk as if searching for prey. These and other observations indicated that these raptors at times hunted at quite low levels within the forest. While these hawks soar frequently and conspicuously above the forest, we observed no attacks on prey from such soaring flights, which appeared to serve mainly as territorial displays. Aldrich and Bole (1937) collected a White Hawk that was attracted by the cries of a wounded Capuchin Monkey (*Cebus* spp.). In addition, White Hawks sometimes responded to acoustical luring at Tikal (Turley 1989). These observations suggest that White Hawks, like many raptors, sometimes use acoustical cues in foraging and may respond opportunistically to situations such as animals in distress.

Some interesting observations of White Hawk behavior were made at Tikal by Gerald Binczik and Susan Booth-Binczik (unpubl. data). While following habituated troops of White-nosed Coatimundis *(Nasua narica)*, these researchers repeatedly observed a White Hawk perch for long periods low in the canopy or sub-canopy near the coatis. It appeared clear that the White Hawk intentionally maintained proximity to these mammals as they foraged along the forest floor, as the raptor was seen to change perches, apparently in order to keep pace with the coatis. Although no prey captures were seen, it seemed likely the hawk(s) followed the coati troop to take advantage of prey startled from cover by these mammals. Similarly, Boinski and Scott (1988) recorded five instances of White Hawks following Red-backed Squirrel Monkey *(Saimiri oerstedi)* troops in Costa Rica, and a pair of White Hawks was found to regularly follow Black-capped Capuchins *(Cebus apella)* in French Guiana, and were observed seven times to capture arboreal snakes disturbed by the monkeys (Zhang and Wang 2000).

The White Hawk has often been described as rather unsuspicious and easily approached (Slud 1964; Smithe 1966; Land 1970; Ridgely and Gwynne 1989), as well as lethargic (Hilty and Brown 1986) or sluggish (Slud 1964). Apparently these adjectives were intended to convey a degree of tameness and to indicate that these hawks tend to remain on a single perch for a long time. In addition, Wetmore (1965) stated that on several occasions he observed White Hawks fly down to perch near at hand "in order to watch me with evident curiosity."

While White Hawks may indeed be more approachable than some other raptors, the adults we followed via telemetry were not tame or easy to approach; caution and stealth were required, as these birds were often nervous and easily flushed. Juveniles, in contrast, were much more approachable. The tameness mentioned by several authors—and the apparent curiosity mentioned by Wetmore—may conceivably relate in part to observations of juveniles, and perhaps may relate also to the penchant this species seems to have for following mammals that act as prey beaters. The terms "lethargy" (Hilty and Brown 1986) and "sluggishness" (Slud 1964) presumably simply describe this species's hunting style, which employs patient scanning from a perch. Nonetheless, while actively hunting, the hawks we followed did not always remain for especially long periods on a given perch. Often after 10–15 minutes on one perch, they moved on.

HABITAT USE

Our reading of the literature gave us the impression that habitat use by White Hawks may vary geographically, with the species showing a penchant for association with large natural or man-made clearings in rain forest in the Guianas, and being more strictly associated with tall mature forest throughout its Mexican and Central American range. In Mexico, all accounts agree that the species occurs in humid evergreen tropical and subtropical forests (Blake 1970, 1977; Binford 1989; Howell and Webb 1995). In southern Veracruz, Lowery and Dalquest (1951) encountered these birds only in "deep, uninhabited jungles" and were of the impression that the hawks avoided passing over open places such as rivers and sand bars. In Guatemala, Land (1970) regarded this species as fairly common in humid forests and plantations of the Caribbean lowlands and Petén, which accords well with our observations. In Belize, Russell (1964) found White Hawks most common in tall rain forest, but also occurring in tall second-growth and semi-open habitats. In Honduras, Monroe (1968) likewise found this species fairly common in lowland rain forests. In Costa Rica, Carriker (1910) characterized it as most common in the more humid lowland areas and absent outside the forest, and Stiles et al. (1989) stated that this raptor prefers forested areas in hilly terrain, agreeing closely with Wetmore (1965) in this regard. In Panama and Colombia, this species has been described as local to fairly common in forest and forest borders in humid lowlands and foothills (Blake 1958; Hilty and Brown 1986; Ridgely and Gwynne 1989).

In short, virtually all accounts from Colombia northward agree that this is mainly a bird of mature, humid lowland forest, though sometimes occurring also along forest edges, around treefall gaps, and occasionally in second-growth and other somewhat open situations. So far as Central American accounts go, Slud's (1964) account is unusual in stating that the species is typical of forest openings, thinned woodland, and clearings with trees; he was of the impression that these birds rarely occurred within extensive forest stands, and then usually in association with some sort of opening. Brown and Amadon's (1968) account, apparently heavily colored by Slud's observations, stated that this species "rarely enters deeply into unbroken rain forest, except perhaps around swampy areas where the forest is somewhat thinned out," and is, we believe, misleading as to the true association of this hawk with mature forest, at least in Central America. In French Guiana, Thiollay (1984, 1985) and colleagues (Jullien and Thiollay 1996) concluded that the White Hawk was associated with large openings in primary forest, particularly natural gaps but also edges of clearings. Thiollay (1985) regarded this species as being more tolerant of deforestation than many other primary forest raptors and categorized it as equally common in primary and successional forest and as occurring to a limited extent in areas of large clearings, plantations, and successional areas with patches of secondary forest.

In sum, published literature is somewhat divided, with many accounts simply regarding this as a forest raptor, and some emphasizing a tendency for it to be associated with clearings, edges, gaps, or successional forest. In our study area, this raptor was closely associated with tall mature forest. Conceivably the species is more prone to use edge and partly deforested habitats in wetter, more equatorial forests, but more quantitative study is needed before one can be certain of such a pattern.

Our six main study territories were amid extensive mature forest deep within Tikal National Park, and three breeding adults and two yearlings that we radio-tagged remained within such habitat. We took data on habitat at 27 perches of the Brecha Sur adult male (during a non-nesting period), 37 perches of the Brecha 0 adult male (while he hunted to supply a large nestling and his mate), 22 perches of the Brecha 0 adult female (while she helped supply the chick near fledging time), 34 perches of the Transect 9 dependent juvenile from the age of 15 months until it dispersed at 22 months, and 72 perches of the Brecha Sur juvenile during a 2.5-month period when it was about a year old. Results regarding habitat use by these five radio-tagged birds are given in Table 10.2.

All sightings of all five individuals were in primary forest (Table 10.2). During his dispersal movements, the Transect 9 subadult was located three times in mature forest fragments outside the park; other than these instances, all localizations of all birds were within the extensive mature forest within the park. None of the radio-tagged hawks was located in association with the several large, human-made clearings near park facilities and Maya ruins, nor along the main road through the park. For four of the hawks, forest at all perches was categorized as Upland, Hillbase, or Transitional Forest—relatively tall, structurally complex, and for the most part closed-canopy forest types. Only one bird was categorized as having used the lower Scrub Swamp Forest type, in 15% of cases (Table 10.2). Data we collected on tree species composition at perch sites verified these forest type designations. These results are in agreement with earlier findings at Tikal, which showed a significant positive correlation between canopy height and the number of White Hawks observed in point counts (Whitacre et al. 1990), indicating that these birds used mainly tall Upland and similar forest types, rather than Low Scrub Swamp Forest.

For the five hawks represented in Table 10.2, perch height averaged 15 m and perch trees 20 m tall. Canopy height in perch vicinities averaged 20 m, again indicating use mainly of tall forest. Perch limbs from 2 to 50 cm in diameter were used, with perch diameter averaging 10 cm, and perch trees ranged in diameter from 17 to 250 cm, averaging 52 cm (Table 10.2). On average, these four hawks (exclusive of the Brecha Sur yearling) were perched above the canopy during 17% of observations (n = 4 individuals, SD = 4), and were beneath some foliage but overlooking an area of the upper canopy surface on average 27% of the time (Table 10.2; SD = 23, n = 4). The birds were perched within the canopy on average 36 ± 21% of the time and in the sub-canopy space 20 ± 19% of the time (n = 4). We never found the birds perched on the ground. The Brecha Sur yearling, based on 72 perch sites, was found 93% of the time in forest 15–25 m tall and 7% of the time in forest 10–15 m tall, reflecting this bird's use of mainly the tall Upland and Transitional Forest types rather than the low-canopied Scrub Swamp Forest.

Selective Use of Treefall Gaps

A number of authors have commented that White Hawks often perch hunt along forest edges, streams, clearings, or treefall gaps (Chapman 1929; Van Tyne 1950; Haverschmidt 1968; ffrench 1973; Ridgely and Gwynne 1989; Stiles et al. 1989). Our data offer at least partial support for the notion that these hawks frequently use treefall gaps within extensive forest. The percentage of sightings in which the hawks were within 11.3 m of a treefall gap ranged from 16.7 to 40.5%, averaging 30.1 ± 10.8% (n = 5 birds). Many of these were gaps caused by single treefalls and were 5 × 10 m or smaller in size, but several were created by multiple treefalls, and a few were as large as 20 by 20–30 m. In no case did we find a focal bird associated with a truly large clearing (100 m or more on a side), although we have observed this casually in slash-and-burn farming areas near the park.

For only one bird were we able to statistically compare the time spent in and near treefall gaps to that expected by chance. This 1-year-old spent more time than expected within treefall gaps and less time than expected near gaps (within 10 m), whereas time spent away from gaps did not differ from that expected (Table 10.3). One possible interpretation is that this bird, detecting a gap nearby, tended to move toward it, but overall, did not consistently seek out gaps. While perched in treefall gaps, the hawk often (48% of the time: n = 19) perched on vines that hung over the gaps, presumably affording a good view of the gap. When not perched at treefall gaps, the hawk tended to frequent areas of relatively open understory, which again presumably afforded it a good view of the forest floor and lower vegetation.

We tested whether the five radio-tagged birds spent more time than expected within 11.3 m of treefall gaps. We failed to find such a pattern, but we consider this result inconclusive. The percentage of sightings in

Table 10.2 Habitat use by five radio-tagged White Hawks in Tikal National Park

Habitat characteristics	Adult male, Brecha Sur	Adult male, Brecha 0	Adult female, Brecha 0	Subadult male, Transect 9	Post-fledged young, Brecha Sur	Mean (n = number of individuals)
Perch Characteristics						
Perch height (m)	18 ± 6 (2–22); n = 27	15 ± 7 (4–30); n = 37	14 ± 5 (8–25); n = 22	13 ± 8 (2–30); n = 34	15 ± 8 (3–36); n = 72	15 ± 2 (n = 5)
Canopy height (m)	19 ± 4	20 ± 3	20 ± 2	21 ± 3	20 m; n = 72	20 ± 1 (n = 5)
Perch diameter (cm)	8 ± 6 (7.5–25); n = 27	11 ± 6 (10–25); n = 25	12 ± 10 (15–23); n = 23	8 ± 8 (10–25); n = 34	10 ± 9	10 ± 2 (n = 5)
Perch tree height (m)	24 ± 5 (13–34); n = 27	17 ± 6 (3–25); n = 37	19 ± 5 (4–50); n = 23	18 ± 7 (1–50); n = 36	20 ± 8 (2.5–46); n = 72	20 ± 3 (n = 5)
Perch tree diameter (cm)	59 ± 38 (17–165); n = 27	59 ± 32 (22–160); n = 28	60 ± 40 (18–165); n = 23	30 ± 24 (4–100); n = 28	50 ± 50 (7–250); n = 72	52 ± 13 (n = 5)
Sightings near truly large clearings or road sides	0 (n = 26)	0 (n = 39)	0 (n = 24)	0 (n = 34)	0 (n = 60)	0 (n = 5)
Sightings in or near tree-fall gaps	34.6%[a] (n = 26)	40.5%[b] (n = 42)	16.7%[c] (n = 24)	20.6%[d] (n = 34)	38.3% (n = 60)	30.1 ± 10.8% (n = 5)
Sightings with grass	7.7% (n = 26)	10.3% (n = 39)	0 (n = 24)	8.8%[c] (n = 34)	— (n = 60)	6.7 ± 4.6% (n = 4)
Primary forest	100% (n = 26)	100% (n = 42)	100% (n = 24)	100% (n = 38)[f]	100% (n = 60)	100% (n = 5)
Forest Types	(n = 22)	(n = 24)	(n = 20)	(n = 35)	(n = 72)	—
Upland Dry	9.1%	8.3%	10%	2.9%	—	7.6 ± 3.2% (n = 4)
Upland Standard	68.2%	20.8%	20%	62.9%	—	43.0 ± 26.2% (n = 4)
Total Upland	77.3%	29.2%	30%	65.8%	31.3%	46.7 ± 23.0 (n = 5)
Hillbase/Sabal	13.6%	8.3%	5%	17.1%	—	11.0 ± 5.4% (n = 4)
Transitional	9.1%	62.5%	65%	17.1%	—	38.4 ± 29.4% (n = 4)
Hillbase + Transitional	22.7%	70.8%	70%	34.2%	53.7%	50.3 ± 21.5% (n = 5)
Scrub Swamp	0	0	0	0	14.9%	3.0 ± 6.7 (n = 5)
Position	(n = 27)	(n = 41)	(n = 23)	(n = 35)	—	—
Above canopy	14.8%	14.6%	17.4%	22.9%	—	17.4 ± 3.9 (n = 4)
Upper canopy surface	59.3%	19.5%	21.7%	5.7%	—	26.6 ± 23.0 (n = 4)
Within canopy	18.5%	17.1%	47.8%	60.0%	—	35.9 ± 21.4 (n = 4)
Sum of last two	77.8%	36.6%	69.5%	65.7%	—	62.4 ± 17.9 (n = 4)
In sub-canopy space[g]	7.4%	48.8%	13.0%	11.4%	—	20.2 ± 19.2 (n = 4)
On ground	0	0	0	0	—	0

[a] Some of these are multiple treefall gaps; sizes as follows: 5 × 8, 8 × 15, 8 × 15, 8 × 15, 8 × 20, 10 × 15, 10 × 50, 20 × 20, 20 × 20, and 20 × 30 m.

[b] 17 gaps; one 4 × 8, five 5 × 8; three 5 × 10, one 7 × 10, one 7 × 12, one 7 × 15, two 8 × 15, one 8 × 20; one 10 × 15, and one 15 × 25 m.

[c] Four treefall gaps, 4 × 6, 4 × 9, 5 × 8, and 10 × 20 m.

[d] Six treefall gaps 2 × 5, 4 × 6, 3 × 7, 5 × 8, 5 × 8, and 5 × 12 m; one disturbed area, possibly burned in the past, 10 × 30 m.

[e] Cases with grass were all tree-fall gaps or clearings.

[f] Primary forest, but one with 10 × 30 m area probably burned in the past.

[g] Includes "at bottom surface of canopy."

Table 10.3 Proximity to treefall gaps in a radio-tagged post-fledged White Hawk, compared with random points within its home range

	Number (%) of observations/points		
	Near gaps	In gaps (within 10 m)	More than 10 m from gaps
White Hawk	19 (32)	4 (7)	37 (62)
Random Points	13 (14)	22 (24)	57 (62)

Notes: Significance: P = 0.003, χ^2 = 11.54, 2 df, MRPP chi-square test.

which hawks were perched within 11.3 m of a treefall gap averaged 30.1 ± 10.8% (range = 16.7–40.5%, n = 5; Table 10.2). Treefall gaps did not seem more frequent within an 11.3 m radius of White Hawk perches than in the forest at large, based on data from our vegetation analyses. However, this analysis may be flawed; our vegetation analysis likely recognized older, more subtle treefall gaps in addition to the fairly recent, obvious treefall gaps recognized during our telemetry efforts with White Hawks. Hence, data on gap incidence may not be comparable in these two data sets. It was our impression, however, that these birds occurred near treefall gaps more frequently than dictated by chance.

We would insert a caution, however. An association with natural treefall gaps is quite different from a tendency to use large anthropogenic clearings or even truly large natural openings. At Tikal, these birds nested throughout extensive stands of tall mature forest. While they may often have hunted at natural treefall gaps within the forest, this does not imply that the species thrives under substantial deforestation or forest alteration.

Two seasons of point counts at Tikal found White Hawks significantly more abundant in mature forest, especially tall Upland Forest, than in the slash-and-burn farming landscape (Whitacre et al. 1992a, 1992b). In 1991 point counts, mean number of individuals detected per 2.5-hour count was far higher in mature forest (0.69 ± 0.13 [SE]) than in the farming landscape (0.17 ± 0.09) (P = 0.004, Row-mean-score test). Our 1989 point counts also detected White Hawks far more often in mature forest (1.19 individuals/count, + 0.25 [SE]) than in the farming landscape (0.07 + 0.07 [SE]) (P = 0.001, Mann-Whitney U-test, U = 55.0, n = 7, 8). Moreover, in the 1991 counts, White Hawk detections were significantly more frequent in Upland Forest than in Bajo (Scrub Swamp) Forest types (P = 0.025, Mann-Whitney U-test, U = 51.5, n = 8, 8), and in Upland Forest than in the farming landscape (P = 0.0008, Mann-Whitney U-test, U = 63, n = 8, 8), but did not differ significantly between Bajo forest and the farming landscape (P = 0.63, Mann-Whitney U-test, U = 36, n = 8, 8). Our ordinations of these same data sets (with data on 20 raptor species and five other bird species) depicted White Hawks as being among the raptors most often associated with mature as opposed to disturbed or successional forest.

BREEDING BIOLOGY AND BEHAVIOR

Nests and Nest Sites

White Hawks showed little selectivity with regard to nest tree species: 13 nests were in nine tree species. One nest was found in each of the following species: Zapotillo *(Pouteria durlandii)*, Silión *(P. amygdalina)*, Jocote Jobo *(Spondias mombin)*, Ramón or Breadnut *(Brosimum alicastrum)*, and Canchán *(Terminalia amazonia)*, while two nest trees were Manax *(Pseudolmedia oxyphyllaria)*, two were Dantos *(Vatairea lundellii)*, and three were Zapotillos *(Pouteria reticulata)*. The same Honduras Mahogany *(Swietenia macrophylla)* was used twice, the second nest being constructed in the same position as the first, which had fallen. Nest trees averaged 53.8 ± 30.9 cm in diameter at breast height (range = 23.5–138 cm, n = 13), and nests averaged 20.3 ± 4.0 m above ground (range = 15–29.5 m, n = 13). Nests were generally in the upper third of the tree, usually within 3 m of the trunk. Most nest trees were slightly emergent above the surrounding canopy. All 13 nests were constructed of dead sticks, and 9 that were measured averaged 42 x 62 cm and 30 cm in overall depth, with nest cups averaging 26 x 31 cm wide and 5.3 cm deep. Intertwined with some nests were vines that grew up the tree, making it virtually impossible to see these nests from the forest floor. At one nest many ants were present.

Few other nests have been described. Chapman (1929) found a nest high in a tall tree in Panama, and a stick nest in Trinidad was built on a bromeliad close to the main trunk, 24 m up in a tree near a jungle clearing (Herklots 1961). A 1-egg clutch in the Western Foundation of Vertebrate Zoology (WFVZ) collection was taken from a nest 18 m up in a Kapok *(Ceiba pentandra)* tree in Trinidad's Caroni marshes.

Despite the evidence just given that White Hawks hunted mainly in tall Upland Forest at Tikal, we found no White Hawk nests in Upland Forest. Rather, all nests were in areas of low-lying Bajo or Transitional Forest, often near areas of Upland Forest: six were amid Bajo Forest, five at the edge of bajos, and one in Transitional Forest. The birds' use of emergent nest trees may help explain why we found nests mainly in forest types of lower stature, as emergent trees are more common in these forest types than in tall Upland Forest. Use of emergent nest trees may help minimize nest predation, as such trees may place the nests above habitual travel routes of monkeys, and may limit access by climbing predators.

Egg and Clutch Size

We climbed to six nests during incubation, and each had a 1-egg clutch. At four other nests, a single chick hatched or was present upon nest discovery. All published clutch-size records of which we are aware are also of 1-egg clutches (Schönwetter 1961; Brown and Amadon

1968; ffrench 1973). Hence, a 1-egg clutch appears typical both at Tikal and throughout the species's range.

Four eggs at Tikal ranged from 54.0 to 62.0 g in mass, averaging 57.0 ± 3.6 g ($n = 4$). These four eggs averaged 54.6 ± 2.9 mm by 44.3 ± 0.25 mm in dimensions. Using Hoyt's (1979) formula, these dimensions yield an estimated fresh egg mass of 58.6 g. Hence the typical 1-egg clutch at Tikal amounts to 8.3% of adult female body mass or, if the heaviest female is included, 7.9% of female mass. Wolfe (1964), describing two 1-egg clutches taken by Smooker in Trinidad, gave measurements of 56.0 by 46.3 mm and 54.2 by 34.0 mm, though the breadth of the latter egg is almost certainly an error and should be 44.0 (L. Kiff, pers. comm.). Finally, Schönwetter (1961, repeated in Brown and Amadon 1968) gave egg dimensions of 53.0–55.7 mm by 42.7–44.0 mm, with two weights of 41.7 and 50.8 g.

Eggs at Tikal were white with reddish brown splotches, except for one that also had a large black spot on the apical end. Wolfe (1964) described Smooker's two Trinidad eggs as resembling poorly marked *Buteo* eggs with sparse pigmentation of light brown. The Trinidad nest mentioned earlier contained one egg, pale bluish white with light brown smears (Herklots 1961). Schönwetter (1961, repeated by Brown and Amadon 1968) also described eggs that were white with a few small irregular brown spots.

Nesting Phenology

We observed nest building in six cases, from 6 February to 27 March. Egg laying occurred from 14 to 31 March at five nests, coinciding with the driest period of the year. Eggs hatched from 31 March to 27 April, late in the dry season, and all five known fledging dates were from 14 June to 31 July, during the early to mid rainy season. The few existing data from elsewhere indicate a similar nesting season. Chapman (1929) found a nest in Panama on 9 March, and incubation was underway at a Trinidad nest on 7 March (Herklots 1961). A Trinidad clutch in the WFVZ collection was taken on 18 April.

Length of Incubation, Nestling, and Post-fledging Dependency Periods

The average incubation period at Tikal was roughly 35–36 days (34–38 days, $n = 3$; one egg hatched after 34–36 days, and two others after 34–38 days). One fledging interval was precisely determined at 65 days. At a second nest, the chick was estimated to be 10 days old upon discovery, and fledged 56 days later, at roughly 66 days. The 1994 Brecha Sur nestling fledged at 58–61 days. At a fourth nest (Brecha 0/4, 1994) the chick fledged at approximately 68–74 days, after two events we regarded as premature fledgings (see Appendix 10.1). A fifth fledging interval of roughly 88 days was considered atypical (see later). In summary, it appears that fledging usually occurs at 60–70 days, with 88 days probably representing an extreme.

VOCALIZATIONS

The typical call, most often heard from soaring birds but also given while perched, has been described as a very harsh, buzzy, buteo-like scream (Brown and Amadon 1968), similar to that of a Red-tailed Hawk *(Buteo jamaicensis)*, but less sharp and not as loud (Andrle 1967), "so hoarse that it sounds asthmatic" (Slud 1964). It has been variously rendered as *shreeeeerr* (Hilty and Brown 1986), *shee-ee-eer* (Smithe 1966), *sheeeeww* or *ssshhhww* (Stiles et al. 1989), or *whii-ii-eehhr* or *hweeiiihrr* (Howell and Webb 1995).

BEHAVIOR OF ADULTS

Incubation Period

Six of the nine main study nests were found during nest construction or incubation. Throughout the breeding season, females continued to add sticks and fresh leaves to nests. Males assumed all hunting duties from the time of egg laying until nestlings were half grown, at which time females began to hunt and bring prey to the nest. During incubation, prey exchanges took place in a tree other than the nest tree, often 30–40 m or more distant. The male would announce his approach with the species's typical harsh whistle, and the incubating female would join him, effecting the prey transfer on a perch. We witnessed no cases of prey caching for later retrieval. During the incubation phase, females rarely if ever ate at the nest, instead consuming prey at or near the exchange site. We never witnessed males incubate, even while females ate. At the 1991 Pajaritos nest the female always covered the egg completely with leaves before slipping silently off the nest to receive prey from the male. This may have hidden the egg from predators, insulated it against chilling or overheating, or both.

Nestling Period

Female White Hawks frequently shaded nestlings and protected them from rain by standing with wings extended over the nest cup. In some cases as the position of the sun changed, the female did so as well, interposing herself and continuing to shade the chick with her outspread wings. Nestlings also moved in under the female at times in an apparent attempt to escape the hot sun. When a female was off the nest, the nestling would often move about, apparently in search of shade. At such times, nestlings sometimes gaped in apparent response to the heat and sometimes vocalized. Adult females also sometimes appeared to be affected by the heat, gaping as they remained in the direct sun, shading the chicks with their wings drooped at the wrist and spread to the sides.

During the nestling phase, of 132 prey items that were observed being brought to nests, in 49% of cases the male brought the prey directly to the nest, while in 51% the female left the nest to receive prey nearby. Females during this phase ate at the nest while feeding the lone nestling. We never witnessed males brooding young. The Pajaritos male behaved differently from other males. He never came all the way to the nest, but rather called the female off the nest and transferred prey to her outside the immediate nest vicinity. At the other nests, when a male landed on or very near the nest to transfer prey to the female, he would often remain perched there for a few minutes before leaving.

BEHAVIOR AND DEVELOPMENT OF NESTLINGS

Nestling Period

Hatchlings were covered in white down except for their shoulders, which sported buffy to reddish brown down. The cere in hatchlings was yellow, the beak black, and iris brown; metatarsi were yellow and talons milky white. Near the end of the second week, the egg tooth disappeared, and the cere began to turn gray. During the third week, pin feathers began to show on the wings, and talons became black. At 21–23 days, young were still covered with white down, without visible tail feathers, and were barely able to stand. At 4–5 weeks, the cere turned gray, and the body was covered with contour feathers; we first observed nestlings feeding themselves at this time. When one nestling was 46 days old, its tail was 5.1 cm in length, and by 50–52 days, a black subterminal band began to develop. At this age, the nestling began to perch on the edge of the nest and flap its wings. At approximately 66 days, this nestling achieved lift for the first time and from then on practiced flapping daily until it fledged at approximately 88 days. Its tail was 9.3 cm long and the black subterminal band was evident when it was last measured, at the approximate age of 66 days.

A "Widowed" Female Fledges a Young Bird

At the Transect 9 nest in 1992, the adult male disappeared when the nestling was roughly 36 days old. The female called for two days, seemingly waiting for the male to deliver prey. Near the end of the second day she brought a large insect to the nestling, and thereafter, she left the nestling unattended while foraging. The nestling lost 40 g during the month prior to fledging. Although frequency of prey deliveries before and after the male disappeared (averaging 1.9 items/day) did not differ significantly (ANOVA, $F_{1,19}$ = 1.390; P = 0.25), we could not document the rate of biomass delivery. This nestling's approximately 88-day nestling period was atypically long in comparison to the other fledging intervals we recorded; this likely reflected suboptimal provisioning by the lone female. The relative roles of males and females in providing for dependent juveniles during their year-and-a-half post-fledging dependency period remain unknown.

Post-fledging Dependency Period

We monitored the fate of five radio-tagged juveniles—four receiving transmitters shortly before or after fledging, and a fifth receiving one approximately 10 months after fledging. Two of the five survived to reach independence 17–19 months after fledging, and the other three died within 2.0–4.5 months after fledging. For the two juveniles that survived to independence, details are as follows.

The 1991 Brecha Sur fledgling was discovered in mid-July 1991 and was trapped and radio-tagged on 17 May 1992, approximately 10 months after fledging (Appendix 10.1). We studied this juvenile's movements intensively for the next 10 weeks, during which it used an area of 46.3 ha (Minimum Convex Polygon [MCP] method). This bird began to perch at progressively lower levels within the forest and began hunting on its own. Although it remained in the area, it began to fly longer distances to the south. The last time we saw this juvenile receive prey from an adult was on 12 December 1992, and we last located it on 20 January 1993, 1.5 km south of the nest. Apparently it had reached independence during this interval and was dispersing. These data indicate a post-fledging dependency of at least 17 months for this individual. This dependent yearling spent most of its time in the upper canopy, where it perched and begged for food. As the summer progressed, this juvenile began to perch at lower levels (regression of perch height on Julian date, P < 0.001, R^2 = 0.30). This decrease in perch height over time suggested the bird may have been learning to forage as the months passed, choosing lower perches in order to better scan the ground and lower vegetation. While during our early observations the bird seemed to be waiting for the adults to bring food, as time passed it began to do some foraging, and we once saw it fly to the ground and capture a lizard.

On 8 July 1992, a nestling was banded and radio-tagged in the Transect 9 nest. This bird fledged on 30 July 1992, at the age of about 88 days, after the disappearance of the adult male several weeks prior. As of 7 July 1993, a year later, this juvenile still remained within 150 m of the nest and was still being fed by the adult female, apparently unaided by any adult male. At the age of 15–22 months, this juvenile was usually found within 50–800 m of the nest. Monitored via telemetry, this juvenile remained dependent until early March 1994, when it precipitously dispersed from the nest vicinity, a week after we saw the female dive at him in what appeared to be an aggressive fashion. Over the next three months this male ranged over a few square kilometers south of the nest, and on 6 July, we found him at least 5 km southwest of his natal nest, in a forested area in the farming landscape south of the park. This young male reached independence and dispersed approximately 22 months after hatching and 19 months after fledging.

Histories of additional fledglings support the above conclusions concerning the prolonged period of juvenile dependency. Even without radiotelemetry, we could often monitor juveniles by means of their frequent vocalizations, their tendency to remain close to their nest, and their recognizably juvenal plumage. The 1993 Naranjal fledgling remained near its nest, surviving at least eight months (until March 1994), while the 1994 Brecha Sur and the 1994 and 1993 Brecha 0 fledglings remained dependent and near the nests as long as they survived—about two, two, and four and a half months, respectively.

The prolonged post-fledging dependency documented here is consistent with findings for several other Neotropical raptors, but represents an extreme for such a modestly sized raptor. All other Neotropical raptors so far shown to have extended dependency periods are larger than the White Hawk, and most appear to have shorter dependency periods than those shown here. For example, Savanna Hawks *(Heterospizias meridionalis)*, roughly 30–40% heavier than the White Hawk, were dependent on adults for 4–7 months after fledging (Mader 1982), and Great Black Hawks *(Buteogallus urubitinga)*, at least 50% heavier than White Hawks, for at least 7–11 months after fledging (see Chapter 11; Mader 1981). For the Ornate Hawk-eagle *(Spizaetus ornatus)*, twice the size of the White Hawk, post-fledging dependency periods were 11–14 months (see Chapter 15), and for the Black Hawk-eagle *(Spizaetus tyrannus)*, half again the size of the White Hawk, post-fledging dependency periods of approximately 10 months appeared typical (see Chapter 14). To our knowledge, the White Hawk is the smallest raptor yet shown to have an extended post-fledging dependency.

SPACING MECHANISMS AND POPULATION DENSITY

Territorial Behavior and Displays

We do not know whether White Hawks defended their entire home ranges against conspecifics, but we did witness some cases of overt territoriality. On 5 January 1994 a White Hawk flying alone above the forest was stooped on by a pair soaring in the area. One of the aggressors perched while the other followed the fleeing bird into the forest canopy. In 1992, as we observed the Baren pair circling over the forest, a neighboring pair of White Hawks circled closer and closer to them. At length the two pairs began stooping at one another, and each of the Baren pair grappled talons with a member of the other pair, both duos of grappling birds simultaneously tumbling out of sight into the canopy. Hurrying to the area, we saw nothing further. ffrench (1973) described similar behavior but interpreted it differently. He saw what he termed a courtship display in which two birds he thought to be mates clutched talons and cartwheeled for hundreds of feet toward the ground. Simmons and Mendelsohn (1993) presented evidence that such cartwheeling flights of raptors most often result from aggressive territorial interactions and are rarely performed among mated pairs.

White Hawks soared frequently and conspicuously above the forest canopy, often in pairs. As suggested by Thiollay (1989b), such soaring likely serves as a territorial proclamation, and we consider this highly conspicuous soaring to be the species's main spacing mechanism among conspecifics. Soaring was most frequent in mid to late morning; such flights began after the morning fog had lifted and the sun had begun to warm the forest. On rainy and overcast mornings, we rarely if ever observed soaring. On sunny mornings, from a single vantage point we often could simultaneously observe two or more pairs of White Hawks soaring over the forest at different points of the compass. These hawks often vocalized while soaring.

In one case, as we observed a pair, the bird we believed to be the male soared very high at one point and then suddenly dropped steeply, joining the female low over the forest. The following day both pair members were seen cutting and carrying nesting material here and copulating. Apart from this dive the only aerial displays we observed were frequent soaring and the instances of aggression noted earlier.

Nest Defense

We observed a number of cases of aggression toward other raptor species, Guatemalan Howler Monkeys *(Alouatta pigra)*, and Spider Monkeys *(Ateles geoffroyi)*. At the Brecha Sur nest on 24 May 1994, with a chick in the nest, the adult pair was seen to stoop at a pair of Great Black Hawks; at the Transect 9 territory in 1992, we observed the White Hawks stooping at Black Hawk-eagles at times. At a Crested Eagle *(Morphnus guianensis)* nest, a White Hawk followed the female eagle, stooping at her, to the eagles' nest tree; here the White Hawk was joined by another White Hawk, presumably its mate, and both soared, calling, above the perched eagle. In a number of cases White Hawks behaved aggressively toward spider and howler monkeys. At the Brecha 0 nest, with a chick in the nest, the female stooped at a spider monkey in a nearby tree. Similarly, in more than one case at the 1991 Pajaritos nest, when spider or howler monkeys drew near the nest, the incubating or brooding female (and in two cases also the male, who was in the vicinity) dropped like a stone, silently, into the understory, and slipped off quietly to some distance, then climbed above the canopy and returned, stooping repeatedly at these primates. The hawks sometimes struck the monkeys, in one case causing a monkey to fall some 10 m to cling, screaming, to a vine below. The female's stealthy departure from the nest no doubt served to keep the nest location secret from these potential nest predators and disruptors, while the ferocious aerial attacks seemed effective in hastening the monkeys' departure. In one case as we climbed a tree near a White Hawk nest containing a nestling, both pair members protested loudly, eventually stooping at the climber—the male once and the female half a dozen times.

Constancy of Territory Occupancy, Use of Alternate Nest Sites

Five territories were monitored for three to four years each, and in four of these, nesting was documented in more than one year. In one case (Naranjal) the hawks nested again in the same tree, rebuilding in the same spot from which their earlier nest had fallen. In one territory (Pajaritos) the second nesting was 150 m from the first, in another territory (Brecha 0/4) it was 630 m from the first, while in a third territory (Brecha Sur) a different tree was used in each of three years, all falling within an area 350 by 130 m. In several other cases in which a nesting effort never fully materialized, we witnessed reconstruction of existing nests, indicating that alternate nests are often used in successive years, but that the hawks also frequently refurbish and reuse nests from earlier years.

Home Range Estimates

We radio-tagged six breeding adults, four fledglings, and one yearling. Twice we placed radios on the 1991 Baren female, but she removed both soon afterward. The radio-tagged 1991 Naranjal female began spending substantial time away from the nest and was difficult to locate soon after her chick fledged. Hence we accumulated adult home range data mainly for three males and one female. We located the 1991 Baren male 55 times, and his estimated breeding season home range (MCP) was 248 ha. An area-observation curve indicated that these 55 localities were probably quite adequate for estimating this male's home range during this interval. The Brecha Sur adult male yielded 27 visual contacts from mid-June 1994 to mid-April 1995, during a time between nesting efforts, without a dependent juvenile. These data yielded home range estimates of 232 ha (MCP) and 335 ha (95% Harmonic Mean [HM]). At the 1993 Brecha 4 nest, we trapped and radio-tagged both breeding adults. Based on 42 locations, we estimated the male's breeding season home range as 290 ha (MCP) and 281 ha (95% HM), though it still slowly increased in size with additional data points. For the adult female we obtained 22 locations, all during the nestling period. All were close to the nest, as this female was still tending a nestling, and yielded an estimated home range of only 50 ha, with the size estimate continuing to grow linearly with each new data point. Other observations also showed that females remained near the nest much of the time while tending a nestling and hence may utilize quite limited areas at this time.

The mean home range for three breeding males was 264 ± 29.4 (SD) ha, which is likely a minimum; the median for these three was 284 ha (averaging HM and MCP estimates for each bird). With larger telemetry samples, and consideration of spatial use by both pair members throughout the annual cycle, a larger home range may very well be indicated. In French Guiana, Thiollay (1989a) estimated the home range for two White Hawks

at 355 and 450 ha, somewhat larger than our estimate, but not sufficiently so to account for the great difference in density estimates between his study and ours. As we discuss later, the difference in apparent density between these two studies seems to have as much to do with the amount of space unoccupied by White Hawks as with home range size.

Inter-nest Spacing and Density of Territorial Pairs

For eight territories near the heart of the park, we measured inter-nest distance among neighboring nests (Fig. 10.1). Mean inter-nest distance was 2266 m (SE = 118 m, range = 1750–2685 m, $n = 7$). Using this inter-nest distance, the Maximum Packed Nest Density (MPND) formula (see Chapter 1) gives an estimate of 4.67 km² of unique space per territorial pair, the square method yields 5.14 km²/pair, and the polygon method 7.14 km²/pair—corresponding to 21.4, 19.5, and 14.0 pairs/100 km² of habitat, respectively. The latter (polygon) estimate seemed quite conservative, as it seemed likely that one or more undiscovered territories may lie within the polygon constructed (Fig. 10.1). Our best density estimate is probably the 5.14 km²/pair given by the square method, for an estimated 19.5 pairs/100 km². A minimal conservative density estimate is 14.0 pairs/100 km², as given by the polygon method. Hence we estimate that the 510 km² of non–Scrub Swamp Forest in Tikal may house as many as 71–99 pairs of White Hawks.

As a partly independent check, spot mapping and nest searches revealed an estimated 4.7 pairs in a 20 km² study plot, or 4.26 km²/pair, for an estimated 23.5 pairs/100 km², which exceeds even our MPND density estimate. This gives us confidence that our methods did not overestimate the actual density of White Hawk pairs.

Geographic Variation in Density of Pairs

Whichever of our nesting density estimates is most accurate, there is little doubt that White Hawks had higher nesting densities in Tikal than were reported in French Guiana by Thiollay (1989a), who estimated a density of six pairs per 100 km²—less than one-third the density we estimated at Tikal. Thiollay estimated a home range size 50% larger than we did. Still, this does not fully account for the difference in density estimates between these two studies. Rather, this disparity represents a striking departure from our estimates of the amount of area *not* occupied by White Hawk territories. Thiollay estimated that 62% of a 42 km² study plot was unoccupied by pairs of White Hawks; in contrast, we estimated that no more than 20% of our 20 km² intensive study plot was unoccupied by White Hawk territories. However, our 264 ha mean (284 ha median) home range estimate implies a potential density of 35.0–37.9 pairs/100 km² if home ranges of this size were closely packed. In contrast, our best estimate of actual density was 19.5 pairs/100 km², only half the potential density suggested by our home

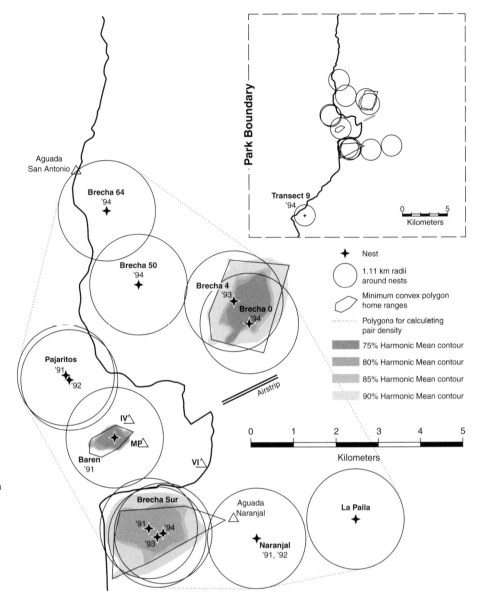

10.1 Tikal National Park boundaries, locations of White Hawk nests studied, with 1.11 km radius buffers, polygon used in estimating density of territorial pairs, and two adult male home ranges indicated. Home ranges shown are 281 ha (Brecha 4 1993) and 335 ha (Brecha Sur 1994), derived by the 95% Harmonic Mean method.

range estimate. In summation, this approach suggests that 20–50% of suitable habitat at Tikal was unoccupied by White Hawks. The strikingly regular spacing among neighboring nests, however (Fig. 10.1), suggests, rather, that suitable habitat at Tikal was essentially filled with White Hawk pairs. We consider it likely that the latter is true, and that our estimates of home range size substantially underestimate the areas actually used by a pair of White Hawks throughout an annual cycle.

The striking difference in density estimates between our study and Thiollay's, though it may in part be due to differences in methods and perhaps in sampling adequacy, is unlikely, we believe, to be completely artifactual. Rather, it seems likely this species does nest at greater densities at Tikal than in French Guiana. Thiollay (1989a) stated that the presence of White Hawk pairs was conditional on the presence of natural openings such as rocky outcrops, rivers, and multiple treefall gaps. In contrast, White Hawk territories at Tikal appeared uni-

formly distributed over areas of Upland and similar forest. Presumably there is something different between Tikal and Thiollay's French Guiana site, in the behavior or ecology of the White Hawk, the nature of the environment, or both. Rainfall at Thiollay's study site is approximately 300–325 cm annually (Tostain et al. 1992)—more than twice the rainfall occurring at Tikal—and we presume that primary forest in French Guiana is taller and structurally more complex than at Tikal. Species richness of many groups of organisms is no doubt substantially higher in French Guiana than at Tikal. How these environmental differences may lead to large differences in White Hawk density between the two sites is unknown. One factor that does *not* appear significant, however, is the species richness of close relatives; while a second species of *Leucopternis* occurred at Thiollay's site, its estimated density was only one pair per 42 km²— far too low, it would seem, to exert a strong influence on the density of White Hawks. It does seem plausible that the greater vertebrate species richness in French Guiana may result in a higher intensity of diffuse food competition by other raptor species, other birds, mammals, and reptiles. Still, the reality of this apparent density difference between the two sites remains in question because of the vastly different methods and sampling intensities achieved by ourselves and Thiollay.

DEMOGRAPHY

Age at First Breeding, Attainment of Adult Plumage

Immature White Hawks of this subspecies are similar to adults but have more black in the primaries, secondaries, and coverts (Blake 1977; Howell and Webb 1995). According to Howell and Webb (1995), the adult plumage is attained with the second prebasic molt at the age of about two years. Breeding females of our study pairs were in typical adult plumage—all white on the wings, except for black on primary tips and in the wrist. Though most breeding males were in adult plumage, the breeding males at two 1991 nests were heavily checkered with black throughout the primaries and secondaries—apparently in subadult plumage. This interpretation was supported by the following observation. In the Baren territory we trapped and banded both breeders in 1991; the male had a good deal of black in his wings. The following year, although we could find no nesting underway, we observed both banded breeders from the previous year, and the male at this time no longer had more black in his wings than is normal for full adults. From this we conclude that he was a second-year bird during his 1991 nesting effort.

Our observations not only suggested that most breeders were more than two years old, but also showed that at least males occasionally breed prior to attaining adult plumage, apparently before they are two years old. This seems rather remarkable considering that the juveniles we monitored via telemetry did not reach independence until 17–19 months after fledging (19–22 months after hatching). If adult plumage is indeed attained at the age of two years, males breeding in juvenal plumage must have done so immediately after achieving independence. This perhaps implies that a high level of foraging proficiency is attained before juveniles become independent and disperse. The age at attainment of adult plumage and at first breeding merit further attention.

Frequency of Nesting, Percentage of Pairs Nesting

A number of observations showed that pairs skipped a year of nesting activity while caring for dependent juveniles from the previous year's nesting. Additional observations showed that when the previous nesting failed or the fledgling died at an early age, adults did not skip a year, but rather nested again the following spring (Ap-

pendix 10.2). Our best estimate is that consecutive successful nesting attempts are begun about two years apart, in the early spring. We detected nesting attempts in 52% (11/21) of territory-years monitored; we regard this as a tentative estimate of the percentage of pairs initiating a nesting effort in a given nesting season.

Productivity and Nest Success

We monitored six territories for two to four years each, for a total of 21 territory-years. For some years at some nests, however, while we did not detect a nesting effort, we were unable to be certain the hawks did not nest undiscovered; hence in one version of our productivity analyses we deleted some territory-years from the sample. We assumed a 1-egg clutch in all cases (see earlier). For eight nests detected before an egg was laid or soon after, results were as follows: of eight eggs laid, six (75%) hatched and four chicks fledged (67% of those hatching); hence 50% of eggs produced a fledgling. Of the four chicks fledging, two died within 2.5 months, one survived at least eight months after fledging, and one survived to reach independence 19 months after fledging. If the fledgling that survived to at least eight months is assumed to have reached independence, then two of four fledglings did so.

If we expand the sample to include chicks that had already hatched when the nest was found, we have eight nestlings in the sample. In addition to those just mentioned, one chick was approximately 10 days old upon discovery, and one was a week prior to fledging; both these additional chicks fledged, for six fledglings in the sample. In this case, 10 eggs produced 8 chicks and 6 fledglings. One of the two latter fledglings died within four to five months, and the other probably survived for several months (though we did not monitor the fledgling, we saw the adults in the nest area frequently the following breeding season, and they gave no evidence of nesting activity). Hence probably three of six fledglings survived to or near independence. One other juvenile, discovered soon after fledging, survived to reach independence 18.5–19.0 months later, for an estimated total of four of seven fledglings surviving to or near independence. Hence our data indicate that about 50% of fledglings survived to reach independence or close to this point.

In 21 territory-years monitored, we detected nesting attempts (an egg laid) in 11 territory-years (52.4%), producing 11 eggs, nine nestlings, and seven fledglings, of which at least four survived to or near independence. Hence, per territory-year, we estimate 0.52 egg laid, 0.43 egg hatched, 0.33 fledgling, and 0.19 fledgling surviving until or close to reaching independence. However, for some of these territory-years, there may have been nesting efforts we failed to detect, such that the results just given may be overly conservative. Another approach is to omit some of the 10 non-nesting territory-years from the sample. In this version we omit one of the apparently non-nesting years for the Pajaritos pair, and all three of the apparently non-nesting years of the Baren pair, where we never detected nesting after 1991. In this scenario we have 17 territory-years resulting in 0.65 egg laid per territory-year, 0.53 egg hatching, 0.41 fledgling produced, and 0.24 fledgling surviving to or near independence. We regard the latter estimates, based on 17 rather than 21 territory-years, as most closely representing actual productivity in our study.

Our best estimate is that 8 eggs/nests produced 4 fledglings (50% success). Our most liberal estimates are that 10 eggs/nests produced 6 fledglings (60%) or 11 eggs/nests produced 7 fledglings (64% success). These success rates are not greatly different from those shown in 10 studies of Red-shouldered Hawks (Crocoll 1994); the five Red-shouldered studies with largest sample sizes showed nest success averaging 62.8% (SD = 8.35, n = 5, range = 53–72%). The Red-shouldered Hawk, similar in size and diet to the White Hawk, may arguably be considered a temperate zone ecological counterpart. Because of differences in clutch size and frequency of nesting per pair, however, productivity per nest and per occupied territory differed greatly between these two species. White Hawks averaged 0.5–0.64 fledgling per nest, or 0.33–0.41 fledgling per territory-year, and about 0.19–0.24 fledgling surviving to independence per territory-year. In contrast, with a mean clutch of 3.1 (n = 5 studies), Red-shouldered Hawks fledged, on average, 1.5 young per nest (SD = 0.26, n = 6, range = 1.1–1.8; Crocoll 1994), and something less than this per territory-year.

Thus, annual productivity per pair appears at least 3.7–4.5 times as great in the Red-shouldered Hawk as in the White Hawk. By the time of independence, the difference is likely greater because of the much longer dependency in the White Hawk. The 50–60% nest success rate we documented for the White Hawk is comparable to that of a number of other tropical accipitrids studied to date (Mader 1982; see other chapters herein).

Sources of Nest Failure and Mortality

The two eggs that failed to hatch did not suffer predation. One suffered embryo death, possibly due to researcher disturbance, while the other may have been infertile; in both cases adults ceased incubating after a longer–than–normal period. Of the two chicks that failed to fledge, one fell during a windstorm at the age of 22 days, and one disappeared at the age of 4–5 days, probably taken by a predator. The adult male White Hawk that disappeared at one nest was assumed to have died, but the cause was unknown.

Parasites and Pests

At times nestlings and adult females on the nest were attended by numerous insects, which in some cases appeared to cause the raptors discomfort. The female at the 1994 Brecha Sur nest was perturbed at times by many small flies; several could often be seen on her beak. Oc-

casionally she used her beak to remove ants crawling on her legs. The nestling here was also perturbed on many occasions (often shaking its head) by flies or bees that crawled on and buzzed around its face in huge numbers. This chick appeared molested at times by the many ants in the nest and sometimes removed them from its legs. The nestling at the 1994 Brecha 0 nest was also sometimes perturbed by many flies or bees.

Falling from nests was a significant cause of death for nestlings of several raptor species at Tikal. Nestlings sometimes moved in response to harassment by insects. Thus it does not seem unreasonable to suppose that harassment by insects may contribute to the incidence of fatal falls. We did not record any instances of White Hawk nestlings suffering botfly infestations.

CONSERVATION

All evidence points to White Hawks at Tikal and the surrounding region being closely associated with tall mature forest. Though one individual and probably more showed a tendency to hunt at treefall gaps at Tikal, this should not be confused with an ability to thrive in areas where the forest has been reduced and fragmented by human activities. A treefall gap—even a multiple-tree gap—is not comparable to a human-made clearing. Two seasons of point counts at Tikal found White Hawks significantly more abundant in the park's mature forest—especially tall Upland Forest—than in the farming landscape nearby. Likewise, ordination of community-wide raptor census results at Tikal portrayed White Hawks as occurring in a portion of "ordination space" corresponding to tall mature forest (D. F. Whitacre et al., unpubl. data).

Apparently the White Hawk is often associated with natural and perhaps anthropogenic forest openings in French Guiana (Thiollay 1984). Throughout the Mexican and Central American portion of the species's range, however, we feel safe in concluding that this species is linked mainly to tall mature forests and will suffer population declines where mature forests are reduced or fragmented. Indeed, Wetmore (1965) noted that in Panama "as land has been cleared for cultivation these birds have disappeared." We believe this is an accurate depiction of the White Hawk's interaction with human-caused deforestation in our study area, and probably in many parts of its range.

HIGHLIGHTS

The only other *Leucopternis* species studied to date is the Gray-backed Hawk (*L. occidentalis*; Vargas 1995), which comprises a superspecies with the White Hawk and has often been considered a subspecies of the latter. Hernán Vargas studied the Gray-backed Hawk in western Ecuador from 1992 to 1994, examining 10 nestings. As in our study, the clutch size was one egg, and the incubation period at one nest was 36 days, within the range (34–38 days) we documented. Young fledged at 56–84 days, averaging 72 days (*n* = 3), similar to our results (60–88 days). Based on 246 identified prey items, the diet (numerical basis) was 59% reptiles, 13% mammals (mainly rodents, *Oryzomys* spp.), 10% birds, 9% crustaceans (freshwater crabs), 6% amphibians, and 3% earthworms. In terms of biomass, reptiles made up 53% of the diet, and mammals 30%. Among the reptiles taken, snakes (*n* = 119) far outnumbered lizards (*n* = 25). While numerous terrestrial snakes were taken, 21 arboreal vine snakes (*Oxybelis* sp.) were taken. Interestingly, among the 25 lizards, *Ameiva* species (*n* = 18) predominated. This terrestrial lizard genus has been mentioned as White Hawk prey by numerous authors. Very likely, the active-search foraging style of these lizards heightens their vulnerability to perch-hunting raptors such as the White Hawk. Amphibians included 12 frogs and 4 caecilians. In sum, a mix of arboreal, terrestrial, fossorial, and aquatic prey were taken. Nests were close to streams, averaging only 76 m from water.

While details of diet differed, no doubt partly because of biological differences between Vargas's well-watered Ecuadorian study site and our relatively dry forest site, the general diet and breeding biology were quite similar in the two studies. One notable difference was that Vargas found these hawks nesting nearly year-round, with six pairs nesting during the rainy season (January–June) and four during the dry season (July–December); this difference occurred despite the fact that in western Ecuador, as at Tikal, there is a single annual rainy season (Vargas 1995).

Perhaps our most notable result at Tikal was our demonstration of a very long period of post-fledging dependency in this handsome raptor, especially for its modest body size. Two radio-tagged juvenile White Hawks were cared for by adults until reaching independence 17–19 months after fledging (19–22 months after hatching). At both sites, the adults did not breed during the year the juvenile remained dependent. These two cases suggested that White Hawk pairs that have successfully fledged a nestling normally skip a year before again attempting to breed. In addition, the period from hatching to fledging (60–70 days, occasionally more) is also long for a raptor of this size.

While it is less surprising to find slow nestling development and prolonged post-fledging dependency in a larger raptor, the interval of dependency demonstrated here is probably the most extreme known for a raptor as small as the White Hawk. This is also one of the smallest raptors thus far known to consistently lay a 1-egg clutch. These traits (small clutch and slow development) are among those often described as typical of a relatively "K-selected" strategy (Pianka 1970), which in turn appears common among tropical forest raptors and other organisms that occur near carrying capacity in relatively stable environments. Beyond this general observation, one wonders whether there may not be something about the White Hawk's foraging ecology that makes it difficult to deliver prey at a faster rate (permitting a larger clutch size or more rapid development) and difficult for fledglings to more rapidly achieve hunting prowess permitting their independence. It is arguable whether these hawks are best regarded as dietary generalists with a penchant for reptiles, or as somewhat specialized reptile hunters. Conceivably, heavy reliance on reptile prey (especially snakes) may prohibit a higher rate of prey delivery to the nest, thereby selecting for a small clutch size and slow nestling development. Also, perhaps this species's largely reptile diet, or its methods of hunting, require an unusual level of experience, contributing to the long period of post-fledging dependency.

Also noteworthy was the apparent greater density of White Hawks at Tikal than that estimated by Thiollay for French Guiana. Better density estimates, however, will be needed from South American localities before we can be sure of the reality of this apparent difference between sites. Finally, there may be geographical variation in the manner in which these hawks are affected by human-caused forest modification and fragmentation—another area for future research.

For further information on this species in other portions of its range, refer to Booth-Binczik et al. 2004, Komar 2003, and Zhang and Wang 2000.

APPENDIX 10.1

Nesting Histories of White Hawk Territories Studied at Tikal, Guatemala

Pajaritos territory. We monitored this territory from 1991 to 1994. Nest 1 (1991) was 22.7 m up in a 53 cm diameter Zapotillo *(Pouteria durlandii)*, 40 m up a modest hill from an area of low-lying Bajo Forest. It was constructed of fresh and old branches, at least partly of Silión *(P. amygdalina)*. We located the nest on 11 March 1991 when the female brought fresh branches; we observed her for the next three days, and on midday of 14 March she had begun incubation. On 15 March we climbed, documenting the 1-egg clutch and nest dimensions. The female abandoned incubation after 44 days. The egg, undamaged, contained a dead embryo; the pair did not re-nest that year. In 1992 a pair nested (nest 5) 29.5 m up in a Danto *(Vaitarea lundellii)* 138 cm in diameter, 150 m from the 1991 nest. As we were unable to trap the adults in 1991, we could not verify whether the same adults were present in 1992. It was clear, however, that these were alternate nest sites within a single territory. This, the tallest White Hawk nest tree we observed, emerged above an area of Bajo Forest. A sizable colony of ants occupied the tree and nest. The egg was laid on 15–17 March 1992 and hatched on 21 April. The nestling disappeared at the age of 3–5 days, probably taken by a predator. In 1993 we detected no nesting in this territory, though we could not be certain the pair did not nest. In spring 1994 we witnessed this pair copulating and carrying sticks but again detected no nesting effort. Though the pair did not nest in either of the historic nest sites, we could not be certain they failed to nest elsewhere, although this was our belief.

Baren territory. On 27 March 1991 we found a nest (nest 2) 22.5 m up in a Jocote Jobo, or a Yellow Mombin (*Spondias mombin*), 84 cm in diameter. The nest was made of fresh and old limbs, at least partly of Malerio (*Aspidosperma* sp.), Copal Colorado (*Cupania belizensis*), and Silión (*P. amygdalina*), with leaves of the latter used as lining. The nest was near the base of a hill cloaked in Upland Forest and was 20 m into a very large area of low-lying Bajo Forest. The single egg, with unknown laying date, produced a nestling that hatched on 2 May and at the age of 22 days was found dead at the base of the tree after an intense thunderstorm the previous night. We searched for nesting efforts in this territory each year from 1992 to 1994, but never again located a nesting effort; in one version of productivity calculations, we delete the latter three territory-years from consideration. The adult female here was radio-tagged twice in 1991, but removed both radios soon afterward. The adult male was also radio-tagged in 1991, yielding 55 relocations.

Naranjal territory. We achieved full monitoring of reproductive activity in this territory from 1991 to 1994. The pair nested in 1991 and 1993, producing a fledgling each time. One survived for at least eight months after fledging, and we believe the other did as well. On 6 June 1991, we found the nest (3) by following a radio-tagged adult female. The nest was 24.2 m up in a 65 cm Honduras Mahogany (*Swietenia macrophylla*) in the center of a large area of Bajo Forest and contained a large nestling, which fledged on 14 June. The nest was constructed on the main trunk where the latter had broken off, and was made of fresh and old branches, at least partly of Honduras Mahogany and Silión. A vine tangle intertwined with the nest, making it nearly invisible from below. We did not monitor this fledgling after it left the nest. In 1992 we conducted extensive observations in the nest vicinity, often seeing the adults but never engaged in reproductive activities; hence we surmise that the 1991 fledgling survived for at least several months, and that the adults did not nest in 1992.

On 5 March 1993 we discovered nest 7 when a pair of White Hawks brought fresh twigs of Marío (*Calophyllum brasiliense*) to the same mahogany tree as in 1991. The old nest had fallen, and a new one had been built in the same position. From 5 to 19 March, the pair continued to rebuild the nest. This nesting produced a juvenile female that, as of March 1994 (eight months after fledging), still remained dependent in the nest vicinity. No hatch date is available. We saw no evidence of nesting in 1994, and the dependent juvenile from 1993 was still several months from reaching independence; it seems safe to conclude the pair did not nest in 1994.

Brecha Sur territory. We fully documented reproductive efforts in this territory from 1991 through late January 1995. On 18 July 1991 we located a fledgling White

Hawk tended by adults. This fledgling remained near (often within 15 m of) a stick nest 16 m up in a young Ramón 30 cm in diameter, on a small hill amid an area of extensive Bajo Forest. A pair of adults was frequently seen in the vicinity, and the nest contained fresh prey remains. As no other raptors were seen in this area, and because the dependent fledgling remained in the vicinity, we concluded this stick nest was the nest from which this juvenile had fledged. After being frequently observed in early spring of 1992, this juvenile was radio-tagged on 17 May 1992 and its behavior studied in detail over a 10-week period during which we took data on 72 perch locations. Based on its subsequent history, we concluded that this juvenile had fledged only weeks before we discovered it on 18 July 1991. The juvenile remained dependent until at least 12 December 1992, when a prey exchange from an adult was last seen; it reached independence between this date and 20 January 1993, at least 17 months after fledging.

In 1992 this pair did not nest, as the juvenile remained dependent throughout the year. On 9 March 1993 we located nest 8 in this same territory, in a Canchán (*Terminalia amazonia*) 300 m southeast of the 1991 nest. The nest was 21 m up in this 71 cm diameter tree and was built out on a limb some distance from the trunk, in a vine tangle. The egg was abandoned when it did not hatch after 50 days of incubation. We found no evidence of re-nesting. In 1994 White Hawks in this territory nested in a third tree, a Manax (*Pseudolmedia oxyphyllaria*) 150 m east of the 1993 nest. The nest, 17 m up in this 28 cm diameter tree, was discovered on 3 March during nest construction and copulation. The egg was laid on 14 March, and we made extensive nest observations from 29 March till the chick fledged on 21 June. We logged 312 hours of observation on 24 days during this period (13 hr/day, two days/wk). The chick hatched 24 April and first left the nest tree on 21 June at the age of 58–61 days. The chick received a radio transmitter on 21 July 1994 and was monitored until its death between 10 August and 12 September, 55 m from the nest tree. It had been eaten by a predator, likely a mammal, but whether it was killed by a predator is unknown. The adult female was banded on 14 June 1994, and the adult male received bands and a radio transmitter on 15 June 1994; from then until 10 April 1995 we made 27 visual contacts and six biangulations on this male. On 31 January 1995 we saw a White Hawk in this territory carrying nesting material; hence we surmise that this pair renewed nesting activity in 1995, some five months after the death of the previous year's fledgling.

Transect 9 territory. We documented all reproduction in this territory in 1992 and 1993. We found nest 6 on 5 March 1992 after observing a pair of White Hawks copulating, calling, and defending their territory against Black Hawk-eagles (*Spizaetus tyrannus*) over a three-week span. We finally located the nest after observing the female carrying branches in her beak to the nest. The nest was 22 m up in a 50.9 cm diameter Danto (*V. lun-*

dellii), and was made of limbs of at least *Aspidosperma* and *Pouteria* species, and lined with leaves of the same. The nest tree was in a zone of Transitional Forest at the base of a small hill. The male nestling, about 23 days old when first observed on 26 May, was estimated to have hatched during the first week of May (approximately 3 May). This chick received a radio transmitter on 8 July and fledged on 30 July at the unusually advanced age of roughly 88 days, after the adult male disappeared on 8 June when the nestling was roughly 36 days old. We assume the adult male had died. Replacing this juvenile's transmitter periodically, we monitored it through 9 March 1994, when it achieved independence and dispersed, and thereafter, obtaining several locations over the four months subsequent to its dispersal. With the aid of telemetry, we gathered habitat use data during 36 visual contacts from the age of 15–22 months. Nearly all observations occurred during the morning hours, from the nest tree to 2 km distant, averaging 500 m from the nest tree. The female did not nest in 1993, when she continued to care for the juvenile. Whether she nested in 1994 after the juvenile dispersed in early March is unknown but seems doubtful. It is also unknown whether she had obtained a new mate by this time, though we saw no evidence that she had.

Brecha 0/4 territory. We documented reproductive performance in this territory in 1993 and 1994. We found nest 9 on 4 May 1993 in an area of Bajo and Transitional Forest. The nest was 20 m up in a Manax 35 cm in diameter and was built in the main fork of the bole, of limbs from at least *Aspidosperma* and *Pouteria* species. A nestling approximately 10 days old was in the nest upon its discovery and fledged 1 July, at approximately 66 days. This female fledgling received a radio transmitter. After it failed in October 1993, we could still always locate her in the nest vicinity by her vocalizations; hence we are certain she died before 13 November 1993. A great number of White Hawk feathers found in the vicinity in mid-October were likely hers. The adult male here received a radio transmitter and was visually located (and habitat data taken) 42 times from 27 May to 8 July 1993, yielding a home range estimate (MCP) of 290 ha. The adult female also received a radio transmitter, which failed prior to the chick's fledging. Her 22 visual localizations during this period (21 June–9 July) thus occurred while she took part in hunting to help supply her own food needs and those of the large chick. These localizations yielded habitat data at perch sites and an estimated 50 ha home range centered on the nest.

The banded pair again nested in 1994. By following the radio-tagged male, we found the nest on 22 February when we witnessed copulation and nest building. The nest was in a Zapotillo *(Pouteria reticulata)* 23.5 cm in diameter and 28 m tall, next to a large area of Bajo Forest, 150 m outside of our 20 km² intensive study plot. The nest was amid a vine tangle and was made of limbs of at least Malerio *Aspidosperma* species. Beginning 24 March, we made frequent daylong observations here, accumulating 131 hours on 12 days. The egg was laid on approximately 12 or 13 March, and the male chick hatched on approximately 13 April. On 10 June, at about 58 days, the chick was found below the tree and returned to the nest by researchers. On 22 June (age approximately 70 days) when we banded the chick, it fledged in response to the climber and was again returned to the nest. The chick fledged definitively around 22–26 June, at the approximate age of 68–74 days. A month later, the fledgling received a radio transmitter; this fledgling was found dead only a few weeks later, cause of death unknown.

La Paila. We discovered this nest on 25 March 1994 when we saw an adult carrying branches to it. The nest was 18 m up in a Zapotillo 43 cm in diameter at breast height and 25 m tall, at the edge of a bajo. It was built of sticks and leaves of at least *Aspidosperma* and *Pouteria* species and was constructed atop an older nest structure. We observed the site periodically from this date through late June. No egg was laid here, and whether this pair nested elsewhere was unknown.

Brecha 50. This nest, within our 20 km² study plot, was discovered on 3 April 1994 after we observed nest building on 16 March. The nest was 19.5 m up in a Zapotillo 42 cm in diameter at breast height and was situated in a fork of five limbs. It was composed at least partly of *Pouteria* limbs and appeared to be an old nest under reconstruction. We observed this site periodically throughout the spring. During April and early May the White Hawks added branches to the nest, but by late May such activity ceased. This nest did not receive an egg, and whether this pair nested elsewhere was unknown.

Brecha 64. We discovered this nest on 14 April 1994. At this time, a juvenile White Hawk, no doubt fledged in 1993, remained dependent on adults. We saw all three birds repeatedly in the vicinity and, in one case, saw an adult transfer prey to the juvenile. The nest was 16 m up in a Zapotillo 36.5 cm in diameter, at the foot of a hill next to a very large bajo. The adults gave no indication of renewed nesting activity that spring.

APPENDIX 10.2
Nesting histories of the six main White Hawk study territories at Tikal

Year	Pajaritos	Baren	Naranjal	Brecha Sur	Transect 9	Brecha 0/4
1991	Nest 1 Egg laid 13/14 Mar, did not hatch (embryo died); incubation abandoned after 44 days, longer than normal period Nest tree: *Pouteria durlandii*	Nest 2 Egg laid prior to 27 Mar; hatched 2 May; chick fell to death in storm on 24 May at 22 days); breeding male apparently a 2 year old Nest tree: *Spondias mombin*	Nest 3 Large nestling when found 6 Jun, fledged 14 Jun; subsequent fate of fledgling uncertain but probably survived several months Nest tree: *Swietenia macrophylla*	Nest 4 Recently fledged young present on 18 Jul discovery of nest; fledged probably in late Jun/ early Jul Nest tree: *Brosimum alicastrum*	No data	No data
1992	Nest 5 Egg = 1 Hatched = yes Fledged = no (nestling disappeared at 3–5 days, probably due to predation) Nest tree: *Vaitarea lundellii*, 150 m from 1991 nest	No nesting at known sites Same banded adults Cannot be certain they did not nest elsewhere in the territory	Much pair activity but no nesting Probably still caring for 1991 fledgling	No nesting; 1991 juvenile still dependent 1991 juv. radio-tagged in May 1992, reached independence between 12 Dec 1992 and 20 Jan 1993	Nest 6 One egg, hatched very roughly 3 May (est. 23 days old on 26 May); fledged 30 Jul, with radio tag; survived to reach independence 9 Mar 1994 Adult male disappeared 8 Jun Nest tree: *Vatairea lundellii*	No data
1993	Activity in same tree as 1992, but no nesting	No nesting at known sites Same banded adults Cannot be certain they did not nest elsewhere in the territory	Nest 7 Egg = 1 Hatched = yes Fledged 1 female, who survived at least to Mar 1994 Nest reconstructed in same *S. macrophylla* as in 1991	Nest 8 Egg laid 17 Mar, did not hatch (addled); abandoned 6 May after 50 days of incubation Nest tree: *Terminalia amazonia*, 300 m SE of 1991 nest	No nesting; 1992 juvenile still dependent	Nest 9 10-day-old chick on discovery on 4 May; Fledged 1 Jul, with radio tag; died by mid-Nov; Both adults radio-tagged Nest tree: *Pseudolmedia oxyphyllaria*
1994	Copulation, nest building in same nest as 1992, but not nesting	No nesting at known sites Same banded adults Cannot be certain they did not nest elsewhere in the territory	No nesting located; probably did not nest; 1993 fledgling still dependent at least to Mar 1994	Nest 10 Egg laid 14 Mar, hatched 24 Apr; fledged 21 Jun; received radio tag 21 Jul; died 10 Aug–12 Sept; found eaten by predator 55 m from nest Nest tree: *Pseudolmedia oxyphyllaria*, 150 m E of 1993 nest	Probably no nesting 1992 juvenile reaches independence and disperses 9 Mar 1994	Egg laid 12–13 Mar, hatched 13–15 Apr; Two premature fledgings 10 and 22 Jun, fledged definitively 22–26 Jun; received radio tag 22 Jul; chick died within 2 months after fledging Same adults as previous year Nest tree: *Pouteria reticulata*
1995	No data	No data	No data	Nest building observed 31 Jan, 5 months after death of 1993 juvenile	No data	No data

11 GREAT BLACK HAWK

Richard P. Gerhardt, Nathaniel E. Seavy, and Ricardo A. Madrid

The "bajos" or seasonal swamp forests of Tikal's low-lying areas support a great diversity of plant and animal life. Scattered emergents tower above the canopy, which is lower here than in other forest types. Small saplings achieve a great density and, together with shrubs, vines, and sedges, often form walls impenetrable to humans. The visitor to this landscape is impressed by the abundance of thorns. The most important survival theme in the bajos, however, is that of water, which tends to occur in feast or famine proportions. During the wet season, bajos are inundated—water flows through arroyo channels and stands waist deep over broad areas, draining away slowly—while in the dry season, deep fissures form in the clay-rich soil. It is this seasonal alternation between inundation and drought that is responsible for the distinctive vegetative composition and stunted structure of this forest.

As one might expect, the assemblage of animals in Bajo Forest also differs from that in Upland Forests nearby. The few year-round water holes existing in Tikal are generally found in bajos. These are visited by a wide array of mammals, including deer, peccaries, Tapirs *(Tapirus bairdii)*, and Jaguars *(Panthera onca)*. Some of the more terrestrial lizards and snakes also abound in bajos, where sunlight reaches the ground through the low, rather open canopy, in treefall gaps, and along the margins of water holes.

One raptor, at least, exhibits a clear preference for bajos at Tikal—Great Black Hawks *(Buteogallus urubitinga)* place their nests above the bajo in emergent trees. Here, late in the dry season, the female incubates her single egg. When the chick hatches, the male (and later the female) brings it a veritable smorgasbord of animal life. Although the Great Black Hawk may have a habitat preference, it is the classical generalist where prey is concerned. Insects, toads, lizards (the bigger, the better), snakes (venomous or not, thin and agile or large and stout, terrestrial or arboreal), active squirrels, sleeping bats and opossums, nestling and adult birds—all are captured and eaten by Great Black Hawks.

Our fieldwork included the first systematic study of the food habits of this species and provided a wealth of new information regarding its breeding biology and behavior. The following pages contain the most comprehensive portrait to date of this handsome hawk.

GEOGRAPHIC DISTRIBUTION AND SYSTEMATICS

The Great Black Hawk is widely distributed from Mexico south to Bolivia, Paraguay, and northern Argentina, and on Trinidad and Tobago (Brown and Amadon 1968; Ferguson-Lees and Christie 2001). These hawks occur from coastal and riverine lowlands up into montane foothills, often to 1000–1500 m in elevation (Monroe 1968; Hilty and Brown 1986; Binford 1989; Stiles et al. 1989; Ridgely and Greenfield 2001), and sometimes to 1800 m (Ridgely and Gwynne 1989; Howell and Webb 1995; Hilty 2003). Two subspecies are recognized: *B. u. ridgwayi* of Central America and Mexico, and the somewhat larger *B. u. urubitinga* of South America. These two forms have sometimes been regarded as distinct species (Ferguson-Lees and Christie 2001).

Until recently, the status of the Great Black Hawk in our study region had remained enigmatic. The first, and apparently only, specimen known for the Department of Petén was collected near Uaxactún, 25 km north of Tikal, in the early 1900s (Van Tyne 1935). Smithe (1966) considered it doubtful that the species nested anywhere near Tikal, and described it as a very uncommon visitor in the park. Since the inception of the Maya Project (Burnham et al. 1988), the species has been regularly observed both in the park and in other areas of the Petén. In 1991, personnel of the Maya Project found the first nest at Tikal (Gerhardt et al. 1993), confirming the species's status as a breeding resident of the area.

MORPHOLOGY

This is a stout raptor, with wings that are broad from front to back, and a relatively short tail. Appendixes 1 and 2 give body mass and linear measurements for *B. u.*

ridgwayi, the subspecies occurring at Tikal. The tarsi are long compared to those of other, similarly sized raptors. Existing data yield a mean male body mass of 1036 ± 165 g (n = 5) and a mean female mass of 1042 ± 254 g (n = 7; Appendix 1) or, if one exceptionally light female (625 g) is omitted, a mean of 1111 ± 192 g (n = 6). In either case, males and females did not differ significantly in weight (P = 0.81, Mann-Whitney U-test, U = 19.0, df = 1, n = 5 males, 7 females), although perhaps they would with a larger sample. Females nominally exceeded males by 0.6–7.2% in body weight, for a (cube root) dimorphism value of 0.2–2.3, but further data are needed to reveal whether this apparent difference is real. The mean Dimorphism Index for wing chord, culmen length, and body mass is 4.0–4.7 (Appendix 2), depending on the body mass used. Hence these raptors appear to be slightly size dimorphic—somewhat less so than most North American buteos (Snyder and Wiley 1976).

The South American subspecies, *B. u. urubitinga*, is larger than *B. u. ridgwayi*. Like several other South American raptors, it varies geographically in size, with birds in southerly, temperate portions of its range being larger than those nearer the equator (Brown and Amadon 1968; Ferguson-Lees and Christie 2001).

Plumage and Soft Parts

Adult Great Black Hawks of the northern subspecies are slaty black with white upper-tail coverts, a broad white band on the upper tail, and white-tipped tail feathers (Plate 11.1). The thighs are black, thinly barred with dusky white. In Tikal birds, the iris is brown and the lores are dark; the bill is black and the legs and feet orangish yellow. The cere has been described as slaty (Brown and Amadon 1968), but in the birds found at Tikal it was orangish yellow; other authors also describe the cere as yellow (Ridgely and Gwynne 1989; Stiles et al. 1989; Howell and Webb 1995).

Great Black Hawks exhibit two basic (post-juvenal) plumages prior to the adult plumage (Howell and Webb 1995). This has been seen in the wild (Dickey and van Rossem 1938) and in captivity, where birds reportedly take 5–6 years to attain full adult plumage (S. Matola, W. Crawford, pers. comm.). It has also been stated that definitive adult plumage is attained at the age of four years (Ferguson-Lees and Christie 2001). At Tikal, an immature Great Black Hawk captured in March had wing and tail feathers representing two age classes. The newer tail feathers exhibited the numerous, contrasting brown and gray bands characteristic of immature plumage. The older wing feathers were buffy, as described for first basic plumage (Howell and Webb 1995), but the newer ones (second basic), while quite dark, were not the black shown by adult Great Black Hawks (third basic plumage; Howell and Webb 1995). Whereas the slaty black adult plumage is striking, the mottled buff and rufous tones of immature plumages create warm and beautiful effects of their own (Plates 11.2 and 11.3).

Tikal chicks were buff colored ventrally and on the thighs, and dark brown dorsally with a dark brown swath on the outer thigh. Their heads were buff with a dark brown, tear-shaped spot on top and a dark brown stripe running through each eye and joining as a band at the back of the head (Plate 11.4). The degree of contrast between the buff and dark brown varied among nestlings. Eyes were dark brown, tarsi and feet yellow, and the cere a slaty blue-gray. Advice for distinguishing Great Black Hawks from Common Black Hawks *(Buteogallus anthracinus)*, especially in juvenal and subadult plumages, is given by Howell (1995) and Howell and Webb (1995).

PREVIOUS RESEARCH

The breeding biology of the Great Black Hawk was previously known only from a handful of serendipitous observations (Brown and Amadon 1968; ffrench 1976; Mader 1981; Thurber et al. 1987), and Hartman (1966) described the egg of *B. u. ridgwayi*, obtained from the oviduct of a collected female. A wide range of prey items has been reported from isolated observations of hunting behavior and prey remains collected beneath roosts or in stomachs (Dickey and van Rossem 1938; Friedmann and Smith 1950; Lowery and Dalquest 1951; Wetmore 1965; Brown and Amadon 1968; Olmos 1990b; Lewis and Timm 1991). Nearly all published dietary accounts are from habitats that, unlike Tikal, have a significant aquatic component throughout the year, such as coastlines, mangroves, and rivers.

RESEARCH IN THE MAYA FOREST

From 1991 to 1994 we studied Great Black Hawks in Tikal National Park. Our studies began in 1991 when we found two active nests. In 1992 we searched for nests, and in 1993 devoted a team full-time to this species, but we had little luck finding nests. In 1994 we had better luck and studied pairs in five territories. In total, we monitored nesting activity during 11 territory-years in five territories, with best data coming from eight of these (see nesting histories in Appendix 11.1). We found nests largely by observing from treetops. Observations of courtship or of hawks carrying nesting material or prey often led to nests. We visited nests every 2–3 days to record nesting phenology, and made behavioral observations from tree platforms approximately 35 m distant. In 1991 we climbed to nests weekly to weigh and measure nestlings, whereas in 1993 and 1994 we avoided climbing to nests until after young fledged or nests failed, except to verify some clutch sizes. We collated unpublished information on clutch size and nesting phenology, using data from egg-set collections of the Western Foundation of Vertebrate Zoology (WFVZ) and the Delaware Museum of Natural History (DMNH). Statistical analyses were conducted using SYSTAT 5.0

(Wilkerson 1990). Except where stated, measures given are means ± 1 SD.

DIET AND HUNTING BEHAVIOR

Great Black Hawks have extremely variable diets and are capable of exploiting a vast array of food types. Published accounts of food items include frogs, lizards (Friedmann and Smith 1950; Wetmore 1965); a snake (Voous 1969); a grebe, Inca Dove (Columbina inca), and a rodent (Dickey and van Rossem 1938); a fish, rodent, and carrion (Mader 1981); nestling Bare-throated Tiger-Herons (Tigrisoma mexicanum: Lewis and Timm 1991); eggs of a Plumbeous Ibis (Theristicus caerulescens: Olmos 1990b); fallen nestlings of egrets (Egretta) and spoonbills (Platalea) (Ferguson-Lees and Christie 2001); carrion, insects, and frogs (Haverschmidt 1962, 1968); nocturnal land crabs (Lowery and Dalquest 1951); and fruits (Bierregaard 1994). Other records also indicate beach scavenging (Todd and Carriker 1922) and extensive feeding on bird eggs and nestlings (Robinson 1994).

We systematically documented prey delivered to three nests at Tikal (Gerhardt et al. 1993 [this article has sometimes been erroneously cited as authored by Brodie and Baness, due to unfortunate pagination in the reprint]; Seavy and Gerhardt 1998), confirming that Great Black Hawks often take a wide variety of prey. Of 130 identified prey items delivered to three nests, 29.2% were lizards, 24.6% snakes, 3.1% either lizards or snakes, 20.8% mammals, 13.9% birds, 6.2% anurans (frogs and toads), and 2.3% insects (Table 11.1).

Reptiles made up the bulk of the diet at Tikal, accounting for nearly 60% of all prey items. Brown Basilisks (Basiliscus vittatus), commonly found near forest edges and openings, made up 70% of the lizards. Wetmore (1965) also commented on the large role played by basilisks in the diet. Arboreal Anolis lizards, presumably captured from forest vegetation, were also observed as prey at Tikal, and at least 12 species of snakes were taken (Table 11.1). The snakes most often identified as prey—the rear-fanged vine snake (Oxybelis sp.) and the Boa Constrictor (Boa constrictor)—are highly and partly arboreal, respectively, but the hawks also captured terrestrial snakes. Most were non-venomous or rear-fanged species, but at least three venomous species were delivered to nests: the Hognosed Pit Viper (Porthidium nasutum), Jumping Pit Viper (Atropoides nummifer), and Eyelash Viper (Bothriechis schlegelii; Table 11.1).

A variety of mammals were also taken, including bats, squirrels, other rodents, and mouse-opossums. Bats delivered to nests were mostly medium sized, weighing roughly 30 g. Three bats identified from prey remains were Jamaican Fruit-eating Bat (Artibeus jamaicensis), a common fruit bat at Tikal, weighing approximately 50 g (Morrison 1983). Rodents in the diet included squirrels (Sciurus spp.), a Big-eared Climbing Rat (Ototylomys phyllotis), and small, unidentified cricetine rodents. Two small opossums—probably Mexican Mouse Opossums (Marmosa mexicana)—were delivered to nests. With the exception of the squirrels, nearly all mammalian prey were nocturnal species, probably vulnerable to predation when encountered during daylight hours. Birds delivered to nests were similarly diverse, including a woodpecker, pigeons, thrushes, and an oriole. While nestlings may have been taken, none of the birds we saw delivered to nests were obvious nestlings. However, two mammals were juveniles—a squirrel (Sciurus sp.) and an unidentified rodent. Anurans and insects were infrequently delivered to nests and difficult to identify. A single Gulf Coast Toad (Bufo valliceps) was found in a nest.

Although observations of foraging were few, the diet at the nest allows us to make some inferences concerning foraging behavior. The variety of prey delivered to our study nests, along with the wide range of prey items and foraging habits reported in the literature, indicate that Great Black Hawks are opportunistic generalists, able to exploit a wide range of prey types and hunting situations. Indeed, no raptor species studied in Tikal had a broader diet than this species (Sutter et al. 2001; see Chapter 23).

That these hawks may often hunt in forest openings or along water holes is suggested by the frequent occurrence of Brown Basilisks in their diet at Tikal; these diurnal lizards are not generally found in tall, unbroken forest (R. Gerhardt, pers. observ.) and are often most abundant along the margins of water bodies (Campbell 1998; Stafford and Meyer 1999; Lee 2000). Some observations of Great Black Hawks hunting supported this conclusion. We watched these raptors move slowly from perch to perch around a small pond, pausing to hunt the shoreline for a few moments from each perch. Wetmore (1926) reported a similar observation in which a pair of birds worked their way along the edge of a lagoon, flying from one mass of reeds to another, at times dropping to the half-submerged vegetation below. In Venezuela, an adult was collected while hunting frogs at a small woodland pool (Friedmann and Smith 1950), and Wetmore (1965) stated that these birds often hunted frogs by perching on low vegetation or on the ground, near small pools and streams, and often in open pastures and fields.

The most detailed hunting observations are from Manu, in Amazonian Peru, where Robinson (1994) found these raptors using several foraging methods that made use of their long legs. They frequently raided the hanging nests of oropendolas and caciques, reaching within to extract 62 nestlings. Great Black Hawks at Manu also frequently walked along river beaches, preying on nestlings of ground-nesting nighthawks; at times they waded in shallow streams, using their wings for support as they reached their talons deep into marsh grass. Finally, Robinson reports several cases of these hawks eating eggs at Hoatzin (Opisthocomus hoazin) nests, as well as one eating an adult Hoatzin and another killing a Roadside Hawk (Rupornis magnirostris) that was mobbing it.

In view of the many records of these hawks feeding on bird eggs and nestlings, it is perhaps not surprising

Table 11.1 Prey items of Great Black Hawks (*Buteogallus urubitinga*) observed at three nests in Tikal National Park[a]

Common name[b]	Scientific name	Caoba and Arroyo Negro nests, 1991 — Number (same as percentage in this case)[c]	Naranjal nest, 1994 — Number[d]	Naranjal nest, 1994 — Percentage (%)[c]	Overall — Total number	Overall — Percentage (%)[c]
Brown Basilisk	*Basiliscus vittatus*	27	2	6.7	29	22.3
Anole	*Anolis* spp.	5	0	—	5	3.9
Helmeted Basilisk	*Corytophanes* sp.	0	3	10.0	3	2.3
Unidentified lizards		1	0	—	1	0.8
Total number of lizards		**(33)**	**(5)**	**(16.7)**	**(38)**	**(29.2)**
Boa Constrictor	*Boa constrictor*	6	0	—	6	4.6
Eyelash Viper	*Bothriechis schlegelii*	1	0	—	1	0.8
Western Indigo Snake	*Drymarchon corais*	3	0	—	3	2.3
Speckled Racer	*Drymobius margaritiferus*	1	0	—	1	0.8
Scarlet Kingsnake	*Lampropeltis triangulum*	*[e]	0	—	*[e]	
Parrot Snake	*Leptophis* sp.	4	0	—	4	3.1
Neotropical Whip Snake	*Masticophis mentovarius*	2	0	—	2	1.5
Green Vine Snake	*Oxybelis fulgidus*	2	0	—	2	1.5
Vine Snake	*Oxybelis* sp.	4	0	—	4	3.1
Hognosed Pit Viper	*Porthidium nasutum*	1	0	—	1	0.8
Jumping Pitviper	*Atropoides nummifer*	1	0	—	1	0.8
Tropical Rat Snake	*Spilotes pullatus*	1	0	—	1	0.8
Berthold's False Fer-de-lance	*Xenodon rabdocephalus*	1	0	—	1	0.8
Snake species		3	2	6.7	5	3.8
Total number of snakes		**(30)**	**(2)**	**(6.7)**	**(32)**	**(24.6)**
Unidentified reptiles		0	4	13.3	4	3.1
Total number of reptiles		**(63)**	**(11)**	**(36.7)**	**(74)**	**(56.9)**
Gulf Coast Toad	*Bufo valliceps*	1	0	—	1	0.8
Unidentified frog or toad		7	0	—	7	5.4
Total number of anurans		**(8)**	**(0)**	**(0)**	**(8)**	**(6.2)**
Tree squirrel	*Sciurus* sp.	3	3	10.0	6	4.6
Cricetid rodent	Cricetidae	3	0	—	3	2.3
Big-eared Climbing Rat	*Ototylomys phyllotis*	0	1	3.3	1	0.8
Total number of rodents		**(6)**	**(4)**	**(13.3)**	**(10)**	**(7.7)**
Jamaican Fruit-eating Bat	*Artibeus jamaicensis*	1	0	—	1	0.8
Unidentified bat		7	1	3.3	8	6.2
Total number of bats		**(8)**	**(1)**	**(3.3)**	**(9)**	**(6.9)**

Common name	Species	No.	No.	%	No.	%
Opossum	*Probably Marmosa*					
	mexicana	1	1	3.3	2	1.5
	Unidentified mammals	0	6	20.0	6	4.6
	Total number of mammals	**(15)**	**(12)**	**(40.0)**	**(27)**	**(20.8)**
Keel-billed Toucan	*Ramphastos sulfuratus*	0	2	6.7	2	1.5
Pigeon	*Columba* sp.	8	0	—	8	6.2
Oriole	*Icterus* sp.	1	0	—	1	0.8
Pale-billed Woodpecker	*Campephilus*					
	guatemalensis	1	0	—	1	0.8
Clay-colored Robin	*Turdus grayi*	0	2	6.7	2	1.5
	Unidentified birds	4	0	—	4	3.1
	Total number of birds	**(14)**	**(4)**	**(13.3)**	**(18)**	**(13.9)**
	Insects	0	3	10.0	3	2.3
	Total identified prey	100	30	100.0	130	100.0
	Unidentified to class	6	14	—	20	—
	Total prey	106	44	—	150	—

[a] Numbers in parentheses are subtotals.
[b] Common names of reptiles and amphibians follow Campbell 1998 and Lee 2000.
[c] Percentage on numerical basis, of prey identified to class.
[d] At the 1994 Naranjal nest, the male brought in 4 mammals, 1 lizard, and 1 snake, whereas the female brought in 1 mammal, 3 lizards, 1 snake, and 2 birds; no size difference was detectable between male and female prey.
[e] One identified from prey remains in nest; not included in numerical calculations.

that several records also exist of carrion feeding. In Colombia, Todd and Carriker (1922) described the diet as being largely of shellfish and crabs washed up on the beach. Haverschmidt (1962) observed a pair feeding on a dead caiman along with many American Black Vultures *(Coragyps atratus)* and Yellow-headed Caracaras *(Milvago chimachima)* and also observed these hawks eating snakes killed by traffic (Haverschmidt 1968). Mader (1981) observed an immature feeding on pig intestines along with American Black Vultures, Turkey Vultures *(Cathartes aura)*, and Southern Crested Caracaras *(Caracara plancus)*. In South America, several authors comment that *B. u. urubitinga* is often seen standing or walking on the ground along riverbanks or on sandbars (Robinson 1994; Ridgely and Greenfield 2001; Hilty 2003); presumably this implies hunting for easy-to-catch or inert prey. Thus, while capable of more rapacious foraging methods, the Great Black Hawk, at some times and places, shows a more scavenging mode of foraging than do any of the other raptors we studied at Tikal.

Prey are apparently captured at all levels of the forest, from ground to canopy. In Trinidad, Trail (1987) observed a Great Black Hawk make a fast, shallow dive at a lek of Guianan Cock-of-the-Rock *(Rupicola rupicola)*, in an unsuccessful capture attempt. Still-hunting from within the canopy may be common, as there were arboreal species represented among each class of prey brought to nests (Table 11.1). The many nocturnal mammals among prey at Tikal—evidently captured by day—and the many reports of feeding on nestlings, bird eggs, and carrion suggest that Great Black Hawks may often invest time in searching for sometimes cryptic but vulnerable and easily captured prey. Their long, nearly bloodless legs presumably enable them to capture venomous snakes with minimal risk, and may aid them in grasp-searching within retreats for nestling birds, hidden lizards, and roosting bats and other mammals. In the parlance of behavioral ecology, such a foraging strategy entails substantial search cost but little pursuit cost—prey may be well hidden, but once found are easy to catch.

HABITAT USE

Great Black Hawks are found in a large variety of habitat types within the tropical and subtropical zones. These include rain forest, savanna with forest patches and gallery forest, forest edge and second growth, marshes, and old pastures with scattered trees (del Hoyo et al. 1994). Many authors describe this species as most often found near water—similar in this regard to the Common Black Hawk—although some authors state that it is less restricted in this regard (Ferguson-Lees and Christie 2001). The Great Black Hawk, however, is clearly less restricted to coastal zones than the Common Black Hawk, and is often the most common black hawk at inland localities (Smithe 1966; Ferguson-Lees and Christie 2001; Ridgely and Greenfield 2001; Hilty 2003). Near water-

ways, Great Black Hawks reportedly occur in fairly dry habitats including dry chaco woodlands, cactus-covered slopes, and pine-oak forest (Brown and Amadon 1968; Ferguson-Lees and Christie 2001). Many authors, however, attribute to this species a fondness for swampy forests, mangroves, and other wetland habitats (e.g., Hilty and Brown 1986; Stiles et al. 1989; Robinson 1994; Howell and Webb 1995), and most previous accounts of the biology of Great Black Hawks have been based on observations of nests or foraging adults closely associated with water (Lowery and Dalquest 1951; Haverschmidt 1962, 1968; Wetmore 1965; Lewis and Timm 1991; Robinson 1994).

The nests we found in Tikal were all situated in primary Bajo (Scrub Swamp) or Transitional Forest types, which are characterized by a low, somewhat open canopy, extremely dense understory, and many emergent trees. Though flooded during the rainy season, these areas were extremely arid during the dry season, when Great Black Hawks laid eggs. Small perennial ponds undoubtedly exist within at least some home ranges, but only one nest was located near a known body of water, a small pond of about 100 m²—and by the end of the dry season, even this pond had nearly vanished. Whereas fish and crustaceans are often mentioned in the literature as important prey items (Brown and Amadon 1968; Bierregaard 1994), there was no aquatic component to the prey we observed at nests in Tikal.

In French Guiana, Thiollay and colleagues (Thiollay 1984, 1989a; Jullien and Thiollay 1996) concluded that within a large area of pristine forest, the occurrence of Great Black Hawks was contingent on the presence of natural openings such as multiple treefall gaps, rock outcrops, or rivers. Prey taken at Tikal also suggested the use of openings and forest edge, but the most conspicuous habitat association was that between Great Black Hawk nests and Bajo Forest and related forest types occurring low in the drainage system. We gathered some habitat use data on a dependent male fledged from the Transect 9 nest in 1994. In 13% of cases (3 of 23), he was found perched in Bajo or Scrub Swamp Forest; 56% (13) in Transitional Forest; 4% (1) in Hillbase Forest; and 26% (6) in Upland Forest. These numbers suggest frequent use of Bajo and Transitional Forest relative to the availability of these forest types.

BREEDING BIOLOGY

Nests and Nest Sites

In Tikal, Great Black Hawks built nests 20–31 m above the forest floor (mean = 25 ± 5 m, *n* = 7), often in association with dense vine tangles (Seavy and Gerhardt 1998). Some nests were located on top of a single large limb 40 cm in diameter, but most were at a two- or three-way fork, supported by living branches 10–15 cm in diameter. One nest was supported only by a thick vine tangle. Nests were positioned well within the tree crown and protected by surrounding foliage of the nest tree and associated vines. Many nest trees and vines were leafless in the dry season, such that nests were rather conspicuous at the onset of breeding but concealed by foliage later in the nesting cycle (Plate 11.5).

The Bajo or Scrub Swamp Forest type in which we found Great Black Hawks nesting is characterized by a low canopy, usually less than 15 m tall, broken by large emergent trees. The hawks placed their nests within these emergent trees: Honduras Mahogany (*Swietenia macrophylla*: two trees), Black Olive *(Bucida buceras)*, Hormigo *(Platymiscium yucatanum)*, Kapok *(Ceiba pentandra)*, and Black Cabbage-bark *(Lonchocarpus castilloi)*. Seven nest trees averaged 29 ± 4 m in height (range = 22–35 m) and 120 ± 81 cm in diameter (range = 66 cm to an astonishing 300 cm). Great Black Hawk nest trees characteristically had little or no crown contact with neighboring trees. Elsewhere, several nests have been observed in mangroves and one each in a pine, a palm, and a Kapok (Brown and Amadon 1968; Haverschmidt 1968; ffrench 1976; Mader 1981).

Previous nest descriptions have been contradictory, with nests described both as deeply cupped (Smithe 1966; Brown and Amadon 1968) and platform-like (Grossman and Hamlet 1964). Six of seven nests observed at Tikal had a platform-like configuration and a seventh had a measurable depression, just 10 cm in depth. Four nests averaged 83 cm in diameter, and three of these averaged 56 cm in external depth. Nests were built of dry sticks and lined with green leafy twigs, which the adults replenished occasionally throughout the incubation and nestling periods.

Egg and Clutch Size

Three nests at Tikal had 1-egg clutches, and two others held a single nestling each. An observation in Mexico by Martin et al. (1954) of four individuals, which they called a family group, may indicate the possibility of a larger clutch size, at least at the northern extreme of the species's range. Clutches of a single egg, however, characterize all of 27 egg sets documented from Venezuela (7 sets), Argentina (18 sets), and Trinidad (2 sets: Swann 1923; Norris 1926; Seavy and Gerhardt 1998, DMNH; WFVZ). Thus, on present evidence, a 1-egg clutch appears the norm throughout the species's range.

For the South American subspecies *B. u. urubitinga*, Girard (1933) described the eggs as quite variable in size and color and gave measurements as 55–65 mm by 50–68 mm. Twelve eggs of this subspecies, from Argentina, averaged 52.1 ± 2.4 mm by 62.4 ± 2.0 mm (Norris 1926), and two from Trinidad measured 47.4 mm by 58.6 mm and 46.9 mm by 57.7 mm (WFVZ). For *B. u. ridgwayi*, an egg ready for laying measured 47.5 mm by 59 mm (Hartman 1966). Based on this latter measurement, Hoyt's (1979) formula predicts a fresh egg and clutch mass of approximately 72.8 g, or about 6.5–7.0% of mean adult female body mass.

Nesting Phenology

Our earliest observation of nesting activity of Great Black Hawks occurred on 23 March, when we observed a pair nest building and copulating (Seavy and Gerhardt 1998). It is likely, however, that some nesting activity begins as early as January and February. Laying dates ranged from late March to early May, and incubation took place entirely within the latter part of the dry season. Nestlings generally hatched in May or June, and young left the nest in late June, July, and early August—early in the rainy season (Seavy and Gerhardt 1998). The timing of nesting thus appeared to be closely linked to seasonal weather patterns. The egg-laying interval was broader than in some of the migratory and insectivorous raptors at Tikal, in which laying was confined to only about a month. Still, Great Black Hawks seemed to nest more synchronously than did the two *Spizaetus* hawk-eagles, in which laying took place over a span of 4–7 months (see Chapters 14, 15).

For Mexico, Central America, and Trinidad, all evidence points to a reproductive phenology similar to that at Tikal, with nesting during the Northern Hemisphere springtime, often beginning during the dry season and concluding during the rainy season. In Panama, an egg ready for laying was obtained from the oviduct of a specimen collected on 17 February (Hartman 1966), and two nests contained young on 17 March and 21 April (Wetmore 1965). In El Salvador a nest on 18 March held a nearly grown nestling (Brown and Amadon 1968), whereas two other chicks fledged around 3 June and in July (Thurber et al. 1987). In Trinidad, eggs were collected on 12 April and 8 May (WFVZ). Two adults collected in Oaxaca, Mexico, on 22 March and 4 April had enlarged gonads (Binford 1989), and a pair in Tamaulipas was observed copulating in June (Martin et al. 1954).

Timing of breeding in Surinam and Venezuela seems different and perhaps demonstrates a broader seasonal nesting window. In Venezuela, Mader (1981) found a nest with an egg on 28 August (this chick fledged in mid-October), while in Surinam, Haverschmidt (1968) reported a nest with a nearly grown chick on 18 October, and a nest with incubation underway on 27 June. In Venezuela, Thomas (1979) reports juveniles seen from July to November, with no details given. Among 12 nests in Argentina, some had clutches each month from August through January (Norris 1926; WFVZ), during the austral spring and summer.

Length of Incubation, Nestling, and Dependency Intervals

One incubation period at Tikal was 40 ± 2 days (Seavy and Gerhardt 1998); to our knowledge this is the only incubation period recorded in the wild. One young made its first flight within the nest tree at the age of 56–61 days (we earlier stated [Seavy and Gerhardt 1998] erroneously that this chick fledged 55 days after hatching), and, we estimate, that it fledged at 62–68 days. Another young fledged at an estimated age of 65–70 days (Seavy and Gerhardt 1998). Apparently the only other published data on the duration of the nestling period are those of Di Giacomo (2000)—the lone chick of one brood left the nest at close to 60 days, and the lone chick at another nest reached the "branching" stage (perched on a limb near the nest, well feathered and seemingly ready to fly), by about 60–66 days (ibid.).

Great Black Hawks at Tikal had a lengthy post-fledging dependency period. We observed an apparent prey delivery to a juvenile that had hatched nearly eight months earlier, and another radio-tagged juvenile remained dependent and within 500 m of the nest to an age of 12 months, when our observations ended; at this time, he showed no sign of approaching independence. Mader (1981) reported an immature, approximately seven months after fledging, perched next to an adult and begging for food. In Argentina, a fledgling was observed periodically near its nest, along with the adult pair, until the approximate age of 7.5 months, about 5.5 months after fledging (Di Giacomo 2000). We tentatively conclude that dependency to an age of at least 12 months is likely.

VOCALIZATIONS

In Tikal, we frequently heard three types of Great Black Hawk vocalizations. During the nest-building period, the birds gave a call composed of a number of short repeated notes, ending with a slightly longer one: *zzi-zzi-zzi-zzi-zzi-zzi-zz-zzzzzzz*. Once nesting had begun, pair members gave a long drawn-out, soft cry: *zzzzzzzzzii-iiiiii*. This was heard occasionally during prey deliveries. A third call, similar to the second but much louder, was given in nest defense.

BEHAVIOR OF ADULTS

Incubation Period

Females conducted most incubation. In five days of observed incubation at the 1994 Naranjal nest during the two weeks prior to hatching, the male incubated on one occasion for half a day; he incubated only one other time, for 4 minutes. Overall, this male spent 9% of his time incubating, 9% in the nest tree, 11% in the vicinity, and 71% of the time out of sight and presumably some distance from the nest (n = 5 days, 57.3 hours of observation). During the same two-week period, the female at this nest spent an increasing amount of time incubating as hatching approached (linear regression, adjusted r^2 = 0.66, $F_{1,3}$ = 8.65, P = 0.06). Overall during this period she spent 68% of the time incubating, 2% standing on the nest, 8% in the nest tree, 9% in the nest vicinity, and 14% of the time out of sight and presumably away from the nest vicinity (n = 5 days, 57.3 hours of observation). The adults at this nest were quite vigilant during incuba-

tion, and for only 5% of the observation time were both adults absent from the nest area. We observed no prey transfers at the nest during the incubation phase, which suggests that both adults met their own foraging needs while off the nest. During these 57 hours of observation, we saw fresh greenery brought to the nest only three times, and only by the male.

Nestling Period

Immediately after a chick hatched, females spent most of their time brooding and standing over it. Males generally delivered prey to a nearby tree, where the female received it to deliver to the nest. At the Naranjal nest, however, the male always delivered prey directly to the female at the nest or nest tree (n = 22). The female then tore off small bits to feed the young. For much of the nestling period, males were away from the nest, presumably hunting. Females carefully protected nestlings from the tropical sun, standing over the chick, with back to the sun and wings partly spread over the nest cup. As the sun changed position, females would also shift, maintaining an effective shading position.

As chicks grew, there was an obvious change in the behavior of adults (Fig. 11.1). Males spent more time near the nest, and females remained away for extended periods, presumably hunting. At the 1994 Naranjal nest, which was observed in detail, this behavioral change was noted 27 days after hatching. Prior to this time the female had never been seen providing prey to the nest, and she had spent only 14% of her time away from the nest; the male captured all 15 prey items brought to the nest prior to this date (Fig. 11.2). After this time the female regularly spent as much as half the day away from the nest and delivered from half to all of the prey items on a given day (Fig. 11.2). From this time (chick day 28) through day 55, when the chick was at the branching stage and our observations ended, the female brought in (and captured, we believe) 76% or 19 of 25 prey items, and the male the remainder. Hence in this case, the bulk of the hunting role passed from male to female during the last month of the nestling period (Fig. 11.2). During this period, the male often provided vigilance near the nest and shaded, brooded, and fed the chick bits of prey—roles that earlier had been performed mainly by the female (Fig. 11.1). As chicks grew older, females also relaxed their assiduous shading behavior. When the Naranjal chick reached the age of six weeks, the female began to respond less and less to the chick's seemingly desperate calls at times when the latter appeared heat stressed; she often perched up to 30 m distant, leaving the nestling exposed to the sun. At such times, the chick went in search of shade on its own.

Prey delivery rates were highest during mid-morning, peaking between 0800 and 1100 at about one delivery every 2.2–2.9 hours (Fig. 11.3). Delivery rates dropped toward midday, to as low as one per 20 hours, and then increased to one every 3–4 hours in the afternoon and early evening. At the 1994 Naranjal nest, during 225.5 hours of observation on 20 days during the nestling

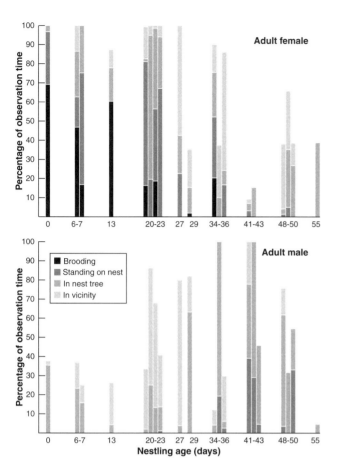

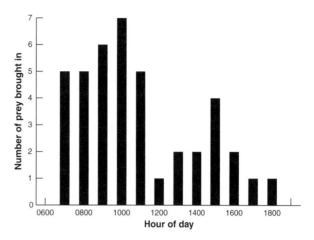

11.3 Time of day adults delivered prey to a Great Black Hawk nest during the nestling period at Tikal: Naranjal nest, 1994. Note: hourly delivery rates are not adjusted for amount of observation time during each hourly period, but observation time was nearly equal for each hourly period.

11.1 Time budgets of a male and female Great Black Hawk during nesting at Tikal, Guatemala: Naranjal nest, 1994.

pair, on 9 June, when their nestling was 36 days old, was twice seen copulating about 200 m from the nest. These were the only copulations observed here in 225 hours of observation on 20 days during the nestling period.

BEHAVIOR AND DEVELOPMENT OF NESTLINGS

Nestling Period

Young were essentially immobile upon hatching, raising their heads only for feeding. The following notes are for the 1994 Naranjal nestling. Before the chick's 34th day after hatching, few detailed notes on its behavior were taken. On day 34, the chick already showed most adult-like behaviors, including preening and shaking its plumage. On this date, the chick vocalized along with adults several times, especially during prey deliveries. On following days, the chick also called when left alone for long periods, and once when a vulture passed overhead. At 34 days, in a stiff breeze, the chick was first noted spreading and flapping its wings; wing flapping was frequent by 42 days. The chick also was first seen eating unassisted at 42 days. At 49 days, the chick was nearly fully feathered and successfully defended itself from the attack of a Collared Forest Falcon (*Micrastur semitorquatus*) while the adults were away. On this day and thereafter, the chick often made small jumps while flapping. At 55 days, the chick had reached the branching stage and clambered around on vines and limbs up to 25 cm from the nest. After a six-day hiatus in observations, the chick was first seen flying at 61 days—from the nest to another limb of the nest tree—a distance of 6 m. We assume this chick fledged (i.e., flew beyond the nest tree) within the next week, at 62–68 days.

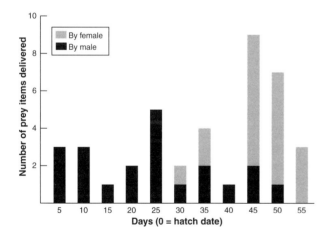

11.2 Changing roles of male and female as hunter at a Great Black Hawk nest at Tikal: Naranjal nest, 1994.

period, we saw the male bring fresh greenery on 14 occasions and the female on seven occasions (together, one sprig/10.7 hr). Thirteen sprigs were believed to be of Silión *(Pouteria amygdalina)*, and eight were believed to be of Honduras Mahogany.

One notable observation involved copulation of an adult pair during the nesting period. The 1994 Naranjal

POST-FLEDGING DEPENDENCY PERIOD

At the 1994 Transect 9 nest, from the time of fledging until the age of one year, we found the juvenile male on average 270 ± 170 m from the nest (n = 24 observations, range = 0–600 m). For this juvenile, mean perch height was 15 ± 4 m (n = 23), in trees averaging 19 ± 5 m tall (n = 23) and 49 ± 32 cm in diameter (n = 22), on perch limbs averaging 9 ± 8 cm thick (n = 22). In 5% of cases (1 of 21), this juvenile was found on a perch protruding above the canopy, in 67% (14) he was just at the upper canopy surface but in a position overlooking an area of canopy, in 19% (4) he was perched within the canopy, and in 9% (2) he was in the upper portion of the sub-canopy space. We had the impression this juvenile often chose emergent perches from which he could survey the surroundings for the approach of an adult with food.

SPACING MECHANISMS AND POPULATION DENSITY

Territorial Behavior and Displays

On clear, sunny mornings these hawks often soared alone or in pairs over the forest. Like Thiollay (1989b), we interpret such soaring flights as functioning mainly in territorial advertisement, and this appears to be the primary mechanism through which pairs space themselves out over the landscape. In one case during the nestling period, a mated pair, during mid-morning, flew about 800 m from their nest and briefly soared together. In Peru, three adults were seen engaged in aerial maneuvers suggesting courtship (Schulenberg and Parker 1981). These birds are reported to dangle their legs while soaring in circles and calling (Howell and Webb 1995).

Nest Defense and Interspecific Interactions

During the nestling period, adults were vigilant near the nest, and several times we watched them drive off potential predators. A male was observed chasing Black-handed Spider Monkeys *(Ateles geoffroyi)* from the area of one nest, a female attacked a group of Plain Chachalacas *(Ortalis vetula)* as they approached another nest, and a pair of Great Black Hawks drove both Black Hawk-eagles *(Spizaetus tyrannus)* and Ornate Hawk-eagles *(S. ornatus)* from the nest area. In contrast, Great Black Hawks repeatedly tolerated the close approach of a pair of Bicolored Hawks *(Accipiter bicolor)* in a situation where nests of the two species were less than 200 m apart.

CONSTANCY OF TERRITORY OCCUPANCY, USE OF ALTERNATE SITES

Nesting territories of Great Black Hawks appeared to remain relatively stable from year to year. When we checked prior-year nesting areas, they generally were occupied by a pair, although incidence of nesting efforts varied annually. One territory was occupied each year from 1991 to 1994, another was studied in 1991 and 1994, and two others were occupied in both 1993 and 1994 (Appendix 11.1). While in one territory the same nest was used for three of the four years we collected data, in other cases alternate nests were constructed in the same tree, or 20 and 350 m from a previous nest. Including all data, consecutive nestings averaged 60 ± 140 m apart (range = 0–350 m, n = 6).

Home Range Estimates

Although we radio-tagged one adult male, we did not obtain enough data to allow estimation of home range area.

Inter-nest Spacing and Density of Territorial Pairs

Figure 11.4 shows locations of the nests we studied at Tikal. The four northerly nests were, we believe, in neighboring territories and yielded a mean inter-nest distance of 4.46 ± 1.09 km (range = 3.48–5.63 km, n = 3). Using this mean inter-nest distance, the Maximum Packed Nest Density (MPND) method (see Chapter 1) yields an estimate of 18.1 km² of unique space per pair, the square method gives 19.9 km²/pair, and the polygon method 21.8 km²/pair (mean of these three = 19.9 km²/pair). These correspond to densities of 5.5, 5.0, and 4.6 pairs/100 km², respectively. Based on this, one can project that Tikal's 576 km² may hold as many as 26–32 pairs; however, it is not clear whether these densities can be extrapolated fairly to the overall mix of forest types within the park. Within the heart of the park, our estimate of 4.6 pairs/100 km² is a proven minimum density, and the range 4.6–5.5 pairs/100 km² probably spans the actual density. In the llanos of Venezuela, the centers of two adjacent territories were about 1.8 km apart (Mader 1981), implying a local density as high as one pair per 3.25 km², and in southeast Peru, Scott Robinson (pers. comm.) made the very similar density estimate of approximately three territorial pairs per 10 km².

DEMOGRAPHY

Age at First Breeding, Attainment of Adult Plumage

In captivity these hawks require 5–6 years to attain adult plumage (S. Matola, W. Crawford, pers. comm.), and the existence of at least two distinct subadult plumages suggests this is true in the wild as well (Dickey and van Rossem 1938). Howell and Webb (1995) and Howell (1995) state that adult plumage is attained with the third prebasic molt, which presumably begins at the age of about three years. Whatever the exact age at which

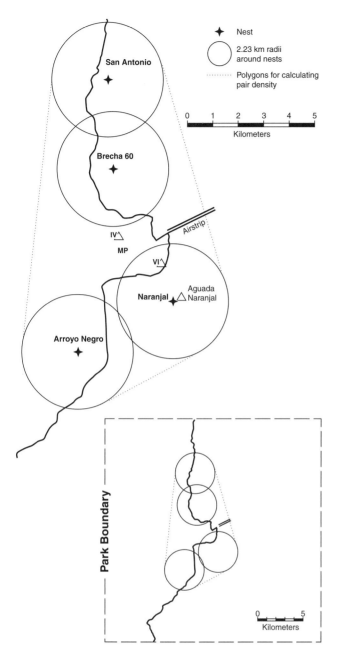

11.4 Locations of Great Black Hawk nests we studied in Tikal National Park, 2.23 km buffers around nests, and polygon used for estimating unique space per pair.

adult plumage is attained, Great Black Hawks probably do not often breed until then. This is suggested both by the clear distinction between the adult and immature plumages and by the fact that birds in immature plumage were not observed soaring above the canopy.

Frequency of Nesting, Percentage of Pairs Nesting

If juvenile dependency to the age of at least one year is typical, we may expect that pairs successfully raising a chick to independence must at best nest every other year. Hence we predict that somewhere in the neighborhood of half of all territorial pairs would initiate nesting attempts in any given year. Using the subset of eight best-documented territory-years, an egg was laid in half of these.

Productivity, Nest Success, and Mortality

Including nests discovered when the chick was near fledging, 10 territory-years produced six eggs and three fledglings (Appendix 11.1), indicating a productivity of 0.27 fledgling/territory-year; this number could be higher if some nestings occurred undetected by us. The best data are obtained by restricting consideration to eight nests discovered before egg laying or early in incubation. Based on these data, productivity and reproductive success of Great Black Hawks at Tikal were low. In three years of study, we observed eight cases of nest building and copulation, half of which apparently did not result in a clutch being laid (Appendix 11.1). Of the five nests in which birds laid eggs, two failed of unknown cause after 15 and 37 days of incubation, and three eggs hatched; two young fledged and the other suffered predation at 5–6 weeks, likely taken by a Tayra *(Eira barbara)* or one of several native cats. Overall reproductive success of laying pairs was three fledglings per six attempts, and productivity was three fledglings in ten territory-years. Among pairs that failed to lay eggs, some may have had dependent young from the previous year's nesting; however, in such a case, one might expect not to observe nest reconstruction and copulation, which we did.

We observed no juvenile mortalities during the post-fledging dependency period, but our small sample precludes conclusions based on this. We observed one instance of adult mortality, when a Boa Constrictor ate a radio-tagged adult male. This male was apparently healthy when last seen on 24 August 1994. Searching via radiotelemetry, we failed to locate him on 31 August and on 12, 20, and 22 September. On 30 September, the male's radio signal was tracked to the stomach of a large boa 865 m from the nest.

CONSERVATION

Some literature portrays this species as quite tolerant of habitat modification, but evidence is contradictory. In Argentina, Olrog (1985) stated that these birds remained common even in areas of extensive forest clearing and could be found nesting on telephone and utility poles. As counterpoint, Lehmann (1970) stated that deforestation had eliminated this species from the Valle del Cauca in Colombia, and a later study (Alvarez-López and Kattan 1995) verified this conclusion. At Tikal, our point counts detected Great Black Hawks significantly more often in primary forest than in the farming landscape (Whitacre et al. 1992a), and Principal Components Analysis ranked this as among the raptor species most associated with primary forest as opposed to heavily altered habitat

(D. Whitacre, unpubl. data). Thus, while this species is apparently resilient to deforestation in some situations, we found no evidence of this in our study area, and the degree of its tolerance in this regard remains unclear. At any rate, Great Black Hawks are not likely to serve as a sensitive indicator of subtle forest disturbance. While wholesale habitat destruction would obviously affect Great Black Hawks along with many other organisms, these raptors' high degree of dietary generalism suggests substantial plasticity, and their habitat use also appears somewhat flexible. Indeed, among our most significant findings is that these hawks are not always as closely associated with aquatic habitats as has often been implied.

Still, the species's frequent use of a specific nesting habitat (at least at Tikal) may increase its vulnerability to anthropogenic habitat changes. The Bajo Forests favored by this species for nesting at Tikal are often among the first areas deforested by subsistence farmers, since soils are deep in these sites, and soil moisture is seasonally higher than in the thin-soiled, rocky Upland Forests. However, large areas of Bajo Forest remain within the huge Maya Biosphere Reserve, which bodes well for this species in our study area. Over a broader region, this species's somewhat generalized habitat selection, diet, and foraging repertoire give reason to hope these handsome birds will long soar over tropical canopies from Mexico to Argentina.

HIGHLIGHTS

Much remains to be learned about Great Black Hawk population ecology. We have not yet determined home range sizes, turnover rates, or the extent of the nonbreeding population. Data now in hand do, however, allow some speculation. Many of this raptor's life history characteristics—including delayed acquisition of adult plumage, probably a delayed onset of breeding, low reproductive effort and success, and an apparently long dependency period—place these raptors toward the "slow" end of the "slow-fast" life history continuum that is conspicuous among birds (Saether 1987, 1988).

While several raptor species at Tikal included snakes in their diet, the Great Black Hawk was the only one of our study species distinguished by both eating and being eaten by the same species of snake—the Boa Constrictor. Given larger samples, we might also find the snake-eating Laughing Falcon (*Herpetotheres cachinnans*), Crested Eagle (*Morphnus guianensis*), and White Hawk (*Leucopternis albicollis*) occasionally involved in similar relationships with the boa and other bird-eating snakes such as the Tropical Rat Snake (*Spilotes pullatus*), Neotropical Bird Snake (*Pseustes poecilinotus*), and the Western Indigo Snake (*Drymarchon corais*).

Our estimate of the density of Great Black Hawks at Tikal, at about 4.6–5.5 pairs/100 km^2, was lower than the 8 pairs/100 km^2 estimated by Thiollay (1989a, 1989b) for a primary forest site in French Guiana. This runs contrary to the pattern shown by most of the raptor species

studied at both sites; for most, our density estimates for Tikal were far higher than those of Thiollay in French Guiana. Differences in sampling methods and intensity make it difficult to know what proportion of these apparent differences are real. For this species, however, it seems likely that densities were indeed higher in French Guiana than at Tikal. Thiollay (1989a, 1989b) concluded that the Great Black Hawk was the most abundant large raptor at his study site; this was certainly not true at Tikal, where the Great Black Hawk ranked among the least numerous raptors (see Chapter 23). Thus, a real difference in density between these two sites seems likely. Also, two territories in Venezuela (Mader 1981) were spaced more closely than any we found at Tikal, again indicating that local densities are at least sometimes higher in northern South America than those we estimated at Tikal.

At Manu, Peru, Scott Robinson's estimate of 3 pairs/10 km^2 scales up to 30 pairs/100 km^2—compared to approximately 5 pairs/100 km^2 at Tikal and 8 pairs/100 km^2 at Thiollay's French Guiana site. We can offer no suggestion as to why nesting density of this species is apparently higher in Venezuela than at Tikal, whereas for most other raptors the reverse appears true. Much evidence, however, indicates that Manu is an extremely productive site, at least in part due to its position in the floodplain of the Rio Manu—perhaps this helps explain this species's apparently higher density there than in French Guiana and Tikal. In any event, confirmation of these tentative regional differences in population density is needed.

Robinson's (1994) observation of frequent Great Black Hawk predation on nestling caciques and oropendolas in Peru stands in contrast to the results of our nest observations at Tikal—whereas a few birds were brought to nests, none were obvious nestlings (although a few juvenile mammals were taken). The simplest explanation may be that these opportunistic raptors readily exploit large colonies of nesting birds within their territories, but territories in Tikal may not have contained such colonies. Furthermore, Robinson was observing cacique and oropendola colonies, which provided a greater opportunity to witness raptor predation on nestlings. A direct comparison of overall diets or diets at the nest would be better able to evaluate whether predation on nestling birds was more important in Peru than in our study site.

Many aspects of the nesting and foraging ecology of the Great Black Hawk deserve further attention. While these hawks apparently preferred Bajo and Transitional Forest as nesting sites, we do not know how they apportioned their foraging efforts within the various habitat types available. Indeed, detailed aspects of habitat selection remain virtually unknown. The fact that we observed no prey exchanges in five days (57 hours) of nest observation during the incubation period is intriguing, as it suggests that the female may have met her own food needs by hunting during this period. This would be unusual in such a large sub-buteonine raptor (Newton 1979), and requires confirmation. However, at the same nest, the male also fed, brooded, and stood watch over the nestling to an unusual degree, once the chick reached the age of four weeks—and at the same time, the adult female captured 75% of the prey brought to the nest. Thus, this species may prove unusual in the degree of equitability in sharing of parental roles during nesting—this would be a worthwhile hypothesis to test in further research. Moreover, because this species shows a highly variable diet, and because size and dispersion of prey items are believed to affect parental role division in raptors (see Chapters 6, 7; Seavy et al. 1998), this widespread raptor may prove an ideal species in which to test hypotheses regarding adult role partitioning in relation to characteristics of the prey base and diet.

For further information on this species in other portions of its range, refer to Beaumont 2005, Carvalho Filho et al. 2006, and Di Giacomo 2000 and 2005.

APPENDIX 11.1

Nesting Histories for Great Black Hawk (*Buteogallus urubitinga*) Territories Studied in Tikal National Park, Guatemala.

Arroyo Negro territory. In this territory, the same nest was used in 1991, 1993, and 1994. The nest was 22 m up in a 27 m Black Olive *(Bucida buceras)* tree, 100 cm in diameter, situated in the wooded bed of the Arroyo Negro, a wet season watercourse. In 1991 one egg was laid and a chick fledged; adults were not banded. In 1993 the adults reconstructed the same nest, but no egg was laid. In 1994 the adults reconstructed the nest and copulated, but again no egg was laid. The nest in 1994 was supported by two limbs 11 and 18 cm thick and partly by a vine tangle, and was 60 cm across and 25 cm in external depth, with a flat upper surface.

In 1994 we observed this nest for 30 hours during the pre-laying period, from 22 March to 13 May. The nest had already been fully constructed, and we witnessed the addition of nesting material only once, on 23 March, when we also observed two copulations, 5–10 seconds in duration. The pair frequently arrived at the nest and manipulated nesting material, and the female often settled into incubation position. However, the pair's visits to the nest grew less and less frequent and finally ceased. Whether the pair nested somewhere else in the territory we could not be certain.

Brecha 60 territory. This territory was studied in 1993 and 1994, the 1994 nest being 20 m from the previous nest. The 1993 nest was 31 m up in a 31.5 m Hormigo *(Platymiscium* sp.) tree, 90 cm in diameter, and the 1994 nest was 22 m up in a huge, 22 m tall, half-dead Kapok *(Ceiba pentandra)*. No egg was laid in 1993; adults were not banded. In 1994, we observed this nest for 30 hours during the pre-laying phase, from 14 March to 27 June. Although this pair did add material to the nest, we did not witness copulation and no egg was laid. Whether the pair nested elsewhere in the territory is not known with certainty, but is doubtful.

Naranjal/Brecha Sur territory. This territory was studied in 1993 and 1994. In 1993 the birds built a nest and laid an egg, but the nest was abandoned after 37 days of incubation. The nest was 28 m up in a 31 m Black Cabbage-bark *(Lonchocarpus castilloi)* 80 cm in diameter. The 1993 nest was 800 m due south of the Naranjal water hole, and the 1994 nest, 30 m up in a 35 m Honduras Mahogany *(Swietenia macrophylla)* 120 cm in diameter, was 200 m south of the previous nest, in arroyo forest. The 1994 nest was amid a vine tangle that helped support it. In 1994, incubation was underway when the nest was discovered on 19 April. The nest was observed on five days (57.33 hours) during the last two weeks of incubation, and the chick hatched on 4 May. Both male and female took part in incubation, for spells of 4–6 hours. Adults did not bring prey to the nest, and both pair members apparently hunted to meet their own needs during their time off the nest. We observed the nest on 20 days (225.6 hours) during the nestling period. From the date of hatching (4 May), we observed the nest frequently over a 55-day span, until 28 June, when the chick made its first flight within the nest tree and our fieldwork ended. In this nest, fully exposed to the sun, the female shaded the nestling carefully, as she did the egg during incubation. The adult male was banded and radio-tagged on 8 June 1994, and the adult female was banded on 31 May 1994. The chick was banded and radio-tagged, and we made 23 visual contacts from 9 August 1994 to 13 March 1995, when the fledgling remained dependent and within 500 m of the nest.

San Antonio territory. This nest was situated near a large, marshy water hole amid extensive mature forest and was studied only in 1994. The nest, 53 cm across and 20 cm in external depth, was 20 m up in a Honduras Mahogany 88 cm in diameter and 25 m tall, in Transitional Forest. In 1994 we observed this nest for 32 hours during the pre-laying phase, from 29 March to 4 May. Although the nest appeared fully constructed, we saw the pair add limbs on eight occasions, and witnessed four copulations of 3, 5, 10, and 20 seconds in duration. The pair was very active in the nest vicinity during this period. During the incubation phase, we observed this nest for 30 hours from 4 May to 15 June. The egg was laid on 4 May, and the female incubated from this date until 15 days later, when the egg was lost to unknown cause. The pair abandoned the area soon after, and no further activity was seen here.

Transect 1 territory. A nest was first discovered here in 1991, 20.5 m up in a 27 m Honduras Mahogany 66 cm in diameter. One egg was laid, and the resulting chick died at the age of 5–6 weeks, apparently killed by a mammal; adults were not banded. This territory was studied again in 1994; the pair used a different nest built in the same tree used in 1991. This nest was discovered with a large nestling, about 60 days old, already present. This chick received a band and radio transmitter and, after fledging, was monitored through 16 May 1995, when it still remained dependent and within 500 m of the nest.

12 ROADSIDE HAWK

Theresa Panasci

The Roadside Hawk *(Rupornis magnirostris)* is the smallest and most common resident buteo-like raptor in the forests of Tikal and among the most common hawks generally in the lowlands of Central and South America. Aptly described as "a fairly typical buteonine generalist" (Bierregaard 1994), this species has borne a host of vernacular names, including Insect Hawk, Gray-tailed Hawk, Large-billed Hawk, and Tropical Broad-winged Hawk (Dickey and van Rossem 1938; Brown and Amadon 1968). The common name now in use is appropriate only to a degree—in addition to occurring frequently along roads and in other open and human-modified habitats, this small, handsome hawk also occurs in mature forest of certain types, which may surprise even many seasoned Neotropical birders.

Often found perched in a lone tree in a clearing, this hawk is generally tame and easy to approach, incessantly vocalizing its petulant-sounding *eeeaah* call when approached (Plates 12.1 and 12.2). This is not a highly aerial species, exhibiting limited soaring and flap gliding, especially over the nesting area early in the breeding season. Most hunting is done from a perch, the hawk making a gliding descent toward prey on the ground or vegetation. These hawks hunt in open areas and also beneath the forest canopy, taking a variety of small vertebrate and invertebrate prey.

GEOGRAPHIC DISTRIBUTION AND SYSTEMATICS

The Roadside Hawk has long been recognized as atypical of the genus *Buteo,* with respect to plumages, molts, osteology, and proportions (Brown and Amadon 1968). First described two centuries ago, by 1844 this hawk was placed in the genus *Rupornis,* which was also considered during the late nineteenth and early twentieth centuries to include Ridgway's Hawk *(Buteo ridgwayi)* and the White-rumped Hawk *(B. leucorrhous:* Hellmayr and Conover 1949; Mayr and Cottrell 1979). Friedmann (1950) and some later authors retained *Rupornis* as a monotypic subgenus of *Buteo* containing *magnirostris,*

and Amadon and Bull (1988) regarded *B. (Rupornis) magnirostris* as the most primitive species of *Buteo,* linking the latter with the sub-buteonines. Johnson and Peeters (1963) presented the case that the Roadside Hawk belongs to a group of closely allied "woodland buteos" also containing the Ridgway's Hawk, Red-shouldered Hawk *(B. lineatus),* Broad-winged Hawk *(B. platypterus),* and Gray Hawk *(B. [Asturina] nitidus).* Amadon (1982) commented that such a group should also include the White-rumped Hawk *(B. leucorrhous),* but not the Gray Hawk.

A molecular study (Riesing et al. 2003) has greatly clarified relationships within *Buteo* and allies, and confirmed the judgments of early taxonomists who believed the Roadside Hawk atypical of *Buteo.* This study found several problems with *Buteo* as formerly defined, and recommended the resurrection of the monotypic genus *Rupornis* for the Roadside Hawk.

The Roadside Hawk is a common and widespread resident throughout its tropical and subtropical range, occurring at lower and middle elevations of Mexico (with occasional Texas records) and Central America south to northern Argentina and Uruguay, including a number of Caribbean islands (AOU 1983). Over most of its range, this species is sympatric with two widely distributed woodland buteos, the Gray Hawk and Short-tailed Hawk *(B. brachyurus),* and in some areas, with the Zone-tailed Hawk *(B. albonotatus).* As forests have been cleared, the Roadside Hawk is said to have expanded to higher elevations, up to 2500 m in northern South America and to nearly 3000 m locally in Peru (Fjeldsa and Krabbe 1990).

Roadside Hawks show ample geographic variation in size and coloration, and as many as 16 subspecies have been recognized (Peters 1931; Hellmayr and Conover 1949), though more recently Brown and Amadon (1968) and del Hoyo et al. (1994) listed only 12 subspecies. Guatemalan birds were attributed by earlier authors (Griscom 1932; Paynter 1955; Smithe 1966) to the subspecies *direptor,* which was not recognized by Brown and Amadon (1968) and del Hoyo et al. (1994). These more recent authors regard *griseocauda* as ranging from northern Mexico to western Panama, including our study area.

MORPHOLOGY

The sexes are similar except in size, females being slightly larger than males. Combining all data for *R. m. griseocauda* gives a mean male weight of 259 g (*n* = 13) and female weight of 284 g (*n* = 15; Appendix 1). Using only Tikal data, the Dimorphism Index (DI) based on (cube root) body mass is 5.3; including weights from elsewhere (Appendix 1), it is 3.1. Linear measurements (Appendix 2) yield a DI of 3.6 for wing chord, 4.0 for tail length, 6.0 for culmen length, and 6.0 for tarsus length.

PREVIOUS RESEARCH

In addition to treatments in field guides and regional works (e.g., Wetmore 1965; Monroe 1968; Oberholser and Kincaid 1974; Rowley 1984), a number of publications describe geographic variation and subspecies in the Roadside Hawk (e.g., Peters and Griscom 1929; Hellmayr and Conover 1949). As this hawk is abundant and easily collected, virtually every publication describing a collection of birds made within its range (e.g., by Chapman, Friedmann, and Wetmore) includes information on this species, and many give brief data on nests, eggs, behavior, diet, and habitat. Among the accounts giving most ecological and behavioral information are Russell 1964, Wetmore 1965, Oberholser and Kincaid 1974, Mader 1981, Stiles and Janzen 1983, Thurber et al. 1987, Beltzer 1990, and Haverschmidt and Mees 1994.

RESEARCH IN THE MAYA FOREST

Initial studies of Roadside Hawks at Tikal were conducted by Vásquez Marroquín and Reyes Moreno (1992), who studied six nestings. A larger body of work was conducted by myself and three field assistants: José M. Castillo, Nehemias Castillo, and Francisco O. Tovar. In 1993 and 1994 we studied this species in an 8.25 km² (5 x 1.65 km) study plot in primary forest near the heart of Tikal National Park, and in an 8 km² (4 x 2 km) study plot in the slash-and-burn farming landscape 10 km south of the park. Studying 32 nestings, our main goal was to compare the species's breeding ecology and success in primary forest and the human-altered landscape. We collected data on habitat affinities, nest site selection, nesting density and success, productivity, prey delivery rates, diet, and foraging behavior. Key results are presented elsewhere (Panasci and Whitacre 2000, 2002) and are summarized here along with additional information. Unless stated otherwise, results are given as means + 1 SD.

DIET AND HUNTING BEHAVIOR

The Roadside Hawk is a dietary generalist, taking a variety of large invertebrates and small vertebrates. Early collectors reported stomach contents ranging from lizards, salamanders, small snakes, mice, and birds to caterpillars, spiders, grasshoppers, and beetles (Dickey and van Rossem 1938; Lowery and Dalquest 1951; Haverschmidt 1962). Haverschmidt (1968) described this species as mainly a lizard eater, also taking frogs, snakes, and large insects. Likewise, Wetmore (1965) characterized the prey as mainly lizards, large Orthoptera, and other insects, "with an occasional small bird, usually a young one, or a mouse." Some other authors have also emphasized the role of lizards in the diet. Other items listed in the diet include bees, ants, centipedes, scorpions, and occasional bats (Ferguson-Lees and Christie 2001).

Previous studies of Roadside Hawk diets have been few and based on small samples. In Argentina, based on 45 prey items from 22 stomachs, Beltzer (1990) documented a diet of insects and spiders (78%, numeric basis), frogs and toads (11%), fish (7%), and rodents (4%), with one grasshopper species comprising 51% of the diet; on a biomass basis, rodents and amphibians no doubt contributed a greater percentage of this sample. Another Argentinian study (Massoia 1988) reported skeletal prey remains found below a perch attributed to the Roadside Hawk. These 19 prey items included 10 Brazilian Guinea Pigs *(Cavia aperea)*, two Lutrine Opossums *(Lutreolina crassicaudata)*, two Cape Hares *(Lepus capensis)*, one large White-eared Opossum *(Didelphis albiventris)*, and four unidentified birds. These remains, in our opinion, are very unlikely to represent prey of Roadside Hawks, and we suspect this perch was used by some larger raptor. For Amazonian Peru, Robinson (1994) gave records of these hawks taking three orthopterans, two cicadas, a snake, and seven lizards, four of the latter being small *Ameiva* lizards taken on the ground. Robinson also cited Groom's (1992) finding that Roadside Hawks ate eggs and nestlings of Sand-coloured Nighthawks *(Chordeiles rupestris)* on riverside beaches at the same Peruvian site.

At six nests studied earlier at man-made openings in the forest around Tikal's Maya ruins, of the 52 identified prey items, 37% were reptiles (13 lizards and 6 snakes), 29% rodents, 15% birds, 17% insects, and 2% frogs; one other item was unidentified (Vásquez Marroquín and Reyes Moreno 1992). On a biomass basis, rodents and birds would probably assume somewhat greater and insects less importance than just indicated. In our larger study at Tikal (Panasci and Whitacre 2000), reptiles and amphibians made up 81.4% of identified prey, with lizards contributing over half of the 140 prey items delivered to nests (Table 12.1). *Anolis* lizards (which are largely arboreal, mostly occurring in the forest interior) were identified only at nests in the primary forest, whereas *Sceloporus* lizards were identified only at nests in the farming landscape. We rarely observed *Sceloporus* species in the interior of tall forest at Tikal, and more often sighted them in forest-edge or other somewhat open habitat; at times, these lizards were very common in slash-and-burn agricultural fields with abundant dead trees and limbs on the ground. Other prey items delivered to nests were insects (9.3%, especially cicadas and grasshoppers), mammals (7.9%, with one squirrel, eight other rodents, and two bats), and birds (1.4%; Table 12.1). Insects, lizards, and frogs typically were delivered to nests intact. In contrast, most mammals were decapitated or partially eaten before delivery, and birds were generally plucked before being brought to the prey exchange site (Panasci and Whitacre 2000).

Nests in Tikal's primary forest and the farming landscape differed significantly in prey composition (Panasci and Whitacre 2000). At forest nests, proportionally more insects and mammals were brought in, while in the farming landscape, amphibians and reptiles played a larger role. This dietary difference likely reflects differences in availability of prey types within the home ranges of the pairs studied, but we have no data on the relative abundance of these prey types in the forest and slash-and-burn habitats. We would want further data before venturing a guess as to whether Roadside Hawk diets in our study area typically differ between mature forest and farming landscapes.

Roadside Hawks most often (40/44 observed cases) used the typical buteonine "still-hunting" technique: hunting from a perch and dropping or gliding down to capture prey (Panasci and Whitacre 2000). In areas of non-continuous canopy cover, as along a dirt road or in man-made clearings, hunting perches were usually 10–25 m up in an emergent tree. In the forest, half of 32 hunts were launched from perches protruding above vegetation, and the other half from perches beneath the canopy, where these hawks perched 8–15 m up to hunt lizards or insects on the ground and vegetation below (Panasci and Whitacre 2000). In three cases we observed Roadside Hawks

drop steeply downward 5 m from a tree limb perch to seize a lizard from the trunk of the perch tree. In four cases hawks in clearings and along forest edge launched hunts from perches 1–5 m high, taking adult cicadas from tree trunks or other vegetation 1.0–2.5 m above ground. Also along a clearing edge, we watched a banded male descend from a 2.5 m high perch, capture a tarantula at the opening of its burrow, remove the spider's legs, and consume it there. Finally, in two cases we observed hawks that were walking on the ground seize a beetle in their talons. Of 44 capture attempts, 84% (39) were successful.

We twice observed these hawks use an aerial style of hunting (Panasci and Whitacre 2000). We observed an unsuccessful aerial pursuit of a Squirrel Cuckoo (*Piaya cayana*) and the capture of a perched Black-cowled Oriole (*Icterus prosthemelas*); the hawk climbed from a 1 m perch to catch the oriole, which was perched approximately 6 m up. Other observers have also occasionally reported Roadside Hawks capturing small birds. For example, Dickey and van Rossem (1938) saw a Roadside Hawk snatch a honeycreeper from a feeding flock. A number of authors have mentioned that local people regard this hawk as a thief of young chickens, as did residents of our study area. While there is no reason to disbelieve this allegation, observations by ornithologists have apparently been few.

Various authors have characterized Roadside Hawks as opportunistic hunters, and our results agree; in several cases we observed these hawks exploit certain prey during brief periods of high availability. For example, after a night of heavy rain a Roadside Hawk was observed

Table 12.1 Number and type of prey brought to nests by Roadside Hawks in primary forest and in a slash-and-burn farming landscape in Petén, Guatemala

Prey type	Primary forest	Slash-and-burn farming landscape
Reptiles	23	57
Spiny Lizards (*Sceloporus* spp.)	0	4
Anoles (*Norops* spp.)	9	0
Helmeted Iguanid (*Corytophanes* spp.)	0	1
Unidentified lizards	11	51
Skinks (Scincidae)	2	0
Snakes	1	1
Amphibians	9	25
Frogs	9	24
Toads	0	1
Mammals	9	2
Spiny Pocket Mouse (*Heteromys* spp.)	4	0
Big-eared Climbing Rat (*Ototylomys phyllotis*)	0	2
Unidentified rats	2	0
Deppe's Squirrel (*Sciurus deppei*)	1	0
Bats (Chiroptera)	2	0
Birds	0	2
Insects	9	4
Cicadas (Homoptera: Cicadidae)	7	1
Grasshoppers (Orthoptera: Acrididae)	2	2
Caterpillars (Lepidoptera)	0	1
Total	**50**	**90**

catching frogs at a newly refilled water hole (Panasci and Whitacre 2000). In two other cases, nocturnal prey were taken during periods of dim lighting, when they may have become active—two bats were delivered to a nest in late afternoon during overcast weather just prior to a rainstorm, and on another overcast afternoon, four juvenile *Heteromys* species were delivered to a nest during a few-hour period (Panasci and Whitacre 2000); these spiny pocket mice are nocturnal and generally remain in burrows by day (Emmons 1990). Dickey and van Rossem (1938) described these hawks as often being attracted to fires, where they exploited injured, dead, or fleeing fire victims; in the process, the hawks sometimes scorched their wing and tail feathers, at times reducing these to stubs. Similarly, in two cases we saw perch-hunting Roadside Hawks catch a small rodent and a grasshopper fleeing agricultural fires. Roadside Hawks also attended army ant swarms to capture small organisms fleeing the ants; we witnessed ant-following behavior on seven occasions. Several authors have noted that these hawks are among the most frequent respondents to squeaking noises made by ornithologists to mimic distressed small animals (Dickey and van Rossem 1938; Wetmore 1944), and this was also our experience at Tikal (D. Whitacre, pers. comm.). Brown and Amadon (1968) reported that these hawks were attracted to noises of nestling birds and would tear open bird nests in search of prey. Finally, a colleague once saw several Roadside Hawks feeding on a termite swarm in Amazonian Colombia (A. Lieberman, pers. comm.); while this may seem unusual, even various eagles (Matthiessen 1972) have been known to feed on fat-rich winged termites (Wiegert and Coleman 1970)—the Häagen-Dazs of the natural world.

Noting this species's somewhat accipiter-like proportions, Johnson and Peeters (1963) raised the question of whether this species may fill an "accipiter niche." At Tikal, we found little evidence of this, if one characterizes an accipiter niche as involving a diet largely of birds and a hunting style involving stealth, ambush, and bursts of speed at close range. Records of birds in the diet are few, and the hunting style at Tikal was typically buteonine.

HABITAT USE

The Roadside Hawk occurs in a wide variety of habitats throughout its range. In at least some areas, it no doubt has increased in abundance along with deforestation and forest fragmentation. It is most often encountered in pastures, agricultural lands, woody successional habitats, forest and river edges, gaps in mature forest canopy, and in woodland and thinned forests. In our study area, however, these hawks occurred also in certain mature forest types restricted to low-lying portions of the topographic gradient. Roadside Hawks seem to occur to some extent in practically all New World tropical lowland habitats except deserts, areas of monotypic row crops, open plains devoid of trees, and dense, closed-canopy forests.

In northeastern Venezuela, in a series of systematic counts, Friedmann and Smith (1950) found this species—along with the Gray Hawk—twice as common in deciduous seasonal forest as in the edge of this forest habitat, and undetected in open savanna. Many authors have noted that in heavy rain forest regions, this species is generally rare or absent in the forest except in openings (Griscom 1932; Wetmore 1943; Bierregaard 1994; Jullien and Thiollay 1996). Wetmore (1957, 1965), however, noted that in Panama these hawks occurred in extensive areas of primary forest, where he believed them to avoid the sub-canopy spaces, perching instead where they were shaded from above but had a view over lower portions of the forest canopy, where he believed them to hunt.

In our study area, the habitat used by Roadside Hawks was largely consistent with indications in the literature. We found these hawks hunting along forest edges, in canopy gaps, in forest clearings, in primary and second-growth forests, and along water bodies. In addition, we observed these hawks hunting beneath the intact canopy of primary forest, which has not to our knowledge been reported before.

Roadside Hawks at Tikal consistently nested (n = 32) in emergent trees protruding above surrounding vegetation (Panasci and Whitacre 2002). In our mature-forest study plot, all nests were in various types of low-canopied Bajo Forest occurring in low-lying areas (i.e., Scrub Swamp and allied forest types); these hawks avoided nesting in the tall, closed-canopy Upland Forest that occupied 90% of the study plot. While other factors may have been involved in this selectivity, the most judicious interpretation, we believe, is that these hawks nested in Bajo Forests to utilize the tall, isolated emergent trees occurring there. Other nests within the park were in isolated trees amid large clearings around Maya ruins and park facilities. Robinson (1994) also reported that Roadside Hawks and numerous other raptors nested in isolated trees in Peru. Isolated and emergent trees, having minimal connection to adjacent canopy, may provide some safety from climbing predators. In the farming landscape near Tikal, 20 Roadside Hawk nests showed no clear association with any particular forest type, and occurred in isolated groups of trees amid cattle pastures or crop fields and woody successional vegetation (Panasci and Whitacre 2002). Presumably hawks in the farming landscape did not need to selectively use any particular habitat in order to have access to isolated or emergent nest trees.

One interesting pattern we noted at Tikal was the following. Both Roadside Hawks and Gray Hawks—a slightly larger species—were common in the farming landscape adjacent to Tikal. In contrast, within the park's primary forests, the Roadside Hawk was a common nester in certain forest types, whereas the Gray Hawk was essentially absent from the park. Despite the superficial resemblance of these two species, apparently some aspect of Gray Hawk habitat requirements is not met in Tikal's forests. We noted some difference in nest placement between Roadside and Gray Hawks in the farming landscape. Three Gray Hawk nests were in larger trees (mean diameter = 143 ± 20 cm, n = 3) than were Roadside

Hawk nests (mean diameter = 25 ± 13 cm, n = 20). Gray Hawk nests were bulky stick nests, larger than those of Roadside Hawks, and all were placed in a fork in the main trunk, though we have seen nests elsewhere placed laterally on large limbs. All three Gray Hawk nests were in topographic depressions, along the edge of mature forest fragments. Elsewhere we have noted Gray Hawk nests in more open habitats in slash-and-burn farming landscapes.

BREEDING BIOLOGY

Nests and Nest Sites

Roadside Hawks at Tikal built flat platform nests (n = 32) loosely constructed of sticks. All 12 forest nests and 12 of 20 slash-and-burn nests were concealed from above or below, or both, by vine tangles. All nests contained some green leaves in a slight depression in the nest center. Nests averaged 34.7 cm by 26.3 cm across (n = 16) and were placed in a branch fork above and away from the main trunk (n = 22), or in a fork of, or against, the main trunk (n = 10). Nesting situations differed somewhat between the mature forest and farming landscapes. In the mature forest, 12 nests averaged 20.3 ± 3.0 m above ground, and nest trees averaged 23.4 ± 3.0 m in height and 54 ± 24 cm in diameter; in the farming landscape, nests averaged only 10.6 ± 3.3 m up in trees averaging 13.0 ± 4.2 m tall and 25 ± 13 cm in diameter (Panasci and Whitacre 2002). In the forest, all nests were in broad-leaved trees, except for one in a palm. In the farming landscape, various tree species were utilized, including a dead snag. In the primary forest, canopy height near the nest averaged 12.6 ± 3.5 m, whereas in the farming landscape, it averaged 3.8 ± 3.2 m; in both cases, nests averaged 7 m above the surrounding canopy (Panasci and Whitacre 2002).

Egg and Clutch Size

Data we collated revealed that clutch size varies geographically in the Roadside Hawk, with clutches near the equator typically being a single egg, and clutches of two predominating at higher latitudes, along with occasional clutches of three. In Venezuela and Surinam, the most common clutch size is one egg, whereas from Costa Rica to Mexico the usual clutch is two, with occasional 1- and 3-egg clutches. Among eight Venezuelan nests, Mader (1981) found six with 1-egg clutches and two with a single nestling each, and Cherrie (1916) reported a 2-egg clutch from the middle Orinoco region. Haverschmidt and Mees (1994) stated that the normal clutch size in Surinam is one egg, and the Penard collection from Surinam contained five nests with one egg each (Hellebrekers 1942). For Colombia, both 1- and 2-egg clutches are reported (Allen 1905; Todd and Carriker 1922), and the same is true for Brazil, much of which lies well south of the equator. In Argentina, de la

Peña (undated) reported a clutch of three and a brood of two, and Salvador (1990) reported five 2-egg clutches and one 1-egg clutch. In northern Argentina, A. Di Giacomo (pers. comm.) found seven 2-egg and two 1-egg clutches. All accounts from Costa Rica northward that make general statements about clutch size assert a normal clutch of two (Smithe 1966; Stiles and Janzen 1983; Howell and Webb 1995). In Mexico, the vast majority of clutches are of two eggs, with occasional 1- and 3-egg clutches (see below).

We analyzed latitudinal patterns in Roadside Hawk clutch size, using data from 127 clutches in the collection of the Western Foundation of Vertebrate Zoology (WFVZ) and 16 published records (Allen 1905; Cherrie 1916; Mader 1981; Haverschmidt and Mees 1994). These 143 clutch records included 106 clutches with two eggs, 31 with one egg, and 6 with three eggs. In Mexico, the mean clutch size was 1.94 ± 0.40 (n = 117). We compared the size of 20 clutches from within 13° of the equator (three from Colombia, two from Costa Rica, one from Guyana, five from Surinam, and nine from Venezuela) with that of 123 clutches from higher latitudes (15–32°)—one from Argentina, two from Brazil, three from Paraguay, and 117 from Mexico. Of the 20 clutches from lower than 13° latitude, all were 1-egg clutches except for a pair of 2-egg clutches from Costa Rica and one from Venezuela; this yields a mean clutch size of 1.15 ± 0.37 (n = 20). Among 123 clutches from higher latitudes, 14 were 1-egg clutches, 103 were 2-egg clutches, and 6 were 3-egg clutches, for a mean clutch size of 1.94 ± 0.40 (n = 123). Including the Tikal data in the latitudinal analysis did not change the outcome, although the mean clutch size for high-latitude nests drops slightly to 1.82 ± 0.48 (n = 151). Among all 143 records, clutch size averaged 1.83 ± 0.48 (n = 143). We compared the size of 20 clutches from within 13° of the equator with that of 123 clutches from higher latitudes (15–32°). For the low-latitude clutches, mean clutch size was 1.15 ± 0.37 (n = 20), whereas for clutches above 13° latitude, mean clutch size was 1.94 ± 0.40 (n = 123), a significant difference (Mann-Whitney U = 315, P = 0.3 × 10^{-11}; regression of clutch size on latitude significant at P = 0.5 × 10^{-9}, with adjusted r^2 = 0.30).

A regression on latitude accounted for 30% of the variance in clutch size (P = 0.5 × 10^{-9}; adjusted r^2 = 0.30). Rather than describing this as a case of latitudinal variation in clutch size, however, it is perhaps most accurate to state that clutches from the Venezuela-Guyana-Surinam region were smaller than clutches from other areas collected to date; latitude is one factor that covaries with clutch size in this sample. At Tikal, we recorded 20 clutches with one egg and 8 with two eggs, for a mean clutch size of 1.29 ± 0.46 (n = 28); there were proportionally more 2-egg clutches (6 of 17) in the slash-and-burn habitat than in the forest (2 of 11), but this was not a statistically significant difference.

Ten eggs from Oaxaca, Mexico, averaged 45.1 mm by 37.1 mm (Rowley 1984). Based on these dimensions,

Hoyt's (1979) formula predicts a mean fresh egg mass of 34.0 g, or 11.9% of mean female body weight. Hence the mean clutch at Tikal (1.29 eggs) amounts to 15.3% of mean female mass, while modal clutches of one and two eggs in different regions represent about 11.9 and 23.8% of mean female mass.

Nesting Phenology

At Tikal, Roadside Hawks laid eggs from late March through May, and nestlings fledged in June and July. Most fledging dates (12 of 15) were between 6 and 26 June, whereas the earliest fledging was noted on 23 May and one chick each fledged on 7 and 15 July. Eggs were laid in the driest period of the year (late dry season), and young fledged during the first month or so of the rainy season. Such timing, with fledging taking place early in the rainy season, is common in Central American birds (Skutch 1950). Insect populations at Tikal and elsewhere in Central American lowlands generally reach their annual peak early in the rainy season (see Chapter 2). The same is true of most small birds, and likely some other vertebrate populations as well (see Chapter 2).

Ample evidence indicates that reproductive timing of Roadside Hawks in Mexico and Belize is similar to that at Tikal (Wetmore 1943; Rowley 1984; Paynter 1955; Russell 1964), with eggs being laid late in the dry season. For 74 freshly laid clutches from northeastern and southeastern Mexico, mean date of collection was 11 April + 13 days (n = 74, WFVZ data). Farther south in Central America, nesting is also reported in the dry season, but it may begin somewhat earlier than in our study area. In Costa Rica, Stiles and Janzen (1983) reported nesting from January or February to April or May, while for El Salvador, Thurber et al. (1987) reported a nest with eggs on 2 February and one with a nestling on 17 February. In Venezuela, Mader (1981) saw courtship in early April (late dry season) and described eight nests—seven with eggs or young hatchlings from 5 May to 10 June, and one with eggs on 11 August, possibly a re-nesting after earlier failure. As the single annual rainy season in Mader's study area occurs in May–December (Mader 1981), his nesting dates fall mainly early in the rainy season, not unlike the timing at Tikal. Dates given by Haverschmidt and Mees (1994) and Hilty and Brown (1986) also support a March–May breeding season in northern South America.

In Tikal, we observed two cases of Roadside Hawks re-nesting after a failed breeding attempt; both pairs rebuilt partially destroyed nests and laid a new clutch. Vásquez Marroquín and Reyes Moreno (1992) also observed one re-nesting, which occurred in June.

Length of Incubation, Nestling, and Post-fledging Dependency Intervals

Vásquez Marroquín and Reyes Moreno (1992) documented 37-day incubation periods for each of two nests in Tikal, and we documented five additional incubation periods. At one nest, eggs hatched after 35 and 37 days of incubation, while at another nest, the eggs hatched after 33 and 35 days of incubation; at a third nest, an egg hatched after 36 days of incubation. Thus, incubation periods ranged from 33 to 37 days, averaging 35.7 ± 1.5 days (n = 7). Nestlings at one nest were hopping to nearby branches at 31 days; both fledged early at 35 days when the nest tree was felled, and both survived. At another nest, the nestling fledged at 38 days.

The duration of post-fledging dependency is unknown. We found no indication, however, that Roadside Hawks have a prolonged post-fledging dependency. Nesting occurred in consecutive years in many territories, in one case involving a marked pair of adults—thus, if prolonged fledgling care takes place, it did not prevent many pairs from nesting in consecutive years. We observed two 1-year-old juveniles perched near nest sites that were successful the previous year, although whether these were their natal nests was not known; no parental care was observed. In these two cases, alternate nest sites within the same territories had young in the nest at the time these juveniles were sighted.

VOCALIZATIONS

Adult Roadside Hawks of both sexes used three distinct vocalizations. The *eeeah* vocalization was a whistled call described as the "complaint call" by Wiley and Wiley (1981) and a "whining whistle" by Brown and Amadon (1968). This call has been aptly described as sounding petulant or angry. Roadside Hawks voiced this call when disturbed or approached by humans. Specific body movements followed this vocalization; hawks would jerk their tails to one side and peer about to locate the source of disturbance. Adults also voiced this whistle when we climbed nest trees.

The second common vocalization was a long *kee kee kee*, ascending at first to a constant pitch, described as the "song" by Wiley and Wiley (1981). This call, somewhat reminiscent of that of a large woodpecker, takes the form of a dry, monotonous series of *kees* repeated at the rate of about five notes per second, for several seconds. This call was given frequently during courtship displays and throughout the breeding season in the presence of mates and intruders. The *eeeah* whistle often preceded the *kee kee kee* or song vocalization when other raptors or conspecifics were within a Roadside Hawk's territory. For example, as one pair's nestling was eaten by a conspecific yearling, they continually gave the whistle call followed by the *kee kee kee* song.

The third vocalization can be described as the begging call. This was a short, low-intensity whistle *(wheet)* usually given in a two-note sequence, with a slight pause between each such pair of notes. Sometimes this two-note call was given only once, but typically it was repeated every 2–3 seconds for a period of time. Young in the nest made this vocalization, as did females who had recently lost nests, in addition to both adults during copulation, prey deliveries, and prey exchanges. Adults sometimes voiced a single *wheet* or three or more in a row, the latter especially during copulation.

BEHAVIOR OF ADULTS

Pre-laying and Laying Period

In March, courtship began and Roadside Hawks became highly vocal. A single male or pair would perform aerial displays while vocalizing continuously. The main display consisted of a male circling above its territory, climbing higher and higher, while loudly voicing the *kee kee kee* vocalization. Conspecifics would respond from the forest with the same call. The male would continue to circle and call, or tuck its wings and swoop down, only to begin circling upward again. Twice we observed interactions in which two pairs of Roadside Hawks interacted in territorial aggression. In both cases, two hawks interlocked talons and fell through the sky in a series of somersaults. In one case the hawks maintained their grip and fell through the canopy; we found them still interlocked, one hanging below a limb and the other above, where they pecked at each other for seven minutes before releasing their grip.

Pairs built a new nest every year, often in a different tree, although we twice observed pairs building new nests in the same tree used the previous year. During nest building, both sexes brought in green leafy twigs. Copulation took place prior to and during the nest-building phase, often during the morning when nest building was most often witnessed. Pairs copulated in the nest tree or a nearby emergent—often one that was later used as a prey exchange site. Before copulation, one pair member gave the begging call, and the mate responded with the same vocalization. The pair would continue to call until the act was completed.

Incubation Period

Females conducted most incubation, with males covering the eggs briefly while the female ate after prey exchanges. The typical prey exchange took place as follows. The male would perch in the prey exchange tree, announcing his arrival with the begging call. Responding with the same call, the female would leave the nest and perch near him in the exchange tree. Both birds continued to call as she approached him and received the prey.

Nestling Period

Early in the nestling period, when only the male hunted, he delivered prey to the exchange site and the female would then take the prey to the nest. Males transferred prey to females talon to talon, and females carried prey to the nest in this fashion, except for cicadas, which were transferred beak to beak and carried by females in the beak. In the nest, the female gave the begging call, and the young responded with a softer version of the same.

Holding the prey in her talons, the female tore off small pieces with her beak and fed the young. As the nestlings grew older—usually when contour feathers began to appear—adult females began to hunt, their roles in this increasing until they brooded the nestlings only at night and during bad weather. At nests with two-nestling broods, females began to hunt about 10 days after the first egg hatched, and both sexes delivered prey to the nest thereafter. At nests with single nestlings, females began hunting at a somewhat later date.

Prey deliveries occurred throughout the day, with 10 deliveries the most observed during a dawn-to-dusk (14.5-hour) nest observation. Delivery rates did not differ significantly between the early nestling period (the first 17 days after hatching; mean = 0.226 prey/hr/chick) and the later nestling period (second 17 days; mean = 0.232 prey/hr/chick), even though females contributed more prey during the second period (Panasci and Whitacre 2002). There was no significant difference between delivery rates at one- and two-nestling nests or between nests in the primary forest (0.37 ± 0.01 prey/nest/hr) and those in the farming landscape (0.44 ± 0.08 prey/nest/hr).

Post-fledging Dependency Period

Parental care of fledglings continued for at least a week after young left the nest, after which we have no information. At nests visited during this time, the fledglings were heard begging throughout the day, in and around the nest tree, the volume of calling apparently correlated with the chicks' hunger level. A parent arriving with prey would voice the begging call from a nearby perch, and fledglings would respond with the same call. We observed both parents bringing prey to fledglings ($n = 5$ nests).

BEHAVIOR AND DEVELOPMENT OF NESTLINGS

Nestlings could hold their heads up for short intervals soon after hatching, and their activity increased with age. Nestlings' bodies were covered in an off-white down, their heads were a creamy beige, and the down around their eyes was gray. The beak tip was black, and the rest of the beak a dirty yellow, as were the cere and legs. Nestling eye color was a smoky gray (Plate 12.3). Within 2–3 days after hatching, longer, gray down feathers covered the heads of nestlings. Within 2–3 weeks, the tips of the brown and cinnamon juvenal plumage could be seen pushing through the skin. At about 30 days, the young were first noted leaving the nest and perching on nearby branches, where they flapped their wings. Nestlings issued a soft begging call when the female was in the nest with a prey item, and sometimes in the absence of any adult.

Siblicide

We suspected that siblicide occurred at two forest nests and observed it clearly at another (Panasci and Whitacre

2002). In the latter incident, the larger of two nestlings attacked the smaller by rearing up and pecking it on top of the head. This occurred throughout one day, until the smaller nestling moved to the edge of the nest and fell 8 m to the ground. At another two-nestling forest nest, food was subsidized daily by humans (wildlife filmmakers) to the brooding female over a 40-day period, and no sibling aggression was documented during this time. Within two days of discontinuation of the food subsidy, the larger nestling began attacking its smaller sibling, which was soon found below the nest tree, vocalizing forcefully while being attacked by a mob of Brown Jays *(Cyanocorax morio)*, which repeatedly pecked its bleeding head. We found no direct evidence of siblicide or sibling aggression in the farming landscape.

SPACING MECHANISMS AND POPULATION DENSITY

Territorial Behavior and Displays

Roadside Hawks tended to defend a more-or-less circular area approximately 150–200 m in diameter. Within each such territory (n = 42) was a core use area usually containing two emergent trees or a nest site. These emergent trees were used for perching, night roosting, and prey exchanges. Roadside Hawks attacked conspecifics (n = 11) or other large raptors (n = 9) that entered the core use area, including Black Hawk-eagles *(Spizaetus tyrannus)*, Ornate Hawk-eagles *(S. ornatus)*, Great Black Hawks *(Buteogallus urubitinga)*, Bicolored Hawks *(Accipiter bicolor)*, and White Hawks *(Leucopternis albicollis)*. Territorial flights varied in intensity but usually began with the resident hawk giving an *eeeah* whistle and a *kee kee kee* call from a perch, then circling upward over the territory, vocalizing intermittently. Hawks used rapid bursts of flapping to climb, interspersed with soaring flight while circling. Once the resident hawk gained peak altitude (about 80–250 m), it would tuck its wings and dive at the intruder. We observed males and females performing such territorial activities together, diving on intruders until the latter withdrew. When we broadcast recordings of Roadside Hawk vocalizations, resident males would sometimes perform the territorial flight in apparent response, at times diving into the forest near us (n = 6). The flight displays described earlier in this chapter, under Pre-laying and Laying Period, probably served territorial as well as courtship functions, and the cases of neighboring pair members falling with interlocked talons attest to the vigor of territorial defense.

Constancy of Territory Occupancy, Use of Alternate Nest Sites

The number and locations of territorial pairs in both study plots were fairly constant between years. Remarkably, while vegetative cover changed dramatically each breeding season in the slash-and-burn plot, locations and

occupancy of Roadside Hawk territories remained quite stable. Limited observations suggested that individual hawks in the forest plot often showed fidelity to traditional nesting areas. One banded male was observed on the same territory for four years, and a banded female was seen on her same territory for three years (R. Thorstrom, pers. comm.). Three males that we banded in 1993 as members of non-nesting territorial pairs were again present on the same territories in 1994, each engaged in nesting. Three other Roadside Hawks (a successfully breeding female and two males) were banded in the forest study plot in 1993, and not observed in 1994 on any territories within the plot; it is not known whether these birds died or relocated.

Home Range Estimates

We estimated the home range of one non-nesting, radio-tagged territorial male that was paired with a female. We obtained 31 locations for this male, throughout the breeding season (19 April–5 July, 1993) and at all times of day. This male remained within 350 m of an emergent roost tree in the center of his core use area. This male's home range was estimated to be 9 ha using the Minimum Convex Polygon (MCP) method, 12 ha using the 85% Harmonic Mean (HM) method, or 14 ha using the 95% HM method; Fig. 12.1).

Inter-nest Spacing and Density of Territorial Pairs

During our two-year study, we recorded 12 nesting efforts in the forest plot (Fig. 12.1). In 1993, the plot supported 13 pairs, 5 of them nesting and 8 non-nesting, territorial pairs. In 1994, this plot held 11 pairs, 7 of them nesting and 4 non-nesting. Hence, nominally, the two-year average was 12 resident pairs (nesting plus territorial) per 825 ha, and breeding density averaged 6 pairs/825 ha. However, a more accurate density is obtained by allocating half territories to pairs situated on the plot boundary. With this method, the density was 10 pairs in 1993 and 9 in 1994, for a mean of 9.5 pairs/8.25 km^2—a density of 1.15 territorial pairs/km^2, equivalent to 86.8 ha/territorial pair.

We documented 20 nesting attempts in the slash-and-burn study plot during our two-year study. In 1993, 10 nesting and 2 non-nesting territorial pairs occupied the plot, while in 1994 there were 10 nesting and 3 non-nesting pairs, for a two-year mean of 12.5 resident pairs/8.0 km^2. Breeding density was the same in both years: 10 nesting pairs/8.0 km^2. Allocating partial territories for pairs on the plot boundary, the density estimate becomes 9 nesting pairs per year, with 2 non-nesting pairs in 1993 and 2.5 in 1994, for a mean density of 11.25 pairs/8 km^2, or 1.41 territorial pairs/km^2—equivalent to 71 ha/pair.

These density estimates suggest either that home ranges were larger than the 14 ha we estimated for one male, or that pairs were spaced farther apart than dictated by home range area, perhaps owing to habitat

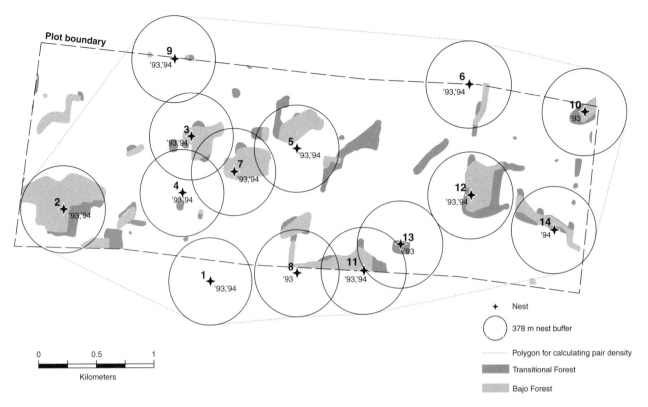

12.1 Space use by Roadside Hawks on an 8.25-km² plot in mature forest in Tikal National Park, Guatemala. Stars represent nests (centers of activity in the case of non-nesting pairs). Around each is drawn a circle of radius 378 m, one-half the mean inter-nest distance. The polygon enclosing all nests was 11.04 km² in area, equivalent to 84.9 ha for each of 13 territories contained therein (see text). Also shown is 85% Harmonic Mean home range (12 ha) of one paired, nonbreeding adult male. Pairs nesting both years = 1, 5; pairs present both years, nesting in 1993 = 9; pairs present both years, nesting in 1994 = 2, 3, 4, 6, 7; pairs only present in 1993, nesting = 8, 10; pairs present both years, non-nesting = 11, 12; pairs present only in 1993, non-nesting = 13; pairs present only in 1994, non-nesting = 14.

heterogeneity. In the primary forest plot, nests were strongly associated with Bajo Forest, and the spacing between areas of this forest type (which were scattered amid a much larger quantity of Upland Forest) seemed to largely dictate the locations of territorial pairs (Panasci and Whitacre 2002). Thus, selective habitat use coupled with habitat heterogeneity appeared to limit nesting density here. In the farming landscape, habitat in nest vicinities did not differ detectably from a random sample, and hawks nested in trees typical of those available.

DEMOGRAPHY

Age of First Breeding, Attainment of Adult Plumage

Juveniles were easily distinguished from adults by their cinnamon-brown plumage and brown iris; the iris was whitish or yellowish in adults. Juveniles attain adult plumage with their first complete (prebasic) molt at about one year (Dickey and van Rossem 1938; Howell and Webb 1995). At Tikal, we observed two individuals that were in adult plumage but maintained the brown iris of young birds, and several museum specimens were re-

ported with the same condition (Dickey and van Rossem 1938; Friedmann 1950); presumably such birds are in their second year. According to Dickey and van Rossem (1938), in El Salvador the annual molt takes place in August and September, perhaps beginning in July; in addition, they reported that some hawks underwent a partial spring body molt. We did not note any adult-plumaged, brown-eyed (second-year) birds nesting. Near the northern extreme of the species's range, Sutton and Pettingill (1942) noted some brown-eyed birds as pairs, believed to be nesting.

Frequency of Nesting, Percentage of Pairs Nesting

Although Roadside Hawk pairs seemed often to nest in consecutive years, a substantial proportion of pairs did not lay eggs in a given year, especially in the mature forest plot. In the forest plot, 5 (39%) of 13 pairs nested in 1993 and 7 (64%) of 11 in 1994, for a 50% rate of nesting overall by territorial pairs. In the farming habitat, 10 (83%) of 12 pairs nested in 1993 and 10 (77%) of 13 in 1994, for an overall nesting rate of 80% by territorial pairs. We do not know why hawks in the primary forest showed a higher incidence of non-nesting than did hawks in the farming landscape. Rates of prey delivery to nests in the two habitats were similar, suggesting that feeding conditions were not radically different (Panasci and Whitacre 2002). One explanation may be that the emergent nest trees preferred for nesting were more limited in number in the mature forest study plot (rising mainly above patches of low-canopied Bajo Forest) than in the farming landscape, where isolated and emergent trees were abundant. Furthermore, interspecific competition for such emergent nest trees may exist—especially in the mature forest, where more raptor species occurred in reasonable abundance than was the case in the farming landscape (Panasci and Whitacre 2002). Several raptor species at Tikal nested mainly in emergent trees, and often in the same low-canopied forest types used heavily by nesting Roadside Hawks. In the farming landscape, suitable nest trees did not seem to be in short supply, as Roadside Hawks the second year of our study never used the same trees as in the first year. In contrast, in two cases in the forest plot, these hawks reused the same nest tree (and once the same nest) during the second year; this may be taken as weak support for the notion that nest trees were more limited in the forest than in the farming plot (Panasci and Whitacre 2002).

Productivity and Nest Success

In our two-year study, productivity and nesting success were low (Panasci and Whitacre 2002). Combining all data, productivity was 0.2 fledgling/territorial pair-year, and only 25% of nesting efforts resulted in one or more fledged young (Table 12.2). Low productivity was in part due to a high incidence (35%) of non-nesting by territorial pairs. In the forest plot, mean productivity for the two years was 0.08 fledgling/territorial pair, and 0.17

nest was successful per nesting attempt. In the slash-and-burn plot, mean productivity was 0.32 fledgling/territorial pair, and 0.30 nest was successful per nesting attempt. In the forest plot, 24 pair-years produced only two fledglings, whereas in the farming plot, 25 pair-years produced eight fledglings (Table 12.2).

Our climbs to some nests to verify clutch size may have affected rates of nest failure—of 13 nests we climbed to, 5 failed, while of 19 nests we did not climb to, 3 failed. While a statistical test (Fisher's exact test, $P = 0.219$) found no significant difference in these proportions, our small sample implies low statistical power (about 30%; a chi-square test on these data has a power of approximately 0.30; power of the Fisher Exact test presumably is similar), and we cannot be confident that no such effect existed. At any rate, it does not seem plausible that the low incidence of *nesting* by territorial pairs could have resulted from our activities. Thus, at least the incidence of nesting, and possibly nest success, was low regardless of any effect caused by us. Vásquez Marroquín and Reyes Moreno (1992) reported two fledglings from six nests at Tikal, for a productivity of 0.30 fledgling per nesting attempt.

Causes of Nesting Failure and Mortality

Two-thirds of nest failures occurred during the egg stage. In the forest plot, 8 of 10 nest failures took place during the egg stage, and in the farming landscape, 8 of 14 did so, the rest failing in the nestling stage. The main known causes of nest failure in the primary forest were predation ($n = 3$) and wind ($n = 2$; Panasci and Whitacre 2002). Based on claw marks on tree trunks and fur caught in tree bark, we suspected some nests were depredated by Tayras (*Eira barbara*), a large, arboreal mustelid roughly the size of a Fisher (*Martes pennanti*). We observed

Table 12.2 Roadside Hawk nesting incidence and success in primary forest and slash-and-burn farming study plots in Petén, Guatemala

Number of:	Primary forest plot[a]		Slash-and-burn plot[b]		Total
	1993	1994	1993	1994	
Non-nesting pairs	8	4	2	3	17
Nesting pairs	5	7	10	10	32
Successful nests	1	1	4	2	8
Nests that failed during egg stage	3	5	4	4	16
Nests that failed during nestling stage	1	1	2	4	8
Fledglings produced	1	1	5	3	10
Eggs documented as laid	5+[c]	8	11+[d]	12+[e]	35+
Estimate of eggs laid[f]	6	8	13	13	40
Estimate of eggs laid[g]	6.29	8	13.58	13.29	41
Territorial pair-years studied	13	11	12	13	49
Fledglings/territorial pair-year	—	—	—	—	0.20

  [a] Study plot 8.25 ha in area.
  [b] Study plot 8.0 ha in area.
  [c] One clutch of 2, three clutches of 1, and one clutch size unknown.
  [d] Three clutches of 2, five clutches of 1, and two clutch sizes unknown.
  [e] Three clutches of 2, six clutches of 1, and one clutch size unknown.
  [f] One egg allocated per unknown clutch.
  [g] 1.29 eggs allocated per unknown clutch.

Tayras near Roadside Hawk nests on three occasions, and we saw these mustelids depredate other bird nests. As noted earlier, we witnessed one case of siblicide in the forest plot and suspected another; no such incidents were noted in the farming landscape. In the forest plot, we also witnessed a case of cannibalism; a second-year bird (adult plumaged with the dark iris of a juvenile) killed and ate a 14-day-old nestling at the nest.

In the farming landscape, the main observed cause of nest failure was human persecution (*n* = 4). People cut down nest trees in two cases, leading to premature fledging in one case and nestling death in the other. Another nestling disappeared from the nest, and we found machete marks on the tree, while at a final nest we saw two boys playing with slingshots in the morning, and by noon, the nest had fallen and the egg was gone.

Parasites and Pests

We did not observe parasitism by botflies (*Philornis* spp.) on Roadside Hawk adults or young, in contrast to the case of Crane Hawks (*Geranospiza caerulescens*; see Chapter 9) and some other raptors that nested in similar habitat. Most Roadside Hawk nests and nest trees were covered with ants. It is unknown whether these raptors were protected from parasites by the presence of ants, as hypothesized for the Laughing Falcon (*Herpetotheres cachinnans*; see Chapter 18).

CONSERVATION

As Roadside Hawks thrive in a wide variety of open and human-modified habitats, it is unlikely that deforestation poses a global threat to the species, and in many areas deforestation has probably led to an increase in this species's population. In at least one case, however, this tolerant, generalist species has suffered a serious local population decline. In El Salvador, these hawks were once common south of the coastal highway in pastures supporting scrubby trees dotted along roadsides and in fence lines (Thurber et al. 1987). In the early 1970s, the trees and fences were removed, pastures converted to cotton fields, and pesticides heavily used. By the 1980s, Roadside Hawks were no longer reported in the area, and Thurber et al. (1987) attribute this decline mainly to habitat loss: conversion of pastures, fence rows, and untended areas to treeless expanses of cotton. It seems plausible, however, that the documented massive use of DDT and other organochlorines on cotton in this area may also have played a role. Martínez-Sánchez (1986) reported that in Nicaragua, Roadside Hawks had been observed with symptoms of poisoning in cotton plantations.

In all likelihood, the effects of deforestation on this species may vary with forest characteristics and climatic regime. Where forests are tall, dense, and wet, these raptors appear to be largely restricted to edges, clearings, and natural gaps; deforestation in this case may lead to increased populations of this species, at least until habitat modification becomes extreme. In drier climates where forests occur as gallery forests or open woodlands of various sorts, deforestation may well have a negative effect on this species. To summarize, while globally this species is in no danger and may sometimes benefit, at least temporarily, from human interventions, in specific cases habitat modification by humans may diminish or eliminate local populations.

Thiollay (1984) reported that tropical raptors in human-modified habitat in South America experienced high mortality due to interactions with humans, and our results agree. The Roadside Hawk is a generalist species that thrives in human-modified environments, and that human persecution often caused nest failure in the farming landscape is not of special concern. However, this result may bode ill for other, less abundant raptor species with lower intrinsic reproductive rates. Reducing persecution should be possible through education and may increase the ability of human-dominated landscapes to retain a maximal proportion of their native raptor faunas.

HIGHLIGHTS

Among our most interesting results were the low productivity and high incidence of non-nesting among territorial pairs (50% non-nesting in the forest plot and 20% in the slash-and-burn plot: 35% overall). The productivity values we documented (0.08 and 0.32 fledgling/territorial pair in the forest and farming landscapes, respectively) were very low for raptor populations not adversely affected by environmental contaminants. We compared our results with values given by Newton (1979, table 23) for temperate zone and tropical-subtropical accipitrids. For these two latitudinal groupings of accipitrids, we calculated species's means for percentage of territorial pairs that nested, nest success, and productivity per territorial pair. Among the temperate zone accipitrids represented in Newton's data (15 species), 84.1% of territorial pairs nested (laid eggs), while in tropical accipitrids (12 African species), 70.3% of territorial pairs nested—significantly different rates (Panasci and Whitacre 2002). Among 15 species of temperate zone accipitrids, productivity averaged 1.23 fledglings/territorial pair, whereas 11 species of tropical accipitrids had productivity averaging 0.51 fledgling/territorial pair. The difference in productivity between temperate and tropical raptors in Newton's sample resulted mainly from the larger mean clutch of temperate raptors (2.8, *n* = 14 species)—nearly twice that of tropical raptors (1.5, *n* = 12 species).

With regard to the percentage of territorial pairs that nested (65% overall), Roadside Hawks resembled the tropical African raptors just cited. However, the productivity rates we documented were lower than all productivity values listed by Newton (1979), except for one based on a small sample size. While the high rate of nonbreeding in those tropical raptors listed by Newton (1979) may have been in part due to extended fledgling care in some species, with a consequent two-year nesting cycle, this was not the case in Roadside Hawks, as individual pairs of Roadside Hawks (including a marked pair) were observed to nest in consecutive years. Nor was the low productivity we observed solely due to low rates of nesting; nest success, averaging 17% in the forest plot and 30% in the farming plot, was lower than in any accipitrid listed in Newton's table 23. In Newton's sample, temperate zone accipitrids averaged 68.3% nest success (SE = 1.75, n = 17), and tropical accipitrids 65.9% success (SE = 3.68, n = 12 [deleting one study with small sample size]), an insignificant difference (Panasci and Whitacre 2002). However, our nest success rates were based on relatively small samples (n = 12 in the forest plot, 20 in the farming plot), and additional data might reveal different mean values.

To summarize, the low productivity rates we observed resulted from a small clutch size, relatively high rates of non-nesting (though not unusually high for tropical raptors, except in the forest plot), and high rates of nest failure. If the productivity values we observed were truly characteristic in the habitats we studied, it would seem these habitats must be "population sinks" for this species (Newton 1979; Pulliam 1988)—habitats absorbing immigrants from elsewhere, but not contributing to net population growth. Many raptors, however, show variable productivity from year to year, often related to food supply (e.g., Korpimäki and Norrdahl 1991; Bednarz 1995), and we speculate that further research would find productivity in our study populations sometimes higher than we documented.

Another notable finding was that Roadside Hawks commonly nested in certain mature forest types at Tikal. While Wetmore (1957, 1965) made a similar finding in Panama, it is not widely appreciated that this raptor sometimes nests abundantly in mature forest.

For further information on this species in other portions of its range, refer to de la Peña 2006, Di Giacomo 2005, Marini et al. 2007, Navarro et al. 2007, Riesing et al. 2003, and Verea et al. 2009.

13 CRESTED EAGLE

David F. Whitacre, Juventino López, and Gregorio López

Using radiotelemetry, we find the female Crested Eagle *(Morphnus guianensis)* four kilometers from the nest where she recently fledged an eaglet. She is perched in the dappled shade of the lower forest canopy. Pinned to the limb by her talons drapes a 1.3 m Tropical Rat Snake *(Spilotes pullatus)*, a handsome jet black creature with creamy speckling. She pulls a bite of meat from the snake, swallows, then pauses to impassively watch a pair of Tayras *(Eira barbara)* in a nearby treefall gap. The Tayras, large members of the weasel tribe, are going in and out of a beehive in a hollow limb that had cracked open when the tree recently blew down. Amid the buzzing throng, the mustelids are feasting on the bees' brood and honey. The Crested Eagle lifts a large foot and scratches her face, pauses, then reaches over her shoulder and preens her scapulars, her lax crown feathers jiggling.

The Crested Eagle, known in Spanish as the Aguila Monera or "monkey-eating eagle," is a mysterious bird that has long captured the imagination of Neotropical raptor enthusiasts. Though much more lightly built, these birds somewhat resemble the Harpy Eagle *(Harpia harpyja)* in plumage (Plates 13.1 and 13.2) and have often been confused with it, especially in our study area, where the most popular field guides do not picture or even mention the Crested Eagle. This superficial resemblance, plus this eagle's large size and apparent rarity, contribute to the fascination with this little-known raptor.

Here we report on the diet, breeding biology, and behavior of two pairs and on spatial use by an adult female and her dependent fledgling. These eagles proved to have a low reproductive rate, protracted juvenile dependency, and a unique diet of nocturnal, arboreal mammals and large snakes. With extensive space demands and a subcanopy hunting style, these are likely among the Neotropical raptors most vulnerable to deforestation.

GEOGRAPHIC DISTRIBUTION AND SYSTEMATICS

The Crested Eagle occurs from Petén, Guatemala, and Belize (and probably southernmost Mexico, see below) southward through humid, forested lowlands to northern Paraguay and Argentina (Brown and Amadon 1968). The northern limit of the species's range is uncertain. In 1981, Ellis and Whaley published the first records for Guatemala, and the first record for Belize was apparently that of Hall (1995); previously, two Honduran records (Monroe 1968) were regarded as the northernmost. Since the 1988 inception of the Maya Project, we have found these eagles to be regular and widespread in Petén, though at low densities. To date there is no fully documented record for Mexico, although the species probably occurs in southernmost Campeche, Quintana Roo, and Chiapas. On 8 April 1992, while censusing raptors from a treetop near the ruins of Calakmul in southeastern Campeche, Jason Sutter and Jorge Montejo Díaz observed a Crested Eagle gliding directly overhead, at 25 m. Their written description and sketch leave little doubt that this sighting, some 45 km within the borders of Mexico, is authentic, apparently the first for Mexico (J. Sutter, pers. comm.). At Bethel, in extreme western Petén, in 1993 we had a long look at a perched Crested Eagle within 5 km of the border with Chiapas, Mexico.

The Crested Eagle has generally been regarded as rare and local in most parts of its range. Except in Panama, few records exist for Central America, and even in South America these birds have nowhere been described as common. Published accounts portray this as mainly a bird of tropical lowlands, occurring up to 500–600 m, with a probable sight record at 1000 m in Colombia (Hilty and Brown 1986). At Tikal, in addition to studying pairs in two territories, we made repeated sightings, indicating the presence of several territories.

MORPHOLOGY

Among the widely distributed Neotropical forest eagles, the Crested Eagle is second in size only to the much heavier Harpy Eagle. Few body mass data are available (Appendix 1). We could find only one published weight for a male—1275 g (Haverschmidt and Mees 1994)—and a captive male yielded an estimated adult weight of 1200 g (Appendix 1). Apparently only one female weight has been published, 1750 g for a female of unstated origin (Brown and Amadon 1968). To this we can add one adult female weight from Tikal; the female at the Naranjal nest, one month after her chick hatched, weighed 1948 g. All other available weight data are from captive birds (Appendix 1). The nestling we studied at Tikal weighed 1630 g on day 72 after hatching and 1697 g on day 92; hence it was judged a female. Additional body mass data are needed.

With a mean female weight of 1850–1975 g, the Crested Eagle is only modestly heavier than the Ornate Hawk-eagle *(Spizaetus ornatus)*, and much lighter than the Harpy Eagle, in which males weigh 4.0–4.8 kg and females 7.6–9.0 kg (Bierregaard 1994). In addition to being more slender, Crested Eagles have proportionally longer tails than the Harpy Eagle. Compared to the Red-tailed Hawk *(Buteo jamaicensis)*, Crested Eagles are about 14% larger in wing chord, 29% larger in culmen length, and about 25% (males) to 50% heavier (females).

Body mass data (Appendix 1) give a tentative Dimorphism Index (DI) value of 12.4, with females (1850 g) weighing about 45% more than males (1275 g). For nine males and six females, Bierregaard (1978) gave the following DI values: 6.7 for wing chord, 6.9 for tail length, 6.9 for tarsus length, 6.6 for middle toe length, and 4.3 for culmen length—similar to the values in Appendix 2, except for culmen length. Mean DI of wing chord, culmen length, and body mass is 7.8. In sum, these birds are moderately size dimorphic, to about the same degree as a Red-tailed Hawk (Snyder and Wiley 1976).

Adult Plumages

The Crested Eagle occurs in a pale and a dark color phase, unrelated to gender and geography. Both phases are depicted in many field guides and in the color photo of Bierregaard (1984). These were thought to be distinct species. The "Guiana Crested Eagle," described from the pale morph, was initially named *Falco guianensis* by Daudin in 1800 and renamed *Morphnus* by Cuvier in 1817. In 1879, Gurney described the "Banded Crested Eagle," *Morphnus taeniatus*, based on the dark morph. This arrangement held until 1949, when Hellmayr and Conover showed that the latter was but a color morph.

PREVIOUS RESEARCH

This magnificent raptor has been little studied. Bierregaard (1984) observed a nest in Brazil over several weeks, and Neil Rettig gathered information on a nest in Guyana, which he has kindly allowed us to include here. Apart from coverage in regional works and field guides, brief accounts exist on distribution (Ellis and Whaley 1981), general description (Lehmann 1943), and eggs (Kiff et al. 1989), and a few observations on hunting and diet (Bierregaard 1984; Julliot 1994; Robinson 1994).

RESEARCH IN THE MAYA FOREST

We studied a nesting in each of two territories—one in Tikal National Park and one in the farming landscape nearby. Nest 1, the "Caoba" nest, several kilometers south of the park, was shown to us by a farmer on 10 June 1994, when incubation was underway. We observed the nest from a platform 24 m up in a tree 35 m distant. The eagles rarely directed their gaze at us. From 10 June to 19 July we observed on 16 days, taking detailed notes during 15 visits totaling 115.7 hours. Observations were 3.2–13.5 hours in length, averaging 7.7 ± 4.55 hours; six periods were from dawn to dusk (11.8 hours or more). Our 39-day observation interval nearly spanned the predicted incubation interval of 48–51 days; hence, we believe the clutch was laid shortly before our first visit. This nesting failed 33–39 days after we began observations, and before the egg(s) hatched; the cause of failure was unknown but may have been predation. The female was incubating on 13 July, but on our next visit (19 July) neither adult was incubating and the pair copulated near the nest. Climbing to the nest on 26 July, we found eggshell fragments. Checking periodically, we found no renewal of nesting during 1994 or 1995. By mid-May 1996, most of the nest had fallen, and fire had passed through this woodlot, within 4 m of the nest tree.

Nest 2, the "Naranjal" nest, was near the center of Tikal National Park (Plate 13.3). Searching for a month in an area where we had repeatedly observed Crested Eagles in previous years, on 8 May 1995 we found the nest with the female incubating. We built a platform 74 m from the nest and on 11 May began dawn-to-dusk observations, usually 13–14 hours in duration, using a spotting scope. We made 979.3 hours of observation on 83 days; observation periods averaged 11.8 hours in duration. We observed this nest during the last three weeks of the incubation period, through the 28 May hatching and until the chick fledged, 14–19 September, at 109–114 days after hatching. We monitored the radio-tagged fledgling until the age of 16 months, when we ended fieldwork (Plate 13.4). We also placed a radio on the adult female and studied her movements over an eight-month period. The radios (model AI-2B, Holohil Systems, Carp, Ontario) weighed 30 g and were affixed via a backpack arrangement using Teflon ribbon.

Pair members were distinguished without difficulty at both nests. At nest 1 the female was a light-morph individual and the male a dark morph. At nest 2 both adults were of the pale morph; here size dimorphism assisted, but the individuals were distinguished mainly by keeping track of both birds' movements during the male's infrequent visits to the nest and by the female's radio tag.

To analyze adult roles, we attributed the position and behavior of the male and female at all times to one of five mutually exclusive categories: incubating/brooding, standing on nest, in nest tree (not on nest), near nest tree (within view of nest tree, usually < 100 m), and away from nest (not within view of nest tree, generally > 100 m).

Time spent standing on the nest but not in incubating posture was regarded as an incubation break.

In documenting the diet, we used only "observed" prey—that is, prey whose delivery we witnessed, that we identified in the nest soon after delivery, or that we observed in the possession of a bird away from the nest (Table 13.1). We included data from prey remains found in or below the nest only when we were sure these were remains of prey we had already observed but not identified, in which case it helped us to reduce the number of observed items that remained unidentified. For observed prey, weight was visually estimated and items placed in categories of 0–50 g, 51–100 g, 101–200 g, 201–400 g, and more than 400 g; in some cases items were weighed. To calculate prey biomass, we used our estimates, as well as mean species weights. For birds, mean weights are from Dunning (1993) and for mammals they are from Emmons (1990), supplemented by our own data. For snakes, in addition to visual estimates and the few prey items we weighed, we derived length-weight relationships from 18 snakes of several species weighed at Tikal; these provided rough weight estimates, based on the estimated length and diameter of snakes we observed as prey. Data were analyzed using SYSTAT 5.0 (Wilkerson 1990) and MRPP chi-square tests (Berry and Mielke 1986). Unless stated, data presented are means ± 1 SD.

DIET AND HUNTING BEHAVIOR

Brown and Amadon (1968) reported the diet as smaller species of monkeys as well as opossums, birds, and reptiles, while Wetmore (1965) reported that in Panama, these eagles are said to feed on the smaller monkeys, large birds, and iguanas. Grossman and Hamlet (1964) gave the diet as opossums, monkeys of the genus *Lagothrix*, young Llamas *(Lama glama)*, and reptiles, while Stiles et al. (1989) reported the diet as snakes, frogs, and mammals up to the size of Kinkajous *(Potos flavus)* or small monkeys, and occasionally birds. Trail (1987) reported one attack at a Guianan Cock-of-the-Rock lek *(Rupicola rupicola)*, and Julliot (1994) reported an instance of predation on a young Red-faced Spider Monkey *(Ateles paniscus)*. Robinson (1994) reported several attacks on primates and the capture of a Common Squirrel Monkey *(Saimiri sciureus)*. A large Tropical Rat Snake is shown in the talons of a Crested Eagle photographed in Belize (Hall 1995).

The only previous quantitative data are those of Bierregaard (1984) for a Brazilian nest: 15 prey items there included 1 frog, 8 mammals, and 6 snakes— 1 Green Anaconda *(Eunectes murinus)*, 3 Emerald Tree Boas *(Corallus caninus)*, and 2 unidentified snakes. The mammals were small, mostly 20–35 cm in head-body length; Bierregaard believed most were rodents or marsupials. Two Kinkajous were also taken. Bierregaard also saw the female attack a flock of Grey-winged Trumpeters *(Psophia crepitans)* and concluded that birds are likely taken at times. At a nest in southern Guyana, Neil

Rettig (unpubl. data) found remains of a squirrel monkey *(Saimiri* sp.), an opossum *(Philander* sp.), a Blue-crowned Motmot *(Momotus momota)*, and a snake.

Nest 1, several kilometers south of Tikal, which failed prior to hatching, yielded only three prey items, all brought in by the male during incubation (Table 13.1). Two were snakes—one of them 70 cm by 2.5 cm and the other larger—and the third was a procyonid mammal, either a Kinkajou or a juvenile White-nosed Coati *(Nasua narica)*. Nest 2, observed through the chick's fledging and beyond, yielded 97 prey items, for a total of 100 between both nests. On the basis of those 97 items identified at least to class, diet at the two nests was 15.0% snakes, 21.0% birds, and 64.0% mammals (Table 13.1). In terms of biomass, the diet was 13% snakes, 10% birds, and 77% mammals.

The eagles took a number of juvenile birds and mammals, suggesting they robbed nests and dens. Of 17 birds taken, at least 6 were nestlings or fledglings, and all five coatis were juveniles. We also suspected a few other mammals (e.g., porcupines) were immature. In some cases it appeared likely the male Crested Eagle raided the same nest more than once, as on 20 June when he brought in two nestling Roadside Hawks *(Rupornis magnirostris)* two hours apart. Again on 14 July, he brought in two birds an hour and a quarter apart that appeared to be of a single species and that we believed were juveniles or nestlings.

Among the 13 snakes taken by the pair at nest 2, at least 11 belonged to three large species: 5 were Boa Constrictors *(Boa constrictor)*, 5 Tropical Rat Snakes, 1 was probably a Western Indigo Snake *(Drymarchon corais)*, and 1 was a green, arboreal species, probably *Leptophis* or *Oxybelis*. Largest was a Boa Constrictor 126 cm long and 8 cm in diameter, estimated to weigh several pounds. While radio-tracking during the post-fledging period, we found the adult female far from the nest, feeding on this snake, probably at the capture site. One of the *Spilotes* snakes was estimated at 130 cm, also a robust snake, though not as heavy bodied as a boa, while the green, arboreal snake, also 130 cm, was quite slender. These snakes were substantially larger than those usually taken by the other main snake-eating raptors at Tikal— the Laughing Falcon *(Herpetotheres cachinnans)*, Great Black Hawk *(Buteogallus urubitinga)*, and White Hawk *(Leucopternis albicollis)*.

All 60 mammals identified to at least family level were partly to totally arboreal, and 84% were mainly nocturnal: squirrels and coatis were the only diurnal mammals among identified prey. Arboreal Mexican Hairy Porcupines *(Sphiggurus mexicanus)*, contributing 9% of total and 14% of mammalian prey, were more frequent than in the diet of any other raptor we studied and comprised most of the rodent portion. Because of the large size of these porcupines, rodents comprised a greater proportion of the diet on a biomass basis (30.6%) than on a numerical basis (14%). These porcupines are highly nocturnal and arboreal in habit (Nowak and Paradiso 1983). In Brazil, Bierregaard (pers. comm.) found an emaciated female

Table 13.1 Diet of Crested Eagles at two nests in and near Tikal National Park, Petén

Prey species	Number of prey caught by Nest 2				Number of prey caught by Nest 1	Number of prey caught, both nests	Percentages on numerical basis (%)	Estimated mean prey mass (g)	Estimated total biomass (g)	Total biomass (%)
	Male	?	Female	Total	Male					
Boa constrictor	1	1	3	5	—	5	—	600[a]	3000	—
Spilotes pullatus	—	3	2	5	—	5	—	250[b]	1250	—
Drymarchon corais	1	—	—	1	—	1	—	190[c]	190	—
Green, arboreal, prob. *Leptophis* or *Oxybelis* sp.	—	—	1	1	—	1	—	165[d]	165	—
Snake species	—	—	1	1	2	3	—	384[e]	1152	—
Total number of snakes	**2**	**4**	**7**	**13**	**2**	**15**	**15.0**	**384[e]**	**5760**	**12.9**
Roadside Hawk (*Rupornis magnirostris*) nestling	2	—	—	2	—	2	—	135[f]	270	—
Juvenile raptor, possibly *Leptodon cayanensis*	1	—	—	1	—	1	—	240[g]	240	—
Mexican Wood Owl (*Strix squamulata*)	1	—	—	1	—	1	—	288	288	—
Juvenile Northern Potoo (*Nyctibius jamaicensis*)	—	1	—	1	—	1	—	185[h]	185	—
White-necked Puffbird (*Notharchus macrorhynchos*)	1	—	—	1	—	1	—	96[i]	96	—
Collared Aracari (*Pteroglossus torquatus*)	—	1	—	1	—	1	—	226[i]	226	—
Keel-billed Toucan (*Ramphastos sulfuratus*)	1	2	—	3	—	3	—	339[i]	1017	—
Masked Tityra (*Tityra semifasciata*)	1	—	—	1	—	1	—	79[i]	79	—
Bird species	7	4	—	11	—	10	—	218[i]	2180	—
Total number of birds	**15**	**6**	**0**	**21**	**—**	**21**	**21.0**	**218**	**4578**	**10.2**
Didelphis marsupialis or *virginiana*	2	1	3	6	—	6	—	1085[k]	6510	—
Didelphidae, probably *Marmosa*	—	1	—	1	—	1	—	70[l]	70	—
Didelphidae, probably *Philander*	1	—	1	2	—	2	—	200[m]	400	—
Caluromys derbianus	23	3	5	31	—	31	—	308[l]	9548	—
Total number of Didelphidae	**26**	**5**	**9**	**40**	**—**	**40**	**40.0**	**413**	**16,520**	**37.0**
Nasua narica, juveniles	3	2	—	5	—	5	—	350[n]	1750	—
Procyonidae (*Nasua* or *Potos*)	—	—	—		1	1	—	1200[o]	1200	—
Total number of Procyonidae	**3**	**2**	**—**	**5**	**1**	**6**	**6.0**	**492**	**2952**	**6.6**
Sciurus (mainly *deppei*)	3	—	1	4	—	4	—	205[p]	820	—
Rodent, probably *Tylomys* sp.	1	—	—	1	—	1	—	240[q]	240	—
Total number of rodents	**7**	**3**	**4**	**14**	**0**	**14**	**14.0**	**975.7**	**13,660**	**30.4**

Table 13.1—*cont.*

Prey species	Number of prey caught by Nest 2				Number of prey caught by Nest 1	Number of prey caught, both nests	Percentages on numerical basis (%)	Estimated mean prey mass (g)	Estimated total biomass (g)	Total biomass (%)
	Male	?	Female	Total	Male					
Unidentified mammals	2	2	—	4	—	4	—	308[s]	1232	—
Total number of mammals	38	12	13	63	1	64	64.0	537	34,368	76.9
Not identified to class	1	2	0	3	—	3	—	—	—	—
Total prey items	56	24	20	100	3	103	—	—	—	—
Total identified to class	55	22	20	97	3	100	100	447	44,707	100

[a] Two boas were estimated at 60 cm and 200–400 g; 300 g was used for these. Based on these and other boa weights, a boa measured as 79 × 5 cm was estimated as weighing 500 g, and one measured at 126 × 8 cm was estimated as 1.3 kg. The mean for these four then is 600 g; same mean is used for the fifth boa.

[b] Two *Spilotes pullatus* were estimated as 130 and 140 cm long, for which we estimate 250 g; the same weight was assigned to the remaining three, for which no size estimate was made.

[c] Measured as 114.3 cm, 190 g.

[d] Estimated as 130 cm and 165 g, the weight of a 122 cm *Masticophis mentovarium* weighed at Tikal.

[e] Mean weight of identified snakes.

[f] Half the adult weight of 270 g.

[g] Half the adult weight of 484 g.

[h] Adult weight (length of remiges suggested near-adult size).

[i] Adult weight.

[j] Mean estimated weight of identified avian prey.

[k] Midpoint of range given for males and females by Emmons 1990.

[l] Midpoint of range given in Emmons 1990.

[m] Low end of weight range given in Emmons 1990; these items appeared smaller than *Caluromys*.

[n] Juvenile coatis at Tikal average 300 g in May, 400 g in June (J. Binczik, pers. comm.); these coatis were all taken at end of May and in early June.

[o] Prey taken 17 June; weight used is halfway between that of juvenile coati in June (400 g) and minimum weight (2 kg) of Kinkajou given by Emmons 1990.

[p] Median weight of *S. deppei*; all observed individuals were believed to be this species.

[q] Median weight of *Tylomys* spp. (Emmons 1990).

[r] Minimum weight given in Emmons 1990.

[s] Weight used for *Caluromys*, the most frequent mammalian prey.

Crested Eagle, too weak to fly, that had her stomach lining impaled by three porcupine quills.

By far the largest part of the diet was provided by the opossums—Didelphidae—contributing 67% of the mammals taken, and 40% (numerical) or 37% (biomass) of the total diet. Of the 40 opossums taken, 31 were Derby's Pale-eared Woolly Opossums *(Caluromys derbianus)*, 2 were probably Gray Four-eyed Opossums *(Philander opossum)*, and 1 was possibly a Mexican Mouse Opossum *(Marmosa mexicana)*. Six opossums were of the larger Common Opossum *(Didelphis marsupialis)*. The raccoon family, Procyonidae, made up 10% of mammalian prey on a numerical basis, and somewhat more on a biomass basis; five were juvenile coatis, and the sixth may have been a Kinkajou.

Crested Eagles brought in nocturnal prey throughout the daylight hours (Fig. 13.1), though less often in early afternoon. Squirrels and most birds are diurnal and coatis largely so. Of 11 identified birds taken, 4 were nestlings or juveniles, and the remaining 7 were hole nesters (Table 13.1), suggesting that they may have been captured by reaching into nests or cavities. The snake *Spilotes* is diurnal (D. Whitacre, pers. observ.), while

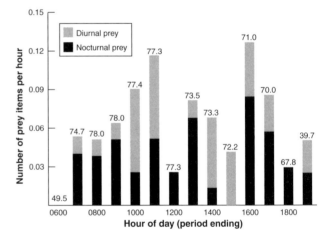

13.1 Prey types brought to nest versus time of day at Naranjal Crested Eagle nest, Tikal, Guatemala. Number at top of bar = total observation time.

the Boa Constrictor is mainly crepuscular to nocturnal (Greene 1983). Our prey delivery data for most hourly periods were based on more than 70 days of observation;

hence the bimodal or trimodal daily pattern of prey deliveries (Fig. 13.1) probably did not result from chance.

Because mammals and snakes of a given species may vary widely in weight, it is not possible to state precisely the mass of such prey items—thus, our estimates of dietary biomass are subject to substantial error (Table 13.1). We calculated a mean overall prey size of just under 450 g. For snakes, we estimated a mean of roughly 384 g and for mammals 537 g; birds in the diet appeared to be smaller on average, approximately 218 g. Most prey were 200–384 g, with substantially larger items taken with fair regularity (Fig. 13.2). The largest prey item may have been the 126 cm by 8 cm boa, with a mass estimated at 1–2 kg.

In view of the heavy predation on arboreal and nocturnal mammals, it is interesting that no bats were recorded in the diet. This is in striking contrast to our results for the Black Hawk-eagle (*Spizaetus tyrannus*, see Chapter 14), which, like the Crested Eagle, took many nocturnal arboreal mammals throughout the day, suggesting similar foraging styles of these two raptors. However, bats (most of which are 50 g or less) may be too small to be profitable prey for Crested Eagles. Also interesting in its absence from the diet was the Central American Agouti *(Dasyprocta punctata)*, an abundant, diurnal, forest-floor inhabitant at Tikal. Adult agoutis, however, at 3.2–4.2 kg (Emmons 1990), probably average quite a bit heavier than the prey items reported here.

We compared prey types brought in by the male and female at the Naranjal nest. Of 97 prey items identified to class (including a few in possession of the adult female away from the nest during the post-fledging period), 55 were captured by the male, 20 apparently by the female, and 22 were already in the nest when we observed them (omitted from consideration here). The male and female caught similar proportions of mammals (69% and 65%, respectively; Table 13.1). The male, however, brought in 15 birds (27%) and the female none. Snakes made up only 3.6% of the prey brought in by the male, but 35% of the female's prey. Based on these data,

prey of the male and female eagle differed significantly in taxonomic composition (MRPP chi-square test: $\chi^2 = 17.3$, df = 2, $P = 0.0001$). If only items brought to the nest are considered, the female's sample is reduced to 14 items, but the male-female difference is still significant (MRPP chi-square test: $\chi^2 = 11.6$, df = 2, $P = 0.003$; including the three items caught by the male of nest 1 did not alter these conclusions). One would want a larger sample of female prey, however, as well as data from additional pairs, before concluding whether the sexes tend to differ in prey selection. One possibility is that the number of young birds taken by the male simply reflects that these items were recorded during the peak nesting season, whereas the female's prey items were recorded later on, when nestlings and fledglings may have been less abundant. Also, it is possible that adults selectively ate certain prey types at their site of capture and brought others to the nest—if so, and if males and females differed in this regard, this could produce inter-sex differences in prey brought to the nest, even if males and females capture the same prey.

We also compared the size of mammalian prey brought to the nest by the male and female at Naranjal (Table 13.2). Only mammals were included in this comparison as it was difficult to estimate size of birds and snakes. Prey size for the male and female eagle did not differ significantly (MRPP chi-square test: $\chi^2 = 2.95$, df = 2, $P = 0.23$). This sample was small, however, and more data might tell a different story, especially if prey captured and consumed away from the nest throughout the nonbreeding season were fully represented among the data; prey found in possession of the female away from the nest during the post-fledging period were among the largest prey documented.

Table 13.2 Number of mammalian prey brought to the nest by male and female Crested Eagles at nest 2, according to size class

Prey size/type	Male	Female
Large		
Porcupine, *Sphiggurus mexicanus*	3	3
Opossum, *Didelphis marsupialis*[a]	2	3
Coati, *Nasua narica*	3	0
	8	6
Medium		
Woolly Opossum, *Caluromys derbianus*	23	5
	23	5
Small		
Probable *Philander opossum*	1	1
Rat species	1	1
Squirrel, *Sciurus* sp.	3	1
	5	2
Total	36	13

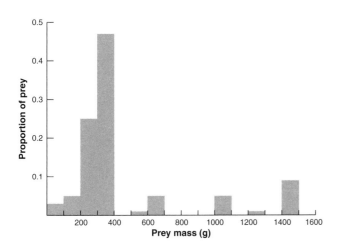

13.2 Estimated prey mass proportion for Crested Eagles at Tikal, based on 100 prey items identified at least to class.

[a] *Didelphis marsupialis* includes *D. virginiana*; we did not distinguish between the two.

Hunting Behavior of an Adult Female

We observed a few definite and a few probable hunting attempts by the nest 2 female. On 15 May, midway through the incubation period, the female left the nest and flew into a nearby tree, where we heard wings beating against foliage and saw two Kinkajous enter a cavity in the trunk. Moments later, the female returned to the nest, having apparently made an unsuccessful attack on one of these mammals. On 14 June, when the nestling was 17 days old, the female launched downward in a steep glide from a tree near the nest, evoking alarm calls from Brown Jays *(Cyanocorax morio)* in the area—though we could not see the quarry, we believed this a capture attempt. On 21 June, we saw a similar flight by the female, again from a tree near the nest tree. On 21 July, the female flew from the nest tree, approximately 75 m and behind another tree, where we heard a crash followed by vocalizations of a squirrel or bird and wing beats of the female in the vegetation. She returned to the nest an hour later, having either made an unsuccessful hunt or already consumed any prey she may have captured. As we followed the adult female via radiotelemetry, we witnessed no clear-cut hunting behavior. In nine cases she had prey in her possession: one Derby's Pale-eared Woolly Opossum, three Mexican Hairy Porcupines, two large Common Opossums, one mammal species, one snake species, and one unidentified prey item. The female was not overly flighty, and we felt our presence made her nervous in only 5 of 28 cases. On occasion we observed her for up to 2.5 hours.

Hunting Behavior of a Dependent Juvenile

We regularly monitored the radio-tagged fledgling, nearly always achieving visual contact. She frequently engaged in capture practice, pouncing on limbs and fruits. In one case she flew upward, seized a fruit, and hung there momentarily. We observed such behavior 3, 5, 7, and 11 months after she fledged. On a few occasions we saw the juvenile make rapid attacks on seemingly agile prey. At the age of eight months, the juvenile was perched on a vine in the sub-canopy space when she suddenly flew downward in an unsuccessful attack; we heard a scuttling sound, as of escaping quarry. At 15 months, the juvenile tried unsuccessfully to seize a grasshopper that flew near her head. This juvenile also demonstrated novel foraging behavior in several cases, searching within cavities, under loose bark, and within epiphytes in rather the manner of a Crane Hawk *(Geranospiza caerulescens)*. At eight months, the juvenile poised, flapping, atop a bromeliad, within which we had the impression she was searching for prey. At the age of 14 months, the juvenile stuck her head into a hole in a tree and came out with a dead leaf in her beak, letting it fall. She then jumped to a perch 1.5 m away but remained interested in the cavity, as a Keel-billed Toucan *(Ramphastos sulfuratus)* called loudly above. The eaglet then returned to the cavity and reached her foot in, but retrieved no prey. At 15 months,

the eaglet walked through the limbs of a Ramón (Breadnut Tree, *Brosimum alicastrum*), as if searching for some small prey item among the bromeliads and bark, and a few minutes later, we observed her searching in similar fashion through another bromeliad-laden Ramón.

These observations of the juvenile searching within a cavity and among loose bark and bromeliads may help clarify the nature of foraging behavior that led to the observed diet. Though this behavior may simply exemplify exploratory behavior by the juvenile, presumably adults also methodically search potential hideouts among the vegetation in order to secure the many nocturnal, arboreal, and cavity-nesting prey animals that appeared in the diet. However, some prey items and a few observed capture attempts also indicate that these eagles made stealthy, rapid attacks on alert, day-active prey. Colombian raptor biologist César Márquez (pers. comm.) once kept a Crested Eagle for a time; this bird seemed to pay a great deal of attention to auditory cues, and Márquez thought that these eagles might hunt partly in this fashion. This, of course, is not unique, as many raptors respond at times to auditory clues such as distress calls of potential quarry.

HABITAT USE

Published accounts describe the Crested Eagle as mainly a bird of lowland primary rain forests, and Jullien and Thiollay (1996) found it to be among the raptors most strongly associated with tall, pristine forest in French Guiana. Our observations agree with this view, but they also show that not only wet rain forests are occupied; the drier, more seasonal forests of northern Petén are also home to these birds. All of our sightings have been in areas with substantial mature forest cover. Our nest 2, within Tikal National Park, was situated in Transitional Forest, at the edge of a small, seasonally flooded bajo area; the surrounding, extensive mature forest included a mix of Upland, Hillbase, and Low Scrub Swamp Forest. A Guyana nest visited by Neil Rettig (pers. comm.) was in an extensive area of primary forest 180 m above sea level, on rolling terrain incised by numerous small streams. Bierregaard's (1984) Brazilian nest was at the well-studied Biological Dynamics of Forest Fragments site near Manaus, where the forest canopy reaches approximately 37 m and rainfall is 2200 mm yearly (Stotz and Bierregaard 1989), nearly a meter more rain than the rainfall at Tikal. Habitat around our nest in the farming landscape is described under Conservation, later in this chapter.

Habitat Use by a Dependent Juvenile

Excluding sightings in the nest tree, we made 23 visual contacts of a radio-tagged juvenile during the year after she fledged. Of these sightings, 65% were in tall Upland Forest, and the rest in Transitional Forest. All sightings were in primary forest, and only 2 of 23 were in the vicinity of substantial clearings or roadways; there was no indication that this juvenile sought out clearings. Mean canopy height near perches was 20–25 m in 10 cases, 15–20 m in 12 cases, and 10–15 m in 1 case. Four (17.4%) of 23 sightings were in the vicinity of treefall gaps; although we made no formal test, this did not suggest to us an affinity for gaps.

Mean height of perches was 18 ± 7 m (n = 23), though the eaglet used perches 4–32 m in height. Perch trees ranged from 8 to 36 m in height, averaging 23.5 ± 6.5 m (n = 22), and averaged 83 cm in diameter (range = 18–162 cm). The juvenile showed a slight tendency to use lower perches after the age of 30 weeks, possibly indicating an increasing tendency to forage. Of 23 perches, 21 were in living trees, the same dead tree was used twice, and a vine was used once. Perch limbs ranged from 3 to 37 cm thick, averaging 10 ± 8 cm (n = 23), and nearly all were horizontal, though the vertical tip of a dead tree was once used. Twice the bird was perched above the canopy surface, 11 times at the canopy surface but overlooking lower areas of canopy, five times within the canopy, three times at the lower canopy boundary, and twice in the sub-canopy space. The juvenile was not especially timid and was relatively easy to observe without apparent disturbance.

Habitat Use by an Adult Female during the Post-fledging Period

On 29 May 1995 we fitted the adult female at the Naranjal nest with a radiotransmitter. From 23 August 1995 through mid-April 1996 (when the radio failed), we made 28 visual contacts with this female, all but 4 of them after the nestling had fledged; two sightings were in the nest tree but only one is included here. We found the female overwhelmingly (81% of locations) in tall, mature Upland Forest—46% of the time in Upland Standard Forest and 35% in Upland Dry Forest. In the remaining cases, she was perched in Transitional Forest (11% of the time), Sabal/Hillbase Forest (4%), and riparian forest (4%). We never observed her in low-canopied Bajo or Scrub Swamp Forest. Understory at perch sites was generally fairly open, as is typical of Upland Forest. Canopy height was 20–25 m in 8 cases, 15–20 m in 12 cases, and 10–15 m in 6 cases. All 28 observations except 1 were in primary forest. Five (18%) of 28 observations were in the vicinity of natural treefall gaps, not enough to convey the impression that she sought out such sites.

We found the female perched from the ground to 35 m up, with perch height averaging 16 m (n = 27). Perch trees ranged from 10 to 35 m tall, averaging 23 ± 7 m (n = 25), and were 18–167 cm in diameter, averaging 57 cm. Perch limbs averaged 8.4 ± 6.8 cm thick (range = 3–35 cm), and all 28 were horizontal except one that was angled; she also perched on two vines, 3 and 12 cm thick. Of 27 perches, in 15% (4 cases) the bird was on a perch protruding above the canopy, in 30% (8) she was at the upper canopy surface, overlooking lower canopy nearby, in 37% (10) within the canopy, in 15% (4) at

the lower canopy margin, and in 4% (1) she was on the ground. Overall, she was within or below the canopy somewhat more frequently (56% of cases) than the juvenile (44%). Neither her mean perch height nor mean distance from the nest changed over the eight months of data collection.

BREEDING BIOLOGY AND BEHAVIOR

Nests and Nest Sites

Nest 1 was 23 m up in a recently dead Marío tree, also known as Antilles Calophyllum (*Calophyllum brasiliense*), 36 m tall and 68 cm in diameter. The nest was centered above the main bole where it divided into three limbs, 24, 30, and 39 cm in diameter, which cradled the nest. The nest was 61 cm in external depth, 97 cm by 148 cm across, with a shallow depression 46 cm by 48 cm across and 10 cm deep. It was built of dead branches averaging approximately 94 (53–152) cm long and about 2.2 (0.8–4.3) cm in diameter. The nest was easily visible from the ground and involved no vine tangle. The nest tree, though in a forest remnant, was not in contact with the surrounding canopy. The nest tree being dead and leafless, this nest was fully exposed to the sun.

Nest 2 was 16.4 m up in a Jobillo, also known as Glassywood tree (*Astronium graveolens*), 34 m tall and 61 cm in diameter. The nest was in a fork of the main trunk, supported by three limbs, 23, 24, and 35 cm in diameter, and was 140 cm by 108 cm across, 70 cm in external depth, and flat on top. It was composed of dead branches averaging about 72 cm in length and 1.1 cm in diameter. The nest was easily visible from the ground and did not involve a vine tangle. Vegetation above the nest provided partial shade, which increased midway through the incubation period as new leaves appeared; a densitometer reading at the nest gave an estimate of 50% cover. Though amid extensive forest, the nest tree was not in contact with the forest canopy at large.

Bierregaard's (1984) Brazilian nest was 28 m up in the lowest fork of a large tree and had a view over the forest canopy to one side. A nest discovered by Neil Rettig (pers. comm.) in Guyana in January 1992 was 24.3 m up in a fork of the main trunk of a 31.1 m Mandio (*Qualea paraensis*) tree. The nest was 1.2 m in diameter, very symmetrical, 60 cm in external depth, and composed of sticks averaging about 1.9 cm thick. It was quite flat on top, with much vegetative debris composing the flat surface, probably as a result of leafy sprigs being brought to the nest.

For these nests, nest tree height averaged 34 ± 3 m (n = 3), nest height averaged 23 ± 5 m (n = 4), nest diameter averaged 122 ± 2 cm (n = 3), and external nest depth averaged 64 ± 6 cm (n = 3).

Egg and Clutch Size

We were not able to verify the clutch size at either Tikal nest, although we suspected nest 2 contained only one egg. In Brazil, Bierregaard (1984) observed a nest with two cream-colored eggs. To our knowledge, the only purported Crested Eagle egg collected in the wild is one from Guyana described originally by Kreuger (1963) and listed also by Schönwetter (1961); however, the identification of this egg is in doubt (Kiff et al. 1989). Four eggs laid by Crested Eagles in captivity were described by Kiff et al. (1989). An egg laid by a captive female from Panama was 64.0 mm by 50.7 mm, with a calculated fresh weight of 90.5 g—substantially larger than three eggs laid by a captive Peruvian female, which were 57.5 by 43.9 mm, 55.5 by 42.7 mm, and 51.8 by 41.2 mm, with a mean calculated fresh weight of 55.0 g. All four eggs were dull white and unmarked, much smaller and differently marked than Kreuger's Guyana egg. Kiff et al. (1989) argue that the eggs of the Peru female were probably atypically small, a phenomenon known to occur in captive birds. Further information on egg and clutch size of this species is clearly needed.

Kiff et al. (1989) used the Peru female's captive body weight (2950 g) to calculate that her 55.0 g mean egg weight amounted to 1.9% of body weight. This female, however, weighed 2226 g when received by the Oklahoma City Zoo, in excellent condition and juvenal plumage (zoo records). With this figure, mean egg weight was calculated to be 2.5% of body weight. If instead the mean of available wild female weights (1850 g) is used, the egg amounts to 3.0% of female body weight. The 95 g egg of the Panama female was estimated at 3.1% of body weight, again using the 2950 g weight of the captive Peru female; if we use instead 1850 g—the mean of wild female weights—this egg weighed 5.1% of mean female mass.

The Peru female, while housed for nine years at the Oklahoma City Zoo, laid at least 25 eggs, and in eight cases laid a second egg 3–10 days (mean = 7.4 days) after the first (Kiff et al. 1989). This timing of second eggs is probably the strongest evidence suggesting that a 2-egg clutch is common for this species, at least in northeast Peru. It remains to be discovered whether clutch or egg size varies geographically. If the modal clutch is two as suggested by available evidence, then a clutch would range from approximately 6 to 10% of female body mass, depending on which egg and female weights are used.

Nesting Phenology

At the first Tikal nest, laying probably occurred shortly before our observations began on 10 June. At nest 2, an egg hatched on 28 May. Based on a predicted incubation period of 48–51 days (see below), this indicates a laying date during the first week of April. Both laying dates fall late in the dry season, and the single nestling hatched at the very end of the average dry season, fledging on or just before 19 September, the height of the rainy season. In Venezuela, Hilty (2003) observed an adult adding green sprigs to a nest in March. Hence, the few data existing for the northern Neotropics suggest nesting commonly begins during the Northern Hemisphere spring, that

is, late in the dry season. In contrast, laying at a Brazilian nest took place at the peak of the rainy season, and hatching occurred at the onset of the dry season (Bierregaard 1984).

Length of Incubation, Nestling, and Dependency Periods

We did not determine the duration of incubation. However, based on an incubation period of approximately 46 days for the Ornate Hawk-eagle (see Chapter 15) and 55–56 days for the Harpy Eagle (Brown 1976b; Rettig 1978; M. Kuhn and C. Sandfort, pers. comm.), and on information in Newton 1979 relating body mass to incubation period, we estimate an incubation period of 48–51 days for the Crested Eagle. The single study nestling fledged at 109–114 days after hatching. Duration of post-fledging dependency is discussed under Behavior and Development of a Nestling, later in this chapter.

VOCALIZATIONS

The eagles gave a variety of vocalizations in the nest vicinity. Female calls included a single-note call resembling the *peech!* call of the Collared Aracari (*Pteroglossus torquatus*), given as she rejoined the male on the nest, her wings twitching in unison with each call; a sharp whistle, given in the male's absence and thought to be a solicitation of his presence; a *flee-flee-flee*, given as she perched on the nest, the male out of view; and a high-pitched, thin, unobtrusive *peee-zhoo*, repeated many times as she observed the male perched nearby with prey he was about to transfer to her. Calls of the male included a very quiet *kluwik* given while he perched with prey he was about to transfer to the female, and a short, loud whistle given, for example, as he left the nest, where the female was also present. Bierregaard (1984) described one call as reminiscent of a bosun's whistle—a shrill, high-pitched *youuu-ree*, the final note brief and ascending. Descriptions by Stiles et al. (1989) agree with those given here.

BEHAVIOR OF ADULTS

A number of authors have noted that these eagles are often seen perched for long periods on high exposed limbs. Though we occasionally observed this, such behavior at Tikal was not frequent enough to make these birds conspicuous or easy to detect.

Incubation Period

The female provided 97.3% of incubation at one nest and all incubation at the other. Between incubation bouts, females would often roll the egg(s) with their beaks. While incubating or standing in the nest, females frequently tugged at nesting material with their beaks, rearranging green sprigs and other material on the nest rim. During the incubation phase, one or both adults were on the nest 97.0 ± 1.1% of the time (n = 2 nests). Eggs were incubated from 39 to 83% of the time, the wide range possibly related to stage of incubation. Females incubated significantly less as incubation progressed, but did not spend less time on the nest; rather, they spent more time standing over and shading the egg(s). Adults at both nests frequently brought leafy twigs to the nest, and less often dead branches.

Nest 1. At this nest, the egg(s) failed to hatch, perhaps destroyed by a predator. On average, one or both adults were on the nest (incubating or standing) 97.7 ± 4.3% of the observation time (n = 16 observation days); on only one day did this figure drop below 95%. If constancy of total nest attendance decreased as the incubation period progressed, this trend was not significant in our sample (regression on Julian date, $F_{1,13}$ = 3.232, P = 0.096, adjusted r^2 = 0.137). The female contributed 97.3 ± 8.3% of total daily presence on the nest (n = 16 days), and 97.3 ± 8.6% of daily incubation (n = 16). She tended the nest nearly constantly, relieved occasionally by the male, in one case for 3.85 hours. During most incubation breaks the female was not relieved by the male and most often remained standing on the nest.

The egg was incubated by one adult or the other during 82.7% of our total observation time and, on a daily basis, 81.0 ± 9.3% of the time on average (n = 15 days, range = 68–100%). If incubation constancy declined during the month of observations, this was not detected in our sample (Fig. 13.3) (regression on Julian date, $F_{1,13}$ = 3.375, P = 0.089, adjusted r^2 = 0.145); however, the amount of time the female spent incubating decreased significantly (Fig. 13.3) (regression, $F_{1,13}$ = 5.93, P = 0.03, adjusted r^2 = 0.26). While the male may have contributed more to incubation as the season progressed, only twice did he provide more than 1% of the day's incubation (8.7 and 34.0%).

The female, on average, incubated 73 ± 71 minutes at a time (n = 51) (using only incubation intervals observed in their entirety). Her incubation stints ranged from 8 to 310 minutes in duration, and median duration was 51 minutes (n = 51). Duration of the female's incubation stints did not change over time (regression on Julian date, P = 0.21). The male incubated eight times during four days, providing 0.3, 0.5, 8.7, and 34.0% of incubation on those days (male incubation periods were 2, 4, 10, 33, 40, 44, and 137 minutes in duration). We observed in their entirety 69 intervals during which neither bird incubated, though in most cases one of the adults was on the nest. These ranged from 1 to 58 minutes in duration with a mean of 17.4 ± 11.4 minutes and a median of 15 minutes; intervals of 2–25 minutes were most common. A scatterplot revealed no trend with stage of incubation.

This nest was quite exposed to the sun and the female often shaded the egg when she stood in the nest during an incubation break, especially during the hot, midday hours. During 22 of 69 incubation breaks, the

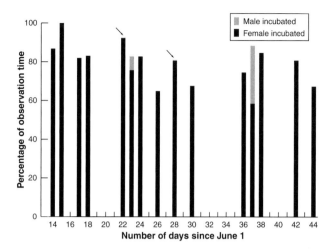

13.3 Incubation behavior of male and female Crested Eagles at nest 1, Tikal. As laying date was unknown, 1 June was used as an arbitrary starting point for graph. Arrows indicate instances when male incubated for a period too short to depict as a bar.

bird attending the nest held one or both wings extended, casting shade on the nest cup and egg. In other cases, although the adult did not hold its wings out, the egg was probably shaded by the bird standing over it or by shadows of tree limbs. It was apparent that the female provided shade when it was most needed, that is, during midday to early afternoon hours of intense sunlight. During these hours, it was uncommon for the egg to go unshaded when not being incubated, except in one case directly after a rain shower. Likewise, on a drizzly, overcast day, we noted shade-giving behavior only once, suggesting that such behavior was elicited by conditions of intense sunlight and heat.

Nest 2. Our observations at this successful nest began 19 days prior to hatching and covered the latter 40% of the presumed 50-day incubation period. We made observations on 13 days during incubation, totaling 102.5 hours and averaging 7.9 hours per observation period. As the male at this nest was never seen to incubate or to brood the nestling, total incubation is documented by examining figures for the female.

During these 13 days, the female spent, on average, 84.6 ± 28.8% of her time on the nest (Fig. 13.4; range = 12.9–100%, n = 13) and 35.2 ± 21.9% of her time incubating (Fig. 13.4; range = 0–72.2%, n = 13). Two dates may be considered outliers, when she spent only 12.9 and 28.3% of her time on the nest; in the other 11 cases she spent more than 88% of her time on the nest. Omitting these two outlier days, on average she spent 96.2 ± 3.7% of her time on the nest (range = 88.8–100%, n = 11) and 39.2 ± 21.2% of the time incubating (range = 0–72.2%, n = 11; Fig. 13.4).

As the incubation period progressed, the female dramatically decreased the percentage of time spent incubating (regression, P = 0.007, adjusted r^2 = 0.452), from a mean of 60% 20 days prior to hatching, to a mean of

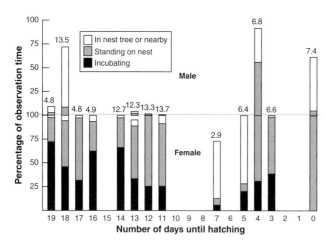

13.4 Total nest attendance (in nest tree or nearby, on nest standing, or incubating/brooding) by male and female Crested Eagles at nest 2, Tikal. Number at top of bar = total observation time, in hours.

10% by hatching day (Fig. 13.5). She did not decrease her total presence on the nest (regression, $F_{1,9} = 1.81$, $P = 0.21$, adjusted $r^2 = 0.075$) (Fig. 13.5), but spent more time standing over the egg, presumably shading it, and less time in incubation posture. (Neither incubation time nor time on the nest bore any relation to the length of observation periods; hence these results appear not to be artifacts of the duration of observation episodes.) At times, 3 hours or more passed without the female incubating, as on 12 May, two weeks prior to hatching. Indeed, her incubation became so seldom that by six days prior to hatching, we concluded that this nesting effort had failed, and then were proved wrong when the egg hatched—this experience should provide a caution to others studying tropical raptors.

Incubation, brooding, and nest vigilance by males. Whereas the male at nest 1 incubated on several occasions, once for as much as 2.28 hours, the male at nest 2 never incubated during our observations. Usually the nest 2 male spent a few seconds to a minute standing on the nest during prey exchanges, and rarely 8–11 minutes. This male remained on the nest more than usual just prior to and around hatching time. On the chick's first day, the male spent 20 minutes on the nest along with the female, and the following day he spent 30 minutes on the nest; four days prior to hatching, he spent slightly more than 2 hours standing on the nest along with the female (Fig. 13.4). If one includes time spent within view of the nest, this male's proximity to the nest was substantially greater on a few occasions. In four cases the male remained near the nest for 64–72% of the observation period (i.e., periods totaling 4.5–9.9 hours each); this occurred on the hatch day, the following day, and on days 4 and 18 prior to hatching. Thereafter, not until day 127 (two weeks after the chick fledged) did the male spend as much as 14% of the observation period in the nest vicinity. We never saw the male brood the nestling. In one

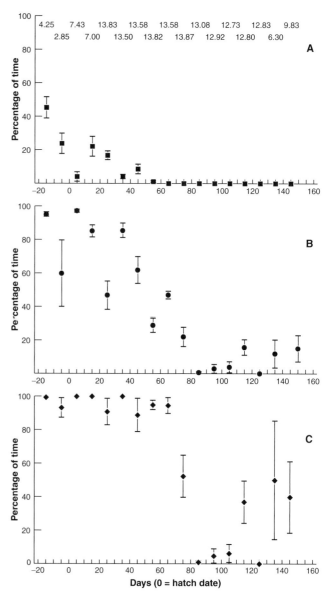

13.5 Time budget of female Crested Eagle during incubation and nestling periods at nest 2 at Tikal. (A) Percentage of time female incubated or brooded. (B) Percentage of time female spent on nest, all activities combined. (C) Percentage of time female spent on or near (< 100 m) nest. Data are means + 1 SE per 10-day period. Number at top of column = total observation time, in hours.

case when the female was absent, the male attempted to feed the 24-day-old nestling, but he seemed inept at tearing off small bits. After a few minutes he abandoned this attempt and flew off.

Nestling Period

The amount of time the nest 2 female spent incubating the egg and brooding the chick declined gradually, beginning prior to hatching (Figs. 13.4, 13.5). The female brooded the nestling very little during its first week, although this was the period when she most diligently stood over the nestling, shading it with wings spread, and feeding it (Fig. 13.5). Brooding by the female was again fairly frequent during the chick's second and third weeks, though decreasing steadily in incidence, and was last noted during the eighth week after hatching.

The female's total daily time on the nest reached its maximum of 90–100% during incubation and up to two weeks after hatching, thereafter declining steadily, and was often zero by 81 days after hatching (Fig. 13.5). As the female spent less time on the nest, she spent progressively more time perched in the nest tree or nearby; this behavior peaked in frequency when the chick was about nine weeks old. The female made a dramatic shift in behavior during a two-week period, from most often being near the nest constantly up to chick age 60–65 days, to often being absent the entire daily observation period by chick age 80 days (Fig. 13.5). During this transitional period, the female often arrived at the nest in the morning and left in the afternoon. At this time, the nestling began to feed itself when the female was absent.

When the female's absence from the nest area reached its maximum at nestling age 80–81 days, she remained away for a week, visited somewhat more frequently for another three weeks, and, after a month had elapsed, once again spent significantly more time in the nest vicinity (Fig. 13.5). At this time she generally visited the nest 1–2 days per week, bringing prey she presumably had caught. When she did visit, she continued to feed the nestling bit by bit and brought foliage to the nest. We began to follow the female via radiotelemetry from the time she moved away from the nest at chick age 81 days. At times we found her more than 4 km from the nest in possession of prey, portions of which she later brought to the juvenile.

Regulation of nestling and female thermal status. The female often appeared to respond rapidly to the thermal environment at the nest, which changed as quickly as a cloud could cross the sun. In some cases she may have responded to vocal cues from the chick, but in other cases she evidently responded directly to weather conditions, for example, when returning quickly to the nest from some distance when rain began. On several days we watched the female alternate between brooding, shading, providing rain shelter, and merely standing on the nest, as conditions changed between hot sun, shade, and rain squalls. As the female changed her position on the nest throughout the day, she frequently positioned herself between the chick and the sun, with her back to the latter and her wings folded or partly spread over the chick.

At times the female seemed to brood and shade the chick at the same time, with her wings spread as she crouched low in a brooding posture. When protecting the chick from rain, the female generally assumed a brooding position. The chick also played an active role, often seeking shade or warmth below the female or moving into areas of natural shade. At times when the nest was under hot sun, we suspected the nestling's insistent vocalizations were spurred by discomfort, and the female provided shading in apparent response. Often when

the nest was bathed in full sun both the nestling and the adult female appeared to suffer discomfort from the heat, drooping their wings at the wrist, gaping, and panting; the same was true of adults while incubating. Shading and apparent heat stress were most common from 1200 to 1300, when the sun was virtually straight overhead (Fig. 13.6); neither shading nor heat stress were noted during early morning or late evening. Cases of apparent heat stress in the female or nestling (n = 21) showed a virtually identical pattern to that seen in Figure 13.6.

The adult female showed evidence of heat stress on seven occasions—once while incubating, five times while shading the nestling during its first 17 days, and once while standing beside the seven-week-old chick. The nestling appeared heat stressed on 14 occasions. Only twice did this occur while the female was shading the nestling; one case occurred while the female was gone for 3 minutes to collect foliage, and one took place as the chick moved out of natural shade to lay in front of the female. One case took place as the female and the eight-week-old chick stood side by side in the sun. Most cases (nine) took place when the chick was 75–101 days old, that is, after the female had largely ceased to remain in the nest vicinity. This highlights the female's success in preventing, for the most part, thermal stress in the nestling up to this point.

The female assiduously covered the nestling during rain. We observed this on nine occasions from nestling age 3 to 59 days. After the female largely abandoned the nest area at nestling age 81 days, the nestling often endured torrential rain on its own; from ages 81 to 122 days, we observed this on 12 occasions. Even then, when the female did visit the nest, often bringing prey, she continued to feed and shade the chick at times, and the chick still sought refuge beneath her. We last noted the female shading the juvenile with wings extended when the chick was 102 days old and had been left alone most of the time for three weeks. In subsequent visits, the fe-

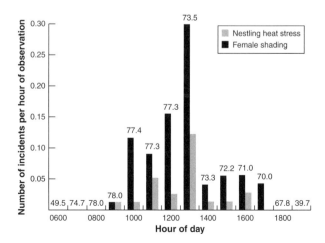

13.6 Daily pattern in which female Crested Eagle at Tikal nest 2 shaded the egg(s) or nestling: frequency of shading events per hour of observation, based on 65 shading events. Number at top of bar = total observation time, in hours.

male still fed the nestling bite-sized morsels at times, at least through chick age 143 days, and stood near it but was not observed to shade it.

Delivery of foliage and sticks to the nest. Eagles at both nests often delivered fresh foliage to the nest, a behavior common in many raptors whose adaptive basis is not well understood. Females made frequent trips to gather green sprigs from nearby trees, and males did so less often. The eagles used their beaks to break off the green sprigs, usually carrying them to the nest in their beaks, but occasionally in their talons. At times an adult tried to secure a sprig from two or three trees before succeeding, and at times they failed and abandoned the effort. Sprigs were placed in the nest cup or on the rim.

The female at nest 2 collected most sprigs from the same few individual trees. Of 134 sprigs which we were able to identify, they included *Aspidosperma* species (60%), Ramón (20%), and *Pouteria* species (14%), with lesser numbers of *Acacia* species, vines, *Rehdera penninervia*, and *Simira salvadorensis*, and, on occasion, green sprigs or leafless limbs of the Glassywood nest tree itself. The prevalence of *Aspidosperma* twigs among those brought in by the female was far higher than the relative abundance of this genus in the forest at large (Schulze and Whitacre 1999). Estimated lengths of sticks and sprigs brought to the nest averaged 30 ± 15 cm ($n = 172$); the vast majority were leafy and 15–45 cm long, but a few leafless sticks and vines were 1 m or longer. Of 183 sticks delivered by the female, 148 were categorized as to condition. Of these, 115 (78%) were green and leafy, and 33 (22%) were bare sticks.

Why many raptors bring greenery to the nest is unclear, and hypotheses range from producing shade for nestlings, ameliorating the nest microclimate through evaporation, or chemically suppressing populations of ectoparasites (Wimberger 1984; Clark 1991; Gwinner 1997), to signaling occupancy of the nest site (Newton 1979). Using time of day as a proxy for the solar/thermal environment at the nest, we tested whether there was a correlation between periods of hot sun on the nest and the rate at which adults delivered green foliage; such a correlation was not evident. In fact, the dip in foliage delivery rate during the midday hours may indicate that the female was reluctant to leave the nestling unshaded even briefly during these hottest hours (Fig. 13.7). The female delivered most foliage during early to mid morning, with a slight resurgence of activity in early evening (Fig. 13.7). This mirrors the general pattern of daily activity for many birds in many environments; thus, it may simply indicate that females brought foliage most often during their hours of greatest activity.

Dividing the nestling period into intervals, we tested whether the female's tendency to deliver green leafy sprigs versus leafless twigs differed during the nesting cycle; we found no difference over time (MRPP chi-square tests, all $P > 0.08$). Delivery of foliage occurred during incubation and was most frequent during the

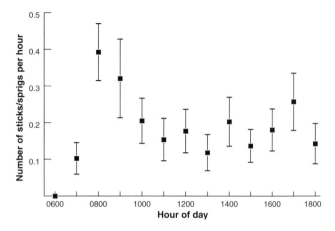

13.7 Daily pattern in which female Crested Eagle at Tikal nest 2 brought foliage and sticks to the nest; data are hourly means over 79 observation days, ± 1 SE.

nestling's first 10 weeks, but it continued occasionally even 4.5 months after the chick had fledged (Fig. 13.8).

We saw the male bring sprigs or sticks to the nest on only 10 of 82 observation days. Of 12 delivered, 8 were leafy, 3 leafless, and 1 unknown. In most cases his delivery rate was one stick or fewer per 13 hours of observation; only once did he bring two items in one day. Most of the male's foliage deliveries were made just before and after the hatching date, when he spent more time near the nest than at any other phase of the nesting cycle.

Our results did not clearly favor any particular hypothesis regarding the adaptive advantage of bringing foliage to the nest, and we consider it likely that multiple advantages are involved. The bimodal form of Figure 13.7 suggests that shading and amelioration of the nest microclimate were not the primary benefits of green foliage in the nest, since female shading behavior and heat-stress incidents were clearly concentrated at midday (Fig. 13.6).

Interactions between pair members. At nest 1, the adults were notably amicable when together in the nest vicinity. They often perched less than a meter apart, on or near the nest, and in one instance, fed side by side on the same snake. In one case the male pecked at the female's head in what appeared to be allopreening, and in another instance they engaged in behavior that might be described as "billing." We observed copulation only after the clutch was destroyed.

We observed more intra-pair strife at nest 2, perhaps simply because we observed this nest over a longer period. During the nestling period, the female sometimes mantled (spread her wings) over the nest upon the male's arrival. The Brazilian female observed by Bierregaard (pers. comm.) also mantled over any prey remains in the nest, upon hearing the male announce his approach. In several cases the nest 2 pair tussled over prey items in the nest, when the male tried to feed from items he had earlier delivered. This male appeared to experience conflicting motivations; at times, after retrieving prey from

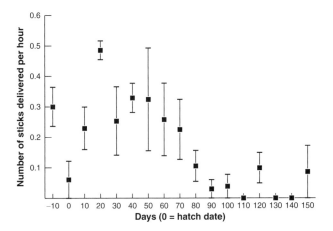

13.8 Seasonal pattern in which female Crested Eagle at Tikal nest 2 brought foliage and sticks to the nest; data are number of sticks/sprigs delivered per hour of observation, one point for each observation day. Curve is LOWESS smooth, default tension, SYSTAT 5.0.

the nest, he flew back to the nest with it, only to repeat the entire sequence.

Prey exchanges. We observed 60 prey exchanges, 57 of them at nest 2. Transfers were mostly from male to female or chick, but a few were from female to chick. The male always transferred prey to the female on the nest or, in one case, within 1 m of it, and in each case the female ate the prey mainly on the nest. The female nearly always announced the male's approach by beginning to call a couple of minutes to a few seconds before he flew in.

The male's routine was almost invariant. Remaining silent, he would fly straight to the nest, where the female (or later, the chick) would seize the prey and mantle over it, calling. After a few seconds, the male would then fly to a higher limb in the nest tree and perch there for a minute or two, calling. He would then perch in one or two other trees increasingly distant from the nest, finally flying out of sight within a few minutes of his arrival. When the female was in a nearby tree, occasionally the male would perch near her before going to the nest, but rather than transferring prey to her where she perched, he always delivered it to the nest. On rare occasions we heard the male announce his approach by calling, but usually we did not hear him call until a moment after he deposited the prey in the nest, when the female or chick was already calling. In contrast, at the Brazilian nest, the male audibly announced his approach before all arrivals at the nest (R. Bierregaard, pers. comm.).

The female and the nestling, upon seizing the prey in their talons, invariably called loudly and mantled over the prey, often continuing for as long as the male remained visible, even a few hundred meters from the nest. They would fall silent and turn their attention to the food item the moment he disappeared from view. The female rarely called when the male was away from the nest, except in nest defense. At times she called softly

while staring fixedly in a certain direction, and we felt certain she knew or suspected the male to be somewhere in that direction, perhaps via the mobbing calls of jays and toucans.

Prey items delivered by the male were usually headless and often largely plucked and half eaten. After delivering prey, the male would occasionally deliver a stick or sprig of foliage to the nest. Late in the nestling period, the female was often perched in a nearby tree when the male delivered prey to the chick at the nest. Often she did not move to the nest to feed the chick until after the male's departure. When the male visited the nest vicinity without bringing prey, the female would vocalize and he would softly answer.

Prey exchanges of Crested Eagles differed in two respects from those of the Ornate Hawk-eagle. First, the female Crested Eagle routinely ate on the nest during the incubation period. Female Black Hawk-eagles also did so, but not female Ornates, which normally ate at some distance from the nest while the male incubated. After nestlings hatched, females of most of the 18 falconiform species we studied at Tikal ate on the nest along with the nestlings. Second, the male Crested Eagle deposited prey at the nest, even when the female was perched in a nearby tree. This was different from the pattern in the Ornate Hawk-eagle, in which males commonly transferred prey to females 30–150 m from the nest.

Roles of parents as providers. At nest 1 the male apparently conducted all hunting for the pair during the incubation phase. He was most often observed near the nest in early morning but absent most of the day. The female rarely left the nest for more than 5–10 minutes; only once did she leave the nest area for 3 hours, while the male incubated. At nest 2 the male was the sole provider from the time of our discovery of the nest (mid-incubation period) until the chick was a month old. On 29 June, when the chick was 32 days old, the female brought in a squirrel, and after this she hunted fairly often; from this date through 19 September, when the young fledged, the female brought in 10 prey items compared to the male's 34.

After the chick fledged, the nest 2 female took on a major role in providing for it, and she may have been the main provider. From 19 September (fledging) through 14 August of the following year, when the juvenile was 14.5 months old, the female brought in 13 of the observed prey items and the male only three. We last saw the male deliver prey one month after the chick fledged; after this time, all 11 observed prey items were delivered by the female. In some cases we knew or suspected she had captured the prey, having observed it in her possession during radio-tracking. While we saw no evidence that she received any of these items from the male, it is possible. In a couple cases we suspected the male had delivered prey to the fledgling. Our impression that the female played the overriding role in providing for this fledgling may have resulted in part from her bearing a radiotransmitter while the male did not. It appeared the juvenile received prey from the adults on average only once weekly.

Prey delivery rates. We analyzed prey delivery rates from the time of hatching through fledging, based on 60 days of nest observations during this 114-day period, averaging 12.77 hours in duration (SE = 0.271, n = 60 days); most sessions were from dawn until dusk. The mean delivery rate was 0.95 item/day (SE = 0.115, n = 60 days). On 21 days we saw no prey delivered, on 25 days we saw one item delivered, on 10 days two items, and on 4 days three items. The highest delivery rate occurred on the hatch day; perhaps the male was stimulated by the hatching event. Apart from this, the days with highest delivery rates occurred between chick age 40 and 85 days.

BEHAVIOR AND DEVELOPMENT OF A NESTLING

The Naranjal nestling, a female, had not hatched as of 26 May and was tiny and weak the morning of 29 May; we assumed she hatched 28 May. This nestling's behavioral development is described in detail elsewhere (Whitacre et al. 2002). We first saw the chick make preening motions on day 16. By day 23 the chick could briefly stand on its legs rather than simply on its tibiotarsi, and took faltering steps, though still without its legs fully extended. On day 24 we first noted stereotyped wing stretching. On day 25 the chick attempted unsuccessfully to eat on its own, and on day 26, when the male arrived with prey, the chick vocalized and tried to seize the prey. On this date, feather shafts were beginning to emerge on the head and wings. On day 37 we first noted the chick standing firmly on straightened legs, and capable of flapping its wings in flight exercise. On day 39, the chick, along with the adult female, extended its wings and vocalized when an American Black Vulture (*Coragyps atratus*) flew overhead.

On day 45 the chick was first noted as wet from rain, the female being absent in the early morning. On day 53, the chick remained standing for 17 minutes and manipulated nest material with its beak; when the female arrived at the nest, the chick spread its wings and erected its feathers in an apparent threat display. After this date, the chick often responded to flying vultures, raptors, and airplanes in this fashion, with plumage erected, wings spread, and beak open. On day 54 the chick stood for 38 minutes but was still unsuccessful at feeding itself. On day 59 the chick stood for more than two hours and again was wet from rain, as the female spent a good portion of the day away; for the first time the chick, with difficulty, tore off and ate a few bites on its own. By day 65, the chick's jumping and flapping were frequent, and we first noted it pounce on and seize nest material.

By day 80 the chick fed on its own more often. By day 81, the adult female began spending most of her time away from the nest, apparently hunting, and periodically brought prey to the nest; the juvenile was now adept at plucking prey items and feeding itself. From this time on, the juvenile weathered frequent heavy rains on its own. At 92 days (28 August), we equipped the chick with a backpack radiotransmitter. The chick's exertions in practice flight and in pouncing on and pecking at nest material and prey remains became increasingly frequent and vigorous, and by day 93 it frequently jumped, flapping, from rim to rim of the nest. At 100 days, it frequently moved nest material around with its beak, and its practice hunting and flapping had reached a fever pitch. As of 13 September (day 108) we had never seen the chick leave the nest or actually fly. On 19 September, the nestling flew around within the nest tree and to other trees; hence it fledged at 109–114 days. At first the fledgling returned to the nest to sleep and receive prey, but within a few days it spent little time in the nest, often flying between trees up to 100 m distant and returning to the nest to receive prey from the adults.

With the aid of radiotelemetry, we followed the fledgling to the age of nearly 16 months. Unlike some other raptors at Tikal, the fledgling called mainly when adults approached, rather than food begging while adults were away from the nest area. At least as late as day 141 (16 October), the female still occasionally fed the chick small pieces of food when she delivered prey.

At 72 days, the chick weighed 1630 g, wing chord was 24.5 cm, total length 45 cm, and wingspan 87 cm; the eyes were dark gray, the feet buffy yellowish, and the cere, facial skin, and beak were black. At 92 days, the chick weighed 1697 g, wing chord was 40 cm, total length 59 cm, and wingspan 114 cm. Elsewhere we present a growth curve and behavioral notes for a captive-reared chick (Whitacre et al. 2002). Developmental mileposts accorded closely between the captive-reared chick and the wild chick we observed. Both were able to stand for some time beginning during their eighth week, and both made their first significant flights at 103–114 days.

The Crested Eagle nestling at Tikal developed at a slower rate than did Ornate Hawk-eagle nestlings, which in turn developed more slowly than did Black Hawk-eagle nestlings. This pattern accords with these species' relative sizes. Black Hawk-eagle chicks first stood up for prolonged periods at five weeks, Ornate chicks during the seventh week, and Crested Eagle chicks at eight weeks. We first noted Black Hawk-eagle chicks flapping vigorously at four weeks, Ornate chicks at five weeks, and the Crested Eagle during the seventh week. The age at which chicks were first able to feed themselves was less variable; this occurred at eight weeks in the Black Hawk-eagle and during the ninth week in the Ornate and the Crested Eagle. A Black Hawk-eagle chick first reached the "branching" stage (walking on limbs near the nest) at 4.1–4.9 weeks; we observed this at 9 weeks in the Ornate Hawk-eagle and 16 weeks in the Crested Eagle. First flights within the nest tree were noted during the eighth week in the Black Hawk-eagle, the tenth week in the Ornate, and at 16 weeks in the Crested Eagle. Fledging took place during the tenth week in the two hawk-eagles and at 16 weeks in the Crested Eagle. These patterns support our conclusions below regarding time to independence in the Crested Eagle.

Post-fledging Dependency Period

We regularly monitored the radio-tagged fledgling from the time of fledging (19 September 1995) until 365 days later (age of 479 days), accumulating 25 visual contacts during this period. At this point, our fieldwork ended. On average, we found the juvenile 190 m from the nest. Although the juvenile frequently returned to the nest or nest vicinity throughout this period, the maximum distance from the nest at which we detected it gradually increased, reaching 600 m only at the age of nearly 15 months—48 weeks after fledging. At the age of 15.75 months, when we ended our monitoring, the juvenile's plumage remained completely white below, it showed no sign of reaching independence, and remained active from the nest to 600 m away (Fig. 13.9). These data indicate a usage area of only 37.5 ha using the Minimum Convex Polygon (MCP) method, or 27 ha by the 85% Harmonic Mean (HM) method, with 1.1 km between the most distant points. Even at the age of nearly 16 months, the hunting attempts we had witnessed were awkward and infrequent, and we saw no sign of the juvenile having captured prey on its own.

Our data suggest that juvenile Crested Eagles require longer to reach independence than do Ornate Hawk-eagles and Black Hawk-eagles, as expected based on their relative sizes. During its first 48 weeks after fledging, the juvenile Crested Eagle on average moved 12–13 m farther from the nest per week, but three Ornate Hawk-eagles increased their maximum distance from the nest by 15–20 m/week during a similar time period (see Chapter 15). When the juvenile Crested Eagle was 16 months old, we had not yet located it more than 600 m from the nest. In contrast, only one of three young female Ornate Hawk-eagles remained within 800 m of the nest at this age; one

dispersed 12 km from the nest at the age of 14 months, and the other regularly moved at least 1.5 km from the nest by the age of 12 months (see Chapter 15). Likewise, a juvenile male Black Hawk-eagle dispersed 20 km from the nest at the age of 12 months (see Chapter 14).

Although the full juvenile dependency period of the Crested Eagle remains unknown, it must be well over 16 months from hatching and could easily be substantially longer. In the Harpy Eagle, Alvarez-Cordero (1996) found that independence is not reached until the young are at least 22.6, and probably closer to 30, months old (i.e., 17–24.4 months after fledging).

SPACING MECHANISMS AND POPULATION DENSITY

Territorial Behavior and Displays

We witnessed no interactions among Crested Eagles other than pair members; hence we cannot comment on intraspecific territoriality or related behavior. Unlike most diurnal raptors at Tikal, these birds neither soared frequently and conspicuously above the canopy, nor voiced loud, frequent vocalizations that carried long distances through the forest. Despite some published references to frequent soaring high above the forest (Wetmore 1965; Hilty and Brown 1986; Bierregaard 1994), we never observed this despite hundreds of hours spent observing nests and censusing raptors from emergent treetops. In this, our observations agree with statements in the literature that the species rarely if ever soars (Ridgely and Gwynne 1989; Robinson 1994; Howell and Webb 1995). Our sightings during point counts over the canopy have been of birds flap-gliding rather purposefully in a straight line, generally only tens of meters above the canopy. At both study nests, adults usually came and went by flying silently, at or below canopy level. However, Julliot (1994) gave a detailed description of a Crested Eagle's capture of a young spider monkey that began with an overflight by the eagle "very high in the sky." Hence it remains unknown how commonly this species may fly and soar high overhead, and conceivably this may vary geographically. This species's lack of daily soaring bouts and of loud vocalizations made it one of the most difficult raptors to detect at Tikal. For locating these eagles and estimating their abundance, playback of conspecific or distressed prey vocalizations may be useful; this method was successful at least once in our fieldwork. We suspect that the primary means of inter-pair spacing may be vocalizations.

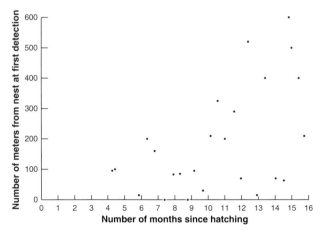

13.9 Rate of movement of dependent juvenile Crested Eagle away from natal nest during first 16 months of life; nest 2, Tikal.

Nest Defense and Interactions with Other Species

We noted 75 incidents in which Crested Eagles interacted with other birds, mammals, or airplanes in nonhunting contexts. Most (39%) were responses to American Black Vultures and Turkey Vultures (*Cathartes aura*);

17% were interactions with Brown Jays, 16% with small passerines, mainly various flycatchers, and 12% with other raptors; 7% were responses to planes and helicopters; 4% were interactions with Black-handed Spider Monkeys *(Ateles geoffroyi)*; 3% were threat responses to something overhead that we could not see; and 1.3% each were interactions with Crested Guans *(Penelope purpurascens)*, toucans, and aracaris.

With Brown Jays, the main interaction was a persistent clamor raised by a retinue of these jays when a Crested Eagle flew through the forest. The jays would eventually lose interest when the eagle remained perched for a time, but they would renew their raucous mobbing whenever the eagle took flight. Toucans and aracaris frequently called when a Crested Eagle flew near, but they appeared less prone than the Brown Jays to persistently mob the eagles. The calling of Brown Jays and toucans often advised us of the eagles' movements through the forest, and the female and nestling clearly paid attention to these calls in anticipating the arrival of the male (or in the case of the nestling, of either adult) at the nest. Flycatchers, most often Social Flycatchers *(Myiozetetes similis)*, also mobbed the female at times, sometimes causing her to change perches.

The most frequent and vigorous defensive responses by the eagles were directed toward American Black and Turkey Vultures. The typical response by the adult female was to lean forward and spread her wings, mantling over the nest and chick. When this display was fully developed, the female erected her plumage and called loudly. By the age of 39 days, the chick responded to vultures and other stimuli in a similar fashion to that just described.

Crested Eagles both harassed and were harassed by other raptors. In one case, a White Hawk followed the female, attacking, to the nest tree, then was joined by its mate as the two soared, calling, above the perched eagle. When a pair of White Hawks soared over the nest, the chick, age 58 days, gave the mantling threat display and called loudly. After the nestling had fledged, the adult female once attacked a pair of White Hawks some 300 m from the nest for several seconds, until they left the area. A pair of Black Hawk-eagles initiated nesting (which failed during incubation) within 200 m of the Crested Eagles' nest, and no aggressive interactions were noted, even when one of the hawk-eagles perched in a nearby tree while the female eagle incubated. During the nestling period, the female eagle drove a Laughing Falcon and a probable Collared Forest Falcon *(Micrastur semitorquatus)* from the nest tree. In one case, she responded strongly when a large raptor soared over the nest; she covered the chick with her wings, calling loudly, until the raptor soared out of sight 3 km away. The adult female was once stooped at by a pair of Great Black Hawks, and the fledgling in one case made several passes at a Hook-billed Kite *(Chondroheirax uncinatus)*, in what we believed to be aggression rather than capture attempts.

Most interactions with mammals were with spider monkeys. When several of these monkeys passed within 15 m of the nest, the incubating female called but did not leave the nest. In another case a group of spider monkeys approached the female, 200 m from the nest, screaming and throwing limbs at her and causing her to change perches. In one case, the female on the nest appeared nervous as she watched a Guatemalan Howler Monkey *(Allouata pigra)* moving nearby, but in other cases, she and howler monkeys appeared oblivious of one another as she moved around, collecting foliage, within 50 m of the primates.

The female eagle showed surprising tolerance when we climbed near the nest. Six days before the egg(s) hatched, we climbed to within 4 m of the female as she perched near the nest tree. She remained calmly perched, flew to the nest, then returned to perch again 5 m from the climber. In another case, having failed to catch her with a *bal-chatri* placed on the nest, we managed to capture her essentially by hand. While Estuardo Hernández was at the nest preparing to lower the unsuccessful trap, the female arrived with prey and perched on the nest less than a meter from him. Estuardo had the presence of mind to place a lasso of cord on the nest. When the female stepped into it, Estuardo pulled the noose tight, capturing her. A female Crested Eagle observed in Brazil was similarly docile and landed in the nest while the observer climbed a tree 5 m away (R. Bierregaard, pers. comm.). Only once did the male at our nest 2 respond aggressively, as a researcher climbed a palm near the nest tree. The male passed near the climber several times but never truly attacked. The docility of these eagles, perching near us as we climbed to the nest, was notably different from the behavior of some other species such as the Ornate Hawk-eagle, which routinely attacked researchers climbing near the nest (see Chapter 15).

Constancy of Territory Occupancy, Use of Alternate Nest Sites

These eagles are year-round residents at Tikal, and they may hold long-term territories. Repeated sightings indicated that certain areas were consistently occupied over the years. For example, the general area where the Naranjal nest was found in 1995 had been the site of several sightings during previous years, and a juvenile Crested Eagle was observed there in 1994, and other juveniles in years since our study.

Home Range Estimates

Via 28 telemetry-aided visual contacts from 23 August 1995 (four weeks before the chick fledged) through mid-April 1996 (30 weeks after the chick fledged), the home range (Fig. 13.10) of the Naranjal adult female was estimated at 24.2 km² (MCP) or 23.5 km² (85% HM). This is probably a conservative estimate of the area she used during this eight-month period; hence we rounded the estimate up to 25 km². During most of this period the female played an important role in provisioning the fledgling. We take this estimate, 25 km², as

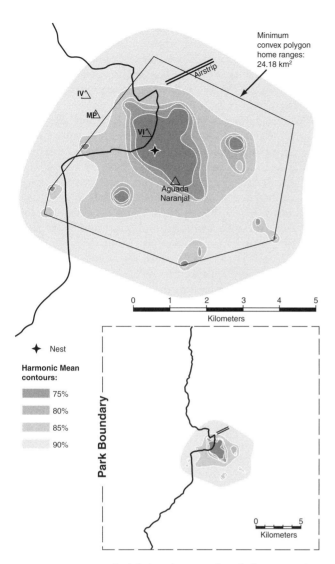

13.10 Home range of adult female Crested Eagle from Aguada Naranjal nest in Tikal National Park. The minimum convex polygon encloses 24.18 km² and the 85% Harmonic Mean is 23.5 km. Taken during the pre- and post-fledgling period.

We had no neighboring nests on which to base estimates of inter-nest distance. The radio-tagged adult female at times was located as much as 3.8–4.0 km from her nest; used as a radius of activity, this may suggest a usage area of roughly 50 km². Using the estimated 25 km² home range of this female and assuming the habitat is saturated with nonoverlapping, contiguous home ranges (bold assumptions), one may arrive at a very tentative maximum density estimate of approximately four pairs/100 km², or 20 pairs in the 510 km² of Tikal National Park that is not Scrub Swamp Forest and hence probably not suitable for nesting of these eagles. However, this is likely an overestimate, for two reasons. First, the 25 km² home range based on only 28 locations of this female may underestimate the true annual home range of a pair. Second, we do not know if pairs fill all available forest habitat; if not, the density suggested here could be a gross overestimate. We favor a roughly 50 km² estimate of usage area, based on the female's locations 4 km from the nest, which yields a rough estimate of 10 or 11 pairs for the park.

During our eight years of study at Tikal, we repeatedly saw Crested Eagles in certain disparate areas and felt confident we knew of at least three and probably four or five territories within the park. If we take into account all information sources, an educated guess would place the number of pairs in the park between 10 and 20, or 2–4 pairs/100 km² of non–Scrub Swamp Forest. Until better information is available, it is prudent to assume the lower end of this range. Thus, one might estimate perhaps 1.75–2.00 pairs/100 km², admittedly a crude estimate. Even if these estimates are accurate, it is possible they apply only in favorable circumstances, and that densities are often lower over large areas. At a French Guiana site, Thiollay (1989a) estimated a density of four individuals (1 or 2 pairs)/100 km², which agrees well with our most cautious estimate.

DEMOGRAPHY

Attainment of Adult Plumage and Age of First Breeding

According to Howell and Webb (1995), adult plumage is attained with the third prebasic molt, that is, during the fourth year. This is in contrast to the Harpy Eagle, which, these authors state, attains adult plumage with the fourth prebasic molt, during the fifth year.

Frequency of Nesting, Percentage of Pairs Nesting

One juvenile showed no sign of reaching independence at 16 months after hatching. Assuming, conservatively, that this juvenile reached independence by the age of 18 months (this could easily take much longer) and adding 50 days for incubation and two months of pre-laying activity, one may predict that the interval

a first approximation of the mean home range size per pair; likely it is conservatively small and we suspect that home ranges may often be twice this size.

Inter-nest Spacing and Density of Territorial Pairs

The Crested Eagle is among the rarest of the 30-odd forest-nesting raptors at Tikal. Of the large forest raptors present or potentially present, the only species more rare here than the Crested Eagle are the Harpy Eagle, the Black Solitary Eagle *(Harpyhaliaetus solitarius)*, if indeed it is present (Beavers et al. 1991; Beavers 1992), and the Black-and-white Hawk-eagle *(Spizaetus melanoleucus)*, for which we made at best a few sightings per year. Still, Crested Eagles are more common in this area than is generally supposed and than suggested by the infrequent casual sightings.

between initiation of nesting efforts may have been at least 22 months. For the Ornate Hawk-eagle, the mean interval from egg laying to egg laying was 23.3 months (see Chapter 15). Given the multiple indications that the young Crested Eagle developed more slowly than did several young Ornate Hawk-eagles, it seems likely that the interval from laying of one successful clutch to the next laying may be longer than 24 months in the Crested Eagle. Thus we expect that fewer than half of territorial pairs initiate a new nesting attempt in any given year.

Productivity and Nest Success

We cannot comment on rates or causes of nesting success or adult mortality. With a 1- or 2-egg clutch, age of first breeding at several years, and with each nesting cycle requiring two years or more, these eagles must have one of the lowest reproductive rates among the raptors we studied at Tikal.

Parasites and Pests

The Naranjal nestling was perturbed by flying insects at times, and occasionally the adult female snapped at these with her beak. At times we noted ants in the nest but did not note any interaction between these and the eagles.

CONSERVATION

One of the great conservation questions concerning Neotropical raptors is their ability to persist in partially deforested landscapes. Our nest 1 was in such a landscape, within a remnant of mature Transitional Forest approximately 200 m by 1 km, amid a patchwork of forest remnants, cornfields, and fallow fields supporting various ages of regrowth. An active cornfield, 550 m by 850 m, reached within 150 m of the nest tree. To quantify habitat in the vicinity, we established 3 km transects in the four cardinal directions from the nest, assigning habitat along each 100 m segment to one of the following classes: primary Upland, Transitional, or Scrub Swamp Forest; active cornfield; and abandoned crop field (regrowth < 10 years old). Based on this sampling, 45% of the area within 3 km of the nest remained in primary forest, 43% was brushy second growth, and 12% was active cornfields. Mean patch size for primary forest was 16.3 ha (n = 14), for young second growth (fallow fields) it was 10.5 ha (n = 17), and for active cornfields, 4.3 ha (n = 8); overall, mean patch size was 10.8 ha. These small patch sizes reflect the fine-grained nature of habitat fragmentation in this landscape. Though composition of the original forest cover is unknown, remaining forest was 41% Upland, 44% Transitional, and 15% Scrub Swamp. The southwest quadrant was still largely covered by mature, mainly Upland, forest, while extensive Scrub Swamp Forest lay to the north, beginning 2 km from the nest and extending north and east from there. This nesting failed before the egg(s) hatched, and we detected no renewal of

nesting here over the next two years, during which time the nest had largely fallen from the tree, and fire had entered the forest fragment to within 4 m of the nest tree. Perhaps we witnessed the last nesting attempt in this nest as this once-forested landscape became increasingly modified by farming.

Witnessing a nesting attempt in this half-deforested landscape tells us little about the species's tolerance of habitat modification, as this may simply have illustrated the attachment of a pair to a traditional site no longer suitable for this species. In our best judgment, these eagles are highly vulnerable to deforestation. They appear to hunt mainly within or below the forest canopy, hence probably will not adapt readily to hunting in more open habitats. Moreover, the diet was dominated by fairly large, arboreal nocturnal mammals, which are mainly forest-associated species and probably decline with deforestation. With regard to the large snakes that figured importantly in these eagles' diets—*Spilotes*, *Drymarchon*, and *Boa*—we know of no data on how these snakes are affected by deforestation. The Crested Eagle was one of several raptor species that Thiollay (1984) concluded was adversely affected by human subsistence hunting. While human hunters may depress populations of important prey species in some situations, in our study area there was little overlap in the species taken by humans and Crested Eagles (see Chapter 15). Shooting of the eagles themselves may be a more serious concern.

Finally, the facts that these eagles use probably 25–50 km² on an annual basis and dwell mainly in the forest interior, combined with an intrinsically low population density and natural rate of increase, suggest that the Crested Eagle may be one of the Neotropical forest raptors most affected by forest loss and alteration. This eagle was listed as "near-threatened" by Collar et al. (1992), with which we concur.

HIGHLIGHTS

The main components of Crested Eagle diets at Tikal—large snakes and medium-sized, arboreal nocturnal mammals—make an intriguing combination. Crested Eagles ate larger snakes than did any other raptor at Tikal. One way to view this dietary combination is the following: For a raptor as large as the Crested Eagle to specialize completely on snakes—as does the much smaller Laughing Falcon—would probably be impractical, as the population densities of the large snakes taken by Crested Eagles are probably too low to allow such a diet. Thus, a would-be predator of large snakes must add something else to its diet in order to have a sufficiently dense prey resource. This may help explain the unique dietary combination of large snakes and medium-sized mammals.

From an ecological standpoint, it is worth noting that snakes are themselves carnivorous. While some snakes eat mainly rodents (herbivores/granivores), others prey heavily on secondary consumers such as birds (and their eggs), lizards, frogs, and other snakes—that is, animals that are themselves carnivorous. Hence, snake-eating raptors occupy some of the highest trophic levels among terrestrial organisms. Among Neotropical forest raptors, the Crested Eagle may well be the most important predator of large snakes. If this is true, these eagles may have some influence on the nature of predator-prey systems that include large snakes as higher-order predators. Overlap for the ophidian (snake) portion of these eagles' diet may be significant not only with certain other raptors but also with large carnivorous mammals such as cats and, especially, with large, snake-eating snakes such as in the genera *Drymarchon* and *Clelia* (Greene 1988). The felid community at one Amazonian site obtained 5.6% of its collective diet from snakes (Emmons 1987).

It is also noteworthy that other raptors comprised at least 19% of the avian prey taken by Crested Eagles and 4% of their total diet (Table 13.1). Throughout most of the Neotropical lowland forest, the Crested Eagle is the second-largest forest raptor. Thus it seems plausible that this eagle may play a significant role as a nest predator of some of the larger birds, especially other raptors—which are generally safe from smaller predators.

To the extent that these eagles prey on smaller predators, one may postulate that their decline or disappearance might give rise to ecological changes that might reverberate through a biotic community. One potential effect is "meso-predator release," a phenomenon in which smaller carnivores, released from predation by larger carnivores, proliferate or are behaviorally released, resulting in higher-than-normal predation on the smaller animals that make up their diets (Latham 1952; Soule et al. 1988; Crooks and Soule 1999), which, in turn, can lead to further ecological effects. This phenomenon has been invoked, for example, as a possible explanation of what seems to be higher-than-normal predation rates on bird nests on Barro Colorado Island, a Central American site where a number of top predators have long been absent (Loiselle and Hoppes 1983).

Notable for their absence or rarity in the Crested Eagle diet were the large game birds—the tinamous and cracids—and this was true also at the Brazilian nest (Bierregaard 1984). In contrast, Ornate Hawk-eagles, both at Tikal and in Brazil (Klein et al. 1988), frequently took large adult birds, including these game birds. In turn, Crested Eagles, like Black Hawk-eagles, seemed to specialize in large part on arboreal mammals, especially nocturnal ones. However, this predominance of nocturnal over diurnal mammals in the diet at Tikal may owe in part to biogeography. Whereas small diurnal primates within the size range of Crested Eagle prey are diverse in South America, none occur at Tikal. In future research, it will be interesting to learn whether Crested Eagles in South America rely more on small diurnal primates and less on nocturnal opossums and porcupines than they did at Tikal; prey records from the Brazil nest suggest perhaps not (Bierregaard 1984), but more data are needed. Finally, our data suggest that male and female Crested Eagles may differ in diet—both in size and in taxonomic composition. This is an intriguing possibility from the standpoint of niche partitioning and the evolution of sexual size dimorphism.

While total nest attendance at our two nests was similar, the percentage of the time adults (mostly females) incubated differed strikingly: 39% of the time at nest 1 and 83% at nest 2. The nest 1 female spent a correspondingly greater proportion of the time standing over and shading the egg(s) than did the nest 2 female. However, nest 1 was in full sun and likely experienced a warmer microclimate than did the partially shaded nest 2, and ambient temperature can affect incubation constancy (Drent 1973). Thus, the difference in incubation behavior between these two nests may have been related to the degree of sun exposure at the nest sites. Variation in incubation behavior between pairs within a raptor species is often great, however, and several pairs must be studied before generalizations can be safely made (Brown 1976; Newton 1979).

Skutch (1976) considered the normal range of incubation constancy to be 60–80% for a variety of small bird species, with a constancy of more than 80% shown mainly by birds that are fed by their mates, such as raptors. In raptors, it has been generalized that adults attend nests 90–100% of the day during incubation and early in the nestling period (Hubert et al. 1995). Seemingly, little is known of how incubation constancy may change during the incubation period (Drent 1973). Skutch (1976) pointed out that while birds may often sit tighter in the face of disturbance as hatching approaches, this does not necessarily translate into more constant incubation overall as hatching approaches—he lists several species in which incubation constancy is just as likely to decrease as to increase as hatching draws near. As mentioned, ambient temperature can affect incubation constancy.

Behavioral roles of raptor pair members change predictably through the course of the nesting season, in a fashion that is generally typical of a genus or a group of genera (Newton 1979). Roles may be largely described by pair members' contributions to hunting, incubation, brooding and feeding of young, and nest defense and vigilance. These specific roles, however, may be regarded as superimposed on a more basic tendency—to remain near the nest, or to travel elsewhere, often to hunt. To describe this continuum of attachment to the nest site, Balgooyen (1976) used the terms "centripety" and "centrifugy," the former referring to adults' attraction to the nest and the latter to their tendency to move away from it. We use the same terminology here.

Seemingly, female centrifugy in Crested Eagles began prior to hatching—as exemplified by a progressive reduction of incubation constancy—and intensified gradually from that time through the nestling period. Time spent by the female incubating the egg and brooding the chick—and total time she spent on the nest—declined gradually, as she spent a correspondingly greater portion of her time remaining vigilant in a nearby tree, until chick ages 60–65 days. At this point the female made a rapid behavioral shift, and by chick age 80 days, she was absent from the nest area most of the time. We documented similar patterns of female centrifugy in Ornate and Black Hawk-eagles at Tikal (see Chapters 14, 15). Hubert et al. (1995) reviewed evidence of similar patterns in other raptors and suggested that increasing activity of chicks was a proximate cue stimulating increased female centrifugy in the Common Buzzard *(Buteo buteo)*.

Because data on density and inter-nest spacing were unavailable due to the limited number of nests observed, it will be of great interest to further explore the population density and spatial use of this species at various sites within their range, as well as methods for detecting these cryptic forest dwellers. As one of the lowest-density Neotropical forest raptors, and with a diet that includes many smaller predators, Crested Eagles should be especially useful as indicator species relative to habitat space and ecological integrity, for reserve planning and management, and for ecological restoration efforts.

For further information on this species in other portions of its range, refer to Burton 2006, Grosselet and Gutierrez Carbonet 2007, Jones et al. 2000, Oversluijs Vasquez and Heymann 2001, and Vargas et al. 2006.

14   BLACK HAWK-EAGLE

*David F. Whitacre, Juventino López, Gregorio López,
Sixto H. Funes, Craig J. Flatten, and Julio A. Madrid*

As we walk along a trail beneath Tikal's dense forest canopy, a clear, ringing call drifts down from on high, *whut, whut-EEEeer . . . whut, whut, whut-EEEeer.* Emerging into a treefall gap, we glance up, finding the anticipated silhouette of a Black Hawk-eagle *(Spizaetus tyrannus)* as it banks against a puff of cumulus high overhead. This handsome, crested hawk-eagle is most often detected in just this fashion—as it soars high above the forest, issuing its distinctive call (Plate 14.1). As with many raptors at Tikal, sightings of these birds perched (Plates 14.2, 14.3) or flying within the forest are rare. Indeed, this hawk-eagle's hunting behavior has rarely been observed. It must be interesting, however, as 70% of the prey items brought to nests at Tikal were bats and other nocturnal mammals delivered during daylight hours. Presumably these hawk-eagles must find such nocturnal prey by searching patiently for their daytime retreats. What clues they use in finding such prey remains a mystery, but one truly strange observation we report here suggests a novel hypothesis (see Highlights, later in this chapter). Dramatic attacks on day-active quarry are also sometimes made. In general, these hawk-eagles appear to prey much more heavily on mammals than does the closely related and broadly sympatric Ornate Hawk-eagle *(Spizaetus ornatus)*, which takes about half mammals and half birds. Although Black Hawk-eagles are widely believed to be more tolerant of deforestation than Ornates, few nests have yet been observed in fragmented landscapes, whereas a number of Ornate Hawk-eagle nests have been observed in modified landscapes, and these often fledged young.

GEOGRAPHIC DISTRIBUTION
AND SYSTEMATICS

The Black Hawk-eagle occurs from southern Tamaulipas and Guerrero, Mexico, to northern Bolivia, Paraguay, and Argentina (Brown and Amadon 1968; AOU 1983; Howell and Webb 1995). The subspecies *S. t. tyrannus* occupies southeastern Brazil and extreme northeastern Argentina, while the smaller *S. t. serus* occurs elsewhere

in the species's range (Brown and Amadon 1968; Bierregaard 1994). Widespread in the lowlands, this hawk-eagle commonly ranges up to elevations of 1000–1300 m and is known to occur up to 2000 or even 3000 m in Guatemala (Vannini 1989a). These hawk-eagles are year-round residents at Tikal, as they likely are throughout their range.

As traditionally conceived, *Spizaetus* is a pantropical genus containing 10 species. It is one of several genera with feathered tarsi—the so-called booted eagles—sometimes considered a tribe, Aquilini. Molecular genetics studies (Helbig et al. 2005) have drastically revised our understanding of relationships among the booted eagles. These studies indicate that while the Aquilini as a whole comprise a monophyletic group, none of the polytypic genera within it *(Spizaetus, Aquila, Hieraaetus)* are monophyletic, requiring revisions of traditional generic boundaries. The aquilines contain three basal clades: Old World *Spizaetus* and *Stephanoaetus*, together forming a sister group to the remainder of the tribe; a Neotropical clade containing the Ornate Hawk-eagle, Black Hawk-eagle, the Black and White Hawk-eagle *(Spizaetus melanoleucus)*, and the Black and Chestnut Hawk-eagle *(S. isidori)*; and an Old World clade consisting of *Aquila* and several allied genera, the boundaries among which are rearranged by this recent work (Helbig et al. 2005). Aquilines as a group diverged roughly 12–15 million years ago, and the Neotropical clade diverged from Old World lineages about 8–11 million years ago (Helbig et al. 2005).

MORPHOLOGY

According to Brown and Amadon (1968) this is one of the more lightly built species of *Spizaetus*. For the subspecies occurring at Tikal *(S. t. serus)*, the largest published set of measurements is that of Blake (1977), based on 11 males and eight females. These, along with smaller data sets from Friedmann (1950) and Bierregaard (1978), are given in Appendix 2. There is some uncertainty regarding size and degree of sexual dimorphism, as these data sets depart substantially on some measurements (Bierregaard's data set is the same as Friedmann's, with four individuals added). Based on Blake's data, values of the Dimorphism Index (DI) are 5.2 for wing chord, 3.5 for tail length, 6.2 for culmen, and 6.9 for tarsus. In contrast, Friedmann's (1950) data give dimorphism values of 14.0 for wing chord, 20.3 for tail, 5.5 for culmen, 9.1 for tarsus, and 8.2 for middle toe. Based on body weight (see below), the DI is 6.7, possibly indicating that the linear dimorphism values given by Blake's data are more representative than those given by Friedmann's slightly smaller data set. Blake (1977) and Friedmann (1950) both give values for unflattened wing chord; hence differences in methods appear not to explain the difference in wing and tail length dimorphism in these two data sets. Conceivably, the discrepancy results from missexed specimens in one or both data sets. Friedmann's specimens were from Mexico, Honduras, Costa Rica, Panama, Surinam, and Brazil; the provenance of Blake's specimens is unstated. Based on Friedmann's data, the wing-tail length ratio for females was 1.14 and for males 1.21, while for Blake's data, the wing-tail ratio was 1.25 for females and 1.23 for males.

Appendix 1 lists all body mass data of which we are aware for *S. t. serus*. Adult females averaged 1115 ± 25 g (n = 5). Omitting one unusually heavy bird, males averaged 911 ± 41 g (n = 3; Appendix 1). Based on the cube root of these weights (Storer 1966), the DI is 6.7.

PREVIOUS RESEARCH

The Black Hawk-eagle has never before been studied in detail. Published information on nesting is sparse: Rangel-Salazar and Enriquez-Rocha (1993) described a Mexican nest from which a juvenile had recently fledged, and listed several prey remains found at the nest. A Panamanian nest attributed to this species held two young in both 1965 and 1968, and five prey items were recorded (Smith 1970). Most other mentions of the species are brief accounts in regional lists and field guides.

RESEARCH IN THE MAYA FOREST

At Tikal we studied five nesting attempts in four territories from 1992 to 1995 and monitored reproductive activity during 12–13 territorial pair-years (13 years if Naranjal territory is included in 1990, when an adult female was first trapped and radio-tagged and was not nesting). Nesting histories and our observation efforts are detailed in Appendix 14.1. Our study methods were similar to those described here in other chapters. We found nests by repeatedly visiting areas of activity, often observing from treetops. Once nests were located, we observed them from nearby tree platforms or the ground. Unless stated, data given are means ± SD.

DIET AND HUNTING BEHAVIOR

At three nestings in two territories, we observed 117 prey items, of which 85 were identified (Table 14.1). The diet was very similar at the three nests. Mammals dominated the diet in each case, comprising 95.3% of identified prey overall, and 93–100% of identified items at individual nests (mean = 95.8 ± 3.7%, n = 3). The balance of the diet was composed of birds—4.7% overall—ranging from 0 to 7.0% at individual nests (mean = 4.2 ± 3.7%, n = 3). Numerically, bats were the predominant prey overall (50.1%) and at all nests, ranging from 42 to 60% of identified prey (mean = 52.3 ± 9.5%, n = 3), followed by tree squirrels, contributing 27.1% of prey overall, and 20–33% of prey at individual nests (mean = 25.8 ± 6.8%, n = 3; Table 14.1). The remainder of the prey was made up of opossums (10.6% overall)—which we believed were mainly Mexican Mouse Opossums (*Marmosa mexicana*)—Mexican Hairy Porcupines (*Sphiggurus mexicanus*, 2.4%), and a single juvenile White-nosed Coati (*Nasua narica*, 1.2%). In terms of biomass, bats contributed an estimated 14% of the diet, squirrels 36%, porcupines 27%, mammals as a whole contributed 93.3% of the diet, and birds the remaining 6.7% (Table 14.2).

One bat taken was the Jamaican Fruit-eating Bat (*Artibeus jamaicensis*), and we believed that many of the bats taken were this species or the congeneric Great Fruit-eating Bat (*A. lituratus*), both of which are common at Tikal. Smaller, unidentified bats were also taken. Among the four birds observed as prey, one was a Mexican Wood Owl (*Strix squamulata*) of unknown age. The absence of reptiles in the diet at the nest was striking. While a few may have been among the unidentified prey brought to nests, substantial numbers were not observed. To these dietary records at nests may be added records of two additional bats and several unidentified other mammals, a headless reptile brought to fledglings by adults (witnessed during telemetry monitoring of the juveniles), and a snake being carried by an adult. In addition, we observed one fledgling on the ground feeding on Common Raccoon (*Procyon lotor*) feces containing crayfish parts and fish scales.

Of those 69 prey items to which it was possible to assign an approximate mass (Table 14.2), 63 items (91%) weighed about 200 g or less, and 40 items (58%) were in the neighborhood of 50 g. Only six prey items were believed to weigh 400 g or more. The maximum weight of prey items is unknown, as Mexican Hairy Porcupines and juvenile coatis are highly variable in weight. Median mass of 69 identified prey items was 50 g, while their average mass was roughly 167 g, heavily influenced by the three largest items (Table 14.2).

We found evidence that males took smaller prey on average than did females (Table 14.3). The two species of squirrels occurring at Tikal were often distinguishable; Deppe's Squirrel (*Sciurus deppei*), at 190–220 g, is half the size of the Yucatán Squirrel (*S. yucatanensis*) (mean = 420 g), and the two differ notably in pelage color (see Chapter 2). In 19 cases we were able to assign species to squirrels observed as Black Hawk-eagle prey (Tables 14.2, 14.3). Males brought in 15 Deppe's Squirrels and

Table 14.1 Diet of Black Hawk-eagles at three nests in Tikal National Park[a]

Prey items	1992 Naranjal		1992 Arroyo Negro	
	Number	Percentage	Number	Percentage
Bats	15	41.7	12	60.0
Sciurus deppei	11	30.6	4	20.0
Sciurus yucatanensis	1	2.8	0	0
Total number of squirrels	**12**	**33.3**	**4**	**20.0**
Marmosa mexicana	4	11.1	4	20.0
Porcupine, *Sphiggurus*	0	0	0	0
Coati	0	0	0	0
Total number of identified mammals	**31**	**86.1**	**20**	**100**
Mammal spp.	3	8.3	0	0
Total number of mammals	**34**	**94.4**	**20**	**100**
Birds	2	5.6	0	0
Unidentified prey	17	—	8	—
Total number of identified prey	**36**	—	**20**	—
Grand total	53	—	28	—

Prey items	1994 Arroyo Negro		All nests combined	
	Number	Percentage	Number	Percentage
Bats	16	55.2	43	50.6
Sciurus deppei	5	17.2	20	23.5
Sciurus yucatanensis	2	6.9	3	3.5
Total number of squirrels	**7**	**24.1**	**23**	**27.1**
Marmosa mexicana	1	3.5	9	10.6
Porcupine, *Sphiggurus*	2	6.9	2	2.4
Coati	1	3.4	1	1.2
Total number of identified mammals	**27**	**93.1**	**78**	**91.8**
Mammal spp.	0	0	3	3.5
Total number of mammals	**27**	**93.1**	**81**	**95.3**
Birds	2	6.9	4	4.7
Unidentified prey	7	—	32	—
Total number of identified prey	**29**	—	**85**	—
Grand total	36	—	117	—

[a] Percentages are based only on prey identified at least to class. Not included here are two bats observed as prey in March 1995 at the Arroyo Negro nest, and prey items observed away from nests (see text). Prey in table were observed at the nest via observations from tree blinds; identifications in some cases were confirmed by prey remains later retrieved from nests.

1 Yucatán Squirrel, while females brought in 2 of each. Although the latter sample is small, these results suggest that males took the smaller squirrels more often than did females. When all mammalian prey were categorized as small or large (with bats, Deppe's Squirrel, and Mexican Mouse Opossums being small and Yucatán Squirrel, Mexican Hairy Porcupines, and coatis being large), males brought in 55 small and 3 large prey, while females delivered 8 small and 3 large prey (Table 14.3), significantly different proportions (MRPP chi-square test; $\chi^2 = 5.61$, $P = 0.02$). Considering the diet in taxonomic terms, males tended to bring in more bats relative to other prey types than did females, although this tendency was not statistically significant (MRPP chi-square test, bats versus all other identified prey; $\chi^2 = 2.1$, $P = 0.15$).

All mammals brought to nests by these hawk-eagles were arboreal, and 71% were nocturnal species, which presumably spend much of the day sleeping in concealed locations. Nocturnal mammals were delivered to nests during every daylight hour (Fig. 14.1), suggesting the hawk-eagles caught them at day roosts. Though captures of these prey were not observed, we believe the hawk-eagles probably searched carefully through the forest canopy or sub-canopy to find sleeping mammals, possibly reaching into tree hollows, vine tangles, or leaf nests. The delivery of an unweaned squirrel pup with eyes still closed also suggests this method of hunting by Black Hawk-eagles.

Aided by radiotelemetry, we observed the hunting behavior of an adult male and female of two different pairs. Some of the resulting observations supported the

Table 14.2 Diet composition based on estimated prey biomass for three Black Hawk-eagle nests studied at Tikal[a]

Prey type	Total	Approximate mass range (g)	Estimated median mass/individual (g)	Total biomass (g)	Percentage of total biomass (%)
Bats (Chiroptera)	43	20–75	50	2150	14.4
Sciurus deppei	20	190–220	205	4100	27.4
Sciurus yucatanensis	3	420	420	1260	8.4
Total number of squirrels	23	—	—	5360	35.8
Marmosa mexicana	9	40–100	70	630	4.2
Sphiggurus mexicanus	2	1400–2600	2000	4000	26.8
Nasua narica	1	1000–2000	1000	1000	6.7
Total number of identified mammals	78	—	—	13,140	87.9
Unidentified mammals	3	—	272[b]	816	5.5
Total number of mammals	81	—	—	13,956	93.3
Birds	4	—	250[c]	1000	6.7
Prey not identified to class	32	—	—	—	—
Total number of identified prey	85	—	—	14,956	100.0
Grand total of prey	117	—	—	—	—

[a] Body masses used for each species are the same as those used in Chapter 15.
[b] Unidentified mammals were assigned the mean mass of squirrels taken.
[c] We assigned 250 g for bird mass because we believe it unlikely that larger birds were often taken; median bird mass may have been lower than this.

Table 14.3 Prey delivered to nests by adult male and female Black Hawk-eagles at Tikal[a]

Prey type	Males		Females	
	Number	Percentage[b]	Number	Percentage[c]
Bats (Chiroptera)	36	56.3	4	33.3
Sciurus deppei	15	23.4	2	16.7
S. yucatanensis	1	1.6	2	16.7
Marmosa mexicana	4	6.3	2	16.7
Sphiggurus mexicanus	2	3.1	0	0.0
Nasua narica (juv.)	0	0.0	1	8.3
Unidentified mammals	3	4.7	0	0.0
Total number of mammals	61	95.3	11	91.7
Birds	3	4.7	1	8.3
Total number identified to class	64	100.0	12	100.0
Grand total	64	100.0	12	100.0

[a] Table includes only items for which it was known which adult brought prey to the nest; thus the total (105) is less than the total number of prey items (117) given in Table 14.1. The adult delivering prey to the nest vicinity was assumed also to have captured it; items transferred by male to female were tabulated as delivered by the male.
[b] Percentage of identified male prey.
[c] Percentage of identified female prey.

hypothesis that these hawk-eagles searched in likely roosting sites of nocturnal mammals, in addition to patiently hunting from a perch. The female at one nest was observed flying among dead fronds hanging beneath the crown of a Botán palm *(Sabal mauritiiformis)* as if searching for prey, and the male of another pair was twice observed to behave in like fashion. Such sites are sometimes used by bats as day roosts (Morrison 1980). One morning while following the same adult male, we observed him spending four hours below the canopy in what appeared to be still-hunting. At one point he descended, landing abruptly in a vine tangle with wings and tail spread, but this apparent attack was unsuccessful. Then, perched 9 m up on a vine below a Botán palm for nearly an hour, he intently watched a Deppe's Squirrel that was calling in the canopy nearby. After the squirrel moved away, the hawk-eagle remained below the canopy, perching in three other trees 5–7 m tall, for two and a half hours. We had the impression that he was hunting throughout this interval.

The only capture by Black Hawk-eagles we witnessed was that of a Yucatán Squirrel by the Naranjal adult female with a six-week-old chick in the nest. From a tree near the nest, she suddenly flew downward among the limbs of a nearby tree and emerged two minutes later with a squirrel in her talons. In three other cases, we saw the adult female at the Arroyo Negro nest attack birds perched or flying near her nest, in what appeared to be capture attempts; all three attacks were launched from a perch. One attack was at a flying bird, while the other two were descending flights into tree foliage, directed at birds that were probably perched. In one case the intended

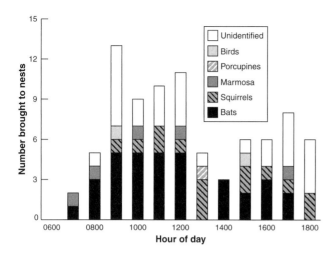

14.1 Type and number of prey items brought to three Black Hawk-eagle nests at Tikal, Guatemala, by time of day.

quarry was a flock of Aztec Parakeets *(Aratinga nana)*. In Peru, Robinson (1994) described an attack on macaws perched in a tree, and a case in which a Black Hawk-eagle dove from the mid-canopy to catch a squirrel 2 m from the ground. He generalized that these hawk-eagles make long (> 50 m) attacks through the forest and stoop downward at prey in open areas. Our observations indicate that additional, often sub-canopy, hunting styles are also used.

The inclusion of many squirrels and occasional birds in the diet at Tikal, and the observations of hunting and attacks on squirrels and birds, show that stealthy hunting and rapid attacks on agile, day-active prey form an important part of this hawk-eagle's foraging repertoire. However, because 70% of the mammals taken at Tikal were nocturnal species, patient searching of their likely daytime roost sites is probably emphasized more than stealthy ambush attacks. It would be interesting to know what clues these raptors use in searching for such roosting prey.

In Quintana Roo, Mexico, Rangel-Salazar and Enriquez-Rocha (1993) found remains of 17 prey items in and below a nest. Fourteen (82%) of these were birds, including seven Keel-billed Toucans *(Ramphastos sulfuratus)* and two Collared Aracaris *(Pteroglossus torquatus)*; three were mammals—two squirrels and a Common Raccoon. The diet portrayed by these prey remains—which resembles the Ornate Hawk-eagle's diet at Tikal more than the Black Hawk-eagle's diet at Tikal—may be somewhat biased, as toucan skulls and beaks are especially durable and conspicuous prey remains. In contrast, the bats that numerically comprised half the Black Hawk-eagle's diet at Tikal were consumed entirely, leaving no trace. While it appears that the Quintana Roo pair may have taken more birds, especially toucans, than did Black Hawk-eagles in our study, this small sample of prey remains from a single nest does not permit firm comparisons with our data from direct observations at several nests. This instance does, however, point out the need for studies of Neotropical raptors in different localities and habitats: geographic variation in diet and other aspects of a species's ecology are expected, and no

study at a single site can provide information applicable throughout a species's range.

Other food items mentioned in the literature include two flycatcher nestlings snatched along with part of the nest by an immature hawk-eagle (Brown and Amadon 1968; Skutch 1960b). Sick (1993) commented that these hawk-eagles prey more on mammals than on birds, that they are said to habitually eat primates, probably including marmosets, and that bats including the Jamaican Fruit-eating Bat had been found several times in stomach contents. Haverschmidt and Mees (1994) noted that a female had opossum fetuses in its stomach *(Didelphidae* spp.), and Albuquerque (1995) mentions a sighting of an individual carrying a *Didelphis* opossum in Brazil. At a Panama nest attributed to this hawk-eagle, Smith (1970) identified two Common Green Iguanas *(Iguana iguana)*, a vine snake *(Oxybelis* sp.), a Black-chested Jay *(Cyanocorax affinis)*, and a Red-tailed Squirrel *(Sciurus granatensis)*. Stiles et al. (1989) state that this species takes "birds up to size of toucans and chachalacas, including nestlings; mammals like squirrels and opossums; large lizards and snakes." In the farming landscape south of Tikal, we heard numerous accounts from farmers of chicken-thieving activities that seemed to implicate the Black Hawk-eagle, and we saw one individual that had reportedly been shot in the act.

HABITAT USE

Published statements about habitat affinities of the Black Hawk-eagle are variable. Several authors describe Black and Ornate Hawk-eagles as occurring in dense lowland forest and do not distinguish between habitats used by the two species. Many authors, however, suggest that the Black Hawk-eagle has a greater preference or tolerance than does the Ornate Hawk-eagle for second-growth, forest edge, and partly deforested landscapes (e.g., Thiollay 1985; Hilty and Brown 1986; Stiles et al. 1989; Jullien and Thiollay 1996). These comments generally come from regions with wetter, taller forest than that of Tikal. However, even at Tikal, some observers have formed a similar impression (J. Madrid, pers. observ.). Such a difference in habitat use by the two species may be real, and there may also be geographic variation in the degree to which Black Hawk-eagles use dense, continuous forest versus disturbed, semi-open, and edge habitats (see Chapter 23). We are, however, skeptical of the more extreme accounts such as that of Slud (1964), who stated that the Black Hawk-eagle is not truly a forest species in Costa Rica and is restricted to second-growth and groves in semi-open areas, often bordering extensive forests. We suspect this impression results more from the greater detectability of the birds in such situations than from actual patterns of habitat use.

Because our studies were mainly conducted in mature forest, we cannot comment conclusively on the relative ability of Black and Ornate Hawk-eagles to thrive in partly deforested habitats. However, our nesting and telemetry studies showed that at Tikal, Black

Hawk-eagles are certainly not absent from mature forest; quite the contrary, these hawk-eagles nested and hunted deep within the heart of the continuously forested park and showed no preference for man-made openings around Maya ruins or park facilities. Nor did they show a preference for the Scrub Swamp Forests that are the lowest-canopied forest type within the park. Rather, all nests were in relatively tall, dense forest of types we deemed Mesic Bajo, Hillbase, and Transitional Forest (see Chapter 2).

All four of the Black Hawk-eagle nests we studied were located in areas of relatively low terrain. Two were in or beside the bed of the Arroyo Negro, a wooded, seasonally flowing drainage course. Forest along the Arroyo Negro is structurally complex, with high, broken canopy, and its deep soils and position low in the drainage system may create a dry season refuge of moister-than-average conditions. The other two nests were in fairly similar Transitional and Hillbase Forest. One was an alternate nest of the Arroyo Negro pair and was located 500 m from the primary nest in the bed of the watercourse, while the remaining nest was in a nearly level area of Hillbase Forest 100 m from the foot of a substantial ridge. Whether the association of these nests with these Transitional Forest types (rather than the more widespread Upland Forest) indicates a preference for such nesting habitat is not clear.

Visual contacts made while following hawk-eagles via telemetry aided in defining their patterns of habitat use. The Arroyo Negro male and the Naranjal female, both having nests along the bed of the Arroyo Negro, did not confine their foraging to this seasonal watercourse but ranged also into other areas of forest. During eight visual contacts, the Arroyo Negro male was observed four times in tall Upland Forest and four times in palm-rich Transitional and Hillbase Forest. In these same eight sightings, this male was perched once at the upper canopy surface, three times within the canopy, three times where the lower canopy surface meets the understory, and once low in the understory. Mean perch height was 12 m (n = 8), ranging from 1 m (a dead log) to 16 m; median height of perch trees was 22 m and median diameter 50 cm. All sightings were in mature forest, with canopy height averaging 20 m.

Two radio-tagged fledglings monitored during their prolonged dependency on adults remained in mature forest at all times. In 14 visual contacts, the juvenile male of the 1992 Arroyo Negro nest remained mainly in tall forest, ranging from Upland through Hillbase, Transitional, and Mesic Bajo Forest types. Median canopy height at 12 perch sites was 21 m, ranging from 5 to 25 m. Of the 14 sightings, this bird was perched at or above the upper canopy surface in four cases, within the canopy five times, and in the understory five times. Perch heights averaged 16 m (n = 12) and varied from the ground to 25 m above; perch trees averaged 22 m tall (range = 6–30 m), and 45 cm in diameter (range = 10–102 cm). Nine perch limbs averaged 5 cm in diameter (range = 2.5–10.0 cm) and were usually living. All 11 sites had some Escobo palm (*Cryosophila stauracantha*) in the understory, in-

dicating that these were neither Dry Upland sites nor low-lying Scrub Swamp sites, but rather were in between these extremes. Ten of 11 sites had young Sabal palms and 7 of 11 had mature palms of this species, indicating, along with other tree species observed, that these were largely Transitional, Mesic Bajo, or Hillbase Forest types. Since most of these contacts with this fledgling hawk-eagle took place within 50–200 m of the nest, these data also characterize the general nest vicinity.

While radio-tracking the juvenile female that fledged from the Naranjal nest in 1992, habitat data were taken during 16 visual contacts. In each case, this juvenile was perched in tall primary forest, not associated with roads or large clearings. This nest was in Mesic Bajo Forest in a narrow swale between two hills cloaked in Upland Forest; 11 visual contacts were made in Upland Forest and three in Transitional Forest (in two cases, forest type was not noted). The juvenile was found perched an average of 19 ± 9 m high (n = 13, range 5–30 m), in trees averaging 24 ± 7 m tall (n = 14, range 9–30 m). In eight cases she was at or above the upper canopy surface, three times within the canopy, and four times in the understory. Perch trees averaged 45 ± 28 cm in diameter (n = 13, range 11–97 cm), and perches were mainly horizontal limbs 1.5–7.5 cm thick (mean = 4.3 ± 2.3 cm, n = 12); eight were living and three dead. Estimated minimum canopy height averaged 18 ± 2.6 m (n = 14), while modal canopy height averaged 21 ± 1.6 m (n = 14) and maximum canopy height averaged 25 m (n = 14). Tree species noted within an 11.3 m radius of the bird's perch supported characterization of most of these sites as Upland Forest, with a few sites being more Hillbase or Transitional in nature. As most of these sightings were 50–150 m from the nest tree, these data characterize the nest environs as well.

Based on our point count results (Whitacre et al. 1992a, 1992b), Black Hawk-eagles appeared similar in abundance at Tikal and two nearby sites (Zotz and Dos Lagunas). In contrast, we failed to detect Black Hawk-eagles in two seasons of fieldwork in the drier, lower forests around the Maya ruins of Calakmul, Campeche, 100 km north of Tikal. At all sites except Calakmul, Black Hawk-eagles were detected more often in point counts than were Ornate Hawk-eagles. Still, our nesting studies suggested that Ornates were in fact more abundant than Black Hawk-eagles at Tikal. Black Hawk-eagles appeared to soar more often, and were more vocal at such times, than Ornate Hawk-Eagles. This highly vocal soaring behavior makes the Black Hawk-eagle more conspicuous than the Ornate, rendering it difficult to ascertain the true relative abundance of the two species based on point counts unadjusted for detection probability.

In point counts, we detected Black Hawk-eagles significantly more often in primary (especially Upland) forest than in the fragmented farming landscape (Whitacre et al. 1992a, 1992b). These hawk-eagles were not entirely absent from the farming landscape, however, and were observed there occasionally. An emaciated Black Hawk-eagle found in the farming landscape south of Tikal died soon afterward in a rescue facility. During studies of the Roadside Hawk *(Rupornis magnirostris)* in the farming landscape (see Chapter 12), Theresa Panasci regularly observed a pair of Black Hawk-eagles. These birds flew between at least two of the larger mature forest remnants but were never seen to perch in the open agricultural lands or young second growth between. While stationed for several months at the Cerro Cahuí Biotopo reserve—an island of a few square kilometers of mature and degraded forest surrounded by largely deforested farming country some 20 km south of Tikal—Julio Madrid heard and saw Black Hawk-eagles soaring overhead virtually daily and observed Ornate Hawk-eagles much less frequently.

BREEDING BIOLOGY AND BEHAVIOR

Nests and Nest Sites

We documented four nests at Tikal. Nest trees were two Ramón or Bread-nut Trees *(Brosimum alicastrum)*, a Black Olive *(Bucida buceras)*, and one *Rehdera penninervia*, from 70 cm to 1.5 m in diameter (mean = 98 ± 35 cm, n = 4), and 28–30 m tall (mean = 30 ± 3 m, n = 4). Nests were placed 23–28 m high (mean = 25.5 m, n = 4). Three of the four nests were 9–12 m out on lateral limbs, while one was 2 m from the bole. All four were built amid vine tangles, and at least two obtained partial support from the vines, as well as being supported by one to five limbs that were mostly 6–11 cm thick. Nests were difficult to see within the vine tangles, and all were at least partly shaded by vegetation overhead. Three of the nest trees were connected with the forest canopy such that Black-handed Spider Monkeys *(Ateles geoffroyi)* could have moved through them without descending to the ground. Though we had assessed the fourth nest tree as "not connected" to the canopy, even here spider monkeys sometimes climbed into the nest tree to harass the hawk-eagles. The four nests averaged 97 ± 9 cm in external diameter (range = 89–115 cm) and 46 cm in external depth (range = 38–56 cm). Nest cups averaged 59 cm in width (range = 41–74 cm) and 5.1 cm in depth (range = 3.4–8 cm).

Compared with 12 Ornate Hawk-eagle nests at Tikal, Black Hawk-eagle nests did not differ significantly in mean or maximum nest dimensions or in height of nests or nest trees (ANOVA; all P > 0.18). Nest placement, however, differed notably. Ornate Hawk-eagle nests were placed in crotches of major limbs, usually 20–30 cm or more in diameter, near the central axis of the tree, rarely more than 6 m from the bole. Black Hawk-eagle nests were placed farther from the center of the tree and were on much smaller limbs (2.5–11.0 cm in diameter), often partly supported by vine tangles. As a result, nests of Black Hawk-eagles were generally within foliage and better concealed than Ornate Hawk-eagle nests, which were in more open areas within the crown and easily visible once discovered.

Egg and Clutch Size

Two Black Hawk-eagle nests at Tikal each contained a single egg, and three others contained single nestlings, suggesting that a 1-egg clutch is typical at Tikal. At the Los Angeles Zoo, two females, probably from Ecuador, each laid an egg and then, 32 and 34 days after these eggs were taken, each laid a second egg (Kiff 1979b). This laying pattern also suggests a normal clutch of one. Clutch size is apparently well documented for only one other member of the genus: the Ornate Hawk-eagle (see Chapter 15). To summarize, our data suggest that a 1-egg clutch is the norm for Black Hawk-eagles at Tikal, as for all *Spizaetus* species studied to date. However, a Panamanian nest attributed by Smith (1970) to this species contained two chicks in both 1965 and 1968; if accurate, this record seems exceptional.

One egg measured at Tikal was 60.2 mm by 50.1 mm, 75 g in weight, and white with brown blotches—similar in size and color to eggs described by Kiff (1979b). Including data from Kiff (1979b), mean egg dimensions are 60.3 ± 2.6 mm (n = 5) by 48.6 ± 1.0 mm (n = 5), and mean egg weight is 76.7 ± 4.3 g (n = 3). Using Hoyt's (1979) formula, these dimensions imply a fresh egg weight of 77.9 g. Thus, the typical 1-egg clutch amounts to 7.0% of adult female body weight.

Nesting Phenology

At three nests, we observed copulation from mid-December through January, in early January, and from late March through much of April, respectively. At one nest where the egg was laid between 15 and 24 February, we witnessed frequent copulations beginning two months prior, when the nest was discovered on 13 December.

No precise laying dates were recorded. To estimate laying dates, we assumed an incubation period of 44 days (see below). Estimated laying dates for five nests were 21 January, 15–24 February, the first half of March, 24 March, and 8 April. The only precisely known hatching date at Tikal was 7 May, while two other hatch dates were estimated as 6 March and 22 May, based on the size of the chick at the time of discovery. Two additional eggs would have hatched around 31 March and 28 April, had they been successful. Observed fledging dates were 6 May, 15 July, and between 12 August and 9 September. Based on a mean nestling period of 10 weeks, two other chicks would have fledged around 9–18 June and 7 July, had they survived.

To summarize, known and estimated laying dates for five nests were between 21 January and 8 April, while five known and estimated hatch dates were between 6 March and 22 May. Three of five fledging dates were between mid-June and mid-July, with the earliest 6 May and the latest between 12 August and 9 September. Copulation at four nests was observed from mid-December through much of April. Based on these observations we conclude that, on average, egg laying at Tikal peaks around early March, hatching around late April to early

May, and fledging around late June to early July. This timing is similar to that of the Ornate Hawk-eagle at Tikal, although the latter species exhibited a broader seasonal range, perhaps simply reflecting the larger sample of nests studied. At Tikal the rainy season usually begins between mid-May and mid-June. Thus, on average, it appears that Black Hawk-eagle eggs at Tikal hatch during the late dry season, and that most chicks are in the nest during at least the first few weeks of the rainy season, and often longer.

Length of Incubation, Nestling, and Post-fledging Dependency Intervals

We did not determine the length of the incubation period at Tikal. For the somewhat larger Ornate Hawk-eagle, we observed an incubation period of 46 days (see Chapter 15). We estimate that incubation in the Black Hawk-eagle is probably 42–44 days, and we used 44 days in describing breeding phenology.

Of three nesting efforts that resulted in fledged chicks, only one was discovered prior to hatching, providing the most precise data on length of the nestling period. The young female at this nest made her first flight from the nest tree at the age of 69 days. One male nestling was about 15 days old at the time of nest discovery and made his first (5 m) flight beyond the nest tree at the age of about 61 days, and more substantial flights at about 65 days. We conclude that a fledging age of 9–10 weeks is typical, though further verification is needed. Fledglings at Tikal reached independence at one year (see below).

VOCALIZATIONS

We recognized two calls of adult males. The first was the species's typical call, often given while soaring above the canopy. This call is well described by E. Willis as *whut, whut, whut-eEEEeeer* (Brown and Amadon 1968). The number of preliminary syllables varies from one to three, and the final *whut-eEEEeeer* may be given a second time. These calls are clear and ringing, and the terminal *eEEEeer* syllable is strongly accented, first rising then falling in pitch and intensity. The males' second call type was a soft, lisping whistle. Though males were usually silent when visiting nests, they occasionally voiced this call on or near the nest, often in apparent response to vocalizations by the nestling or adult female. Once we observed a male voice a quiet *pit pit* while perched 1 m from his mate, just prior to copulation.

We recognized three calls of adult females. Call number one was given during copulation. Call number two might be termed a food-begging call. It was similar to the whistling *hui hui hui hui hui* begging call of the juvenile and was used in the same context, when the male approached the nest with prey. At such times, the female voiced this call loudly and persistently, generally until the male moved out of sight. This call, whether given by a female or nestling, seemed to motivate males to leave

the vicinity. The third call was given by both males and females in nest defense and was described in field notes as *pi-pii pit pit pi-pii, pi-pii pit pit pi-piii.* Also during nest defense, both males and females were noted a few times to voice a protracted *huiiiiiio,* which rose and then fell in pitch.

We recognized three vocalizations given by nestlings. The first was a loud *hui hui hui hui* food-begging call, often given insistently for a prolonged period when the adult male arrived with prey. When males arrived at the nest with prey, older juveniles would aggressively seize the prey, mantling over it with quivering, extended wings and head feathers erect, and loudly and continually voice this call until the male left. After juveniles reached the age of one month and were able to feed themselves, they voiced a second type of call when the male arrived at the nest without prey. In such cases, males remained on the nest longer (up to 13 minutes) than they did when delivering prey. This call was also voiced at times by juveniles when the female was on the nest, again, when nestlings no longer required the female's assistance in feeding. Our interpretation was that the call stemmed from the chick's desire to have the female leave the nest. Nestling call number three may be described as *pipipipi,* repeated many times, and was heard in contexts of self-defense, as when we climbed to a nest to measure the chick, band it, and mount a radio transmitter.

BEHAVIOR OF ADULTS

Pre-laying and Laying Phase

In the Guatemalan highlands, on 18 February, Jay Vaninni (1989a) observed "a complex courtship display flight involving contact and roll-overs"—the only published account of courtship flight in this species of which we are aware. We observed no courtship displays at Tikal; if these occurred, they may have taken place prior to the beginning of our seasonal observations.

We observed the behavior of Black Hawk-eagles during the pre-laying period at two nests—the Bejucal nest in 1992 and Arroyo Negro in 1995. The Bejucal nest was found on 25 March and observed through 21 April, when we concluded it had been abandoned without an egg having been laid. The Arroyo Negro 1995 nesting effort was detected on 13 December 1994 when the radio-tagged male and the banded female were observed copulating, and the female carried sticks to the nest used in 1992.

Behavior at both nests was much the same. The birds were present during at least half our visits. At such times, the female often remained within 100 m of the nest and spent occasional periods on the nest, often manipulating nest material, sometimes joined there by the male. We observed 11 copulations: 7 on the nest and the other 4 within 100 m of the nest. All lasted 8 seconds except one that lasted 16 seconds. At both nests, the females brought in nesting material, at least some of it green foliage. We did not note males carrying nesting material, but our observations were too few to be certain that they didn't. Pair members, especially the females, were fairly vocal during these observations. Twice the Arroyo Negro male was noted soaring and calling above the nest vicinity, and the Bejucal male once was seen dropping from a great height, straight to the female, and then copulating with her. On 11 January the Arroyo Negro pair copulated three times in 50 minutes, and on 26 January they did so twice in 25 minutes.

Incubation Phase

We made observations during the incubation phase at three nests. On average, these three females incubated 97.7 ± 1.9% of the time and remained in the nest vicinity the remainder of the time, occasionally bringing green foliage to the nest during brief incubation breaks. Males were away from the nest vicinity on average 89.9 ± 12.9% of the time ($n = 3$). For two males, these figures were 95.2 and 99.2%, while the third male was away only 75.2% of the time. On one occasion the latter male remained on the nest beside the female for an hour after a prey exchange, and on another occasion he spent more than an hour perched in a nearby tree. Males were not observed to incubate at any of these nests. At all three nests, males delivered prey directly to the nest, and females consumed it there, perhaps obviating the need for the male to take a turn incubating while females fed. In this, these hawk-eagles differed from Ornate Hawk-eagles, in which the female usually ate at a location away from the nest, and the male often incubated or shaded the egg until she returned. In Black Hawk-eagles, we no longer witnessed copulations once incubation had begun.

Nestling Phase

Females. Our best observations during the nestling period came from the 1992 Naranjal nest, which we observed from before hatching until after the female chick fledged, and the 1994 Arroyo Negro nest, where the male chick was about two weeks old when we discovered the nest. The progression of adult female behavior was similar at the two nests, but it differed in the rate of change (Figs. 14.2, 14.3). At both nests, the time spent by females on the nest gradually decreased and was effectively zero by 63–68 days after hatching. While this gradual decrease in nest attendance began nearly immediately after hatching at the Naranjal nest (Fig. 14.2), it occurred somewhat later and more abruptly at the Arroyo Negro nest (Fig. 14.3).

Females tended their young chicks closely, protecting them from sun and rain and feeding them bites of food, sometimes more than 100 in a session. During the nestlings' first month, females left the nest only a few minutes at a time, often bringing a sprig of foliage on their return. During the first week after hatching, females brooded or shaded nestlings nearly constantly, sometimes achieving both at the same time by spreading one or both wings over the nest cup while settled

in a brooding position over or next to the chick. By the end of the first week, females shaded chicks more than they brooded them. Females often appeared heat stressed while they shaded the nestling. They would stand with their back to the sun, gaping as they drooped their spread wings at the wrist, appearing very wilted to observers. Females shaded nestlings frequently until the latter reached the age of one month, then rapidly decreased such shading behavior. Once nestlings aged six weeks, females shaded them only rarely.

As their nestlings grew, females showed an overall slow but steady decline in attentiveness to them and in time spent at the nest (Figs. 14.2, 14.3). For example, during the first week after hatching, the Naranjal female spent 20–30% of her time standing over and feeding her chick. This degree of attendance was very constant through 42 days, then tapered off to nothing by the age of 60 days (Fig. 14.2). Beginning at about day 5, the female spent increasing amounts of time on the nest but not close to the chick. This activity reached 30% of the female's time budget by nestling age ten days and tapered off from ages two weeks to one month. As this activity declined, the female spent accordingly more time off the nest, perched in nearby trees. The female's total percentage of time on the nest was nearly 100% at the time of hatching, dropping to 90% by day 10, 80% by nestling age three weeks, 60% by one month, 35–40% by age six weeks, and decreasing from 5% to zero during the chick's tenth week (Fig. 14.2). The remainder of the female's time budget—time spent perched within sight of the nest, and time spent out of sight—increased proportionately as time spent on the nest decreased. By the time the nestling aged six weeks, the female began to hunt and to provide prey, and by day 65 she was absent almost constantly, presumably spending much of her time hunting. At the 1994 Arroyo Negro nest, where the chick was about two weeks old at the time of nest discovery, the adult female's behavior was similar to that at the Naranjal nest (Fig. 14.3). Female behavior during the latter part of the nesting cycle at the 1992 Arroyo Negro nest, which was found when the chick aged about six weeks, was also similar. The Naranjal female often carried scraps of prey remains and feathers away from the nest, dropping these twice each from perches 30 and 60 m away, and in five other cases flew off with them in various directions.

Males. Behavior of adult males during the nestling phase was consistent at the three nests observed. Males were nearly always away from the nest, presumably hunting, and rarely spent as much as 1.5% of the observation period on or near the nest. With rare exceptions, males made only brief visits to nests, mainly to transfer a prey item to the female or nestling. Usually, they remained only 10 seconds to a minute or two on the nest. After prey transfers, males often left straightaway, but at times they lingered for 2–5 minutes in a nearby tree, sometimes calling softly. Females and nestlings typically seized the prey and mantled over it, and the males invariably jumped to the other side of the nest the moment the transfer was

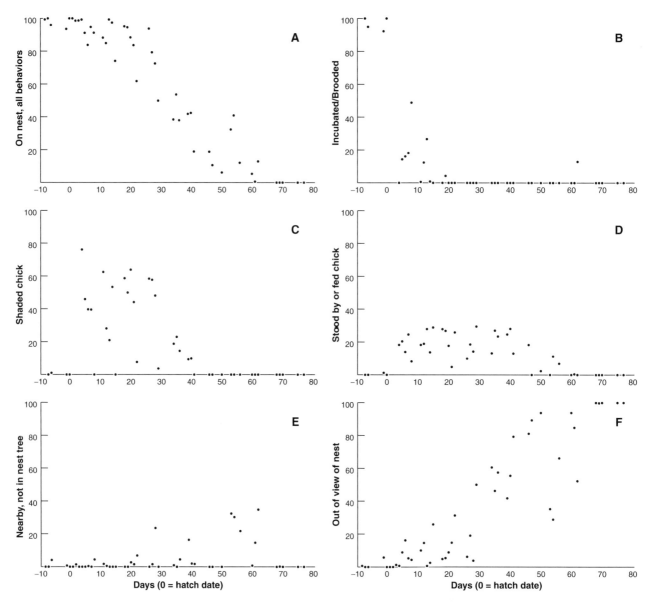

14.2 Percentage of time spent in various behaviors by the adult female Black Hawk-eagle at the 1992 Naranjal nest, beginning ten days prior to hatching. The chick fledged at 69 days after hatching.

made. Adult females and nestlings called persistently as long as the adult male remained in sight, and this calling seemed to encourage the male to move away from the nest. Adult males often flew off at treetop height, but sometimes they circled upward, calling, before moving off at some height. When adult females were off the nest, males occasionally lingered on the nest 5 minutes or more, sometimes manipulating nesting material.

Late Nestling and Early Fledgling Phase

Females at two nests (Naranjal 1990, Arroyo Negro 1994) first brought in prey they apparently had caught when their nestlings were six weeks old. At a third nest where the hatching date was not precisely known (Arroyo Negro 1992), we first observed this when the nestling was be-

lieved to be roughly 61 days old. At the first two nests, by nestling age 65–70 days (i.e., fledging age), adult females effectively abandoned the nest area, thereafter visiting it only briefly to deliver prey to the fledgling. At one of these nests, as the female ceased her vigilance near the nest during the nestling's tenth week—around the time of its first flight beyond the nest tree—the male twice spent 3-hour periods near the nest. At the other nests, we did not make detailed observations so far into the fledging period, so we cannot say whether such an increase in male vigilance around fledging time was typical. After chicks fledged, adults usually brought prey to the nest, and the fledgling, often perched within 30–100 m, upon hearing their vocalizations would hurry to receive it there. At other times a fledgling, apparently hungry, would linger at the nest, begging persistently, while both adults were away.

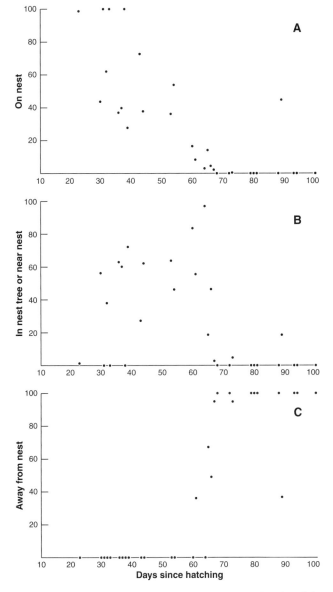

14.3 Percentage of time spent in various behaviors by the adult female Black Hawk-eagle at the Arroyo Negro nest, beginning at nest discovery at estimated chick age 15 days. The chick fledged at about 61 days.

Addition of Twigs and Live Foliage to Nests

Adults, especially females, periodically brought dead sticks and sprigs of green foliage to nests. In several hours of observation during the pre-laying period at the 1992 Bejucal nest, we twice saw the female bring green sprigs to the nest; she frequently manipulated sticks in the nest with her beak, as did the male on occasion. During the incubation period, in 55.25 hours of observation at two nests, we witnessed greenery brought five times, that is, once every 11 hours.

Most deliveries of greenery and sticks were observed during the nestling period. At the 1992 Naranjal nest, from eight days before hatching to the time of fledging, we saw the female bring foliage or sticks to the nest 79

times, and the male once. Delivery of sticks and foliage was frequent from the time of hatching, reaching its peak frequency at the age of one month, and decreasing to zero after nestling age 56 days (Fig. 14.4). The female brought material to the nest most actively during mid-morning and briefly in late evening, and sporadically during the afternoon (Fig. 14.5). We had the impression that green foliage predominated over dead twigs during the hot midday hours (Fig. 14.5), but this tendency was not significant (MRPP chi-square tests, grouping the hours into early, middle, and late, $P > 0.12$). Of the 80 sticks brought in, 26 (33%) were believed to be Malerio (*Aspidosperma* spp.), 22 (28%) Ramón, 2 (2.5%) Chicozapote (*Manilkara zapota*), 1 (1.3%) Zapotillo (*Pouteria* sp.), 3 (3.7%) vines, and 26 (33%) were unidentified; they ranged from approximately 10 to 35 cm in length. Sixty-two (78%) of the 80 were sprigs of living foliage, and in several cases we watched the female break these from

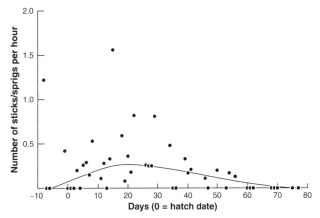

14.4 Hourly rate at which sticks and green foliage were brought to the 1992 Naranjal Black Hawk-eagle nest at Tikal, from ten days prior to hatching, until fledging at 69 days after hatching. Curve is produced by SYSTAT LOWESS procedure, stiffness = 0.5.

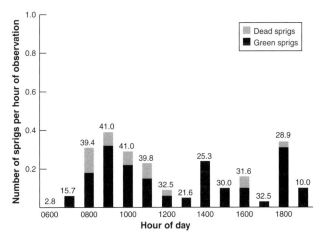

14.5 Rate at which sticks and green foliage were brought to the 1992 Naranjal Black Hawk-eagle nest throughout the day, from eight days prior to hatching until fledging at about 69 days; based on 80 sticks/foliage sprigs brought to the nest. Numbers above bars indicate number of hours of observation for each time period.

trees with her beak as she perched. In other cases she tried unsuccessfully to do so. Observation time at the 1992 and 1994 Arroyo Negro nests was much less, but females at both also brought foliage to the nest during our nestling period observations.

BEHAVIOR AND DEVELOPMENT OF NESTLINGS

Nestling Period

Hatchlings and young nestlings were covered in white down. A known-age female had black feathers emerging on her wings by day 11, and by day 34 she was cloaked in black plumage, with white visible only on her head (Plate 14.4). One female nestling 44 days after hatching weighed 1101 g, or 99% of the mean adult female weight, and at day 57, she weighed 1145 g, more than the average adult female. One male weighed 875 g at the age of roughly 81 days, 20 days after his first flight from the nest tree, and another male weighed 955 g at an estimated age of 66 days. Both these males were in the apparent weight range of adult males.

Our best data on behavioral development of nestlings come from a female of known hatch date. During her first week she spent most of her time sleeping, often brooded by the female. She raised her head only long enough to receive bits of food from the female. By day 4, she moved in beneath the female at times, pecked briefly at foliage in the nest, and stood for two seconds before falling. During her second week, the chick was notably more mobile, occasionally fluttering her wings and standing briefly or shuffling about. By days 19–21, feather emergence was substantial and the chick was first noted as preening, which was frequent thereafter. By this time the chick was much more active, frequently standing, sometimes for up to two minutes, and taking a few steps before falling. She often spread her wings, sometimes flapping, and occasionally jumped as she moved about. On day 21 the chick twice joined the adult female in calling when the latter received prey from the male.

On day 26 the chick remained standing, walking, and picking at nest material for nine minutes, and on day 27 she was first noted flapping vigorously. During her fourth week, she often pecked at leaves and sticks in the nest and continued to jump and flap while walking about the nest. On day 28, she was first noted mantling over prey delivered by the male, and on the same day, when spider monkeys moved within 2 m of the nest, she gave a threat display, extending her wings, gaping widely, and calling. By 28 days, when the nestling was well feathered and very active, walking and flapping back and forth in the nest, we still had not seen her feed on her own. The nestling was first seen clambering out of the nest onto nearby limbs and vines on day 29, and by day 34, she did this frequently, usually remaining for 10 minutes or so. By day 34, she remained standing most of the time, and we first observed her feed herself successfully, though the female continued to feed her as well. Though the adult female was increasingly off the nest, the chick continued to seek shade beneath her when she was on the nest. At other times, the chick appeared to seek areas shaded by vegetation overhead.

During the chick's sixth week we observed rapid transitions in her behavior and that of the adult female. When the male arrived with prey in the female's absence, the chick quickly mantled over it, calling until the male withdrew from the area. The female began to shade the chick less often during this period, and the chick at times showed apparent heat stress, walking about with her beak open, sometimes calling softly. When the female arrived in the afternoon of day 39, the chick rapidly moved beneath her, but the female moved away, leaving her sprawled in the sun. Minutes later, when the female settled down with wings spread in her usual chick-shading posture, the nestling quickly stood up and gave an apparent threat display, with wings extended and plumage erect, gaping widely as she called. Apparently at this stage of the chick's development, the adult female's spreadwing shading posture was perceived as a threat display, eliciting in turn a threat display from the nestling. The female still fed and shaded the nestling at times, although during this week, the chick was left to stand alone in the rain rather than covered by the female as before.

On day 40, the female for the first time brought prey she apparently had caught, visiting the nest only to leave it there. The nestling still had difficulty feeding herself but eventually succeeded. During the chick's eighth week, she frequently had tugs-of-war with the female over prey; when the adult female won, she invariably proceeded to feed the chick. On days 42 and 49, the nestling gave a well-developed threat display at a researcher who placed a trap on the nest. On day 54, the chick was first noted to fly from one rim of the nest to the other. Gaping as if in heat stress, she spent substantial time climbing in limbs 1–2 m from the nest, possibly in search of shade. She also seized a limb above the nest with her talons, pecking at it as if it were quarry. On day 56 she was frenetically active, climbing about 3 m above and 7 m laterally from the nest, tugging at limbs with her beak, and again jumping and flapping about on the nest. She was first described as flying among limbs near the nest on this day and was still fed at times by the female.

On day 61, the nestling made three flights of at least 3 m within the nest tree, and on day 68, she made many short flights in the nest tree, mostly within 3–4 m of the nest. On day 69, she made her first flight from the nest tree during an intermission in our observations; with the aid of radiotelemetry, we found her 10 m up in a tree 75 m from the nest. Over the next week, she often visited the nest and nest tree, but she also used various trees 5–20 m from the nest, changing perches frequently. She perched as high as 27 m, apparently able to maintain and increase her altitude within the forest, whether by flying or hopping upward.

The male nestling of the 1994 Arroyo Negro nest was about 15 days old when the nest was found; hence ages given below are approximate. By the age of about 23 days, when observations began, this male chick was standing and walking about on the nest. By about day 30 he spent a good deal of time standing and preened and fed himself unaided, though he was still often fed by the female. From day 31 on, he practiced flapping. From the age of five weeks on, we noted him occasionally tugging or pecking at green foliage in the nest. By about 44 days, he often moved to the extremities of the nest. On day 53 we first noted short hopping flights in which he flew up to 0.5 m above the nest to perch on a vine. From this day on, he made frequent short flights of this nature and frequently climbed about in the vines near the nest, often remaining perched for a couple hours 1–3 m from the nest. Also on this day, we first saw him practice "capturing" epiphytes and sticks in the nest, pouncing on and seizing them with his talons. On days 60 and 61 his short flights within the nest tree were very frequent, and on day 61 we first saw him fly outside the nest tree, to a tree 5 m away. On days 65–73, he made frequent flights between trees up to 45 m from the nest tree, but he still spent substantial intervals on the nest. From days 79 to 89, he spent much less time on the nest and usually returned there only to receive prey brought by the adults. By estimated day 89, he moved as much as 100 m from the nest.

When the 1992 Arroyo Negro nest was found, the male nestling was estimated to be six weeks old. The sequence of behavioral development in this chick was similar to that of the two chicks just described, although the dates of behavioral mileposts were not accurately known. This chick fledged between the estimated ages of 82 and 110 days.

Post-fledging Dependency Period

We monitored three radio-tagged fledglings during the post-fledging dependency period. All three spent most of their time within 50–150 m of the nest and continued to return to the nest to receive prey from adults. One fledgling male reached independence at the age of 12 months when it abruptly dispersed 20 km from the nest. Another fledgling male was found dead at the age of 5.7 months, eaten by some predator 45 m from the nest tree. A fledgling female likewise was found eaten at eight months after hatching, 60 m from her natal nest. These two fledglings had shown no indication of dispersing or reaching independence by the time of their deaths. We conclude that a post-fledging dependency period of one year from the time of hatching is typical. Though we noted that adult males delivered prey to fledglings in some cases, we do not know the relative contribution of males and females in provisioning fledglings.

SPACING MECHANISMS AND POPULATION DENSITY

Territorial Behavior and Displays

We commonly saw lone Black Hawk-eagles, and sometimes pairs, soaring in circles high above the forest, often while voicing their distinctive call. This typically took place in mid to late morning, when soaring conditions were good, but we also observed this behavior in the afternoon. We never saw any apparent hunting flights launched from a soar, but occasionally we saw hawk-eagles dive steeply downward to the nest or to the mate; in other cases we could not establish the context of the dive. Though it is possible these birds hunt to some extent while soaring high above the forest, we interpret these characteristically vocal flights as primarily serving a territorial display function. Such aerial display appeared to take place throughout the year.

Nest Defense

We saw no aggressive flights directed against conspecifics and few interactions with other raptors. Though females remained in nearly constant vigilance until their chicks were about six weeks old, they did no more than protest vocally at disturbances around the nest. We were never attacked by an adult of either sex while climbing to nests, a claim that can scarcely be made about the Ornate Hawk-eagle.

Far from attacking other species near their nests, these hawk-eagles were mercilessly harassed by spider monkeys. This was noted once at the Naranjal nest but was observed frequently at both the 1992 and 1994 Arroyo Negro nests, where these monkeys often passed nearby. At the Arroyo Negro site we saw spider monkeys harass the chick or adults on five days at each nest in 1992 and 1994, sometimes repeatedly in one day. Up to six spider monkeys would enter the nest tree or nearby trees, two or more of them approaching to within 1–2 m of the nest, often shaking limbs and calling and directing these menacing actions toward the chick. In response, the chick would face them with a threat display: wings spread to the sides, all plumage erected, gaping widely and calling. When the female was nearby, she sometimes flew to the nest or a nearby perch in response to the chick's calls, or if already perched nearby, she simply remained perched. She often would call, but we never saw her or the male make a flight toward the monkeys to drive them away. Several times when neither adult was nearby, the chick protested the monkeys' advances alone. The single occurrence observed at the Naranjal nest was similar, with the female merely calling from her nearby perch.

A dramatic incident occurred one morning when the 1994 Arroyo Negro nestling was about 60 days old. Three spider monkeys arrived at the nest and began to attack the nestling, who responded with a typical threat display, calling loudly. The monkeys persisted in their attack, eventually throwing the nestling from the nest; he caught some vines below the nest with his talons and remained hanging there. The monkeys then walked about on the nest and shook it as they broke off all or most of the green limbs that had shaded it. After sitting in the nest for a bit, they departed, 12 minutes after the episode began. The nestling managed shortly to climb back into the nest. Thereafter, the nest, which had received substantial shade, was largely exposed to the sun.

Nestlings at all three nests commonly responded when American Black Vultures *(Coragyps atratus)* or Turkey Vultures *(Cathartes aura)* flew over the nest vicinity. Typically, the juvenile would give a threat display with wings spread and call loudly for up to two minutes. Adult females at times also appeared alarmed when vultures flew over the nest area and would call in response. The only interactions with other raptors we observed were a case in which a Roadside Hawk followed and

mobbed an adult male, and another in which a small unidentified raptor harassed an adult female for 15 minutes. In the latter case, the female changed perches a few times to avoid this pest, which then finally moved off. Black Hawk-eagles were frequently harassed by Brown Jays *(C. morio)*, and sometimes by Keel-billed Toucans and other small birds. Only once did we see an adult attack one of these persecutors; in this case, a female darted at a Brown Jay that had been mobbing her loudly, causing the jay to move to a more distant perch, where it continued its heckling.

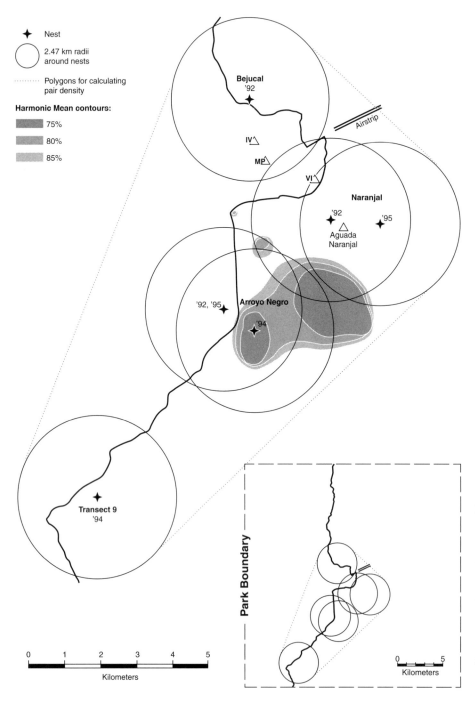

14.6 Nest sites, density estimation polygon, and a home range of Black Hawk-eagles studied in Tikal National Park, 1992–95. Circles around nests have radii of one-half the mean inter-nest distance (2.47 km). The polygon shown encloses 113 km², for an estimated 28.3 km² of space per pair. Also shown is the 95% Harmonic Mean home range (12.9 km²) of the Arroyo Negro 1994 adult male during seven months of the post-fledging dependency period. Nesting histories for each territory are given in Appendix 14.1.

Constancy of Territory Occupancy and Use of Alternate Nest Sites

The four territories we studied were occupied by pairs of Black Hawk-eagles each year. Pairs reused prior nests during some years and occupied alternate nests during others. In 1993, the Naranjal pair used the same nest as in 1992, and in 1995 they had built and occupied a new nest 1.34 km away (Fig. 14.6). The Arroyo Negro pair used the same nest in 1992 and 1995 and a different nest 500 m away in 1994 (Fig. 14.6). The Transect 9 pair was observed copulating in 1995 at the same nest where this female had laid an egg in 1994 (Fig. 14.6).

Home Range Size

Using radiotelemetry, we estimated home ranges for an adult male and an adult female, though both were likely underestimates of the area used because of the small number of relocations obtained. We estimated the home range of the Arroyo Negro adult male during the period when this pair had a dependent fledgling in 1994–95 (Fig. 14.6). Between 9 August 1994 (two months after the young fledged) and 1 March 1995 (8.5 months after the fledging date), this adult male used a Minimum Convex Polygon (MCP) area of 9.65 km^2, and a 95% Harmonic Mean (HM) area of 12.9 km^2. Based on only 19 relocations, these home range estimates are likely smaller than the actual home range of this bird. In 1990, we studied the movements of a non-nesting adult female that had been trapped away from any nest and radio-tagged and later was proved (via her band) to be the female of the Naranjal pair in 1992. She occupied largely the same area in both years. During 11 locations from 8 June to 13 July 1990, she used at least 3.4 km^2, clearly an underestimate of her actual home range.

Inter-nest Spacing and Density of Territorial Pairs

We believe the northern three of our study pairs were nearest neighbors, but it was less certain that the southernmost (Transect 9) pair was a nearest neighbor to these (Fig. 14.6). Based on the northern group of nests, mean inter-nest distance was 4328 ± 173 m (n = 2), and, based on all four nests, it was 4953 ± 1091 m (n = 3). Three methods were used to estimate the density of territorial pairs: the Maximum Packed Nest Density (MPND), square, and polygon methods (see Chapter 1). The two inter-nest distances given here yield MPND estimates of 17.0 and 22.3 km^2/pair, and square estimates of 18.7 and 24.5 km^2/pair.

We applied the polygon method of estimating pair density in three ways. First, using the inter-nest distance from the northern group of nests (4328 m), we constructed a polygon for only those territories; the area of this polygon was 63.6 km^2, or 21.2 km^2/pair. Second, using the mean inter-nest distance derived from all four territories (4953 m) to enclose all four territories, we determined the resulting polygon to be 113.0 km^2 in area,

or 28.3 km^2/pair (Fig. 14.6). Finally, the smaller inter-nest distance was used to enclose all four territories, yielding a polygon of 100.2 km^2, or 25.0 km^2/pair.

In sum, our estimates of space per pair ranged from 17.0 to 28.3 km^2. As we have no way to know which of these estimates is most accurate, we propose one pair/22.5 km^2, the approximate midpoint of this range, as our best estimate of pair density. Based on only three inter-nest distances, these estimates are clearly tentative. Still, extrapolating based on the extremes of 17.0 and 28.3 km^2/pair and Tikal National Park's roughly 510 km^2 of non–Scrub Swamp Forest area, we can estimate that probably 18–30 pairs of Black Hawk-eagles may hold territories in the park, with the best estimate probably about 4 or 5 pairs per 100 km^2 of acceptable habitat, or 22–25 pairs for the park.

The density estimate just given (one pair/20.0–22.5 km^2) is only half that implied by the 9.7–12.9 km^2 minimum home range estimated for the Arroyo Negro male. While this male's home range was likely underestimated owing to the small number of telemetry locations, the large discrepancy between his home range estimate and our density estimates may also indicate that some areas of forest were unoccupied by pairs; that is, pairs were far enough apart that adjacent home ranges were not contiguous. In accord with this idea, in French Guiana, Thiollay (1989a) believed that pairs of this species were widely spaced, with substantial unoccupied forest between. Moreover, the fact that our study nests were in Transitional, Hillbase, and Arroyo Forest may indicate that the distribution of these forest types affected nest spacing; our sample of nest sites was too small to test this hypothesis.

Another piece of evidence supported our estimate of one pair per roughly 20 km^2. In 1994 we tried to enumerate all pairs of several raptor species in a 20 km^2 study plot of Upland Forest with small inclusions of Bajo and Hillbase Forest. We repeatedly observed Black Hawk-eagles in only one corner of the plot—this was likely the territory of the Bejucal pair we studied in 1992 and 1993. It appeared that only one pair occurred on this plot, for a density estimate of one pair per 20 km^2 or more of forest.

Our density estimates, ranging from 3 to 5 pairs/100 km^2, were far higher than the one pair/100 km^2 estimated by Thiollay (1989a) for Black Hawk-eagles at a French Guiana site. In addition, our measurements of inter-nest spacing suggested that Black Hawk-eagles occurred at lower densities than did Ornate Hawk-eagles at Tikal. Black Hawk-eagle nests were significantly farther apart (mean = 4.33–4.95 km) than were 10 nearest-neighbor nests of Ornate Hawk-eagles (mean = 3.13 ± 0.62 km) (Mann-Whitney U test, P = 0.05).

DEMOGRAPHY

Frequency of Nesting, Percentage of Pairs Nesting

Black Hawk-eagle pairs typically nested every other year, unless a nest failed or the fledgling died at an early age, in which case the pair normally nested the following year. At the 1993 Arroyo Negro nest, when the previous year's fledgling reached independence by 10 May, well into the nesting season, there was no new nesting activity. This pair nested again in 1994, with an egg laid on about 6 March. The resulting fledgling died by late August 1994, and the pair nested again in 1995, laying between 15 and 24 February. Hence, this pair skipped a year when it had a dependent fledgling from the previous year, but it did not skip a year following the early death of a fledgling. At the Naranjal territory in 1993, following the death of the previous year's fledgling by 4 January, the pair initiated nesting behavior in March and April.

In several instances when we did not detect a nesting effort in a given territory, we could not devote sufficient search effort to be certain that the pair did not nest at an unknown site. However, for five years in two territories, we had adequate data to evaluate frequency of nesting. In each territory, there were three nesting efforts in five years. Thus, in this modest sample, a nesting effort took place in 6 (60%) of 10 territory-years.

Productivity and Nest Success

Our data permit only tentative statements regarding nesting success and productivity. In 10 territory-years, two pairs laid five eggs, four of which hatched. All four chicks fledged. Two fledglings died at the ages of 5.7 and 8 months, and the remaining two reached independence at the age of one year. In this sample, 2 of 10 territory-years resulted in a fledgling that survived to independence, an impressively low reproductive rate.

Sources of Nest Failure and Mortality

Our data permit few comments on sources of nest failure and adult mortality. Two fledglings that died at 5.7 and 8 months were eaten, but whether they were killed by predators or scavenged after death is unknown. The adult female at Naranjal was found dead of some cause other than predation. An emaciated adult found in the farming landscape, which quickly died in an animal rescue facility, was believed by the veterinarian there to have succumbed to an agricultural toxin, but this is unverified.

Parasites and Pests

In the three nestlings we handled, we did not note botflies, which were sometimes seen in raptor chicks at Tikal.

CONSERVATION

There can be little doubt that deforestation is the most serious threat to these hawk-eagles. As discussed earlier, many observers believe these eagles to be more tolerant of habitat modification than the Ornate Hawk-eagle. While this may be true, we found no evidence that these birds were common in heavily deforested areas. In our study area both of these hawk-eagles were distinctly less common in the agricultural landscape than in areas of extensive primary forest, and we believe that both will probably persist only in areas retaining substantial forest cover. Interestingly, we have found several instances of Ornate Hawk-eagles nesting, often successfully, in forest remnants in farming landscapes, whereas we have yet to learn of a Black Hawk-eagle nest in similar circumstances. Whether this signifies a real difference, and if so, whether it results from lower population densities or greater habitat or nest site selectivity of Black than Ornate Hawk-eagles, is not known. That Black Hawk-eagle nests at Tikal were farther apart than those of Ornate Hawk-eagles suggests lower population densities in the former species, but this requires verification. We suspect that both species can tolerate some degree of deforestation and habitat fragmentation, so long as adequate areas of mature forest or old second growth remain, but the limits of their tolerance in this regard remain to be explored.

It is likely that mortality due to shooting also affects these raptors, especially in human-dominated landscapes. Although their nests are, in our experience, much better hidden than those of Ornate Hawk-eagles, Black Hawk-eagles themselves are the more conspicuous of these two species, owing to their more frequent and more vocal soaring. This, and the reputation they have as chicken thieves—deserved or not—may lead to significant attrition by shooting.

HIGHLIGHTS

With a diet of 95% mammals, the Black Hawk-eagle had one of the most mammal-dominated diets of the 20 raptor species we studied at Tikal. Three other raptor species at Tikal overlapped in diet to a potentially meaningful degree with this hawk-eagle. Possibly the most significant overlap was with the Collared Forest Falcon (*Micrastur semitorquatus*), mammals accounting for 46% of its diet, including Deppe's Squirrel (24.7%), Yucatán Squirrel (6.5%), and fruit-eating bats *Artibeus* and other bats (9.5%; see Chapter 17)—all frequent prey of Black Hawk-eagles. The diet of the Black-and-white Owl (*Strix nigrolineata*), which preys heavily on bats, also overlapped substantially (see Chapter 22); 53 of 73 Black-and-white Owl pellets contained bat remains, including 13 Jamaican Fruit-eating Bats, a species also taken by Black Hawk-eagles. Ornate Hawk-eagles, with a diet of half birds and half mammals, relied far less on mammals than did Black Hawk-eagles; however, because squirrels made up 40% of the Ornate's diet, there was substantial dietary overlap between these two hawk-eagles. Crested Eagles (*Morphnus guianensis*), like Black Hawk-eagles, preyed heavily on nocturnal arboreal mammals, which made up 70% of their diet. Overlap with mammals taken by Black Hawk-eagles was minimal, however. Crested Eagles mainly took larger mammals, especially Derby's Pale-eared Woolly Opossums *(Caluromys derbianus)*, but also porcupines, coatis, and *Didelphis* opossums; squirrels formed only 4.7% of the Crested Eagle diet, and no bats were observed as prey.

The absence of reptiles among prey identified at Black Hawk-eagle nests is striking. We observed these hawk-eagles with reptilian prey on only two occasions, both occurring away from nests. Unless reptiles were frequently eaten away from the nest, it seems likely these hawk-eagles encountered far more snakes and lizards than they captured; this, in turn, suggests they often either failed to detect or elected not to attack these prey types. If so, one must wonder what is the selective advantage of such behavior. Presumably the answer must involve gains in efficiency by specializing on certain prey types while avoiding others.

The Black Hawk-eagle—with its 1-egg clutch, prolonged juvenile dependency, and consequent pattern of nesting only every other year—provides another example of a tropical forest raptor showing slow reproduction and prolonged investment in a single young at a time. Perhaps our most intriguing findings, however, were the species's novel diet and attendant mysteries regarding this hawk-eagle's foraging behavior.

Black Hawk-eagles somehow located and captured nocturnal arboreal mammals throughout the daylight hours, presumably by searching for them in their daytime retreats. This behavior is apparently not unique to Tikal, since opossums and bats are mentioned as prey a number of times in the literature. While this suggests that a penchant for preying on nocturnal mammals may be a consistent feature of the Black Hawk-eagle's foraging ecology, further research at additional localities is needed to verify whether this is a consistent pattern throughout the species's range.

This diet of nocturnal mammals poses the question of how these raptors locate their prey. Are they able to find bats and opossums only when these mammals stir and move about, or can the hawk-eagles locate these mammals visually or aurally even when the latter are nearly motionless and perhaps hidden from view? Our observation of a juvenile Black Hawk-eagle eating Common Raccoon feces may rank as the most bizarre observation during our studies at Tikal. How might this visually oriented predator recognize raccoon scats as edible? Prevailing wisdom has long held that most birds have a very poor sense of smell, but this is no longer believed true (Bang and Wenzel 1985; Waldvogel 1989; Wenzel 1990). Thus, it is perhaps not heretical to suggest the value of investigating whether these hawk-eagles use olfactory clues in searching during the day for their hidden nocturnal prey.

For further information on this species in other portions of its range, refer to Burton 2006, Canuto 2008, and Naveda-Rodríguez 2007.

APPENDIX 14.1

Reproductive Histories of Four Black Hawk-Eagle Territories at Tikal, Guatemala, and Summary of our Study Efforts at Each

Naranjal territory. In 1990 we caught, banded, and radio-tagged an adult female in this territory and tracked her movements via telemetry from 8 June to 13 July. She did not lead us to a nest and appeared not to be nesting at the time. We made a tentative minimum estimate of her spatial use via 13 telemetry localizations. On 29 April 1992, we found the nest of this same female, containing one egg. A female nestling hatched 7 May and fledged 15 July at 69 days. We conducted nest observations, often from dawn to dusk, totaling 389 hours on 47 days from 29 April to 23 July (mean = 8.28 ± 3.54 hr/day). This fledgling was found dead on 4 January 1993, sixty meters from the nest, apparently due to predation. She was still dependent on adults at this time, 242 days after hatching and 173 after fledging.

In 1993, this pair was observed copulating and renewing nesting behavior in March and April, but the adult female (the same banded individual as in 1992) died in mid-April of some cause other than predation. There was no further nesting activity that year. In 1994, a pair was present in the same area. The female, of course, was new, but we could not verify whether the male was the same individual as earlier. This pair apparently did not nest in 1994. In 1995, we discovered a new nest built 1.34 km east of the previous nest. Incubation was begun but the nest failed; whether it failed before or after hatching was unknown. The nest remained inactive two months later. In 1996, we checked the nests used in 1992 and 1995 and observed no activity. However, our observations were insufficient to rule out the possibility that nesting occurred elsewhere in the territory.

Arroyo Negro territory. A nest (site A) was discovered here on 3 July 1992, containing a male nestling estimated to be 40–45 days old, for an estimated hatching date of roughly 22 May. We observed this nest on 20 occasions from 3 July until 12 August, when the nestling was estimated to be roughly 82 days old. The nestling was banded and radio-tagged on 21 July at the approximate age of 60 days; it weighed 955 g then. It fledged between 12 August and 9 September at the estimated age of 82–110 days. Aided by telemetry, we frequently monitored this fledgling's movements thereafter. Between 4 April and 10 May 1993, this fledgling dispersed to an area 19.7 km from its natal nest and reached independence at the age of 12 months. The adult pair at Arroyo Negro apparently did not nest in 1993.

In 1994, we discovered a second nest (site B) on 21 March, 400 m north of nest A. The nest contained a male chick about 15 days old; the hatching date for this bird was estimated to be 6 March. This nestling fledged on 6 May at roughly 61 days after hatching. We conducted 32 nest observations totaling 302.5 hours during the nestling period, between 21 March and 14 June; 15 sessions were 11–13 hours in duration, 11 were 6–7 hours, four were 3–4 hours, and two were 1 hour or less. The banded, radio-tagged fledgling was found dead on 23 August 1994, forty-five meters from the nest, 109 days after fledging and about 170 days after hatching. This juvenile remained dependent on the adults up until its death.

In 1995 the same banded pair of adults nested, again in nest A. We observed them copulating and carrying sticks to the nest on 13 December 1994 and on 4, 11, and 26 January 1995. The single egg was laid sometime between 15 and 24 February 1995. The chick fledged and was monitored until reaching independence in late May or early June of 1996. In 1996 we detected no activity in either of the two known nests; we believed the pair did not nest that year, as the 1995 fledgling remained dependent during the nesting season.

Transect 9 territory. We first located a nest here, containing one egg, on 15 March 1994. This nest failed during incubation, and no evidence of re-nesting was observed. From 15 March to 27 April we conducted nine nest observations totaling 46.55 hours. In 1995, we witnessed copulation on 4 January, 400 m from the 1994 nest tree. We believed that the pair was not using the 1994 nest tree, but no further visits were made after this date. In 1996 we detected no nesting at the 1994 site, but our observations were not extensive enough to rule out the possibility of nesting elsewhere in the territory.

Bejucal territory. We found a nest here on 25 March 1992, during the pre-laying period, and observed it on 11 days from 25 March to 21 April. An adult pair was present sporadically, most often in the morning. The female spent much time on the nest, bringing and manipulating nest material, and several copulations were observed. Our observations ended 21 April, when we concluded that the nesting effort had been abandoned prior to laying. Details of these events are given earlier in the chapter under Pre-laying and Laying Phase. In 1993 we observed a pair in the area, but no nesting effort was apparent in March. In 1994, a pair was frequently observed in the area, but it was not known if nesting occurred. In 1995, a pair was observed in the general area, but again no detailed observations were made. In 1996, the earlier nest was gone, and no activity was witnessed during our limited observations in the vicinity.

15 ORNATE HAWK-EAGLE

David F. Whitacre, Julio A. Madrid, Héctor D. Madrid, Rodolfo Cruz, Craig J. Flatten, and Sixto H. Funes

The Ornate Hawk-eagle *(Spizaetus ornatus)* is aptly named, as this regal bird is as exotic in appearance as the Maya kings and priests of ancient Tikal, who adorned themselves with ornate costumes and plumed head-dresses. With its tall, black crest, rich chestnut face and neck, and boldly black and white barred underparts, this strikingly handsome raptor combines the robust body size of a Red-tailed Hawk *(Buteo jamaicensis)*, the long tail and agility of a Northern Goshawk *(Accipiter gentilis)*, and the powerful feet of an eagle (Plate 15.1). Ornate Hawk-eagles are formidably armed predators that use stealthy, short-range ambush to capture mostly medium-sized birds and mammals. Males commonly take prey such as squirrels and toucan-sized birds, as well as birds up to the size of a Great Tinamou *(Tinamus major)*, whereas females sometimes take larger prey, including Great Curassows *(Crax rubra)*, Ocellated Turkeys *(Meleagris ocellata)*, and young White-nosed Coatis *(Nasua narica)*. These hawk-eagles are forest birds, moving nimbly through the canopy despite their large size. They soar frequently above the forest (Plate 15.2), often while issuing a distinctive call, and consequently can be observed with some frequency. Their movements are often announced by the raucous mobbing of Brown Jays *(Cyanocorax morio)*, which facilitates locating these raptors in the field.

GEOGRAPHIC DISTRIBUTION AND SYSTEMATICS

The Ornate Hawk-eagle ranges from Tamaulipas and Jalisco, Mexico, southward throughout humid, tropical-forested lowlands and foothills to Bolivia, Paraguay, and northern Argentina, occurring also on Trinidad, Tobago, and Isla Coiba off Panama. The nominate subspecies *S. o. ornatus* ranges from eastern Colombia throughout the South American range, while *S. o. vicarius* occurs from western Colombia and Ecuador to the northern extent of the species's range (Brown and Amadon 1968; Bierregaard 1994). Ornate Hawk-eagles are known mainly from sea level up to 1500–1800 m (Monroe 1968; Hilty and Brown 1986; Howell and Webb 1995), with occasional records up to 3000 m (Stiles et al. 1989). This hawk-eagle is a year-round resident at Tikal.

MORPHOLOGY

Appendix 2 gives linear measurements for *S. o. vicarius*, the subspecies occurring at Tikal. The robust capture apparatus of these birds is illustrated by the following measurements. Middle toe length (without talon) averaged 49.1 mm in males and 53.4 mm in females, while middle talon length averaged 24.6 mm in males and 26.9 mm in females. Hind talons averaged 36.7 mm in males and 39.1 mm in females. In Tikal birds, foot span from the tip of the central talon to that of the hallux talon was 132 mm for an adult male, and 139 mm for an adult female. Based on these data, males have larger feet relative to body size than do females—as is common in raptors (Appendix 2).

Wingspan averaged 120 cm for six Tikal females and 105 cm for two males. The sexes do not overlap in wing chord and overlap only slightly in tail length, culmen length, tarsus, and middle toe length (Appendix 2). The ratio of tail length to wing chord (mean of both sexes) is 0.734 (Appendix 2), compared to 0.852 in the Black Hawk-eagle *(S. tyrannus)* and 0.866 in the Crested Eagle *(Morphnus guianensis)*—reflecting the proportionally longer tails of the latter two species.

Combining all available body mass data, adult males weigh 1028 ± 72.4 g (n = 4) and adult females 1452 ± 100.8 g (n = 11; Appendix 1)—about 40% more. Juveniles were slightly lighter, males averaging 984.5 ± 55.1 g (n = 4) and females 1353 ± 130.6 g (n = 6, range = 1200–1582 g). Dimorphism Index (DI) values are as follows: 11.5 for body mass (Appendix 1), 10.9 for wing length, 10.3 for tail, 5.5 for tarsus, 8.4 for middle toe, and 12.8 for culmen length (Appendix 2; Bierregaard 1978). The mean of DI values for wing length, culmen, and body weight is 11.7, exceeding that for all North American *Buteo* species, and equivalent to that of the Gyrfalcon *(Falco rusticolus)* and Merlin *(F. columbarius;* Snyder and Wiley 1976).

PREVIOUS RESEARCH

Current knowledge of this species's biology was summarized by Brown and Amadon (1968), Bierregaard (1994), and Ferguson-Lees and Christie (2001). Previously, limited observations of nesting biology and behavior were made at two nests in Guatemala (Lyon and Kuhnigk 1985) and one in Brazil (Klein et al. 1988), while Robinson (1994) reported on diet and hunting behavior at a Peruvian site. Our initial results at Tikal were reported by Flatten et al. (1990) and Madrid et al. (1991, 1992).

RESEARCH IN THE MAYA FOREST

From 1989 to 1996, we studied the breeding biology of Ornate Hawk-eagles in and near Tikal National Park. We studied 10 territories within the park and took limited data on three territories in the farming landscape nearby (Appendix 15.1). In one of our 10 study territories (La Paila), we never observed a nesting attempt, and, therefore, we omit this site from analyses of reproduction. For the remaining nine territories, we documented reproduction during 45 territory-years, allowing estimation of population-level reproductive rates.

We found nests by searching the forest on foot while listening for vocalizations and scanning likely trees, by observing hawk-eagle activity above the canopy from emergent trees and Maya temples, and by radio-tagging an adult and subsequently following it to a nest. Nests outside the park were brought to our attention by others or discovered by local members of our research team.

We trapped 11 adults (9 females and 2 males), fitting them with unique combinations of one or two color-anodized, numbered aluminum locking bands. We banded 17 juveniles shortly before or after fledging—11 females, 5 males, and 1 of unknown sex. Many juveniles and adults were fitted with radio transmitters, mostly from Holohil Systems of Carp, Ontario. Most radios were mounted as backpacks, using Teflon ribbon joined by cotton thread over the sternum; in some cases we confirmed that these eventually fell off. Backpack radios weighed 18–22 g, well under 3% of body weight. Some fledglings were tagged with 12 g tail-mounted transmitters. Adults were captured by use of *bal-chatri* or noose carpet, nearly always on the nest during the nestling stage, and fledglings were captured by similar methods, on or near the nest. Most radios had life spans of 18–24 months, necessitating periodic retrapping of adults and juveniles to replace failing radios. Retrapping birds was often difficult or impossible, and in many cases this led to radios failing prior to juveniles reaching independence. We were often able to monitor dependent juveniles even without active radios, as they remained near their nest and were often highly vocal.

We made extensive observations at many nests, nearly always from a platform built in a nearby tree. Nest observations took place mostly from dawn to dusk, though were briefer in some cases. Behavioral data were taken ad libitum, noting every behavior of consequence by adults or chicks. In most cases we determined laying dates by climbing to nests, though in some cases laying was inferred from incubation behavior. Hatching was generally inferred from behavior of adults and chicks and confirmed in some cases by climbing to nests. We considered fledging date to be that when the nestling first flew outside the nest tree. The end of post-fledging dependency was recorded as the date when the juvenile began pronounced dispersal movements away from its natal area. This was determined mainly for radio-tagged juveniles, supplemented by observations of juveniles without radios. Values given are means ± 1 SD unless otherwise stated.

DIET AND HUNTING BEHAVIOR

These hawk-eagles are bold predators that prey mainly on birds and mammals. A penchant for preying on large cracids is implied by common names such as Gavilán Cojo litero (Guan Hawk: A. Caso, pers. comm.) and Curassow Hawk (Russell 1964), and several published records support this tendency. Russell (1964) observed an Ornate Hawk-eagle that had captured a male Great Curassow, and Olrog (1985) mentioned two Ornates caught in mist nests, apparently attracted to entangled Chestnut-winged Chachalacas *(Ortalis garrula)*. Ornate Hawk-eagles have also been observed to capture a Little Blue Heron *(Egretta caerulea*: Van Tyne 1935) and a Ringed Kingfisher *(Megaceryle torquata*: Wetmore 1965). In Trinidad, these raptors are known to farmers as chicken thieves (ffrench 1973), and in Venezuela, Friedmann and Smith (1955) related that one raided a farm, carrying off half-grown chickens—they also collected one that had fed on a full-grown "guan" ([*sic*]: Rufous-vented Chachalaca, *Ortalis ruficauda)*.

In El Salvador, a female collected by Dickey and van Rossem (1938) had fed on a monkey and apparently also on an American Black Vulture *(Coragyps atratus)* attracted to the monkey carcass—their account implies that they themselves had killed the monkey several days prior. Stiles et al. (1989) described the diet as medium-sized mammals, birds up to guan size, large lizards, and snakes. In Belize, Ornate Hawk-eagles have been observed trying to catch a snake in a river (Russell 1964) and, in two cases, feeding on large Common Green Iguanas *(Iguana iguana)* along rivers (Clinton-Eitniear et al. 1991). At Tikal, Lyon and Kuhnigk (1985) reported six prey items: a young tinamou, a Plain Chachalaca *(Ortalis vetula)*, a young Crested Guan *(Penelope purpurascens)*, a Gray-headed Dove *(Leptotila plumbeiceps)*, a leaf-nosed bat (Phyllostomatidae), and an unidentified bird.

The most substantial previous dietary data are those of Klein et al. (1988) from near Manaus, Brazil. Of 49 prey items (mostly from remains collected at the nest), 63% were birds, 33% mammals, and 4% reptiles. The 31 birds included 13 cracids, 11 tinamous, and 2 macaws, and the

16 mammals included 13 dasyproctid rodents (agoutis or acouchies). Prey recorded at Manu, Peru, included two wood-rats *(Proecomys* sp.), a rat-sized rodent, three Purple Gallinules *(Porphyrula martinica)*, a Gray-necked Wood Rail *(Aramides cajanea)*, a colubrid snake, a Common Squirrel Monkey *(Saimiri sciureus)*, and two Saddleback Tamarins *(Saguinus fuscicollis:* Robinson 1994). At the same site, unsuccessful attacks were witnessed on oropendolas, caciques, Pale-winged Trumpeters *(Psophia leucoptera)*, monkey troops, macaws, parakeets, Hoatzins *(Opisthocomus hoazin)*, and chickens (ibid.). Finally, Sutton and Pettingill (1942) observed an individual carrying a squirrel-sized mammal and an attack on a quail in Mexico. Brown and Amadon (1968) mention mammals such as Kinkajous *(Potus flavus)* and Cacomistles *(Bassariscus sumichrasti)* as prey.

Diet at Tikal

We observed 408 diet items at Tikal and based diet calculations on the 325 items identified at least to class. Our prey data (Table 15.1) are based only on items observed brought to nests, adjusted as follows for prey remains found in or, rarely, below nests. When a prey remain was identified at the nest in conjunction with nest observations, it was added to the diet tabulation if not already observed, and its identity was assigned to one of the unidentified items observed earlier that day or the previous day. Prey remains collected when nest observations were not underway were omitted. Among 325 identified prey items, 56.3% (183) were birds and 43.7% (142) mammals, with no other classes noted. On a biomass basis, birds contributed 69.8% of the diet and mammals 30.2%. Although reptiles presumably must be taken occasionally in Tikal as elsewhere, we observed none in the diet.

Of 183 birds delivered to nests, most frequent was the Keel-billed Toucan *(Ramphastos sulfuratus)*, comprising 25.2% of identified birds, followed by the Plain Chachalaca (14.6% of identified birds), and Great Tinamou (8.9%; Table 15.1). The bird groups most often taken were galliforms (contributing 55.5% of avian biomass), tinamous (21.4% of avian biomass), toucans (13.5%), parrots (3.2%), and pigeons and doves (3.0%). Together, these five groups comprised 89.4% of identified birds in the diet and 97% of avian biomass. Galliforms and tinamous together contributed an estimated 77% of avian biomass and 36.1% of all prey biomass (Table 15.1).

Among the 131 identified mammals in the diet, most frequently taken were Yucatán Squirrel *(Sciurus yucatanensis)* and Deppe's Squirrels *(S. deppei)*, together contributing 87.8% of identified mammalian prey and 78.9% of mammalian prey biomass. Small opossums contributed 3.1% of identified mammals, bats 3.8%, young procyonid White-nosed Coatis *(Nasua narica)* 3.1%, arboreal porcupines 0.8%, and Central American Agoutis *(Dasyprocta punctata:* large, terrestrial, diurnal rodents) 1.5%. Because of their large size, some of the latter groups contributed more to diet biomass than their frequency in the diet suggests; coatis made up an esti-

mated 7.9% of mammalian biomass, porcupines 3.9%, and agoutis 8.3%.

Prey items ranged in size from 50 g bats to adult Great Curassows weighing an estimated 4.1 kg. Overall, prey items identified to species had a mean weight of 537 ± 44 g (n = 254) and median weight of 420 g. Seventy percent of prey items were between 150 and 450 g in mass, with very few items less than 100 g, and a modest number were from 600 g to 4 kg (Fig. 15.1). Items of 1 kg or larger contributed nearly half the diet biomass (Fig. 15.1).

Birds in the diet averaged significantly heavier (mean = 695 ± 929 g, n = 123) than did mammals (mean = 388 ± 307 g, n = 131).[1] However, median weights of birds (339 g) and mammals (420 g) did not differ significantly.[2] The larger mean size of avian than mammalian prey resulted in part from a few very large avian prey. The birds most frequently taken—Keel-billed Toucans, Plain Chachalacas, and Great Tinamous—weighed approximately 320, 560, and 1050 g, respectively, whereas squirrels—the most frequent mammalian prey—weighed 200–400 g.

Hunting Behavior

A few accounts of Ornate Hawk-eagle hunting behavior have been published. While studying lekking of Guianan Cocks-of-the-Rock *(Rupicola rupicola)*, Trail (1987) witnessed 56 raptor attacks on these birds, including 8 by Ornate Hawk-eagles, resulting in two captures. The hawk-eagles made fast, shallow dives into the lek area, capturing quarry on or near the ground. After attacks by Ornate Hawk-eagles, the Cocks-of-the-Rock remained wary for hours, spooking repeatedly with no apparent cause; the birds seemed to regard these raptors as especially dangerous predators. In Peru, Robinson (1994) described several hunts and generalized that Ornate Hawk-eagles attacked from concealed perches in the canopy or along edges, making short (< 50 m) horizontal

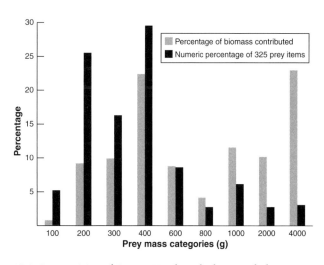

15.1 Composition of Ornate Hawk-eagle diet at Tikal, Guatemala, by prey size classes.

Table 15.1 Observed diet of Ornate Hawk-eagles at nests in and near Tikal National Park[a]

Common name	Scientific name	Number in observed diet	Percentage of diet, numerical basis[b]	Estimated mass per item (g)[c]	Percentage of diet, biomass basis[b]
A. Diet by prey species					
Birds					
Great Tinamou	*Tinamus major*	11	3.38	1052	6.35
Little Tinamou	*Crypturellus soui*	2	0.62	216.5	0.24
Thicket Tinamou	*Crypturellus cinnamomeus*	1	0.31	419	0.23
Slaty-breasted Tinamou	*Crypturellus boucardi*	2	0.62	443	0.49
Tinamou species	—	6	1.85	831.9[d]	2.74
Plain Chachalaca	*Ortalis vetula*	18	5.54	563	5.56
Crested Guan	*Penelope purpurascens*	2	0.62	2060	2.26
Great Curassow	*Crax rubra*	7[e]	2.15	4133	15.88
Ocellated Turkey	*Meleagris ocellata*	2[f]	0.62	2000[g]	2.17
Spotted Wood-Quail	*Odontophorus guttatus*	1	0.31	304	0.17
Gray-necked Wood-Rail	*Aramides cajanea*	1	0.31	397	0.22
Scaled Pigeon	*Columba speciosa*	1	0.31	243.5	0.13
Ruddy Quail-Dove	*Geotrygon montana*	1	0.31	115	0.06
Pigeon/dove species	probably mainly *Columba*, *Leptotila*, and/or *Geotrygon*	13	4.00	171.5[h]	1.22
Non-Amazona parrots	*Pionus* or *Pionopsitta*	1	0.31	198[i]	0.11
Red-lored Parrot	*Amazona autumnalis*	1	0.31	416	0.23
Mcaly Parrot	*Amazona farinosa*	1	0.31	610	0.33
Parrots	*Amazona* spp. (probably including *albifrons*)	2	0.62	410.7[j]	0.45
Parrot species	one or more of the above	2	0.62	325.6[k]	0.36
Guatemalan Screech-owl	*Megascops guatemalae*	1	0.31	107	0.06
Emerald Toucanet	*Aulacorhynchus prasinus*	1	0.31	154.5	0.09
Collared Aracari	*Pteroglossus torquatus*	4	1.23	226	0.50
Keel-billed Toucan	*Ramphastos sulfuratus*	31	9.54	339	5.77
Woodpecker species	genus?	1	0.31	114[l]	0.06
Pale-billed Woodpecker	*Campephilus guatemalensis*	1	0.31	242	0.13
Brown Jay	*Cyanocorax morio*	6	1.85	204	0.67
Montezuma Oropendola	*Psarocolius montezuma*	2	0.62	324	0.36
Great-tailed Grackle or Groove-billed Ani	*Quiscalus mexicanus* or *Crotophaga sulcirostris*	1	0.31	139.2[m]	0.08
Unidentified birds	—	60	18.46	694.8	22.88
Mammals					
Opossums, believed to be Mexican Mouse Opossums	believed to be *Marmosa mexicana*	4	1.23	70	0.15
Large, fruit-eating bats	*Artibeus jamaicensis* and *A.* sp.	2	0.62	50[n]	0.06
Bat species	—	3	0.92	50[o]	0.08
White-nosed Coati	*Nasua narica*	4[p]	1.23	1,000[q]	2.19
Deppe's Squirrel	*Sciurus deppei*	3	0.92	205	0.34
Yucatán Squirrel	*Sciurus yucatanensis*	6	1.85	420[r]	1.38
Deppe's or Yucatán Squirrel	*S. deppei* and/or *S. yucatanensis*	106	32.62	348.3	20.27
Mexican Hairy Porcupine	*Sphiggurus mexicanus*	1	0.31	2000	1.10
Central American Agouti	*Dasyprocta punctata*	2	0.62	2100[s]	2.30
Unidentified mammals	—	11	3.38	387.7	2.34
Prey unidentified to Class	—	83	—	—	—
Total number of prey		**408**	—	—	—
Total number of prey identified to class		**325**	**100.00**	**560.6[t]**	**100.00**
B. Diet by taxonomic group					
Tinamous	—	22	6.77	831.9	10.04
Galliformes	—	30	9.23	1581.3	26.04
Gruiformes (rails)	—	1	0.31	397	0.22
Pigeons and doves	—	15	4.61	172.5	1.42
Parrots	—	7	2.15	385.2	1.48
Owls	—	1	0.31	107	0.06
Toucans/toucanets	—	36	11.08	321.3	6.35
Woodpeckers	—	2	0.62	178	0.20
Passeriformes and Cuculiformes	—	9	2.77	223.5	1.10
Total number of identified birds		**123**	**37.85**	**694.8**	**46.91**

Table 15.1—*cont.*

Common name	Scientific name	Number in observed diet	Percentage of diet, numerical basis[b]	Estimated mass per item (g)[c]	Percentage of diet, biomass basis[b]
Unidentified birds	—	60	18.46	694.8	22.88
Total number of birds		**183**	**56.31**	**694.8**	**69.79**
Didelphidae	—	4	1.23	70	0.15
Chiroptera	—	5	1.54	50	0.14
Procyonidae	—	4	1.23	1000	2.19
Squirrels	—	115	35.38	348.3	21.99
Large, arboreal, nocturnal rodents	—	1	0.31	2,000	1.10
Large, terrestrial, diurnal rodents	—	2	0.62	2,100	2.30
Total number of identified mammals	—	131	40.31	387.7	27.87
Unidentified mammals	—	11	3.38	387.7	2.34
Total number of mammals		**142**	**43.69**	**387.7**	**30.22**
Prey unidentified to class	—	83	—	—	—
Total number of prey	—	408	—	—	—
Total number of prey identified to class		**325**	**100.00**	**560.62[t]**	**100.00**

[a] Dietary data are mainly from nests in 10 territories in Tikal National Park, with a few data from nests in three additional territories nearby. Number of prey items by year were as follows: 45 items in 1989; 57 in 1990; 85 in 1991; 64 in 1992; 38 in 1993; 31 in 1994; 3 in 1995; 1 in 1996; and 1 in year unknown. Only prey items observed when brought to nests or identified in nests soon after delivery are included here; prey identified from remains in nests were included only as discussed in text.

[b] Percentages are based on prey identified to class.

[c] Except as stated in other footnotes, prey body mass data for birds is from Dunning 1993 and Smithe 1966; mammal weights from Emmons 1990 and our own data.

[d] Weighted average of identified tinamous in diet.

[e] Includes 2 adult females, 2 adult males, 1 juvenile, and 2 sex unstated.

[f] Includes 1 juvenile.

[g] For juvenile, we used a weight of 1 kg; for the other, we used mean female weight of 3 kg (Porter 1994).

[h] Unweighted *mean of Columba speciosa, C. nigrirostris, Leptotila verreauxi, L. plumbeiceps, and Geotrygon montana—the most common large doves at Tikal.*

[i] Mean of *Pionus menstruus* and *Pionopsitta haematotis.*

[j] Unweighted mean of *Amazona farinosa, A. autumnalis,* and *A. albifrons.*

[k] Unweighted mean of the 5 parrots listed.

[l] Unweighted mean of the 6 most common woodpeckers at Tikal.

[m] Mean of male weights for these 2 species.

[n] Weight from Morrison 1983.

[o] Assumed to be the same size as *Artibeus,* as they appeared large.

[p] All juveniles.

[q] Based on adult weight of 4.5 kg (Emmons 1990), we estimated 1 kg for juveniles.

[r] Based on specimen we weighed at Tikal.

[s] Based on 3.2–4.2 kg range in Emmons 1990, one eaten on ground over several days by a female at the presumed point of capture was assumed to weigh 3.2 kg; one brought to nest, apparently by male, was assumed to weigh 1 kg, as this is approximately the maximum weight males were able to carry.

[t] 560.6 = mean individual weight, frequency weighted, including estimated unknowns; 616.9 = mean weight of prey species (once each), including unknowns; 660.1 = mean weight of prey species (once each), omitting unknowns.

attacks at quarry in trees or diving downward at quarry on the ground or water. This description is consistent with hunting behavior we observed at Tikal. The description by Stiles et al. (1989) is also apt: "hunts mostly inside forest, perching at medium heights to scan for prey, flying silently and swiftly between perches."

We observed 13 prey-capture attempts by Ornate Hawk-eagles—4 by adult females, 3 by adult males, and 6 by dependent juveniles. Three (23%) of the 13 were successful. In 3 cases, the hawk-eagle made contact with the prey, which escaped, and in 7 cases, the hawk-eagle got within 0.3–1.0 m of the prey, which escaped. Of the 13 hunts, 8 took place in tall, primary Upland Forest, 4 occurred in a grove of Black Olive *(Bucida buceras)* trees emergent above low-canopied

Scrub Swamp Forest, and 1 took place in "tintal"—low Scrub Swamp Forest dominated by Campeche Wood *(Haematoxylon campechianum).* The three prey items captured were a Great Tinamou and a young coati (both captured by adult females) and a Keel-billed Toucan (taken by a dependent juvenile). The 10 failed attacks were on a Great Tinamou (1), an adult female Great Curassow (1), Crested Guan (2), Keel-billed Toucan (1), Melodious Blackbird *(Dives dives:* 1), a large dove (1), a coati (1), and squirrels *(S. deppei* and/or *S. yucatanensis:* 2).

Ten attacks took place in the morning and three in the afternoon. In four cases the quarry was on the ground at the moment of attack, in four cases it was 4–7 m up in vegetation, vocalizing, and in three cases the quarry was in flight, 15–20 m above ground. Most attacks were launched at quarry less than 20 m from the hawk-eagle (Fig. 15.2). Of 12 hunts in which the angle of the attack flight was recorded, 11 were at a descending angle toward quarry (at least 4 of them vertical or steeply downward), and 1 was approximately horizontal, directed at a toucan in flight. In eight cases we noted whether the raptor flapped at some point during the attack or glided throughout; half of the attacks were of each type. Even in flapping attacks, the hawk-eagles usually or always glided during the final approach. The three successful attacks were made from 8, 10, and 20 m, steeply downward, and without flapping. We were told by a park guard of a fourteenth (unsuccessful) attack on an Ocellated Turkey; its inclusion lowers the success rate to 21.4%.

In addition to these attacks, we witnessed many cases in which a hawk-eagle appeared to be stalking quarry. When we approached a radio-tagged hawk-eagle, we often found it perched below the canopy, quite close to (sometimes within 10 m of) appropriate quarry such as tinamous, guans, or curassows. On two occasions when we inadvertently flushed a Great Tinamou in such a situation, the hawk-eagle immediately attacked (once successfully), suggesting that it had been intent on this quarry prior to our approach.

Our observations suggest that these hawk-eagles often use a hunting strategy of stealthy approach to within striking range and then dashing out of cover to attack unsuspecting quarry at close quarters—essentially a "stalk-and-ambush" strategy. In contrast, the few attacks we witnessed on flying birds were launched from a perch as the quarry flew near, and had the appearance of spur-of-the-moment, opportunistic attacks. We never saw Ornate Hawk-eagles attack prey from soaring flight above the forest. At all times when we knew or suspected Ornate Hawk-eagles to be actively hunting, they perched and moved within or below the forest canopy.

While hunting, these hawk-eagles were plagued by a clamorous retinue, especially of Brown Jays, but also of Keel-billed Toucans and various flycatchers. We quickly learned that when the hue and cry of several Brown

Jays went up, it paid to dash in and search, as this often signaled the movement of a raptor through the area. While many raptor species were heralded in this way, Ornate Hawk-eagles were announced in this manner especially often. We assume that hawk-eagle prey species such as squirrels, tinamous, and guans have long since learned the same lesson, and it would seem that Ornate Hawk-eagles, in order to take their quarry by surprise, may need to avoid announcement by Brown Jays for substantial periods of time. When a hawk-eagle remained perched for some time, the Brown Jays eventually would go on about their business. Thus, these hawk-eagles may be forced into an even more patient, stealthy hunting style than is dictated by the direct responses of their quarry. On the other hand, it is also possible that birds are sometimes captured while mobbing the hawk-eagles—we sometimes saw mobbing toucans in what seemed dangerous proximity to a perched Ornate, and these toucans occurred abundantly in the diet of these hawk-eagles.

Attacks utilizing stealthy approach and ambush are typified by the following examples. At the Tintal nest in 1991 the adult female caught a young coati. The female, who was brooding the nestling in early morning, flew at 0612 to a nearby perch, then changed perches at 0630 to a tree 20 m from our observation blind. Five minutes later a band of coatis passed 6 m from the observation tree, and the female attacked a small one near the rear of the troop as it moved over vegetation 5 m above ground. She seized it by the neck and remained in that position, waiting 7 minutes until it was dead before beginning to pluck it. It was our impression that the female had heard this band of coatis approaching and had positioned herself to intercept them. In another case, as we approached an adult male Ornate we were radio-tracking, we flushed a Great Tinamou from the ground in an area where many fruits of Tzol *(Blomia prisca)* had fallen from a fruiting tree. When the tinamou flushed, the hawk-eagle, which had been perched nearby at a height of 6 m, attacked in a descending, flapping pursuit that was unsuccessful but caused a feather to fall from the quarry. In a similar case, we flushed a Great Tinamou from the ground, and the adult female hawk-eagle we were tracking captured it by attacking from a perch 15 m above ground and 10 m to one side of the tinamou. In both cases we thought it likely that the hawk-eagles were already aware of the quarry, probably stalking it, and had been waiting for the appropriate moment to launch an attack when we inadvertently forced the issue.

Attacks on birds in flight often seemed more opportunistic and hastily executed than the attacks just described. Moments after delivering a squirrel to the nest, the 1991 Tintal male launched downward from the nest in an unsuccessful attack on a flying dove; the rapid shift from one activity to another suggested this attack was opportunistic in nature. In 1990, the Tintal male chased a Keel-billed Toucan 20 m across an open space, 15 m above the forest canopy, in a level, twisting tail-chase that ended when the toucan escaped into cover.

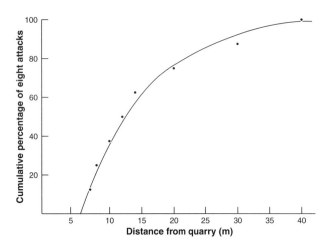

15.2 Distance between Ornate Hawk-eagle and quarry at moment attack was launched, from observations at Tikal.

Manner of Feeding on Very Large Prey

In several cases hawk-eagles consumed very large prey items on the ground over a period of days. It was apparent that these items were too heavy for the raptor to carry up into a tree and too large to be consumed in a single day. On 2 August 1995, for example, we located an adult female in the same area where her radio signal had placed her the previous day. She had a full crop, and, upon searching the area, we found on the ground 12 m from her, less than one-fourth of a Central American Agouti, estimated to have weighed 3.2 kg when intact. The agouti had been dead long enough that it was beginning to smell. Similarly, on 30 May 1995, we found the same hawk-eagle perched below the canopy, 6 m above ground, while some 8 m away on the forest floor was a dead Great Curassow. Half of this bird, mainly the breast region, had been eaten, while the head and legs remained intact. One radio-tagged female at the age of 14 months was observed on the ground beside a partly eaten female Great Curassow. Maggots were present on the carcass, and part of the breast had been eaten. Later the same day, we again found the hawk-eagle feeding on this carcass. The following day she was again perched on the ground here, and believed to be feeding on the curassow, of which little remained. By this time, the carcass smelled strongly and was covered with maggots and bluebottle flies.

Ornate Hawk-eagles could not always fly with the largest prey items they caught, as illustrated by the following. A male hawk-eagle was observed on the ground with a Great Tinamou in his talons, only the head having been eaten. We hoped to catch the male by flushing him off the prey item and placing it in a noose-covered wire envelope to which the male hopefully would return. As we gave chase, we observed that the male had difficulty flying with quarry of this weight. He labored up to a perch, but flying onward, he lost altitude until he was again nearly on the ground, then once again he labored up to a higher perch, and again lost altitude as he flew on. In other cases as well, we saw males eating Great Tinamous on the forest floor. These tinamous average nearly the same weight as an adult male Ornate Hawk-eagle (Table 15.1).

Some Comments on the Diet

We never observed Ornate Hawk-eagles taking Black-handed Spider Monkeys *(Ateles geoffroyi)* or Guatemalan Howler Monkeys *(Allouata pigra)*, even small juveniles. While small enough to be within the most frequent size range of Ornate Hawk-eagle prey, these young primates cling to adults, which may make it dangerous for these raptors to attempt such predation. Closer to the equator, a number of smaller, squirrel-sized primates occur, and they have been documented as prey of this hawk-eagle. Also of note was the low incidence of agoutis in the diet documented at Tikal. It is possible that these large rodents figure more prominently in the diet than indicated

by prey brought to the nest; these rodents are so large that they may often be eaten at the site of capture, as was apparently the case for the few agoutis we documented as prey. It seems likely that during the nonbreeding season, Ornates may take agoutis more often than suggested by the diet at the nest—especially females, which hunt for themselves during the nonbreeding season. All of the especially large prey items we documented—including Great Curassows, Ocellated Turkeys, and agoutis—were taken by females and were documented when we observed females eating them in the forest, presumably at the capture site. The largest items we witnessed males take were Great Tinamous, estimated to weigh 1050 g.

Black Hawk-eagles and Crested Eagles at Tikal took many nocturnal mammals, whereas Ornate Hawk-eagles took very few. Ornate Hawk-eagles appeared to focus more on detecting day-active quarry, rather than searching for nocturnal animals in their daytime retreats, as these other two raptors apparently did. The fact that Ornate Hawk-eagles at Tikal took very few bats, unlike the Black Hawk-eagle, and no snakes, unlike the Crested Eagle, constitutes a marked divergence of diets among these similarly sized raptors. Ornate Hawk-eagles must surely have observed large snakes and lizards more often than suggested by these items' total absence from this raptor's observed diet at Tikal. It seems an inescapable conclusion that these hawk-eagles opted not to attack many suitably sized reptiles that they must have encountered. Reptiles are also poorly represented in this hawk-eagle's diet elsewhere (Klein et al. 1988; Robinson 1994).

HABITAT USE

Most authors regard Ornate Hawk-eagles as denizens of dense lowland foothills, rain forests, and moist tropical forest, and several authors state that Ornates inhabit extensive tracts of primary forest more often than the Black Hawk-eagle (e.g., Bierregaard 1994; Ferguson-Lees and Christie 2001). In contrast, Slud (1964) commented that these birds occur mainly along forest edge, and he believed they did not penetrate deeply into forest "except perhaps along natural breaks." Brown and Amadon (1968) repeated Slud's comments, describing these birds as forest eagles that prefer the presence of some open areas. Hilty and Brown (1986) also suggested that Ornate Hawk-eagles are usually found near forest openings or in broken hill forest rather than deep inside unbroken forest. At Tikal, we found no evidence supporting the idea that these birds frequent mainly forest edge or areas of forest with natural openings. Rather, these hawk-eagles occurred widely throughout large areas of unbroken mature forest within the park.

Nesting Habitat

Ornate Hawk-eagles at Tikal nested in a variety of mature forest situations, but nearly always where a particularly large tree protruded above the surrounding canopy. In a previous report (Flatten et al. 1990), we emphasized that nests known to us at that time were all in ecotonal situations, mainly between taller and lower forest types but also between forest and crop fields, and near hills. As more nests were discovered in later years, this pattern did not hold true, as some nests were also found within areas of more homogeneous Upland Forest type. Still, most nests were in emergent trees protruding above areas of low-canopied Scrub Swamp Forest types, but close to tall Upland Forest where these hawk-eagles mainly hunted (see next section). The proximity of some nests to cultivated areas is unlikely to reflect a preference for such partly deforested areas; rather, it may simply reflect persistent use of traditional sites in the face of agricultural expansion.

Hunting Habitat

Ornate Hawk-eagles showed a preference for hunting in tall Upland Forest and largely avoided the low, dense Scrub Swamp Forests while hunting. They hunted mainly within and beneath the forest canopy. These conclusions were supported by monitoring 16 individuals via radiotelemetry and by our observation of several prey capture attempts. We believe the infrequent use of Scrub Swamp Forest by radio-tagged birds may be explained by the dense sub-canopy vegetation of this habitat, which probably inhibits the hawk-eagles' ability to stalk and capture prey below the canopy. At sites where we found Ornate Hawk-eagles perched and, we believed, hunting, they were often in areas of relatively open understory, such as hilltops where Dry Upland Forest provided an open understory or along old logging roads that provided tunnel-like openings below the canopy. Such areas of open understory may have facilitated the ability of these large raptors to hunt beneath the canopy. Occasionally, however, birds were attacked in flight above the canopy. At the Tintal nest, for example, tall Black Olive trees created a grove-like situation above the low canopy of Scrub Swamp Forest. Here a toucan was taken as it crossed a gap between emergent trees along a seasonal watercourse.

Pairs that nested in emergent trees above Low Scrub Swamp Forest often commuted to adjacent areas of tall Upland Forest to hunt. For example, during prey deliveries at the Tintal nest, amid Low Scrub Swamp Forest, the adult male usually approached the nest from a hill to the west covered with Upland Forest or from a grove of emergent Black Olive trees in the same direction—apparently he hunted mainly in those habitats. We saw no evidence that Ornates sought out man-made openings such as road margins or plazas maintained around Maya ruins.

Habitat Use During Visual Contacts

We gathered habitat data on 16 radio-tagged hawk-eagles when visual contact was made. Here we summarize results for the Arroyo Negro male during a period when he provided all or most food for a dependent juvenile. The following results are based on 24 visual contacts from 22 June 1993 to 12 July 1994.[3] We located this male 490–2480 m from the nest ($n = 24$), median distance of 925 m. In 11 cases he was resting or eating, in 10 cases we believed he was hunting, and in 3 cases he flushed before we could evaluate his activity. His distance from the nest did not differ with activity, that is, whether we deemed him to be hunting versus resting/eating.[4] We found this male perched most often within the forest canopy (12 of 24 cases), once at the canopy's lower boundary, in 5 cases at the upper canopy surface with a view over some portion of the canopy, and in 4 cases on a perch protruding above the canopy. Only once did we find him perched in the understory, and once on the ground, eating a Great Tinamou. On average, he was perched 13 ± 6 m from the ground (range = 0–30 m, $n = 23$). Perch trees averaged 21 ± 6 m tall (range = 12–30 m, $n = 20$), and 40 ± 28 cm in diameter at breast height (dbh) (range = 9–138 cm, $n = 22$); all perch trees (and one vine) were living. Perch limbs averaged 11 ± 6 cm in diameter (range = 3–25 cm, $n = 23$). This male's habitat use typified that of other radio-tagged Ornate Hawk-eagles in that he mainly used the taller forest types rather than the Low Scrub Swamp Forest. Of 23 visual contacts, 13 (57%) were made in tall Upland and Hillbase Forest types, while 7 (30%) were in Transitional Forest and 3 (13%) in Scrub Swamp Forest. Mean forest height in the vicinity was 10–15 m in 8 cases, 15–20 m in 12 cases, and 20–25 m in 3 cases.

BREEDING BIOLOGY AND BEHAVIOR

Nests and Nest Sites

Nesting of the Ornate Hawk-eagle was previously studied by Lyon and Kuhnigk (1985) at two nests in Tikal and by Klein et al. (1988) at a nest in Brazil. A number of other nests have been discovered with minimal data recorded; these are listed later under Nesting Phenology.

We documented nest and nest site characteristics at 14 nests in Tikal and 3 others nearby. Nest trees were Honduras Mahogany (*Swietenia macrophylla*: 6 individuals), Kapok (*Ceiba pentandra*: 4), Black Olive (2), one each of *Ficus* species, Florida Fishpoison Tree (*Piscidia piscipula*), Spanish Cedar (*Cedrela odorata*), Silión (*Pouteria amygdalina*), and Antilles Calophyllum (*Calophyllum brasiliense*). With the exception of one Black Olive and the Antilles Calophyllum tree, all nest trees were alive. Nests ranged in height from 16 to 30 m, averaging 22.9 ± 3.8 m ($n = 17$), while nest trees ranged from 21 to 40 m in height, averaging 29.9 ± 4.8 m ($n = 17$). These were quite large trees, averaging 107 ± 33 cm dbh ($n = 17$); the largest was 178 cm in diameter. All nest trees were large emergents having minimal contact between their crowns and the surrounding forest canopy.

Nests averaged 2.9 m from the main trunk ($n = 17$). Seven nests were against or in a fork of the main trunk, while four were within 3.5 m of the trunk; all but one were within 6 m of the trunk. Supporting limbs numbered two to five and ranged from 7.5 to 63.0 cm in thickness, averaging 28.1 ± 12.8 cm (n = 49 limbs, 16 nests). Nests averaged 102 ± 14 cm in outside diameter (n = 16) and 49 ± 11 cm in outside depth (n = 16). Nest cups averaged 44 ± 8 cm across (n = 4) and 22 ± 13 cm deep (n = 4). The sticks of which nests were built averaged about 65 cm in length and 1.5 cm in diameter (n = 17 nests). Adults (mainly females) added leafy branches to the nest cups throughout the incubation and nestling periods. Though adults sometimes removed such foliage from nests once it had dried, it more often became incorporated into the nest structure.

All nests were easily visible from the ground, situated as they were against the main trunk or in major forks where foliage was sparse. Only 1 of 16 nests had a vine tangle in the immediate vicinity; this nest was in a dead tree that had become overgrown with vines. The two nests in dead trees were exposed to the sky, while all others had partial cover provided by foliage overhead. In no case was vegetation close above the nest; nests were generally partly shaded but poorly concealed from above.

In several details of nest placement, Ornate Hawk-eagles differed markedly from Black Hawk-eagles (see Chapter 14). Ornate nests were mainly on large, open limbs, near the main axis of the tree, and easily visible. In contrast, Black Hawk-eagle nests were on smaller limbs, closer to the periphery of tree crowns and among foliage where they were difficult to see, often amid dense vine tangles. Ornate Hawk-eagle nests were similar in placement to Crested Eagle nests (see Chapter 13).

Egg and Clutch Size

One egg at Tikal was white with a pale, yellow-brown, mottled wash and fine, light brown speckling. This resembles an egg described earlier at Tikal, which was white with small, faint brownish red blotches (Lyon and Kuhnigk 1985). To our knowledge, the only published egg description is that of Kiff and Cunningham (1980), of two eggs laid by a captive female, probably from Guyana—these were unmarked and bright bluish white when laid, rapidly fading to a much paler bluish white. Another captive pair, probably also from Guyana, laid more than two-dozen very pale blue, unspotted eggs, over a 10-year period (D. Mancini, pers. comm.).

The 26 nestings we studied at Tikal all had 1-egg clutches or broods of one. A 1-egg clutch is also the norm for other members of *Spizaetus* for which data exist (Brown and Amadon 1968). To our knowledge, the only published egg measurements are those of Kiff and Cunningham (1980), on the two eggs laid by a captive female from Guyana. These eggs, each from a 1-egg clutch, were 57.7 mm by 44.2 mm and 58.2 mm by 43.4 mm. At Tikal we measured two eggs as 61.8 mm by 42.4 mm (68.0 g) and 63.2 mm by 51.5 mm (87.5 g); these two eggs averaged 77.8 g, although Hoyt's (1979)

formula predicts a fresh mass of 75.5 g. Hence, the normal 1-egg clutch amounts to 5.2–5.4% of adult female body mass. The interval required to replace a lost egg is suggested by the following. On 29 January 1992, the Brecha Norte pair lost their egg 41 days after laying. Another egg was laid on 2–7 March, 32–37 days after the loss of the first egg. When the egg was taken from a captive pair, re-laying also occurred after about one month, on average (D. Mancini, pers. comm.).

Nesting Phenology

Ornate Hawk-eagles nested with a distinct seasonality at Tikal, but their laying dates spanned a longer period than those of any other raptor for which we had a similar sample size. We recorded or inferred 23 laying dates, 20 hatching dates, and 13 fledging dates (Fig. 15.3, Appendix 15.2). Pairs laid eggs during each month from November to May, with 83% of egg laying occurring from January to April (Fig. 15.3). Eggs hatched in December and from February to July, with 90% of hatching taking place from February through June. Fledglings left the nest from April to October, with 85% of fledging occurring from May to August. Egg laying took place throughout the dry season and ended prior to the onset of rains in June. Most eggs hatched during the latter two-thirds of the dry season, with only 20% of eggs hatching at the onset of the rainy season or soon thereafter. Young fledged mainly during the first two-thirds of the rainy season. Juveniles varied substantially in the age at which they reached independence, which may have affected the date of subsequent nesting efforts. The nestling period of Ornate Hawk-eagles overlapped broadly with the period of most bird nesting at Tikal (see Chapter 2). That these hawk-eagles showed a defined nesting season at Tikal is supported by observations of their re-nesting behavior after failed attempts. Three nesting attempts that failed between January and March were almost immediately followed by the laying of another egg. In contrast, three

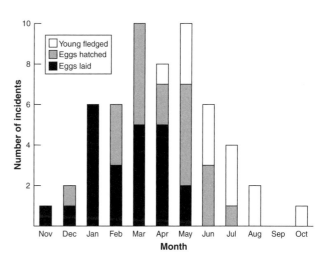

15.3 Nesting phenology of Ornate Hawk-eagles at Tikal, based on 26 nestings.

Table 15.2 Duration of interval from hatching to fledging in Ornate Hawk-eagles

Nest and year	Days from hatching to fledging[a]	Midpoint of estimate	Sex of nestling
Tintal 1989	82–84	83	Male
Naranjal 1990	92–93	92.5	Male
Naranjal 1992	70	70	Male
Punta de Pista 1993	82–85	83.5	Male
Punta de Pista 1991	722	72	Female[b]
Chikintikal 1992	60–69	64.5	Female
Arroyo Negro 1993	84–90	87	Female
Balastrera 1994	63–76	70.5	Female
Punta de Pista 1994	80–87	83.5	Female
Brecha Norte 1993	64–72	68	?

Notes: Minimum duration: mean = 74.9 days (n = 10, range = 60–92)
Maximum duration: mean = 79.8 days (n = 10, range = 69–93)
Best estimate: mean of midpoints = 77.5 ± 9.6 days (n = 10).
Same, median = (72 ± 83)/2 = 77.5

[a] Range of days indicates the possible range that resulted from not having observed each nest every day.
[b] First flight, at 66 days, was accidental; fully fledged at 72 days of age.

nesting attempts that failed between April and July were not followed by another laying until January–March of the following year.[5] Information from elsewhere in Central America and Mexico indicates a nesting season similar to what we documented at Tikal.[6]

Length of Incubation, Nestling, and Post-fledging Dependency Periods

Approximate length of the incubation period was determined for nine eggs. The range 43–48 days included all nine of these eggs. Incubation was most often 44–46 days in duration, with six of nine eggs falling in this interval. Many eggs produced by a captive pair, probably from Guyana, pipped after a mean of 44.5 days of incubation at 99–99.5 °F (D. Mancini, pers. comm.).

Precisely known nestling periods of 10 Ornate Hawk-eagles ranged from 60 to 93 days (10–13 weeks) from the time of hatching until the chick left the nest, although a few nestling periods may have been as brief as 65 days (Table 15.2). Our best estimate of the median duration of the nestling period is 77.5 days. Duration of the nestling period seemed to vary with the circumstances afforded by the nest tree—chicks appeared to remain in trees with large, spreading crowns longer than in smaller nest trees, from which earlier fledging appeared to be the rule. Fledged juveniles remained dependent on adults, on average, to the age of 15.2 months (12.4 months after fledging), but this was quite variable: see Post-fledging Dependency Period later in this chapter.

VOCALIZATIONS

Adult Males

Vocalizations given by Ornate Hawk-eagles were generally high-pitched, thin, piping whistles, varying in the number and duration of notes and their rate of repetition. The most commonly heard call, which we deemed the species's "typical" or "advertisement" call, was often given by male hawk-eagles as they soared over the forest. Hilty and Brown (1986) rendered it as a loud, whistled *whit, wheéeuuu, whep, whep, whep, whep,* with the first note sometimes faint or lacking, and the second note being highest in pitch, most emphasized, and down slurred. The placement of the emphasized, down-slurred note at the beginning of the call rather than the end distinguishes the Ornate's call from that of the Black Hawk-eagle. This call is variable and may lack the drawn-out phrase, in this case resembling *hui . . . pi-pi-pi, hui . . . pi-pi-pi-pia, hui . . . pi-pip* and other variants (Howell and Webb 1995). When voiced during soaring, this call draws attention to the calling bird and probably aids in territorial proclamation.

Adult males arriving with prey near a nest voiced a different call we termed the "arrival call": *pitpit-pitpit-pitpit-pitpit,* and so on. This or a similar call was also given at times by males as they incubated. When delivering prey to dependent fledglings, adult males sometimes voiced a third call resembling the fledglings' food solicitation call (see Nestlings).

Adult Females

Among the most frequently heard female vocalizations was a call we termed the "begging call," as it was given when we believed females were hungry and calling to the male for food, and when males approached the nest vicinity. This call was given softly or loudly, and in variable tempo. It may be represented as *hui-hui-hui-hui* with more rapidly repeated syllabic pairs interspersed at times, such as *huihui . . . huihui.* Sometimes when they remained off the nest, eating, while the male incubated, females voiced what appeared to be the same call but in a long-drawn-out series, *huihuihuihuihuihuihui.* At other times when the male arrived at the nest with prey, these female begging calls gave way to a similar but more variable, more excited-sounding call: *fliyi-fliyi, pipio, pio, pio,* or *hui-hui-fliyi-fliyi.*

Females used a distinct call while being mobbed by Social Flycatchers *(Myiozetetes similis)* and other small birds and when responding near the nest to certain other species perceived as threatening, such as other raptors soaring overhead. The notes comprising this call were the sharpest notes we heard females give. The call was always brief, composed of one to five syllables, *fli-flii-fli-fliii*, or sometimes varied as *fli-fli-fli-flio-flio*. During the pre-laying and copulation phase, we heard some adult females at the nest give vocalizations resembling the "advertisement call" most often heard from males in flight: *hui . . . hui . . . pi-pi-pi-pi, hui . . . hui . . . pi-pi-pi-pi*. Generally both pair members vocalized while copulating. The nature of the copulation vocalization was distinctive enough that experienced observers could tell by hearing it that copulation was occurring, but we lack a phonetic representation of it.

Nestlings

When nestlings appeared contented, as after eating or while being preened by the female, they voiced soft calls. When distressed—for example, when suffering discomfort from the heat—nestlings voiced a chirping call: *chipchop-chipchop-chipchip-chop*, and so on, given at the rate of three phrases per second, sometimes for a minute or more. When an adult arrived at the nest with food, nestlings voiced a food solicitation call: *huiii-huii-huiii-huii, hui-hui-hui, hui-huiii-huiii-huiii*, and so on. This or a similar call was also given in a defensive-aggressive context, as when another raptor flew near the nest.

Other Notes on Vocalizations

Land (1970) stated that one of this hawk-eagle's calls resembles the gobble of a young domestic turkey. We did not hear any calls that fit this description. Slud (1964) stated that Ornate Hawk-eagles, while hunting, sometimes make a cry "that could be taken for the snarling growl of a 100-pound cat" and recounted a case in which two guans on the ground screamed frantically each time a hawk-eagle, which was perched above them, uttered this catlike cry. Kilham (1978) described extremely loud alarm calls given by a Crested Guan under attack by an Ornate Hawk-eagle at Tikal, and likened these in volume to those of a howler monkey. He also heard guttural sounds and growls that made him wonder if the guan was under attack by a jaguar. Kilham appears to attribute these catlike calls to the guan and noted that Russell (1964) described a female Great Curassow with young as making a "threatening, mammal-like snarl, similar to that of an angry dog" during a distraction display. We never witnessed Ornate Hawk-eagles giving loud, catlike calls as described by Slud. In view of these observations, it appears that clarification is needed as to whether the hawk-eagles or their cracid quarry are responsible for such calls.

BEHAVIOR OF ADULTS

Adults generally used three to five large, emergent trees within some 50 m of the nest as favorite perches. The hawk-eagles often removed prey remains and sometimes hawk-eagle feathers from the nest, dropping them in flight or from one of these perches (see also Klein et al. 1988). On rare occasions we witnessed hawk-eagles drinking from the Arroyo Negro—a small stream flowing beneath the forest canopy and reduced to small, disjunct pools during much of the year.

Pre-laying Period

Courtship and nest reconstruction often took place over at least one to two months prior to egg laying. We were unable to determine with certainty how far in advance of egg laying this behavior began. We suspect, however, that initial visits to the nest may begin as soon as the previous year's juvenile has either died or reached independence, which often occurred as much as four months prior to egg laying.

Occasionally we saw pair members soaring together. On 20 November 1991 at the Naranjal nest, four months after their previous young had reached independence, the adult pair was observed soaring together some 300 m above the nest area in late morning (the male's identity was verified by his radio signal). They circled repeatedly, soaring high over the nest area, and then soared off until out of sight. At times, hawk-eagles soared to high altitudes, greater than 300 m. During the incubation and nestling phases, adult males were often seen soaring over the nest vicinity and other parts of their territory while the female remained close to the nest.

A number of authors have described Ornate Hawk-eagles as performing aerial maneuvers they interpreted as courtship displays. These have been described as individuals diving from high in the air to above the forest canopy, sometimes then completing a loop (Slud 1964); soaring together, a pair member rolling over, the pair sometimes touching talons (ffrench 1973); short, "butterflylike" wing flutters while soaring (Hilty and Brown 1986); and climbing flight with "deep, floppy wingbeats" followed by stoops with closed wings, "almost somersaulting at times" (Howell and Webb 1995). Such displays did not appear common at Tikal, but we did witness similar behavior in at least one case, by a male hawk-eagle 200 m above the Naranjal nest tree.

During several weeks prior to laying, pair members, especially females, were often visible in the nest vicinity. Males provided all or nearly all food for females during this period. Females spent much of their time in the nest tree or within 100 m, especially in the morning, when most nest building took place. At times both pair members loitered on the nest together, rearranging nest material and vocalizing softly. Early in the pre-laying phase, females were often absent from nest vicinities during the afternoon, but as laying drew near, they were present more frequently. Nests were often reused, typically after substantial reconstruction. Early in the reconstruction process, mainly dead sticks were brought to the nest, often haphazardly deposited at first, then carefully arranged later, mostly by females.

Adult behavior in the nest vicinity during the pre-laying period was documented during 156 hours of observation at five nests belonging to four pairs (mean = 31.2 ± 16.5 hours of observation, n = 5 nests). Females remained in nest vicinities on average 97.2 ± 2.3% of the time (n = 5), while males were present 30.4% of the time on average and much more sporadically (SD = 23.5%, n = 5). Only rarely was a male near a nest in the absence of the female (mean = 2.8 ± 2.3% of the time, n = 5). When males lingered near the nest during this phase, they often stood on the nest for modest periods and manipulated nest material. It was during this period that we witnessed most cases of males bringing sticks and green foliage to the nest.

Males and females very often stood briefly on the nest immediately prior to copulation, which generally lasted 6–12 seconds. The frequency of copulation in the nest vicinity during the pre-laying period was documented during 204 hours of observation at six nests, during which 60 copulations were witnessed. Copulation frequency averaged 0.34 ± 0.16 per hour of observation overall (n = 6 nests), and during periods when both adults were visible near the nest throughout, it averaged 2.18 ± 1.73 per hour (n = 5 nests). We never witnessed copulation after the egg was laid, except when an egg was lost and the pair began a new nesting attempt.

Typical copulation sequences occurred as follows. During the pre-laying period, females often perched near the nest and vocalized repeatedly for 10–30 minutes. Males often responded to these calls by perching a few meters away and calling quietly in return. Immediately upon the male's arrival, the female would often enter the nest and begin manipulating nest material with her beak. The male would intently observe the female, then fly or jump to the nest, at which time the female would respond by hopping to a nearby perch, where the pair would then copulate. Most copulations occurred in the nest tree on a horizontal limb 50 cm to 6 m from the nest. In some cases we witnessed copulation 100 m from the nest in habitual perch trees, and we noted one copulation 350 m from a nest.

Females conducted most nest reconstruction, bringing in 92 sticks compared to 12 brought in by males during one subset of our observations. Occasionally, however, males played a major role in bringing nest material and spent a good deal of time in the nest and nest tree. For example, during early morning of 13 March 1992 at the Naranjal nest (during the pre-laying period), the male spent 39% of a 3.75-hour observation period on the nest, 53% elsewhere in the nest tree, and the remainder nearby, before leaving the area in late morning. This male made nine brief visits to the nest, and was the only bird observed adding sticks and manipulating

nest material that morning. The female, however, was also present in the vicinity, and three copulations were observed. During the pre-laying period, we observed this male bringing sticks to the nest during four observation days, but never again after the egg was laid. This agrees with our impression that nest visitation and the delivery and manipulation of nesting material by males are behaviorally associated with copulation.

Addition of Greenery to the Nest

Once nest reconstruction was complete and the egg laid, nearly all material brought to the nest during the rest of the nesting cycle was sprigs of live foliage from trees, the rest being leafless sticks. Most foliage was brought by females, but males occasionally delivered foliage throughout the nesting cycle. Females often gathered foliage repeatedly from the same individual trees. These were invariably large nearby trees whose open architecture appeared to facilitate perching. Foliage was collected from Black Olive, Honduras Mahogany, Silión, Antilles Calophyllum, Black Cabbage-bark (*Lonchocarpus castilloi*), other tree species, and occasionally from vines. We could not determine whether the hawk-eagles were selective with regard to the species of foliage collected. On occasion, females broke off and then dropped one or more branches before bringing them to the nest, but whether such sprigs were intentionally discarded was not clear.

In collecting greenery, adults most often used the beak to grasp and break off the branch, though they often used their feet to bend or hold down the limb, and their wings and tail for support against the perch or adjacent limbs.

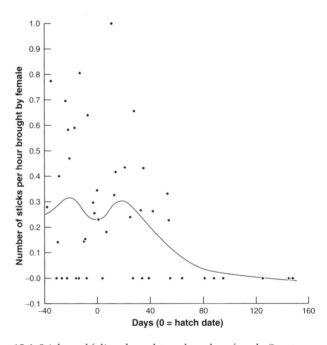

15.4 Sticks and foliage brought per hour by a female Ornate Hawk-eagle during the incubation and nestling periods at Tikal. Curve is SYSTAT LOWESS smooth, tension = 0.35.

Also, hawk-eagles sometimes grasped a branch with one or both feet and hung below, flapping, for up to several seconds, to break it off. Females usually carried greenery in their beaks, though occasionally in their talons. At one representative nest, foliage was delivered most often during incubation and before the chick reached the age of six weeks (Fig. 15.4).

Possible functions of greenery in the nest are discussed in Chapter 13. Here it bears noting that we were impressed by a possible shade-producing function of greenery in Ornate Hawk-eagle nests. While incubating the egg and tending the chick, females often rearranged nesting material—principally sprigs of greenery—that were within reach. Frequently they took sprigs from the nest bowl and placed them on the rim, between the sun and the nest bowl, effectively deepening the bowl by building up its perimeter, which cast shade across the bowl. The extent of this behavior was such that often only the female's head was visible as she incubated or brooded. Likewise, the nestling was often hidden from view as it lay in a deep nest bowl produced in this manner.

Incubation Period

Males provided all prey during the incubation period, and females conducted most incubation. Eggs were rarely left uncovered; females took incubation breaks 1–68 minutes in duration, but most often 6–14 minutes (Table 15.3). During breaks, females often remained standing on or near the nest, frequently preening, and often brought in branches—mostly sprigs of green foliage—to the nest. Females delivered an average of 0.25 ± 0.11 branch per hour to nests (range = 0.19–0.42, n = 5 nests totaling 236.4 hours of observation), 92% of these being sprigs of green foliage (Table 15.3). During midday incubation breaks, females tended to remain standing on the nest, shading the egg, while in the evening when nests received shade, they often left the nest during incubation breaks.

During the incubation phase, males generally visited the nest vicinity only when bringing prey. They transferred prey to females on the nest rim, on a limb of a nearby tree, or in a tree up to 300 m distant. We never observed adults eating on the nest during the incubation period. As the female ate in a nearby tree, the male would often incubate, although pairs varied in this regard. The Naranjal male incubated during each of four observed prey exchanges, whereas some other males incubated after only one out of three or four prey transfers (Table 15.4). Duration of incubation stints by males varied from 20 minutes to 4 hours, averaging 99 minutes (Table 15.4). While incubating, males took infrequent breaks of 3–10 minutes' duration, during which they stood in the nest and preened. The length of time males remained incubating appeared to depend on the female: in general, males ceased incubating only when the female returned to the nest to resume incubation, at which time the male would step off the nest and quickly leave the area.

When we climbed to nests that contained an egg, we nearly always found the egg covered by green foliage. It appeared that females intentionally covered the egg before leaving the nest, though we were unable to observe in sufficient detail to verify this. In one case when the female remained incubating until the climber was 1 m away, the egg was not covered, perhaps because she departed in a hurry. Skutch (1976) described similar egg-covering behavior in a variety of birds, including the Bald Eagle *(Haliaeetus leucocephalus).*

Nestling Period

Roles of parents. From the moment the egg was laid, females began spending essentially all their time on the nest or in the nest tree, but surprisingly, from almost the same moment, females also began a process of "centrifugy" (Balgooyen 1976), that is, of progressively maintaining less strict proximity to the egg, chick, and nest (Fig. 15.5). At first this decline in proximity to the egg and nest occurred slowly, and incubation constancy was generally above 90% during the first half of the incubation period.

Once an egg hatched, the female at first would brood the nestling a great deal, but she soon began to stand over and shade more than brood it (Fig. 15.6). By the time the chick was 7–10 days old, time spent in close brooding by the female generally dropped to 20–30%. Initially, as females spent less time brooding and shad-

Table 15.3 Duration of incubation breaks[a] by females (without being spelled by males) and number of limbs brought per hour

Nest	Extreme durations (minutes)	Most common durations (minutes)	Number of observation periods	Observation time (hours)	Number of limbs brought in by female per hour[b]
Tintal 1991	1–59	12–15	14	112.00	—
Chikintikal 1992	1–68	6–15	16	44.33	0.135
Naranjal 1992	1–48	6–13	20	111.33	0.305
Temple V 1991	1–36	6–12	9	24.00	0.417
Brecha Norte 1991—1	1–62	6–13	8	20.68	0.193
Brecha Norte 1991—2	1–58	6–14	11	36.08	0.194

[a] Generally females remained standing (often preening) on or very near the nest during these breaks.
[b] 92% were sprigs of green foliage, the rest bare sticks.

Table 15.4 Behavior of males in nest areas during the incubation period

Nest	Visits to nest	Number of times male brought prey	Number of times male incubated	Duration of incubation (minutes)
Tintal 1991	4	4	3	45, 50, 84
Punta de Pista 1991	4	3	1	87
Chikintikal 1991	4	1+	0	—
Naranjal 1992	4	4	4	20, 171, 201, 242
Temple V 1991	3	1	1	36
Brecha Norte 1991—1	3	3	3	55, 56, 120
Brecha Norte 1991—2	3	3	1	123
Total	25	76+% of time	52% of time	mean = 99 minutes

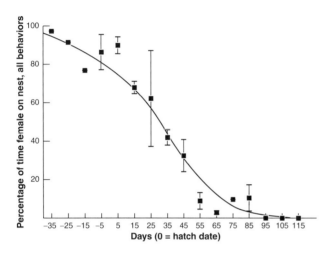

15.5 Percentage of time female Ornate Hawk-eagles spent on the nest (all behaviors combined) during incubation, nestling, and early fledging periods. Means ± SE per 10-day period for two nests. Curve is SYSTAT DWLS, tension = 1/n. Although based on only two nests, this figure typifies our study nests.

ing, they spent correspondingly more time perched on the nest, in the nest tree, or within sight nearby. Shading behavior by females peaked two to five weeks after hatching (Fig. 15.6). Total female presence on the nest (brooding, shading, feeding and preening the young, and simply standing guard on the nest) on average dropped to 50% by the time chicks reached the ages of 3–6 weeks, and declined to essentially zero when chicks were 50–85 days old (Fig. 15.5).

Generally around six to seven weeks after the chick hatched, centrifugy by females became pronounced as they spent ever greater periods of time away from the nest vicinity—presumably engaged in hunting—and began to bring in prey we believe they had captured. For example, at the 1992/93 Arroyo Negro nest, when the chick was six weeks old, we saw the female attack a squirrel near the nest. Later the same day, she left the nest area, returning 40 minutes later with half a squirrel that she may have caught. During several weeks in the late nestling and early fledgling periods, females became temporary but significant providers of food to the young, after which the adult male's role as the primary provider was again resumed. For example, at the Punta de Pista nest in 1991, eight days prior to the chick's fledging, the female was first observed delivering prey we believed she

captured. During the first three weeks after the chick's fledging, this female delivered (and, we believe, captured) 8 of the 11 prey items documented, and during the fourth through sixth weeks after fledging, she was responsible for only one of nine observed prey deliveries.

Although males often incubated eggs while females ate, only females brooded and fed nestlings. Females also played a greater role than males in nest defense, but this may have owed simply to their more frequent presence near the nest. At times when a male was present but not the female, males defended vigorously against intruders.

We recognized four stages of female behavior during the nestling period, distinguished mainly by the frequency with which the female shaded, brooded, and fed the young, and the proportion of her time spent on the nest, in the nest tree, nearby, or away from the nest area. Although these stages were concrete enough to merit recognition, transitions between them were gradual, and the age of the nestling when female behavior progressed to the various stages was somewhat variable.

In stage 1, from hatching until the chick was about 17–21 days old, the female remained on the nest nearly constantly, leaving only briefly to collect foliage and receive prey from the male. Though females never ate on the nest during the incubation phase, they normally did so once the chick had hatched. Brooding during stage 1 was frequent, and when not in contact with the chick, the female usually stood over or beside it. Stage 2 began when females began to perch for short periods off the nest, remaining vigilant, always within 30 m of the nest, and always within view of it. Females left the nest during this stage only when it was not raining and when intense sun was not on the nest. The third stage was marked by the tendency of the female to remain off the nest for an hour or more at a time, often in trees as much as 450 m from the nest. This stage was reached when the nestling—usually about six weeks old (mean = 41.2 ± 3.9 days, $n = 6$, range = 37–46 days)—was well covered with feathers and had begun to manifest aggressive reactions to vultures, other large birds in flight, and sometimes airplanes. During stage 3, the female continued to spend much time on the nest. She fed the nestling, brought greenery, and still shaded and brooded the chick frequently. The fourth stage was reached when the female remained absent from the nest vicinity for hours at a time, and began to capture and bring prey to the nest—when chicks were 54–76 days old.

Occasionally a male visited the nest without prey, perching there or in a nearby tree for up to 25 minutes. Males generally moved within or below the forest canopy when bringing prey to the nest, but after a prey delivery, they sometimes circled up high over the canopy, calling, before soaring away. Males rarely brought foliage to the nest during the nestling period, but sometimes they spent considerable time within view of the nest. At Naranjal in 1992, the male rarely visited the nest vicinity after the chick hatched, and virtually never after the chick's third week (Fig. 15.7A). In contrast, the Punta de Pista male in 1991 showed a pronounced tendency to spend more time in the nest area when the chick was 9–15 weeks old (Fig. 15.7B). This coincided with the period when the female completed her progressive decrease of nest attendance and was absent nearly constantly from the nest vicinity.

Shade-giving behavior by females. Females appeared to adjust their chick-brooding and shading behavior to conditions at the nest, including degree of cloudiness, rainfall, and degree of shading from the sun provided by vegetation. When temperatures were chill, females brooded, and when the sun beat down, they shaded the chick. When rain squalls came, females covered the nestling, sometimes hastening back from some distance to do so. Females used various postures to shade nestlings from sun and rain. In the early morning, when the sun was at a low angle, a female would provide shade simply by lying beside the nestling, her body interposed between sun and chick. As the sun rose higher, she most often stood over the nest bowl, back toward the sun, with wings partly to fully extended. The degree of wing extension appeared to correlate closely with the intensity of sun or rain on the nest: when the sun disappeared behind a cloud or when rain subsided, females closed their wings, extending them again when the sun emerged or showers renewed. At times females gaped, appearing uncomfortably warm as they maintained their shading vigil against intense sun.

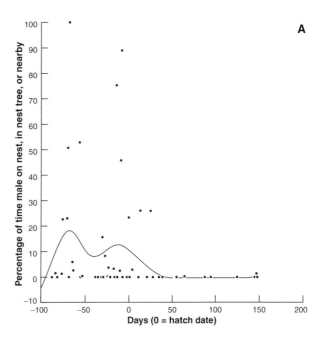

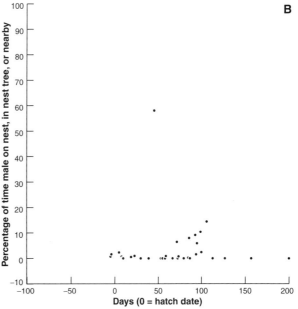

15.7 Percentage of time male Ornate Hawk-eagles spent on the nest, in the nest tree, or nearby (within view, usually < 50 m from nest): (A) Naranjal 1992. (B) Punta de Pista 1991. Curves are SYSTAT LOWESS, tension = 1/n.

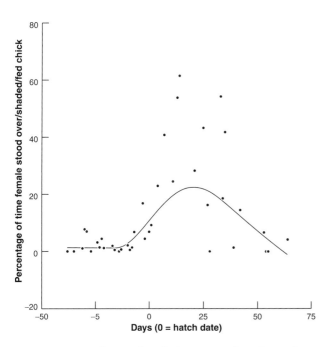

15.6 Percentage of time a female Ornate Hawk-eagle stood over egg/nestling during incubation and nestling periods. ("Standing over" includes shading, preening, feeding: i.e., time on nest in close proximity to egg/nestling, but not incubating/brooding.) Curve is SYSTAT LOWESS, tension = 0.35.

Prey deliveries and exchanges. During the nestling period, males rarely visited the nest vicinity except to bring prey. Often a female anticipated her mate's arrival by staring intently in a certain direction and quietly voicing the *hui-hui-hui* begging call. The clamor of Brown Jays and Keel-billed Toucans appeared to assist adults in tracking one another's whereabouts; many times as we watched a female or chick on the nest, it would turn a keen eye toward a distant clamor of Brown Jays or toucans, sometimes calling quietly. Often, soon after, the male would advertise his approach from that direction with a *pit-pit* call (see Vocalizations earlier in this chapter), voicing this for three to five seconds at a time, repeatedly over a period of several minutes. Females often flew off to receive prey on a perch 100–200 m from the nest, though they sometimes received it on the nest or as much as 400 m distant. When the male delivered prey to the female some distance from the nest, he often left the area immediately afterward. In contrast, when he brought prey directly to the nest, the pair would at times remain perched there together for up to 5 minutes.

Prey exchanges between perched adults were usually accomplished foot to foot, with the female grasping the prey with one foot and taking it from the male's grasp. Occasionally, the male departed just before the female's arrival, and the prey was left on a perch or the nest, where she soon found it. Prey delivered by the male was sometimes headless, sometimes partly or fully plucked, and at times the front half (the breast portion, in birds) had been consumed. Females seemed more likely to deliver intact prey items than were males. When prey were brought to the nest with the head intact, the female would often discard the bill (e.g., from a toucan) or the skull (from a mammal) away from the nest. Removal of prey remnants from the nest might be adaptive for various reasons, perhaps chief among them that odors from prey scraps, especially any dropping to the forest floor, might alert potential nest predators such as mammalian Carnivora to the presence of a nest. When males delivered prey directly to nests, whether before or after the nestling fledged, they normally remained on the nest for less than a minute, and often for only a few seconds. Though we never made a definite observation of prey-caching behavior, in some cases we believed that a female retrieved prey left earlier by the male at a nearby cache site.

Post-fledging Interaction

Between breeding seasons, when adults often had dependent juveniles, we saw no evidence of interactions between mated pairs. For example, during many months when we radio-tracked both members of the Naranjal pair, we never found them together. Normally, males seemed to be the sole providers for fledglings; however, there were exceptions. At one point, for example, the Naranjal male ceased to feed the juvenile for a period of three to four weeks, and the female took over this task. We speculate

that the juvenile's persistent begging calls may have encouraged the female to increase her role in provisioning.

BEHAVIOR AND DEVELOPMENT OF NESTLINGS

Nestling Period

During the first week after hatching (0–7 days), nestlings spent most of their time sleeping, raising their heads only to receive bits of food from the female, who brooded and shaded them nearly constantly. During the second week, nestlings moved about on their tarsi more often. During week 3, nestlings voiced the begging call when females arrived at the nest, and were first noted preening. They shuffled about more successfully on their tarsi, arriving at times on the nest rim, and were capable of a few small, unsteady steps on straightened legs. On day 16 a nestling tried to feed itself, without success, and by the end of the third week, it defecated over the nest rim. By the end of the fourth week, some nestlings stood for periods of up to a minute, and on day 29 one remained standing for six minutes. During week 5, nestlings manipulated sticks in the nest and frequently gave food-begging calls. They walked about on their feet rather than their tarsi and were first noted flapping their wings. During this period, when monkeys approached the nest, a nestling called in alarm. Nestlings at this age were still covered and shaded a good deal by females. During the sixth week, nestlings were increasingly mobile, making small jumps while flapping in place, though some were still unable to remain standing for more than 5–10 minutes. At this age, nestlings became aggressive when receiving prey from adults. With head low and plumage erect, they would pounce on the prey item and mantle over it, quivering their wings or flapping shallowly, and begging loudly. They also behaved this way when in possession of food when a vulture or other raptor flew low over the nest. The adult male usually left the nest quickly after prey deliveries, and the nestling typically called and would not eat until the adult withdrew completely from the nest vicinity. When an adult remained perched within 100 m, the nestling remained agitated, eating only in short spells and calling repeatedly.

During the seventh week, nestlings were still unable to feed themselves, despite their attempts. Though still brooded at times by females, they remained standing much of the time and walked about the nest; still, they often sprawled in the nest center, especially during the hours of most intense sun. During week 8, nestlings remained standing most of the day; they played with sticks and flapped in place. On day 52, one nestling managed to eat a few bites unaided. When researchers climbed to nests, nestlings at this age responded with a threat display, spreading their wings, calling, and gaping. By week 9, chicks became noticeably more independent. Most were able to feed themselves, and many began to leave the nest to perch up to 6 m away on nearby limbs. Flapping in place was frequent.

During the tenth week, several nestlings were first seen flying among perches within the nest tree, and several made their first flights outside the nest tree. By this age, many nestlings greatly decreased their time spent at the nest, returning there mainly to receive prey and eat. During the eleventh week, some young were seen making mock attacks on fruits or branches of the nest tree, pouncing on them or seizing them from below in short, steep, upward flights. By week 12 (78–84 days), the juvenal plumage appeared complete. Young continued making practice attacks on branches or fruits and received prey on the nest or in nearby trees (Plate 15.3).

Post-fledging Dependency Period

Activities of fledglings at first remained centered on the nest tree as adults continued to deliver food to the nest or to nearby trees. By two weeks after fledging, juveniles often spent the night in a tree, up to 100 m distant from the nest. Initially, juveniles ate on the nest or elsewhere in the nest tree, but increasingly they fed in other nearby trees as time progressed, often using grass-like clumps of epiphytes as feeding platforms. Juveniles typically perched high in emergent trees, voicing a food-begging call for protracted periods. When resting with a full crop, however, they were more often perched quietly in a shaded location within or beneath the canopy. Later, as juveniles began to make hunting attempts, we often found them perched within or below the canopy.

We plotted distance from the nest versus age for three fledgling males (Fig. 15.8) and five females (Fig. 15.9). The three males rarely wandered more than 50–100 m from the nest until past the age of 5 months. By 11 months old, the male fledged at Naranjal in 1992 had only once been found more than 350 m from the nest, whereas the 1989 Tintal fledgling, from the age of 9 months to his death at 13.3 months, often ranged up to 500 m, but never beyond, during our observations. Among five females, three rarely moved more than 300 m from the nest until the age of six months (Fig. 15.9). By nine months, two were found on average 100–300 m from the nest, while two others averaged 500 m from the nest. Three females dramatically increased their wandering at the age of 12–14 months and were often found from 900 m to 3 km from the nest, but they were also still found near the nest at times. Two of these three females abruptly dispersed at the ages of 14.00–14.25 months, while the third retreated to the nest vicinity for a time and then finally dispersed at the age of 20.0–20.4 months (Fig. 15.9).

At some point during their dependency, juveniles began to soar occasionally over the forest. For the 1989 male Tintal fledgling, we first observed this at 11 months after hatching, then again at 13 months, when he soared 100 m above his natal nest. The 1991 female fledgling from Arroyo Negro was first noted soaring above the forest at 11.75 months, and at about 13 months she soared extensively over the vicinity of her nest; during the next six weeks she achieved independence and dispersed (Appendix 15.3).

We monitored 13 fledglings (9 of them radio-tagged: Plate 15.4) to determine their behavior, spatial use, survival, and time to independence. We were able to monitor four female juveniles until they reached independence and dispersed, on average, at 15.2 ± 3.4 months after hatching (range = 12.5–20.4 months) and 12.4 ± 3.6 months after fledging (range = 8.5–17.4 months;

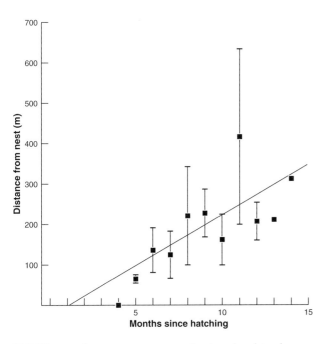

15.8 Distance from nest versus months since hatching for three juvenile male Ornate Hawk-Eagles at Tikal. Monthly means for three birds ± SE.

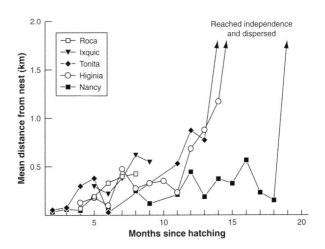

15.9 Distance from nest versus age for five juvenile female Ornate Hawk-eagles at Tikal. Shown is the monthly mean for each individual. Month 6 = month 6.0 to before month 7, etc. Tonita, Higinia, and Nancy reached independence and dispersed at the end of their respective lines.

Table 15.5). Ample evidence indicated that no fledglings reached independence at less than 11.5 months. Histories of the remaining nine individuals were as follows. For one female without a radio, we have no data beyond the age of five months. The remaining eight individuals (five with radios, three without) were monitored, on average, to the age of 11.5 ± 1.5 months (range = 9.5–13.3; Table 15.5). At this point we lost contact with these juveniles when we could no longer locate them because their transmitters had died, they had dispersed beyond radio range (n = 5 birds), or for juveniles without radio transmitters, because we could no longer locate them by their vocalizations (n = 3 birds). During this period some of these individuals no doubt reached independence and dispersed, while others may have perished prior to reaching independence.

Development of Hunting Skills

For many months, juvenile Ornate Hawk-eagles remained within a small area of forest, where adults brought them food. During this time they were presumably exposed to many forest creatures, including appropriate quarry. Early prey capture attempts are probably in large part opportunistic and driven in part by playful curiosity. As greater experience was gained, the transition to true, adult-like hunting was made as the hawk-eagle began to create capture opportunities for itself by patiently stalking quarry until arriving, undetected, within striking range.

We first observed juveniles make prey capture attempts at the ages of 12–13 months, although earlier attempts may have been made, unobserved by us. The 1993 fledgling female from the Arroyo Negro territory made what we believed to be her first capture at 13.5

months, when she captured a Great Curassow 600 m from her nest. Within the next two weeks, she reached independence and dispersed to an area 8 km away. For another young female—the 1991 Arroyo Negro fledgling—the first hunts we observed were attacks on a squirrel and a bird on 21 and 28 April 1992, when she was about 12 months old. On the latter date she was observed eating a bird we believe she caught. This female reached independence 2.5 months later, at the age of about 14.5 months. At the age of 12.25 months, we believe one juvenile female (the 1991 Punta de Pista fledgling) was stalking a guan that we inadvertently flushed 12 m from her. She reached independence and dispersed two months later, at the age of 14.25 months. Further details on these three juveniles are given in Appendix 15.3.[7]

SPACING MECHANISMS AND POPULATION DENSITY

Territorial Behavior and Displays

Like most diurnal raptors at Tikal, Ornate Hawk-eagles frequently soared high over the forest, at times calling. We believe such over flights are the primary mechanism through which these birds advertise their territorial claim. A case of cannibalism may also have represented territorial behavior; see Highlights later in the chapter.

Nest Defense and Interspecific Interactions

As long as we did not climb the nest tree, Ornate Hawk-eagles generally paid us little attention. They occasionally perched in an observation tree along with us, and

Table 15.5 Fates of Ornate Hawk-eagle fledglings

Year	Territory	Sex	Individual	Age at last observation (months)	Time from fledging to last observation (months)	Dispersed?	Age at dispersal (months)
1989	Tintal	M	Toky	13.3	10.5	No[a,b]	—
1990	Naranjal	M	Toky II	11.0	8.2	Yes[c]	—
1992	Naranjal	M	Condorito	10.5	8.25	No[d]	—
1993	Punta de Pista	M	Rudi	10.0	8.0	No[e]	—
1990	Brecha Norte	F	Roca	12.5 (approx.)	8.5 (approx.)	Yes[c]	12.5 (approx.)
1990	Temple IV	F	Ixquíc	13.0	10.25	?[e]	—
1991	Punta de Pista	F	Tonita	14.25	12.0	Yes[c]	14.25
1991	Arroyo Negro	F	Higinia	14.0 (approx.)	12.0 (approx.)	Yes[c]	14.0 (approx.)
1992	Chikintikal	F	Sac-Nicté	13.0	11.0	?[f]	—
1993	Arroyo Negro	F	Nancy	20.0–20.4	17.0–17.4	Yes[c]	20.0–20.4
1994	Punta de Pista	F	Silvia	5.0	3.0	?[f]	—
1994	Balastrera	F	Pati	12.0 (approx.)	10.0 (approx.)	?[f]	—
1993	Brecha Norte	?	Percy	9.5	7.0–7.5	?[f]	—
Mean				12.2 ± 3.44 (n = 13)	9.7 ± 3.28 (n = 13)		

[a] Died.
[b] Eaten by duo of subadult conspecifics.
[c] Definitely reached independence and dispersed.
[d] 95% certainty killed and eaten by a human working in the forest as a xatero (palm frond gatherer).
[e] Radio failed, fate unknown.
[f] Not radio-tagged, but monitored without radio.

sometimes they visited the nest when a climber was within a meter or two. Aggressiveness toward researchers, however, varied among adult females and appeared greatest during the egg and early nestling stages. The Punta de Pista female would often attack observers while they were climbing to the observation blind, although once they were hidden there, she ignored them. In contrast, the Tintal female never attacked researchers as they climbed to the observation platform, which was a similar distance from the nest. When we climbed to active nests, adult hawk-eagles often attacked repeatedly—up to 30 times or more—sometimes striking climbers with their talons. Even though helmet, goggles, and thick jacket were required equipment, climbers sustained a number of lacerations in the course of the study. In their aggressiveness toward researchers climbing near the nest, Ornate Hawk-eagles greatly eclipsed the Crested Eagle and Black Hawk-eagle (see Chapters 13, 14).

Adult females were sometimes harassed mercilessly by small birds, and at times this appeared to cause a female to cut short her forays in search of foliage to add to the nest. Likewise, fledgling hawk-eagles were frequently mobbed by Brown Jays and other small birds, which appeared at times to cause the raptors to change perches. The species most often eliciting a defensive reaction in adult hawk-eagles, as well as older nestlings and fledglings, were the Turkey Vulture *(Cathartes aura)* and American Black Vulture. Even when a vulture soared high overhead, hawk-eagles often grew agitated and gave a threat display or called or did both. We were impressed by the vigor of these responses, which seemed disproportionate to the threat posed by the vultures. American Black Vultures, however, are known to behave in a predatory fashion (Buckley 1999) and could be predators on Ornate Hawk-eagle eggs or young chicks.

The mammal most often eliciting a strong visible reaction from these hawk-eagles was the Black-handed Spider Monkey. On occasion, females dove at spider monkeys that moved near the nest, sometimes striking them, as the following cases illustrate. On 21 April 1990 at the Temple IV nest, a pair of spider monkeys approached to within 5 m of the nest, eliciting a defensive response in the nestling. The adult female hawk-eagle, perched near the nest, called and flew to a tree 75 m distant. She then returned and, with a gliding approach, struck a talon blow to the head of the monkey nearest the nest. Screeching loudly, the monkeys fled the area. In another case, the Arroyo Negro female struck a spider monkey that approached the nest, causing the primate to fall 10 m to a lower limb.

During May 1991 when there was a nestling in the Tintal nest, we spotted a Tayra *(Eira barbara)*—a partly arboreal member of the weasel family, roughly the size of a Fisher *(Martes pennanti)*—on the ground near the observation tree. The Tayra began to climb a tree 5 m from the tree in which the observer was stationed, and 65 m from the hawk-eagle nest. When the Tayra had climbed about 10 m, the adult female hawk-eagle perched near the nest tree attacked, hitting the mammal twice in

the head and knocking it from the tree. The hawk-eagle perched nearby, and 20 minutes later the Tayra began to climb the tree in which the female was perched. Again the female attacked it, knocking it from the tree. The Tayra screamed and left the area.

Constancy of Territory Occupancy, Use of Alternate Nest Sites

Some Ornate Hawk-eagle territories were relatively stable, in that territories were occupied in much the same areas from year to year. Moreover, the same banded adults were often present and paired during multiple years.[8] We observed a number of traditional nest sites fall into disuse, however, as well as shifts in territory boundaries and ownership. At the Naranjal nest, after two successful nestings in 1990 and 1992 and a 1994 attempt that failed prior to egg laying, no evidence of nesting was observed through 1998—except for a few new branches that appeared periodically in the nest. We were unable to determine whether nesting took place elsewhere in this territory, or whether the territory went unoccupied or was divided among neighboring territories. Similarly, the Tintal nest site, where nesting took place in 1989 and 1991, was never again used, nor did we detect nesting activity in the vicinity during periodic visits. The two Balastrera nests, situated 1.7 and 2.0 km from the Tintal site, were first discovered in 1992 and were thereafter used in turn through the end of our fieldwork in 1996; it is possible that these were alternate nest sites within the Tintal territory. It seems more likely, however, that they represented a shift of territorial boundaries, as these sites were unusually far from the Tintal nest to be alternate nest sites within the same territory, yet unusually close to be occupied by a neighboring pair.

The best-documented cases of changes in territory boundaries and occupancy involved the Temple IV/V, Chikintikal, and Arroyo Negro territories (Fig. 15.10). In 1990, a nesting female—"Maya"—was banded at the Temple IV nest and produced a fledgling, which probably reached independence in March 1991. In October 1991, Maya was observed copulating—and later laying an egg in March 1992—at a nest 2.87 km to the west-northwest of her Temple IV nest (Fig. 15.10). This second nest was in an area where we had not previously known of a territory, and we named this the Chikintikal territory.

This movement may have been prompted by the fact that early in 1991, while Maya's mate fed their recent offspring and Maya had a rest period between nestings, another adult female, which we banded and gave the name Invasora, along with an unbanded male, had occupied a nest (Temple V nest) 700 m southeast of Maya's Temple IV nest (Fig. 15.10). This new female laid an egg in early November 1991, but the chick died in early March. Throughout this period, this radio-tagged female used much the same area that Maya had used when nesting at Temple IV. In November 1992, some 200 days after the death of her chick at the Temple V nest, this second female, Invasora, was observed at the Arroyo Negro nest,

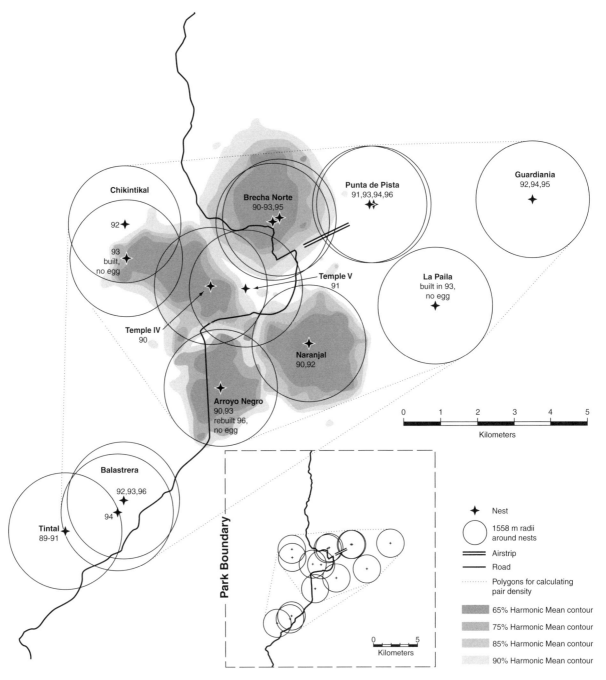

15.10 Ornate Hawk-eagle nest locations and polygons used for calculating nesting density at Tikal. The outer polygon, including the Balastrera territory, is based on a mean inter-nest distance of 3.116 km and yields 12.0 km² per pair. The inner polygon approximates the situation omitting the Balastrera territory; mean inter-nest distance was 2.96, yielding 9.87 km² per pair. Also included is a representation of movements of two adult female Ornate Hawk-Eagles among three neighboring territories, and their nesting histories.

copulating with a male that was not banded until later (Fig. 15.10). Here, in the same nest used by an unbanded pair that fledged a young in 1991, Invasora and her mate nested late in 1992 and early in 1993.

Meanwhile, to our knowledge, there was no nesting in the Temple IV/V area in 1992. Maya remained in the Chikintikal area until 1996, when she showed a tendency to reoccupy the Temple IV/V area. After copulating and nest building in 1993/94, we detected no nest-

ings by Maya, but in 1995 we trapped and radio-tagged her at a Crested Eagle nest quite close to the Naranjal Ornate Hawk-eagle nest. After Invasora's successful nesting at Arroyo Negro in 1992, we detected no further nestings there, although the pair was apparently present in 1995. In 1999 Robin Bjork (pers. comm.) found an active nest in the Temple V vicinity, but we were unable to determine whether this pair included any banded individuals.

To summarize this confusing account, hawk-eagles of three territories—Arroyo Negro, Temple IV/V, and Chikintikal—interacted in such a fashion that territory boundaries and occupancy shifted several times (Fig. 15.10). In fact, it is possible that no more than two of these three territories were active at a given time. In particular, it is possible that the Temple IV/V "territory" is an area that is often divided or exchanged between the Chikintikal territory and the Arroyo Negro territory.

We suggest that this dynamic situation may have resulted in part from two factors. First, the home ranges we documented were substantially larger than the exclusive area per pair implied by the inter-nest spacing we observed. That is, nests were close together relative to the area used by pairs, which may have led to boundary adjustments. Second, on average, only half the territorial pairs in the population are nesting (have an egg or young chick) in a given year, due to the two-year duration of the nesting cycle. We speculate that during the year-long period in which a pair has a dependent fledgling, they may be vulnerable to invasion by other adults who are at the point of initiating a nesting attempt. Perhaps in these formidably armed bird predators, territory occupants sometimes readjust their patterns of spatial use rather than risk outright aggressive contests.

Home Range Estimates

We estimated home range area for 16 Ornate Hawk-eagles, including 2 adult males, 4 adult females, a subadult male and female, 3 juvenile males, and 5 juvenile females. Here we summarize results for the six adults.

Adult males. We estimated home ranges for two adult males banded as breeders: the male banded at the Naranjal nest in 1990, and the male banded at the Arroyo Negro nest in 1993 (Table 15.6). For the Naranjal male, we used 108 relocations taken during a 16-month period spanning the last four months of dependency of one fledgling, the following six months during which there was no nesting effort, and throughout the subsequent nesting effort nearly to the date of fledging. These data yielded a 100% Minimum Convex Polygon (MCP) estimate of 18.8 km² (Table 15.6). Using only the 59 relocations from one nesting effort,[9] the MCP estimate was 14.2 km². For the Arroyo Negro male[10] the MCP home range estimate was 9.6 km² (Table 15.6), though this is likely an underestimate.[11] To summarize, for these two males our best home range estimate is 10–14 km², with up to 19 km² used during two consecutive breeding cycles.

Adult females. During incubation and the first half of the nestling period, adult females rarely strayed far from the nest and had exceedingly small home ranges, estimated at 46 ha (Table 15.6). Including the time up to fledging, one female's home range was estimated as 3.0 km² (Table 15.6). This is still a relatively small area, even though females often hunted during the latter part

of the nestling period. The only female home ranges that helped shed light on the total area used by a pair of hawk-eagles were those embracing periods when females were between nesting efforts—either with a dependent fledgling or after the loss of a fledgling. We measured such home ranges for three females: MCP home range estimates averaged 19.5 ± 5.94 km² (Table 15.6). To summarize, though radio-tagged females used no more than 3 km² during the incubation and early nestling periods, they used an average of 11–19 km² during the period of fledgling dependency, depending on the method of area calculation used. Considering both male and female data, mean home range estimates ranged from 10 to 14 up to 19.5 km².

Inter-nest Spacing and Density of Territorial Pairs

Table 15.7 lists distances between Ornate Hawk-eagle nests we studied at Tikal. Nearly all of these nests were repeatedly active at the same time as their neighbors and thus are safely regarded as distinct territories. Our best estimate of mean inter-nest distances omits the Tintal and Balastrera nests, as these two (possibly alternate nests within the same territory) were probably not nearest neighbors to the remaining nests (Fig. 15.10). This gives a mean inter-nest distance of 2.96 ± 0.53 km (n = 7). Using the polygon method (see Chapter 2), this yields an estimated 9.87 km² of unique space per pair (Fig. 15.10), which we consider our best estimate of Ornate Hawk-eagle pair density in our study area.[12] A less-preferred version includes the Balastrera territory, giving a mean inter-nest distance of 3.116 ± 0.663 km (n = 8); the resulting 107.9 km² polygon yields an estimate of 11.99 km²/pair.[13] However, it is likely that a tenth, undiscovered territory also occurred within this polygon, in which case space per pair is estimated as 10.79 km². Thus, our best density estimate is 10 pairs of Ornate Hawk-eagles per 100 km² of similar habitat (Fig. 15.10), with the roughly 510 km² of Tikal National Park that is not Low Scrub Swamp Forest providing space for about 50 pairs of Ornate Hawk-eagles. It is likely, however, that other factors may adjust this number downward, and this may best be viewed as a rough maximum density in this region.[14]

Meshing home range results with inter-nest distances. To summarize our telemetry results, two males used 10–14 km², or up to 19 km² if two consecutive breeding cycles are considered together, and three females used an average of 19.5 km² during the period of fledgling dependency. In contrast, our best estimate of pair density was one pair per 10 km². Various conclusions may be advanced in response to these results. First, they suggest these hawk-eagles are at local carrying capacity, since there seems to be no unoccupied space between pairs. Second, one might conclude that this is good Ornate Hawk-eagle habitat, again, because all forest other

Table 15.6 Home ranges estimated for radio-tagged adult male and female Ornate Hawk-eagles in Tikal National Park

Nest	Year	Individual	Number of relocations	Dates	Time span (months)	100% MCP[a]	95% HM[a]	85% HM[a]	Comments
Adult males									
Naranjal	1991/92	Felipe	108	21 Mar 1991 to 24 Jul 1992	16	18.8	14.8	8.3	Major parts of two nesting cycles
Naranjal	1992	Felipe	59[b]	1 Feb 1992 to 24 Jul 1992	6	14.2	—	—	From early nest reconstruction nearly to fledging
Arroyo Negro	1993/94	Saddam	42	22 Jun 1993 to 2 Jun 1994	11.5	9.6[c]	9.5[c]	5.7[c]	With dependent fledgling
Adult females									
Naranjal	1990/91	Esperanza	64	20 Oct 1990 to 31 May 1991	8	13.2	11.3	7.77	With dependent fledgling
Brecha Norte	1990/91	Ruca	87	15 Aug 1990 to 18 Jul 1991	11	20.3	18.1	11.6	7 months with dependent juvenile, plus 4 months after juvenile dispersed, and again during pre-nesting period
Temple IV/V	1990/91	Maya	48	18 Aug 1990 to 20 Jul 1991	11	25.0	26.1	14.2	Same as above
Chikintikal	1991/92	Maya	52	31 Oct 1991 to 27 Jul 1992	9	3.0	1.7	0.78	Copulation through fledging
Temple IV/V	1991/92	Invasora	33	30 Oct 1991 to 18 Jun 1992	8	0.46	0.08	0.03	Copulation through loss of nestling

Note: MCP = Minimum Convex Polygon; HM = Harmonic Mean.
[a] Unless indicated, all home range estimates are in square kilometers and are asymptotic, hence valid estimates of area.
[b] These 59 points are a subset of the 108 points above, and include one nesting cycle only.
[c] Home range area still increasing with last relocations; therefore, home range size probably underestimated.

Table 15.7 Inter-nest distances among nearest-neighbor Ornate Hawk-eagle territories in Tikal National Park, 1989–96[a]

Territories	Inter-nest distance[b] (km)
Guardiania to La Paila	3.89
La Paila to Punta de Pista	3.20 (3.17, 3.23)
Punta de Pista to Brecha Norte	2.69 (2.74, 2.84, 2.55, 2.64)
Brecha Norte to Temple IV/V	2.26 (2.42, 2.61, 1.93, 2.09)
Temple IV/V to Chikintikal	3.27 (2.87, 2.80, 3.74, 3.67)
Temple IV/V to Naranjal	2.69 (3.10, 2.27)
Naranjal to Arroyo Negro	2.72
Preferred mean[c]	**2.96 ± 0.53 (n = 7)**
Arroyo Negro to Balastrera[d]	4.21 (4.03, 4.39)
Balastrera to Tintal[e]	1.86 (1.70, 2.01)
Alternate mean[f]	**3.12 ± 0.66 (n = 8)**

[a] Using the minimum spanning-tree method (Gower and Ross 1969; Selas 1997); n nests yield $n - 1$ inter-nest distances; each nest is joined to its nearest neighbor until all nests are joined to at least one other.
[b] Mean distance is given for territories with multiple nest sites.
[c] Preferred version, omitting last two distances.
[d] Omitted in preferred version: Balastrera likely not a nearest neighbor to Arroyo Negro.
[e] Omitted in both versions: Tintal only active in 1991, possibly an alternate site within Balastrera territory.
[f] Less-preferred version: including Balastrera, which was probably not a nearest neighbor of Arroyo Negro.

than extensive areas of Low Scrub Swamp appeared to be occupied by pairs. Third, individuals moved over areas larger than the average amount of exclusive space per pair, indicating substantial spatial overlap among neighbors.

Such spatial overlap among neighboring pairs may be facilitated by the fact that the nesting cycle is two years in length. Because of this species's two-year nesting cycle, at any given time only slightly more than half the pairs should be engaged in a new nesting attempt, while most others tend juveniles during their protracted dependency. If all nestings were successful and if success of different pairs were uncorrelated with particular years, then about half of pairs should initiate new nesting attempts each year. However, because of nest failures and juvenile deaths prior to independence, somewhat more than half of pairs should initiate nests each year.

Therefore, mean distance between simultaneous, new nesting attempts must be greater than the mean distance between all nest sites. Presumably, the food demands of a pair with a large, rapidly growing nestling may be greater than those of a pair with a dependent fledgling. Thus, if inter-nest spacing is adjusted to food demands, it may be adjusted to the average food demands of a mixture of current nesting attempts and pairs caring for dependent fledglings/subadults. Stated another way, perhaps the fact that nearly half of pairs are non-nesting or tending a subadult at any given time may allow high pair density relative to average home range size.

DEMOGRAPHY

Age of First Breeding, Attainment of Adult Plumage

At 371 days after hatching, a juvenile female had not molted any of her original wing and tail feathers, but she showed evidence of molt in other areas. Feathers of her minor wing coverts and wrist, for example, were all new, as were a few feathers on her back. On the sides of her breast were new, black-barred feathers in this previously white area.

We have no direct information on age of first breeding in Ornate Hawk-eagles. According to Howell and Webb (1995), Ornates reach adult plumage at the age of about two years. Presumably, they do not commonly breed before that age, but whether there is substantial additional delay before breeding is unknown. The apparent low turnover of breeders suggests that adult mortality is low and territory occupancy and fidelity high—factors that translate into few breeding opportunities for young birds. This in turn suggests that breeding in young birds may often be deferred for some time. That two birds in subadult plumage were seen together a number of times and ate a juvenile conspecific suggests that birds of this age may at times form a pair bond and even compete for a territorial space. In other raptors, breeding by individuals in immature or subadult plumage occasionally occurs, especially during years of high prey abundance, when marginal areas may become viable territories (Newton 1979). Whether this occurs in Ornate Hawk-eagles remains unknown.

Frequency of Nesting, Percentage of Pairs Nesting

Because of prolonged juvenile dependency, Ornate Hawk-eagles typically nested only every other year unless the prior nestling/fledgling was lost at an early stage. Consequently, the year after a successful nesting, adult males were engaged in caring for a dependent juvenile, and adult females appeared to generally have a respite between nesting efforts, although they helped provision the juvenile in some cases.

In numerous cases, hawk-eagles began nest reconstruction and copulation, but then, for reasons not apparent to us, they did not lay an egg in any of the known nests. Depending on our certainty of the adequacy of our coverage, either these were regarded as cases when nesting did not occur, or these territory-years were omitted from calculations if we believed it possible the eagles nested undetected. To estimate the productivity of our study population, a conservative approach would be to delete any territory-years in which we were not 100% certain whether nesting took place. However, this would create a bias toward over-estimating productivity; some years when nesting did not take place would probably be omitted from calculations, while years when nesting did occur would less likely be omitted, as we were relatively unlikely to miss discovering a nesting attempt. Hence, we have estimated productivity under the scenario we consider most likely, involving 45 territory-years during which we believe our monitoring was adequate—and also a more conservative scenario described here, involving 43 territory-years. Another potential source of bias is that beginning a territory's data series with its year of discovery biases productivity upward, as nesting must have been underway for us to find and include the territory in our sample—territories were unlikely to be newly discovered in years when adults did not nest or were caring for a dependent yearling. To remove this potential bias, we also calculated productivity while deleting the first year of data for each territory.

Nine territories studied in Tikal National Park are considered here.[15] These nine territories had known nesting histories during 3, 3, 4, 5, 4–6 (Temple IV/V), 6, 6, 5–7 (Naranjal), and 7 years, totaling 43–47 territory-years (Appendix 15.1). At Temple IV/V we were not certain, in two years, whether nesting occurred at an unknown site, while at Naranjal, we were not certain during two years whether the area was still an active territory, as no nesting had been detected for some time. We favor using 45 territory-years as the basis for calculations: in this version we regarded the Naranjal territory as no longer active after 1994, but retained Temple IV/V during 1994–95 in the analysis, as we were reasonably confident of our knowledge of its nesting status.

In 45 territory-years, we detected and studied 26 nesting efforts (Appendix 15.1). Thus, a nesting attempt occurred (an egg was laid) in 58% of territory-years. If, rather, we consider our monitoring adequate in only 43 territory years (omitting Temple IV/V during 1995–96 from the sample), then the 26 documented nestings amount to a nesting effort in 60.5% of territory-years. Finally, if we delete the first year of data for each territory, we have 36 territory-years, in which 18 eggs were laid; in this case, an egg was laid in 50.0% of territory-years. This latter approach, however, seemed excessively conservative and unlikely to be more accurate than the alternatives. We favor the first result given, with nesting efforts occurring in 58% of territory-years.

Productivity and Nest Success

Of 16 eggs laid in nests already under observation, 13 (81%) hatched and 10 (77%) of 13 chicks fledged. Eighteen eggs were found before hatching and can be used to calculate survival rate from hatching to fledging.[16] Fifteen (83%) of these hatched and 11 (73%) hatchlings survived to fledging age; nearly all of these were known to have fledged, while one or two that were monitored up until fledging age but not thereafter are included here as having fledged. Of 16 eggs fully documented, 10 (62.5%) produced fledglings. Using our best estimates for each vital rate, however, we estimate that 59.6% of eggs produced fledglings. A compromise is to average these rates, for a 61% survival rate from laying to fledging.

Juveniles that fledged appeared to have a high probability of surviving to independence, which occurred, on average, 12.4 months after fledging. Of 13 fledglings monitored, 3 were known to have died, at 8, 8.25, and 10.5 months after fledging (Table 15.5). At least four individuals reached independence, and we believe that several of the remaining six fledglings did so as well.[17] Seven months after fledging, 12 (92%) of 13 fledglings remained alive, for a monthly survivorship rate of 0.99. At this rate, one can predict that about 87% of fledglings should reach independence, or about 11 of the 13 fledglings we monitored; our best estimate is that 10 of the 13 fledglings actually reached independence.

In sum, our best estimate is that 61% of eggs laid produced a fledgling and 53% an independent subadult. Given that an egg was laid in 50–58% of territory-years, we estimate that, on average, each territory produced 0.31–0.35 fledgling per year and 0.265–0.308 independent subadult per year. Given our estimate of 10 territories per 100 km², about 3.1–3.5 fledglings and 2.6–3.1 independent subadults were produced yearly in each 100 km² of suitable habitat.

Causes of Nesting Failure and Mortality

Three nest failures occurred during incubation and two during the nestling period. In addition to predation on eggs and nestlings, chance events played a role in nest failure. One nest tree was blown over in a windstorm, destroying the nest and egg. Chicks also sometimes fell from nests with no apparent external cause. At the 1991 Tintal nest, the 51-day-old chick fell to its death as we watched. We suspect the same fate befell the chick of the Guardiania nest in 1994, whose remains were found 10 m from the base of the nest tree. Another chick that fell while climbing through limbs during the "branching" phase converted the fall into its first, faltering flight.

Rates of Turnover of Territorial Adults

The only adult mortality we documented occurred in June 1991 at the Tintal nest, where the adult male convulsed and died during handling by a researcher. Resightings of banded adults gave the impression that adult mortality rates were low, and in no case did we suspect that a banded adult had died.[18] Rather, failure to verify the presence of known individuals generally resulted from failure to locate a nesting attempt.

Evidence for a Floating Population of Nonbreeders

We did find limited evidence for the existence of non-territorial, subadult birds. A subadult male was captured and radio-tagged on 31 July 1992 at the Arroyo Negro nest. While his origin was unknown, his subsequent movements centered on the Chikintikal territory. He was not a previous fledgling from the Arroyo Negro nest (where we caught him) because, by plumage, he was about the same age as the most recent (1991) fledgling female from this nest, who reached independence two weeks before this male's capture. During the 21-month period from 3 August 1992 to 16 May 1994, we collected 59 locality points on this subadult male. He ranged over an area of 80 km² (MCP) or 61 km² (95% Harmonic Mean)—a much larger area than that used by any other radio-tagged raptor at Tikal. His pattern of movement did not change noticeably during this period. We do not know whether this male was representative of subadults in general, in ranging over a much larger area than the typical adult home range. While ranging over a large area may increase the likelihood of detecting a vacancy in the breeding population, it may also imply that this male moved about in response to territorial adults he encountered.

Parasites and Pests

We noted no cases of botfly infestations among the 17 Ornate Hawk-eagle nestlings we handled. This suggests a lower rate of botfly infestation than in some other raptors at Tikal. Adults and nestlings were sometimes annoyed by flying insects, and ants were present at times in some nests.

CONSERVATION

In El Salvador, Thurber et al. (1987) judged this species "certainly extirpated from the lowlands" but with repeated sightings in an area of remaining montane forest, and Komar (1998) regarded it as possibly extirpated from El Salvador. In southern Brazil, Albuquerque (1986) called this bird rare, implying a population decline since the turn of the century, and Belton (1984) judged the species probably extinct from Brazil's southernmost state, Rio Grande do Sul. Albuquerque (1995), however, found it still present in remnants of Brazil's much-reduced coastal rain forests, though likely in small numbers. Olrog (1985) believed this species had decreased seriously in northeast Argentina since 1965, when, he states, it was common there. Slud (1964) implied that this hawk-eagle had been eliminated, or largely so, from Costa Rica's central plateau, and Stiles (1985) regarded the species as rare in Costa Rica, indicating that it was listed as Endangered there. These examples illustrate that populations of this species have apparently declined in many areas where deforestation has reduced or eliminated favorable habitat.

Although Ornate Hawk-eagles are primarily occupants of extensive areas of tall, mature forest, we and others have observed a number of pairs nesting in partly deforested areas. We made limited observations on three nests in partly deforested farming landscapes south of Tikal (San Pancho, Chechenal, and La Mula nests). The San Pancho nest, which fledged a young in 1990, was in an area of karst where each of the steep, haystack-like hills retained forest cover, while the swales between were largely claimed by slash-and-burn farming. The nest tree here was at the edge of the forest but less than 50 m from a cornfield. The La Mula nest, which held (and probably fledged) a large nestling in 1994 and 1996, was in an area where more than half the land surface was still covered by mature forest, though with small slash-and-burn fields replacing more of the forest each year. During both years, this nest was at the very edge of a hilltop forest remnant of substantial size, adjacent to a cornfield. The Chechenal nest was also in a partly deforested landscape with increasing amounts of slash-and-burn farming in the area. This pair raised (and probably fledged) a nestling in 1993 and 1995. The nest tree, in a swale 100 m wide between two hills, stood amid corn and bean fields; the forest immediately surrounding the nest had been cleared, probably at least five years prior. In this area, most remaining forest was also in the form of small patches on hilltops.

These instances indicate that Ornate Hawk-eagles are capable of nesting successfully in areas experiencing some degree of deforestation. It remains to be demonstrated, however, exactly how much deforestation and habitat fragmentation these birds can tolerate. Based on the high rates of territory occupancy and nest site fidelity we observed at Tikal, we suspect that the presence of nesting pairs in such agricultural frontier areas may simply reflect the tenacity of established pairs to traditional nest sites in the face of habitat change. Also, in the cases cited, the nests we observed were all near large remaining blocks of forest that may have served as sources of immigrants and as foraging areas. Hence, these instances may reveal little about the species's ability to persist in more thoroughly deforested landscapes, or farther from large reservoirs of intact forest.

Our observations in Central America and southern Mexico indicate that farming landscapes along the forest frontier tend to become severely deforested over time. This does not lead one to be highly optimistic about the future of Ornate Hawk-eagles in the average farming landscape. However, if deforestation can be arrested while perhaps half the landscape or more remains in mature forest, including some large forest fragments many tens of square kilometers in area, it seems likely that these and other forest raptors might persist. Much research is still needed on this topic.

Nest trees of Ornate Hawk-eagles were among the largest trees in the forest. We have extensive data on the size-class distribution of trees at Tikal, and the frequent use of very large trees by these hawk-eagles leaves no doubt that they showed selectivity toward such large trees from among those available. Nearly half the nest trees (8 of 17) belonged to the three species most highly prized by commercial loggers in this region—Honduras Mahogany, Spanish Cedar, and Antilles Calophyllum. Though we do not know whether nest sites are limiting to Ornate Hawk-eagle populations, it is clear that not just any large tree is suitable as a nest site—nest trees typically were towering individuals whose crowns were largely disjunct from the forest canopy. It seems possible that selective logging of large, canopy-emergent trees could affect populations or productivity of these hawk-eagles by limiting the abundance of suitable nest trees.

Though habitat modification is probably the largest threat to populations of most Neotropical forest raptors, for some species—especially the larger ones—hunting by humans can present an important threat, through both direct shooting and indirect effects. Thiollay (1984) presented a modicum of evidence that hunting in pristine forest can adversely affect Neotropical forest raptors. He believed such effects to result partly from the impact of human hunting on prey species shared by some raptors and humans. Prey competition with humans is especially likely in raptors that prey on large cracids, which are often among the species most heavily hunted by people. In our study area, Ornate Hawk-eagles overlapped in diet with human hunters more than any other forest raptor. Prey species that were also taken by human hunters comprised 49.4% of the biomass of Ornate Hawk-eagle diets at Tikal (unpubl. data). Excluding unidentified prey, overlap between Ornates and humans in our study area (Schoener's [1970] Index of Symmetric Overlap) ranged from 4.7 to 12.9%, depending on which of four studies of human hunters is used as the basis of comparison (mean overlap = 8.3 ± 3.4%, n = 4 studies of human hunters). Overlap was mainly for game birds (cracids and tinamous), with a small degree of overlap involving coatis and agoutis.

Direct shooting is a major mortality source for Harpy Eagles *(Harpia harpyja)* and may, in some cases, be a more important conservation issue for that species than is habitat alteration (Alvarez-Cordero 1996). Owing to their size and conspicuousness, Ornate Hawk-eagles are likely among the Neotropical forest raptors for which shooting can be a serious threat. We know of at least two and probably three cases in which these raptors were shot or trapped at Tikal (in one or two cases, the bird was eaten). Traylor (1941) also mentions an Ornate Hawk-eagle shot by a local person in the northern Yucatán Peninsula, and Iñigo-Elias et al. (1987) reported one shot by a farmer in Veracruz, Mexico.

One anecdote regarding food competition between humans and these stately hawk-eagles bears telling. In 1991 or just prior, high-ranking personnel of a Guatemalan conservation agency were on an inspection tour along the logging road that penetrates north from Tikal into the remote heart of the Maya Biosphere Reserve. While carefully straddling the deep ruts in their Suzuki jeep, they came upon an Ornate Hawk-eagle in the road, clutching an Ocellated Turkey it had killed moments earlier. After savoring for a moment this lucky encounter, these officials were seized by inspiration. Dashing ahead, arms waving, they scared the hawk-eagle from the carcass and stole it. Traveling on to their destination at Dos Lagunas, they made a grand meal of the pilfered booty (E. L. Duende, pers. comm.).

HIGHLIGHTS

One of our most interesting observations was a case of cannibalism among subadult hawk-eagles (Plate 15.5). This involved the male offspring from the Tintal nest, which fledged in early June 1989 and was monitored via telemetry until his death 10.5 months later (13.3 months after hatching), at which point he was in subadult plumage. Up until 15 July 1990, this radio-tagged yearling had ranged mainly south of the nest. On 15 July, however, we found a subadult female Ornate Hawk-eagle consuming his radio-tagged carcass 225 m north of his natal nest, while another subadult male perched with a full crop 100 m away. It appeared that these two subadults had formed a pair bond, and we believe they killed the radio-tagged yearling. The male of this pair was in nearly full adult plumage and believed to be about three years old, while the female was beginning her molt into adult plumage and thought to be about two years old. During previous weeks, we had detected this duo more than once in this general area.

Several instances of cannibalism are known in other rain forest raptors, for example, the Crowned Hawk-eagle *(Stephanoaetus coronatus)* of Africa (J. Skorupa, pers. comm.). Cannibalism among tropical forest raptors may be evidence of saturated habitat in which it is difficult to find a breeding space unoccupied by conspecifics. Similarly, while cannibalism was not cited, fatal territorial battles among conspecifics were the leading source

of adult mortality among Golden Eagles *(Aquila chrysaetos)* in the Swiss Alps (Haller 1996).

While some biologists emphasize the opportunistic nature of raptor predation, and tend to describe many raptors as dietary generalists, the diet of the Ornate Hawk-eagle presents a serious challenge to any who would describe this species's prey selection as "generalized." Very few records exist of Ornate Hawk-eagles taking reptiles, and among 325 prey items identified at Tikal, we did not record a single reptile. While no doubt these are taken at times, it seems clear that the Ornate Hawk-eagles that we studied at Tikal must have passed up many opportunities to prey on large lizards and snakes. Whatever the basis for such dietary selectivity (e.g., interspecific competition leading to dietary specialization, or some other basis), this nearly total avoidance of reptiles in the diet appears to be strong evidence for dietary selectivity. Moreover, while the Crested Eagle and Black Hawk-eagle at Tikal (see Chapters 13, 14) took mainly nocturnal mammals, capturing them throughout the day, the mammalian portion of the Ornate Hawk-eagle's diet was predominantly day-active mammals, mainly squirrels. This implies different foraging methods among these three large raptors, with the Ornate Hawk-eagle emphasizing stealthy search, stalking, and ambush of active mammals, rather than opportunistic searching of daytime retreats for nocturnal mammals. Indeed, many observations indicated the importance of stealth in the hunting style of these hawk-eagles. They often appeared to engage in prolonged stalking, or prolonged periods waiting near prospective quarry for favorable conditions, before launching an attack.

Ornate Hawk-eagles sometimes took very large prey—sometimes too large for them to carry in flight. In several cases, they consumed such large prey items on the forest floor, apparently at the capture site, over a period of days during which the hawk-eagle remained perched near the carcass. Whether they defended such carcasses from other predators or scavengers is not known.

Keel-billed Toucans often mobbed Ornate Hawk-eagles and, in turn, were often eaten by the hawk-eagles. Whether these hawk-eagles take advantage of the toucans' mobbing behavior, capturing them in the act, remains to be learned.

For further information on this species in other portions of its range, refer to Burton 2006, Canuto 2008, Carlos and Girão 2006, ffrench 2004, Giudice 2007, Greeney et al. 2004, and Naveda-Rodríguez 2007.

Reproductive histories of Ornate Hawk-eagle territories studied in Tikal National Park

Year	Arroyo Negro	Balastrera	Brecha Norte	Chikintikal	Guardiania	Naranjal	Punta de Pista	Temple IV/V	Tintal
1989	—	—	—	—	—	—	—	—	Egg laid about 26 Apr; male (Toky) fledged 28 Aug
1990	—	—	Fledged female (Roca) by 2 Jul; very approx. dates = laying 25–27 Feb, hatching 7–9 Apr	—	—	Egg laid 27 May, hatched 8 Jul; male (Toky II) fledged 8 Oct; adult female (Esperanza) banded	—	Egg laid approx. early Jan, hatched mid-Feb; female (Ixquic) fledged 12 May; adult female (Maya) banded in May	No nesting; Toky eaten 15 Jul by duo of immature conspecifics
1991	Nest discovered 5 Sep, w/ juvenile female (Higinia) approx. 4 months old; egg laid about mid-Mar; fledged 1; adult pair unbanded	—	Roca dispersed in Mar; copulation seen in Mar but no nesting until Dec; egg laid about 18 Dec, failed before hatching	Temple IV female (Maya) moves here; probably no egg laid, or else failed; copulation seen late Oct	—	Toky II last located 13 Jun, probably dispersed; no nesting	Egg laid 5–6 Mar, hatched 21–22 Apr; female (Tonita) fledged by 27 Jun; adult female (Centella) banded	Ixquic hunting some by 20 Mar, probably reached independence at that time; new adult female (Invasora) nests 700 m from Maya's 1990 nest, lays egg 5–6 Nov, hatches 16–26 Dec; chick dies 2–5 Mar 1992 (before fledging)	Copulation seen early Mar, egg laid 19 Mar, hatched 6 May; chick fell to death 6 Jun. Adult male dies during handling in Jun
1992	Higinia dispersed 16 Jul. New female (Invasora) observed copulating late Nov; subadult male (Posho) caught here 31 Jul 1992, his origin unknown	Nested, young fledgling first seen 21 Sep; later fate unknown	Copulation seen late Feb, egg laid 2–9 Mar, failed 16–20 Apr, prior to hatching; no further nesting attempt	Egg laid 18–19 Mar, hatched 6–8 May; female (Sac-Nicté) fledged early Jul	Nest found 2 Jun, w/ 3-month-old nestling; fledging success uncertain	Copulation seen in Feb, Mar; egg laid 4–5 Apr, hatched 19–21 May; male (Condorito) fledged 30 Jul	No nesting; Tonita dispersed 30 Jun, monitored until Apr 1994	No nesting; Maya nesting at Chikintikal, Invasora has moved to Arroyo Negro; this territory now unoccupied?	No further activity ever seen here

APPENDIX 15.1—cont.

Year	Arroyo Negro	Balastrera	Brecha Norte	Chikintikal	Guardiania	Naranjal	Punta de Pista	Temple IV/V	Tintal
1993	Invasora laid egg about 1 Jan; hatched 16–18 Feb; juvenile female (Nancy) fledged 12–18 May	Egg laid before 1 Jun, hatched 23–27 Jun; chick disappeared 7–22 Jul (before fledging)	Egg laid early Jan, hatched 16–18 Feb; chick died by 18 Mar, before fledging; re-nested, egg laid 20–25 Apr, hatched 8 Jun; male (Percy) fledged 11–19 Aug	Last contact with Sac-Nicté 9 Jun; probably reached independence then; copulation seen Oct–Dec	Apparently no nesting but few data collected	Condorito last located Apr/May, almost certainly caught and eaten by xatero (palm collector); nest reconstruction by 2 Dec	Alternate nest found 6 Jan, egg laid 15–19 Jan, hatched 4 Mar; male (Rudi) fledged 25–28 May; apparently died between 21 Oct and 29 Dec; copulation seen 29 Dec	No nesting; territory unoccupied?	—
1994	Nancy dispersed 13–30 Oct	Egg laid before 17 Feb (alternate nest), hatched 29–31 Mar; female (Pati) fledged 2–13 Jun	Percy last monitored 20 Mar; no nesting activity	Copulation seen in Feb; no nesting seen, but may have moved to an unknown site	Chick hatched about 7 Mar; failed by 20 Apr, before fledging; no further nesting	Pre-nesting behavior but ended, perhaps due to conflict with Great Black Hawks nesting nearby; no nesting in known site; nest has fallen	Copulation seen early Jan; egg laid 12–18 Jan, hatched 7–14 Mar; female (Silvia) fledged 24 May	No nesting; territory unoccupied?	—
1995	No nesting	Probably no nesting, Pati believed still dependent	Egg laid mid-Apr, survived to fledging age, probably fledged	No nesting at known sites; Maya's movements suggest she did not nest at all	Egg laid early Apr; nest tree fell over by 22 May, before hatch; no further nesting seen	No nesting in historic site; none detected elsewhere	Silvia believed still alive, may have reached independence; egg laid, probably in late Oct (see 1996)	—	—
1996	Nest rebuilt by 8 May, no further activity; no nesting in historic nests, probably none at all	Egg laid around late Feb, hatched around early Apr; female (Gayita) fledged	1995 fledgling believed still alive; no nesting activity	No nesting detected; probably no nesting at all	—	No nesting in historic site; none detected elsewhere; status as an active territory questionable in 1995 and 1996	Nest found in late May, w/ 5–6 month-old fledgling (see 1995)	—	—

Note: — = an absence of data.

APPENDIX 15.2

Breeding chronology of Ornate Hawk-eagles in Tikal National Park

Territory	Nest number	Date found[a]	Date eggs laid	Date hatched	Fledging date	Fledgling last seen[b]
Brecha Norte	1	19 Jun 1990	est. 25–27 Feb 1990[c]	est. 7–9 Apr 1990[c]	ca. 2 Jul 1990	19 Mar 1991[d]
	2	10 Oct 1991	17–20 Dec 1991	no[e]	na	na
	2	same	2–9 Mar 1992	no[e]	na	na
	2	4 Jan 1993	14–15 Jan	16–18 Feb 1993	na[d]	na
	1	7 Apr 1993	20–25 Apr 1993	8 Jun 1993	11–19 Aug 1993	20 Mar 1994[d]
	1	18 Apr 1995	11–18 Apr 1995	24–29 May 1995[c]	probably, date?	?
Temple IV	1	8 Apr 1990	est. 6–7 Jan 1990[c]	est. 15–17 Feb 1990[c]	12 May 1990	20 Mar 1991[d]
Temple V	1	25 Oct 1991	5–6 Nov 1991	16–26 Dec 1991	no[7]	na
Chikintikal	1	23 Oct 1991	17–19 Mar 1992	6–8 May 1992	15 Jul 1992	9 Jun 1993[d]
	2	28 Oct 1993	no[g]	na	na	na
Punta de Pista	1	30 Jan 1991	5–6 Mar 1991	21–23 Apr 1991	27 Jun 1991[h]	30 Jun 1992[i]
	2	23 Dec 1992	15–19 Jan 1993	4 Mar 1993	25–28 May 1993	Dec 1994[l]
	2	21 Dec 1993	11–18 Jan 1994	7–14 Mar 1994	24 May 1994	21 Jul 1994[d]
Naranjal	1	24 Apr 1990	27–28 May 1990	7–8 Jul 1990	8 Oct 1990	13 Jun 1991[d]
	1	1 Feb 1992	31 Mar–5 Apr 1992	19–21 May 1992	30 Jul 1992	6 Apr 1993[i]
	1	21 Dec 1993	no[g]	na	na	na
La Paila	1	20 May 992	no[k]	na	na	na
	2	13 Apr 1994	no[k]	na	na	?[l]
Guardiania	1	2 Jun 1992	est. mid-Jan 1992[k]	est. 1 Mar 1992[k]	> 2 Jun 1994	na
	1	17 Mar 994	est. 21 Jan 1993[k]	est. 7 Mar 1994[k]	no[f]	na
	1	30 Mar 995	30 Mar–11 Apr 1995	no[e]	na	na
Arroyo Negro	1	5 Sep 1991	est. late Mar 1991[k]	est. 5 May 1991[k]	< 5 Sep 1991	31 Jul 1992[i]
	1	23 Nov 992	23 Dec–14 Jan 1993	16–18 Feb 1993	12–18 May 1993	30 Oct 1994[d]
Tintal	1	5 Apr 1989	24–29 Apr 1989	3–6 Jun 1989	26–28 Aug 1989	15 Jul 1990[i, m]
	1	7 Mar 1991	14–20 Mar 1991	4–6 May 1991	no[f]	na
Balastrera	1	21 Sep 1993	?	?	< 21 Sep 1992	28 Jan 1993[d]
	1	1 Jun 1993	est. 10 May 1993[k]	26 Jun 1993	no[f]	na
	2	17 Feb 1994	est. 11–13 Feb 1994[k]	29–31 Mar 1994	2–13 Jun 1994	4 Apr 1995[d]
	1	7 May 1996	est. late Feb 1996[k]	est. 7 Apr 1996[k]	yes, date?	?

Note: na = not applicable.
[a] Date this nesting effort was detected.
[b] Date last seen while still dependent on adults.
[c] Date estimated based on length of incubation or nestling period.
[d] Fate unknown.
[e] Nest failed during incubation period.
[f] Nest failed during nestling period.
[g] Pair built nest but did not lay eggs.
[h] Accidental "fledging."
[i] Dispersed from natal area.
[j] Died prior to dispersal.
[k] Nest found but not occupied; possibly a failed nesting attempt from 1993.
[l] Last checked soon after fledging.
[m] Eaten by duo of subadult conspecifics.

231

APPENDIX 15.3

Achievement of Independence by Three Juvenile Ornate Hawk-Eagles

The female Higinia was estimated to be four months old when discovered at the Arroyo Negro nest on 5 September 1991. Monitored extensively by radiotelemetry and visually, she reached independence and dispersed 10.5 months later, in late July 1992, at the age of about 14.5 months. During the interval prior to dispersal, we made 379 hours of observations near this nest. The first definite hunts that we observed by this bird were attacks on a squirrel and a bird on 21 and 28 April 1992, when she was an estimated 12 months old. We believe she had caught a bird we observed her eating on the latter date. A possible hunt was also seen 3 weeks earlier. We first observed her soar above the forest on 21 and 23 April at roughly 11.7 months, and on 27 May (approximately 13 months old) she soared extensively over the nest area. At this time we had trouble locating her radio signal, at least in part due to more frequent and extensive movements away from her nest. We last located her on 16 July, at an estimated 14.5 months old, and we believe she reached independence and dispersed during this period. During the last month of contact, we made several observations in which it appeared she was hunting birds.

The female Nancy hatched at the Arroyo Negro nest on 16–18 February 1993, and fledged between 12 and 18 May 1993, at the age of 86–91 days. From the time of her fledging until she dispersed beyond radio range at 20.4 months, we made 79 telemetry relocations, nearly always achieving visual contact. We did not monitor this bird frequently during her first year, but we documented her ranging from the nest to 200 m distant. From the age of one year until she dispersed, we monitored her frequently; she often returned to the nest vicinity, but she also spent a great deal of time 200–700 m from the nest and was once noted 900 m from the nest. We first noted what we believed to be hunting behavior on 30 May 1994 (at the age of 15.5 months) and again on 15 and 27 June. By late June 1994, when this female was 16.2 months old, she was returning to the nest vicinity less frequently, and we believed it likely she was capturing some of her own food. Her plumage at this time was beginning to resemble that of an adult, with chestnut feathers predominating in her face and neck. During this period she began to frequent a certain area 400–500 m from the nest, and she subsequently dispersed out of radio range in this direction during late October 1994, at the age of 20.0–20.4 months.

The female Tonita hatched 21–23 April 1991 at the Punta de Pista nest, and she first flew outside the nest tree on 27 June at the age of 66 days, when she accidentally fell and fluttered to a nearby tree. She returned to the nest tree and remained there on 1 and 2 July. Not until 3 July did we see her fly between the nest tree and several nearby trees. It appears, therefore, that a more legitimate fledging age for her is 72 days. On 1 May 1992 (age 12.25 months) we believed she was stalking a Crested Guan that we inadvertently flushed 12 m from where she was perched and 2 km from the nest. Three days later we observed her twice attacking dead limbs in what appeared to be the same sort of practice attacks we observed in nestlings prior to them leaving the nest tree. She reached independence and dispersed two months later, at the age of 14.25 months.

Chapter 15 Notes

1. t test; $P = 0.001$, t = 3.484, DF = 147, $n = 254$.
2. Mann-Whitney $U = 872$; $P = 0.25$; chi-square approximation = 1.311, df = 1, $n = 254$; this test assumes two-thirds of unidentified squirrels were *S. yucatanensis* and one-third were *S. deppei*, the same ratio observed in identified squirrels.
3. All were made on separate days except in one case when two contacts were made 2.5 hours apart.
4. Mann-Whitney $U = 61.0$, chi-square approximation = 0.692, df = 1, $P = 0.41$.
5. For example, the Brecha Norte pair laid an egg in December 1991, but this nest failed in January 1992. A second egg was laid in March, but the nest failed again in April. This pair did not lay again to our knowledge until January 1993. This egg hatched in February, but the nestling perished in mid-March. Another egg was then laid in April, resulting in a fledged chick in August. At the Guardiania territory, a nestling perished in April 1994, and this pair was not observed laying again until March or April of 1995. At the Tintal territory, a dependent juvenile was killed in July 1990 prior to reaching independence. This pair waited about eight months before laying again in March 1991. Similarly, at the Balastrera territory, a nestling perished in June 1993, and another egg was not laid until the following February.
6. A female taken in the Yucatán Peninsula on 10 May was incubating (Paynter 1955). At a nest in Panama, adults were visible around the nest during the dry season (from the end of December onward), with a nestling visible at the onset of the rainy season in late April and May, and fledging around the end of May, early in the rainy season (Brown and Amadon 1968). Another Panama nest held a nearly grown young in early October (Willis and Eisenmann 1979), and Russell (1964) cited a Belizian nest found in April that had been occupied only a few days earlier by a pair of Great Black-hawks *(Buteogallus urubitinga)*. Stiles et al. (1989) mentioned a nest containing one well-grown young in Costa Rica during April and May. In Oaxaca, Mexico, Binford (1989) reported a nest with one young in mid-June, and a female in Chiapas, Mexico, spent much time at a nest in April (Brown and Amadon 1968).
7. While most juveniles dispersed and reached independence while still in recognizably juvenile plumage, one female (Ixquic, fledged in 1990 from Temple IV/V territory) was in nearly adult plumage when she reached independence; small areas of white remained on her neck.
8. For example, the Brecha Norte female, who was banded in 1990, remained in this territory yearly through 1995, making six nesting attempts at two nest sites. Similarly, the Punta de Pista female, banded in 1991, remained on her territory through 1994, making three nesting attempts. Both adults at the Naranjal territory, banded in 1990, remained on this territory at least through spring of 1994, nesting there twice.
9. From early during nest reconstruction to fledging date.
10. Forty-two relocations during an 11.5 month period in which a fledgling remained dependent.
11. The cumulative area/observation curve for this male was still climbing with additional data points.
12. This inter-nest distance yields an MPND (see Chapter 2) estimate of one pair per 7.97 km^2 and a square estimate of one pair per 8.76 km^2.
13. This inter-nest distance gives an MPND estimate of one pair per 8.83 km^2 and a square estimate of one pair per 9.71 km^2.

14. The two inter-nest distances given above yield MPND estimates of 8.8 and 8.0 km² per pair, and the square method yields 9.7 and 8.8 km² per pair. While these estimates may reflect local spacing in areas of high density, we consider the 10–12 km² derived by the Polygon Method to be more conservative and defensible.

15. The La Paila territory is omitted because, in four years of monitoring, we never witnessed an egg laid there.

16. And in 16 cases, the nest was under frequent observation from before the egg was laid; in 8 other cases, the nesting event was discovered when there was already a nestling or fledgling.

17. We lost radio contact with several of these through radio failure or dispersal.

18. The adult female banded as a breeder in 1990 at the Brecha Norte nest remained alive and on this territory at least through 1996. The adult female banded at the Temple IV nest in 1990 moved to the Chikintikal territory in 1992, where she remained alive at least through 1996. A breeding female banded as an adult in 1991 at the Temple V territory moved to Arroyo Negro in 1993 but was not definitely identified thereafter. The female banded as a breeder at the Punta de Pista nest in 1991 was present on this territory at least to 1994. The female banded as a breeder at the Naranjal nest in 1990 remained on this territory at least to 1994. The adult male banded as a breeder at Naranjal in 1990 was present at least to 1994, and the breeding male banded at Arroyo Negro in 1993 was present at least to 1995.

16 BARRED FOREST FALCON

Russell K. Thorstrom

More often heard than seen, the Barred Forest Falcon *(Micrastur ruficollis)* is an elusive dweller beneath the forest canopy. Once the typical call is recognized, however, this species has proved ubiquitous in tall mature forest at Tikal and neighboring areas. These birds are most active in the very early morning, when the call, reminiscent of a miniature dog barking in the forest, is given repeatedly, often by both pair members, beginning well before dawn. Barred Forest Falcons call at this time from a high tree perch, later descending to the sub-canopy and understory, where they spend most of their time. Except during their early morning and late evening calling, these birds are secretive and difficult to detect. They are generally silent and relatively inactive during the afternoon, when they often hunt and rest on low perches beneath the canopy.

Forest falcons belong to the genus *Micrastur*, containing at least seven species occurring in Neotropical forests. Members of this genus are rather accipiter-like in proportions, being short winged, long tailed, and long legged, and have a slight, harrier-like facial disc—a combination of traits unique to this group (Plates 16.1, 16.2). The short wings and long tail are believed to enhance maneuverability within dense vegetation, while the facial ruff presumably lends auditory aid in localizing prey, as do similar structures in owls and harriers. Such an auditory aid may be valuable in the dim light of the forest understory—especially given the crepuscular habits of these ghostly forest dwellers. Forest falcons are also rather accipiter-like in behavior, sitting motionless for periods and then suddenly dashing out to capture prey. Sometimes they can be seen moving from perch to perch while hunting, or perching motionless at the periphery of an army ant swarm (Plate 16.3), awaiting an opportunity to seize an insect or lizard fleeing the ants. That new species continue to be discovered and described (e.g., Schwartz 1972; Whittaker 2002) attests to the secretive behavior of these birds.

GEOGRAPHIC DISTRIBUTION AND SYSTEMATICS

The Barred Forest Falcon ranges from Mexico's northernmost humid tropics to northern Argentina, Para-

guay, and Bolivia, and eastward through Brazil and the Guianas (Brown and Amadon 1968; Bierregaard 1994). This species occurs from lowland forests up to subtropical and montane forests at 2500 m and occasionally higher (Bierregaard 1994). Six subspecies are normally recognized, varying in size, amount of barring, and coloration, which is also highly variable within subspecies, some showing both rufous and gray morphs (Bierregaard 1994; Ferguson-Lees and Christie 2001). The subspecies we studied at Tikal, *M. r. guerilla*, is the northernmost now widely recognized, ranging from Mexico to Nicaragua (Bierregaard 1994). Some authors, however, continue to recognize *M. r. oaxacae* as a distinct form in the Sierra Madre del Sur of Oaxaca and Guerrero, Mexico—an area of high biotic endemism (Howell and Webb 1995).

MORPHOLOGY

The sexes were difficult to distinguish in the field, and we normally relied on behavior for this purpose. In the hand, males at Tikal were more grayish overall than females and had fine, evenly spaced gray barring on a pale grayish white breast and abdomen. Females were browner overall, with a brown-sepia back, nape, throat, and neck, and ventral barring wider than in males. Beaks were blackish toward the tip and sooty black at the base, while the cere and sides of the gape were orange-yellow. The iris was dark brown in immatures, changing to smoky gray or buffy yellow in adults. The head and upper parts of adults were variable in color, tinged with gray or rufous, and underparts were heavily barred with dark brown or gray. Tails were dark blackish gray with three or four narrow white bars. Birds in juvenal plumage were more brownish than adults and varied from white to cinnamon on the underparts, with brown ventral barring or almost no barring; unbarred juveniles resembled adult Collared Forest Falcons *(M. semitorquatus)* in appearance. On the breast and abdomen, juvenile females had thick, evenly spaced brown barring on a light brown ground color, whereas juvenile males had narrower dark barring (Plate 16.4).

Males at Tikal averaged 168 g (n = 13), and females 238 g (n = 17; Appendix 1). Males thus weighed about 10 g less than a female Sharp-shinned Hawk *(Accipiter striatus)*, and females about 60 g more. The Dimorphism Index for (cube root) body mass had a value of 11.4. Based on 18 adult females and 13 adult males at Tikal, linear dimensions (Appendix 2) yielded the following dimorphism values: 6.6 for wing length, 6.1 for tail length, 9.3 for culmen, and 8.1 for hallux length. Females were significantly larger than males in mass and in length of beak, hallux, wing, tail, middle toe, and foot span (t-test, all $P < 0.05$). The mean dimorphism value for mass, culmen, and wing chord was 9.1, slightly greater than that for the Northern Goshawk *(Accipiter gentilis*: Snyder and Wiley 1976). Thus, these raptors are moderately size dimorphic. Additional measurements are given by Friedmann (1950), Brown and Amadon (1968), Blake (1977), and Bierregaard (1978).

PREVIOUS RESEARCH

No detailed studies had been conducted on this species prior to our work, and the nest, eggs, and young went undescribed until we found the first nests in 1988 (Thorstrom et al. 1990). Most published information is in the form of brief notes describing hunting behavior (Smith 1969; Willis 1976; Trail 1987; Rappole et al. 1989), association with ant swarms (Willis et al. 1983; Mays 1985), and recognition of Lined Forest Falcon *(M. gilvicollis)* and Cryptic Forest Falcon *(M. mintoni)* as distinct species (Schwartz 1972; Whittaker 2002).

RESEARCH IN THE MAYA FOREST

We studied Barred Forest Falcons in Tikal National Park from 1988 to 1996, monitoring 20 territories. We searched the forest daily from March to July to document activity of potential breeding pairs. We focused our efforts during the early morning and evening hours of greatest activity, often beginning well before first light. We followed pairs by sight and sound until a nest was confirmed by the birds' behavior. When possible, we climbed to nest cavities to document eggs or young, sometimes using a flashlight (Plate 16.5). Observations at nests totaled more than 1800 hours, mostly during 1989–92, with incidental observations from 1993 to 1995 and during four weeks in 1996.

For each nest, we determined laying date, clutch size, hatching success (by periodic checks of nest cavities), and fledging success. We considered the incubation period to extend from the laying of the penultimate egg (usually the second) to hatching of the same. We measured eggs to the nearest 0.1 mm with Vernier calipers and egg and nestling mass to the nearest gram with a 100 g Pesola spring scale. We marked nestlings with food coloring on the crown soon after hatching to allow individual recognition. We measured the length of the tarsus, culmen (front of cere to beak tip), middle toe, and

foot span with dial calipers to the nearest 0.1 mm, and tail and wing chord to the nearest millimeter, using a wing ruler.

We placed radio transmitters (216 kHz, Holohil Systems) on several fledglings and adults to document home range sizes and natal dispersal. We used 1.2 g leg-mounted transmitters on two fledglings, 3.1–4.0 g tail mounts on two fledglings and 15 adults, and 6 g backpack mounts on two fledglings and five adults. We conducted radiotelemetry on foot, generally narrowing a bird's location down to one or two trees, then tying the location to known features via a pace and compass map or occasionally by GPS. Names of forest types follow Schulze and Whitacre (1999), as summarized in Chapter 2. Unless stated, measurements given are means ± 1 SD.

DIET AND HUNTING BEHAVIOR

Barred Forest Falcons at Tikal used several hunting techniques, which we categorized as (1) still-hunting, (2) active perch hunting, (3) "flush" and (4) "scratch" hunting, (5) nest robbing, and (6) ant swarm still-hunting. At times several of these hunting styles were observed during a given foraging trip. During our radio-tracking, the forest falcons hunted mainly within mature Upland Forest, usually less than 10 m from the ground, but occasionally at higher levels in the forest. Owing to the secretive nature and wariness of these birds, we could not quantify the relative frequency with which they used the hunting methods just listed. Rather, hunting observations resulted from opportunistic sightings while radio-tracking.

In still-hunting, the forest falcons scanned the surrounding area from a perch for many minutes, dashing out to attack prey via direct flight or a shallowly descending attack. We observed forest falcons using this technique to catch lizards *(Anolis* spp.), as well as birds at ground-level water feeders in the park. Active perch hunting was the most frequently observed hunting style and was often used by males provisioning females and nestlings. In this hunting mode, the forest falcons changed perches frequently, scanning for several minutes from each perch before moving on. Forest falcons using this hunting style were extremely active, never remaining on a single perch for long. The average time interval from the moment we spotted a perched, hunting male during radio-tracking to the time he changed perches was 5.7 ± 0.8 (SE) minutes (n = 37, range 1–23 minutes), and males often covered a large area before capturing prey. We observed forest falcons catching lizards *(Ameiva* and *Anolis* spp.) and birds in this manner. *Ameiva* lizards were captured on the ground, and *Anolis* species were plucked from vegetation or the ground.

In flush hunting, the forest falcons used their feet to cling to and shake branches and leaves, apparently attempting to flush prey. Twice we observed forest falcons catching *Anolis* lizards in this fashion. In scratch hunting, the forest falcons used their feet, chicken-like, to move forest-floor leaf litter, in order to dislodge or flush prey hidden there—mainly arthropods and lizards. To rob bird nests, the forest falcons presumably used their acute hearing to locate nestlings that were hidden within hanging nests. Twice we observed an adult male hanging from a pendant flycatcher nest and thrusting his foot through the entrance. Nestling passerines—naked or slightly feathered and unable to fly—were observed frequently as prey remains at forest falcon nests.

Occasionally these raptors were observed at army ant swarms, mainly of *Eciton* species. Sallying to the ground from perches normally less than 2 m up, the forest falcons captured insects and lizards fleeing the ants. From 18 February to 10 August 1989, four radio-tagged adult males were located at nine ant swarms, remaining there a total of 323 minutes (range = 4–169 minutes) during 1386 minutes of observation; this amounts to 23% of their time during these observation periods. Altogether, these adult males were found attending ant swarms 8.4% of the time during 273.2 hours of radio-tracking in 1989. During two observations at ant swarms, one male was observed catching two *Anolis* lizards, two beetles, one cockroach, and two unidentified arthropods (possibly spiders) in seven attempts. We did not see Barred Forest Falcons capture birds at any of the nine ant swarms attended by these raptors. Three fledglings 10 weeks old were located twice at ant swarms, spending 13% of their time (80/620 minutes during three observation periods) attending ant swarms. Two of the fledglings were observed catching a katydid and three unidentified insects, and we observed one unsuccessful attack on a bird. Forest falcons attending ant swarms sometimes (four of eight cases) were observed scratching in the leaf litter, apparently to flush fleeing arthropods and lizards that had taken shelter. At these times, the forest falcons' long legs seemed advantageous, as did their long tails, which they used as a balancing aid. Similar behavior of Barred Forest Falcons at ant swarms was reported by Willis et al. (1983).

Adults and fledglings attended ant swarms fairly frequently at Tikal, yet we did not observe this foraging method as often as one might expect based on its frequent mention in the literature (Brown and Amadon 1968; Willis et al. 1983; Bierregaard 1994). It seems possible that ant swarm attendance by these forest falcons may be somewhat overrepresented in the literature relative to its true occurrence, simply because these cryptic raptors are not often observed except at special hunting opportunities such as those provided by ant swarms, which are famous for the spectacular birding opportunities they provide.

Prey types most often mentioned for Barred Forest Falcons include lizards, birds, small mammals, and large insects (e.g., Brown and Amadon 1968). From South American specimens, Schwartz (1972) reported stomach contents including nestling birds, a crab, a mouse opossum (*Marmosa* sp.), a large beetle, and various arthropods. In Peru, Robinson (1994) gave records of Barred Forest Falcons capturing two *Ameiva* lizards, attending army ant swarms twice, attacking understory bird

flocks, and regularly attacking birds in mist nets. For four individuals, he reported stomach contents that included three small passerines, two small mammals, one beetle, miscellaneous insect parts, and a nematode head.

At Tikal, we observed 600 prey items delivered to females, nestlings, and fledglings from 1988 to 1991. Nearly one-third (195) of the items were unidentified, especially late in the nestling period, when forest falcons flew secretively into nest cavities without calling the mate out to receive prey. We consider it unlikely that the unidentified prey items were very different from those identified; hence we report prey composition based on the percentage of identified items. On a numerical basis, reptiles were the predominant prey, accounting for 249 of 405 identified items or 61.5% of the diet, followed by 89 birds (22.0% of the diet), 33 insects (8.1%), 24 mammals (5.9%), and 10 amphibians (2.5%; Thorstrom 2000).

The most detailed portrait of the diet was provided by 1989 and 1991 nest observations, during which we identified 267 of 380 items delivered. Of the 267 identified prey items, 64.0% (171) were lizards, 19.5% (52) were birds, 7.9% (21) arthropods, 4.1% (11) amphibians (frogs), 3.3% (9) mammals, and 1.1% (5) snakes (Table 16.1). Considering the diet in terms of biomass reduces the importance of lizards and increases that of birds and mammals; on this basis, lizards comprised 31.3% of the diet, birds 54.6%, mammals 9.7%, frogs 2.8%, snakes 1.0%, and insects 0.5% (Table 16.1). Lizards included 57 small *Anolis* species, 21 large *Anolis* species, 28 teiids (presumably all *Ameiva* spp.), 11 *Laemanctus* species, 5 *Corythophanes* species, and 49 unidentified lizards. The snakes included one coral snake or mimic. Arthropods consisted of eight cockroaches and 13 other items including spiders and beetles. Birds included five Blue-crowned Motmots (*Momotus momota*), two flycatchers (Tyrannidae), two Emerald Toucanets (*Aulacorhynchus prasinus*), and one each of the following: dove (*Leptotila* sp.), woodcreeper (*Dendrocincla* sp.), Spot-breasted Wren (*Thryothorus maculipectus*), and Clay-colored Robin (*Turdus grayi*). Avian prey ranged in size from an unidentified warbler (*Dendroica* sp.) at approximately 9 g and Spot-breasted Wren at about 15 g to the *Leptotila* dove at approximately 155–170 g. Mammals included seven rodents, one bat, and one unidentified mammal; the rodents were believed to include *Heteromys* and *Oryzomys* species (Table 16.1).

In 1989 and 1991, the same banded female was seen eating small fruits of the Tzol tree (*Blomia prisca*) on the ground during incubation, about one to two weeks after the egg-laying period (Thorstrom 1996a). This female ate the fruits for approximately five minutes, picking them off the ground with her beak.

Rates of prey deliveries to nests peaked during the fourth and fifth weeks after nestlings hatched, when food requirements of the broods presumably were maximal (Fig. 16.1). At this time, female forest falcons contributed to provisioning of broods and brought in many small songbirds, especially nestlings and fledglings. After the young forest falcons fledged, prey delivery rates

Table 16.1 Diet of Barred Forest Falcons at the nest during 1989 and 1991, Tikal National Park

Prey type	Prey		Biomass	
	Number	Percentage (%)	Gram	Percentage (%)
Insects	**21**	**7.9**	**31.5**	**0.5**
Blattaria (cockroaches)	8	3.0	12.0	0.2
Unidentified insects	13	4.9	19.5	0.33
Amphibians	**11**	**4.1**	**165.0**	**2.8**
Unidentified frogs	11	4.1	165.0	2.79
Lizards	**171**	**64.0**	**1848.1**	**31.3**
Anolis spp.	57	21.3	222.3	3.76
Anolis spp., large	21	7.8	289.8	4.9
Ameiva spp.	28	10.5	700.0	11.84
Laemanctus sp.	11	4.1	165.0	2.79
Corythophanes sp.	5	1.9	226.0	3.82
Unidentified lizards	49	18.4	245.0	4.15
Snakes	**3**	**1.1**	**60.0**	**1.0**
Birds	**52**	**19.5**	**3230**	**54.6**
Gray-fronted Dove (*Leptotila rufaxilla*)	1	0.4	160.0	2.71
Blue-crowned Motmot (*Momotus momota*)	5	1.9	665.0	11.25
Emerald Toucanet (*Aulacorhynchus prasinus*)	2	0.7	300.0	5.08
Dendrocolaptidae	1	0.4	35.0	0.59
Tyrannidae	2	0.7	30.0	0.51
Troglodytidae	1	0.4	15.0	0.25
Muscicapidae	1	0.4	75.0	1.27
Unidentified birds	39	14.6	1950.0	32.99
Mammals	**9**	**3.3**	**576**	**9.7**
Unidentified mouse	1	0.4	50.0	0.85
Unidentified rat	6	2.2	456.0	7.72
Unidentified bat	1	0.4	20.0	0.34
Unidentified mammal	1	0.4	50.0	0.85
Total number of identified prey	**267**	**1.0**	**5910.6**	**99.9**
Unidentified prey	**113**			
Total	**380**			

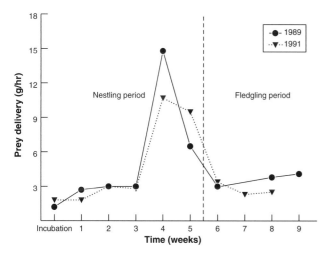

16.1 Rates of prey delivery to Barred Forest Falcon nests at Tikal, Guatemala, during two years of study, in grams per hour. Based on five nests in 1989 and one nest in 1991.

decreased to a level similar to that during the incubation period (Fig. 16.1). The proportional contribution of lizards and birds to the diet at the nest changed through the forest falcons' nesting period (based on 1989–91 data: chi-square test; $\chi^2 = 11.8$, 3 df, $P < 0.01$). The frequency

and biomass of lizards in the diet peaked during incubation, making up 72 (74.2%) of 97 identified prey items, and 66.4% of biomass, and decreased thereafter. In contrast, few birds were brought in during the falcons' incubation period, and birds contributed most during the nestling period, making up 25 (28.7%) of 87 identified prey items and 47.7% of prey biomass at this time.

HABITAT USE

Barred Forest Falcons at Tikal used mainly mature Upland Forest. Their territories, however, often contained several patches of Bajo and Transitional Forest types within a matrix of Upland Forest. Radiotelemetry and associated sightings showed that the forest falcons visited these different forest types but spent most of their time in Upland Forest. Nesting occurred nearly exclusively in Upland Forest; only 1 (1.4%) of 70 nests was in Transitional Forest, and none were in second-growth or modified forest. We rarely observed Barred Forest Falcons in the slash-and-burn farming landscape near the park. It was our impression that in such mosaic habitats, these forest falcons are restricted mainly to the interior of mature forest remnants. We detected Barred Forest

Falcons significantly less often in the farming landscape than in Tikal's mature forest when listening for their calls during predawn point counts (Whitacre et al. 1992a, 1992b).

We sampled vegetation surrounding six Barred Forest Falcon nests and five Collared Forest Falcon nests, allowing us to compare nesting habitat of the two species; results are given in Chapter 17. In brief, Collared Forest Falcons nested mainly in Transitional or Hillbase Forest types with dense understory, whereas Barred Forest Falcons nested almost exclusively in tall Upland Forest with closed canopy and fairly open understory (Enquist et al. 1992). Barred Forest Falcons also nested in areas of taller forest canopy than did Collared Forest Falcons (see Table 17.3). In general, forest in the vicinity of Barred Forest Falcon nests was characterized by abundant aroids, dense canopy cover, few vines, sloping topography, and canopy height of 17–32 m (Enquist et al. 1992; see Table 17.3).

BREEDING BIOLOGY

Nests and Nest Sites

As in most falconids, Barred Forest Falcons did not build a nest structure, nor did they bring greenery to nests. All nests were in tree cavities and had a substrate of decayed wood. Barred Forest Falcons are "secondary" cavity nesters, using naturally occurring cavities or those excavated by other species (Plates 16.6, 16.7). Ninety-four percent (66 of 70) of nesting attempts were in naturally occurring cavities, nearly all of which appeared to have developed where a limb had broken off or the tree's heartwood decayed. Only four nesting attempts were in holes apparently excavated by woodpeckers. Sixty-six nesting attempts (94%) were in cavities with only one entrance; the other four nestings (6%) were in a site with two entrances, used in four consecutive years.

Only large mature trees provided nesting cavities for Barred Forest Falcons. We documented 70 nesting attempts in 39 trees of 15 species. Species most often used were Spanish Cedar (*Cedrela odorata*), with 20 (28.6%) of 70 nestings, and Cantemó (*Acacia dolichostachia*) and Ramón Colorado (*Trophis racemosa*), with 6 nestings (8.6%) each. Nests in Spanish Cedar produced 23 (31%) of the 75 fledglings we documented. Nest cavities averaged 17.4 ± 4.2 m above ground (range = 10–30 m, n = 65), and 80.9 ± 58.8 cm in depth (range = 3–200 cm, n = 34). Nest trees averaged 94.8 ± 40.5 cm in diameter at breast height (range = 30–190 cm, n = 39) and were significantly larger in diameter than a random selection of forest trees at Tikal (Enquist et al. 1992). Though nests in tree cavities are clearly the norm for this species, Baker et al. (2000a) observed a pair nesting in a cliff cavity not far from Tikal. Among 70 Barred Forest Falcon nesting attempts, nesting success was significantly different between live and dead trees (MRPP chi-square = 8.10, P = 0.004). Most nests (33 of 51) in live trees were successful, and most (14 of 19) in dead trees failed.

Roost Trees

Using radiotelemetry, we located 27 trees used as night roosts by five males. Three species made up 89% (24) of the 27 roost trees: Ramón (Breadnut Tree, *Brosimum alicastrum*) 41%, Ramón Colorado 30%, and Spanish Cedar 19%. Roost trees were smaller on average than were nest trees, averaging 36.0 cm in diameter (range = 17–82 cm) (t-test, t = 3.9, df = 21, P < 0.001), compared to 95 cm for nest trees.

Egg and Clutch Size

Of 57 first clutches, 21 were of two eggs and 36 were of three eggs (mean = 2.6 eggs). Two replacement clutches were of two eggs, for an overall mean clutch size of 2.6 eggs. Mean egg dimensions were 43.8 ± 1.9 mm by 34.2 ± 1.1 mm (n = 30), and eggs averaged 28.0 ± 1.7 g in weight (n = 30), or 11.8% of mean female body mass. The modal clutch of three eggs was equivalent to 35.3% of mean adult female body mass, and the mean clutch of 2.6 eggs was equivalent to 30.6%. Hoyt's (1979) formula for predicting fresh egg mass gave precisely the same results.

Laying and hatching interval and sequence. We determined the interval between laying of consecutive eggs for two 2-egg and two 3-egg clutches (n = 6 inter-egg intervals). In five cases the subsequent egg was laid two days after the first, while in the sixth case the last egg of a 3-egg clutch was laid three days after the second. We could not determine when incubation began in relation to clutch completion, but we determined the hatching sequence for several nests. At two nests the first- and second-laid eggs hatched less than 24 hours apart. In one 2-egg nest, both eggs were pipping simultaneously. In two 3-egg clutches, first- and second-laid eggs hatched less than 24 hours apart, and the third eggs hatched less than 24 hours after the second.

Relaying after nest failure. We never witnessed a second nesting attempt following a successful nesting. In two cases, however, females laid a second clutch after losing the first during incubation. In one of these cases, incubation was underway on 8 May 1989, and by 20 May the nest had failed, the eggs apparently destroyed by a predator. On 31 May, the first egg of a 2-egg replacement clutch was laid in an alternate nest 300–400 m from the failed nest. This nest hatched two young on 9 July but failed owing to predation on 22 July, when most nests were fledging young. The other re-nesting occurred in 1990, with the second attempt occurring 150 m from the first. One egg of the replacement clutch hatched on 8 July, but the nest failed two weeks later when the nestling disappeared.

In two other cases, pairs whose nests failed during incubation investigated cavities thereafter but did not lay eggs. In the first of these pairs we observed cavity investigation one to three weeks after the loss of their first clutch. At a second territory, the male lost his mate and the clutch to predation at the end of May and acquired a new female within four weeks. During July and August we twice observed this male courtship-feeding the female. This pair was still investigating cavities as of 7 August, two months after our latest documented laying date.

Nesting Phenology

Barred Forest Falcons were quite synchronous in nesting, with most first clutches laid from mid-April to mid-May and the mean laying date consistently in the first few days of May—just prior to the average onset of the wet season (Fig. 16.2). Earliest recorded laying dates were 13 April 1988, 16 April 1989, 15 April 1990, 15 April 1991, and 22 April 1992. Mean laying date for first clutches was 2 May $\pm$ 2.1 days and did not vary across years ($F_{2,18}$ = 0.29, P > 0.10); 31 first clutches were laid between 13 April and 14 May. Most hatching occurred from mid-May to mid-June (20 May–17 June, n = 21 clutches)—a period which brackets the average onset of the rainy season. Peak hatching occurred during the first week of June (Fig. 16.2), and mean hatching date (6 June $\pm$ 2.5 days) did not vary among years ($F_{2,11}$ = 1.7; P > 0.10). Fledging dates spanned 40 days (25 June–3 August), with a mean fledging date of 15 July $\pm$ 10.4 days (n = 28), during a period of usually reduced rainfall during the rainy season (Fig. 16.2).

The breeding season of Barred Forest Falcons at Tikal began early in the dry season, during late February and early March, and extended through the beginning of the wet season to August and September (peak of the wet season). Barred Forest Falcons had nestlings and fledglings early during the rainy season, a time when many songbirds in this region nest (Skutch 1950), and likely a time of maximal availability of fledgling songbirds as prey. This is also a time of maximal insect abundance (see Chapter 2). The breeding cycle lasted nearly 20 weeks from courtship to the dispersal of young.

Length of Incubation, Nestling, and Post-fledging Dependency Periods

The incubation period from laying of the penultimate egg to hatching of the same was 33–35 days (n = 6 nests), and fledging occurred 35–44 days after hatching (mean = 38.3 $\pm$ 3.3 days, n = 13). Juveniles dispersed from their parents' territories four to seven weeks after fledging, presumably reaching independence at that time.

VOCALIZATIONS

Except when rain inhibited calling, these raptors appeared to call each morning throughout the year. Following the Collared Forest Falcon, the Barred Forest Falcon was usually the second common diurnal bird to begin calling in the predawn darkness. During the breeding season, calling began, on average, 22.9 $\pm$ 6.9 minutes before official sunrise (n = 48). The "normal" call ("dawn" call or "advertisement" call), very similar to the barking of a small dog, consists of a single, accented note resembling the word *ark*, repeated in a measured, deliberate cadence, in series that can last for 15 minutes or more. This call, given by both sexes, may be regarded as true song, as it seems to be associated with territorial defense and maintenance of the pair bond (Marler and Hamilton 1966).

During agonistic interactions between neighboring males along territorial boundaries, two other calls were heard. Often one bird voiced a rapid, swelling, "agonistic chitter," and the other responded with a rapid "clucking." A fourth call—a "mewing cluck" or "chicken-like cackle"—was given by females soliciting food from their mates. A final call, given by recent fledglings, was a repetitive, single-note call that sounded similar to the normal adult call but was weaker, lower in pitch, and with notes that were longer in duration.

During incubation and early in the nestling period, females rarely left the nest to call, nor did they call within the nest cavity except to quietly respond to the male's vocalizations. They began the normal dawn calling once they began to roost outside the nest cavity, when the chicks reached the age of two weeks and females ceased brooding them at night.

BEHAVIOR OF ADULTS

Pre-laying and Laying Period

We observed courtship activity from February to July. Typically, courtship interactions began when a male vocalized from a roost before sunrise, and a conspecific,

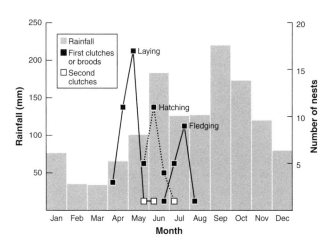

16.2 Breeding phenology of Barred Forest Falcons in Tikal National Park, in relation to rainfall seasonality.

usually the mate, responded. Normally, the male then moved toward a potential nest tree, calling and sometimes diving near the tree, rapidly repeating a loud dawn call. The female would then move to the vocalizing male, sometimes calling as she went. Usually the male would fly to the potential nest cavity, calling loudly. After the female arrived, the male would either fly in and out of the cavity in an apparent nest cavity display, or would wait nearby until she entered the cavity. During such cavity inspections, males remained in the cavity 2.5 ± 1.7 minutes on average (range = 1–7 minutes, $n = 11$).

If the female entered the cavity, the male would either move off to check another potential nest site and call the female over, or would fly off to hunt. Once the female appeared interested in a site, she would remain in the vicinity, guarding the cavity. Occasionally females chased away other cavity-nesting birds such as parrots, woodpeckers, and toucans. During 350 hours of observation during this period, the location of the female was known for 93 hours, of which 11.4 hours (12.3%) was spent inside the prospective nest cavity and 81.6 hours (87.7%) outside but within 50 m of the eventual nest cavity. On returning to the area, the male would voice soft barks. If the female was present, she would fly to the male, voicing the chicken-like cackle associated with food solicitation. If the male had food, the female would fly to him and take it. During such exchanges, females were aggressive and vocal.

After a prey exchange, the male was usually allowed to mate; 25 (33%) of 76 observed copulations occurred immediately after prey exchanges. The birds were silent during copulation, which lasted three to seven seconds (mean = 5.1 ± 1.5 seconds, $n = 76$). The earliest observed copulation attempt occurred on 21 February 1989—seven weeks prior to egg laying—when a female shook a male off her back after apparently receiving a dove from him 20 minutes earlier. The earliest successful copulation was noted on 28 February 1989, six weeks before egg laying. All observed copulations ($n = 76$) occurred between 0526 and 0944, averaging 0653. In one case we observed three copulations in one minute, and in another case eight during five hours, both occurring two weeks prior to egg laying. Occasionally, copulations occurred after the pair left their roosts early in the morning but before they reached the prospective nest tree (5 of 76 observed copulations).

After copulating, the male would move away from the nest, often to hunt, while the female ate and cleaned her beak and feet. She would then fly into the nest cavity, remaining within from one minute to three and a half hours as the laying date neared. Generally, such courtship interactions, along with voicing of the barking song, lasted until mid-morning, when members of the pair moved off in different directions.

By early April, as laying neared, females remained increasingly near nest sites, spending most of their time in "pre-laying lethargy" (Newton 1979), while males provisioned them. At this time, females remained within 25 m of the nest hole and, during 33 hours of nest observation, spent 51% of the time inside and 49% outside the nest cavity. Once incubation had begun, females repelled copulation attempts by knocking the males off with their wings or flying away.

Incubation Period

Only females incubated, while males hunted and provided nearly all food. After receiving a prey item from the male, females generally ate on a limb within 15–40 m of the nest tree, returning to the nest cavity soon after. During 123 prey deliveries during incubation, only seven times did males enter the nest cavity after transferring prey to the female; they spent an average of 8.3 ± 6.7 minutes within (range 1–20 minutes, $n = 7$). During the incubation phase, males were known to be in the nest vicinity for 29.1 hours, and two individuals spent a total of 58 minutes (3.3% of this time) inside the cavities during these seven incidents. While it is possible males settled briefly on eggs during these rare visits within the nest cavities, I believe it most likely they were mainly verifying the status of the eggs. Similar behavior is characteristic of North American accipiters (Snyder and Wiley 1976) and the European Sparrowhawk (*Accipiter nisus*: Newton 1979). Male falcons (genus *Falco*), in contrast, often play a role in incubation, at least covering the eggs while the female eats (Newton 1979). In 600 hours of nest observation during the incubation period, we knew the female's precise location during 381 hours. Of this 381 hours, females spent 91.3% of the time inside and 8.7% outside the nest cavity. During many periods outside the nest cavity, females perched in view at the cavity entrance, observing their surroundings. During the incubation period, females usually ate on a limb within 15–40 m of the nest tree.

Nestling Period

When an adult male arrived in the nest area, he would voice "soft barks," alerting the female of his arrival. The female would respond instantly by flying out of the nest. Usually, males transferred prey to the female 10–40 m from the nest, accompanied by the female's excited "chicken-like cackle." Occasionally, the male delivered food directly to the nest when the female did not leave the cavity or was not in the vicinity. Food exchanges were brief and the female returned directly to the nest, apparently to feed and brood the young. We did not observe males feeding young; they only dropped off food and quickly left.

Roles of adults in provisioning young. In 1989 and 1991 we determined the sex of forest falcons delivering 380 prey items to nests during the pre-laying, incubation, nestling, and fledgling periods. Adult males delivered (and captured) 93% (39) of 42 prey items during the pre-laying period, 98% (123) of 125 items during incubation, 81% (113) of 140 items during the nestling period, and

66% (48) of 73 items during the fledgling period. Females began to hunt and bring in prey during the second week of the nestling period and continued doing so into the fledgling phase. However, nearly all prey deliveries after the second week after fledging were by males. Taking the breeding season as a whole (pre-laying through the fledgling period), males delivered more prey items and biomass than did females; of 267 identified prey delivered to nests in 1989 and 1991, five males brought in 3819 g (75.7%) and five females delivered 1227 g (24.3%) of prey biomass.

Prey transfers, preparation, and caching. Prey caching by Barred Forest Falcons near the nest was rare; we observed only one instance, and possibly a second. Late one evening during the nestling period, a female was attempting to deliver a partially eaten Emerald Toucanet to the nestlings, but was interrupted by a Keel-billed Toucan (*Ramphastos sulfuratus*) that flew near as she approached the nest. Rather than enter the nest, the forest falcon flew to an understory *Cryosophila* palm 15 m from the nest and cached the bird in the center of the palm, 5 m above ground, then flew out of view.

Adults brought birds to the nest usually headless and sometimes plucked, whereas lizards were usually brought in whole. Prey transfers among adults were conducted foot to foot while perched on a limb. Fledglings also received prey foot to foot—on the ground during the first few weeks after fledging, then on tree limbs as they gained experience.

BEHAVIOR AND DEVELOPMENT OF NESTLINGS

Nestling Period

Recently hatched chicks had white natal down, closed eyes, shiny white nails, yellow-orange legs and cere, and short, pale yellow beaks that were deep and strongly laterally compressed (Plate 16.8). Young uttered a soft, high-pitched scream—*screee*—when we handled them. By the second and third days after hatching, eyes were open and nestlings held their heads up. They had brown-black iris and blue-black pupils.

By one week, nestlings had lost their egg tooth, and the tops of their feet were changing to a yellow-green. At 8–10 days, primaries were beginning to emerge from sheaths, and nestlings began displaying death-feigning behavior when we climbed to nests. Nestlings displayed this behavior even before we grasped them to remove them from the nest. They would sense our presence, lie down on their sides, and remain motionless, appearing dead to the observer at the nest entrance. When touched, they either remained still or moved abruptly. This appeared to be an anti-predation behavior.

Nestlings were able to partially sit on their tarsi by 10 days. By 13–15 days, they could stand on their tarsi and had lost the down from the facial area. The base of the culmen and the lower mandible were changing to a brownish black, facial skin was changing to an olive green, and the talons were turning gray. Adult females still roosted nightly in the nest cavity. During the day, females began making short hunting trips, spending up to an hour at a time away from the nest. By 16–18 days, the nestlings' primaries, secondaries, and upper primary coverts had grown rapidly, and tail feathers and contour feathers along the sides of the chest and back had emerged from their sheaths. At this age, the nestlings could stand on their feet.

By 20–22 days, tarsal and toe length (*n* = 5) did not differ significantly from those of adults (*n* = 25). The iris changed to a chocolate brown and facial skin to a dark olive green, while pupils remained blue-black. At this time, the adult forest falcons left intact prey items in the nest for the young to eat, and females spent more time than before hunting and resting away from the nest, and began to roost away from the nest as well. The adult male delivered food directly to the nest if the female did not respond to his call. By the fourth week after hatching (28–30 days) the nestlings were nearly covered with body feathers except on the head, back, and tibiotarsi.

Fledging Period

Young fledged, on average, at the age of 5.5 weeks (38.3± 3.3 vdays, range = 35–44 days, *n* = 13). At five weeks, young were fully feathered and near fledging, though their tails were less than half grown even at fledging. Young females at this time weighed less than adult females; one to two days before fledging, they averaged 202.6 ± 16.8 g (*n* = 10), whereas adult females averaged 219 ± 9.1 g (*n* = 4) (*t*-test, *t* = 2.18, 11 df, *P* = 0.025). In contrast, fledgling males were heavier than adult males; one to two days prior to fledging, young males weighed up to 188.0 g and averaged 180.0 ± 8.0 g (*n* = 7), while adult males averaged 166.0 ± 6.1 g (*n* = 7)—93% of the mean mass of fledgling males (*t*-test, *t* = –3.57, 9 df, *P* = 0.003). Fledgling males, however, were significantly smaller than adult males in most linear dimensions— exposed culmen (88.9%), hallux (87.9%), tail (46%), and wing chord (80%)—but not in foot span or length of middle toe. In females also, fledglings did not differ from adults in foot span and length of middle toe, but they were significantly smaller than adults in body weight (85.2%), culmen length (88.8%), hallux length (87.6%), tail length (44.0%), and wing chord (74.8%).

Three fledglings were weighed and measured immediately prior to fledging and again three to four weeks later; they had reached near-adult weights and measurements by the time of the second handling. Two males examined at 37 and 38 days and again at 66 and 68 days, respectively, both weighed 109% of mean adult male mass initially, and 102 and 103% on the second occasion. In wing chord they were 83 and 84% of adult values initially, and 98 and 99% on second capture; in tail length they were 55 and 44% of adult values initially and 96 and 95% later; and for culmen length, they were

96 and 97% of adult values initially and both 99% later. A female handled at 36 and 57 days weighed 97% of mean adult female weight initially and 94% later, and had 77% of adult wing chord initially and 93% later, 40% of adult tail length initially and 92% later, and 88% of adult culmen length initially and 95% later.

Post-fledging Dependency Period

From 1989 to 1995 we banded 57 nestlings. We fitted several fledglings with radio transmitters and observed their behavior and movements during the initial stages of the fledgling period. When first leaving the nest cavity, young fluttered from the cavity entrance to a nearby perch in the understory, often about 10 m above ground. No young returned to a nest cavity after fledging.

During the first week after fledging, young usually remained within 50 m of the nest tree. They generally remained apart and flew weakly or ran on the ground or along a fallen tree when changing perches. Adults arriving with food vocalized the normal "bark" call, and fledglings responded with the soft, rapid *screee* and bark calls. Adults delivered prey to fledglings on the ground or on a small vine or fallen trunk close to the ground. By two weeks after fledging, young tended to remain closer together and often chased one another on the ground and through the understory. Such play may help the young develop hunting skills. When an adult arrived with food, the fledglings flew to it, and the first to arrive often received the prey item. Young that did not receive food would watch, vocalizing *screee*, and wait for siblings to finish, then would eat any food remaining.

In three cases, we found fledgling forest falcons in association with army ant swarms three to four weeks after fledging. The young usually perched 1–2 m above the swarming ants. On several occasions the young forest falcons caught insects fleeing the ants (cockroaches, crickets, and beetles). Several times young were seen chasing small birds at ant swarms. By three weeks after fledging (ages of 63–72 days) young chased any small animal that moved on the ground and were observed catching small lizards (*Anolis* spp.). By four weeks, all fledglings that were trapped (*n* = 3) had hardened tail feathers and flew proficiently, darting through the lower canopy.

Natal dispersal. Fledglings slowly moved around in their natal areas week by week until dispersing from their parents' territories four to seven weeks after fledging. At one territory, two young were fed by the adult male up to six weeks after fledging, remaining in an area approximately 250 m south of the nest tree and vocalizing from morning roosts in a manner similar to adults, but more softly. After week 6, the radio-tagged fledgling could not be located; apparently this bird and its nestmate had dispersed. One fledgling male, at the age of about five to six weeks, moved 500 m from its nest and remained in that area for at least four months. This area

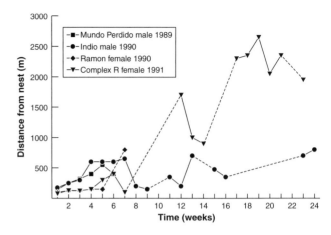

16.3 Distance from nests versus weeks since fledging for four radio-tagged Barred Forest Falcon fledglings at Tikal.

was between two breeding territories in dense Bajo or Scrub Swamp Forest, a forest type not favored by nesting Barred Forest Falcons. After four months the transmitter failed and this male could no longer be located; hence we do not know whether this was the extent of his dispersal movement. One young female dispersed seven weeks after fledging and remained 1.9 km to the northeast in an area of Bajo Forest for at least three months until the transmitter failed (Fig. 16.3).

Of 57 nestlings banded from 1989 to 1995, only 1 was subsequently observed as a breeder; a female banded as a nestling in 1995 was observed in 1996 occupying a nest site and fledging two young 4 km southeast of her natal nest. While many young may have dispersed beyond the vicinity of the 20 nesting territories we studied, this incident demonstrates that not all young dispersed beyond our study area. Moreover, the several-month residencies of two juveniles mentioned earlier 0.5 and 1.9 km from their natal areas further suggest that additional juveniles may have settled relatively nearby.

SPACING MECHANISMS
AND POPULATION DENSITY

Territorial Behavior and Displays

Unlike most other raptors at Tikal, these secretive birds did not display by soaring above the canopy; rather, they used loud, far-carrying vocalizations as a territorial signal. They vocalized at dawn every morning except during rain. Forest falcons responded to vocalizations of neighboring conspecifics by using the typical "barking" call. Occasionally intruders were detected calling within a pair's territory, eliciting "aggressive" vocalizations from the territorial male: these were faster and louder than the normal call. We heard such "aggressive" calls frequently during the courtship and early incubation periods. Intruders appeared to be unpaired forest falcons, possibly searching for territorial vacancies. No intruders were observed to take over a territory. In one instance during the incubation period, two neighboring, radio-tagged males had a territorial encounter during one of our radio-tracking sessions. This conflict occurred at the boundary between the Complex R and Temple V territories and intensified to aggressive vocalizations and chases lasting for 10 minutes until the Temple V male flew in the direction of his nest with the Complex R male following a short distance.

It appeared that home ranges and territories were essentially equivalent in this species, as individuals appeared to defend their entire home range. This is evidenced by the instance just recounted, as well as by other observations of breeding season agonistic behavior between neighboring territory holders and between these and floaters not in possession of a territory.

Although we observed no known cases of extra-pair copulations, we cannot be sure they did not take place. In one instance, a radio-tagged breeding male was tracked and observed in the vicinity of a neighboring nest, about 500 m from his own. The female at this neighboring site, who was incubating at the time, was observed feeding on a lizard moments after this male passed below her nest. Although we suspected this male had delivered the lizard, we could not be certain. In May 1989 we observed another incident of movement by a radio-tagged male into neighboring territories. The female of the Caoba Trail pair was killed on the nest during incubation. Her mate responded by making what seemed to be efforts to locate her and presumably to deliver food to her. Via telemetry, we followed this male as he crossed two neighboring territories 2.4 km west of his nest, calling constantly, apparently in search of his mate. He called much of the morning and until mid-afternoon. By July, this widowed male had paired with another female, and we observed them visiting cavities approximately 200 m west of the failed nest site.

Nest Defense, Interspecific Interactions

Although we observed few cases of nest defense, observations revealed occasional strife with other birds over nest cavities. In April 1989, prior to egg laying, a female Barred Forest Falcon was observed chasing a Keel-billed Toucan from the cavity entrance repeatedly over several consecutive days. In April 1991, at a nest that had been used by a pair of Barred Forest Falcons for two consecutive years, Mealy Parrots (*Amazona farinosa*) laid two eggs. While the parrots occupied this nest, the forest falcons carried on courtship nearby. On 10 May, the forest falcons had reoccupied the cavity and begun laying eggs. The parrot eggs had disappeared. We do not know if the failure of this parrot nest was due to predation or harassment by the Barred Forest Falcons; we witnessed no interactions between the raptors and the parrots. When researchers climbed to nest cavities, females sometimes attacked in nest defense, grasping the climber's hand and leg with their talons, and raking them across the hand on several occasions.

Constancy of Occupancy of Territories, Use of Alternate Sites

Forest falcon pairs were on territories every year based on 85 territory-years of data from 16 territories having at least three years of nesting attempts each—hence the territory occupancy rate was 100%. However, the birds did not nest every year (see Frequency of Nesting, Percentage of Pairs Nesting later in this chapter). For breeding adults banded in 1989, mean residence time for males on territories during the seven-year study was 6.1 ± 1.5 years (n = 7) and for females 5.0 ± 2.8 years (n = 7). These residence times greatly exceed those of three small temperate zone raptors (see Thorstrom 2001).

Reuse of the same nesting cavities was frequent. Of 70 nesting attempts by pairs whose earlier nest sites were known, 44 (63%) were in the same cavity, while 26 (37%) were in alternate cavities within the same territory. During this seven-year study, 11 breeding pairs occupied only one nest tree each, four used two different nest trees each, four utilized three nest trees each, and two occupied four different nest trees. Pairs that nested successfully tended to remain at the same nesting site, while pairs that failed tended to switch to another nest site within the same territory. (Of 22 pairs nesting successfully, 5 changed nest sites the next year, whereas of 18 nesting unsuccessfully, 12 changed nest sites the following year; chi square = 7.63, P = 0.005.) Breeding pairs that used several alternate nest sites raised fewer young than did pairs that continuously used the same nest cavity (chi-square test, χ^2 = 30.00, 3 df, P < 0.001).

Home Range Estimates

Using radiotelemetry, we estimated home range areas for a number of individuals. For home range determination, we used the Minimum Convex Polygon (MCP) and Harmonic Mean (HM) methods. We employed both 100% MCP home ranges (using all detections) and 95% MCP home ranges, which omit outlying records. Home range estimation is sensitive to sample size (Schoener 1981), and size estimates increase asymptotically with increasing number of relocations (Beckoff and Mech 1984). For breeding male Barred Forest Falcons, the asymptote was reached for the MCP home range estimate at 60 to 80 relocations, beyond which additional locations did not result in a larger estimate.

The 100% MCP home range for one nonbreeding male was 74.8 ha, and for 11 breeding males it ranged from 76.2 to 174.8 ha, averaging 122.8 ± 30.2 ha. This nonbreeding male was unpaired and did not frequent a nest site. We monitored him during one breeding season when he was radio-tagged; he was not detected during subsequent years. On average, 5.5% of a home range overlapped with one or more neighboring home ranges. For the same 11 breeding males, the 95% MCP home range estimate averaged 97.8 ± 23.9 ha (range = 62.3–137.7 ha), and the 85% HM home range estimate averaged 114.6 ± 28.8 ha (range = 71.8–164.8 ha), with 2.8%

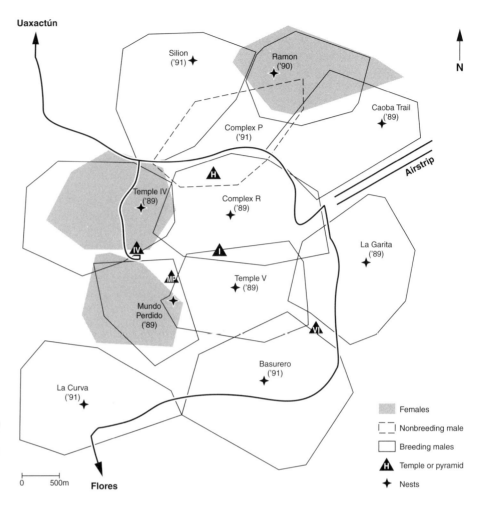

16.4 Estimated home ranges (100% Minimum Convex Polygon) of radio-tagged adult male Barred Forest Falcons during the breeding season at Tikal. Results are given for 10 breeding males and one nonbreeding male.

overlap among home ranges. For three breeding females late in the nestling period, the average 100% MCP home range estimate was 75 ± 11.3 ha (range = 67.0–88.0 ha).

Figure 16.4, depicting 100% MCP home ranges for males, was constructed from data collected during the 1989 through 1991 breeding seasons. We believe this figure accurately represents the average year-to-year distribution of breeding Barred Forest Falcons in the area we studied and in surrounding areas of Upland and similar forest. In this most-intensively studied portion of the park, Barred Forest Falcon territories essentially filled Upland and similar forest types. The areas around Temple IV and Mundo Perdido were man-made clearings, which the forest falcons avoided. The home range of the one nonbreeding male overlapped those of breeding males by 52% using the 100% MCP method and 55% under the 85% HM method. This nonbreeding male appeared to occupy an area between breeding male territories, and overlapped more in territory with neighboring breeding males than did breeding males with one another (Fig. 16.4).

A notable sequence of events took place in 1992 in the well-established Complex R territory. The banded female here disappeared in April, as did the banded male of the Caoba Trail territory, adjacent to the east. The radio-tagged Complex R male then moved into the Caoba Trail territory and paired with the unattended female there. He then occupied the eastern part of his old territory and shifted most of his range into the area used by the previous Caoba Trail male; he ranged over an area nearly the same size as his former territory, with nearly 20% overlap into his 1989 and 1991 home range areas. Telemetry showed that in 1992, this male ranged in basically the same area used by the previous Caoba Trail male, as revealed by telemetry in 1989. This male's departure from the Complex R territory left open an undefended area, and an unbanded pair established a nest at Group F, 150 m south of the old Complex R nest, and within the historic home range of the original Complex R male. We never radio-tagged the new Group F pair, but in 1996 an immature-plumaged female here nested 50 m east of the nest used by the Complex R pair from 1988 to 1992. These two cases illustrate that different individuals occupying a given area sequentially over time at least sometimes used space in a similar fashion.

Inter-nest Spacing and Density of Territorial Pairs

Using the minimum inter-nest distance method, distances between occupied neighboring nests during five years of study averaged 1.04 ± 0.27 km (range = 0.38

to 1.55 km, n = 19 [n = 19 inter-nest distances among 20 territories; based on 39 nest trees used in 70 nesting attempts]). Using the Maximum Packed Nest Density (MPND) method (see Chapter 1), this yields an estimate of 0.98 km² of exclusive space per pair, whereas the square method indicates 1.08 km²/pair. Home range estimates based on radiotelemetry gave similar results. Subtracting the mean overlap value of 5.5% from the mean 122.8 ha home range (100% MCP) for 11 breeding males gives an estimate of 116.7 ha of exclusive area per pair, and subtracting the mean overlap of 2.8% from the mean 114.6 ha home range estimate (85% HM) gives 111.4 ha of estimated exclusive area per pair. Thus, on average, Barred Forest Falcon pairs in our study area used 1.0–1.1 km² of exclusive space per pair.

DEMOGRAPHY

Attainment of Adult Plumage and Eye Color

Barred Forest Falcons presumably molt into definitive adult plumage at their second full molt, at the age of roughly two years, but this requires verification. This statement is based largely on the fact that a female banded as a fledgling in 1995 was studied as a breeder the following year, when she retained her characteristically juvenile brown barring and white collar at least at the onset of the breeding season and probably throughout. Two other observations also indicate that the juvenal plumage is retained at least through the first year. In March 1990, an 11 month old (banded as a nestling in 1989) was observed in juvenal plumage, and in April 1989 another immature-plumaged bird was seen. Although its exact age was unknown, it presumably hatched during the spring of the previous year. Among the many females trapped, three of those in adult plumage retained the dark brown iris of immatures, indicating that definitive plumage is achieved before adult iris color, as in some *Accipiter* species (Newton 1986; Palmer 1988).

Age of First Breeding

Among 70 nesting attempts in 98 territory-years, we found one 12-month-old female breeding, and two other breeding females that were probably two years old, detailed as follows. In 1996, a female banded as a nestling in 1995 was located on a territory 2.5 km from her natal nest, paired with a male in adult plumage. This female, still in juvenile plumage, raised two young. A female in adult plumage but with dark brown iris raised at least one young in 1989, and an adult-plumaged female with dark eyes was captured while breeding at another territory; as outlined, these birds were probably in their second year. Apart from these instances, all other breeding females had adult plumage and eye color (light brown to buffy yellow iris). The normal age at first breeding for females is likely three years or older. We saw no evidence of males breeding in juvenal plumage or with subadult

eye color; thus first breeding in males probably also normally occurs during year 3 or beyond.

Frequency of Nesting, Percentage of Pairs Nesting

Individual pairs often nested in consecutive years. However, although all study territories were occupied each year, not every pair attempted to nest each year. In 98 territory-years at 20 territories, we observed nesting attempts in 70 territory-years (71.4%; Table 16.2). For some purposes, we focus here on 85 territory-years of data from sixteen territories for which we had at least 3 years of known nesting attempts: three territories for 3 years, two for 4 years, four for 5 years, one for 6 years, and six for 7 years. Nesting attempts were known to have occurred in 63 territory-years (74.1%; Table 16.3). If instead we calculate nesting frequency separately for each territory, pairs nested on average in 73.2 ± 21% of years (range = 50–100%, n = 16 pairs; Table 16.3). It was sometimes difficult to be certain that nesting did not take place in a given instance, but any error of this type that may exist here is very small. The percentage of territorial pairs nesting in a given year (71.4–74.1%) was similar to that in the tropical Puerto Rican Sharp-shinned Hawk (74%: Delannoy and Cruz 1988), and to the mean rate of nesting among 12 species of tropical and subtropical raptors (mean = 70.3% of pairs nesting), which showed a lower rate of nesting than did 15 temperate zone species (mean = 84.1%; see Chapter 12).

In the 27% of territory-years in which pairs failed to breed, the adults simply skipped a year of nesting, and then nested again in a subsequent year. In such cases, the adult pairs were observed together, inspecting potential nest sites early in the breeding season, but did not lay eggs. Some pairs did not breed for two or three years, and then resumed breeding, always with the same mate throughout. We detected no synchrony of nonbreeding among pairs within the study population, nor any correlation between the occurrence of such breeding hiatuses and previous nesting success. We do not know why pairs failed to breed at times, but suggest that this probably relates to some breeding mechanism peculiar to tropical environments where birds have long life spans.

Productivity and Nest Success

From 1988 through 1995 (98 territory-years), we documented 70 nesting attempts (Table 16.2). Details were known for 59 nests; in these, 154 eggs were laid, for a mean clutch size of 2.6 (Table 16.2). In the 59 nests (n = 154 eggs), 97 eggs (63%) hatched, and 75 (77%) of the hatchlings fledged. Overall, 1.07 young fledged per breeding attempt and 0.77 per territorial pair-year (Table 16.2). Nest success (percentage of nests fledging one or more young) for 68 fully documented nestings was 54.4%. Nest success was 56.1% (37/66 nests) for first nests, and neither of two re-nestings was successful.

Table 16.2 Reproductive success of Barred Forest Falcons in Tikal National Park, 1988–95

Year	Territorial pairs (number)	Nesting attempts (number)	Nests with confirmed clutches (number)	Eggs in confirmed clutches (number)	Mean clutch size	Eggs hatched (number)	Young fledged (number)
1988	4	4	4	10	2.5	2	*
1989	8	8	7	20	2.9	11	8
1990	9	8	6	17	2.8	12	11
1991	12	12	10	27	2.8	12	12
1992	14	6	5	14	2.8	10	9
1993	14	11	9	23	2.6	16	10
1994	18	12	10	25	2.5	20	11
1995	19	9	8	18	2.3	14	14
Total	98	70	59	154	2.6	97 (n = 63)	75 (n = 77)

Year	Fledglings/ breeding attempts (number)	Fledglings/ territorial pair (number)	Fledglings/ successful pair (number)	Breeding success rate/ territorial pair (%)[a]	Reproductive success rate(%)[b]
1988	*	*	*	*	* (n = 0/2)
1989	1.0	1.0	2.7	38	38 (n = 3/8)
1990	1.4	1.2	2.8	44	50 (n = 4/8)
1991	1.0	1.0	2.0	50	50 (n = 6/12)
1992	1.5	0.6	1.8	36	50 (n = 5/6)
1993	0.9	0.7	1.7	43	55 (n = 6/11)
1994	0.9	0.6	1.8	33	50 (n = 6/12)
1995	1.6	0.7	2.0	37	78 (n = 7/9)
Total	1.07 (n = 75/70)	0.77 (n = 75/98)	2.0 (n = 75/37)	38 (n = 37/98)	54 (n = 37/68)

Note: * In 1988 the outcome of two nests that contained nestlings was not documented and the other two failed (see Thorstrom 1990).

[a] Breeding success rate is proportion of nests with at least one young fledged per cumulative territorial pair.

[b] Reproductive success rate is proportion of nests with at least one young fledged per fully documented nesting attempt.

Annual hatching success did not differ significantly among years from 1989 to 1993 (chi-square test, χ^2 = 1.7, 4 df, $P > 0.5$), nor did reproductive success (Table 16.2). Overall, 75.6% of hatchlings fledged (Table 16.2), and this proportion also did not vary between years (chi-square test, χ^2 = 0.3, 3 df, $P > 0.1$). From the age at which chicks could be sexed (approximately three weeks) to fledging time (approximately five weeks), we did not detect a mortality differential between male and female chicks; of 35 chicks that fledged, 19 (54%) were males and 16 (46%) females (chi-square test, χ^2 = 5.25, 3 df, $P > 0.1$).

Some pairs/territories produced more young than did others. One territory produced fledglings in five of six nesting attempts, totaling 14 fledglings, while four territories were successful each year nesting was attempted (two to four nesting attempts each; Table 16.3). In contrast, two territories never produced young in two and three nesting attempts (Table 16.3), and a pair in one territory attempted to nest seven times and successfully produced young only once. Of the Barred Forest Falcons that attempted to breed (27 males and 27 females), 30% (8) of the females and 22% (6) of the males produced no young, and 19% of the breeders produced more than half of all fledglings (54%). This is similar to results for the

European Sparrowhawk *(Accipiter nisus)*, in which 20% of females produced 50% of all fledglings (Newton 1986), and the Eastern Screech-owl *(Megascops asio)*, in which 21% of females produced 50% of offspring (Gehlbach 1994). Newton (1989) suggested that chance events were responsible for most of the variation in lifetime reproductive success among individuals.

Survival Rates and Adult Turnover

We trapped and banded 27 adult males and 27 adult females and each year, we determined the identity of adults present in each territory to the extent possible. Turnover of adults on nesting territories was low, and fidelity to sites and mates was 100%—the only mate changes that occurred were replacements of females known to have been killed and two males that disappeared and presumably died.

In 1989, seven adult males and seven adult females were banded on their nesting territories early in the breeding season. At the end of the 1995 breeding season (i.e., after six years and seven breeding seasons had elapsed), five of the same males and four of the females remained on these territories (Table 16.4). Assuming all the missing individuals perished, this indicates a 94.5%

Table 16.3 Territorial occupancy, nest-site fidelity, nesting success, and productivity for 16 Barred Forest Falcon territories, Tikal National Park, 1988–95

Territory	Years pair on territory (number)	Nesting attempts (number)	Years of nesting attempts/years pair on territory	Nest success, number (%)	Young fledged (number)
Complex R (1988)[a]	5	4	0.80	2 (50)	5[a]
Temple V (1988)	7	7	1.00	1 (14)	3
Mundo Perdido (1988)[a]	7	6	0.86	2 (33)	4[a]
Caoba Trail (1988)	7	7	1.00	5 (71)	8[a]
Temple IV (1989)	7	4	0.57	4 (100)	10
La Garita (1989)	7	5	0.71	2 (40)	4
Ramon (1989)	7	6	0.86	5 (83)	14
Brecha 60 (1990)	6	4	0.67	2 (50)	5
Basurero (1991)	5	2	0.40	0 (0)	0
La Curva (1991)	5	2	0.40	2 (100)	3
Silión (1991)	5	3	0.60	1 (33)	2
Group F (1992)	4	4	1.00	4 (100)	7
Transecto 5 (1992)	4	2	0.50	2 (100)	1
Brecha 18 (1993)	3	2	0.67	1 (50)	2
Brecha 64 (1993)	3	2	0.67	1 (50)	1
Bejucal (1993)	3	3	1.00	0 (0)	0
Total	85	63	0.73 ± 0.21	34	69 (54%)

[a] Territory excluded from 1988 calculations because outcome of nesting was uncertain.

Table 16.4 Survivorship of adult Barred Forest Falcons breeding in Tikal National Park from 1989 through 1995[a]

Study period	Female (n = 7)	Male (n = 7)
1989 courtship period	7	7
1989 breeding season	5	7
1990 breeding season	5	7
1991 breeding season	5	7
1992 breeding season	5	5
1993 breeding season	5	5
1994 breeding season	4	5
1995 breeding season	4	5

[a] Numerical values are the number of individuals remaining alive, of the birds banded in 1989. All birds were banded as adults (at least second year) at the start of the 1989 breeding season. Hence, the minimum age of the oldest birds at the study's end was 9 years.

annual survivorship rate for males and a 91.1% annual survivorship rate for females over this six-year span. If, however, most mortality took place during the breeding season (see below), it may be more accurate to calculate survivorship over a seven-year period, in which case the annual survival rate of males was 95.3% and for females was 92.3%; we favor these figures. Many individuals remained alive in the last year of our fieldwork, having occupied the same territories throughout. Some were believed to be at least nine years old at that time, since they were presumably at least three years old during the first year of study.

Nearly all turnover of adult females resulted from predation during the incubation and nestling stages, when two females were killed and a third disappeared. In one case, the female's wing feathers and a foot were found in the nest cavity and at the base of the nest tree, indicating predation by a mammal. Another instance involved a radio-tagged female. Her radio signal issued from within the base of the hollow nest tree and eventually descended into the root system and then disappeared; we presumed she was eaten by a snake. A third female, who disappeared during the incubation stage, probably also suffered predation. We observed her mate arrive in the nest area with food, vocalizing loudly for 2–3 hours during several days, with no response from the female. On climbing to the nest we found the eggs destroyed, with no trace of the female.

We documented no details of male mortality, but two males disappeared from territories they had previously occupied for several years; presumably they had died. Recruitment of new breeders in our study territories occurred only when a member of a pair disappeared, and in one case, when loss of opposite-sex members of two neighboring pairs resulted in formation of a pair by the surviving widowed neighbors, leaving a vacant territory that was then filled by unbanded adults.

Causes of Nesting Failure and Mortality

The main cause of nest failure was predation, which accounted for at least 30 (91%) of 33 nest failures. In decreasing importance were egg predation (18 instances, or 55% of all nest failures), nestling predation (9 cases, or 27% of nest failures), and predation on adult females during incubation and brooding (3 cases, or 9% of nest failures). In three cases (9%), the cause of nest failure was unknown, though one of these, late in the nestling period, was likely also due to predation. We observed no nest failures due to desertion or starvation, but parasites may have played a role in two deaths (see below). Nesting success (64.7%) was significantly higher in live than in dead trees (26.3%) (chi-square test, $\chi^2 = 5.1$, 1 df, $P < 0.05$). Number of young fledged was not significantly related to the diameter of the nest tree, but higher nest cavities (> 16.6 m) showed a near-significant tendency to fledge more young than did nests closer to the ground (t-test, $t = -1.4$; 24 df, $P = 0.08$).

For fledglings, it seemed clear that the first few days after leaving the nest were the time of greatest vulnerability to predators. Two to three days after fledging, one young was killed by a Common Gray Fox *(Urocyon cinereoargenteus)* and another by a Boa Constrictor *(Boa c. imperator)*, while the remains of a third were discovered within 25 m of a nest tree. A fourth young was found dead of unknown cause four days after fledging, 25 m from the nest tree. Two young from another nest disappeared one week after fledging, probably also killed by predators.

Parasites and Pests

By two weeks or so after hatching, many nestlings had small dipteran larvae (thought to be botflies) attached beneath the epidermis on their face and toes, and to primaries that were still in blood sheaths. Only two nestlings appeared detrimentally affected by these larvae; one had an enlarged middle ear region and the other a large protuberance affecting the left eye. The first-described young disappeared from its nest a week after these observations, and the other was killed by a Common Gray Fox two days after fledging. It seems plausible that these dipteran larvae may have contributed to the death of these two young, perhaps making them more vulnerable to predation.

CONSERVATION

The dependence of Barred Forest Falcons on mature Upland Forest in our study area (see Habitat Use earlier in this chapter) suggests that this species is adversely affected by deforestation. Moreover, these birds used a territory about 1 km^2 in area. Many of the forest fragments remaining in the farming landscape were smaller than this, likely limiting these birds' ability to thrive and reproduce in the human-altered landscape. Moreover, nest cavities could be a limiting resource, especially where large trees are subject to harvest. That nearly a third of nests were in Spanish Cedar is significant, as this is an important species for the logging industry in northern Central America. Continued logging of this valued lumber species could limit the availability of nest sites, for forest falcons and other wildlife.

Barred Forest Falcons apparently differ geographically in habitat use, and thus also in vulnerability to anthropogenic habitat change. While in our study area this species was restricted to tall mature forest, in Amazonia, Barred Forest Falcons are said to occur mainly in second growth, forest edge, tidal swamp forest, and gallery forest (Bierregaard 1994) with the similar Lined Forest Falcon *(M. gilvicollis)* largely replacing the Barred Forest Falcon in moist primary forest. Bierregaard (1994) also states that the Barred Forest Falcon is much less common in areas where it co-occurs with the Lined Forest Falcon. Similar conclusions were reached by Jullien and Thiollay (1996) and Schwartz (1972), who found Lined Forest Falcons restricted to moist forest and Barred Forest Falcons extending into drier forests.

This apparent regional difference in use of primary versus successional or modified forest suggests that deforestation and forest alteration may be more detrimental to Barred Forest Falcons in our study area than in much of South America. However, deforestation in South America is likely to impact Lined Forest Falcons, Slaty-backed Forest Falcons *(M. mirandollei)*, and other congeners as much as it does Barred Forest Falcons in our study area. The Barred Forest Falcon's wide geographic range, modest spatial requirements, high population density, and presumably minimal vulnerability to shooting, all bode well for its future survival. Still, widespread deforestation should be expected to diminish populations of this species throughout its range.

HIGHLIGHTS

Barred Forest Falcons showed a breeding strategy similar to that of other tropical raptors studied to date, resulting in relatively low nesting productivity. Compared with temperate zone raptors of similar size, Barred Forest Falcons laid large eggs relative to female body mass; laid smaller clutches; had longer incubation, nestling, and post-fledging dependency periods; and a high rate of nest failure, all contributing to low productivity. (Post-fledging dependency of Barred Forest Falcon was 28–49 days [exactly 42 days for two young]; for temperate zone raptors, published values are 26 days for Sharp-shinned Hawk, 40–53 days for Cooper's Hawk *[Accipiter cooperi]* [twice the size of the Barred Forest Falcon], 21–28 days for European Sparrowhawk, and 14–21 days for Merlin *[F. columbarius]*.) Young Barred Forest Falcons reached independence five to seven weeks after fledging, a much shorter dependency period than is shown by some tropical forest raptors, as expected for their small body size. Females provided all incubation, a pattern more typical of the genus *Accipiter* than of most falconids studied to date.

In Tikal, Barred Forest Falcons are year-round residents with stable territories from year to year. These forest falcons showed longer tenancy on territories than did similar-sized temperate zone raptors—this probably results both from greater survival rates and from their tendency to remain permanently on a territory. Because suitable habitat at Tikal appeared saturated with territorial pairs, territories likely are difficult to acquire, which may inhibit birds from searching for vacancies or switching territories. Barred Forest Falcon pairs showed higher fidelity to mates and territories (100%) than did European Sparrowhawks (50–57%: Newton 1989) and Eurasian Kestrels *(Falco tinnunculus,* 71–76%: Village 1990). Mate replacements in this forest falcon species occurred only upon the death of a mate, unlike the case in these two temperate zone raptors. Nesting in 71–74% of territory-years, Barred Forest Falcons at Tikal fit the pattern of less frequent nesting attempts in tropical than in temperate zone raptors described by Panasci and Whitacre (2002).

While nest predation is often the largest cause of nesting mortality in small songbirds (Ricklefs 1969a; Martin 1995), it is generally thought to be less important in raptors (Newton 1979). This cavity-nesting forest falcon, however, was greatly affected by nest predation, which accounted for at least 91% of 33 nest failures. This is among the highest rates of nest predation recorded among raptors. Among six raptor species reviewed by Newton (1979: Table 34), only one approached this level of nest predation—in the Red-footed Falcon *(Falco vespertinus)*, predation accounted for 70% of 57 nest failures. In the other five species, nest predation caused from 0 to 21% of nest losses.

Barred Forest Falcons were long-lived, with annual survival rates estimated at 95.3% for adult males and 92.3% for adult females; several were at least nine years old at the end of the study. This is higher than annual survivorship estimates of 66–71% for four temperate raptors of similar body size: Eurasian Kestrel (Village 1990), Eastern Screech-owl (Gehlbach 1994), European Sparrowhawk (Newton 1989), and Merlin (James et al. 1989; Warkentin et al. 1991). All known cases of female mortality took place during the nesting season, implying that nesting presented a significant risk to female forest falcons.

One interesting pattern is that densities of Barred Forest Falcons, and of the most common *Micrastur* species at a given site, vary regionally within the Neotropics. Barred Forest Falcons in Tikal occurred at a density of about one territorial pair per square kilometer, giving this species the highest population density of any diurnal forest raptor at Tikal. At many South American sites, however, the Barred Forest Falcon is not the most abundant *Micrastur* species. Thus, to compare localities, it is probably most meaningful to compare the densities of the most abundant *Micrastur* species at each. Estimated densities of the most abundant *Micrastur* species in Manaus, Brazil (Lined Forest Falcon: 1–2 pairs/km² [Klein and Bierregaard 1988; Bierregaard 1994]) and Manú, Peru (Barred Forest Falcon: 1.5 to 2 pr/km²), ranged from about the same density we determined at

Tikal, to about twice that density. In French Guiana, Lined Forest Falcon (slightly larger than Barred Forest Falcon) was the most common *Micrastur*—with density estimated at one pair or fewer per 3 km² (Thiollay 1989a)—one-third the density of Barred Forest Falcon at Tikal and one-third to one-sixth of the Peru and Brazil estimates.

Taken at face value, these numbers suggest that the most common forest falcon species at Manaus and Manú occur at up to twice the density as pairs as do Barred Forest Falcons at Tikal, and up to six times the density in French Guiana. Conceivably, differences in primary productivity of these forests might produce such density differences among the most common *Micrastur* species at each site. Another factor that might affect *Micrastur* densities is the number of similar species (including congeners) occurring together, among which food resources must be divided. However, along with *Micrastur* densities apparently being higher at Manaus and Manú than at Tikal, these sites also have a higher species richness of other forest falcon species, of other raptor species, and potential food competitors among other taxa. Thus, higher *Micrastur* densities at these sites than at Tikal are not expected on this basis. Moreover, the apparently sixfold lower density in Venezuela than at the Peruvian and Brazilian sites seems surprising. If such density differences are real, the identity of the underlying factors is an intriguing question. Confirmation of these density figures, however, is probably needed at all the South American sites, before attempting to interpret apparent differences in density. The Manaus and Manú density estimates are remarkably similar, perhaps suggesting that these two estimates are robust. Thiollay's (1989a, 1989b) study, in contrast, was focused mainly on detecting raptors soaring over the canopy, and his density estimates for these secretive sub-canopy dwellers were tentative at best. Only further study can verify the extent of geographic variation in population densities of *Micrastur* species.

For further information on this species in other portions of its range, refer to Di Giacomo 2005, Marini et al. 2007, Nunnery and Welford 2002, Poulin et al. 2001, and Röhe and Pinassi Antunes 2008.

17   COLLARED FOREST FALCON

Russell K. Thorstrom

The Collared Forest Falcon *(Micrastur semitorquatus)* is an elusive, ghostlike presence that calls from the deep recesses of the forest in predawn darkness or after dusk. Rarely seen, this forest falcon utters its hollow, resonating, *ow* or *ahr* call persistently from hidden perches high in the forest canopy. These calls, with an almost human quality, carry great distances through the silent forest at these times of day. Weighing 560–940 g, this striking raptor is the largest member of the falconid genus *Micrastur*. The facial disc—a characteristic of the genus presumably used as an auditory aid in hunting—lends an owlish appearance, and the long, arched tail with deeply rounded tip comprises fully half the bird's total length (Plates 17.1, 17.2).

This is a bold and fearless predator, using stealth and ambush to achieve surprise. It uses its long tail, long legs, and short wings in highly maneuverable flight through the canopy and runs with amazing agility through limbs and thickets on the ground. Though most prey are modest in size, on occasion these falcons take prey as large as young Crested Guans *(Penelope purpurascens)*, young Great Curassows *(Crax rubra)*, or an adult female Ocellated Turkey *(Agriocharis ocellata)*, the latter weighing 2.7–3.2 kg—impressively large prey for a raptor of this size. The short, rounded wings of this genus, coupled with their long tail and legs, represent a case of morphological convergence with the genus *Accipiter*, no doubt based on their common reliance on forest interior habitats.

These secretive, hole-nesting forest falcons occupy a variety of tropical and subtropical forest habitats, from extensive stands of mature forest to gallery forest, forest edge, and second growth, and seem to prefer areas with dense understory. Nests at Tikal were in some of the largest trees in the forest—old-growth behemoths of Honduras Mahogany *(Swietenia macrophylla)* and Spanish Cedar *(Cedrela odorata)*—posing the question of whether selective logging of these prized hardwood giants affects populations of this raptor.

GEOGRAPHIC DISTRIBUTION AND SYSTEMATICS

Collared Forest Falcons range from the Tropic of Cancer on both of Mexico's coastal plains (and with sightings in south Texas: Wheeler and Clark 1995) southward in tropical and subtropical forested regions to eastern Bolivia, northern Argentina, and Paraguay (Brown and Amadon 1968). These birds range from sea level up to 1800 or 1950 m (Ridgely and Gwynne 1989; Howell and Webb 1995). Two subspecies are recognized: *M. s. naso* occurring from Mexico south through Central America and west of the Andes in Colombia, Ecuador, and extreme northwest Peru; and *M. s. semitorquatus* occurring east of the Andes (Friedmann 1950; Brown and Amadon 1968; del Hoyo 1994). This species is a year-round resident at Tikal and no doubt elsewhere in its range.

MORPHOLOGY

Body weights and linear measurements from Tikal and elsewhere are given in Appendixes 1 and 2, and additional measurements are given by Blake (1977). Four adult males at Tikal averaged 587 g and six females averaged 869 g (Appendix 1). At this body mass, a male Collared Forest Falcon is roughly 5% heavier than a female Cooper's Hawk *(Accipiter cooperii)*, and a female Collared is about the same mass as a male Northern Goshawk *(A. gentilis)*.

Sexual Size Dimorphism

Six females at Tikal were significantly larger than four males in mass and in dimensions of beak, tail, and wing chord (*t*-test, all $P < 0.005$), but not in foot span. Based on Tikal birds, Dimorphism Index (DI) values were 13.1 for body mass, 9.2 for wing chord, 11.7 for tail length, and 6.2 for culmen length (Appendixes 1, 2). Dimorphism values for published data sets are somewhat smaller, at least with regard to wing and tail length (Appendix 2). However, Wetmore (1965) stated that many museum specimens are mistakenly sexed, obscuring the true degree of size dimorphism. Thus, we believe our dimorphism estimates from Tikal are the best available, since we were confident of our sexing, and all birds were from the same population.

Based on body weight, Collared Forest Falcons are slightly more size dimorphic than the Cooper's Hawk, and slightly less so than the Gyrfalcon *(Falco rusticolus)* and Northern Harrier *(Circus cyaneus;* see Table 1.2). Taking the mean dimorphism values for weight, wing chord, and culmen, however, the Collared Forest Falcon's 9.5 DI is only moderately large; by this measure, this forest falcon appears slightly more dimorphic than the Northern Goshawk but less so than several other North American raptors (see Appendix 1 and Table 1.2).

Ecomorphology

It has often been noted that the relatively short, rounded wings and long tail of *Micrastur* species represent convergent evolution with *Accipiter* species, both groups being adapted to dashing flight through the cluttered airspace of forest and thickets. However, the extremely short-winged, long-tailed morphology represented by the Collared Forest Falcon is worth emphasizing. Peeters (1963) showed that Collared Forest Falcons have proportionally much longer tails than the Northern Goshawk and Cooper's Hawk, with the tail slightly exceeding the wing chord in *M. semitorquatus,* and wing chord exceeding tail length by 18–40% in these two accipiters. Indeed, no other North American raptor has such a relatively long tail and short wings as this forest falcon. These features seem to reflect a degree of adaptation for pursuit through dense vegetation that is more extreme than in many accipiters. We noted also that tail feathers were relatively flexible, which may reduce breakage while pursuing prey through heavy cover. Wing feathers may also be unusually flexible (see Hunting Behavior later in this chapter).

Plumage Characteristics

This species is generally regarded as showing three color morphs: a light morph (white below), a tawny morph (pale to rich tawny or buffy below), and a dark morph (essentially all dull black or brownish slate, with some white barring on the flanks and tail: Brown and Amadon 1968; del Hoyo et al. 1994). Birds of all three types were at Tikal, though the dark morph appeared uncommon, as elsewhere (del Hoyo et al. 1994; Howell and Webb 1995). Our research involved mainly light and a few tawny birds; only one dark-morph individual was seen.

We noted no plumage differences between the sexes, and size was not reliable for sexing birds in the field except when they were together. Thus, we distinguished the sexes mainly by nesting behavior. The cere and bare facial skin of adults were olive green, and tarsi and feet olive to yellowish green. Fledglings were heavily barred below with brown and had dark chocolate brown eyes, in contrast to the lighter brown eyes of adults (Plate 17.3).

PREVIOUS RESEARCH

No detailed study of this species has previously been made. Published accounts include notes on hunting behavior (Smith 1969; Willis 1976; Trail 1987; Rappole et al. 1989; Robinson 1994), first egg and nest descriptions for the genus (Wetmore 1974; Mader 1979), association with ant swarms (Willis et al. 1983; Mays 1985), and a second description of nest, eggs, and young (Thorstrom et al. 1990). Peeters (1963) discussed the species's morphology and the behavior of captive birds.

RESEARCH IN THE MAYA FOREST

At Tikal we studied five territories in the park and collected fewer data on a sixth pair in the farming landscape nearby. From 1988 to 1993, we monitored five territories for a total of 13 territory-years; nesting histories are given in Appendix 17.1. Study methods were as described in Chapter 16 for the Barred Forest Falcon *(M. ruficollis).* We searched the forest and known territories daily from February to August, to document nesting activity of territorial pairs. We observed nests from the ground or from tree platforms, using 7–10X binoculars at 25–50 m.

Most prey items were recorded when delivered to nests, but some were documented away from nests during radio-tracking. We included only prey we observed delivered to nests or in possession of adults, and omitted prey found in nests (Snyder and Wiley 1976; Wiley and Wiley 1981; Marti 1987). We could not identify amphibians and insects to species, and assigned these to higher taxonomic categories. Considering together prey of both forest falcon species resulted in 37 prey categories, which were used in all niche breadth and overlap calculations. *Anolis* lizards were classed as "small" (< 20 cm total length) and "large" (> 20 cm). Mammal weights follow Emmons and Feer (1997), bird weights come from Smithe (1966) and Dunning (1993), and reptile weights were taken in the field.

Food Niche Breadth (FNB) for each forest falcon species was calculated using Levins's (1968) equation: FNB = $1/\Sigma p_i^2$, where p_i is the proportion of the i th prey category in the diet. To compare raptors with different numbers of prey categories, a standardized niche breadth value (FNBS) was also calculated as follows: FNBS = $(\text{FNB}-1)/(n-1)$, where n is the number of potential prey categories (Levins 1968). Niche overlap based on dietary frequency data (not biomass) was calculated using Schoener's (1970) index of symmetrical overlap: Overlap = $1-(1/2)(\Sigma|p_ij-p_ik|)$, where p_i is the proportion of the ith prey category for species j and k.

At five nests, we sampled tree species composition and several aspects of vegetation structure. Five 0.04 ha (tenth-acre) circles were sampled at each nest tree—one centered on the nest tree and one adjacent in each cardinal direction. Identical sampling was conducted at six Barred Forest Falcon nests.

We radio-tagged eight adults (four females and four males) and three fledglings, using radios from Holohil Systems, Canada. Adults were fitted with either a backpack arrangement (16 g) or a tail mount (8 g). One female and two male fledglings received 5 g tarsi-mounted transmitters one week prior to fledging; they were trapped six weeks after fledging, tarsal mounts were removed, and tail-mounted radios attached. Adult females were trapped on *bal-chatris* (Berger and Mueller 1959) baited with a chick or rat, placed on the ground (n = 4) or in a tree (n = 1). Two adult males were trapped at the nest entrance using wire hoop traps (Thorstrom 1996b) and two using a *bal-chatri* on the ground and near the nest entrance. Three fledglings were trapped using a modified *bal-chatri*/noose carpet with prey remains placed within. Home range estimates used the 100% Minimum Convex Polygon (MCP) method and the 85% Harmonic Mean (HM) method.

DIET AND HUNTING BEHAVIOR

Prey items previously reported for the Collared Forest Falcon include chachalacas (*Ortalis* spp.) mammals, lizards, snakes, and insects (Brown and Amadon 1968), and predominantly birds (Robinson 1994). From 1990 to 1992, we documented 223 prey items delivered to females, nestlings, and fledglings at Tikal (Table 17.1). In addition, 36 items were delivered to fledglings at one nest by a third adult, an "extra" male (see below) that brought in mostly toucans—we present the diet calculated with and without this male's contribution. Without the contribution of the extra adult, based on 171 identified prey items, the diet on a numerical basis was 46.2% mammals (79 items), 34.5% birds (59), 18.7% reptiles (13 lizards and 19 snakes), and 0.6% amphibians (one frog). On a biomass basis, the diet was 37.3% mammals, 53.8% birds, 8.9% reptiles, and 0.05% amphibians. We presume the 52 unidentified prey items resembled those identified. When the extra male's contribution is included, the diet (206 identified items) becomes 43.7% birds (90 individuals), 40.3% mammals (83), 15.5% reptiles (32), and 0.5% amphibians (one frog).

Prey ranged in size from a frog (20 g or less) to an adult female Ocellated Turkey—probably in the neighborhood of 2.7–3.2 kg (Smithe 1966). The turkey was captured by a breeding adult female who then fed on the carcass for two days until she was able to carry the remainder to the nest. Other large prey items included a young Crested Guan. The Great Curassows we documented among prey were young ones up to one-quarter grown, probably still accompanied by adults.

Of the 13 lizards taken, 12 were iguanids (*Corythophanes* spp.), among the largest lizards occurring at Tikal. Squirrels dominated the mammalian portion of the diet, contributing 89% of all mammalian biomass and 33% of total prey biomass; the 79 identified mammals included 42 Deppe's Squirrels (*Sciurus deppei*, 190–220 g) and 11 Yucatán Squirrels (*S. yucatanensis*, 400 g). A surprising number of bats were taken, including two fruit bats (*Artibeus* sp.) and 14 unidentified. The remaining mammals included seven rat-sized rodents including the Hispid Cotton Rat (*Sigmodon hispidus*) and two mice believed to be spiny pocket mice (*Heteromys* spp.).

In addition to the notably large birds listed here, a great variety of medium-sized birds were taken (Table 17.1). Among the 59 birds identified as prey were nine Collared Aracaris (*Pteroglossus torquatus*), seven Plain Chachalacas (*Ortalis vetula*), seven young Great Curassows, six Keel-billed Toucans (*Ramphastos sulfuratus*), four woodcreepers (*Dendrocincla* spp.), three tinamous (*Crypturellus* and/or *Tinamus* spp.), and three Brown Jays (*Cyanocorax morio*). Other notable avian prey included a Mexican Wood Owl (*Strix squamulata*; Table 17.1).

The mean weight of observed prey items (mostly captured by males, but some by females) was 239 ± 19 (SE) g (n = 171); excluding the Ocellated Turkey, the mean was 220 ± 11 (SE) g (n = 170). The median prey weight, with or without the turkey, was 205 g, the same as that of Deppe's Squirrel—the most frequent item in the observed diet. This amounts to 35% of the average weight of a male Collared Forest Falcon. The mean weight of individual lizards was 44 g, snakes very roughly 160 g, mammals 186 g, and birds 373 g. Omitting the Ocellated Turkey, birds in the diet averaged 328 g, with a median weight of 311 g and represented the largest prey taken by these forest falcons, perhaps because of the great diversity of avian prey species within the appropriate size range. By comparison, few species of diurnal mammals occur at Tikal, the most notable being the two species of squirrels. The most frequent prey were birds and mammals weighing 200–400 g (Table 17.1).

In the Aguadita territory in 1990, a male that was not one of the banded pair members began delivering prey to the two young four weeks after they fledged. We observed this adult deliver 36 prey items by the eleventh week after fledging: 27 Keel-billed Toucans, 2 Collared Aracaris, 2 unidentified birds, 4 Deppe's Squirrels (*S. deppei*), and 1 unidentified prey item. In terms of biomass, this extra adult delivered 12.6 kg of prey during this eight-week period.

Based on 37 prey categories—mostly prey identified to species, but in some cases genera or higher categories—the smaller Barred Forest Falcon had a diet dominated by lizards—especially *Anolis* species—and as a result had a narrower dietary niche breadth than did the Collared Forest Falcon, which in turn took a greater diversity of birds and mammals (Table 17.2). Including all prey identified at least to genus, dietary overlap between these two forest falcons was 0.49 (Table 17.2). Dietary overlap of Collared Forest Falcons with some other large raptors is probably more meaningful than that between these very differently sized congeners—Collared Forest Falcon prey items averaged 10 times more massive (239 g) than Barred Forest Falcon prey (24 g: Table 17.2).

Hunting Behavior

These secretive raptors frequent the dense undergrowth and lower canopy of evergreen and deciduous forests, as well as tall second growth, edge situations, and semi-open habitats where sufficient cover exists. They hunt mainly in dense vegetation and attack prey from a concealed perch, pursuing it in flight and by running, hop-

Table 17.1 Prey brought by Collared Forest Falcons to nest at Tikal[a]

Prey type	Prey		Weight (g)	Biomass	
	Number	Percentage		Gram	Percentage
Amphibians	**1**	**0.58**	—	**20**	**0.05**
Unidentified frog	1	0.58	20	20	0.05
Lizards	**13**	**7.60**	—	**577.4**	**1.42**
Corytophanes spp.	12	7.02	45.2	542.4	1.33
Unidentified lizards	1	0.58	35	35	0.09
Snakes	**19**	**11.11**	—	**3040**	**7.46**
Unidentified snakes	19	11.11	160	3040	7.46
Birds	**59**	**34.50**	—	**21,910**	**53.77**
Tinamidae	3	1.75	440	1320	3.24
Cracidae					
Crested Guan (*Penelope purpurascens*)	1	0.58	600	600	1.47
Great Currasow (*Crax rubra*)	7	4.09	500	3500	8.59
Plain Chachalaca (*Ortalis vetula*)	7	4.09	563	3941	9.67
Phasianidae					
Ocellated Turkey (*Agriocharis ocellata*)	1	0.58	3000	3000	7.36
Spotted Wood Quail (*Odontophorus guttatus*)	1	0.58	300	300	0.74
Strigidae					
Mexican Wood Owl (*Strix squamulata*)	1	0.58	240	240	0.59
Momotidae					
Blue-crowned Motmot (*Momotus momota*)	1	0.58	133	133	0.33
Ramphastidae					
Keel-billed Toucan (*Ramphastos sulfuratus*)	6	3.51	350	2100	5.15
Collared Aracari (*Pteroglossus torquatus*)	9	5.26	220	1980	4.86
Emerald Toucanet (*Aulacorhynchus prasinus*)	1	0.58	150	150	0.37
Picidae					
Golden-fronted Woodpecker (*Melanerpes aurifrons*)	1	0.58	81	81	0.20
Chestnut-colored Woodpecker (*Celeus castaneus*)	1	0.58	85	85	0.21
Dendrococlaptidae	4	2.34	37	148	0.36
Corvidae					
Brown Jay (*Cyanocorax morio*)	3	1.75	200	600	1.47
Unidentified birds	12	7.02	311	3732	9.16
Mammals	**79**	**46.20**	—	**15,192**	**37.29**
Yucatán Squirrel (*Sciurus yucatanensis*)	11	6.53	400	4400	10.80
Deppe's Squirrel (*Sciurus deppei*)	42	24.56	205	8610	21.13
Artibeus sp.	2	1.17	50	100	0.25
Unidentified mouse	2	1.17	76	152	0.37
Unidentified rat	7	4.09	150	1050	2.58
Unidentified bat	14	8.19	50	700	1.72
Unidentified mammal	1	0.58	180	180	0.44
Total number of identified	**171**	**100.00**	—	**40,739.4**	**100.00**
Unidentified prey	**52**	—	—	—	—
Total	**223**				

[a] Most prey items were captured by males, but some by females. Only observed prey are included here; prey remains are not listed.

ping, and fluttering through limbs or on the ground. Radio-tracking helped us observe hunting behavior, but even so, we were able to witness few captures because of the secretive nature of these birds. Adult males often still hunted, remaining on a perch for several minutes before flying, often a long distance, to another perch or making a capture attempt. We had the impression that these forest falcons hunted partly through their sense of hearing, perhaps aided by the pronounced facial disc, an anatomical feature which they share with owls and harriers (Rice 1982).

We noted the following examples of hunting. Once we observed an adult male pursue a small Deppe's Squirrel around a tree, capture it, and fly to the ground, killing

Table 17.2 Food niche breadth, dietary overlap, and estimated mean weight of prey of Barred and Collared Forest Falcons brought to the nest, Tikal

Food niche parameters	Barred Forest Falcon	Collared Forest Falcon
Total number of identified prey items	267	171
Mammal species identified in diet	3	6
Bird species identified in diet	7	15
Lizard species identified in diet	5	1
Mean weight (g) of prey[a]	23.7 ± 2.5 (267)	238.9 ± 18.9 (170)
Mean weight (g) of avian prey[a]	62.1 ± 15.3 (52)	373.4 ± 49.5 (59)
Mean weight (g) of main prey types other than birds[a]	lizards: 13.8 (122)	mammals: 179 ± 12.5 (78)
Food Niche Breadth[b]	7.9	13.8
Standardized Food Niche Breadth[b]	0.33	0.49
Dietary overlap[c]	0.49	—

[a] Calculations are based on 37 prey categories at the species, genus, and family level. Values are means ± SE (n).
[b] Levins's (1968) formula.
[c] Schoener's (1970) index of symmetrical overlap.

it there. On another occasion, a field technician radio-tracking an adult male flushed a large iguanid lizard, which then ran toward the forest falcon. The falcon turned to see the cause of the commotion and swooped down, seizing the fleeing lizard with its feet. We made several observations of radio-tagged adult males and females chasing Plain Chachalacas on the ground and through trees, by running after them. A female Collared Forest Falcon 13 weeks after fledging caught a Keel-billed Toucan while chasing it around a tree trunk; she used her wings and feet to maneuver around the tree more quickly than the flying toucan, and captured it from behind, presumably by grasping it with her feet.

In Costa Rica, Gary Stiles (pers. comm.) witnessed a prey capture that vividly demonstrated this species's ability to move through dense vegetation in pursuit of prey. In an old treefall gap with a seemingly impenetrable thicket on one side, Stiles flushed a *Leptotila* dove, which flew up from the ground into this thicket, approximately 8 m from the observer. Immediately a Collared Forest Falcon (which may have been watching the dove) appeared from behind the observer and "flew directly and seemingly suicidally at the thicket, dived into it and seemed to be thrusting itself through using both wings and feet," reappeared at the other side with the dead dove in one foot, and silently flew off low over the ground.

In Peru, Robinson (1994) recorded Collared Forest Falcons unsuccessfully attacking Yellow-rumped Caciques (*Cacicus cela*: seven attacks on flying caciques and five on perched individuals) and Chestnut-headed Oropendolas (*Psarocolius wagleri*) at their colonies. The one successful attack reported by Robinson (1994) was on an adult Spix's Guan (*Penelope jacquacu*).

A mixed contingent of birds including Brown Jays, Keel-billed Toucans, Montezuma Oropendolas (*Psarocolius montezuma*), Collared Aracaris, ant-tanagers (*Habia* spp.), and chachalacas frequently harassed Collared Forest-Falcons at Tikal. While tracking radio-tagged forest falcons, we often observed chachalacas, Brown Jays, and toucans voice alarm calls as the forest falcon showed pursuit intention movements, that is, started forward but did not take flight.

In Panama, Smith (1969) described behavior of the Collared Forest Falcon unlike any we observed. In three cases at different localities, he observed individuals perched 1–2 m up, voicing a "rapid, whining *keer, keer, keer . . .* given on an ascending scale." In each case, the falcon was mobbed by many small birds, perhaps in response to the falcon's vocalizations, and in two cases the falcon darted unsuccessfully after the mobbing birds. Smith conjectured that forest falcons used these calls as a hunting ploy, to elicit mobbing and thus lure birds within capture range. He suggested that the ventriloquial qualities of the *Micrastur's* calls rendered the forest falcon's location difficult for the songbirds to track, allowing the raptors to achieve surprise in attacking mobbing birds at close quarters. We never witnessed such behavior at Tikal, nor did we hear vocalizations like these. It seemed apparent, however, that individual forest falcons showed differences in hunting behavior, a case in point being the frequent captures of Keel-billed Toucans by the "satellite" male mentioned earlier. This male often delivered two toucans a day for nearly a month, whereas all other males typically brought in squirrels. As toucans often mobbed Collared Forest Falcons and other raptors, it is conceivable that the satellite male captured them during their mobbing actions, as suggested by Smith (1969).

HABITAT USE

Our observations of habitat use by this species accord well with published statements. Collared Forest Falcons occupy a range of forest types from mangroves and tropical deciduous forest to rain forest, occurring also in gallery forest and tall second growth. They appear to be partial to forest types and sections of the forest with relatively dense vegetation in the lower strata, through which they easily move.

Differences in the geographic ranges of Collared and Barred Forest Falcons illustrate the contrasting habitat preferences of the two. Collared Forest Falcons occur throughout the semiarid Yucatán Peninsula, whereas the Barred Forest Falcon is restricted to the moister basal third of the peninsula (Lopez Ornat et al. 1989; Howell and Webb 1995). Similarly, Collared Forest Falcons occur continuously along Mexico's relatively dry Pacific coastal plains as far north as Sinaloa, with the Barred Forest Falcon being much more restricted in distribution along that coast. These patterns illustrate that the Barred Forest Falcon is restricted to moister forest types, whereas the Collared Forest Falcon, in addition to occurring in these moist forest types, extends widely into areas of lower, drier, more deciduous forest. A similar pattern is indicated for Costa Rica (Stiles et al. 1989).

Collared Forest Falcons at Tikal spent most of their time 5–20 m above ground, from the lower sub-canopy to mid-canopy level. They frequently used those forest types with the densest sub-canopy vegetation, that is, Transitional, Hillbase, and Scrub Swamp Forests (see Chapter 2). On occasion we observed these falcons in dense vegetation in man-made openings near park facilities and along forest edges where chachalacas and other medium-sized birds were common.

Nesting Habitat

Vegetation sampling at several nests each of Barred and Collared Forest Falcons allowed us to compare nesting habitat of the two species. Barred Forest Falcon nests were mainly located amid tall Upland Forest on gentle hill slopes with shallow soil, whereas Collared nests were found mainly in Transitional or Hillbase Forest, on level sites with deep, clay-rich soil, near the foot of a gentle rise (Table 17.3). Barred Forest Falcon nests were in areas of tall dense canopy with relatively open understory, many aroids, and few vines. In contrast, Collared Forest Falcon nests were in areas of relatively low canopy and dense understory with many vines, especially small-diameter vines (Table 17.3). Maximum canopy height around Barred Forest Falcon nests (mean = 27.3 ± 2.9 m, $n = 6$) was significantly taller than at Collared Forest Falcon nests (mean = 22.6 ± 1.6 m, $n = 5$; Table 17.3) (Mann-Whitney U-test, $P = 0.018$), and minimum canopy height likewise averaged higher at Barred nests (17.4 m) than at Collared nests (14.0 m; Table 17.3) (Mann-Whitney U-test, $P = 0.028$). Collared Forest Falcons that we radio-tracked also showed a predilection for hunting in Transitional, Hillbase, and Scrub Swamp Forest zones, where the dense understory may have aided their undetected approach of quarry, facilitating their partly cursorial pursuit style.

BREEDING BIOLOGY

Nests and Nest Sites

Collared Forest Falcons did not build nests or bring foliage to the nest. All nests ($n = 7$) were in tree cavities, with the eggs laid on a substrate of rotten wood chips (Plate 17.4). Nest cavities were 15.7–24.0 m above ground (mean = 19.9 ± 3.1 m, $n = 6$) in trees ranging in diameter from 90 cm to an astonishing 314 cm (mean = 167 ± 91 cm, $n = 5$). Four nests had two or more entrances, and two nests had a single entrance; the only previously reported nest also had two entrances (Mader 1979). All nests were in naturally occurring cavities resulting from a broken limb or heartwood decay; none were in cavities excavated by other birds. Cavity entrances averaged 56 ± 37 cm by 39 ± 13 cm (range = 16–130 cm by 22–62 cm), and the mean depth of nest cavities was 37 ± 19 cm (range = 23–150 cm).

Only quite large cavities were used and it appeared that only very large, old trees provided suitable cavities. With females weighing nearly 900 g, Collared Forest Falcons are large birds and probably require a relatively large nest cavity. Collared Forest Falcons used Spanish Cedar trees in four of the nine nesting attempts studied; the same tree was used in three years by one pair (Cedro pair, 1988, 1989, and 1992), and another tree by a second pair—La Curva, 1991. A Honduras Mahogany was used

Table 17.3 Comparative characteristics of Barred Forest Falcon and Collared Forest Falcon nest environs at Tikal[a]

Feature	Collared Forest Falcon	Barred Forest Falcon
Small vines	Very abundant	Scarce
Large vines	Moderately abundant	Scarce
Understory density	Dense	Open to moderately open
Canopy cover	Moderately dense	Very dense
Aroids	Few	More abundant
Topography	Level to gently sloping	Sloping
Mean canopy height[b]	17.8 m (12–21 m; $n = 5$)	17.7 m (17–32 m; $n = 6$)
Height of tallest canopy[c]	22.6 ± 1.6 m ($n = 5$)	27.3 ± 2.9 m ($n = 6$)
Height of lowest canopy[d]	14.0 m ($n = 5$)	17.4 m ($n = 6$)
Soil depth[e]	110 cm ($n = 4$)	40 cm ($n = 4$)
Forest types	Mature Transitional and Hillbase Forest	Mature Upland Forest

[a] Based on quantitative sampling at 5 nests of Collared Forest Falcon and 6 of Barred Forest Falcon, from Enquist et al. 1992, unpubl. ms.

[b] Not significantly different: Mann-Whitney U test; $P = 0.46$.

[c] Significantly different: Mann-Whitney U test; $P = 0.018$.

[d] Significantly different: Mann-Whitney U test; $P = 0.028$.

[e] Significantly different: Mann-Whitney U test; $P = 0.04$.

twice, a Black Cabbage-bark *(Lonchocarpus castilloi)* once, a Yellow Mombin *(Spondias mombin)* once, and a Chicozapote *(Manilkara zapota)* once. These falcons nested in trees that were far larger than expected given the distribution of tree diameters present in the forest (B. Enquist et al. 1992). In addition, from among available trees of suitable size (> 52 cm in diameter), Collared Forest Falcons demonstrated positive selection for Spanish Cedar and for no other species as nest trees (ibid.). It is possible, however, that this selectivity was also based on tree size, as many of the largest trees in the forest were Spanish Cedars.

Egg and Clutch Size

Six nests had 2-egg clutches, one had a 3-egg clutch, and one a 1-egg clutch, for a mean (and modal) clutch size of 2.0 ± 0.5 ($n = 8$). Mean egg dimensions were 56.4 ± 2.0 mm by 43.6 ± 1.5 mm ($n = 9$). The average egg mass was 53.4 ± 2.6 g ($n = 9$), or 6.1% of the 869 g mean female body mass; hence the modal and mean 2-egg clutch is equivalent to 12.3% of mean female mass. Using Hoyt's (1979) formula, estimated fresh egg mass was 58.6 g, or 6.7% of mean female body mass; thus, the modal and mean 2-egg clutch would equal 13.5% of mean female mass. All eggs were subelliptical and dark reddish brown with tan and dark brown spots, similar to published descriptions (Wetmore 1974; Thorstrom et al. 1990).

Nesting Phenology

Adults visited prospective nest cavities from January through March, and by early March females spent most of their time in "pre-laying lethargy" (Newton 1979). Eggs ($n = 7$ clutches) were laid in February and March, during the early to mid dry season, with one exception (see below), and hatched late in the dry season, from 16 April to 16 May, with the average hatch date being 23 April ± 5.0 days ($n = 6$). The mean date of fledging was 13 June ± 9.3 days ($n = 8$), at the onset of the rainy season. One atypically late clutch was laid by the Cedro pair in May 1992. This late laying apparently resulted from usurpation of a traditional nest site by nesting American Black Vultures (*Coragyps atratus*; see Nest Defense and Interspecific Interactions later in this chapter).

Length of Incubation, Nestling, and Post-fledging Dependency Periods

At Tikal, our data indicate an incubation period of 46–48 days, a nestling period of 46–56 days, and post-fledging dependency of 6–11 weeks. At the 1989 Cedro nest the female had been incubating for 40 days when the nest was depredated. At another nest, an egg hatched after 46–48 days of incubation. One young fledged at 46 days, but most fledged 50–56 days after hatching (mean = 50.0 ± 4.2 days, $n = 3$ males and 2 females).

One female apparently reached independence and dispersed six weeks after fledging, and one young male was independent or nearly so by seven weeks after fledging, when he was killed by a raptor. This male's female sibling was still attended by an apparently nonparental adult 11 weeks after fledging, although she also caught a toucan at this age. It appears safe to conclude that these forest falcons generally reach independence within a few months after fledging. That they did not show a prolonged dependency period is supported by the fact that pairs often nested in consecutive years, even when successful the previous year. For example, the Aguadita pair nested in 1990, 1991, and 1992. Long incubation, nestling, and post-fledging periods extended the breeding cycle of the Collared Forest Falcon to nearly 28 weeks, which is protracted in comparison to that of similarly sized temperate raptors such as the Northern Goshawk, at 20–21 weeks (Reynolds and Wight 1978; Palmer 1988).

VOCALIZATIONS

Throughout the breeding season, Collared Forest Falcons vocalized each morning except during rain. Males and females began calling before sunrise and continued until mid-morning. These forest falcons were generally the first diurnal birds to begin calling in the morning, on average 36.2 ± 10.1 minutes before official sunrise ($n = 39$ calling bouts of six individuals throughout the nesting period). On moonlit nights Collared Forest Falcons would occasionally vocalize between 0100 and 0400 in the morning. These ritual predawn and early morning vocalizations probably serve mainly to advertise territorial occupancy and maintain spacing among neighbors.

The typical call (advertisement call or song), given by both male and female, was a series of mellow, prolonged *ahr* or *ow* notes, three or four given in a series over a span of about eight seconds, followed by another series after a 20–30-second pause. The most common arrangement was a four-note call consisting of two consecutive *ahr* notes, a two-second pause, then the last two notes. In addition, females at times voiced a rapid tempo call featuring 10–30 rapid *ahr* notes with increasing tempo, starting after the normal three- to four-note call; this female call spanned 10 seconds.

Early morning calling was often conducted from a tall perch, probably near the night roost. Pair members generally roosted 100–300 m apart and 50–400 m from the prospective nest tree. Typically, vocalizations began when either a male or a female forest falcon called from a roost and the mate responded. Normally, pair members vocalized alternately from their respective perches. Occasionally a distant neighboring or unpaired bird would respond to the vocalizations of the territorial pair, as revealed by behavior of a radio-tagged, unpaired female in 1989 who sometimes responded vocally to calling by neighboring pairs.

Young Collared Forest Falcons voiced soft *scree* calls during the early nestling period, at the ages of one to three weeks. As they approached fledging age, and for several weeks after fledging, young voiced single-note *ahr* calls, ranging in intensity from soft to loud, in the context of food begging. Adults delivering prey to fledglings would give a soft *ahr* call as they attempted to locate the young, and young responded with rapid, single-note *ahr* calls. Upon receiving prey, fledglings gave a rapid, excited *ahr* call. A sibling not receiving food often continued to softly voice food-begging *ahr* calls.

Low-frequency notes such as those used by Collared Forest Falcons are believed advantageous for long-range communication by forest birds, because such low frequencies suffer less attenuation by forest vegetation than do higher-pitched calls (Morton 1975; Wiley and Richards 1982). Calling in the still predawn hours resulted in the falcons' calls being audible over great distances (see Territorial Behavior and Displays in this chapter). Diurnal species tend to concentrate their long-range communications in the first hours of daylight and to a lesser extent in the evening (Henwood and Fabrick 1979), presumably at least in part because conditions for sound propagation and reception are good at these times. Since the daily temperature cycle tends to lag behind the solar cycle, atmospheric conditions for long-range sound propagation should usually be more favorable at sunrise than at sunset (Wiley and Richards 1982). In addition to physical sources of signal attenuation and degradation (e.g., wind, largely tied to the sun's heating action), biotic sources of signal interference at Tikal increased dramatically soon after the typical period of the Collared Forest Falcons' morning calling—as many bird species, cicadas, and Guatemalan Howler Monkeys *(Alouatta pigra)* all began to contribute to ambient noise levels.

BEHAVIOR OF ADULTS

Pre-laying and Laying Period

We observed courtship activities in January, February, and March. After early morning vocalizations, a pair during the pre-laying phase would often move to the potential nest site. Normally, the male flew into the nest cavity and began voicing soft, continuous *ahr* calls. He would remain in the cavity until the female entered, or would exit and then the female would enter. When the pair were together in the nest vicinity, they continually voiced soft *ahr* calls. When the male left the area, the female typically remained in the nest vicinity, calling softly and apparently guarding the cavity.

Prey exchanges during this period took the following form. As the male approached the nest vicinity, he voiced soft *ahr* notes while 50–100 m away. After responding with soft *ahr* calls, the female would fly to the vocalizing male. If the female did not respond, the male would call every 30–60 seconds, moving closer to the nest site until the female called back. When the female reached the male, she responded with faster and louder, soft *ahr* vocalizations. If the male arrived with food, an aggressive and vocal food exchange occurred.

We observed only two copulations of these secretive birds, both occurring after prey exchanges, and before the female ate. Copulations were silent and lasted five to eight seconds. After the copulations, the male flew to the nest briefly or left the area. Generally, morning courtship ceased by two to three hours after dawn, and evening courtship was observed during the last half hour or so before the falcons went to roost.

Incubation Period

Females performed all incubation, and males provided essentially all food until late in the nestling period, when females began hunting and delivering prey to the young. During the incubation period, the female left the nest every day for 5–50 minutes in early morning and late evening to call, voicing the normal song and the rapid tempo call. We located several nests by concentrating on the call given by the female in the morning.

Nestling Period

Roles of adults as providers. Total prey items delivered per day ranged from none during incubation to six during the nestling period. During incubation and the first weeks of the nestling period, males provided nearly all prey. Females began to hunt and bring in prey when young were 17–21 days old (*n* = 3 nests). Taking the nesting cycle as a whole and omitting the contribution of the non-pair male at one nest, males delivered 66% of observed prey biomass (11,430 g) and females 34% (5960 g).

Prey transfers, preparation, and caching. A male arriving near the nest with prey would alert the female by calling softly. The pair then often called back and forth for a minute or so. The female usually flew toward the male and seized the prey. Prey transfers were made from foot to foot on a perch and sometimes were very aggressive, with the bird in possession of the prey appearing reluctant to relinquish it. Prey items, especially mammals, were usually headless and partially eaten when delivered by males. Females usually delivered prey intact except for very large items.

We witnessed few instances of prey caching, but in several cases, males delivered prey items containing maggots to the nest, suggesting they had been cached and later retrieved. I observed the following instance of prey caching on 29 June 1991. At 0524, I began radio-tracking the adult male of the La Curva nest, who was tending two recently fledged young. I located the male about half a kilometer from the nest. At 0624, the male's signal fluctuated as if he were engaged in some activity, and I glimpsed a silhouette of the male running on the ground. At 0635 the male flew to a slanted perch, looked around, and hopped down the slanted tree. He grabbed what turned out to be a large chunk of meat from a crotch of the tree, dropped to the ground, and began running northward with the food item clutched in one foot. I then inspected the site from which he had retrieved the item, and collected feathers apparently from this cached prey item.

Post-fledging Dependency Period

At both nests where we monitored young birds after fledging, the male was their main provider. Females at these two nests rarely delivered food during this period and mostly remained away from the nest vicinity. The adult female at the 1991 La Curva nest stopped delivering food to the young during the second week after fledging, and the male became the sole provider thereafter. At the Aguadita nest, during the middle of the post-fledging period an extra (presumably nonparental), unbanded adult male forest falcon began delivering prey to the two fledglings. Both parents stopped delivering food to the young when this unbanded male took over the provisioning role; he performed all prey deliveries subsequently observed, sometimes delivering two Keel-billed Toucans per day. In the fifth week after fledging—when young were 84–91 days old—the extra male accompanied (led?) both young to an area 2.5 km south of the nest, and by week 7 they ranged to 3.5 km from the nest. During week 8 after fledging (105–112 days old), the fledgling male was found dead, apparently killed by a raptor. The fledgling female was still attended by this extra adult male until week 11 after fledging (126–133 days old), when we lost radio contact with her: we assumed she had reached independence when she was observed catching a toucan 11 weeks after fledging. The significance of this apparently nonparental adult's behavior in providing for these fledglings was unknown. One possibility is that his behavior was oriented toward forming a pair bond with the young female; it would be interesting indeed to know who killed the young male. On the other hand, many birds have occasionally been seen caring for juveniles that are not their own offspring—even of other species—a phenomenon that results when juvenile begging triggers adult feeding responses, sometimes in situations that must be regarded as accidental (Shy 1982; Frumkin 1994).

**BEHAVIOR AND DEVELOPMENT
OF NESTLINGS**

The six hatchlings we examined were similar to Barred Forest Falcon hatchlings (see Chapter 16) but larger. They had white natal down, closed eyes, shiny white nails, pale yellow legs and cere, and beaks that were short, deep, laterally compressed, and yellowish white. By two to three days after hatching, eyes were open and the nestlings held their heads up. They had brownish black iris and blue-black pupils. By days 7–14, the cere, feet, and toes were yellowish white and nails light gray; the egg tooth was still faintly present. During their third week, the young actively defended themselves when disturbed, lying on their back or side and reaching out in efforts to foot the researcher's intruding hand. The facial skin around the eyes was turning yellow-green, the upper and lower mandibles were becoming light gray at the base, and primaries were emerging.

During the chicks' fourth week, tail feathers were emerging, the mandibles were turning grayish black near the cere, and legs and feet were nearly adult in size. Natal down between the eyes and beak had disappeared, the iris was a dark chocolate brown, and the facial skin was turning greenish olive. Breast and auricular feathers began emerging and wing coverts grew rapidly. During the fifth week, nails were turning black, body feathers were developing, natal down was disappearing, and mandibles were changing to brownish yellow toward the tip. Facial skin was a dark greenish olive, the cere was yellowish green, and the legs and tarsi were yellowish white. Nestlings at this age begged for food without any aural stimulus from the adults. During the sixth week, down feathers had mostly disappeared, and juvenile contour feathers nearly covered the body. Tail feathers were one-third developed and chest feathers began showing barred coloration. At the ages of six to seven weeks, tail feathers were half grown and wings two-thirds grown, and few natal down feathers remained. Mandibles were dark brownish black. Young males ($n = 5$) had tawnier chests than did females ($n = 3$).

Post-fledging Dependency Period

At the moment of fledging, with tail and wing feathers far from fully grown, young fluttered clumsily from the nest cavity to a nearby branch or tree. No young returned to nest cavities after fledging. Fledglings remained perched about 10 m high at first, except during feedings, when they dropped to the ground to receive prey.

Adults arriving with food voiced a soft *ahr* call, to which fledglings would respond with similar but louder food solicitation calls. The adult usually delivered food to the nearest fledgling early in the fledgling period, and to the one quickest to respond later in the fledgling period. Adults deposited prey on the ground or on fallen trees early in the fledgling period, whereas later the young would intercept the approaching adult, take the food item, and fly away from its sibling. The unfed young would continue to beg or would search for the sibling that had received prey. A hungry fledgling encountering a sibling with food would perch nearby, beg for food, and watch the feeding young, occasionally attempting to steal the food item, usually without success. Sometimes after the first fledgling finished eating, it would carry off the food item and cache it, preventing its sibling from eating the remainder. More often, the waiting young would commandeer the prey remnant after the first young stopped eating.

During the first week after fledging, young remained within 50 m of the nest. Normally, fledglings remained separated by some 10–50 m and flew weakly or ran on the ground or limbs from one perch to another. As night approached, the young hopped and fluttered to the upper canopy to spend the night. In the morning they would vocalize briefly from their high roost before dropping to the ground to play and wait for food. Early in the fledgling period, young spent a good deal of time on the ground—perhaps half the time in some cases. At two weeks after fledging (age 63–70 days), young remained together and often chased one another on the ground and through vines and trees. As time passed, the fledglings spent less time on the ground.

By three weeks after fledging (age 70–77 days), young started moving farther from their nest sites and were capable of longer flights; they often traveled 200 m or more to intercept an adult bringing food. During this period we once found fledglings at an army ant swarm, where they watched and unsuccessfully pursued insects and lizards for two hours during an eight-hour observation period, their behavior similar to descriptions by Willis et al. (1983), Mays (1985), Thorstrom (1993), and Robinson (1994). By the fourth week after fledging (age 77–84 days), young were extremely active in playing and chasing and watched other animals attentively. In several cases we observed fledglings chasing a Nine-banded Armadillo (*Dasypus novemcinctus*), evidently a form of play or hunting practice.

SPACING MECHANISMS AND POPULATION DENSITY

Territorial Behavior and Displays

Collared Forest Falcons did not soar above the canopy as do many Neotropical forest raptors, and there is little question that their loud calls are their principal means of locating and communicating with mates, neighbors, and other conspecifics—for example, in proclaiming territories and maintaining spacing among neighbors. Collared Forest Falcon vocalizations often carried well over a kilometer—a much greater distance than did the higher-pitched and lower-volume calls of Barred Forest Falcons. The relative audibility of these two species at a distance is in keeping with their relative home range sizes, with the more distantly audible Collared Forest Falcon using much larger home ranges than the smaller Barred Forest Falcon. Among animal sounds at Tikal, only those of the howler monkey and Laughing Falcon (*Herpetotheres cachinnans*) carried farther than those of the Collared Forest Falcon. We witnessed no territorial disputes among Collared Forest Falcons.

Nest Defense and Interspecific Interactions

In the Cedro territory, in February 1992, American Black Vultures nested in the cavity the forest falcons had used in two previous years. We removed the vultures' egg in March when the forest falcons were engaged in courtship near this nest. The falcons then occupied the site, laying two eggs between 19 and 21 May. This nesting attempt was two to three months later than the normal laying period of Collared Forest Falcons at Tikal; it seems probable the vultures' use of the cavity prevented the falcons' laying here until we intervened. Apparently

this forest falcon pair did not have an alternate nest site available, and they waited for their historic nest cavity to be vacant. In Ecuador, a pair of Collared Forest Falcons was observed usurping a nesting cavity in use by Great Green Macaws (*Ara ambigua*: López-Lanús 2000). Over a four-month period, the forest falcons attempted to usurp the cavity, attacking the macaws 180 times. They succeeded when the female forest falcon entered the cavity and removed the macaw chick, dropping it to the ground. The forest falcons immediately occupied the cavity and appeared to be incubating eggs within a month (ibid.). These two instances suggest that the large cavities favored by Collared Forest Falcons may at times be in short supply.

Some biologists are concerned that the recent spread of Africanized Honeybees (*Apis mellifera* hybrid) throughout the Neotropics may result in displacement of cavity-nesting birds from nest sites. In one case, honeybees—presumably Africanized—established a hive in a cavity 1–2 m from a Collared Forest Falcon nest. When we attempted to climb to the forest falcon nest, the tree's movement caused the bees to attack us, forcing us to abandon efforts to inspect this nest. The bees also entered the forest falcon nest cavity on this occasion, but the nest went on to produce fledglings.

One female (Aguadita territory) was notably aggressive, always attacking researchers when they climbed to the nest. She was also very bold in remaining on the nest, and in one case, I could not check the status of the eggs/chicks until I removed her from the nest by hand; this also occurred at one Barred Forest Falcon nest.

Constancy of Territory Occupancy, Use of Alternate Nest Sites

These forest falcons were year-round residents at Tikal, and they showed a high degree of fidelity to nesting territories and mates. We observed no cases of territory switching and only one case of adult turnover, when the male of the Curva pair was found dead and the surviving female later re-paired. Two territories with known histories of five and three years were occupied each year (Table 17.4). Four study pairs were observed during more than two years, but only two of these (Aguadita and Cedro) nested in two or more years: both of these pairs used two alternate nest sites (Table 17.4).

Home Range Estimates

Using radiotelemetry, we estimated home ranges for four adult females and three adult males (Table 17.5); three fledglings were also radio-tracked during the post-fledging period. All radio-tracking of adults took place between March and August—hence all home ranges, including those of nonbreeders, were described for the breeding season. Nonbreeders had no dependent young during the time they were radio-tracked. For most breeding and nonbreeding adults the asymptotic home range estimate (Minimum Convex Polygon [MCP]) was

Table 17.4 History of breeding attempts by Collared Forest Falcons at Tikal, 1988–93

	Nesting territories[a]									
	Cedro (1988)		Aguadita (1990)		La Curva (1991)		El Caoba (1991)		Chico (1993)	
Nesting year	NA	Fled	NA	Fled	NA	Fled	NA	Fled	NA	Fled
1988	1	?	—	—	—	—	—	—	—	—
1989	1	0	—	—	—	—	—	—	—	—
1990	0	0	1	2	—	—	—	—	—	—
1991	0	0	1	1	1	2	1	0	—	—
1992	1	0	1	2	[b]	—	—	—	—	—
1993	[b]		[b]	—	—	—	—	—	1	1
Total	3	?	3	5	1	2	1	0	1	1

Notes: — = lack of data, not absence of a nesting attempt; NA = number of nesting attempts; ? = the outcome in 1988 was not observed.

[a] Date in parentheses is the year nest was first located. Fled is the number of young fledged.

[b] Pair present but insufficiently studied to know whether nesting occurred.

reached at 70–90 relocations, but a few home range estimates were still expanding after 100 localizations.

Figure 17.1 depicts the MCP home ranges we estimated for adult Collared Forest Falcons. There was little overlap in home ranges of neighbors, even when estimated in different years; this suggests fairly stable, complementary spatial use patterns. The spatial distribution of home ranges suggested that suitable habitat was largely filled with abutting home ranges or territories, with little unused space between neighbors. Collared Forest Falcons used much larger home ranges than did the smaller Barred Forest Falcon. The home range of the Aguadita Collared Forest Falcon pair, for example, encompassed territories of at least six pairs of Barred Forest Falcons.

Two breeding males had estimated home ranges of 7.1 and 11.8 km² (MCP), for a mean of 9.4 km². One of these males (La Curva/Naranjal) showed a larger home range during four months as a nonbreeder than during seven months as a breeder (Table 17.5). Our only home range estimate for a breeding female was 11 km² (MCP) or 13.5 km² (85% Harmonic Mean [HM])—similar to the largest home range estimate for a breeding male—whereas three nonbreeding females had much smaller estimated home ranges, with the two best estimates averaging 5.5 km² (MCP; Table 17.5). During incubation and the first part of the nestling period, females were closely tied to nest vicinities. The breeding Aguadita female moved around a good deal during the early part of the fledgling period, which was included in our radio-tracking, and this may account for her estimated home range being larger than those of three nonbreeding females. Averaging all three breeding adult home ranges gives an average home range estimate of 10.0 km² (MCP) or 10.7 km² (85% HM)—our best estimates of the mean annual home range of a breeding pair.

Other individuals monitored by telemetry included an adult-plumaged female trapped in May 1989 ("Airport female" in Table 17.5), which ranged over a large area and appeared to be a nonbreeding floater without a clear territory. Her signal disappeared one month later when either the radio failed or she moved beyond radio

contact. In 1991 an adult male (not listed in Table 17.5) was radio-tagged at a nest several kilometers south of the park, near the village of Caoba. He was difficult to follow, as he traveled large distances in this farming landscape, and after a time his signal disappeared. When we later examined the nest, it contained three addled eggs, apparently abandoned. We believed it likely that one or both adults of this pair were killed by humans during this nesting attempt, resulting in abandonment of the eggs (see Conservation later in this chapter).

Size of home range no doubt varies among pairs and over time, depending on habitat and other factors including density of prey and possibly of conspecifics. As an overall estimate of home range size, however, we regard the estimate of 10–10.7 km² given here as appropriate; this accords well with the mean nesting density we calculated, of one pair per 9.22 km² (see below).

Inter-nest Spacing and Density of Territorial Pairs

Mean inter-nest distance among four neighboring, occupied nests was 3070 ± 550 m (range = 2520–3615 m, $n = 3$; Fig. 17.2). A fifth nest was only about 2110 m from one of the four, but its location was not as precisely established; hence it is omitted from the above calculations. This inter-nest distance yields a Maximum Packed Nest Density (MPND; see Chapter 2) estimate of 8.57 km² of exclusive space per territorial pair, whereas the square method gives an estimated 9.42 km² per pair, and the polygon method 9.22 km² per pair (Fig. 17.2). Considering these and the home range estimates given earlier, we estimate an average density of approximately 10 territorial pairs of Collared Forest Falcons per 100 km² of equivalent habitat in our study area.

In South America, density estimates have been made for this forest falcon at two sites. In French Guiana, in an area with four *Micrastur* species, Thiollay (1989b) estimated densities on the order of five territorial pairs of Collared Forest Falcons per 100 km², that is, half the density we estimated at Tikal. At a gross, distributional level, four or five *Micrastur* species supposedly occur at or near Manu

Table 17.5 Home range estimates (hectares) for Collared Forest Falcon adults studied via radiotelemetry at Tikal[a]

Nesting territory (year)	Number of locations	Reached asymptote	Dates of data collection	Breeding individual		Nonbreeding individual	
				Minimum Convex Polygon	85% Harmonic Mean	Minimum Convex Polygon ±SD	85% Harmonic Mean ±SD
Females							
Airport (1989) nonbreeder	41	No	May–Jun	—	—	229	212
Cedro (1990) nonbreeder	88	Yes	March–May	—	—	572	579
Pucté (1991) nonbreeder	26	No	March–Jun	—	—	535	617
Mean of 2 nonbreeding females above				—	—	554 ± 26	598 ± 28
Aguadita (1990) breeder	96	Yes	April–Aug	1098	1346	—	—
Males							
La Curva (Naranjal) (1990) nonbreeder	89	Yes	May–Aug	—	—	885	783
Aguadita (1990) breeder	158	Yes	May–Aug	1176	1215	—	—
La Curva (1991) breeder	94	Yes	May–Nov	713	635	—	—
Mean of 2 breeding males above				944 ± 328	925 ± 411	—	—

[a] The following fledglings were followed by telemetry, though home range estimates were not made: Aguadita 1990-a, June–August; Aguadita 1990-b, June–July; La Curva 1991, June–November.

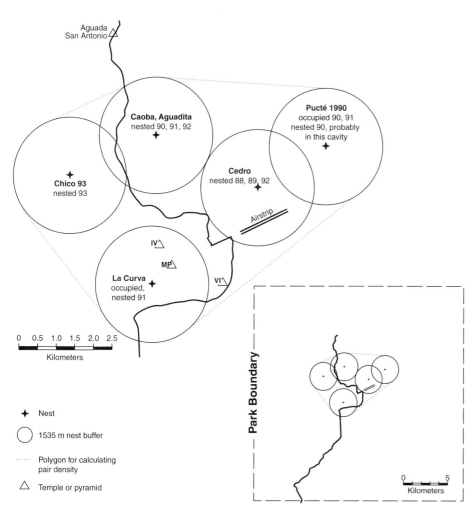

17.1 Collared Forest Falcon nest locations buffered by half the mean inter-nest distance, and polygons used in estimating pair density, Tikal.

National Park in Amazonian Peru; however, Terborgh et al. (1990), Robinson (1994), and Robinson and Terborgh (1997) report only three species—any additional species, if present, must be rare. Terborgh et al. (1990) and Robinson and Terborgh (1997) estimated a density of 0.1 to 0.25 pair of Collared Forest Falcons per square kilometer—that is, from the same density as at Tikal, to a density about 2.5 times as great. These South American density estimates, however, were based on less intensive study than ours.

DEMOGRAPHY

Frequency of Nesting, Percentage of Pairs Nesting

We monitored five territorial pairs a total of 10–13 territory-years, during which nine nesting attempts were made, for an overall nesting frequency of 69–90% of territory-years (Table 17.4). Ten territory-years were well studied such that all nestings were detected; in an additional three territory-years, our coverage was not complete, although we suspected nesting did not occur. The two territories with the longest known histories were the Cedro territory (five years) and the Aguadita territory (three years). Both were occupied each year. The Aguadita pair nested in each of three years, and the Cedro pair in three of five years, for a mean nesting frequency of 80% for these two territories. Birds that had failed in one year

tended not to breed the following year, whereas birds that were successful tended to breed continuously from one year to the next (e.g., Aguadita pair; see Table 17.4).

Productivity and Nest Success

In eight fully documented nesting attempts, 16 eggs were laid, for a mean clutch of 2.0 (Table 17.6). Nine (56%) of 16 eggs hatched, and eight (89%) of these fledged (five males and three females), for an average of 1.0 fledgling per breeding attempt and 0.8 fledgling per territory-year. Five (63%) of eight nesting attempts fledged one or more young (Table 17.6). We did not observe re-nesting when first attempts failed during the egg or nestling stage.

Collared Forest Falcons had a breeding success rate of 63% (5/8) and productivity of 1.0 fledgling per breeding attempt and 0.8 fledgling per territory-year. Several similar-sized north temperate raptors show at least twice this level of breeding productivity: Northern Goshawks fledged 1.7–2.5 young per nesting attempt, Prairie Falcons *(Falco mexicanus)* 1.9–3.1, Peregrine Falcons *(F. peregrinus)* 1.8–2.7, and Cooper's Hawks 1.7–3.5 young per nesting attempt (Palmer 1988).

Causes of Nesting Failure and Mortality

The main cause of nest failure was predation on eggs (n = 2 nests), probably by mammalian or avian predators. In addition, at the Aguadita nest, one of two nestlings disappeared in 1991 from unknown cause. At the Cedro nest, the lone four-week-old nestling died in the nest when the adults stopped delivering prey; they appeared to have abandoned the young. This may have been caused by researcher activity, as we had trapped the adult male shortly before he ceased bringing prey to the young.

We documented two deaths of recent fledglings. In July 1990, one of the two fledglings from the Aguadita nest was found dead some 2 km from the nest during his eighth week after fledging (age of 105–112 days). We suspect he was killed by a raptor, as feathers lying beside him had been plucked. In November 1991, a radio-tagged fledgling male from the La Curva territory was found dead of unknown cause, about 3 km from the nest, 16 weeks after fledging. We speculate that he may have died while making the transition to independence. His sibling, a female, had dispersed seven weeks after fledging. At fledging, this young male was smaller and, at 400 g, some 200 g lighter than were two other fledgling males; he seemed poorly developed behaviorally and weak in flight, spending most of his time on the ground. He appeared to have fledged in a malnourished state, which may have contributed to slow development and subsequent death, especially if the adult male ceased prey deliveries and the young male did not capture prey; the adult male here also died during this time period, but which of the two died first is unknown. This was the only case of adult mortality we documented.

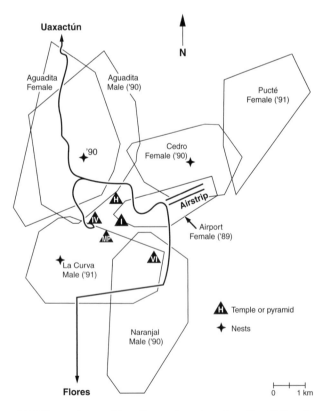

17.2 Home ranges of breeding male and nonbreeding Collared Forest Falcons in Tikal National Park, Guatemala, using the 100% Minimum Convex Polygon method. (Nonbreeding home ranges were those for the Pucté female in 1991, the Cedro female in 1990, the Airport female in 1989, and the Naranjal male in 1990; all others are breeding home ranges.)

Parasites and Pests

We observed no instances of botfly infestations of nestlings.

Table 17.6 Reproductive parameters for Collared Forest-Falcons at Tikal, 1988–93

Year	Territorial pairs (number)	Nesting attempts (number)	Eggs/clutches (number)	Eggs hatched (number)	Young fledged (number)	Fledglings/ breeding attempt (number)
1988[c]	1[c]	1[c]	?	1[c]	?	?
1989	1	1	2/1	0	0	0/1
1990	2	1	2/1	2	2	2/1
1991	4	3	7/3	4	3	3/3
1992	2	2	4/2	2	2	2/2
1993	1	1	1/1	1	1	1/1
Total	11	9	16/8	9/16 (56%)	8/9 (89%)	1.0 (n = 8/8)

Year	Fledglings/ territorial pair (number)	Fledglings/ successful nesting (number)	Breeding success rate/territorial pair (number)[a]	Reproductive success rate/ nesting attempt (number)[b]
1988[c]	?	?	?	?
1989	0/1	0/0	0/1	0/1
1990	2/2	2/1	1/2	1/1
1991	3/4	3/2	2/4	2/3
1992	2/2	2/1	1/2	1/2
1993	1/1	1/1	1/1	1/1
Total	0.8 (n = 8/10)	1.6 (n = 8/5)	5/10 (50%)	5/8 (63%)

[a] Breeding success rate is proportion of territory-years producing at least one fledgling.
[b] Reproductive success rate is proportion of nesting attempts that resulted in at least one young fledged.
[c] The 1988 nesting attempt is excluded from most calculations, as the clutch size and outcome were unknown.

CONSERVATION

This species's ability to use a variety of forest types, including second growth, may bode well for its persistence in the face of anthropogenic habitat change. However, Collared Forest Falcons at Tikal utilized very large, old-growth trees for nesting, presumably because the large natural cavities favored as nest sites were provided by such trees. Moreover, several nests were in tree species important to the commercial logging industry. This raptor thus presents an interesting contrast; while it is able to forage in areas of dense understory, including edge and second-growth habitats, unusually large trees seem important for nesting. Hence, modified habitat that might otherwise be suitable for this species may not provide nesting opportunities.

Six of nine nesting attempts took place in prized commercial tree species. Whether populations of Collared Forest Falcons are affected by the selective removal of large, nesting trees remains unknown. The incidents cited earlier involving American Black Vultures and Great Green Macaws suggest that nesting cavities for these forest falcons may sometimes be limited. As researchers focus on the conservation utility of logged forests, this and other species that appear reliant on cavities in large trees of high commercial value (e.g., Honduras Mahogany and Spanish Cedar) would seem worthy of special attention. It remains to be seen what degree of flexibility in nest site use these birds may demonstrate in areas with few large, old trees. It would be interesting to know whether other nest sites, such as epiphytes or stick nests of other raptors, would be used in some

situations. In addition, efforts to manage and protect large mature trees in human-altered and second-growth habitats might increase the chances of these birds nesting successfully in such modified habitats. It may also be worth investigating whether these falcons would nest in artificial structures simulating natural cavities, as a means of increasing reproduction in certain situations.

Finally, this is likely one of the Neotropical forest raptors most prone to direct conflict with rural humans over the issue of who shall eat the poultry. In Costa Rica, Gary Stiles (pers. comm.) witnessed two attacks by Collared Forest Falcons on poultry, including an adult rooster, and Stiles found this to be the main raptor species referred to as "gavilán pollero"—chicken hawk—by rural Costa Ricans. Villagers in our study area also recognized this as a poultry thief often persecuted by rural residents. The failure of a nest in the farming landscape near Tikal likely resulted from human disturbance or shooting of one or both adults.

HIGHLIGHTS

The overriding impression left by this forest falcon is one of mystery. Though its presence is indicated daily by its eerie, human-like calls in the silent predawn blackness, sightings are occasional at best. Decidedly crepuscular in habit, these secretive birds maintained spacing via their loud predawn, late evening, and sometimes nocturnal vocalizations.

Also remarkable is this forest falcon's extremely long-tailed, long-legged morphology, and its penchant for moving through the densest understory or on the ground, as much by foot as by wing. In form and motion, Collared Forest Falcons somewhat resemble the chachalacas they often take as prey. Prey taken were distributed nearly equally between mammals (37–46%), mainly squirrels, and birds (34–54%), especially toucans and chachalacas. Another remarkable feature is the species's occasional capture of huge prey items in relation to its own body size.

With males at nearly 600 g and females nearly 900 g, this was the largest raptor we studied at Tikal in which fledglings reached independence within several weeks of fledging; all larger raptors showed extended post-fledging dependency, resulting in nesting cycles longer than 12 months. The case we witnessed of an apparent helper at the nest begs further study; helpers have been documented in a few other raptor species (Mader 1975; Bednarz 1987), but not previously in a tropical forest raptor.

These secretive raptors nested in large natural cavities in the largest trees in the forest, often in commercially valuable species such as Spanish Cedar and Honduras Mahogany, and anecdotal evidence suggests that suitable cavities may sometimes be in short supply. Whether logging affects populations of Collared Forest Falcons would be an intriguing question for further research.

For further information on this species in other portions of its range, refer to Carrara et al. 2007, Di Giacomo 2005, Gerhardt 2004, Guedes 1993, López-Lanús 2000, and Vallejos et al. 2008.

APPENDIX 17.1

Nesting Histories of Collared Forest Falcon Territories Studied at Tikal

At Tikal we studied mainly five Collared Forest Falcon territories within the park, and collected fewer data on a sixth pair in the farming landscape nearby.

1. Cedro territory. Nest tree: Spanish Cedar *(Cedrela odorata)*

This territory was located in 1988 and studied from 1988 to 1992; it was occupied each year, and nesting attempts took place in 1988, 1989, and 1992. In 1990 and 1991 the pair was present and the female visited the nest site but did not lay eggs. During four consecutive years from 1989 to 1992, this pair did not nest successfully. A nestling was produced only in 1988; though our fieldwork ended before it fledged, we suspect it did fledge. In 1989 this pair laid two eggs, which suffered predation prior to hatching (40 days after laying), and in 1992 the pair produced one nestling, which died, apparently of starvation. We banded this adult female in 1990, and she was associated with the same nest in 1991 and 1992. The male here was trapped and banded in 1992.

2. Aguadita territory. Nest trees: Yellow Mombin *(Spondias mombin)* and Honduras Mahogany *(Swietenia macrophylla)*

This territory was discovered in April 1990 when we trapped and radio-tagged a female with a brood patch; she led us to the nest, where she was incubating two eggs. The male was banded and radio-tagged later in the year. We studied this pair from 1990 to 1992, and nesting attempts were made in each of these years. In 1990 the pair fledged two young. In 1991, they switched to a nest site 600 m north of the original site and fledged one young; a second nestling disappeared during the first week after hatching. In 1992 this pair fledged two young at the same nest site used in 1991.

3. La Curva territory. Nest tree: Spanish Cedar

This territory was studied only during a 1991 nesting attempt. The adult male was radio-tagged as a nonbreeder in 1990 near the Naranjal water hole—not far outside his 1991 breeding home range. The female at this territory was never banded. This pair fledged two young in 1991, both banded. The juvenile female dispersed seven weeks after fledging. The young male, which fledged at subnormal size, was radio-tagged; in November, approximately 16 weeks after fledging, he was found dead about 3 km from the nest, of unknown cause; we speculate that he died during the transition to independence. The adult male was also found dead of an unknown cause in November 1991, about 4.0 km northeast of the nest site; which one died first was unknown. In 1992 we observed a pair checking a potential nest cavity 1.0 km north of the 1991 nest site; we believed this 1992 pair to be within the same (La Curva) territory.

4. Pucté territory. Nest tree: Pucté *(Bucida buceras)*

We studied this territory in 1990 and 1991. In 1990 we observed two fledglings, fed by a dark-morph adult male during August, but we did not find the nest tree at that time. In March 1991, in the same area, we trapped and radio-tagged an adult female near a Pucté tree with a cavity that showed signs of a previous year's nesting—probably the cavity from which the 1990 young fledged. This female apparently did not nest in 1991. Because uncertainties remained surrounding the nesting history of this territory, it was omitted from Tables 17.4 and 17.6 and from calculations of reproductive performance.

5. Chico territory. Nest tree: Chicozapote *(Manilkara zapota)*

This territory, in the extreme southwest corner of our 20 km² intensive study plot, was studied only in 1993, when this pair fledged one young from a clutch of undetermined size.

6. El Caoba territory. Nest tree: Black Cabbage-bark *(Lonchocarpus castilloi)*

The five territories just listed were all within Tikal National Park. In addition, we discovered this nest in 1991, several kilometers south of the park in the farming landscape near the village of El Caoba. We trapped and radioed the adult male on 12 March 1991 and followed him for a week before his signal disappeared. We believed it likely he was shot, which may have contributed to the failure of this nest, where the unhatched clutch of three eggs was abandoned in April.

18   LAUGHING FALCON

Margaret N. Parker, Angel M. Enamorado, and Mario Lima

Among the most distinctive sounds of the New World tropics, and a trademark of Tikal, is the remarkable duetting of Laughing Falcons *(Herpetotheres cachinnans)*. Early mornings and late evenings are punctuated by these unmistakable calls echoing from an emergent treetop. The Laughing Falcon's vocal repertoire is expansive, including calls resembling shouting, chuckling, and hysterical laughter. Laughing Falcons often duet with their mates in long, crescendoing calls, and the resulting *waaaah–koh* song has earned these birds the local name "guaco."

Although they are adept woodland hunters, Laughing Falcons are most often seen perch hunting from conspicuous perches in open areas (Plate 18.1). In the forests of Tikal, we saw them most often high in the branches of dead or emergent trees. From such conspicuous perches, these boldly marked falconids have earned a place in Latin American folklore and a deserved reputation as dedicated snake hunters. From 1990 to 1992 we studied Laughing Falcon food habits and nesting behavior at Tikal, as well as their vocal repertoire, habitat, and spatial use.

GEOGRAPHIC DISTRIBUTION AND SYSTEMATICS

The Laughing Falcon ranges throughout tropical lowlands from Sonora and Tamaulipas, Mexico, south to eastern Bolivia, northern Argentina, Paraguay, and southern Brazil (Brown and Amadon 1968). The nature of geographic variation in this species remains contentious. As many as six subspecies have been described, but we follow Hellmayr and Conover's (1949) conclusion that only two races are distinguishable. *Herpetotheres c. cachinnans* occurs throughout the species's range except in the humid Pacific coastal strip from eastern Panama to extreme northern Peru, where the smaller, darker race *H. c. fulvescens* occurs (Brown and Amadon 1968; Ferguson-Lees and Christie 2001). There is much individual variation in size; examining specimens from throughout the species's range, Hellmayr and Conover (1949) found

the largest individuals were from western Mexico and Nicaragua. Most common in lowlands, these falconids have been recorded up to 1800 m in Honduras (Monroe 1968) and rarely to 2400 m in Colombia (Hilty and Brown 1986).

Ridgway (1873, 1875) was the first to demonstrate osteological differences between the Accipitridae and Falconidae, and the first to bring together "such aberrant genera as *Polyborus, Micrastur,* and *Herpetotheres*" (Jollie 1977). While generic relationships within the Falconidae have long been attended by uncertainty (Becker 1987; Griffiths 1999), recent studies based on both anatomy and molecular genetics have led to an improved understanding of falconid relationships (Kemp and Crowe 1990; Griffiths 1999; Braun and Holznagel 2002; Griffiths et al. 2004).

It has long been suspected that *Herpetotheres* and *Micrastur* are more closely related to one another than to the other genera of the family. Molecular analyses support this conclusion and depict the clade containing these two genera as basal to the remainder of the family. The genus *Herpetotheres* has no close living relative, and whether its lineage was ever more diverse is not known. While it is clear that *Herpetotheres* and *Micrastur* are allied, the relationships among component species are not yet fully understood (M. Braun, pers. comm.).

MORPHOLOGY AND PLUMAGE

The Laughing Falcon has an accipiter-like flight profile, with short, rounded wings and a moderately long tail. It lacks the tomial tooth on the upper mandible characteristic of the genus *Falco*. The legs and toes are rather short and stout, a trait common to many snake-eating raptors (Bierregaard 1994), and are covered with small, rough, more or less hexagonal and slightly concave scales, which may function as an adaptation to withstand bites from venomous snakes (Brown and Amadon 1968). The buffy white breast and head are broken by a black mask that covers the eyes and malar area and wraps around the back of the head. The masked plumage of this stocky, large-headed falconid is unique among the Falconidae, most resembling that of certain caracaras (Plate 18.2). The back and wings are dark brown, and the long tail is banded chocolate and cream.

Laughing Falcons are not noticeably sexually dimorphic, and we most often distinguished the sexes by the presence of a brood patch in females during nesting. At Tikal, two adult males weighed 643 and 645 g and one adult female weighed 668 g. The best weight data from any one area appear to be those of Haverschmidt and Mees (1994) for Surinam, where seven males averaged 570 g and three females averaged 623 g (9% more), suggesting a slight degree of size dimorphism. Combining all available data (Appendix 1), males averaged 601 g (n = 13) and females 675 g (n = 8)—12% more. Thus, males weigh about the same as a male Peregrine Falcon *(Falco peregrinus)*, and females perhaps 10% more.

Appendix 2 gives linear measurements for 29 males and 27 females from Nicaragua to Peru, and 23 males and 13 females from Mexico, Guatemala, and Honduras. These data give little indication of any size dimorphism in this species (Appendix 2). In summary, this species may be slightly size-dimorphic, but larger samples may be necessary to determine this conclusively.

PREVIOUS RESEARCH

The Laughing Falcon had been subject to no formal ecological study prior to our work in Tikal. Laughing Falcons are known as snake and lizard hunters in Latin American folklore and in the anecdotes of explorers, ornithologists, and oologists (Sheffler and van Rossem 1944; Brodkorb 1948; Lowery and Dahlquest 1951; Wolfe 1954, 1957, 1959; Skutch 1954, 1960a, 1969; Marchant 1960; Haverschmidt 1962). Although the Laughing Falcon is a relatively common raptor, often seen along roads or near villages throughout much of the humid, forested Neotropics, little is known about its ecology, except that it is a cavity-nesting snake eater. Myths abound regarding many raptors, and Laughing Falcons have earned a reputation as "a bird of many accomplishments, the most universally known of which is the ability to forecast rain" (Sheffler and van Rossem 1944).

RESEARCH IN THE MAYA FOREST

We studied 23 nestings of Laughing Falcons in 12 territories over the three years of this study. Seven territories were within Tikal National Park and five in the slash-and-burn farming landscape south of the park. We located nests mainly by listening for duetting pairs. Duets occurred throughout the nesting season, but most regularly during the preincubation phase in February and early March. We listened for duets at dawn and dusk, and if we heard them from the same location over a period of days, we searched there for a nest. As we approached a nest tree, the female or pair typically exhibited anxious,

defensive behavior by perching in or near the nest tree, flaring their wings, and voicing short vocalizations reminiscent of a hysterical laugh.

When we found an active nest, we climbed the tree if possible, confirming the presence of an egg or chick, and measured and weighed it with Vernier calipers and a Pesola spring scale. With nestlings, we measured tarsal length and width, hallux and culmen length, wing chord, body length, and tail length, on average every four days in 1990, every six days in 1991, and every eight days in 1992. Young were banded with size 7-A locking aluminum bands at the approximate age of 45–50 days, within about 10 days prior to fledging. We examined nestlings for ectoparasites, and if these were present, we described them and noted their number, taking a sample if possible. We noted the presence and absence of ants in the nest tree, collected samples, and categorized their abundance. We measured nest height by lowering a cord from the nest, and nest tree and canopy height using a Haga altimeter (Forestry Suppliers, Jackson, MS). Forest within a 40 m radius of the nest tree was described, and trees within this radius were counted, they were identified where possible, and measurements were recorded of those with a diameter at breast height (dbh) larger than 10 cm.

Using binoculars and spotting scope, we observed nests from blinds on the ground or in trees within 50 m of the nest. Observation periods in 1990 averaged six hours, and in 1991 and 1992 were from dawn to dusk. When an entire day of observation was not possible, observations from dawn to 1200 from one day were paired with those from 1200 to dusk the following day. The position of adults was denoted using the following seven categories: on the nest, within 5, 10, 25, 50, or 100 m of the nest, and more than 100 m from the nest or out of view.

Chronology of Nest Observations

In 1990 we studied nests in each of four territories in Tikal National Park and casually observed several pairs in the nearby farming landscape. We observed the Casita nest daily from 21 March (during incubation) through 13 May (two days after the nestling died, likely a result of botfly [possibly *Philornis* sp.] infestation). After this, the adults did not return regularly to the nest tree. Observations at this nest totaled 342 hours. We observed the Bajo nest for 87.5 hours from 6 April to 1 May, until the chick disappeared, likely also the result of severe botfly infestation. We observed the San Antonio nest for 93 hours from 18 April (when the chick was five to six days old) until 11 June, when the chick fledged at 59 days; it did not return to the nest thereafter. One month after fledging, this young was observed approximately 6 km north of the nest, in the company of both adults. We located and climbed to the Arroyo Negro nest on 30 April, within a day after the chick hatched. We observed this nest for 58 hours before the young fledged on 20 June at age 55 days, not returning to the nest thereafter.

In 1991, we studied seven nesting attempts within the park and five in nearby farming areas, conducting more than 2000 hours of observation at six nests. In 1992 we studied three nests within Tikal and four in the farming landscape.

Prey Identification

During nest observations, we identified prey items to the extent possible as adults delivered them to the nest. We collected prey remains when we climbed to nests, but we found very few remains in nests and no pellets or remains below nest trees. For reptiles delivered to nests, we noted head shape, ventral and dorsal coloration as well as any pattern, and estimated length and diameter, in centimeters. Several of the observers conducting nest observations were able to identify many reptiles to genus or species. In addition, we used books (Alvarez del Toro 1971; Peters and Orejas-Miranda 1986; Campbell and Lamar 1989; Campbell and Vannini 1989) and previously captured reptiles to aid in identification and to become familiar with the local herpetofauna. In many cases, species were unequivocally identified, either in the hand or—for some readily identifiable species—through visual observation. In other cases, visual identifications were inconclusive, so categorization was necessary.

Snakes recognized to species were categorized as venomous or nonvenomous and as mainly terrestrial or arboreal, based on our own observations, published knowledge, and consultation with herpetologists (E. Smith, pers. comm.). Only pitvipers (Viperidae) and coral snakes (*Micrurus* spp.) were recorded as venomous; the many rear-fanged species were categorized as harmless. When snakes were not recognized to species, if they were very slender and green, we categorized them as arboreal, and if they were thick bodied and of various other color patterns, we regarded them as terrestrial. We believe that our error in these categorizations was minimal.

To give observers experience in estimating length and diameter of snakes brought to nests, a snake was held at nest level while observers watching from the normal observation post calibrated its known dimensions against the tree or nest site. Also, we collected, weighed, and measured 23 snakes and 43 lizards captured or found dead on roads, photographed many, and traced their outlines for practice estimating dimensions and body mass. Because of the great variation in mass to length ratios among individuals and species—especially of snakes—we were unable to estimate dietary biomass.

Radiotelemetry

To equip birds with radio transmitters, we trapped them using a *bal-chatri* trap near the nest or in a favored hunting area. We tested rodents, lizards, and dead and rubber snakes as trapping bait, but only live snakes were used successfully. In 1990 we trapped three adults within the park (one pair and one male) during the nestling period and fitted them with radio transmitters. We followed

these birds as they foraged, and when this was not possible, we estimated their locations via biangulation approximately every two days over a period of 110 days spanning much of the nestling and early post-fledging periods. We trapped the adult male of the San Antonio pair on 16 May 1990, affixed a 5.5 g radio to his tail, and monitored him for 197 hours over a 78-day period. We conducted biangulation in part from three platforms constructed in emergent trees. The 1990 fledgling from this nest was also radio-tagged within a week before fledging, using a 3.5 g tarsal-mounted transmitter that remained active for eight days after fledging. After this we could often locate this bird by tracking the radio-tagged adult male who continued to feed the fledgling until our departure a month after the fledging date.

We trapped the adult male of the Casita pair on 5 May 1990, when the chick was 31 days old (six days before the chick died), attached a 5.5 g backpack transmitter, and conducted 592 hours of radio-tracking over a 90-day period, obtaining 74 localizations. We trapped the female of this pair on 15 April 1990 and affixed a 5.5 g radio as a tail mount. We tracked her for 470 hours over 95 days, gathering 91 location points. Although she lost the radio when she molted the tail feathers to which the transmitter was attached on 21 July, we continued to gather data on her movements via her conspicuous vocalizations and frequent proximity to her radio-tagged mate. In 1992 we radio-tagged the adult male of the Casita pair late in the nestling period and collected 61 location points from 28 June to 31 July. In 1991, we estimated home ranges for falcon pairs in the farming landscape via visual observations (without use of radiotelemetry).

Statistical Analyses

All statistical analyses were performed using SYSTAT (Wilkerson 1990). Growth curves were fitted using the SYSTAT NONLIN module, with 20 iterations. We tape-recorded pairs duetting and, for some individuals, recorded several types of vocalizations ($n = 16$ recordings). We documented the timing and duration of duets during nest observations.

GENERAL BEHAVIOR

Laughing Falcons are not very aerial and are usually seen in flight only when moving directly between perches. On their short, rounded wings, these stocky raptors fly with rapid, shallow, stiff wing beats alternating with short glides. Although Wetmore (1965) reported occasional soaring, most authors state that these falconids typically do not soar, and our observations agree.

DIET AND HUNTING BEHAVIOR

One of our main objectives was to compare the diet and hunting behavior of these reptile specialists in the mature forests of Tikal and in the agricultural lands adjacent to the park. We documented the diet primarily by observing at nests as adults brought prey to their mates or nestlings. Our attempts to follow birds as they foraged were rarely successful, although several hunting sessions and one capture episode provided insights into their behavior.

Among 767 Laughing Falcon prey items observed during the study, snakes (86%) and lizards (8%) made up the vast majority (Table 18.1). In the primary forest of Tikal, Laughing Falcons fed exclusively on snakes (99.5% of the diet), whereas in the farming landscape they expanded their diets to include a number of lizards, several rodents, two birds, and even several fish (Table 18.1). Among four nests in the farming landscape, each studied for two years, one nest (Sal Petén) exhibited a more diverse diet than did the other three. This nest was in the most fragmented and degraded habitat, at the edge of a small lake. Among this pair's 146 observed prey items (1991 and 1992), the diet consisted of 70% snakes (102), 12% lizards (18), 12% rodents (18), 4% fish (6), and 1% birds (2). In contrast, at the other three farming landscape nests together, 214 prey items were exclusively snakes (176, or 82%) and lizards (38, or 18%). Among 663 snakes observed as Laughing Falcon prey, we distinguished at least 20 species. Among the 58 lizards observed as prey, at least seven species were represented.

At nests in both mature forest and farming landscape, the temporal pattern of prey deliveries was bimodal, with highest delivery rates in early to mid morning and in late afternoon/evening. Many raptors at Tikal showed a similar bimodal pattern of prey deliveries, which may result simply from daily activity patterns of the raptors. However, this pattern may also reflect greater snake activity during the cooler morning and evening hours than during the hours of most intense heat. In addition, Laughing Falcons in the park's mature forest made morning prey deliveries slightly later on average (1115) than at nests in the farming landscape (1019).[1] This may reflect the fact that the largely deforested farming landscape warms up earlier in the day than does the forest interior (M. Parker, pers. observ.); we suspect that activity of many snake species may be curtailed by intense sun and heat earlier in the day in the farming landscape than in the forest.

In both mature forest and farming landscape habitats, the types of snakes captured by Laughing Falcons showed an association with time of day. Snakes that were deemed mainly terrestrial in habit—for example, coral snakes *(Micrurus* sp.), Fer-de-lance *(Bothrops asper)*, and Western Indigo snakes *(Drymarchon corais)*—were captured most frequently during the morning hours. In contrast, snakes that were at least partly arboreal—for example, *Leptophis* species, *Oxybelis* species, and *Boa constrictor*—were captured throughout the day.

For snakes in the diet (n = 556 that could be categorized), we compared the contribution of arboreal versus terrestrial and venomous versus nonvenomous snakes to Laughing Falcon diets in the primary forest and the farming landscape (Table 18.2). There was a near-significant tendency for arboreal snakes to comprise a greater proportion (53.3%) of the diet in the forest than in the farming landscape (45.4%), with the remainder of the categorized snakes being terrestrial in habit.[2] In the farming landscape, mean estimated prey length (33.8 cm) was significantly less than that (42.9 cm) in the mature forest,[3] indicating that shorter (presumably largely terrestrial) snakes and other smaller vertebrates were taken more often in the human-modified habitat than in mature forest. The proportion of venomous snakes in the diet in the farming landscape (23.1%) was significantly greater than in the mature forest (12.1%);[4] many of these were Fer-de-lance, a dangerously venomous pitviper much feared by local people.

The degree of dietary plasticity we observed in this seemingly specialized snake hunter suggests a greater behavioral flexibility than previously suspected. In mature forest, Laughing Falcons acted as snake specialists, with 385 snakes and two lizards brought to nests.[5] The fact that lizards, rodents, birds, and even fish were added to the diet in the farming landscape suggests that the falcons perceived important differences between prey resources of the intact and modified forest habitats. The expansion of diet breadth in the farming landscape presumably indicates that from the falcons' perspective, the profitability of preying on these additional items, compared to snakes alone, differed between these two habitats. Whether this shift in diet has to do with lower absolute abundance of snakes in the farming landscape, or more to do with the relative abundance, availability,

Table 18.1 Diet of Laughing Falcons in Tikal's mature forest and in adjacent farming landscape

Prey type	Mature forest		Farming landscape		Total	
	Number	Percentage	Number	Percentage	Number	Percentage
Snakes	385	99.5	278	73.2	663	86.4
Lizards	2	0.5	56	14.7	58	7.6
(Reptile subtotal)	387	100.0	334	87.9	721	94.0
Rodents	0	—	18	4.7	18	2.3
Fish	0	—	6	1.6	6	0.8
Birds	0	—	2	0.5	2	0.3
Unidentified	0	—	20	5.3	20	2.6
Total	387	100.0	380	100.0	767	100.0

Table 18.2 Proportion of arboreal versus terrestrial snakes and venomous versus nonvenomous snakes delivered to Laughing Falcon nests in mature forest and adjacent farming landscape[a]

Predominant habit of snakes	Mature forest		Farming landscape		Overall	
	Number	Percentage[a]	Number	Percentage[a]	Number	Percentage[a]
Arboreal	163	53.3	114	45.4	277	49.8
Terrestrial, nonvenomous	105	34.4	79	31.5	184	33.1
Terrestrial, venomous	37	12.1	58	23.1	95	17.1
(Terrestrial subtotal)	(142)	(46.6)	(137)	(54.6)	(279)	(50.2)
Unknown	80	27	107	—	—	—
Total number classified as to activity height and venom	305	—	251	—	556	—
Total	385	—	278	—	663	—

[a] Percentages are based on the total number that were classified as to activity height and venom.

species composition, and size distributions of these different prey types in the two habitats, is unknown.

We have no information allowing comparison of population densities, biomass, or species composition of reptile assemblages within Tikal's mature forest and in agricultural areas bordering the park. In view of the substantial deforestation in the farming landscape, it seems likely that arboreal and forest-dependent reptiles must be reduced in abundance and species richness there. On the other hand, some reptiles no doubt are favored by deforestation, and we observed, for example, that spiny lizards (*Sceloporus* species) were often conspicuously abundant in areas where slash had been felled for farming—much more so than within stands of tall mature forest.

Foraging behavior. We observed Laughing Falcons still hunting in all forest strata, from within the understory less than 1 m above ground to 8 m above the canopy in emergent trees, as well as from snags or other tree perches in open areas such as pastures. We assume that these perch-hunting raptors found prey largely through visual detection of prey movements. This assumption seems to receive some support from the fact that these raptors often perched motionless for periods of several hours. Two common genera of diurnal arboreal snakes at Tikal—*Leptophis* (parrot snakes) and *Oxybelis* (vine snakes)—show remarkable crypsis, not only in body form and coloration but also in behavior. Both of these snakes show locomotory specializations that increase their visual resemblance to vines swaying in the wind—no doubt enhancing the effectiveness of their camouflage while moving through above-ground vegetation (Fleishman 1985). We speculate that the Laughing Falcon has been a source of natural selection leading to such crypsis in these snake genera. Whether these falcons also possess the ability to search out snakes that are hidden—inactive among vegetation—remains unknown.

Methods of prey capture. Only once did we witness a snake capture, when an adult female Laughing Falcon attacked a Fer-de-lance basking on a dirt road at 0700 one morning. The falcon approached the snake on foot, with wings outspread and low. The falcon then hopped

over the snake several times, passing approximately 1 m above it, causing it to turn toward the bird repeatedly to maintain a defensive posture. After landing, the bird swept her primaries along the ground with wings outstretched. Immediately following her fourth pass over the snake, she landed and reached out her foot, pulled the snake beneath her, and quickly dispatched it with a bite to the neck. It seemed that the falcon offered its wings as a target, distracting the snake's attention from more critical areas of the body, and that the repeated jumping over the snake might have enabled the bird to position the snake for safer capture. Wetmore (1965) also heard descriptions of similar capture behavior, though he stated that only one wing was outstretched by the Laughing Falcon perched on the ground near its snake quarry.

Observations by Haverschmidt (1968) indicate that the terrestrial hopping attack we witnessed is not the only capture mode employed. Haverschmidt observed nine prey captures and stated that "the great force with which the bird pounced downward was always very striking. The actual contact on the ground was clearly audible." His description implies that snakes are often seized straightaway, without the "dancing" attack we witnessed. He also indicated that several such captures were made amid thick undergrowth.

In the case we witnessed, the snake was killed at the time of capture. Often, however, male Laughing Falcons brought live snakes to the female or chick. When this was the case, the female would either kill the snake with a bite to the neck or swallow the entire snake alive. Venomous and nonvenomous snakes were brought to the nest with heads usually attached, but no venomous snake was seen delivered alive. Both venomous and nonvenomous snakes were also delivered at times without heads, but this was an unusual occurrence, in contrast to Skutch's observations in Costa Rica (Skutch 1983).

HABITAT USE

Throughout their range, Laughing Falcons occupy an array of wooded habitats from humid rain forest to palm savanna, forest edge, and arid thorn forest. They are generally regarded as most common in semi-open habitats and are stated by various authors to occur seldom if at all within unbroken forest (Brown and Amadon 1968; Bierregaard 1994). The degree to which these falcons are restricted to naturally open or human-modified forest seems to vary geographically. In South American localities with tall, dense, humid forest, Laughing Falcons are said to be rare or absent within primary forest and are restricted to forest edge and areas with clearings or second growth (Thiollay 1985; Bierregaard 1994; Jullien and Thiollay 1996). In contrast, in Colima, Mexico, Schaldach (1963) found this falcon abundant, showing "a decided preference for heavy virgin stands of Tropical Deciduous Forest" but also occurring in dense, tall thorn forest. Rather than this pattern indicating that the birds' habitat preference varies geographically, we suggest it may simply reflect the way in which forest characteristics vary regionally. Mature forest in some areas (e.g., dryish, outer tropical sites) may resemble human-altered forest in other areas (wetter, more equatorial sites) more closely than it does mature forest at the latter sites. At Tikal, where vegetation and climate are intermediate between those of the dry deciduous and thorn forests of Colima and the wetter forests of Amazonia and the Guianas, we found Laughing Falcons to occupy both primary forest and the successional patchwork of the farming landscape.

We observed these falcons most often at the edge of a forest gap, opening, or road, but this was likely due, in part, to the ease with which they could be seen in these sites, compared to the forest interior. Within Tikal's mature forest, six nest sites were in Upland Forest and two in areas of low-lying Bajo or Scrub Swamp Forest, and radio-tagged falcons ranged through both forest types. Outside the park, nests were located in forest fragments 0.5–2.0 ha in size, often at the edge of cleared land.

Five of the 13 nest sites we studied were situated within 500 m of permanent bodies of water, three of them in the farming landscape. The pairs within these territories regularly hunted from perches along the margins of these water bodies. Frog-eating snakes, including Speckled Racers (*Drymobius margaritiferus*), Cat-eyed Snakes (*Leptodeira septentrionalis*), and *Leptophis* spp., were the most commonly observed items among prey taken by Laughing Falcons from these foraging areas (17 of 20 items observed as prey). At the Sal Petén nest, which was along the shore of a lake in heavily modified habitat, the pair was seen perch hunting at the edge of the lake during 71% (75/106) of foraging observations, and occasionally they captured fish. At the San Antonio nest, located near a small marshy lake surrounded by mature forest within the park, the birds were observed foraging along the shore during 17% (24/141) of foraging observations.

BREEDING BIOLOGY AND BEHAVIOR

Nests and Nest Sites

All 13 nest sites that we discovered (used in 23 nesting efforts) were in naturally occurring cavities or depressions in a tree, and nearly half (six) were associated with epiphytes. The falcons displayed no nest-building behavior, laying their single egg on the available substrate—usually decayed wood and bark. Nest sites ranged from fully enclosed cavities in which the birds disappeared into a hole and dropped down to a nesting chamber within the hollow trunk, to depressions in a crotch or a broken limb—and included the ends of broken-off trunks, with or without a surrounding wall of vertically projecting wood. One nest was in a trampled area in the center of a coarse, grasslike epiphyte. Four of the five nest sites in the farming landscape were surrounded by epiphytic cacti, which proved to be an excellent deterrent to climbing biologists and may also lend protection against other climbing mammals. This cactus was not found near nests within Tikal. Laughing Falcon nests have been found in association with epiphytic cacti in other areas as well (Robbins and Wiedenfeld 1982).

Laughing Falcons elsewhere have been found occasionally to nest in stick nests of other raptors (Marchant 1960) and often in epiphytes (Wolfe 1959). Of 25 Mexican clutches in the Western Foundation for Vertebrate Zoology (WFVZ) collection, 9 were associated with epiphytes.[6] Although the original description of cliff nesting in this species (Sheffler and van Rossem 1944) was apparently in error (Wolfe 1954), another egg in the WFVZ collection, taken from a Sonoran cliff nest by Sheffler in 1949, is correctly identified (R. Corado, pers. comm.)—moreover, the adult female was collected, further confirming the identification.

Within Tikal National Park nine of the nests were in living trees and four were in dead trees; in the farming landscape, all five nests were in live trees. Nest heights averaged 21 m (range = 16–31 m). Nest trees at Tikal included three Honduras Mahogany (*Swietenia macrophylla*), two Spanish Cedar (*Cedrela odorata*), and one each of Black Olive (*Bucida buceras*), *Ficus* species, Kapok (*Ceiba pentandra*), Eggfruit Tree (*Pouteria campechiana*), Zapote Mamey (*P. sapota*), Chicozapote (*Manilkara zapota*), Florida Fishpoison Tree (*Piscidia piscipula*), and Bombacaceae species. In both the park and the farming landscape, nests were located in the tallest tree within the local area and were isolated by either height above, or distance from, adjacent trees. Nests averaged 4.3 m (range 2.0–7.5 m) above the adjacent canopy, in trees that averaged 25.7 m in height (range = 16–34 m) and ranged from 50 cm to more than 1 m in diameter at breast height.

Egg and Clutch Size

In the northern portion of this species's range, the normal clutch size is one, with a single 2-egg clutch reported from Mexico—of 25 Mexican clutches in the WFVZ collection, one was a 2-egg clutch and the rest 1-egg clutches. All clutches and broods we observed at Tikal were likewise of one. In Argentina, de la Peña (1983) reported two 2-egg clutches from the same nest, one of them a replacement clutch laid after he collected the first. Whether a 2-egg clutch is frequent in any part of the species's range remains to be demonstrated.[7]

Table 18.3 Measurements of Laughing Falcon eggs

Locality	Length (mm)	Breadth (mm)	Measured mass (g)	Estimated mass (g)	Source
Horqueta, Paraguay	56.5	45.6	—	64.3	Wolfe 1938
San Luís Potosí, Mexico	58.0	44.6	—	63.1	Congreve 1954
Veracruz, Mexico	62.4 ± 2.2 (n = 3)	48.9 ± 1.4 (n = 3)	—	81.6	Wolfe 1964
Tikal, Guatemala	58.8 ± 2.8 (n = 3)	47.0 ± 1.7 (n = 3)	58[a], 72[a], 76[b]	71.0	This study
Mean of the above	59.8 ± 3.0 (n = 8)	47.3 ± 2.0 (n = 8)	68.7 ± 9.5	73.1 (n = 8)	

[a] Weighed within two weeks of hatching.
[b] Addled egg.

The first legitimate egg reported for this species was from Horqueta, Paraguay (Wolfe 1938). Egg dimensions and mass from Tikal and elsewhere appear in Table 18.3. All three eggs examined closely at Tikal were maroon, mottled with brown splotches. Combining all data (Table 18.3), Hoyt's (1979) formula predicts a mean egg mass of 73.1 g, or 10.8% of the mean 675 g female body mass. Egg size appears to vary regionally (perhaps in accordance with body size) as three eggs from Veracruz, Mexico, had a predicted fresh mass of 81.6 g, whereas three from Tikal had a predicted fresh mass of 71.0 g (Table 18.3).

Nesting Phenology

Compared with other raptors at Tikal, Laughing Falcons nested relatively early. During the three years of this study, the rains at Tikal tapered off during January, and Laughing Falcons laid eggs during late February and early March—the early to mid dry season.[8] Eggs hatched in early to mid-April,[9] during the earliest signs of the coming rainy season, and fledging often took place in June,[10] early in the rainy season. One pair that lost their first egg to predation laid a replacement egg 21 days later.

Length of Incubation, Nestling, and Post-fledging Dependency Periods

Two incubation intervals were estimated at 45 days, and two fledging intervals were approximately 55 and 59 days. The duration of post-fledging dependency is not fully known, but it appears substantial. In those cases where we monitored radio-tagged adults or young during the first month after fledging, adults attended and fed fledglings, often a few to several kilometers from the nest. Our monitoring did not extend long enough beyond fledging to witness the juveniles reaching independence. In several cases, however, juveniles were observed in company with their parents at the age of one year. At the Sal Petén nest during the 1992 breeding season, the previous year's fledgling (determined by bands) was repeatedly observed in company with the breeding pair. In addition, on four occasions, three adult-plumaged Laughing Falcons were seen duetting together, and we believed these cases to involve an adult pair and their young of the previous year. On 12 March 1992 (i.e., at the very outset of the nesting season) a colleague observed a pair of Laughing Falcons in company with a juvenile at Aguada Bejucal, near one of our nest sites (M. Vásquez, pers. comm.). Finally, we once observed a banded 1-year-old in the company of two other Laughing Falcons within several kilometers of that bird's natal area. We never observed these juveniles receive prey from adults, and whether this occurs remains unknown. In sum, the duration of post-fledging dependency in the Laughing Falcon remains unknown, though some intriguing observations suggest the possibility that association with parents, if not dependency, may at least sometimes be prolonged. However, if young did show prolonged post-fledging dependency, this did not prevent adults from breeding each year.

VOCALIZATIONS

We recognized five elements in the vocalizations of Laughing Falcons, some of which compared quite well to descriptions of vocalizations of the Laughing Kookaburra (*Dacelo novaeguineae*: Reyer and Schmidl 1988), a masked, reptile-eating, tropical kingfisher. The five elements we recognized were denoted as *wah*, *koah*, laugh, cackle, and chuckle components. Both male and female Laughing Falcons used all five elements in various intra-pair interactions.

Paired Laughing Falcons engaged in ritualized vocal and visual displays, often duetting near their nest sites at dawn and dusk. They assumed erect postures, usually less than 2 m apart, and called in a stereotyped fashion. These duets typically lasted from 2 to 12 minutes and began when one bird initiated a loud, slow, monosyllabic *wah* call and was soon joined by its mate uttering a *koah* call in syncopation. These calls built in volume and tempo, crescendoing in alternating *wah-koah* calls that were audible for several kilometers.

The *wah* element was a short or long wail that, when repeated and increased in tempo, was often a precursor to the *koah* call (given by the mate), and initiated a duet. *Wah* was also given as a solo call, usually from the nest vicinity. The *koah* element was "a short, guttural sound, which can be given alone, often repeatedly, but also as an introduction to the laugh song" (Reyer and Schmidl 1988). When this call preceded a laugh call, it sounded

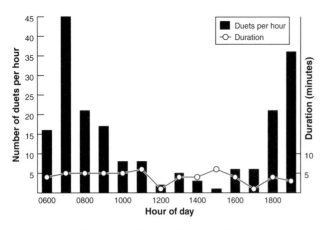

18.1 Temporal pattern and mean duration of duetting by Laughing Falcons, based on 195 duetting bouts.

shorter and screamlike. When given alone, it had a softer sound and apparently then did not indicate any threat to the bird (see below). The *wah* and *koah* calls composed the main elements of the duet and were interchanged in syncopation.

The Laughing Falcon's laugh call comprised a series of loud, cascading *ha ha ha* sounds, which often incorporated some of the other vocal elements listed here. When given in defensive contexts, the laugh sounded hysterical, and when no apparent threat was present, it had a more musical sound. Laughing Falcons used the laugh call while driving various potential predators from the nest vicinity (see Nest Defense, later in this chapter) and while the egg at the 1991 Casita nest was eaten by Tayras (*Eira barbara*).

The Laughing Falcon's cackle was "a harsh, loud, repetitive sound, similar to the '*ha ha*' component of the laugh song but with a short '*a*'" (Reyer and Schmidl 1988 after Parry 1968). This call sometimes preceded a full laugh call. The cackle was used by adults defending the nest. This loud, hysterical-sounding call was often accompanied by a spreading of the wings and tail in defensive display, and often developed into a laugh call that could last for half an hour. The chuckle call was a soft, rolling cackle, sometimes lower in tone; it was heard only in adults and rarely developed into another type of call. It seemed to indicate that all was well. When pairs foraged, they often used a soft chuckle or *koah* that was audible to observers within only about 20 m, and seemingly functioned as a contact call. The female frequently used the chuckle while she tended the nest, especially when the male approached with prey. The chuckle often alerted observers to the fact that the male was approaching and the female would soon leave to intercept a food delivery.

The daily pattern of duetting is indicated in Figure 18.1. Duets were most common at dawn or soon after and again at dusk. Duets lasted from 1 to 12 minutes, averaging 4.7 minutes in duration. Fully developed duetting decreased in frequency during the first several weeks of incubation, although when the male brought food to the

nest thereafter, the pair often gave abbreviated duets less than 1 minute in duration (mean = 18 seconds, range = 4–62 seconds, $n = 48$).

BEHAVIOR OF ADULTS

Where percentages are given in this section, they are based on pooled data from all six 1991 nests.

Pre-laying and Laying Period

Early in the pre-laying period, Laughing Falcons often visited the potential nest site two or three times daily, occasionally roosting in the nest tree. After an early-morning session of duetting in or near the nest tree, they often perched for up to two hours, vocalizing softly, after which the male would fly off, presumably to hunt. The female would remain within 50–100 m of the nest tree, or at times would also leave to hunt. By 1700 or 1800, pair members generally returned to the nest vicinity, often duetting for several minutes prior to roosting in the area.

Incubation Period

Throughout the incubation phase, female Laughing Falcons stayed on or near the nest virtually the entire time, while males delivered food. Males were never observed incubating. During incubation breaks, females would leave the nest and perch 5–7 m away, often for 15–20 minutes, preening. Generally the male would leave during the morning and return with a snake, perching perhaps 50 m from the nest. The male would call softly, and the female would leave the nest and approach him. Seizing the prey, the female would often call loudly for a time and eventually begin eating. The male meanwhile often flew to the nest tree and perched there for 10–15 minutes before leaving the area.

Nestling Period

During the chicks' first few days, females brooded them nearly constantly and fed them small morsels. Females performed all brooding of the young and during the bulk of the nestling period remained largely on the nest (45% of the time) or within 5 m of the nest (21%). Females began to leave the nest area when the chick was about 20 days old. At this point, the female typically left for short periods (< 20 minutes), gradually increasing her time away until her foraging time equaled that of the male when the chick was about 35 days old.

The male would transfer prey to the waiting female in the nest vicinity, and she would then take it to the nest where the nestling would seize it. As chicks grew larger, females sometimes delivered snakes alive, which gave the chicks experience footing and biting living prey. Like all falconids we studied, these raptors did not bring greenery or any other nest material to the nest.

BEHAVIOR AND DEVELOPMENT OF NESTLINGS

Nestling Period

Hatchlings were covered in unusually dense down, cream colored except for the black mask, which was present at hatching. Weak and moving little at first, by two weeks the chicks were more mobile, and by three weeks they could be heard vocalizing softly in apparent food begging. Chicks remained quiet in the nest when they were young but after about 20 days, they voiced the cackle call during handling. By the age of two weeks, chicks voiced the *wah* call softly from the nest when the adult pair duetted, and at about 40 days they began to duet with their parents. From the age of about six weeks, chicks sometimes moved 1–2 m from the nest, yet initially they always returned to it and began hopping practice flights.

Chicks at 10 nests were periodically measured and weighed, beginning soon after hatching until about the age of 55 days (an estimated 10 days prior to fledging), at which point we stopped handling them to avoid premature fledging. Of all parameters measured, body mass and tarsal length best reflected growth rates (Fig. 18.2). By examining residual plots, we assessed the fit of several nonlinear growth models (Gompertzian, logistic, modified von Bertalanffy, and allometric curves: German and Meyers 1989, following Lebeau et al. 1986). Data were best fit by allometric curves: $Y = a * t^c$, where Y = mass and t = age. The individual chick measured most often (16 times) and regularly had the best fit to the allometric model (corrected r^2 = 0.97), and this was the only model to fit the group as a whole (corrected r^2 = 0.45).[11] Multiple measures taken on one individual were corrected using Bonferroni adjustments.

Post-fledging Dependency Period

Juvenal plumage at Tikal was similar to that of adults, except that juveniles had a more reddish cast dorsally (Plate 18.3). Chicks fledged with a slightly shorter and less-banded tail than that of adults (Haverschmidt 1948; Brodkorb 1948; M. Parker, pers. observ.). During the first week after fledging, young birds left the immediate vicinity of the nest tree but often returned to within 25–40 m of it. Within one to two weeks, fledglings traveled up to 200 m from the nest, moving 10–15 m from adults to eat alone after receiving prey. After the third week, fledglings moved farther afield and by the fourth and fifth week often had moved a kilometer or more from the nest tree and continued to be fed by adults at this distance from the nest. In cases in which we monitored radio-tagged adults or young during the first month after fledging, both pair members of three observed pairs attended and fed fledglings, often several kilometers from the nest. In 1990, one fledgling, 64 days after fledging, was found 6 km north of its nest with both adults in attendance. Throughout our post-fledging observations, fledglings joined their parents in vocal duetting sessions.

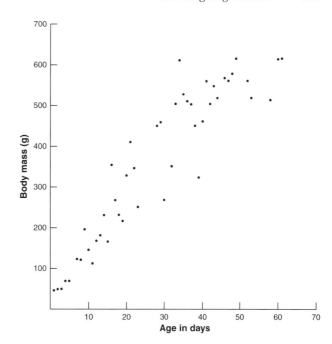

18.2 Body mass growth curves for eight Laughing Falcon nestlings.

SPACING MECHANISMS AND POPULATION DENSITY

Territorial Behavior and Displays

The degree to which neighboring pairs overlapped in home range use was unknown. We witnessed no instances of overt intraspecific aggression. Our impression, however, based on acoustic surveys, mapping of calling pairs, and visual observations, was that relatively little spatial overlap occurred among neighboring pairs. The following observations seem to support this notion of minimal spatial overlap among neighbors. When the Casita pair abandoned their 1990 and 1991 nesting efforts because of nestling death and egg predation, respectively, the adults began duetting at the assumed perimeter of their territory immediately after abandoning their nesting attempt. In both cases, this behavior lasted for several weeks, and it may have been a type of territorial patrol and boundary assessment after being confined to a relatively small area during nesting.

Laughing Falcons did not use soaring over-flights as a means of territorial display, as do many diurnal Neotropical raptors. With little doubt, the main spacing mechanisms of Laughing Falcons are their loud, far-carrying vocalizations and the conspicuous visual signal provided by their bold, contrasting facial markings as they perch prominently on an emergent, open perch. Along with calls of the Guatemalan Howler Monkey *(Aloutta pigra)*, the duetting calls of the Laughing Falcon were the animal sounds at Tikal audible at the greatest distance. These calls were audible for a few kilometers and further under some circumstances, especially at dawn and dusk when conditions were best for such long-distance signaling (see Chapters 9 and 17). Such effective long-range signals may be related to the fact that these raptors appear to use home ranges that are relatively large for their body size.

Nest Defense

Nest defense was manifested by agitated use of the cackle vocalization. This loud call was often accompanied by wings and tail spread in defensive display, for example, when researchers approached to climb the nest tree. Laughing Falcons also used the laugh call while driving various potential predators from the nest vicinity, including Black-handed Spider Monkeys *(Ateles geoffroyi)*, American Black Vultures *(Coragyps atratus)*, Brown Jays *(Cyanocorax morio)*, Keel-billed Toucans *(Ramphastos sulfuratus)*, Plumbeous Kites *(Ictinia plumbea)*, and White-nosed Coatis *(Nasua narica)*.

Constancy of Territory Occupancy, Use of Alternate Nest Sites

Territory location and occupancy appeared largely stable from year to year, not only in the park but also in the rapidly changing farming landscape, where, of five nests studied in 1990, four were again occupied in 1991. Of four nests studied within the park in 1990, three were reoccupied by the same banded pairs in 1991, and at least the Casita pair remained intact in 1992 as well. Pairs often used the same nest cavity for multiple years, including sites where they failed to fledge young for three consecutive years. However, in a few cases, banded pairs switched to alternate nest sites as much as 1.6 km apart but apparently within the same territory.

Home Range Estimates

Within primary forest. In 1990, the radio-tagged Casita pair maintained a home range that included several human-made clearings around the main ruins, in addition to large areas of surrounding forest (Fig. 18.3). While this pair occasionally used clearings around Tikal's Maya temples, more than 85% of recorded locations were in the forest away from these openings. Using the Minimum Convex Polygon (MCP) method, we estimated this male's breeding season range at 25.3 km² (*n* = 74 locations), and the female's at 14.6 km² (*n* = 91). These are relatively large home range estimates for a raptor of this body size (Newton 1979), but the nature of habitat within this home range may help explain this result. Tikal's partially cleared central ruins lay in the center of this home range and were only lightly used by the falcons; inclusion of these apparently nonfavored areas may have led to a larger-than-average home range. In 1992, we retrapped the pair, attached new transmitters, and monitored their movements over a three-month period. They were then using a nest site 1.6 km from their previous nest tree, which had collapsed—yet they hunted in the same general area as before.

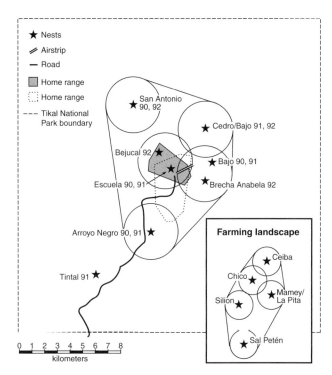

18.3 Locations of Laughing Falcon nests and home ranges studied in and near Tikal National Park, Guatemala, 1990–92. Nests in farming landscape (all 1991) were several kilometers south of the park and are drawn to the same scale as nests in the park. Radii of circles around nests are one-half the mean inter-nest distance in the two situations: for nests in the park, radii = 2155 m; in the farming landscape, radii = 1116 m. Polygons indicate areas used in calculating pair density: 103.24 km² in park, 27.22 km² in farming landscape.

In 1992, the radio-tagged adult male of the Casita pair, during the latter part of the nestling period, often hunted within 500 m of the nest. Based on 35 locations late in the nesting and early in the post-fledging period, we estimated this male's 85% Harmonic Mean (HM) home range as 3.2 km². Later in the post-fledgling period, the fledgling moved 1.5 km east of the nest and was frequently in the company of the adult male. Overall, the 100% MCP home range estimate for the adult male during the late nestling and post-fledging period was 6.7 km² (*n* = 61 locations; Fig. 18.3).

The male of the San Antonio pair, amid continuous forest, had a 1990 home range (MCP) estimated at 6.5 km², but owing to the inaccessibility of this male's home range, more than 40% (*n* = 41) of our attempts to locate him were unsuccessful, as we could not detect a radio signal. This suggests that he ranged farther than indicated by our 6.5 km² figure. The nestling at this nest was fitted with a transmitter seven days prior to fledging, and we followed the fledgling for four weeks before the radio stopped functioning. Eight weeks after fledging, we located this juvenile 6 km north of its nest and within several meters of its parents—further suggesting a large home range for this pair.

Of these various home range estimates, it is difficult to know which may be closest to a typical value for our study area. Estimates of 6.5 and 6.7 km² for two breeding males may well be underestimates, whereas the 25.2 and 14.6 km² estimates for a breeding male and female may be atypically large, mainly because much scarcely used habitat was included within this pair's home ranges. However, the latter two estimates correspond fairly closely to our best density estimate for Laughing Falcons within the park, which was one pair per 21–22 km² (see below). Hence we tentatively accept this range (21–22 km²) as realistic within Tikal, while recognizing that home ranges as small as 7 km² may be used in some cases.

In the farming landscape. Outside the park, pairs were observed more easily in the open habitat, and we constructed rough home range estimates by following these birds without radiotelemetry. In the more forested portions of the farming landscape, home range estimates were approximately 9 km², but they declined to roughly 4 km² in the most degraded, intensively farmed habitat. Although our fieldwork did not allow a rigorous comparison of home range sizes in the forest and farming landscapes, it is safe to conclude that home ranges were generally smaller in the farming landscape.

This difference in home range size between mature forest and the farming landscape was accompanied by differences in the way Laughing Falcons used their ranges. Birds foraging in the patchwork habitats of the farming landscape tended to hunt at a limited number of sites, often from the same few perches. These heavily used sites were usually near water, and frog-eating snakes were often captured. Hence these Laughing Falcons appeared to exploit certain—probably prey-rich—areas within their home ranges more intensively than they did other areas. Within the forests of Tikal, though Laughing Falcons hunted along water sources at times, they did not seem to concentrate their foraging activities so clearly at these sites, and appeared to spread their hunting efforts more evenly over their home ranges.

Inter-nest Spacing and Density of Territorial Pairs

Spacing and density in primary forest. Five pairs we studied within the park were thought to be neighboring pairs, and a sixth was probably not a nearest neighbor to any of these (Fig. 18.3). Using the minimum spanning tree method (Gower and Ross 1969; Selas 1997), we calculated mean distance between neighboring nests in three ways. First, we used only nests simultaneously active in a given year, for 1990, 1991, and 1992. The mean inter-nest distance for these three years was 4.27 ± 0.22 km (n = 3, based on two, three, and four inter-nest distances per year; Table 18.4). This yields a Maximum Packed Nest Density (MPND) estimate of 16.6 ± 1.7 km²/pair (n = 3), a square estimate of 18.3 ± 1.9 km²/pair (n = 3), and a mean polygon estimate of 21.9 ± 3.6 km²/pair (n = 3; Table 18.4).

Second, when the average nest site within each of five nearest-neighbor territories is plotted (regardless of simultaneity of nesting), this approach yields a mean inter-nest distance of 4.50 ± 0.95 km, for an MPND estimate of one pair per 18.4 km², a square estimate of one pair per 20.3 km², and a polygon estimate of one pair per 21.5 km² (Table 18.4). A final, and probably too conservative method, is to regard all six territories as nearest neighbors, which gives a mean inter-nest distance of 4.70 ± 0.94 km (n = 5), yielding an MPND estimate of 20.1 km²/pair, a square estimate of 22.1 km²/pair, and a polygon estimate of 26.0 km²/pair (Table 18.4). As we most trust the first two scenarios, and the polygon method, our best estimate is a density of one pair per 21.9 km², or about 4.6 pairs per 100 km² (Table 18.4). The third scenario, estimating one pair per 26 km², may be taken as a minimum density estimate for Tikal.

Spacing and density in the farming landscape. In the farming landscape, we studied five pairs that made nesting attempts in 1991. Four were certainly nearest neighbors, and a fifth may also have been (Fig. 18.3). Inter-nest distances were 1.84, 1.89, 2.19, and 3.01 km, the latter datum involving the nest of questionable neighbor status. Including all four measurements, mean inter-nest distance was 2.23 ± 0.54 km (n = 4), for an MPND estimate of 4.53 km²/pair, a square estimate of 4.97 km², and a polygon estimate of 5.43 km²/pair (Table 18.4). If the outlying nest is omitted, mean inter-nest distance was 1.97 ± 0.19 km, for an MPND estimate of 3.53 km²/pair, a square estimate of 3.88 km²/pair, and a polygon estimate of 4.00 km²/pair (Table 18.4). For our purposes, it is more conservative to include the outlying nest, yielding a polygon estimate of one pair per 5.43 km², or as many as 18 pairs per 100 km² in similar habitat.

Comparative nest spacing in forest and farming landscapes. Nests in the farming landscape were significantly closer together than those in the primary forest.[12] In the farming landscape, nests averaged 2.23 ± 0.54 km apart (n = 4 inter-nest distances). In the primary forest, when all territories are taken together, nests averaged 4.50 ± 0.95 km apart (n = 4), whereas nests simultaneously active averaged 4.27 ± 0.22 km apart (n = 3 yearly means; Table 18.4). Hence by all measures, nests within the park were about twice as far apart as nests in the farming landscape, and there seems little doubt that pair densities were considerably higher in the farming landscape than in the primary forest (Fig. 18.3). This accords with our observation of smaller home ranges (roughly 4–9 km²) in the farming landscape than in Tikal's mature forests (7–22 km²).

Two aspects of Laughing Falcon behavior support the notion that these raptors used large areas compared with the areas other raptors of similar body size use. First, the fact that alternate nest sites used by a single marked pair were as much as 1.6 km apart, with little change in foraging arena, implies a large home range. Second, fledglings

Table 18.4 Inter-nest spacing and density of territorial pairs of Laughing Falcons at Tikal[a]

Method, source data	Estimated mean inter-nest distance	In primary forests of Tikal National Park, estimated unique space per territorial pair (km²/pair)			
		MPND	Square	Polygon	Best
Based on all six pairs shown inside park in Fig. 18.3.					
Simultaneously active nests[a]	4.27 ± 0.22 km (n = 3 years)[a]	16.6 ± 1.7 (n = 3)	18.3 ± 1.9 (n = 3)	21.9 ± 3.6 (n = 3)	21.9
Regardless of simultaneity of nesting, six territories[b]	4.70 ± 0.94 km (n = 5)	20.1	22.1	26.0	26.0
Based on the northernmost group of five nests inside park in Fig. 18.3.					
Regardless of simultaneity of nesting, five territories[c]	4.50 ± 0.95 (n = 4)	18.42	20.25	21.55	21.55
In the farming landscape					
Five nests simultaneously active in 1991	1.84, 1.89, 2.19, 3.01 mean = 2.23 ± 0.54 (n = 4)	4.53	4.97	5.43	5.43
Same, deleting southernmost nest as a possible outlier	mean = 1.97 ± 0.19 (n = 3)	3.53	3.88	4.00	4.00

[a] Three yearly mean inter-nest distances (1990, 1991, 1992), based on 2, 3, and 4 inter-nest distances per year.
[b] Using mean nest locations within each of six territories in Fig. 18.3.
[c] Using mean nest locations within northernmost five territories in Fig. 18.3.

and adults wandered in company over large areas—up to 6 km from the nest—again implying large home ranges.

In Venezuela, Mader (1981) found four pairs of Laughing Falcons inhabiting an area of about 13.4 km² of mixed palm savanna and forest, implying about 3.35 km² of space per pair. This is similar to the density estimated by Scott Robinson (pers. comm.) for a highly productive floodplain forest at Manu, Peru: three to five pairs of Laughing Falcons per 10 km² of forest, or roughly 2.0–3.3 km² of space per pair. Our estimate for the farming landscape—one pair per 4.0–5.4 km²—is not greatly different from the above estimates, which lends further confidence that densities in favorable habitat may often be in the range of one pair per 3–5 km². In French Guiana, Thiollay (1989a) gave no density estimates for this species, which was essentially absent from mature forest in his study area. Our estimated density for Tikal's mature forest—one pair per 21–22 km²—is 4–10 times lower than these other values; we cannot guess, at this point, whether such a low density is typical in mature forest habitats in Guatemala or elsewhere.

DEMOGRAPHY

Age of First Breeding, Attainment of Adult Plumage

In several cases subadults were seen in company with their parents. This suggests that subadults often—perhaps usually—do not breed.

Frequency of Nesting, Percentage of Pairs Nesting

Within the protected forests of Tikal, our Laughing Falcon study pairs nested every year, with the possible ex-

ception of one pair in 14 territory-years, yielding a 7% non-nesting rate. Non-nesting was far more common in the farming landscape. In 1991, our year of most intensive study, none of six pairs within the park failed to nest, while five pairs in the farming landscape failed to nest in four of nine pair-years over a two-year period. Even if some of the latter had attempted to nest and failed, it seems that the incidence of non-nesting was higher in the farming landscape than within the park. However, it may have been easier to detect non-nesting pairs in the farming landscape than inside the park, where duetting of pair members was a principal aid in detecting these falcons amid the extensive forest; thus, the observed difference between these two habitats in incidence of non-nesting may be partly artifactual. We do not know why the incidence of non-nesting was higher among pairs in the farming landscape than among forest pairs. However, one possibility is that while suitable hunting habitat and prey were available in the farming landscape, nest sites were limited because of felling and burning of remaining trees, resulting in a relative scarcity of nest sites.

Productivity and Nest Success

Of 22 pair-years monitored, 14 produced a fledgling, for 0.64 fledgling per territory-year. Including one re-nesting, 14 (61%) of 23 nesting attempts during the three years fledged young successfully. In the mature forest, 8 (57%) of 14 nests were successful, and in the farming landscape, 6 (67%) of 9 nesting efforts were successful—not a significant difference between these two habitats.[13] Predation accounted for 5 (56%) of 9 nest failures. Predators responsible were a family of Tayras at one nest, unknown predators at three nests, and an avian predator at the fifth nest. In one of these cases, the nestling was heavily infested with botflies, which may have facilitated predation. Two pairs (22%) abandoned eggs—one, an addled egg after 61 days of incubation and one, during a second nesting attempt, after 33 days of incubation, apparently due to disturbance by a nearby colony of Africanized Honeybees (*Apis mellifera* hybrid). At one of the farming landscape nests (Chico), the male was shot on 11 April, during the incubation period, and the female continued to incubate, leaving the nest to hunt about once every two days. Within 15 days of the male's death, another male was delivering prey to the brooding female, and this pair remained together after the nest was abandoned on 17 May, possibly in response to intensive burning of the surrounding fields. The status of the egg or chick at the time of nest abandonment was not known. In 1990, one chick fell from its nest and was found to have a heavy ectoparasite load (likely *Philornis* sp.), a factor that may have contributed to its fall. Hence, causes of nesting failure may be summarized as: predation 56%, Africanized Honeybees 11%, botflies 11%, human disturbance 11%, and addled egg 11%.

Parasites and Pests: Laughing Falcons, Botflies, and Ants

In 1990 we observed four nests intensively, two of which produced a fledgling each, and two of which failed. Both successful nests contained ants (*Pseudomyrmex* sp.), which were present each time we climbed to the nests. The two nestlings in trees with ants had no visible botflies or other conspicuous ectoparasites. In contrast, the two chicks that failed to fledge had visible ectoparasites, and neither of these nests contained ants. One of these chicks was found dead at the bottom of the nest tree in mid-May; on necropsy, we found this bird had seven botfly larvae, three of them severely impacting the left eye and ear canal. This nestling's feather development was markedly retarded compared to similar-aged healthy chicks. The other nestling whose nest was not shared by ants harbored at least five botfly larvae, and this chick was taken from its nest by an avian predator.

Having noted the suggestive pattern in 1990, we modified our 1991 study plan to test for a relationship between the presence of ants in nest trees, incidence of parasitism, and nesting success. Of 12 nesting attempts in 1991, 7 were successful, 6 of these being associated with hymenopteran colonies: 5 with ant colonies and 1 with a hive of (presumably Africanized) Honeybees (Table 18.5). Of the 5 unsuccessful nesting attempts, 3 trees contained ant colonies; at one of these, the egg failed to hatch, at one the chick suffered predation, and one was abandoned when surrounding fields were burned. The other 2 failed attempts were by 1 pair that shared a tree with a honeybee colony; the first egg suffered predation, and the second nesting was abandoned apparently owing to harassment by the bees (Table 18.5). For both years considered together, there was a near-significant tendency for successful fledging to be associated with presence of Hymenoptera,[14] especially with ants.[15]

Ectoparasites, mainly botflies (probably *Philornis* sp.), were also found on chicks that fledged, but no more than three per chick.

Thus, nesting in association with ant colonies may have increased the fledging success of Laughing Falcons. This effect, if real, may have at least two mechanisms. Ants swarmed over and stung researchers when we climbed to Laughing Falcon nests and may have some deterrent effect toward potential nest predators. Ants may also have decreased rates of botfly parasitism. We often noted ants crawling on Laughing Falcon chicks, and they did not appear to bother or harm the chicks. We assume that ants foraged within the Laughing Falcon nests they occupied, and if ectoparasite eggs or larvae were available on a chick or in the nest, their consumption by ants would likely be beneficial to the birds.

Botflies are conspicuous insects that deposit their eggs or larvae on host nestlings. The larvae burrow into the chick and eventually exit to pupate at the bottom of the nest. Smith (1968) found that in Chestnut-headed Oropendolas (*Psarocolius wagleri*) and Yellow-rumped Caciques (*Cacicus cela*), chicks with more than seven botfly larvae usually died, and in Puerto Rican Sharp-shinned Hawks (*Accipiter striatus venator*), nestlings parasitized by *Philornis* botflies had mortality rates nearly four times those of un-parasitized nestlings (Delannoy and Cruz 1991). Among Puerto Rican Sharp-shinned Hawk nestlings, infestation with seven or more botfly larvae was inevitably fatal, but as few as two larvae sometimes killed a nestling. Well-nourished chicks can typically withstand a few larvae, but larger numbers of botfly larvae would likely slow growth and development, rendering a chick more susceptible to disease or predation. Larval parasites can burrow into and damage critical anatomy, as evidenced by larvae in the ear canal and eye of the Casita chick in 1990. In Puerto Rican Sharp-shinned Hawks, larvae occurring on the nestlings' throat or near the eyes were quickly fatal, regardless of their number (Delannoy and Cruz 1991).

Smith (1968) found that two icterids were protected to some extent from botfly infestation when the birds nested near colonies of bees or wasps that preyed on or repelled botflies. Similarly, we suspect that Laughing Falcons nesting in trees with colonies of certain kinds of ants may suffer less nesting mortality from botfly infestation than do pairs nesting without ants. Whether the ants benefit in turn, by scavenging the falcons' prey remains, is not known; however, prey remains were notably scarce in these nests.

CONSERVATION

In general, Laughing Falcons appear to be fairly resilient to partial deforestation and forest modification. This is illustrated by Thurber et al.'s (1987) finding that these raptors were fairly common in El Salvador, where little natural forest remained and where many other rap-

Table 18.5 Histories of Laughing Falcon pairs/territories studied at Tikal

Territory	Year	Territory active	Laid egg	Egg hatch	Chick fledge	Ants	Bees	Cause of failure	Comments
Nests within Tikal National Park									
1. Casita	1990	Yes	Yes	Yes	No	No	No	Botflies	Fell to death, probably owing to botfly infestation
	1991	Yes	Yes	No	No	No	Yes	Predation	Day 28 of incubation, Tayras ate egg
	1991	Re-nest[a]	Yes	No	No	No	Yes	Bees	Abandoned day 33 of incubation, due to Africanized Honeybees
(Bejucal)[b]	1992	Yes	Yes	Yes	Yes	?	?		—
2. Bajo	1990	Yes	Yes	Yes	No	Yes	No	Avian predator	Probably facilitated by botflies
	1991	Yes	Yes	No	No	?	No	Failure to hatch	Abandoned after 61 days of incubation; addled egg
(Brecha Anabela)[c]	1992	Yes	Yes	Yes	No	?	?	Predation of chick	—
3. San Antonio	1990	Yes	Yes	Yes	Yes	Yes	No		—
	1991	Yes	Yes	Yes	Yes	Yes	No		—
	1992	Yes	Yes	Yes	Yes	?	?		—
4. Arroyo Negro	1990	Yes	Yes	Yes	Yes	Yes	No		—
5. Tintal	1991	Yes	Yes	Yes	Yes	Yes	No		—
6. Cedro	1991	Yes	Yes	Yes	Yes	Yes	No		—
7. Ceibal[d]	1991	Yes	Yes	Yes	Yes	No	No		—
Nests in farming landscape									
8. Ceiba	1991	Yes	Yes	Yes	Yes	Yes	No		—
	1992	Yes	Yes	Yes	Yes	?	?		—
9. Chico	1991	Yes	Yes	No	No	Yes	No	Field burning caused abandonment	—
10. Sillion Sur	1992	Yes	Yes	Yes	Yes	?	?		—
	1991	Yes	Yes	No	No	Yes	No	Predation	Failed 30 Jun, 34 days into observations
11. Abejas	1992	Yes	Yes	Yes	Yes	?	?		—
	1991	Yes	Yes	Yes	Yes	No	Yes		—
12. Sal Petén	1991	Yes	Yes	Yes	Yes	Yes	No		—
	1992	Yes	?[e]	?	No	?	?	Predation	—

[a] Pair re-nested 21 days after Tayras ate first egg.
[b] Same banded pair of adults, 1990–92; Bejucal site is alternate site used by Casita pair.
[c] Brecha Anabela is alternate nest site in Bajo territory.
[d] The Ceibal nest was at the boundary of Tikal National Park, adjacent to the farming landscape.
[e] We assume an egg was laid in this case.

tors had declined or disappeared. They found Laughing Falcons even in open country with only a few widely scattered trees. Presumably their snake-dominated diet may allow these birds to occupy quite open habitat so long as prey are sufficiently numerous and perch and nest sites are available. Still, one suspects that with extreme habitat conversion, these birds would ultimately decline in population, if for no other reason than nest site shortage.

With their diet consisting almost exclusively of snakes, Laughing Falcons feed at an unusually high trophic level, even for a raptor. We know of no evidence that this species suffered reproductive problems during the height of the DDT era, but eggs collected in southeastern Mexico between 1952 and 1969 had residues of at least five organochlorine compounds and reportedly showed eggshell thinning relative to the pre-DDT era (Iñigo-Elias et al. 1991). Their extraordinarily high trophic position may make these falconids of interest in continuing studies of environmental contamination.

Our comparative studies of Laughing Falcons in mature forest and the agricultural landscape did not lay to rest any questions concerning these birds' relative success in these two habitats, but they raised several. First, falcons in the farming landscape appeared to use smaller home ranges and to occur at higher densities than in the primary forest nearby. This increased density in the agricultural landscape, in conjunction with higher rates of non-nesting, may indicate something about the density and spatial distribution of prey in the two habitats. Laughing Falcons in the farming landscape habitually hunted at a limited number of sites, often along water holes. Whether or not total prey abundance was higher in the farming landscape, perhaps prey were more concentrated in certain small areas, resulting in a patchier or more concentrated distribution of search effort by the Laughing Falcons, and in smaller home ranges than in mature forest.

Although we detected no difference in nesting success between the two habitats, the incidence of non-nesting among pairs in the farming landscape was substantially higher than among forest pairs. This species poses intriguing questions for further study, among them that of possible source-sink dynamics among differing habitats. If, in further research, non-nesting proves more frequent in agricultural situations, it will be key to learn whether fewer pairs there attempt to nest, or whether they more often fail because of disturbances such as prevalent burning prior to planting. This burning typically takes place when Laughing Falcons are incubating or have young chicks—potentially vulnerable points during the nesting cycle.

To summarize, we believe that partial deforestation represents less of a threat to this species than to many other forest raptors we studied at Tikal and conceivably may temporarily lead to an increase in populations in some cases. During the early phases of deforestation by slash-and-burn farmers, many dead snags are created, which may offer nest sites for these cavity nesters, as

well as abundant hunting perches. However, when such snags eventually fall they are not replaced, and it is common for such farming landscapes to become increasingly denuded of trees. Nest trees and hunting perches probably become limiting to Laughing Falcon populations in the more thoroughly deforested landscapes. Moreover, even if deforestation has a transient beneficial effect on Laughing Falcon populations in some areas, at other sites—particularly, in drier areas—these raptors seem more reliant on mature forest, as Schaldach (1963), for example, found in Colima, Mexico. In such cases, deforestation might have a more negative impact on Laughing Falcon populations than it does in moister areas.

HIGHLIGHTS

It is scarcely possible to overstate the uniqueness of this handsome raptor, which has no close relative or ecological counterpart. While several Old World raptors specialize on eating snakes—for example, snake eagles (*Circaetus* spp.) and serpent eagles (*Spilornis* spp.)—these are all accipitrids, and thus have a very different phylogenetic heritage than the Laughing Falcon. And while a number of Neotropical raptors eat many snakes, the Laughing Falcon, based on present knowledge, is the species most narrowly restricted to an ophidian diet. Interestingly, some of our study pairs in the farming landscape took more types of non-snake prey than did pairs in the mature forest. Much of this diet/habitat correlation, however, was provided by a single falcon pair, and further research will be needed before concluding whether a tendency toward a broader diet is typical in partially deforested habitats.

Since many raptors at Tikal showed a 1-egg clutch, the Laughing Falcon was not unusual in this regard. However, this species provides the only known example of a modal 1-egg clutch among the more than 60 species included in the Falconidae. Whether the 1-egg clutch is related to a diet of snakes is unknown and worth further attention. Our data and those of others suggest that home range size and spacing of nesting pairs can vary markedly in this species; further study is needed to identify more precisely the pattern of spatial use at any given site. Tentative indications, however, are that at least in primary forest, these falconids may use a relatively large area for a raptor of their body size. If this holds true in further study, it will be of interest to investigate whether, across various raptor species, large home ranges are correlated with the proportion of snakes in the diet. Another intriguing observation was that adults were sometimes accompanied by a fledgling from the previous year. We do not know, however, whether subadults received parental care (i.e., food)—this topic awaits further study.

Another of our results that merits further research is the apparent beneficial effect of ants in Laughing Falcon nests. Botflies led to nestling mortality in some cases and were suspected of contributing to mortality in others. The apparent effect (statistically near significant) of ant presence on nesting success is suggestive in this regard. Perhaps ants reduce mortality by minimizing botfly infestation or reduce the frequency of nest predation.

For further information on this species in other portions of its range, refer to de la Peña 2006, Di Giacomo 2005, DuVal et al. 2006, and Gerhardt 2004.

Chapter 18 Notes

1. Mann-Whitney U-test, $P = 0.002$.
2. MRPP chi-square test, $\chi^2 = 3.54$, $P = 0.06$.
3. t-test, $P = 0.001$.
4. MRPP chi-square test, $\chi^2 = 11.7$, $P = 0.0006$.
5. The two lizards were brought to a nest along the boundary between the park's mature forest and the adjacent farming landscape.
6. Fourteen of these clutches were in natural tree cavities, four in epiphytes, three in tree crotches with epiphytes, one in a crotch with no epiphyte mentioned, two amid cacti atop palmetto stubs, and one in a niche in a cliff.
7. For instances of 2-egg clutches in *Herpetotheres*, a request to a Neotropical raptor interest group Listserver produced evidence of two 2-chick broods: one in Veracruz, Mexico (Sergio Aguilar, pers. comm.), and one within 300 km of Asunción, Paraguay (Fernando Feás, pers. comm.).
8. Based on known hatching dates and an estimated 45-day incubation period, three eggs were estimated as laid on 18 February, 26–27 February, and 15 March; a fourth egg was laid on roughly 21 February (14–28 February). For the first three, mean

laying date was 1 March; if the fourth case is included, mean laying date is 27 February.

9. Three known hatching dates were 4 April, 12–13 April, and 29 April (mean = 15 April, $n = 3$); a fourth egg hatched 1–14 April—using the midpoint (7 April), the mean of four hatch dates is 13 April.

10. Two observed fledging dates were 11 and 20 June (mean = 15.5 June, $n = 2$); based on the observed 57-day (55–59-day) fledging period, a third chick would have fledged about 31 May, and a fourth about 27 May–10 June, had these survived to fledging.

11. *Editor's note:* For growth-rate analyses of Chapter 23, a separate curve-fitting exercise was conducted with these data, using TableCurve 2D software (SYSTAT Software, 2002), in order to estimate the growth constant, K, for the steepest segment of the growth curve. The logistic growth constant, K, equals four times the growth rate slope at the point of inflection, when slope is expressed as the proportion of the asymptotic body mass achieved per unit time (Ricklefs 1967, pers. comm.). Omitting three outlying data points, 10 curvilinear equations gave good fits (adjusted r^2 ranging from 0.932 to 0.935), with mean slope of 0.0285 ± 0.0013, roughly corresponding to a logistic K of 0.1139. The logistic equation (adjusted $r^2 = 0.934$) gave a slope of 0.0299, for a logistic K value of 0.1196.

12. Mann-Whitney U-test, $P < 0.025$. Two versions of this test: one version included all nest sites, regardless of simultaneity of nesting efforts, whereas the other version used only distances between nests with simultaneous nesting efforts, both versions gave the same result.

13. MRPP chi-square test, $\chi^2 = 0.20$, $P = 0.66$.
14. Fisher's exact test, $P = 0.12$.
15. Fisher's exact test, $P = 0.09$.

19 BAT FALCON

Margaret N. Parker and David F. Whitacre

Few sounds are as evocative of the humid Neotropical lowlands as the piercing voice of a Bat Falcon *(Falco rufigularis)* as it sweeps in to perch on some bare limb high above the canopy. And there are few sights more dramatic than one of these compact, crisply colored falcons launching itself from a tall, bromeliad-spangled perch, in a breath-taking stoop toward a bird crossing a canopy gap below. For the naturalist fond of observing predators in action, we heartily recommend these handsome falcons (Plate 19.1).

This hole-nesting falcon is a swift and highly aerial hunter of the interface between sky and vegetation. Habitually perching on dead snags or tall exposed limbs, these falcons capture prey almost exclusively in flight. Hunts vary from long, near-vertical stoops to low flights calculated to startle prey from the forest canopy or low ground cover. The only common *Falco* of Neotropical forests, these small falcons are often quite vocal, making them conspicuous and easily observed. A wide spectrum of habitats is used, from unbroken expanses of mature forest to pastures dotted with small forest remnants (Plate 19.2). These falcons are not particularly wary of humans, and at times they live in close proximity to people, even perching on TV antennas while hunting over villages or urban edges.

GEOGRAPHIC DISTRIBUTION AND SYSTEMATICS

Bat Falcons occur from southern Tamaulipas and Sonora, Mexico, to northern Bolivia, Paraguay, and Argentina, as well as the island of Trinidad (Brown and Amadon 1968). Most common in humid, forested lowlands, these falcons range up to 1500 m in Guatemala (Land 1970) and Mexico (D. Whitacre, pers. observ.). Within the genus *Falco*, Fox (1977) argues that Bat Falcons are most closely allied with the two other Neotropical falcons—the Orange-breasted Falcon *(F. deiroleucus)* and Aplomado Falcon *(F. femoralis)*—the three of them comprising, perhaps along with the New Zealand Falcon *(F. novaeseelandiae)*, a southern radiation within the genus. Cade (1982)

favored this interpretation, pointing out that all four share a relatively laterally compressed, keeled beak and tails that are dark gray to black, interrupted by narrow white bars, an arrangement unique among the falcons.

MORPHOLOGY AND PLUMAGE

The Bat Falcon is a solidly built falcon, similar in proportions to the Merlin *(F. columbarius)*, Peregrine Falcon *(F. peregrinus)*, and Teita Falcon *(F. fasciinucha)*. As in these other bird-eating falcons, the wings at rest end approximately at the tail tip. We surmise that morphologically, the Bat Falcon, like the other falcons just mentioned, is especially well adapted for swiftly accelerating stoops.

The Bat Falcon is among the most size-dimorphic of falcons (Appendixes 1, 2). Fourteen males averaged 130 g (108–148 g) in weight, and nine females 206 g (177–242 g: Haverschmidt and Mees 1994). Females weigh nearly the same as a female Merlin, while males weigh a bit more than a female American Kestrel *(F. sparverius)*. Values of the Dimorphism Index for (cube root) body mass, wing chord, and culmen length are 15.3, 15.2, and 23.4, averaging 18.0,[1] which exceeds the equivalent mean for the Peregrine Falcon (15.7), the most size-dimorphic North American falcon (Snyder and Wiley 1976). In several linear dimensions, the female is 9.5–26.5% larger than the male,[2] while in body mass, the female is 59% heavier than the male. In light of the species's highly bird-dominated diet, this high degree of sexual size dimorphism accords well with the pattern prevalent among raptors, in which species taking agile prey, especially birds, show the greatest size dimorphism (Newton 1979; Temeles 1985).

Cade (1982) predicted that the Bat Falcon will prove to be among the fastest members of the genus in level, flapping flight. In watching Bat Falcons fly and hunt, Whitacre et al. (unpubl. ms.) gained the impression that these small falcons are heavily wing loaded, as an aid in rapid flight and perhaps in acceleration or final velocity achieved during stoops. Using data from Cade (1982), these researchers showed that the Bat Falcon is more heavily wing loaded than the average *Falco* species of its body size. However, verification of whether Bat Falcons do indeed accelerate rapidly in a stoop or reach particularly high velocities awaits direct measurement.

In northern Central America and Mexico, fledgling Bat Falcons are readily distinguished from adults. The throat, chin, and auricular regions, which are largely white in adults, are uniformly colored salmon, cinnamon, or orange in juveniles. In addition, juvenile Bat Falcons have a few heavy, blackish bars on the rufous undertail coverts, lacking in adults (Friedmann 1950; Barreto 1968; Brown and Amadon 1968); this feature also occurs in adult Orange-breasted Falcons (Blake 1977). When perched together, male and female Bat Falcons are easily distinguished by size. Behavior and voice also aid in distinguishing the sexes, especially during nesting, when the female is invariably present near the nest and often voices the food solicitation wail. The male appears periodically with food, often using high-pitched *pip* notes while near the female.

PREVIOUS RESEARCH

The Bat Falcon is one of the best-studied Neotropical raptors, having been studied three times prior to our work at Tikal. William Beebe (1950) first studied a pair and their progeny in a Venezuelan cloud forest throughout one breeding season. Beebe had an immense pair of binoculars set on a tripod, and the falcons' eyrie was near his lab window, allowing him to observe minute details of these falcons' daily lives. The second study also took place in Venezuela, where Monte Kirven (1976) produced a doctoral dissertation on the diet and hunting behavior of two pairs in a lowland pasture setting. Finally, Gary Falxa, Devora Sharp, and David Whitacre studied 29 pairs in the lowlands of southern Mexico during 1977 and 1978 (Cade 1982; Falxa et al., unpubl. ms.; Sharp et al., unpubl. ms.; Whitacre et al., unpubl. ms.). In addition, a few brief notes report on diet and hunting behavior (Chavez-Ramirez and Enkerlin 1991; Seijas 1996). For aspects of breeding biology and behavior, we rely heavily on the results of Falxa, Sharp, and Whitacre, hereafter often denoted simply as the "Mexican study."

RESEARCH IN THE MAYA FOREST

We studied Bat Falcons at Tikal in 1991, concentrating on two pairs nesting among the central ruins (Temple IV and Temple I pairs) and two pairs nesting elsewhere in the park (Bajo and Camino pairs). Our studies focused on diet, hunting behavior, and foraging and nesting role division between pair members. Observations began the last week of February and intensified during March and April, totaling more than 1500 hours by the end of the nesting period. We typically observed nests from dawn to dusk, every three or four days at each nest. Observations were continuous and aided by the use of binoculars and a spotting scope.

The Temple I nest was in a small cavity in the vertical limestone facade of Tikal's Temple I and was observed from the top of a small ruin 60 m away. The Temple IV nest was similarly located in a hole on the west facade of Temple IV and was observed from an upper landing of the temple, 7 m below. This close proximity did not disturb the falcons, which were accustomed to tourists climbing there. From this landing we collected pellets and prey remains nearly every day of the breeding season. The Bajo nest was 19.5 m up in the trunk of a dead Black Olive tree *(Bucida buceras)*, in a cavity formed by a broken limb and was observed from a ground blind 40 m away. The Camino nest was 21 m up in a dead tree in a cavity in a termite nest, most likely formed by trogons—it was observed from a tree blind slightly above nest level, 34 m distant. This nest failed between 11 and 13 April, prior to hatching.

We took data on prey deliveries and the adult falcons' excursions away from the nest. At the Temple I and Temple IV nests, views over the canopy were largely unobstructed, allowing us to observe 541 complete foraging trips. For these trips we recorded time, duration, sex of the falcon, prey type, success of the hunting excursion, and the habitat in or over which the falcons hunted. We also recorded the height of hunting flights, using the following categories: less than 2 m above ground, within the forest, within the canopy, at the canopy's upper contour, and aerial. We also noted the position from which attacks occurred: soaring, gliding, perch hunting, flapping, strenuous flapping, or gleaning from trees. We analyzed only those excursions that began and ended within our view, omitting those during which the falcon was out of our sight for more than three minutes. This left us with 256 usable hunting excursions from the Temple IV nest and 263 from the Temple I nest.

To document prey used by Bat Falcons during the breeding season, prey items were identified as they were brought to the nest or via remains collected there. Observers became familiar with avian prey species at the Temple IV nest, aided by the proximity of our observation site. Through observations, collection of feathers as they were plucked, and use of field guides, we often identified avian prey to species at the time of observation. Feathers and body parts collected were used to identify unobserved prey and to confirm visual identifications. Avian remains were identified by comparison with study skins at the Museum of Vertebrate Zoology, University of California, Berkeley. Of 83 bird species identified as prey, 34 (41%) were identified from remains as well as visually, 17 (21%) only via remains, and 32 (38%) only visually (Appendix 19.1). Bird weights were taken from published sources, mainly Dunning (1993). For prey items identified only to family or other category, for example, "hummingbird," we assigned the mean mass for all species of that category identified among prey items at Tikal. Items not identified at least to taxonomic class (*n* = 16) were deleted from biomass estimates.

To estimate insect masses, we caught and weighed specimens of the following: Odonata (Anisoptera: dragonflies,

n = 7), Lepidoptera (n = 4), Coleoptera (Hydrophilidae, Scarabaeidae, Buprestidae, Elateridae, n = 13), Homoptera (Cicadidae, n = 2), Diptera (Tabanidae, n = 16), Orthoptera (katydids, Tettigoniidae, n = 3, and grasshoppers, Acrididae, n = 13), and Hymenoptera (Apidae: bumblebees, Bombinae, n = 3). Insects were weighed using a 10 g Pesola scale, whose accuracy was tested against an Ohaus GA-110 electronic balance. The Pesola performed well,[3] underestimating true weight by 5–6%—thus, we did not correct the field Pesola weights.

We collected dragonfly and lepidopteran remains below the Temple IV nest and from the nest cavity and perch and cache sites after the young fledged. Wings were typically intact, allowing identification. Dragonfly species were differentiated and enumerated (though not identified) by examining wing venation patterns. Lepidopterans were identified to species by Nicholas Haddad, who was studying butterflies at Tikal at the time. We assumed lizards in the diet to be arboreal species. Their masses were estimated using weights of 19 *Anolis* lizards caught and weighed at Tikal. Bat remains collected below the Temple IV and Bajo nests were identified, and masses estimated, using Eisenberg 1981.

Statistical analyses utilized SYSTAT (Wilkerson 1990) and consisted of paired and unpaired t-tests, Kruskal-Wallis tests, and Mann-Whitney U tests. P values of 0.05 or less were considered significant.

DIET AND HUNTING BEHAVIOR

The Bat Falcon is not a bat specialist, as its common name implies. The makeup of the diet depends somewhat on local hunting opportunities but is most often dominated by birds in terms of biomass and by insects numerically. Bat Falcons at Tikal preyed on a wide variety of insects and birds, with a few bats and lizards also taken. Prey items ranged in size from 0.3 g horseflies to a 220 g White-fronted Amazon Parrot (*Amazona albifrons*). The four study pairs took at least 197 prey species among the 1438 prey items documented (Table 19.1). Below the Temple IV nest we collected remains representing 21 species (78 individuals) of butterflies and 47 individual moths. More than 800 dragonflies of several species were taken by the four falcon pairs. Remains of four bats were collected beneath two nests; two were identified as Jamaican Fruit-eating Bats (*Artibeus jamaicensis*).

We identified 83 bird species among prey items at Tikal, from five orders and 21 families (Appendix 19.1). Of 309 birds documented in the diet, at least 45% were highly aerial species: 70 hummingbirds and 69 swallows. These results are similar to those found in Mexico (Whitacre et al., unpubl. ms.) and in Beebe's (1950) Venezuelan cloud forest study. In each case, birds of aerial or canopy habits and those associated with openings and edges predominated in the diet, with forest interior and understory species only occasionally taken. In Bat Falcon diets at Tikal, best represented by number of species

were the flycatchers, with 13 species. Ten species each of warblers, emberizid finches, and tanagers occurred in the diet. Several bird species that dwell mainly beneath the forest canopy were also taken, including White-breasted Wood-Wren (*Henicorhina leucosticta*), Todymotmot (*Hylomanes momotula*), and Plain Xenops (*Xenops minutus*). Presumably these were taken when they crossed gaps or otherwise left dense forest cover, or were snatched from foliage at the edge of clearings or in "canyons" in the canopy surface.

At Tikal, as in Mexico, during the incubation and early nestling periods, male Bat Falcons brought predominantly birds to the nest to feed the female and nestlings. As the nestlings grew larger, the falcons relied increasingly on insects, with both adults capturing them at frequent intervals for themselves and to feed the young. At many nests in Mexico and at one Tikal nest (the Bajo site), dragonflies were the insects most frequently taken. At this nest, dragonflies comprised 94% of the insects brought to the nest, and an estimated 89% of the dietary biomass over the nesting season. At two other Tikal nests, dragonflies comprised only 26–43% of insect prey (numerically), and butterflies, beetles, spiders, and large bees assumed greater importance. Spiders may have been taken from webs or vegetation or while ballooning.

In studies to date, Bat Falcon diets varied from almost solely birds or bats to include many insects (Table 19.2). On a numerical basis, the percentage of vertebrates in the diet (mainly birds and bats but with occasional lizards, frogs, snakes, and mice) ranged from 16% (Mexico) and 23% (Tikal) up to 85 and 91% in Venezuelan lowlands and cloud forest, respectively, the remainder of the diet being made up of insects—85% in Mexico, 77% at Tikal, and 15 and 9% in Venezuelan lowlands and cloud forest. In terms of biomass, however, the results are quite different since insects are much smaller than most vertebrate prey items. On a biomass basis, the estimated importance of birds varied from 32% of the diet in Venezuelan lowlands to 61% at Tikal, 65% in Mexico, and 85% in Venezuelan cloud forest. Bats made up 10% of prey biomass in the cloud forest, 14% in Mexico, and only 2% at Tikal; in Venezuelan lowlands bats comprised 64% of diet biomass, but this number is probably not representative, being based on only two falcon pairs over a short period. The total contribution of vertebrates on a biomass basis was more consistent between studies, varying from a low of 66% at Tikal to 83% in Mexico and 96% in both Venezuelan cloud forest and lowlands.

Some portion of these apparent dietary differences among studies is likely a result of differences in methods, especially in the average mass of insect prey used in calculations. For example, Whitacre et al. (unpubl. ms.) used 0.65 g as the estimated weight of dragonflies and other insects, and 2 g for cicadas. Insect weights we used at Tikal are shown in Table 19.1, and they were uniformly much greater than the weights used in the Mexican study. Insect weights at Tikal were based on larger samples than in the Mexican study, and our verification

Table 19.1 Diet of Bat Falcons at four nests at Tikal

Prey category	Estimated mass (g)	Bajo nest (number)	Temple I nest (number)	Temple IV nest (number)	Camino rest (number)	Total (number)	Percentage of total prey (numeric)	Total mass (g)	Percentage of total prey (biomass)
Dragonfly	2.2	738	37	61	3	839	—	1845.8	—
Butterfly	2.4	18	11	27	1	57	—	136.8	—
Beetle	4.05	3	9	29	0	41	—	166.05	—
Grasshopper	7.1	1	3	3	1	8	—	56.8	—
Cicada	5.1	9	7	7	0	23	—	117.3	—
Bumblebee	5.5	10	2	42	0	54	—	297.0	—
Spider	0.3	3	0	18	0	21	—	6.3	—
Katydid	4.9	0	0	6	0	6	—	29.4	—
Horsefly	0.3	0	0	2	0	2	—	0.6	—
Unidentified arthropods	2.2	0	14	39	1	54	—	118.8	—
Total arthropods	**Midpoint of mass ranges below**	**782**	**83**	**234**	**6**	**1105**	**76.8%**	**2774.85**	**33.9%**
Bird, 15.5–20.0 g	17.75 g	4	1	7	2	14	—	248.5	—
Bird, > 20–30 g	25 g	2	3	4	1	9	—	229.5	—
Bird, > 30–40 g	35 g	3	1	3	0	7	—	248.5	—
Bird, > 40–70 g	55 g	0	0	1	1	2	—	111.0	—
Bird, > 70 g	75 g	1	4	0	0	4	—	300.0	—
Unidentified	32.3	1	22	11	1	35	—	1130.5	—
Total number of birds > 15 g	—	**11**	**31**	**26**	**5**	**71**	—	**2359.0**	—
Hummingbird	6.8	16	19	35	0	70	—	476.0	—
Swallow	15.5	0	50	19	0	69	—	1069.5	—
Bird, 6.5–11.0 g	8.75 g	11	6	16	1	34	—	297.5	—
Bird, 11.1–15.0 g	13.05 g	6	7	8	2	23	—	300.15	—
Unidentified bird < 15 g	10.9	4	16	21	1	42	—	457.8	—
Total number of birds < 15 g	—	**37**	**98**	**99**	**4**	**238**	—	**2600.95**	—
Total number of birds		—	—	—	—	**309**	**21.5%**	**4959.95**	**60.5%**
Lizard	7.6	0	2	1	0	3	—	22.8	—
Bat	35.8	2	0	3	0	5	—	179.0	—
Unidentified vertebrate	—	0	6	9	1	16	—	259.2	—
Total number of non-bird and unidentified vertebrates	—	**2**	**8**	**13**	**1**	**24**	**1.7%**	**461.0**	**5.6%**
Total number of vertebrates	—	—	—	—	—	**333**	**23.2%**	**5420.95**	**66.1%**
Total number of prey	—	**832**	**220**	**372**	**16**	**1438**	**100.0%**	**8195.8**	**100.0%**

Table 19.2 Diet at the nest of Bat Falcons in four studies (%)[a]

Prey type	Tikal	Mexico	Venezuela, cloud forest	Venezuela, lowlands
Numeric diet				
Birds	20.6	12.0	74.8	37.8
Bats	0.4	2.7	14.7	47.4
Other vertebrates	2.2	0.8[2]	1.8	0.0
Total vertebrates	23.2	15.5	91.3	85.2
Arthropods	76.8	84.5	8.7	14.8
Biomass diet				
Birds	60.5	65.0	85.0	32.0
Bats	2.2	14.0	10.0	64.0
Other vertebrates	3.4	4.0[b]	< 1.0	0.0
Total vertebrates	66.1	83.0	96.0	96.0
Arthropods	33.9	17.0	4.0	4.0

[a] Sample sizes: Tikal, n = 1405 identified prey items; Mexico, n = 925 identified prey items; Venezuelan cloud forest, n = 218 identified prey items; Venezuelan lowlands, n = 135 identified prey items.
[b] Unidentified vertebrates.

of our scale's accuracy (see earlier) gave no reason to question our weights. It seems unlikely that the mass of individual insects taken actually differed greatly between Tikal and southeast Mexico; use of a common set of insect weights in both studies diminishes the apparent difference in the percentage of the diet composed of insects and vertebrates in the two localities (Table 19.2). Still, there is no doubt that differences in diet occurred between studies. The low use of bats at Tikal is unquestionable, and it also seems likely that falcons in the Venezuelan cloud forest took fewer insects than did those in Tikal and Mexico. The Mexican and Venezuelan lowland studies were conducted mainly in human-altered habitats including extensive pasture, and this may have led to some dietary contrasts with Tikal. However, the Venezuelan cloud forest site, like Tikal, was primarily mature forest; hence the difference in use of insects versus vertebrate prey between these two sites was not due to differences in degree of forest modification.

It is clear that local prey availability affects what these opportunistic falcons prey on, often resulting in pronounced dietary differences between neighboring pairs. At Tikal, a response to abundant, vulnerable prey was evidenced when the falcon pair at Temple I preyed heavily for several days on young Ridgway's Rough-winged Swallows (Stelgidopteryx ridgwayi) fledging from a nearby colony. This falcon pair captured 22 swallows in three successive days and at least 49 swallows during the breeding season. Likewise, the Bajo pair preyed far more heavily on dragonflies than did the other three pairs, and these insects were obviously more abundant in the vicinity of this nest than at the other three territories.

Male and Female Roles at Tikal

Males at Tikal, on average, delivered slightly larger prey (7.6 ± 0.30 g [SE], n = 875) than did females (6.6 ± 0.53 g [SE], n = 502).[4] This pattern resulted from the fact that

during the incubation and early nesting periods, when the male was the main provider, vertebrate prey played a larger role than during the latter part of the nestling phase, when both adults brought in an increasing proportion of insect prey. Although females took the few largest prey items, falcons of both sexes took all categories of avian and insect prey types and sizes. Males captured prey weighing from 0.5 to 165.0 g and females from 0.5 to 185.0 g. The Temple I female was observed with one prey item estimated at 220 g, the largest recorded in this study. Capture success was similar for both sexes when complete foraging excursions were considered. At the Temple I and IV nests (pooled), males were successful in 71.9% of foraging excursions and females in 70.4% of excursions.

Overall, males played a larger role as providers than did females. Of the 543 timed foraging excursions, 353 (65.0%) were performed by males and 190 (34.3%) by females.[5] Of the 1423 prey items delivered by adults during observations at the four nests, males delivered 875 items (61.5%) and females 520 (36.5%).[6] All females contributed less prey biomass than their respective mates.[7]

Hunting Behavior

Hunts were launched from perches or from soaring flight, sometimes from substantial altitudes—especially when adequate winds appeared to facilitate soaring. These falcons also used a hunting strategy that has been described as a "contouring" flight in which the birds cruise at high speed, low over vegetation, with the apparent hope of startling prey into flight at close quarters (Whitacre et al., unpubl. ms.). Quarry is sometimes flushed intentionally from the forest canopy or ground cover, but with rare exceptions it is caught while in open air, and we never saw these falcons catch prey on the ground.

Bat Falcons, like several other falcons studied, generally rely heavily on the element of surprise in capturing prey, and this sometimes involves the use of concealing cover during an attack. At Tikal, where we often used motorcycles for transportation, we were twice unnerved by a Bat Falcon speeding head-on toward us at eye level as we rode along. In both cases the falcon passed within arm's reach, apparently intent on quarry behind us. Presumably we were being used as cover to achieve surprise of quarry we had passed and perhaps flushed from roadside vegetation.

A variety of habitats may be used in hunting, from open pastures (usually with tall snags used as hunting perches) to closed-canopy forest. In 526 timed hunting excursions, 126 (24%) were foraging flights over open areas (mowed plazas around Maya ruins) and 400 (76%) were over intact forest. Of 541 hunting excursions (not all timed), 68 (12.5%) were low (< 2 m) over the ground, 62 (11.4%) were in open areas, at variable heights, 38 (7%) were contouring flights low over the forest canopy, 129 (23.9%) were directed at the forest canopy, 123 (22.7%) were within the forest, 97 (18.0%) were aerial (well above the canopy or ground), and 24 (4.4%) were of unknown nature. Hunts within the forest were cases in which the bird dove into deep clefts in the canopy's upper surface, or passed at high speed through gaps in treetop foliage. Bat Falcons did not, to our knowledge, actually drop beneath the closed canopy of the forest. Bat Falcons at Tikal hunted mainly over the forest, and nearly all of the hunting deemed to take place in open areas occurred within view of the Temple I nest as this pair hunted swallows from a nearby colony over short-cropped lawn.

In more than 900 hunts observed in southern Mexico (Whitacre et al., unpubl. ms.), Bat Falcons were observed only once to snatch prey from the foliage of a tree. In contrast, Bat Falcons in Tikal used a novel foraging behavior on 17 occasions. In these cases, the falcons flew into dense canopy foliage, striking a limb or clump of leaves with their extended feet, and then fell away, often repeating this behavior, up to 11 times on one occasion. This foraging tactic, observed most often during a several-week period at the Temple II nest, was used to dislodge large bees that perched on the undersides of leaves. The falcons captured the bees in flight after dislodging them in this manner. The falcons did not appear to be simply striking random bits of foliage, but rather seemed to target areas where bees were sheltered in the shade of thick leaves. Bat Falcons also captured these bees in direct flight while perch hunting and, as with other stinging insects, the falcons held the bee in their talons and removed the stinger with their beak before eating the insect on the wing or delivering it to the nest.

Of the 541 hunting excursions observed at Tikal, 386 (71.3%) resulted in prey capture, 154 (28.4%) were unsuccessful, and 1 (0.3%) was of unknown outcome. Success rates ranged from 62% in hunting excursions low over the ground (n = 24) to 77% in contour flights over the forest canopy (n = 31), to 99% in aerial excursions (n = 73). These success rates were judged on a per-hunting excursion basis, which sometimes included attacks on multiple prey items per flight. This approach gave higher success rates than methods used by Whitacre et al. (unpubl. ms.) and in most other analyses of raptor hunting success, which have usually quantified success rates per attack on prey, rather than per hunting excursion (e.g., Rudebeck 1950, 1951; Cresswell 1996). In Mexico, success per attack on prey was 59.6% overall (n = 778), but this was heavily influenced by the high success rate of insect hunts (success rate = 64.3%, n = 661). Of 50 bat hunts, 24% were successful, and success in bird hunts was between 8.3 and 15.2% (Whitacre et al., unpubl. ms.).

Foraging flights at Tikal ranged in duration from 3 to 360 seconds and averaged 52.9 ± 3.2 seconds (SE) for the Temple I pair and 79.5 ± 4.1 seconds (SE) for the Temple IV pair. Duration of foraging flights differed significantly between the sexes,[8] with females spending a mean of 54.1 ± 4.6 seconds (SE) and males 71.9 ± 4.7 seconds (SE) per hunting excursion. Females spent less time per foraging bout than did males at all times of day except at dusk (1800–2000), when females on average spent nearly twice as long away from the nest as did males. These numbers refer only to flights originating near the nest and within our view and reflect the habitats used by the falcons during these flights, which we felt encompassed the majority or entirety of their territories.

HABITAT USE

Bat Falcons occupy habitats ranging from mature humid tropical lowland and subtropical montane forest to pastures with scattered forest remnants. At Tikal and elsewhere (Beebe 1950), Bat Falcons do nest amid extensive forest, but it remains unclear whether, in such situations, they are often tied to local differences in canopy height created by natural openings or topographic features. Bat Falcon densities in extensive areas of mature forest often appear lower than those in areas of agricultural frontier or riparian situations, but they have never been adequately measured. In many cases, both in mature forest and in agricultural mosaics, nesting Bat Falcons appear sparsely and irregularly distributed, and we felt this to be the case in the farming landscape immediately south of Tikal, where we never found these birds to be common. It remains unknown whether Bat Falcon abundance in such situations is limited by availability of nesting cavities, hunting perches, prey abundance, or some other factor.

Bat Falcons appear to reach highest densities in mosaic areas containing some mature forest, some young successional forest, and an abundance of nest and perch sites in the form of dead snags (Sharp et al., unpubl. ms.). Such conditions often occur along meandering lowland rivers and, at least temporarily, along the agricultural frontier, as slash-and-burn farmers create a landscape of active and abandoned fields interspersed with forest remnants (Paynter 1955; Iñigo-Elias 1993; D. Whitacre, pers. observ.). Seemingly, the most favored situations are those having ample second growth interspersed with mature forest and abundant nesting and hunting snags. Riverine situations seem especially favored. Sharp et al. (unpubl. ms.) speculated that this apparent preference for river-edge situations is based on a combination of factors: aggregation of insectivorous birds and bats foraging low over such rivers; exposure of avian and bat prey over the water, far from cover; juxtaposition of tall riverbank snag perches with the low water surface, creating a height advantage for the perched falcon; and the abundance of young successional vegetation, which allows the falcon a height advantage and provides high densities of avian prey.

Bat Falcons do sometimes nest in areas of extensive pasture, but Sharp et al. (unpubl. ms.) found that males of such pairs often commuted more than 1 km to hunt at remnant forest patches, remaining away from the nest for hours at a time. These researchers concluded that, perhaps with rare exceptions, Bat Falcons need some forest or woodland areas in which to hunt.

BREEDING BIOLOGY

Nests and Nest Sites

As is true of all *Falco* species, Bat Falcons do not build nests but simply make a scrape in the substrate of a suitable site. Bat Falcons nest in natural cavities in trees or cliffs and occasionally in cavities of man-made origin, including the Maya temples of Tikal. Falxa et al. (unpubl. ms.) observed 18 nestings in cavities in dead trees, three in live trees, three on cliffs, two on spathes of palm fronds, and one attempted nesting in the girders of a steel highway bridge.[9] Two of the four nests we studied at Tikal were in holes in vertical surfaces of the upper portions of limestone Maya temples, one was in a cavity formed by a broken branch in the trunk of a dead tree, and one in an old termite nest in a dead tree, within a cavity probably excavated by trogons. The two tree nests at Tikal were 19.5 and 21 m above ground, while the two temple nests were at least 40 m above ground in sites protruding above the surrounding forest canopy.

Egg and Clutch Size

Each of four clutches at Tikal contained three eggs. Twenty-five museum egg sets and two other clutches, all from southern and eastern Mexico, showed a mean and modal clutch size of 3.0 eggs (two clutches of two, 23 clutches of three, and two clutches of four: Falxa et al., unpubl. ms.). Falxa et al. (unpubl. ms.) observed asynchronous hatching and concluded that incubation began before clutches were complete.

We did not measure any eggs. Schönwetter (1961) gave mean dimensions of 12 eggs as 40.6 mm by 31.7 mm, and other published measurements are similar (Wetmore 1965; ffrench 1973; Rowley 1984). Based on Schönwetter's data, Hoyt's (1979) formula yields a predicted mass of 22.3 g per egg, or 10.8% of the mean 206 g female body mass. Hence the mean 3-egg clutch amounts to approximately 32.5% of mean female body mass.

Nesting Phenology

By the end of February, prey exchanges had begun between pair members at each of the four Tikal study nests, and territories were actively defended by both sexes. Estimated laying dates for these four nests were during the second and third weeks of March, and hatching occurred approximately 30 days later. Three young per nest fledged from the Temple I, Temple IV, and Bajo nests during late May and early June; the Camino nest was abandoned during incubation, apparently as a result of nest site damage and predation. Timing of nesting was similar in southern Mexico, with eggs laid in March and April and young fledging from mid-May to late June (Falxa et al., unpubl. ms.). Other sources indicate similar timing throughout Central America and northern South America (Bierregaard 1994).

This timing, with most fledging occurring from mid-May to mid-June at the onset of the rainy season, may be advantageous for several reasons. First, many songbirds that breed in the United States and Canada are present early during this period, either wintering or in migration, and were important prey both at Tikal and in Mexico. Hence, this food resource is present during Bat Falcon egg formation, incubation, and feeding of nestlings; migrants are largely absent by late May. Second, many resident bird species also nest at the onset of the rainy season, such that their populations, including many inexperienced young, presumably reach their annual peak about the time young Bat Falcons fledge. Finally, many insect populations also reach their annual peak during the first weeks of the rainy season (see Chapter 2), affording a plentiful supply of easily caught prey during the nestling period and as the young falcons learn to hunt.

Length of Incubation, Nestling, and Post-fledging Dependency Periods

Apparently no data exist on the precise hatching and fledging intervals of the Bat Falcon. Falcons in this size range typically have incubation and nestling periods of about 30 days each (Kirven 1976; Cade 1982). No research has yet determined at what age fledglings reach independence, though they have been observed being fed by adults for at least 4–12 weeks after fledging (Beebe 1950; Falxa et al., unpubl. ms.). It is clear that young Bat Falcons achieve independence prior to the onset of the following breeding season and sometimes attempt to breed during their first year (Falxa et al., unpubl. ms.). Further details are given under Post-fledging Dependency Period, later in this chapter.

VOCALIZATIONS

While lone Bat Falcons are generally silent, members of mated pairs are often highly vocal. Researchers in

Mexico described the vocal repertoire in some detail (G. Falxa, D. Whitacre, D. Ukrain-Sharp., pers. comm.).

Kew-kew-kew. This is the characteristic call of the Bat Falcon, familiar to any Neotropical birder. Haverschmidt (1962) described this call as identical to that of the Hobby *(Falco subbuteo)*. It is used by both sexes and in many contexts, but most commonly in connection with locomotion and also during nest defense. At times, pair members utter this call each time they land after a flight.

Slow kew kew kew. G. Falxa, D. Whitacre, and D. Ukrain-Sharp (pers. comm.) described this call as being more protracted and quieter than the usual *kew kew kew* vocalization. It is used in a variety of contexts, usually while pair members are near each other in the vicinity of the nest. Sometimes both pair members used this call while perched, one bird often answering the other's call. Frequently, one bird pips while the other voices this call (ibid.).

Wail. The wail call, which might also be termed a whine or begging call, is a protracted, high-pitched wail with a quavering quality. G. Falxa, D. Whitacre, and D. Ukrain-Sharp (pers. comm.) found this call to be used exclusively by adult females and by nestlings/fledglings—in both cases it served a food-begging function. In addition, adult females sometimes voiced this call when bringing prey to fledglings. This call often appears effective in motivating adults to hunt and bring prey (adult males in the case of a female voicing it: ibid.).

Pipping and chittering. Bat Falcons use a variety of calls described by G. Falxa, D. Whitacre, and D. Ukrain-Sharp (pers. comm.) as "pips" and "chitters." These are emphatic, clipped syllables, uttered singly and in groups of two, three, or many, and varying in tempo and pitch. Chitters are rolling series of notes repeated at least 10 times per second, giving a drum-roll effect. Pipping calls are quiet and used mainly when pair members are close together. Adult males in Mexico used pipping vocalizations five times as often as females, and pairs during the early- and pre-nesting phases used pipping calls more commonly than did pairs with young (ibid.). Pipping tends to occur in different behavioral contexts, depending on the breeding status of pairs. During courtship and nest site selection, both pair members use pip vocalizations mainly in nest site–related contexts, followed in frequency by pair-related and locomotory contexts. At eyries where nesting was underway, pipping was heard mainly in the context of prey transfers, followed in frequency by nest-related contexts (ibid.).

BEHAVIOR OF ADULTS

Pre-laying and Laying Period

In one study in Mexico, cavity investigation and supplemental feeding of females by males began at least 35 days prior to the onset of incubation (G. Falxa, D. Whitacre, D. Ukrain-Sharp., pers. comm.). Copulation often followed immediately after cavity investigation by one pair member or the other; males often flew directly from a prospective nest cavity to mount the female (ibid.). Several flight displays were also observed in the Mexican study, including mutual flights by pair members and rapid flights or stoops past the nest site (ibid.). Frequent stoops by males at non-prey species also appeared to serve a display function (ibid.). Females often used an apparent copulation-solicitation display in which the head plumage is compressed while the head is craned forward and the bill often pointed downward (ibid.).

Incubation Period

Bat Falcons at Tikal showed a more equal division of sex roles during nesting than did Bat Falcons in Mexico and compared to many other falcons. Females at Tikal were frequently off the nest while males incubated for periods up to six hours. This in itself is not highly unusual, as males often provide 10–25% of incubation, sometimes more, in other *Falco* species (see Chapter 20). However, that Tikal females actively foraged during all phases of the breeding cycle is somewhat unusual, and that they often delivered prey to incubating males at two nests is quite unusual among falconiforms. The Temple IV female delivered 16.4% of the prey biomass brought to that nest during incubation, and the Temple I female delivered 24.9% of prey biomass delivered there during incubation. The Bajo female was not observed to bring prey to the nest during incubation; this pair subsisted mainly on dragonflies.

In the Mexican study (Falxa et al., unpubl. ms.), male Bat Falcons performed essentially all hunting for themselves, their mate, and the nestlings, from some time prior to egg laying until the nestlings were partly grown—the norm in most raptor species. Even prior to egg laying, females spent most of their time perched near the prospective nest cavity, periodically receiving prey from the male. From this time until they began to aid in hunting when chicks were half grown, females rarely left the nest vicinity (ibid.). In the Mexican study, females performed all incubation duties and left the nest infrequently and briefly, usually only when called by the male to receive prey but sometimes to hunt dragonflies nearby or to perch and preen. In the early days after hatching, females continued to brood the young steadily, leaving the nest only long enough to eat prey brought by the male and to bring prey to the young (ibid.). The researchers in the Mexican study never observed a male taking a turn in brooding nestlings.

Nestling and Fledgling Periods

Prey transfers, preparation, and caching. Details of prey exchanges and caching behavior were described by Falxa et al. (unpubl. ms.) and resembled those of other *Falco* species. Most prey transfers from male to female took place on perches near the nest site from beak to beak, with one or both birds often effecting a horizontally inclined, head-low posture during the exchange. Males announced their approach with prey using the *kew kew kew* or slow *kew kew kew* call, and females typically voiced the food solicitation wail throughout the exchange. When a female did not leave the nest to receive prey, the male often cached it, and females also cached prey received from the male. Both sexes retrieved cached items to eat, to feed the young, or in the case of males, to offer the female. The most frequent cache sites were bromeliads on limbs in or near the nest tree (ibid.).

The female as hunter. In the Mexican study, as the young grew larger, adult females began to augment the hunting efforts of their mates (Falxa et al., unpubl. ms.). This activity on the part of females may have been facilitated by the abundance of dragonflies and/or cicadas in eyrie vicinities. Though one female was observed hunting dragonflies as early as the twelfth day of incubation, most began to hunt when nestlings were two to three weeks old (ibid.). After the young were out of the nest and learning to fly, the role of adult females as providers increased, approaching that of the male in some cases.

Females at Tikal played a larger role in provisioning young than in the Mexican study and sometimes provisioned incubating males, an unusual observation among falconiforms. The Bajo female, though she did not bring prey to the incubating male, delivered 41.9% of the prey biomass to the nest during the nestling period, while the Temple IV and Temple I females delivered 23.7 and 36.6% of prey biomass during this period. By the fledgling phase, the Bajo, Temple IV, and Temple I females had contributed 33.6, 40.8, and 44.6%, respectively, of the total prey biomass delivered to these nests.

This greater equitability of male and female roles at Tikal than in other Bat Falcon studies may be related to the apparently greater reliance of the falcons at Tikal on insects than in other studies. As argued by Seavy et al. (1997b), reliance on small prey such as insects makes it less efficient for males to perform all provisioning duties, because repeated travel to the nest with such small items is not energetically efficient. More equitable division of incubation and brooding duties, with adults largely feeding themselves during stints away from the nest, is predicted under such circumstances.

Prey delivery rates. Rates of prey biomass delivery to nests differed among pairs at Tikal. During timed foraging excursions, the Temple I pair delivered 12.5 g of prey per minute, versus 5.7 g/min for the Temple IV pair. However, these rates are per hunting excursion (mostly < 3 minutes in duration) and do not reflect daily delivery rates. The higher rate for the Temple I pair owes in part to the fact that they preyed heavily on Ridgway's Rough-winged Swallows, which nested in the same temple as the falcons.

Birds accounted for most of the prey biomass during incubation at three of four Tikal nests, but as the nestling phase progressed, the falcons relied more heavily on insects. The Temple IV pair delivered prey of about the same mass throughout the breeding season, while the Temple I pair delivered their heaviest prey items during the nestling period. Falcons of the Bajo pair exploited dragonflies at an increasing rate throughout the nestling and fledgling periods, resulting in a steady decline in the mean mass of prey items throughout the breeding season. During the nestling period, this pair delivered prey items 12 times more frequently than during incubation, often at a rate of one dragonfly every 30 seconds. Dragonflies comprised an estimated 89% of this pair's prey biomass over the entire nesting season. The Temple I and Temple IV pairs showed much greater variability, switching prey types frequently in apparent response to changing patterns of prey availability. It appeared that bumblebees, various species of butterflies and moths, grasshoppers, cicadas, and fledgling swallows each provided prey of ephemeral abundance, and each was hunted almost exclusively during their respective times of peak availability.

BEHAVIOR AND DEVELOPMENT OF NESTLINGS

Nestling Period

Ten nestlings examined at four Mexican eyries had white or pale grayish white down, and nestlings up to two weeks old had pale gray to chartreuse ceres (Falxa et al., unpubl. ms.). At fledging, juveniles' ceres and orbital rings varied from pale lemon yellow to dull chartreuse or dull blue-gray. All fledglings had yellow feet and legs, though paler than those of adults. In adults, cere and orbital rings were bright yellow while feet and legs were often brighter, ranging from yellow to orange (ibid.).

Post-fledging Dependency Period

During the fledgling period at the Tikal nests, an adult usually accompanied the young while they flew for short periods near the nest. Both sexes accompanied the fledglings during these initial flights and aggressively drove off potential predators. The young birds roosted within 50 m of the nest up to 40 days after fledging, beyond which we do not have information. Beebe (1950) and Falxa et al. (unpubl. ms.) give further details concerning behavior during the post-fledging period.

SPACING MECHANISMS AND POPULATION DENSITY

Territorial Behavior and Displays

The main territorial display in Bat Falcons appeared to be the simple act of perching prominently. With its bold pattern of black helmet, blackish chest, and contrasting white throat, a Bat Falcon on a tall open perch is recognizable at a great distance. Similarly, Bat Falcons often voice their typical *kew kew kew* call loudly and repeatedly throughout the day, which must be a conspicuous territorial signal to conspecifics.

Interspecific Interactions

At Tikal, Bat Falcons drove at least nine species of larger raptors from nesting territories, sometimes pursuing them more than 2 km, with repeated, vigorous stoops. We observed attacks on Swallow-tailed Kite *(Elanoides forficatus)*, Plumbeous Kite *(Ictinia plumbea)*, Crane Hawk *(Geranospiza caerulescens)*, White Hawk *(Leucopternis albicollis)*, Great Black Hawk *(Buteogallus urubitinga)*, Roadside Hawk *(Rupornis magnirostris)*, Ornate Hawk-eagle *(Spizaetus ornatus)*, Black Hawk-eagle *(S. tyrannus)*, and Orange-breasted Falcon *(F. deiroleucus)*. Six other species of large birds were also attacked—American Black Vulture *(Coragyps atratus)*, Turkey Vulture *(Cathartes aura)*, Plain Chachalaca *(Ortalis vetula)*, Red-lored Amazon *(Amazona autumnalis)*, Brown Jay *(Cyanocorax morio)*, and Montezuma Oropendola *(Psarocolius montezuma)*. Additional species chased from nest environs in Mexico are listed by Falxa et al. (unpubl. ms.).

Constancy of Territory Occupancy, Use of Alternate Nest Sites

Bat Falcons are generally regarded as living in pairs (Beebe 1950; Haverschmidt and Mees 1994), and in Mexico, Falxa et al. (unpubl. ms.) gathered limited evidence that pairs were present near nests during the nonbreeding season. Casual observations at Tikal also indicated that falcon pairs may remain in association during the nonbreeding season and continue to frequent or periodically visit nest sites. The Mexican study also found that the pair bond can exist independent of fidelity to a nest site; some pairs had no apparent attachment to a fixed area and were seemingly in the process of searching together for a nest site. Some nest sites are occupied repeatedly over the years, and while no data exist for marked birds, it seems likely that pair members often remain together for multiple years.

Although Beebe (1950) observed the July movement of seven Bat Falcons through a mountain pass in Venezuela, and Monroe (1968) noted that nonbreeding falcons appeared on islands off the Honduran coast, migrations in which a Bat Falcon population seasonally deserts any part of its breeding range have not been reported. We conclude that this species typically remains as long-term mated pairs, resident on a nesting territory throughout the year.

Inter-nest Spacing and Density of Territorial Pairs

No home range estimates have been made for this species, at Tikal or elsewhere. We obtained only approximate data on nearest-neighbor distances among pairs at Tikal. In 1991, the Temple I and Temple IV pairs nested approximately 760 m apart. The distance from the Temple I pair to the Bajo pair was roughly 4.5 km, and from the Temple IV pair to the Camino pair roughly 3.5–4.0 km. In addition, we observed a non-nesting pair roughly 2 km north of the Camino pair. We believed these five pairs were nearest neighbors, yielding a rough average of 2.75 ± 1.7 km among neighboring eyrie sites ($n = 4$). Though two pairs at Tikal nested less than 1 km apart, on average, nesting pairs were spaced much farther apart than this, both in the mature forest of the park and in the farming landscape to the south, where pairs seemed surprisingly uncommon and patchy in occurrence. In Mexico, Sharp et al. (unpubl. ms.) found a mean distance of 1.5 km (range = 1.3–1.9 km, $n = 4$) between nearest-neighbor pairs in areas of homogeneous favorable habitat with a probable excess of nest sites. They took this to be an estimate of the minimum spacing between nests in favorable conditions, corresponding to a density of one pair per 2 km², and concluded that Bat Falcons often occur at lower but rarely higher densities than this.

DEMOGRAPHY

Age of First Breeding, Attainment of Adult Plumage

Bat Falcons reach definitive plumage with their first full molt, beginning at the age of about one year. Of 18 pairs scrutinized closely by Falxa et al. (unpubl. ms.), in 14 pairs both members were in adult plumage, that is, two years or older. Three pairs were composed of an adult male and a 1-year-old female, while one pair was of two 1-year-olds; none of these bred successfully. This pattern coincides with that seen in many temperate zone raptors, in which subadult females occur as pair members more often than do subadult males (Newton 1979). Though apparently breeding usually begins after attainment of adult plumage, it is not known whether first breeding is often deferred beyond the age of two years.

Frequency of Nesting, Percentage of Pairs Nesting

Bat Falcons nested on Temple I and Temple IV nearly every year of the Maya Project (1988–1997). Although we noted one incident of an apparently non-nesting pair at Tikal, our impression was that most territorial pairs nested in a given year. While we have no data on banded adults, it seems certain that a given pair often nests in consecutive years.

Productivity and Nest Success

In Mexico, 10 of 15 study pairs successfully raised young, producing 2.5 fledglings per successful nesting and 1.67 per nesting attempt (Falxa et al., unpubl. ms.). Three of our four study pairs at Tikal fledged young in 1991, for a mean of 2.25 young fledged per nest and three per successful nest. We have no other formal data on nesting success at Tikal, but casual observations from 1988 to 1997 suggested that reproductive success was similar to the rates given here during this time.

Causes of Nesting Failure and Mortality

The single nesting failure we noted at Tikal occurred during incubation, apparently due to predation. Falxa et al. (unpubl. ms.) also suspected predation in at least one nest failure.

CONSERVATION

Bat Falcons are generally little disturbed by human presence per se, and where adequate hunting and nesting habitat exists, these falcons demonstrate an ability to thrive in close proximity to humans. At Tikal, the Bat Falcons that nested yearly on Temple IV were quite unheeding of human presence. We have heard of at least two instances in which children have thrown grasshoppers or other insects up in the air, with Bat Falcons swooping down to take these offerings before they hit the ground. And Whitacre et al. (unpubl. ms.) observed three individuals regularly hunting bats over a small town, perching on television antennas over the rooftops.

Bat Falcons tolerate and may even benefit, at least temporarily, from the conversion of pristine forest into a landscape mosaic including pastures and young second growth, along with substantial mature forest. Bat Falcons often reach high densities in such landscapes, as noted by Paynter (1955), but this is generally a transitory condition; such landscapes tend to become increasingly deforested over time, with pastures eventually claiming most of the land surface (D. Whitacre, pers. observ.). Once pastures dominate the landscape, suitable nest sites and hunting perches may become limiting, since large trees are not replaced as favored nesting snags succumb to decay and repeated charring during the yearly burning cycle. We have witnessed the demise of several nesting trees as these snags fell after repeated burning; Bat Falcon pairs disappeared from these sites (D. Whitacre, pers. observ.). Moreover, whether such pasture-dominated landscapes often retain adequate prey for Bat Falcons, especially vertebrate prey, is unknown. Finally, while Bat Falcons are often common along the agricultural frontier, it is not known whether this abundance owes, at least in part, to the proximity of large remaining areas of mature forest, both as a "food factory" and as a possible source of immigrant Bat Falcons.

These falcons' tolerance of habitat alteration, combined with the species's abundance, modest space requirements, and wide geographic distribution, bodes well for the species's future. However, a number of cases can be cited in which this species has virtually disappeared from portions of its range in apparent response to human activities, and this trend can be expected to continue where deforestation proceeds to an extreme degree. Stiles (1985) believed that populations in Costa Rica had declined greatly in the previous 15–20 years and proposed listing it as endangered there. He suggested that shooting and pesticides may have played a role in this decline, and we would add that general levels of habitat alteration likely also contributed, especially where large acreages of forest have been converted to monocultures of sugarcane, oil palm, cattle pastures, and low groves of sun coffee.

Likewise, in 1977, Sharp et al. (unpubl. ms.) searched large areas of southeastern Mexico within the historic range of this species before finding a population large enough to study in the coastal plains of Chiapas and Tabasco. Though they found pairs farther north in the forested Sierra Madre Oriental, they found none on the gulf coastal plains of Veracruz and Tamaulipas, despite many nesting records from that area from the 1950s and earlier. In El Salvador, Bat Falcons have been extirpated or nearly so (Thurber et al. 1987). In both El Salvador and northeast Mexico, it is difficult to guess the relative effects of various factors, as both regions have suffered a combination of deforestation, conversion to intensive agriculture and pastures, and heavy use of DDT and other organochlorines from the 1950s through the 1970s, especially on cotton.

DDT use may well have led to the disappearance of Bat Falcons from certain areas. By 1980, for example, this species essentially disappeared as a breeder from the Pacific coastal plain of Guatemala (J. Vannini 1989a, 1989b), even though suitable Bat Falcon habitat remains in this once-forested area. Cotton farming here covered large acreages from the 1950s to the late 1970s, and prodigious quantities of DDT and a host of other organochlorine insecticides were sprayed on cotton in this region until about 1980 (Murray 1994). Research that might have revealed any impacts of these insecticides on Bat Falcons or other raptors was not conducted here during this time. In southeast Mexico, however, eggshell thinning and high levels of DDE in egg lipids were demonstrated in Bat Falcon and Aplomado Falcon eggs collected between 1954 and 1967 (Kiff et al. 1980). As early as 1970, Cattle Egrets (Bubulcus ibis) on Guatemala's Pacific coast showed dramatic evidence of eggshell thinning in cotton-growing regions and high DDE levels in the eggs (S. Herman, unpubl. data). It is probably safe to conclude that on Guatemala's Pacific coast and perhaps in portions of Mexico, pesticides contributed to the virtual disappearance of Bat Falcons. However, the decreasing reliance on organochlorine insecticides globally since the 1970s suggests that the worst of this threat is over. In the future, the extensive clearing and domestication of once-forested landscapes will probably do more to reduce the geographic distribution and global population of this handsome falcon than any other single factor.

HIGHLIGHTS

The most notable result of our Bat Falcon studies at Tikal was the relatively equitable division of male and female roles in the four pairs studied. These pairs differed in behavior from most *Falco* species and previously studied Bat Falcons, in that they shared incubation, brooding, and foraging roles more equally than is often observed among falcons, especially in such a highly dimorphic member of the genus. Males played relatively large roles in incubation, and females at two nests brought prey to incubating males, an unusual observation among falcons and among falconiforms in general. Female Aplomado Falcons are also known to feed males (Hector 1986), as are female Red-necked Falcons (*F. chicquera*: Olwagen 1984). However, pair members in both these species frequently hunt cooperatively, which Bat Falcons rarely do, making our observation of females feeding males all the more interesting.

Other factors that may facilitate foraging by female Bat Falcons during nesting are the tropical climate and abundance of prey—especially insect prey—in the nest vicinity. Warm ambient temperatures seemingly allowed females to remain off the eggs and young chicks for short periods without ill effects in order to hunt near the nest; for these cavity nesters, the hot sun is presumably less of a problem for eggs and chicks than for open-nesting raptors. Moreover, availability of prey near nests allowed females to hunt while guarding eggs and chicks against predation.

Bat Falcons at Tikal appeared to rely more on insects in our observations than in other studies to date, although differences in study methods may compound the differences in these estimates among studies. If the greater reliance on insects at Tikal is robust, it may be that the more equitable sharing of nesting roles between males and females at Tikal is promoted by this reliance on relatively small prey.

Differences in habitat surrounding the four Tikal nests may help explain differences in diet and parental roles observed among these pairs. The Bajo pair, nesting amid Scrub Swamp Forest, fed less on birds than other pairs and relied heavily on dragonflies throughout the nesting cycle. Bird species richness is certainly lower in Scrub Swamp than in Upland Forest (Whitacre et al., unpubl. data), and conceivably, prey birds were less abundant or harder to catch in this low, dense habitat, leading to greater reliance of the falcons on dragonflies. The Bajo female played a larger foraging role than females nesting in Upland Forest, suggesting that this female increased her foraging role to compensate for poorer foraging prospects of her mate in this habitat. Studying American Kestrels in California, Rudolph (1982) documented a similar instance: in one pair occupying a low-quality territory, pair members shared incubation more equally than usual, and the female hunted more than expected, apparently compensating for poor habitat quality.

For further information on this species in other portions of its range, refer to Carvalho et al. 2007 and Di Giacomo 2000 and 2005.

APPENDIX 19.1

Bird species among Bat Falcon prey at Tikal, from observations and prey remains at four nests

Family Scientific name	Common name	Means of identification	Mean body weights (g) from Dunning 1993
Psittacidae			
Aratinga astec	Aztec Parakeet	O & C	76.9
Amazona albifrons	White-fronted Amazon	C	206.0
Apodidae			
Chaetura vauxi	Vaux's Swift	C	17.1
Panyptila cayennensis	Lesser Swallow-tailed Swift	O & C	21.1
Trochilidae			
Phaethornis superciliosus	Long-tailed Hermit	C	6.0
Phaethornis longuemareus	Little Hermit	C	3.0
Phaeochroa cuvierii	Scaly-breasted Hummingbird	C	8.9
Florisuga mellivora	White-necked Jacobin	O	7.4
Chlorostilbon canivetii	Canivet's Emerald	O & C	2.5
Heliothryx barroti	Purple-crowned Fairy	O & C	5.5
Momotidae			
Hylomanes momotula	Tody-motmot	C	29.3
Picidae			
Melanerpes pygmaeus	Yucatán Woodpecker	C	38.6
Melanerpes aurifrons	Golden-fronted Woodpecker	O & C	80.9
Furnariidae			
Xenops minutus	Plain Xenops	O	10.6
Thamnophilidae			
Thamnophilus doliatus	Barred Antshrike	O & C	27.9
Microrhopias quixensis	Dot-winged Antwren	O & C	7.9
Tyrannidae			
Myiopagis viridicata	Greenish Elaenia	C	12.3
Leptopogon amaurocephalus	Sepia-capped Leptopogon	O	11.7
Todirostrum cinereum	Common Tody-Flycatcher	C	6.4
Platyrinchus cancrominus	Stub-tailed Spadebill	C	—
Myiobius sulphureipygius	Sulphur-rumped Flycatcher	O & C	—
Empidonax sp.	*Empidonax* sp.	O & C	10.3
Attila spadiceus	Bright-rumped Attila	C	39.1
Myiarchus yucatanensis	Yucatán Flycatcher	C	21.4
Myiarchus tuberculifer	Dusky-capped Flycatcher	O	19.9
Myiarchus tyrannulus	Brown-crested Flycatcher	C	43.8
Pitangus sulphuratus	Great Kiskadee	O	61.0
Myiozetetes similis	Social Flycatcher	O & C	28.0
Tyrannus melancholicus	Tropical Kingbird	O	37.55
Genera incertae sedis			
Schiffornis turdinus	Thrush-like Shiffornis	O	31.7
Pachyramphus cinnamomeus	Cinnamon Becard	O	20.3
Pachyramphus major	Gray-collared Becard	O	25.2
Tityra semifasciata	Masked Tityra	O	79.3
Tityra inquisitor	Black-crowned Tityra	O	43.3
Cotingidae			
Cotinga amabilis	Lovely Cotinga	O	71.5
Pipridae			
Manacus candei	White-collared Manakin	O & C	19.9
Pipra mentalis	Red-capped Manakin	O & C	15.0
Vireonidae			
Vireo griseus	White-eyed Vireo	O & C	11.4
Vireo pallens	Mangrove Vireo	O	11.6
Hylophilus ochraceiceps	Tawny-crowned Greenlet	O	10.4
Hylophilus decurtatus	Lesser Greenlet	O	8.8
Vireolanius pulchellus	Green Shrike-Vireo	O	24.0

Family Scientific name	Common name	Means of identification	Mean body weights (g) from Dunning 1993
Hirundinidae			
Progne chalybea	Gray-breasted Martin	C	42.9
Stelgidopteryx serripennis ridgwayi	Ridgway's Rough-winged Swallow	O & C	15.9
Hirundo rustica	Barn Swallow	O & C	16.0
Troglodytidae			
Thryothorus maculipectus	Spot-breasted Wren	O	14.85
Thryothorus ludovicianus	White-browed Wren	O & C	18.7
Uropsila leucogastra	White-bellied Wren	O	9.05
Henicorhina leucosticta	White-breasted Wood Wren	O	15.7
Sylviidae			
Ramphocaenus melanurus	Long-billed Gnatwren	O & C	9.7
Polioptila plumbea	Tropical Gnatcatcher	O	6.0
Mimidae			
Dumetella carolinensis	Gray Catbird	O & C	36.9
Parulidae			
Vermivora peregrina	Tennessee Warbler	O & C	10.0
Vermivora celata	Orange-crowned Warbler	O	9.0
Parula pitiayumi	Tropical Parula	O	6.9
Dendroica petechia	American Yellow Warbler	O	9.5
Dendroica pensylvanica	Chestnut-sided Warbler	O & C	9.6
Dendroica magnolia	Magnolia Warbler	C	8.7
Dendroica coronata	Yellow-rumped Warbler	C	12.5
Mniotilta varia	Black-and-white Warbler	O & C	10.8
Setophaga ruticilla	American Redstart	O & C	8.3
Wilsonia citrina	Hooded Warbler	O	10.5
Thraupidae			
Eucometis penicillata	Gray-headed Tanager	O	27.0
Habia fuscicauda	Red-throated Ant-tanager	O & C	38.5
Thraupis episcopus	Blue-gray Tanager	O & C	35.0
Thraupis abbas	Yellow-winged Tanager	O & C	45.0
Euphonia affinis	Scrub Euphonia	O & C	10.0
Euphonia elegantissima	Elegant Euphonia	O & C	15.0
Euphonia gouldi	Olive-backed Euphonia	O	14.0
Chlorophanes spiza	Green Honeycreeper	C	19.0
Cyanerpes lucidus	Shining Honeycreeper	O & C	11.4
Cyanerpes cyaneus	Red-legged Honeycreeper	O & C	14.0
Emberizidae			
Volatinia jacarina	Blue-black Grassquit	O	9.7
Sporophila torqueola	White-collared Seedeater	O	8.7
Oryzoborus funereus	Thick-billed Seed-Finch	O & C	—
Tiaris olivacea	Yellow-faced Grassquit	O & C	8.9
Arremon aurantiirostris	Orange-billed Sparrow	O	34.5
Arremonops rufivirgatus	Olive Sparrow	O & C	23.6
Saltator atriceps	Black-headed Saltator	O	79.7
Guiraca caerulea	Blue Grosbeak	O	28.4
Passerina amoena	Lazuli Bunting	O & C	15.5
Passerina cyanea	Indigo Bunting	O & C	14.5
Icteridae			
Dives dives	Melodious Blackbird	C	96.2

Note: O = visual observation; C = remains collected.

Chapter 19 Notes

1. A lesser disparity in culmen length, using data from Friedmann (1950), still yields a mean value of 16.4% for these three measures.

2. The Dimorphism Index equals 8.3 for tarsus length, 18.9–23.4 for beak measurements, 12.9 for middle toe length, 12.7 for middle talon length, and 9.1 for hind talon length (Bierregaard 1978).

3. Pesola weight $= -0.002 \pm 0.946$ Ohaus weight; $r^2 = 1.0$, df $= 1, 77$; $P < 1.0 \times 10^{-15}$ ($n = 79$).

4. Kruskal-Wallis statistic $= 18.2$, $P < 0.005$.

5. Gender of the foraging bird was unknown in four cases.

6. In 28 cases we could not determine the sex of the bird delivering prey.

7. Only three nests were used in this analysis, as the Camino nest failed prior to hatching.

8. $t = 16.8$, df $= 518$, un-paired t- test, $P < 0.014$.

9. An additional nest on a palm spathe was observed in the Selva Lacandona of Chiapas, Mexico, by E. Iñigo-Elias (pers. comm.).

20 ORANGE-BREASTED FALCON

Aaron J. Baker, David F. Whitacre, and Oscar A. Aguirre

Nesting on large cliffs or emergent trees, typically in areas of primary forest, the Orange-breasted Falcon *(Falco deiroleucus)* is among the least known of the world's 39 falcon species. It is generally regarded as rare to very rare and spottily distributed; the reasons for this apparent rarity and localized distribution are poorly understood. Somewhat smaller than a Peregrine Falcon *(F. peregrinus)*, these falcons hunt above the forest canopy for birds and bats and are among the few falcons worldwide that are closely associated with moist tropical forests (Cade 1982). Here we report the most extensive biological studies to date on this enigmatic falcon.

GEOGRAPHIC DISTRIBUTION AND SYSTEMATICS

The Orange-breasted Falcon occurs locally in New World tropical forests from Guatemala, Belize, and perhaps southern Mexico, to northern Argentina, Paraguay, and Bolivia (Brown and Amadon 1968; Cade 1982). This falcon's presence as a breeding species in northern Central America has been known since at least 1958 (Smithe 1966), but to our knowledge, only the pair nesting on the Maya temples of Tikal National Park is well documented. From 1979 to 1983, Mike Arnold and Peter Jenny searched for additional pairs in Guatemala, Belize, and Ecuador, locating four nest sites in Guatemala, four in Ecuador, and none in Belize (Jenny and Cade 1986; M. Arnold, pers. comm.). The first published sight record in Belize was apparently that of Haney (1983). Elsewhere in Mesoamerica, and indeed throughout its range, details of the distribution and population status of this species have remained largely unknown. We recently reviewed historic and recent evidence of this falcon's occurrence in Mesoamerica (Baker et al. 2000b). With the exception of Guatemala and Belize and four falcon pairs discovered in eastern Panama since the year 2000, all adequately documented Mesoamerican records were more than 20 years old at the time of this writing. Apart from these recent Panamanian nests, to our knowledge the only nesting records known for Mesoamerica have been in Guatemala and Belize (Baker et al. 2000b).

One of the great mysteries concerning this falcon's distribution is whether it does or did nest in Mexico; we think it is likely that a small Mexican breeding population awaits discovery. The only widely known specimen record for Mexico is a single record from Tecolutla, in central Veracruz (Friedmann et al. 1950), but Howell and Webb (1995) thought this locality might be in error. A second specimen, in Chicago's Field Museum (185,927), has been widely missed by those interested in this species. This specimen, which is correctly identified, was collected on 30 April 1947 at Palma Real, Ocosingo, Chiapas, Mexico (D. Willard, pers. comm.). One published sighting in Mexico gives adequate descriptive detail (Hardy et al. 1975), and several other sightings have been reported. Some time prior to 1973, the late Miguel Alvarez del Toro, one of Mexico's premier ornithologists, made a January sighting amid the spectacular cliffs of the Canyon del Sumidero just outside Tuxtla Gutierrez, Chiapas. Several sightings have been reported from Veracruz, including one by W. J. Schaldach in 1972 (unpubl. ms.) and one of a bird in flight by Iñigo et al. (1989). In addition, Col. L. R. Wolfe made a sighting prior to 1977 between Villahermosa and the Tabasco coast (pers. comm. to D. Hector). González-García (1993) lists the species as resident in the Montes Azules Biosphere Reserve of the Selva Lacandona, in eastern Chiapas, Mexico, but gives no details. No doubt the Montes Azules/Selva Lacandona region, not far from active eyries in Guatemala, and with extensive, intact humid forest, is the portion of Mexico most likely to hold breeding pairs.

For Honduras, apparently a single 1937 specimen exists (Monroe 1968); for Nicaragua two specimens exist— one taken in 1962 by Howell (1972) and one listed by Salvin and Godman (1904) as from Matagalpa. To our knowledge, there are no records for El Salvador. The few specimen records for Costa Rica and Panama date from the nineteenth century, and even in 1910, Carriker regarded this species "only a very rare straggler" at best. Stiles et al. (1989) stated that no definite record was known for Costa Rica in the previous 30 years. Likewise for Panama, Ridgely and Gwynne (1989) stated that no definite evidence existed for the twentieth century, with possible sightings in 1957 and 1970. Griscom's (1932) account of nesting in "cathedral and church towers and belfries in the heart of towns and cities" in Nicaragua and western Panama is likely based on confusion with the Bat Falcon *(Falco rufigularis)*, as pointed out by other authors (Wetmore 1965; Howell 1972). A published sight record from Costa Rica (Herman and Hedstron 1990) is also, we believe, open to doubt.

In South America, the species is regarded as rare and local in Ecuador (Ridgely and Greenfield 2001) and very rare and local in Colombia and Venezuela (Hilty and Brown1986; Hilty et al. 2003). In Ecuador, Ridgely and Greenfield (2001) estimated the country-wide population as certainly no more than a few-dozen pairs, perhaps fewer, and in French Guiana, Thiollay (1989a) found this falcon to be probably the rarest, most patchily distributed raptor. The species's density in other South American nations appears no greater than indicated here. This falcon's known and probable distribution is further discussed under Conservation, later in this chapter, and the species's systematic position is discussed in Chapter 19.

MORPHOLOGY AND PLUMAGE

This is a medium sized, highly dimorphic falcon, with a robust, stocky build similar to that of a Peregrine, the wings at rest reaching roughly to the tail tip (Plate 20.1); the ratio of wing chord to tail length is 2.03, essentially the same as in North American Peregrines (data in Palmer 1988). The largest available sets of linear measurements (from Blake 1977 and Bierregaard 1978) are summarized in Appendix 2. Values of the Dimorphism Index based on the above (those from Bierregaard 1978 are in parentheses) are 13.5 (15.7) for wing chord, 14.2 (19.6) for tail, 19.4 (19.5) for culmen length, and 12.4 (13.6) for tarsus.[1]

Cade (1982) gives the weight range for males as 330–360 g and for females as 550–650 g, pointing out that the largest females weigh twice as much as the smallest males. Few body weights have been published, but a female from Guatemala weighed 654 g (Smithe and Paynter 1963), while Haverschmidt (1963), for specimens from Surinam, gives two male weights as 338 and 340 g and two female weights as 595 and 620 g. Based on all these data, females average approximately 605 g and males 339 g (Appendix 1), giving a (cube root) body mass dimorphism value of 20.2. As an overall indicator of the magnitude of dimorphism, the mean of dimorphism values for body mass, bill length, and wing chord has often been used (Snyder and Wiley 1976). For the Orange-breasted Falcon, this value is 18.1, exceeding that for all North American raptors except the Sharp-shinned Hawk *(Accipiter striatus:* Snyder and Wiley 1976).

Cade (1982) contended that the Orange-breasted Falcon is the most formidably armed of falcons relative to body size, with proportionally huge feet, long toes, heavy talons, and a massive, laterally compressed, deep beak. We tested this by comparing this species with Bat Falcon, Aplomado Falcon *(Falco f. femoralis)*, Prairie Falcon *(F. mexicanus)*, Peregrine Falcon *(F. p. anatum)*, Gyrfalcon *(F. rusticolus)*, and American Kestrel *(F. s. sparverius)* (data from Bierregaard 1978). The ratio of culmen chord to total body length and to wing and tail length was greater in *F. deiroleucus* than in the other six species. The same was true for upper mandible depth, length of middle toe, and length of middle and hind talons in relation to total, wing, and tail length. Hence, Cade's impressions are borne out in this cursory analysis.

In a morphometric analysis of essentially all *Falco* species (38 species and several subspecies), Kemp and Crowe (1994) found that two axes captured much of the morphological variation present: an axis reflecting overall body size and especially foot size, and an axis reflecting tail length and, to a lesser extent, wing length. Female Orange-breasted Falcons, with short wings and tail and large feet, clustered close to male Peregrine Falcons of two subspecies and the closely allied Barbary Falcon *(F. pelegrinoides)*, while male Orange-breasted Falcons clustered close to female Taita Falcons *(F. fasciinucha)*, female Oriental Hobbies *(F. severus)*, and female Bat Falcons. Collectively, these two groups comprised the shortest winged, shortest tailed, and largest footed of the 38 falcon species analyzed (ibid.).

In flight, Orange-breasted Falcons appear heavily wing loaded, demonstrated by observations that Orange-breasted Falcons in breeding chambers cannot climb in flight as readily as can Peregrines in a similarly confined space. Moreover, they land at higher air speeds than Peregrines, with an audible thump (R. Berry, pers. comm.). Both of these features are as expected in a highly wing-loaded species. In a regression analysis of estimated wing loading versus body mass for several falcon species, this species does indeed have higher-than-average wing loading for a falcon of its size (see Chapter 19). Heavy wing loading is correlated with relatively fast characteristic flight speeds (e.g., "maximum range speed," "minimum power speed": Pennycuick 1975, 1989). Cade (1982) predicted that this species will prove to be among the fastest members of the genus in level, flapping flight, and we speculate that the same may be true of their stooping speeds. In addition to relatively high wing loading, Cade (1982) states that this species has especially stiff flight feathers that he interprets as an adaptation for rapid flight and swift acceleration.

Identification

The extreme similarity of the Orange-breasted Falcon to the much more common Bat Falcon, coupled with inadequate descriptions in many field guides, routinely leads to confusion of these two species in the field (Ridgely and Gwynne 1989; A. Baker, pers. observ.); hence many reported sightings are questionable. Advice for distinguishing these species was given by Howell and Whittaker (1995) and Whittaker (1996). We would emphasize these authors' admonition that the best way to learn to identify Orange-breasted Falcons is to first gain extensive experience with Bat Falcons, and we suggest special attention to vocalizations.

PREVIOUS RESEARCH

Only a handful of publications has addressed the ecology of the Orange-breasted Falcon (Boyce 1980; Cade 1982; Jenny and Cade 1986; Baker 1998; Baker et al. 1992, 2000b). The validity of earlier egg descriptions was disputed by Boyce and Kiff (1981), and egg descriptions were given by Kiff (1988). Distributional records were given by Whittaker (1996). Apart from these, most literature treatments are brief accounts in field guides or regional works.

RESEARCH IN THE MAYA FOREST

From 1991 to 1997 we studied the ecology, behavior, and population status of this species at 19 nest sites in Guatemala and Belize. Elsewhere we have published our findings with regard to this species's population status and conservation prospects in Mesoamerica (Baker et al. 2000b). Here we summarize those findings and present for the first time our results with regard to diet, hunting, breeding biology, and behavior.

We studied Orange-breasted Falcons in northern Petén, Guatemala, and throughout Belize. Elevations in Petén range from 100 to 600 m; topography is gently rolling, broken in a few areas by steep, jumbled karst terrain and, rarely, by large escarpments with limestone cliffs. Belize is similar in topography, but it includes the Maya Mountains, a range that has extensive karst, reaches to 1120 m in elevation, and is transected by several major river systems, which in places has formed abundant tall cliffs (Plate 20.2). This cliff-laden range remains largely forested in Belize but has lost much forest cover where it extends into Guatemala. Vegetation in Belize is similar to that described earlier (see Chapter 2), though rainfall is higher in some areas, leading to differences in the semi-deciduous tropical forest cover. At higher elevations in the Maya Mountains, Caribbean Pine *(Pinus caribaea)* forests dominate a landscape interspersed with areas of broadleaf riparian vegetation.

In 1991 we began collecting data on productivity and occupancy of the six Orange-breasted Falcon nesting sites known to us at that time, and in 1992 we began searching for additional pairs in Guatemala and Belize. By 1994, we had located six nest sites in Guatemala and 13 in Belize. Each year from 1992 to 1997 we visited most previously documented nest sites and searched for additional pairs. To determine occupancy and productivity, we visited nest sites once or twice each year, typically early in the season during courtship or incubation (February–March), and again just after young typically fledged (June–July). Four of the 19 nest sites were omitted from most analyses of site occupancy and productivity; two of the four were monitored only one year owing to difficulty of access, and two were occupied for only one year.

In 1996 we conducted behavioral observations at two sites in Guatemala and two in Belize; unless otherwise noted, all behavioral data presented here are from these observations. We conducted observations in periods of four consecutive days, from predawn to post-dusk (approximately 13 hours 50 minutes per day). We made three rotations through the four sites, for a total of 48 days of observation (661 hours 15 minutes) during the nesting season. Two of us (A. J. B. and O. A. A.) conducted observations, each on a half-day shift. We observed eyries from 50 to 100 m distant. At two sites we observed from small platforms constructed in tree crowns in front of the nest cliffs; at one site we observed from the rim of a large limestone sinkhole in which the nest was located; and at one site, from atop a Maya temple opposite the temple housing the eyrie. We collected behavioral data continuously, using focal animal sampling (Altmann 1974), aided by binoculars (10 × 40), spotting scope (25 × 60), wristwatch, compass, and notebook.

We divide the nesting cycle into courtship, incubation, nestling, and fledgling stages. For time-activity budgets, "nest ledge" is defined as present on the actual nesting ledge; "present in area" is within 100 m of, but not on, the nest ledge; and "absent from area" is more than 100 m from the nest ledge. Birds "absent from area" were always out of view and often 500 m or more from the nest ledge. We estimated the size of observed prey items, using 50 g categories. We define a hunting attempt as a pursuit of an individual prey item, regardless of the duration or number of dives by the falcon or evasions by the prey. This is consistent with the definition of a "hunt" or "attack" in similar studies (Rudebeck 1950, 1951; Dekker 1980; Cresswell 1996).

To analyze site occupancy and nesting success in relation to habitat modification, we grouped nest sites into two categories: mature forest and habitat mosaic. Mature forest nest sites ($n = 7$) were those surrounded by mature forest for at least a 5 km radius. Habitat mosaic nest sites ($n = 8$) were those that had some deforestation or forest modification (successional forest, agriculture, or pasture) but occupying less than 50% of the area within a 5 km radius of the nest, the remainder being mature forest. To assess differences in reproduction and occupancy between the two habitat categories, data for all years were combined.

To test for trends over time in the annual proportion of nest sites occupied and the proportion of pairs producing one or more young, we used binary logistic regression (Minitab 1996; Pampel 2000). To test for trends in yearly mean number of young fledged by successful pairs, we used ordinal logistic regression (Pampel 2000). In analyzing the percentage of sites occupied annually, we omitted the year in which a site was discovered. Most statistical comparisons used one-way analysis of variance (ANOVA) and *t*- tests performed in Minitab 11 (Minitab 1996), and multiple random permutation procedure (MRPP) chi-square tests (Berry and Mielke 1986); logistic regressions also employed Minitab 11. All *P* values of 0.05 or less were considered statistically significant; values given are means ± 1 SD.

DIET AND HUNTING BEHAVIOR

Table 20.1 lists all bird species identified as prey in our study and by Jenny and Cade (1986). Most species were identified from remains collected on nest ledges and below perches. In 1996 we observed 105 prey items brought to nests. Of these, 90 (85.7%) were birds and 15 (14.3%) were bats. Of the 90 birds, 82 were placed into estimated 50 g size categories. Figure 20.1 gives a histogram of prey weights for identified prey species and prey items observed in 1996. In numerical terms, 25–75 g prey items were most frequent, and more prey species fell into this size class than any other. In terms of biomass, items less than 25 g contributed an estimated 3.3% of prey biomass, 25–75 g items contributed 26.6%, 75–125 g items made up 5.5%, 125–175 g items 20%, and items of 175–225 g 44.4% of prey (Fig. 20.1). The mean size of identified prey species (85 g) and mean estimated size of observed prey items (90 g) amounted to about 25–26% of adult male body mass. Half of total prey biomass, however, was contributed by prey items estimated to weigh at least 175 g—50% or more of adult male body mass (Fig. 20.1).

Among the prey most often taken were doves and small parrots. Both of these groups were conspicuous in their morning commutes above the canopy, which likely made them available to these predators of birds in flight. Also commonly taken were highly aerial species such as White-collared Swifts (*Streptoprocne zonaris*). Medium-sized birds frequenting treetops were often taken, including tityras, becards, kingbirds, toucanets, and woodpeckers. Several trogons and motmots were also taken, no doubt when these sub-canopy dwellers were exposed in treetops or flying across forest gaps. Large flying insects also played a minor role in the diet and were probably important to juveniles learning to hunt. Nearly all fledglings were observed pursuing and, at times, capturing and eating flying insects. We only once observed an adult feeding on insects; during a brief period of high availability of a large beetle species, an adult male that had recently lost his mate captured and consumed on the wing 23 individuals in 16 minutes.

We observed three hunting techniques: (1) perch-initiated diving pursuits, (2) perch-initiated level to climbing pursuits, and (3) flight-initiated diving pursuits. Perch-initiated dives typically began from a cliff-top tree and were usually non-flapping stoops from near vertical to an approximately 45° angle downward at quarry flying below the perched falcon but above the forest canopy. Birds under attack typically escaped by diving into the forest where the falcons did not follow. Diving attacks initiated from flight were similar but were started from soaring flight about 100–500 m above the forest canopy. Hunting attempts initiated from flight were often directed at higher flying quarry, at times resulting in multiple dives and escape maneuvers before the quarry reached the safety of the canopy. Perch-initiated level to climbing pursuits were directed at high-flying prey species, especially swifts, which the falcons appeared to force upward until the prey made an attempt at a downward escape, at which point the falcon would dive in pursuit. Level to climbing pursuits typically involved multiple dives and climbs, often lasting 5–10 minutes and covering several hundred meters.

In 1996 we observed 208 hunting attempts (124 by males, 81 by females, and 3 cooperative), all in the vicinity of nest sites. Of these, 199 were unsuccessful, and 9 (4.3%) resulted in prey capture. Nearly all (99.1%) of the observed hunting attempts were initiated from tree perches atop cliffs and most were diving pursuits (Table 20.2). We observed three cooperative hunts by pair members—two initiated from perches and one from flight; all three were initiated by the male diving at prey flying above the canopy. The male's diving pursuit forced the prey to make an evasive maneuver upward into the path of the female, who had begun her dive just after the male.

The mean number of hunts observed per day in the vicinity of eyries varied by breeding stage. Four females together averaged 0.5 ± 1 hunt per day in the vicinity of nest sites during 12 days of the courtship stage, 1.25 ± 1.57 hunts per day during 16 days of incubation, and 2.87 ± 4.38 during 16 days of the nestling stage. During the same days, four males averaged 1.67 ± 1.87 hunts per day in the

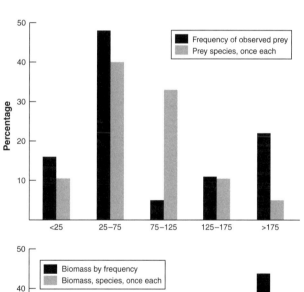

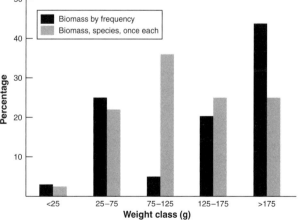

20.1 Frequency distribution of Orange-breasted Falcon prey weights for prey items observed at nests (n = 83) and identified prey species (n = 36).

Table 20.1 Species and weight of Orange-breasted Falcon prey in Guatemala and Belize

Species	Weight (g)[a]	This study	Jenny and Cade 1986
Killdeer (*Charadrius vociferus*)	92.1/101		*
Lesser Yellowlegs (*Tringa flavipes*)	81.0		*
Spotted Sandpiper (*Actitis macularia*)	40.4	*	
Pectoral Sandpiper (*Calidris melanotos*)	97.8/65.0		*
Scaled Pigeon (*Columba speciosa*)	262/225	*	*
Mourning Dove (*Zenaida macroura*)	123/115		*
Ruddy Ground Dove (*Columbina talpacoti*)	46.5	*	*
Gray-headed Dove (*Leptotila plumbeiceps*)	170		*
Blue Ground Dove (*Claravis pretiosa*)	67.3	*	
Ruddy Quail-Dove (*Geotrygon montana*)	115	*	
Olive-throated Parakeet (*Aratinga nana*)	76.9	*	*
Brown-hooded Parrot (*Pionopsitta haematotis*)	149		*
White-crowned Parrot (*Pionus senilis*)	212	*	
Pauraque (*Nyctidromus albicollis*)	53.2	*	
White-collared Swift (*Streptoprocne zonaris*)	110	*	*
Lesser Swallow-tailed Swift (*Panyptila cayennensis*)	21.1		*
Citreoline Trogon (*Trogon citreolus*)	83.1	*	
Violaceous Trogon (*Trogon violaceus*)	51.5	*	*
Slaty-tailed Trogon (*Trogon massena*)	141.0	*	
Blue-crowned Motmot (*Momotus momota*)	133.0	*	
Emerald Toucanet (*Aulacorhynchus prasinus*)	160/149	*	
Acorn Woodpecker (*Melanerpes formicivorus*)	82.9/78.1	*	
Golden-olive Woodpecker (*Piculus rubiginosus*)	56.2/55.4	*	
Barred Woodcreeper (*Dendrocolaptes certhia*)	64.2	*	
Eastern Kingbird (*Tyrannus tyrannus*)	43.6	*	
Rose-throated Becard (*Pachyramphus aglaiae*)	30		*
Masked Tityra (*Tityra semifasciata*)	79.3	*	*
Black-crowned Tityra (*Tityra inquisitor*)	43.3	*	*
Green Jay (*Cyanocorax yncas*)	78.5	*	
Swainson's Thrush (*Catharus ustulatus*)	30.8		*
Gray-breasted Martin (*Progne chalybea*)	42.9		*
Ridgway's Rough-winged Swallow (*Stelgidopteryx serripennis*)	15.9		*
Tropical Mockingbird (*Mimus gilvus*)	58.4	*	
Red-legged Honeycreeper (*Cyanerpes cyaneus*)	14.0		*
Botteri's Sparrow (*Aimophila botterii*)	19.9		*
Rose-breasted Grosbeak (*Pheucticus ludovicianus*)	45.6		*
Melodious Blackbird (*Dives dives*)	96.2	*	

Note: Asterisk (*) = species recorded as prey in each study.

[a] Male/female or both; most weights are from Dunning 1993.

Table 20.2 Frequencies of three types of hunting behavior used by male and female Orange-breasted Falcons in nest vicinities in Guatemala and Belize

Type of attack	Males	Females	Cooperative	Total
Perch-initiated dive (number)	77 (62.1%)	58 (71.6%)	2 (67%)	137 (65.9%)
Perch-initiated climb or level pursuit (number)	46 (37.1%)	22 (27.2%)	1 (33%)	69 (33.2%)
Flight-initiated dive (number)	1 (0.8%)	1 (1.2%)	0	2 (0.9%)
Total	124 (59.6%)	81 (39.0%)	3 (1.4%)	208 (100%)

vicinity of nest sites during courtship, 4.69 ± 5.79 during incubation, and 1.5 ± 2.28 during the nestling stage.

For 205 hunts we estimated the distance flown during the hunting attempt to the nearest 25 m. For all males combined, estimated hunting flight distance averaged 225 ± 270 m (*n* = 126) and for females, 165 ± 190 m (*n* = 79). No differences were found in mean length of hunting flights between the sexes or among individual females[2] or males.[3] Both males and females hunted most actively in the morning, with a brief secondary peak in hunting near dusk (Fig. 20.2B).

HABITAT USE

These falcons may be found in a variety of mature forest types, from humid evergreen lowland forest, to the subhumid, sub-deciduous forest of our study area, to subtropical forest up to at least 1100 m (Cabot and Serrano 1986), 1300 m (Stiles et al. 1989), 1700 m (Meyer de Schauensee and Phelps 1978), or even 2400 m in elevation (Hilty and Brown1986). We regard the statement that habitat is "primarily open forest and forest edge" (AOU 1983) as somewhat misleading. Slud's (1964) com-

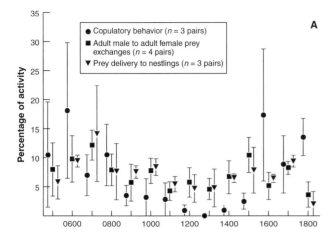

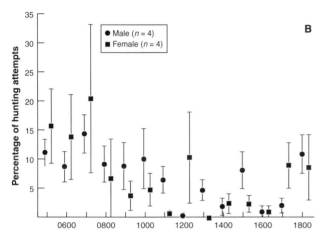

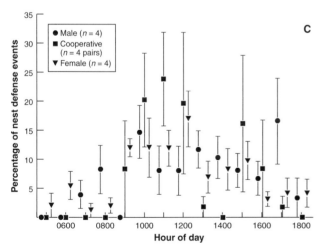

20.2 Daily pattern of Orange-breasted Falcons' (A) copulatory behavior, adult male to adult female prey exchanges, and prey deliveries to nestlings; (B) hunting behavior; and (C) territorial events. Data are means ± 1 SE.

ment that in Costa Rica this species prefers the very wet, cloud-forested middle elevations also seems inaccurate. While most authors have regarded this as primarily a forest species, a few have described broader habitat tolerances including more open situations. Boyce (1980) suggested these falcons may prefer ecotones, "especially where complex mature forests are interrupted by savan-

nas, clearings, high cliffs, or rivers"; he speculated that these falcons capitalize on high prey diversity and densities provided by the "edge effect."

Sightings are often made along lowland rivers in extensively forested areas (Haverschmidt 1968; Whittaker 1996), which may, in part, have given rise to the frequent mention of "forest edge" in habitat characterizations. However, two factors are important in this context. First, rivers are prime routes for human travel; hence the frequency of sightings there may have little to do with the falcons' relative use of these sites versus sites in the more remote hinterlands back from the river's edge (Howell and Whittaker 1995). Second, a river through a forest provides the falcons with a tall perch over a broad, low, open area—hence providing favorable hunting sites, given this species's hunting style. Yet it may be the surrounding forest that serves as a "food factory," producing prey to be hunted along the river margins. In this sense, "forest edge" sightings along jungle rivers must be distinguished from other situations that may be deemed "forest edge," such as areas with extensive deforestation.

Several authors have proposed that this species relies largely on primary forest and is sensitive to habitat modification (Cade 1982; Thiollay 1985; Ellis and Smith 1986; Jenny and Cade 1986). Visiting a number of sites in Ecuador where specimens had been collected in the past, Jenny (Jenny and Cade 1986) found most to be largely deforested and apparently also without Orange-breasted Falcons. Jenny and Cade (1986) found Orange-breasted Falcons only in sites with extensive mature forest, and in one case, deforestation between Jenny's visits had removed a nest tree and surrounding forest, apparently causing the pair to move to a relatively uncut portion of the same ridge some 2 km away (P. Jenny, unpubl. ms.).

In our study, eyries in partly modified habitat fledged significantly more young per successful pair than did pairs amid pristine forest (Table 20.3; Baker et al. 2000b). This finding may lend support to the notion that this species prefers ecotones, especially where mature forests border other habitat types. However, it is important to note that the habitat mosaic sites we studied were dominated by mature forest. Hence, the increased falcon productivity we documented for the partially altered habitat does not imply that more heavily modified landscapes are also favorable to this species. While this species may not be absolutely dependent on pristine tropical forest habitat, the fact that no nest sites have been found in entirely—or even predominantly—deforested areas leads us to conclude that these falcons have an affinity for unmodified to slightly modified tropical forest habitat. In a study of raptor habitat selection along a successional and habitat modification gradient, Jullien and Thiollay (1996) characterized this species as relying on mature forest to the same extent as other mature forest species including the Harpy Eagle *(Harpia harpyja)* and Barred Forest Falcon *(Micrastur ruficollis)*. In sum, we favor Hilty and Brown's (1986) characterization of this as a "forest-based" species; though often seen perched along forest edges such as rivers, these birds, throughout most of their range, appear to be strongly associated with tracts of mature forest.

In Argentina, Olrog (1985) states that the typical habitat of this species is not wet forest but rather savanna woodlands; he indicates it may also occur at the edge of wet forest. In the South American Chaco, Short (1975) ascribes this falcon a broader habitat amplitude than does Olrog, calling it a bird of forested mountain slopes, forest edges and clearings, and savannas. In southern Brazil, Albuquerque (1986) mentions several localities of sight records, all in pampas regions with gallery forest along streams and cliffs. These observations may indicate regional differences in habitat use by this falcon, with birds at these southerly localities perhaps more often occupying open habitats than is typical farther north.

BREEDING BIOLOGY AND BEHAVIOR

Nests and Nest Sites

In Guatemala and Belize, 18 of 19 nest sites were on cliffs and one was in an emergent Cohune Palm *(Orbignya cohune)*. All sites were in extensive mature forest (*n* = 11) or habitat mosaics dominated by mature forest (*n* = 7). On cliffs, the adults made their scrapes and laid eggs in sites ranging from exposed ledges to deep potholes, typically near the center of the cliff (Plate 20.3). The single palm nest (to our knowledge the first documented nesting in a tree in Central America and the first known case of nesting in a palm; see Boyce and Kiff 1981) was in a tree about 20 m tall that emerged from a small patch of 10 m tall successional forest surrounded by extensive mature forest. Here, the falcons nested on a horizontal leaf spathe. Fifteen nest sites were in river valleys or on escarpments in the immediate vicinity of water, usually a river or creek. Nesting cliffs averaged 125 m in height (*n* = 18; Table 20.4).

In eastern Ecuador, Jenny (unpubl. ms.) found 3 pairs of Orange-breasted Falcons with fledged young nesting beneath epiphytes in the main crotches of large emergent trees. In the same area, D. Pearson (pers. comm.) had previously documented nesting in a large Kapok *(Ceiba pentandra)*. Jenny concluded that nesting in the spacious crotches of huge emergent trees was frequent in Amazonian Ecuador (Jenny and Cade 1986). Tree nesting presumably may occur elsewhere as well, and it would be worth publishing any observed instances. Observations of this species well within the Amazon Basin (Whittaker 1996), where cliffs are rare, may suggest that tree nesting is common there. However, all other instances of nesting known to us for South America involve cliff nesting, which is known for at least Ecuador (F. Ortíz, pers. comm.), French Guiana (Thiollay 1989a), Peru, and Bolivia (J. O'Neill, pers. comm.). Thiollay (1989a) stated that in French Guiana the species had never been found breeding except on cliffs.

All three of Jenny's Ecuadorian eyrie sites and the one described by D. Pearson were on ridges overlooking large expanses of forest. He felt that the position on ridges was significant. We found that these falcons hunted largely from cliff top perches and concluded that a perch elevated above surrounding terrain is advantageous in this hunting style. Hence we agree with Jenny that these falcons may favor nest sites whose terrain or vegetation offers the falcons an elevated position from which to hunt. This would apply not only to cliffs and knolls but also to the forested river-edge situations in which these falcons are often observed (Whittaker 1996).

Egg and Clutch Size

Table 20.5 summarizes clutch and brood sizes and the number of recently fledged nestlings observed in the wild in our study and by Jenny and Cade (1986). As shown by Boyce and Kiff (1981), the original description of eggs attributed to this species (Coltart 1951) was no doubt in error. Kiff (1988) described 10 eggs laid by a captive female taken as a nestling from Petén, Guatemala. These eggs averaged 49.09 (46.71–52.99) mm by 38.96 (37.07–39.92) mm; in addition, Jenny and Cade (1986) measured one Guatemalan egg in the wild at 51.5 mm by 39.0 mm. Using Hoyt's (1979) formula, Kiff's mean measurements give a predicted mass of 40.8 g per egg, or 97.9 g for the mean 2.4-egg clutch. Hence an egg amounts to 6.5% of mean female body mass, and the mean 2.4-egg clutch amounts to 15.7% of female body mass. A color photo of the eggs is included in Kiff 1988.

Nesting Phenology

The breeding season in our study area began with courtship in January and February. Egg laying peaked in late March and early April (mid dry season; Fig. 20.3), hatching was concentrated in May (dry to rainy season transition), and young fledged mainly in June and July (early rainy season; Fig. 20.3). Young of many bird species in Central America fledge early in the rainy season, presumably to take advantage of high prey abundance at that time (Skutch 1950). Reproductive timing of Orange-breasted Falcons resulted in young fledging at a time of year when abundance of their avian prey, particularly young, inexperienced birds, is probably maximal.

Length of Incubation, Nestling, and Post-fledging Dependency Periods

Although we did not document the precise length of any hatching or nestling intervals, falcons of this size typically have an incubation period of approximately 30 days, and young fledge at the approximate age of 40 days (Jenny and Cade 1986). We do not know how long fledglings received care from adults, but we assume they reach independence well before the subsequent breeding season. A given eyrie was often the site of nesting in consecutive years, with no indication that pairs skipped a year after a successful nesting.

Table 20.3 Orange-breasted Falcon site occupancy and reproductive performance at mature forest and habitat mosaic sites, 1992–97

Parameter	Mature forest sites	Habitat mosaic sites	P
Site occupancy	73.3% (n = 35)	77.8% (n = 36)	0.727[a]
Pairs fledging ≥ 1 young	40.7% (n = 27)	62.5% (n = 32)	0.098[b]
Fledglings/successful pair (mean ± SD)	1.36 ± 0.67 (n = 11)	2.11 ± 0.68 (n = 18)	0.004[c]

[a] MRPP χ^2 = 0.12. χ^2
[b] MRPP χ^2 = 2.73.
[c] t_{27} = 2.89

Table 20.4 Physical characteristics of Orange-breasted Falcon nest sites in Guatemala and Belize

Variable	Mean ± SD	Range	n
Cliff height	125 ± 89 m	35–400 m	18
Cliff width	243 ± 145 m	15–600 m	16
Nest to brink[a]	72 ± 54 m	10–200 m	14
Nest to left[b]	109 ± 71 m	3–235 m	13
Nearest neighbor[c]	7.6 ± 6.1 km	1.7–19.2 km	17

[a] Distance from nest ledge to top of cliff.
[b] Distance from nest ledge to left edge of cliff.
[c] Distance to nearest other Orange-breasted Falcon nest site, omitting two outliers with nearest-neighbor distances of 74 and 53.5 km.

Table 20.5 Summary of Orange-breasted Falcon clutch and brood sizes, and number of recently fledged young

Variable	Mean ± SD	Mode	Range	n
Clutch size	2.4 ± 0.8	2	1–4	12[a]
Brood size	1.9 ± 0.7	2	1–3	18[b]
Fledgling number	1.8 ± 0.7	2	1–3	35[c]

[a] n = 10 (this study) and 2 (Jenny and Cade 1986).
[b] n = 14 (this study) and 4 (Jenny and Cade 1986).
[c] n = 30 (this study) and 5 (Jenny and Cade 1986).

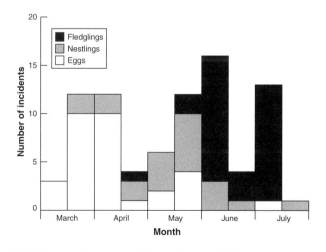

20.3 Nesting phenology of Orange-breasted Falcons in Guatemala and Belize. Each bar represents a two-week interval. Sample sizes for pairs with eggs are 29 (this study) and 2 (Jenny and Cade 1986); sample sizes for pairs with nestlings are 16 (this study) and 5 (Jenny and Cade 1986). In any given year, each nest is represented once for each stage of nesting (egg, nestling, fledgling) and is moved to the subsequent category upon first such observation.

VOCALIZATIONS

We recognized eight vocalizations that we considered distinct. These are described and their behavioral contexts summarized in Table 20.6. To our knowledge, the vocal repertoire of Orange-breasted Falcons is shared most closely with the Bat Falcon (Kirven 1976; Falxa et al. unpubl. ms.; D. Whitacre, pers. observ.). The behavioral context of the chip vocalization appears similar to that of the Peregrine Falcon's *eechip*, Gyrfalcon's *chup*, Prairie Falcon's *kuduchip* (Wrege and Cade 1977), and Merlin's *(F. columbarius)* tic (Feldsine and Oliphant 1985). Behavioral contexts of the rapid *kak* are similar to those of defensive calls of other *Falco* species, as are those of the wail and chitter (given by females and males, respectively, during copulation), and beg (given by females and juveniles, mainly in a food solicitation context; Wrege and Cade 1977; Cade 1982). The numerous behavioral contexts of the slow *kak*—a less intense version of the defensive rapid *kak*—appear similar to those of less intense versions of defensive calls in other falcon species (Nelson 1970; Cade 1982; Feldsine and Oliphant 1985).

BEHAVIOR OF ADULTS

Pre-laying and Laying Period

We observed several behaviors that appeared to be courtship related: mutual vocalizing, ledge displays, perch supplanting, and flight displays. The falcons were highly vocal at all stages of breeding. Mutual calling of the chip vocalization (Table 20.6) appeared to have a significant role in courtship. All pairs observed just prior to egg laying spent considerable time calling back and forth while perched, sometimes occupying several hours in this way.

Individual male and female ledge displays were typified by a horizontal body posture; walking in small circles; scraping with the feet; picking with the beak at leaves, rocks, and twigs; and shallow bowing directed away from or parallel with the cliff. Individual ledge displays were one to five minutes in duration. In mutual ledge displays, both adults remained in a horizontal body posture facing one another and made shallow bows accompanied by vocalizations: females voiced a low chip, and males uttered all three types of chip vocalizations (Table 20.6). Mutual ledge displays were typically less than one minute in duration. Females occasionally (< once daily) displaced males from their perches. As males usually left the nest vicinity when displaced, such female behavior appeared to stimulate the male to hunt.

Females used the low chip vocalization, and rarely the beg, prior to displacing males.

We observed two types of flight displays in courting males: "power flying" and "power diving" (Nelson 1970). Power flying consisted of a strong flapping flight back and forth in front of a cliff face, typically near the top of the cliff. Rarely we observed males rolling approximately 90° to either side or undulating up and down during power flying, as they did when pursuing prey. Males voiced the slow *kak* (Table 20.6) periodically during and after power flying. Power diving consisted of a 45° to vertical dive in front of the cliff face, not directed at prey, and at times accompanied by corkscrew maneuvers. Power diving usually culminated with the slow *kak* (Table 20.6). Pairs were also observed soaring together up to several hundred meters above ground in the vicinity of nest sites during the courtship period—behavior that possibly serves in courtship or territorial advertisement.

Copulation was always preceded by both adults vocalizing from different perches. In all cases males and females vocalized chip calls back and forth; at times one or both used the slow *kak* and rarely the female used the beg call (Table 20.6). Copulation often followed a prey exchange or a male ledge display. A few seconds prior to mounting, usually when the male was in flight toward her, the female bent forward to a near-horizontal position with her tail slightly fanned. During copulation females

Table 20.6 Descriptions and behavioral contexts of vocalizations of adult male and female Orange-breasted Falcons

Vocalization	Description	Adult female behavioral context	Adult male behavioral context
Rapid kak	Single notes of relatively high frequency and short inter-call interval, repeated 2–30 times	Territorial encounters	Territorial encounters
Slow kak	Single notes of relatively low frequency and long inter-call interval, repeated 2–30 times	Arrival in the vicinity of the nest site following an absence, after feeding and about to return to the nest ledge to incubate or brood, and rarely during ledge displays and territorial encounters	Arrival in the vicinity of the nest site following an absence, after copulation, during flight displays, and rarely during ledge displays and territorial encounters
Low chip	Single note of relatively low frequency, often repeated 2–50 times with 1–20 sec inter-call intervals	Preceding and following nearly all behavior	Preceding and following nearly all behavior
High chip	Single note of relatively high frequency, rarely repeated, often preceded and followed by the low chip	None	Preceding and following copulation, prey exchanges, and during ledge displays
Double chip	Low chip followed immediately (< 1 sec inter-call interval) by a squeaky version of the high chip	None	Prior to prey exchanges and copulation, and rarely during ledge displays
Wail	Relatively high-frequency wailing notes with a short (< 1 sec) inter-call interval	During copulation	None
Chitter	Relatively high-frequency single notes with a very short (< 0.5 sec) inter-call interval	None	During copulation
Beg	Relatively high-frequency and long inter-call interval (0.5–2.0 sec) wailing notes, at times culminating in a brief series of higher frequency notes	Prey solicitation	None

remained tipped forward to an angle of approximately 45°, wings held slightly open, and the head was held up in line with the body while the tail was angled up and to one side. Males usually mounted by resting their tarsi on the female's back, with talons balled into a fist or with the foot open and the hallux apparently slipped under the female's humeri. Males flapped throughout mounting, balancing in an upright posture with the neck extended in a downward curve and tail angled downward, at first moving in time with wing flapping before culminating in a brief, stationary tail press. Periodically throughout copulation females wailed and males chittered (Table 20.6). After copulation, males nearly always circled the area a few times vocalizing the slow *kak* (Table 20.6) and either returned to perch next to the female or elsewhere near the eyrie or left the area.

During the pre-laying period, we observed 56 copulations by two pairs in 12 days (4.7 ± 2 per day). When incubation began, copulations became rare, though early in the incubation period at one site we observed five copulations in four days (0.8 ± 1.3 per day). We did not witness copulation later in the nesting cycle. More than 95% (55) of 58 observed copulations occurred on cliff-top trees, with only 2 occurring on nest ledges and 1 on a non-nest ledge. Copulation duration for all pairs combined was 9 ± 2.6 seconds (n = 31). Copulation was most frequent in the morning and evening (Fig. 20.2A).

Courtship behavior was similar to that of other *Falco* species. We observed ledge displays, perch displacing, and flight displays nearly identical to those documented for Peregrine Falcons, Gyrfalcons, Prairie Falcons, and Merlins (Enderson 1964; Nelson 1970; Wrege and Cade 1977; Feldsine and Oliphant 1985; Clum and Cade 1994).

Parental Roles and Activity Budgets throughout Nesting

Males showed similar daily activity patterns during the courtship and incubation phases; they spent 46% of their time in the nest area, 43–47% absent from the nest area, 5–6% of the time on nest ledges, and during 2–5% of the time, male location was unknown (Fig. 20.4A). During the nestling stage males spent more time (57.1%) absent from the nest area; time present in the nest area (42.4%) did not change much, and little time (0.3%) was spent on nest ledges (Fig. 20.4A). Throughout the fledgling stage, males spent 99.9% of the time absent from the nest area, returning only briefly to deliver prey (Fig. 20.4A).

In contrast, females spent very little time absent from the nest area during courtship and incubation (2 and 3%, respectively), and increased time spent on nest ledges from 18% during courtship to 85% during incubation (Fig. 20.4B). During the nestling stage females decreased time on nest ledges to 42.9%, increased time absent from the nest area to 7.7%, and spent the remainder in the nest area (Fig. 20.4B). Like males, females were absent from the nest area 99% of the time during the fledgling stage, returning only to deliver prey (Fig. 20.4B).

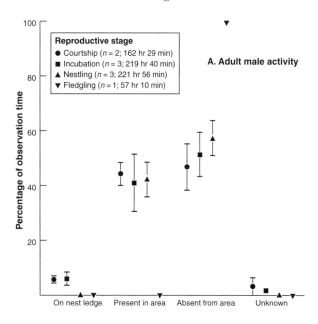

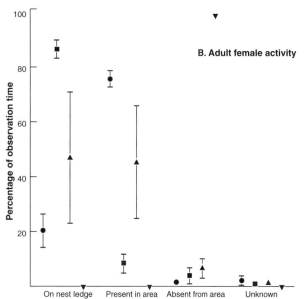

20.4 Time-activity patterns of (A) male and (B) female Orange-breasted Falcons as percentage of observation time during courtship, incubation, nestling, and fledgling stages of reproduction. Data are means ± 1 SE.

While present in the nest area, both males and females spent the majority of time loafing and preening. Little time was spent pursuing prey or defending the nest area, as most attacks on prey and nest defense events were brief (< 1 minute). During the courtship period, both adults participated in preparing the scrape. At all nests, the female performed nearly all incubation, rolling the eggs often, and shading them with her body and partly spread wings. During this period males entered nest ledges only when the female was away from the ledge, eating or performing other maintenance activities. Males appeared clumsy when incubating, often stepping on eggs and continually readjusting their position on the eggs.

Prey Transfers, Preparation, and Caching

We observed 95 prey deliveries from adult males to females. Combining data for all pairs, the mean number of prey deliveries by males per day during the courtship stage was 2.0 ± 0.77 ($n = 12$), during the incubation stage 2.1 ± 0.9 ($n = 16$), and during the nestling stage 2.9 ± 0.7 ($n = 16$). Four types and locations of prey exchanges were observed. Bill-to-bill transfers on tree perches ($n = 23$) and nest ledges ($n = 22$), and foot-to-foot aerial transfers ($n = 19$) were the most commonly observed types, each accounting for approximately 20% of all prey exchanges. On occasion, a male held prey in his beak and the female grasped it with her foot. During prey transfers, males typically assumed a horizontal posture as in other falcons and voiced chip calls; females gave beg calls (Table 20.6). Prey transfers, like hunting attempts (Fig. 20.2A and 20.2B), were most frequent in early to mid morning and in the late afternoon to evening hours.

We observed adults delivering prey to nestlings 63 times during 16 days (3.9 ± 1.1 per day) at three nests that held one, three, and an unknown number of nestlings. Of 63 prey items delivered to nestlings, 57 (90.5%) were brought to the ledge by females and 6 (9.5%) by males; most of the items delivered by females were believed to have been captured by their mates. Adults flying to the eyrie with prey often advertised their approach with the slow *kak*; nestlings always voiced a soft beg during prey deliveries.

We observed falcons caching prey 24 times (adult females = 18, adult males = 5, juveniles = 1) in 1996, mostly in epiphyte clumps on trees ($n = 15$) but also on small cliff ledges ($n = 9$). In previous years we also observed falcons cache prey in crotches of tree limbs. All prey cached by the female had been delivered to her by the male, and all prey cached by the male had been offered to and refused by the female while she incubated or brooded.

BEHAVIOR AND DEVELOPMENT OF NESTLINGS

Nestling and fledgling Orange-breasted Falcons behaved like young of other *Falco* species (Sherrod 1983). Downy nestlings, when not being brooded by an adult, were observed sleeping, awkwardly walking around the scrape, and picking at pebbles and other debris on the nest ledge. Older, feathered nestlings spent much of the day perched on the edge of the ledge, flapping in place, preening, vocalizing the rapid *kak* at passing intruders, and periodically begging when the adults were within view. When adults approached the ledge with food, both young and older nestlings consistently begged.

Fledglings spent considerable time vocalizing and pursuing flying insects, airborne vegetation, birds of all sizes, each other, and their parents. The beg vocalization, accompanied by wing fluttering, was given almost constantly any time adults were near the nest. Playful

hunting and its probable function in learning to capture prey has been well documented and discussed for young Peregrines (Parker 1979; Sherrod 1983).

SPACING MECHANISMS AND POPULATION DENSITY

Territorial Behavior and Displays

As in the Bat Falcon, prominent perching on tall, bare limbs, and frequent vocalizations in the eyrie vicinity probably help these falcons advertise their presence to conspecifics. We did not observe these birds making regular, soaring over-flights of their home ranges, as did many forest raptors at Tikal, seemingly as a territorial display. However, it seems likely that the normal flight activities may contribute toward territorial advertisement. We witnessed only one case in which Orange-breasted Falcons drove off a conspecific.

Interspecific Interactions and Nest Defense

Two expressions of nest defense behavior were observed in eyrie vicinities: defensive vocalizations only, and pursuit flights with or without vocalizations. During vocal nest defense, the rapid *kak* (Table 20.6) was given 1–50 times, usually until the intruder left the area. During defensive pursuit flights a rapid or slow *kak* may or may not be given prior to flight, accompanied by the falcon flying in above the intruder and making short dives, often striking its back and usually voicing the rapid *kak*.

We observed nest defense behavior on 220 occasions; in 133 cases the falcons only vocalized, while in 87 cases they made attack flights with or without defensive vocalizations (Table 20.7). Females averaged 1.1 ± 1.8 nest defense attack flights per day and 2.1 ± 3.4 instances of vocal nest defense without attack flights ($n = 44$ days, all females pooled). Males averaged 0.7 ± 1.3 attack flights daily and 0.6 ± 0.9 instances of vocal defense without attack flights ($n = 44$ days, all males pooled). Attack flights involving both pair members averaged 0.3 ± 1 per day, and instances of both sexes simultaneously giving nest defense vocalizations averaged 0.4 ± 0.6 per day ($n = 44$ days). Females engaged in nest defense more than twice as often as males, perhaps because they were near the nest a greater proportion of the time. Males, however, showed a greater tendency to engage in attack flights than did females, which most often gave defensive vocalizations only (Table 20.7).[4] Jenny and Cade (1986) also noted that while females often initiated attacks in nest defense, males seemed more ardent in their attacks.

Thirteen species of raptors and one corvid elicited nest defense responses in 1996. American Black Vultures (*Coragyps atratus*) and Turkey Vultures (*Cathartes aura*) elicited most nest defense behavior—65.4 and 25.4% of all responses, respectively. Roadside Hawks (*Rupornis magnirostris*) elicited five responses and White Hawks (*Leucopternis albicollis*) four, whereas three species

Table 20.7 Types and frequencies of Orange-breasted Falcon nest defense behavior observed in the vicinity of nest sites

Type of nest defense behavior	Male	Female	Both pair members	Total
Vocalization only (number)	26 (45.6%)	91 (64.5%)	16 (72.7%)	133 (60.4%)
Attack flight only (number)	2 (3.5%)	4 (2.8%)	0 (0.0%)	6 (3.2%)
Vocalization and flight (number)	29 (50.9%)	46 (32.6%)	6 (27.3%)	81 (36.8%)
Total number	57 (25.9%)	141 (64.1%)	22 (10%)	220 (100%)

elicited responses twice each: Hook-billed Kite *(Chondroheirax uncinatus)*, Swallow-tailed Kite *(Elanoides forficatus)*, and Black-and-white Hawk-eagle *(Spizaetus melanoleucus)*. Species eliciting one response each were the Ornate Hawk-eagle *(Spizaetus ornatus)*, Great Black Hawk *(Buteograllus urubitinga)*, Crane Hawk *(Geranospiza caerulescens)*, Plumbeous Kite *(Ictinia plumbea)*, Collared Forest Falcon *(Micrastur semitorquatus)*, Orange-breasted Falcon, Brown Jay *(Cyanocorax morio)*, and three unidentified species. In addition to these species, in prior years we observed Orange-breasted Falcons driving away Bat Falcons, Short-tailed Hawks *(Buteo brachyurus)*, Red-tailed Hawks *(B. jamaicensis)*, Black Hawk-eagles *(S. tyrannus)*, King Vultures *(Sarcoramphus papa)*, and Osprey *(Pandion haliaetus)*. While most defensive behavior we witnessed was directed toward raptors and vultures, Jenny and Cade (1986) also observed vocal defensive pursuits directed at several species of large parrots *(Amazona* sp.), oropendolas, Keel-billed Toucans *(Ramphastos sulfuratus)*, and Plain Chachalacas *(Ortalis vetula)*.

Incidence of nest defense behavior (Fig. 20.2C) was lowest in the morning and by far most frequent during midday, when both pair members often responded together. The high frequency of nest defense during midday is likely related to the prevalence of rising thermals at that time and the associated soaring behavior of the two vultures that together accounted for 89% of defensive events. On average, nest defense began when intruders were an estimated 75 ± 74 m from nests, but the modal distance of intruders when the falcons began to respond was closer to 25 m.

Constancy of Territory Occupancy, Use of Alternate Nest Sites

In all, we monitored 19 eyrie sites for a total of 70 site-years, of which 54 (77%) had pairs in attendance. Yearly mean occupancy rate was 76.0 ± 6.6% (n = 6 years), whether all sites are included or only the nine sites monitored for each of six years. In cases in which we found sites unoccupied, there is a small chance that the falcons occupied undiscovered alternate sites in the same territory, but we believe that the probability that we missed a nesting effort was quite small. We knew of only one territory where the falcons used two alternate nesting sites, and these were only a few hundred meters apart.

Home Range Estimates

We made no estimates of home range size. However, one male that we radio-tagged frequently foraged eastward from the eyrie a distance of roughly 10 km. Notably, this male had limited hunting opportunities from atop the nest cliff, which was in a sinkhole.

Inter-nest Spacing and Density of Territorial Pairs

Nesting density in our study area appeared to be largely controlled by the occurrence of large cliffs on which these falcons nested. Though the single pair of palm-nesting falcons we studied indicates that tree nesting does occur in this area, we believe it is infrequent. From 1988 to 1996 we had 10–50 personnel engaged yearly in field studies of raptors in this area; were tree nesting by Orange-breasted Falcons common in this area, we would expect to have encountered more than the single instance we did. At least within the 576 km² of Tikal National Park, this falcon is not a widely dispersed breeder.

However, palms may yet prove to be a more frequent nesting substrate than shown to date. Since the end of our formal study, in the spring of 1998 a colleague observed three fledgling Orange-breasted Falcons in northwestern Petén in an area without cliffs but with many Corozo palms (J. López, pers. comm.), and in the spring of 2000, we found an additional pair nesting in a Corozo palm in the humid forests of northwestern Petén (O. Aguirre, pers. observ.). If nesting in palms or other trees turns out to be frequent, this would cause us to adjust our perceptions as to nesting densities and distribution and regional and global population size in this species.

While at times pairs nested as close together as 1.7 km, they averaged 7.6 km apart within local high-density enclaves in Belize (n = 17; Table 20.4)—the highest nesting density yet recorded for the species. This inter-nest spacing implies roughly 60 km² of unique space per pair within such high-density enclaves. Regionally, known densities were much lower than implied by this inter-nest spacing. The 13 known nest sites in the 23,000 km² nation of Belize yield an estimated one pair per 1770 km², whereas density appeared far lower in Petén, Guatemala, where we knew of only six eyries in the 21,000 km² area of the Maya Biosphere Reserve. No doubt additional eyries exist in both nations, such that actual regional densities are higher than what is implied by these minimum density figures.

Thiollay (1989a) found this falcon to be probably the rarest, most patchily distributed raptor in French Guiana, where it so far is known only as a cliff nester. The scattered inselbergs there provide very limited breeding sites, and Thiollay estimated a maximum density of less than one pair per 3000 km². Although tree nesting appears frequent in eastern Ecuador, even there, the species is considered rare and local (Ridgely and Greenfield 2001).

DEMOGRAPHY

Age of First Breeding, Attainment of Adult Plumage

As in other falcons, Orange-breasted Falcons in captivity attained adult plumage with their first prebasic molt, beginning at the age of about one year (T. Cade, pers. comm.). We assume that as in other large falcons, these falcons usually begin breeding at the age of two years or later. The age at which they usually enter the breeding population is of course dependent on the dynamics of the population in question and is unknown.

Frequency of Nesting, Percentage of Pairs Nesting

Pairs at a number of cliffs nested in consecutive years. Although we had only one banded adult, we presume that in many cases the same individuals nested in consecutive years.

Productivity and Nest Success

We found no significant trends over time in the annual percentage of sites occupied, percentage of pairs fledging one or more young, or the number of young fledged at successful sites (Table 20.8); this was true using data from all sites and restricting analyses to those nine sites monitored yearly from 1992 to 1997. Overall, of 59 pair-years in which eggs were laid, 31 (52.5%) resulted in one or more young being fledged, and successful nesting pairs on average fledged 1.83 fledglings/successful nest (n = 29). Productivity, based on 70 eyrie-years over a six-year period, was 56 fledglings, or 0.80 fledgling/territory-year. Using only those nine sites monitored for at least five years, productivity was also 0.80 fledgling/territory-year.

Causes of Nesting Failure and Adult Mortality

In one case we found an adult female dead on the eyrie ledge; cause of death was unknown.

Parasites and Pests

Because we normally did not climb to nests, we obtained little information on ectoparasites of nestlings. Three nestlings removed by The Peregrine Fund for captive breeding were each infested with botflies—one so heavily that we doubt it would have survived without removal of these parasites.

CONSERVATION

One cannot discount the possibility that DDT or other organochlorine insecticides have affected populations of

Table 20.8 Annual reproductive success and eyrie occupancy rates of Orange-breasted Falcons in Guatemala and Belize, 1992–97

Parameter	Sample	1992	n	1993	n
Site occupancy[a]	All sites	66.7%	6	77.8%	9
	Subsample[b]	66.7%	6	77.8%	9
Fledging ≥ 1 young[c]	All sites	80.0%	5	62.5%	8
	Subsample[b]	80.0%	5	57.1%	7
Fledglings/successful pair (mean ± SD)[c]	All sites	1.75 ± 0.96	4	1.8 ± 1.1	5
	Subsample[b]	1.75 ± 0.96	4	2.0 ± 1.15	4

Parameter	1994	n	1995	n	1996	n	1997	n
Site occupancy[a]	72.7%	11	73.3%	15	80%	15	85.7%	14
	77.8%	9	77.8%	9	77.8%	9	77.8%	9
Fledging ≥ 1 young[c]	41.7%	12	45.5%	11	63.6%	11	41.7%	12
	42.9%	7	42.9%	7	42.9%	7	42.9%	7
Fledglings/successful pair (mean ± SD)[c]	1.8 ± 1.1	5	1.8 ± 1.1	5	1.57 ± 0.53	7	2.2 ± 1.1	5
	2.0 ± 0.0	3	2.0 ± 0.0	3	1.67 ± 0.58	3	2.33 ± 1.15	3

[a] The year in which a site was discovered was omitted from analyses of site occupancy.
[b] Subsample refers to those 9 sites monitored continuously from 1993–97.
[c] All tests for trends over years used logistic regression and were nonsignificant, $P > 0.05$.

this species in the past. Two eggshell fragments from an Orange-breasted Falcon nest at Tikal in 1979 contained 180 ppm DDE and 50 ppm heptachlor epoxide on a lipid weight basis (Jenny and Cade 1986)—levels that correspond to about 10% eggshell thinning in Peregrines (ibid.). Nonetheless, it seems quite unlikely that any pesticide effects account for the apparent rarity of this species throughout most of its range; rather, the species seems to be intrinsically rare. With little doubt, the main conservation challenge for this species is deforestation and perhaps other forms of habitat modification by humans, as detailed earlier under Habitat Use.

Demographic Trends in Our Study Population

Falcons in our sample from Guatemala and Belize appeared to maintain steady reproductive and occupancy rates during the six-year study period, suggesting a stable local population (Baker et al. 2000b). However, it is unknown whether these results can be validly extrapolated to the regional population as a whole, nor is the size of the regional population known. Historic information on the species's distribution in Mesoamerica is too sketchy to allow analysis of any possible changes in distribution that may have occurred. However, the antiquity of most records perhaps suggests the species's population may have contracted in at least some portions of its Mesoamerican range.

Extent and Locale of Populations

The distribution and population size of this falcon remain largely shrouded in mystery. Any area in the Neotropics where humid lowland to lower montane forests occur in conjunction with large cliffs should be considered potential nesting habitat. To some extent, occurrence of suitable cliffs coincides with the occurrence of limestone karst geology. Moreover, sites with unusual prey abundance, such as large colonies of *Streptoprocne* swifts, deserve special attention as possible attractants to nesting pairs of this falcon.

In 1999 and 2000, The Peregrine Fund personnel made a concerted effort to discover any cliff-nesting Orange-breasted Falcon population that may exist in Central America south of Guatemala and Belize. A search for occupied nesting cliffs was conducted from Honduras to eastern Panama, as detailed by Thorstrom et al. (2002). Sixty-six potential nesting cliffs were surveyed for occupancy from the ground, 55 were inspected by fixed-wing aircraft, and 262 were inspected by helicopter (ibid.). The great majority of seemingly suitable cliffs in conjunction with mature forest were found in Honduras and Panama; these were predominantly in remote, mountainous areas surveyed mainly by helicopter. In Honduras and El Salvador, numerous large cliffs were located, but in many cases, little or no forest remained in the vicinity. Nicaragua possessed huge tracts of mature forest in its Caribbean lowlands, but no cliffs were seen in over-flights there, whereas "cliffs" in Costa Rica were virtually all

steep, vegetation-cloaked volcanic slopes probably unsuitable for falcon nesting. From El Salvador and Honduras through Costa Rica, no Orange-breasted Falcons were located. These researchers felt confident they had discovered the majority of potential nesting cliffs in these four nations, and that no falcons escaped their notice on the cliffs they visited on the ground.

In Panama, 200 cliffs were surveyed from the air, producing Orange-breasted Falcon sightings at one cliff and suspected sightings at two others. Follow-up visits on the ground verified the presence of four pairs of Orange-breasted Falcons nesting on limestone cliffs in remote, forested mountainous regions of Darien, near the Colombian border. These have been monitored by The Peregrine Fund since 2000 (A. Muela, pers. comm.).

These recent efforts lead to the following conclusions. Eastern Panama harbors at least a small number of cliff-nesting Orange-breasted Falcons, but it appears unlikely that a substantial number of cliff-nesting pairs exists between Guatemala/Belize and eastern Panama. In visits to 18 cliffs in southeastern Honduras (the Sierra del Warunta and Montanas de Colon) during April and May 1999 in search of this species, none were detected (Anderson 1999) and additional searching of 60 cliffs here by helicopter later the same breeding season also produced no sightings (Thorstrom et al. 2002). To conclude, our earlier speculation that Orange-breasted Falcons in Belize and Guatemala constitute a disjunct northerly population (Baker et al. 2000b) appears largely substantiated. Tree-nesting pairs might exist anywhere in suitable forest habitat in Central America, but this remains speculative except for the few instances of tree nesting discovered in our Guatemalan study area.

As the bulk of the species's geographic range occurs in South America, presumably the bulk of its population also resides there. However, the species's population status remains poorly known there as well, and it has not yet been found anywhere at a density rivaling what we documented in Belize. Believing this species to prefer terrain of some elevation or slope, Cade (1982) speculated that it may be absent from the central Amazon Basin. If the species were solely reliant on cliff nesting, its absence as a nester from the central Amazon region would be largely ensured. However, observations at several localities in the central Amazon Basin (Whittaker 1996) suggest this falcon is widespread there. Presumably falcons there nest in trees, though this remains to be demonstrated.

Population Tendency and Status

Ramos (1986) speculated that these falcons were extinct or at least declining in Mexico. We regard their status in Mexico as too poorly known to be certain of this, but some decline there is reasonable, based on the amount of deforestation that has occurred. Stiles (1985) indicated that the species is listed as endangered in Costa Rica and is perhaps locally extinct (Stiles et al. 1989). Globally, the species was regarded as near threatened by Collar et al. (1992) but was not listed by Stattersfield et al. (1998); we recommend a globally vulnerable or near-threatened listing for this species. This species's patchy distribution and apparent rarity in most regions, as well as its heavy reliance on large cliffs and mature forest, suggest that it is intrinsically vulnerable to human habitat alteration or to any other factor that significantly reduces the species's reproduction or survival. Further studies on abundance, status, or ecology anywhere within its range would be valuable to the species's conservation prospects.

We consider it likely that Guatemala and Belize are home to fewer than 50 breeding pairs of Orange-breasted Falcons. It seems prudent to treat this northern enclave as disjunct from the rest of the species's breeding range. Continued vigilance will be necessary to ensure the well being of this northernmost population segment.

HIGHLIGHTS

The 4.3% hunting success rate we observed for Orange-breasted Falcons near nests is among the lowest documented for any raptor (Rudebeck 1950, 1951; Parker 1979; Temeles 1985) and, in particular, is lower than that documented during the breeding season in other *Falco* species (Cade 1982; Roalkvam 1985; Jenkins 2000). Most prey were captured away from the observable vicinity of nest sites, however, and it is possible that success rates were higher in hunts farther afield. Studies of nesting Peregrine Falcons, Gyrfalcons, Aplomado Falcons, and Merlins have revealed hunting success rates from 15 to 93% in eyrie vicinities (Parker 1979; Treleaven 1980; Bird and Aubry 1982; S. K. Sherrod in Cade 1982; Hector 1986; Dekker 1988; A. P. Martin in Ratcliffe 1993; Jenkins 2000); and Roalkvam (1985), in his review of Peregrine Falcon hunting efficiency, found a range of 7–83% attack success rates for all ages at all times of year. Several authors agree that Peregrine Falcons hunt at variable levels of intensity, presumably resulting at times in atypically low success rates (Treleaven 1961; Dekker 1980; Ratcliffe 1993; Jenkins 2000). The low hunting efficiency we observed for Orange-breasted Falcons could result from a high proportion of less-serious hunting attempts in eyrie vicinities, but we have no reason to believe this is so—in general, it is thought that nesting falcons hunt at high intensity, to meet high food demands at this time.

Perhaps most directly comparable are the hunting success rates of other cliff-nesting falcons in tropical localities. Breeding Peregrine Falcons *(F. p. minor)* in tropical and temperate South Africa averaged a 12.7% success rate in 251 hunts at nest sites (Jenkins 2000), and cliff-nesting Taita Falcons in Zimbabwe had a 21% success rate in 29 hunts (Hartley 2000). Taita Falcons, though only 50–65% as massive as Orange-breasted Falcons, are ecologically similar, being tropical cliff nesters with similarly stout, short-tailed, small-winged body proportions—these two species have the heaviest wing loading for their body size of 32 falcon species compared (see Chapter 19; see also Hartley 2000). The 4.3% hunting success rate for Orange-breasted Falcons in eyrie vicini-

ties was substantially lower than that recorded in these other studies of tropical cliff-nesting falcons.

While in eyrie vicinities, Orange-breasted Falcons initiated more hunts from perches (99.1% for both sexes combined, $n = 208$) than have been shown in most studies of other falcons, which launched proportionally more hunts from flight. In five studies, nesting Peregrine Falcons perch hunted, on average, in 66% of hunts (range = 58–75: Jenkins 2000), while urban-nesting Merlins initiated 58% ($n = 80$) of hunting attempts from perches (Sodhi et al. 1991). When hunting away from eyrie vicinities, Orange-breasted Falcons may have launched more hunts from flight than we observed near eyries. Still, this dramatically higher proportion of perch hunts by Orange-breasted Falcons compared to other falcons in nest vicinities suggests a greater propensity for perch hunting in this than in other falcons studied to date.

A tendency to hunt from perches may reflect relatively expensive flight in Orange-breasted Falcons, resulting from flight morphology adapted to hunting birds in flight in the tropical forest environment. The ability to achieve high speeds in a dive may be especially important to falcons that must catch avian prey briefly exposed over dense forest. While acceleration in a dive should be more rapid in falcons with light wing loading (Andersson and Norberg 1981), heavily wing-loaded falcons should achieve greater maximum dive velocities than falcons with lighter wing loading (ibid.; Tucker 1998). Orange-breasted Falcons have heavy wing loading for falcons of their size (see Chapter 19). However, because the energetic cost of flight also increases with wing loading (Pennycuick 1975, 1989), it may be advantageous for these falcons to minimize time spent in flight and thus to hunt frequently from perches. Jenkins (1995) found such a result in comparing sympatric Peregrine Falcons and Lanner Falcons *(Falco biarmicus)* in southern Africa; being more heavily wing loaded, Peregrine Falcons were predicted to have higher flight speeds, higher flight costs, and lesser soaring capabilities than the more lightly wing-loaded Lanner Falcons. Observations confirmed that Peregrine Falcons hunted more often from perches, spent less time flying, and when flying, flapped more, compared to Lanner Falcons in similar situations. However, Peregrine Falcons were more agile and more capable of catching birds in flight than Lanner Falcons were. Heavy wing loading was viewed as one of a suite of adaptations making Peregrine Falcons highly effective at catching birds in flight, though with a flight-energy cost that makes perch hunting more important to them than to falcons less specialized for capturing birds in flight. We hypothesize that the same argument applies to the Orange-breasted Falcon.

The opportunity to perch hunt from a commanding cliff top may also be important to these falcons in providing a height advantage over prospective quarry while avoiding costly searching and climbing flight. In South Africa, Jenkins (2000) evaluated the importance of cliff height and cliff elevation above surrounding terrain, for hunting styles and success of Peregrine Falcons. The initial height advantage of the falcon over quarry at the beginning of an attack was the factor most strongly associated with hunting success, such that attacks from taller cliffs yielded higher success than those from shorter cliffs. In contrast, elevation of the cliff base over surrounding terrain had no effect on hunting success, which Jenkins interprets as follows. Elevation of the cliff base over surrounding terrain is determined mainly by the length of the talus slope below a cliff; however, the vertical extent of a talus slope often does not add to the vertical air space through which a falcon can accelerate or maneuver, and thus adds little to the falcon's hunting opportunities (but see Gainzarain et al. 2000 for a contrary view). Tall cliffs provide a height advantage over some quarry at no energetic cost to the falcons. Since it is mainly terminal velocity, and not acceleration, that scales up with wing loading, and since very long stoops (up to 1 km or more) are required in order to approach terminal diving velocity in near-vertical stoops (Tucker 1998), very tall nesting and hunting cliffs may be especially important for heavily wing-loaded falcons such as the Orange-breasted Falcon. However, it is not known whether falcons in the wild often accelerate to diving velocities approaching their aerodynamic maxima; a trained Gyrfalcon limited its diving speed to 52–58 m/sec, though it could have reached higher speeds had it not intentionally increased drag via wing positioning and possibly other postural adjustments (Tucker et al. 1998).

Another factor promoting perch hunting may be the following: a falcon soaring over a tract of forest might cause potential quarry to stay beneath the canopy until it had passed. Conversely, a perched falcon may be less conspicuous and hence more easily achieve surprise. The importance of the element of surprise in hunting by falcons has been shown by Cresswell (1996), who found higher capture rates for Peregrine Falcons when they achieved surprise. The Orange-breasted Falcon's proportionally large feet and beak may also be adaptations for hunting above dense forest (Cade 1982), aiding the falcons in holding onto and quickly dispatching prey before it can escape into forest cover. A rapid coup de grâce may be especially important when handling powerfully biting prey such as parrots and some bats, which might be capable of injuring a falcon's toe.

Cooperative hunting has been observed in at least seven *Falco* and five accipitrid species (Bent 1938; Cade 1982; Ellis et al. 1993), but this was uncommon in Orange-breasted Falcons, with only three cooperative hunts among 208 observed. Considering the low hunting success rate we documented and the higher success rates afforded by cooperative hunting in other species (Roalkvam 1985; Hector 1986), we were surprised to observe such infrequent cooperative hunting in this species. However, most falcons that regularly hunt as pairs occupy more open habitats, where one bird may cut off aerial escape routes while the other chases avian prey from scattered trees or shrubs (Hector 1986); such a strategy may be ill-suited to densely forested habitats.

Cade (1982) and Jenny and Cade (1986) suggested that Orange-breasted Falcons were decidedly crepuscular, if not partly nocturnal, in hunting behavior. However, several falcon species are known to be quite crepuscular at times, and a few even to hunt by moonlight on occasion. We found no evidence that Orange-breasted Falcons are more crepuscular than many other falcons. Although bats were captured, the earliest hunting behavior we observed was at dawn and the latest at dusk. Nearly all hunting attempts we observed were accompanied by vocalizations, and we camped many nights within hearing distance of several nest sites and never heard the falcons vocalize before dawn or after dark.

Orange-breasted Falcons show greater size dimorphism than most other raptors, with males weighing less than 60% as much as females. Although the selective pressures leading to reversed sexual size dimorphism remain unresolved, it is interesting to note some behavioral correlates of the degree of dimorphism. That these falcons are the most heavily armed member of the genus and also among the most size dimorphic appears consistent with Cade's (1960, 1982) hypothesis that reversed sexual size dimorphism in raptors results from social relations between male and female. If a clear-cut social dominance is necessary in dangerously armed species in order to maintain a smoothly functioning pair bond, it might follow that the most formidably armed species would show the most extreme differences in body size. We are skeptical, however, that the dominance hypothesis provides an adequate explanation for patterns of size dimorphism in raptors.

Cade (1982) also suggested that extreme size dimorphism may result in male falcons being unable to adequately warm the eggs or young, which may promote extreme partitioning of parental roles. Compared to other *Falco* species, male Orange-breasted Falcons showed a low frequency of nest ledge attendance and consistently awkward attempts at incubation and brooding. During the incubation period, males spent only 5% of their daylight time, versus females' 85%, on the nest ledge, and females alone seemed to incubate at night. Nest site attendance by males appeared to occur only as relief to the female while she ate or engaged in other maintenance activities. These results are similar to, but more extreme than those for most other highly dimorphic falcons. Male Peregrine Falcons were found to spend 12–33% of daylight hours in incubation (Cade 1960; Herbert and Herbert 1965; Enderson et al. 1972; Cramp and Simmons 1980; Hustler 1983; Carlier 1993; Ratcliffe 1993) and male Gyrfalcons spent 17–24% of the time incubating (Poole and Bromley 1988; Clum and Cade 1994). Even more extreme, Nelson (1970) estimated that at mid-incubation, male Peregrines incubated 30–50% of daylight hours.

Male Orange-breasted Falcons also contributed little to brooding young; females brooded 42.9% of the time while males scarcely brooded at all (0.32% of the time). Again, this pattern appears more extreme than in other highly dimorphic falcons. Male Gyrfalcons brooded 5–25% of daylight hours (Jenkins 1978; Poole and Bromley 1988; Clum and Cade 1994) and male Peregrine Falcons approximately 5–10% during the first 15 days of the nestling stage (Carlier 1993) or virtually not at all (Nelson 1970; Enderson et al. 1972; Hovis et al. 1985).

The Orange-breasted Falcon, with a bird-dominated diet and among the most size dimorphic of raptors, fits the well-known relationship between degree of dimorphism and reliance on agile, especially avian, prey (Newton 1979). Moreover, with the very low hunting success rate we found, this falcon conforms with the relationship between low hunting success and high degree of size dimorphism found by Temeles (1985). Temeles's "prey vulnerability" hypothesis takes into account that prey attributes such as agility and use of dense habitat may render prey less available, resulting in lowered hunting success of some raptors. The extremely low hunting success rate we found, if it proves characteristic for this species, may relate to the fact that these falcons hunt mainly birds flying over tropical forest—a resource that may have inherently low availability or vulnerability. Peregrine Falcons in South Africa had highest hunting success over habitats of sparse to moderate cover, and lowest success over dense cover (Jenkins 2000). Conceivably, low hunting success is a consistent feature for large falcons hunting birds in flight over dense tropical forest.

The ecologically similar Peregrine Falcon shows greatest density and nesting productivity at mid to high latitudes, and lowest density and productivity in the tropics (Jenkins 1991, 2000; Jenkins and Hockey 2001). In addition, nesting cliffs used by Peregrine Falcons in the tropics were taller, on average, than those at higher latitudes; as taller cliffs are believed to provide better hunting opportunities, this pattern suggests that physical site characteristics optimal for hunting are more important to Peregrine Falcons in the tropics than elsewhere, perhaps compensating for lower prey availability in the tropics (Jenkins 1991, 2000; Jenkins and Hockey 2001). Jenkins (1991, 2000) and Jenkins and Hockey (2001) speculate that lower nesting productivity and use of mainly tall

cliffs by Peregrine Falcons in the tropics are symptomatic of generally lower availability of avian prey in the tropics than at higher latitudes. They hypothesize that prey are more available to nesting Peregrine Falcons at higher latitudes because avian breeding seasons are more temporally compressed at higher latitudes than in the tropics, producing a larger pulse of avian prey during falcon nesting. However, other factors such as larger avian clutch sizes at higher latitudes could also lead to such a latitudinal trend in availability of birds as prey.

The low hunting success rate we found calls for verification by further study. In addition, it would be instructive to compare rates of prey biomass delivery to nests with those of other falcons. If the low hunting success we found is typical of this species, and if biomass delivery rates are unusually low, these factors might help explain the apparent patchiness and rarity of this falcon throughout its range, and might have conservation implications. Perhaps, as suggested for Peregrine Falcons in the tropics (Jenkins 1991, 2000; Jenkins and Hockey 2001), only the most favorable hunting situations, in terms of prey abundance and vulnerability, and in terms of the hunting advantages provided by tall cliffs, permit Orange-breasted Falcons to thrive. If this is true, then subtle changes in the abundance or species composition of available prey might convert acceptable sites into sites unfavorable for these beautiful and dramatic birds.

For further information on this species in other portions of its range, refer to Berry et al. 2010.

Chapter 20 Notes

1. In addition, from a slightly smaller sample, Bierregaard (1978) gave the following dimorphism values: 14.7 for middle toe length, 8.2 for upper mandible depth, 12.1 for upper mandible width, 16.8 for middle talon length, and 23.8 for hind talon length.
2. ANOVA, $F_{3,75}$ = 1.344, $P > 0.05$.
3. ANOVA, $F_{3,113}$ = 1.451, $P > 0.05$.
4. MRPP chi-square test; χ^2 = 5.98, P = 0.014.

21 MEXICAN WOOD OWL

Richard P. Gerhardt and Dawn M. Gerhardt

We were in the ancient forest before sunrise hoping to find the diurnal roosts of a pair of Mexican Wood Owls *(Strix squamulata)*. If we could pinpoint these roosts, we could come back at dusk with a trap, capture the birds, and put radio transmitters on them. While we strained our ears for the last, muffled contact calls by which the pair of owls notified one another of their respective whereabouts, the day shift was coming alive.

Blue-crowned Motmots *(Momotus momota)*, handsome birds with racket-tipped tails, were giving the call described by their name. A distant Collared Forest Falcon *(Micrastur semitorquatus)* had finished his customary predawn calling bout, and a nearer Barred Forest Falcon *(M. ruficollis)* had just begun his routine, the higher-pitched, more frequent notes a sped-up version of his larger relative's call. Family groups of Plain Chachalacas *(Ortalis vetula)* seemed to vie with neighboring clans for some sort of daily decibel award. The voices of Crested Guans *(Penelope purpurascens)*, from the tops of trees 35 m in height and Great Curassows *(Crax rubra)* from nearer the ground served as a prelude to the gobbling of Ocellated Turkeys *(Agriocharis ocellata)*. By the time the parrots and Montezuma's Oropendolas *(Psaracolius montezuma)* joined the chorus, we knew that the owls must be at their roosts. As we strained our ears for those last owl calls, the air around us was shattered by the bellowing, from just above our heads, of the loudest terrestrial animal, a Guatemalan Howler Monkey *(Aloutta pigra)*.

This is how Dawn remembers her first morning in the forest at Tikal, and thus began our research on the Mexican Wood Owl. We caught that pair of owls the same evening. The radio transmitters allowed us to find the nest once the female began incubating, to locate the male's daytime roosts and follow his nightly activities, and to find regurgitated pellets through which we could study the owls' food habits. During the next three years, we found 21 active nests of this common owl and learned a great deal about its ecology in the primary forests of Tikal.

GEOGRAPHIC DISTRIBUTION AND SYSTEMATICS

Developments in molecular genetics have revised our understanding of phylogenies within Strigidae, and both the Mexican Wood Owl and the Black-and-white Owl *(S. nigrolineata;* see Chapter 22) are now recognized in the genus *Strix.*

The Mexican Wood Owl is a rather common and widespread wood owl of the Central American and northwest South American tropics and subtropics; its range extends from Nuevo León and Sonora, Mexico, to Panama and northernmost Colombia (König and Weick 2008). It has also been reported from Hidalgo County, Texas (Lasley et al. 1988). This has been described as the most abundant of the larger tropical owls in Mexico (Blake 1953) and probably the most common forest owl of the humid lowlands of Honduras (Monroe 1968). Indeed, from Mexico to Panama, the Mexican Wood Owl is uniformly regarded as fairly common to common (Dickey and van Rossem 1938; Slud 1960, 1964; Russell 1964; Wetmore 1965; Binford 1989; Ridgely and Gwynne 1989; Stiles et al. 1989; Howell and Webb 1995) or even abundant (Schaldach 1963). Three subspecies are generally recognized (König and Weick 2008), with the subspecies occurring at Tikal being *S. s. centralis,* which ranges from southeast Mexico to northwest Colombia.

MORPHOLOGY AND PLUMAGE

Mexican Wood Owls are medium-sized owls, lacking "ear" tufts. As adults, they are dark brown with small white spots on the crown, nape, back, and wings. The breast and belly are white but heavily streaked vertically with deep brown. Noticeable white crescents on either side of the beak offset dark, chocolate brown eyes. The feet and cere are dull yellow and the beak gray-green (Plates 21.1, 21.2).

Published reports on Mexican Wood Owls list adult body masses of 176 to 248 g (Burton 1973) and 235 to 305 g (Voous 1989). These references do not list the sources for these data; presumably they come from a small number of museum skins and captive individuals and represent unknown portions of the species's range. In Tikal, males averaged 240 g in weight (range 220–256 g, *n* = 7), and females 335 g (range 308–385 g, *n* = 11; Appendix 1). Thus, male Mexican Wood Owls weighed about the same as a male Long-eared Owl *(Asio otus)*, whereas female Mexican Wood Owls weighed much more than a female Long-eared Owl—about the same as a female Northern Hawk-Owl *(Surnia ulula:* Snyder and Wiley 1976).

With a Dimorphism Index (DI) value of 12.1 (Appendix 1), Mexican Wood Owls exhibit the most pronounced body mass dimorphism yet documented among owls (Earhart and Johnson 1970; Andersson and Norberg 1981; Mueller 1986), with the possible exception of the European form of Tengmalm's Owl *(Aegolius funereus)* (Korpimäki 1986; Lundberg 1986). Mexican Wood Owls in Tikal also exhibited pronounced dimorphism with respect to flattened wing length (DI = 5.4) with females' flattened wing length (average 24.6 cm, *n* = 12) being significantly longer than males' (average 23.3 cm, *n* = 8), and with respect to tail length (DI = 6.2; females 15.0 cm, *n* = 11; males 14.1 cm, *n* = 7; Gerhardt and Gerhardt 1997). We expand on this theme under Highlights, at the end of this chapter.

PREVIOUS RESEARCH

As was the case with most raptor species that breed in Tikal, Mexican Wood Owls were essentially unstudied prior to the Maya Project. Collecting for museums had provided what little was known regarding the distribution, morphology, and food habits of this species. A single reference to a nest was available (Belcher and Smooker 1936), and no published information existed regarding ecology, demography, home range size, and other life history traits.

RESEARCH IN THE MAYA FOREST

The research we conducted at Tikal from 1989 through 1992 represents the first systematic study of this common owl. The initial step was a calling survey (Gerhardt 1991b), which assessed the responsiveness of Mexican Wood Owls to playbacks of recorded conspecific calls and examined the relationship between response rate and various environmental parameters. Subsequently, we began a study of the owls' breeding biology, which provided much of the information presented here (Gerhardt 1991a; Gerhardt et al. 1994a, 1994b). We conducted this research in the mature Upland Forest surrounding the main ruins of Tikal. Since Mexican Wood Owls can be found in other forest types in the park and in a variety of natural and human-altered habitats in other parts of the species's range, some of our results from Tikal may not be applicable to the species as a whole.

DIET AND HUNTING BEHAVIOR

The most common method of studying owl diets is via pellet analysis. Owls tend to swallow their prey either entirely or in large bites without plucking or rejecting any portion. Although owls lack a crop, they nonetheless form indigestible portions into pellets, which are then regurgitated. Such matter tends to be more intact than in the pellets of falconiforms, and fur, feathers, bones, and other hard parts can aid the scientist in identifying the prey eaten. This method, however, is not without bias. Whereas certain hard, chitinous parts such as head capsules, beetle elytra, and grasshopper legs frequently appear in pellets, many soft insect parts are completely digested. Therefore, arthropods and other invertebrates tend to be underrepresented in pellets relative to their true occurrence in the diet.

We collected pellets from beneath the diurnal roosts of several Mexican Wood Owls, primarily males. We found very few pellets in nest vicinities but did collect uneaten prey remains during nest checks. We quickly learned that finding pellets is much more difficult in the tropics than in more temperate regions. Pellets of Common Barn Owls *(Tyto alba)* can last for years in the shelter of a barn, and Long-eared Owls, which often roost in winter in large numbers and near the ground, can leave behind large accumulations of pellets that remain intact from one year to the next. Though Mexican Wood Owls did not roost especially high during the breeding season, it was nonetheless a rare pellet that reached the ground intact. Palm fronds, vines, and other vegetation either caught entire pellets or caused them to break apart so that only scattered pieces reached the forest floor. When a pellet did reach the ground intact, its half-life could be measured in minutes, not days, as ants, termites, and reduviid bugs carried it off piece by piece.

Mexican Wood Owls in our study area ate mostly insects and rodents. Of the 52 pellets (or parts thereof) that we collected, 23 (44%) contained only insect parts, and all but 1 (98%) contained some insect exoskeletal material. Insects found most commonly in pellets were scarabaeid beetles and acridid grasshoppers, with cockroaches, snout beetles (Curculionidae), and long-horned beetles (Cerambycidae) found less frequently. These insects were generally represented by intact legs; beetle elytra were typically broken into small bits. Twenty-six (50%) of the pellets contained bones or fur of rodents, most of these belonging to Fulvous Rice Rats *(Oryzomys fulvescens)* and Hispid Cotton Rats *(Sigmodon hispidus)*. Lizard mandibles were found in two pellets, and the jaw of a small bat in one. Whereas insects were found in nearly all pellets, vertebrates were represented in 56% of pellets. A rough estimate, based on these data, would be that invertebrates contributed 60% of the biomass of the diet of these owls, and vertebrates the remainder.

We found remains of nine prey in Mexican Wood Owl nests. These included two cockroaches, a large snout beetle, a treefrog *(Hyla* sp.), a ranid frog, a lizard *(Anolis* sp.), a Big-eared Climbing Rat *(Ototylomys phyllotis)*, a rice rat, and the feathers of a small bird. We observed Mexican Wood Owls capturing or carrying a katydid and two rice rats.

HABITAT USE

Mexican Wood Owls typically hunted in dense mature forest and exclusively at night. Low perches were commonly used, and most of the rodents captured were terrestrial species. One male, whose home range included human habitations, spent time hunting rats near kitchen facilities.

Our Mexican Wood Owl research was conducted almost exclusively in Upland Forest at Tikal, but the species was present in other forest types within the park. Indeed, we have encountered these owls in a great many, often human-modified forest habitats and have even watched an individual perched, hunting, on a 3 m pole in the middle of a slash-and-burn cornfield bordered by forest (D. Whitacre, pers. comm.). Further research is needed to determine the relative density and productivity of Mexican Wood Owls in the varied habitats in which they are found.

BREEDING BIOLOGY AND BEHAVIOR

Nests and Nest Sites

We located 21 active Mexican Wood Owl nests—seven in 1990, six in 1991, and eight in 1992 (Bonilla et al. 1992; Gerhardt et al. 1994b). All were in cavities in live trees. Most cavities (n = 18) were in the trunk and were formed by decay following a branch tearing from the main trunk (Plate 21.3). One cavity was formed by the breaking of the trunk itself; this nest was partly shaded both by a living branch of the nest tree and by the large leaves of a *Philodendron* plant. Two other nests were in the main crotch of the respective nest trees. One of these was a depression only 10 cm deep, but it, too, was overhung by a *Philodendron*. Mean dimensions of cavity entrances were 17.2 cm by 32.3 cm (n = 13), and the mean depth of cavities was 51.8 cm (range = 10–250 cm, n = 13). The 21 nests were placed in 10 tree species, with Ramón or Bread-nut Tree *(Brosimum alicastrum)* and Allspice *(Pimenta dioica)* used nine and four times, respectively. Mean nest height was 12.2 m (range = 8.4–17.5 m, n = 21).

Egg and Clutch Size

We found clutches of two and three eggs (mean = 2.2, n = 18). All eggs were non-glossy, off-white, and elliptical, being only slightly longer (mean = 42.1 mm) than wide (mean = 36.0 mm, n = 24). Mean egg mass was 27.7 g (n = 24), amounting to 8.3% of mean female body mass, and 2- and 3-egg clutches representing 16.5 and 24.8% of the mean 335 g female body mass, respectively. Hoyt's (1979) formula, using K_w = 0.55 (determined from data given by Schönwetter 1964), gives a predicted fresh egg mass of 30.0 g, or 9.0% of the mean female body mass at Tikal, with the 2.2-egg average clutch amounting to 19.7% of female body mass.

Nesting Phenology

Mexican Wood Owls in this population were quite synchronous in their nesting. Almost all eggs were laid during the last half of March, and hatching occurred in mid to late April. Fledging occurred near the end of May, just prior to the onset of the rainy season. Whatever other factors may influence the timing of nesting in this species, nest site selection no doubt plays a role—many of the cavities used for nesting were filled with water once the rainy season arrived. We documented two apparent renesting attempts following initial failures; one resulted in failure, but the other was successful and extended much later into the season than did all other nests.

Length of Incubation, Nestling, and Post-fledging Dependency Periods

Incubation periods were 28–30 days long (estimate from four eggs). Three young fledged at the ages of 27–29 days, and another at 32–33 days. Fledglings remained within their natal home range until at least three months after fledging, roosting with and being fed by one or both parents. We believe young reached independence and dispersed shortly thereafter (by four months after fledging). At about this time, the first prebasic molt yielded owls that looked like adults, and family groups were not observed after August. These durations for incubation, nestling, and post-fledging dependency are quite similar to those of the Spotted Owl *(Strix occidentalis:* Forsman et al. 1984), a temperate species with approximately twice the body mass as the Mexican Wood Owl. Compared to the Long-eared Owl (Marks et al. 1994), a temperate owl of similar mass to the Mexican Wood Owl, Mexican Wood Owls had slightly longer incubation periods, longer nestling periods (and did not exhibit "branching" behavior to the extent that Long-eared Owls did), and longer post-fledging dependency periods. Thus, the entire breeding period was somewhat protracted relative to that of a northern owl species of similar size.

VOCALIZATIONS

We heard three distinct vocalizations by Mexican Wood Owls in our study area. Males frequently and females occasionally uttered a three- to six-note call. This served as a territorial vocalization and could easily be elicited from both sexes throughout the breeding season (Gerhardt 1991b) by broadcasting a recorded call. This territorial call was consistent among individuals in having three primary hoots, each separated by nearly one second. One, two, or three additional, more muffled notes preceded or followed these primary hoots in the calls of most individuals. The presence and number of these subsidiary notes were consistent for the individual and facilitated our recognizing individuals by their calls. Miller (1963) discovered that the voice box of the Mexican Wood Owl is more enlarged and specialized than that of other owls investigated. This enables the Mexican Wood Owl to produce an especially low-pitched note for a bird of its size. The calls of females were higher pitched than those of males.

The vocalization given most often by the female was a catlike yowl, used as a food solicitation call. The frequent use of this call near the nest was the best clue to the location of the nest tree. The third call, a soft, rather high-pitched hiss, was given only by the young. It, too, seemed to be used in food begging.

The territorial call was described by Wetmore (1968) as *bru bru* and *bu bu bu* and by Ridgely (1976) as *keeoow-eeyo* or *cowooawoo*. These descriptions from Panama do not accurately suggest the calls of Mexican Wood Owls in Tikal, which are more consistent with the "guttural hoots" described by Blake (1953) from Mexico. It is likely that there exists some variation in the primary call across the range of the species and that Mexican Wood Owls have a more varied repertoire than what we heard.

We exposed Mexican Wood Owls to broadcasts of recordings of conspecific calls during a six-week period roughly spanning their nesting season (Gerhardt 1991b). Their responses were counted, and the relationship between calling frequency and parameters such as weather, light intensity, and time of night was evaluated. The owls were quite responsive to broadcast of conspecific calls throughout the study period. Wind was the only factor, of those tested, that affected the potential to hear an owl call. The high rate of response (40%) we detected under a variety of conditions suggests that broadcast of recorded calls can be used as a census tool for this species and highlights the territorial nature of Mexican Wood Owls.

We recorded the territorial calls of 10 male Mexican Wood Owls and analyzed them to determine whether we could recognize individuals by voice (unpubl. data). We generated sonograms and measured various parameters of each call, including number and placement of the muffled, subsidiary notes, and starting and ending pitches of the primary notes. We then had someone else measure these same parameters on other (not previously measured) calls as a blind test of the chosen parameters. Referring to the recorded measurements, this person was able to correctly assign the individual's identity to all recordings using only five parameters. We also overlaid individual notes on one another to derive a similarity index and generated a correlation matrix (Gaunt et al. 1994) to compare similarity of notes of one individual with those of another using the program SIGNAL (Beeman 1993). To these correlations, we applied a Multiple Response Permutation Procedure (Slauson et al. 1991), to examine the effect of comparison type (self or other) on correlation coefficients. Self-correlations were found to be significantly greater than those between different owls. Thus, individual Mexican Wood Owls were recog-

nizable to humans listening in the field, to humans using the metrics of their sonograms, and by statistical analysis. Undoubtedly, Mexican Wood Owls are themselves able to recognize individual conspecifics by voice. They likely rely on vocalizations for imparting and receiving a great deal of information associated with establishing and maintaining territories; finding, courting, and maintaining contact with their mates; choosing nest sites; finding food; and performing other activities.

BEHAVIOR OF ADULTS

We visited 274 different diurnal roosts on 407 occasions. The typical perch was a horizontal branch or vine in a dense section of forest. Mean diurnal roost height was 5.3 m (range = 0.5–18 m, n = 274). Cavities were never used for daytime roosting. Mexican Wood Owls often spent the day within 2 m of the forest floor, particularly on the hottest days. When not on nests, females tended to roost with their mates, and, later, family groups roosted together. Members of a pair or family often roosted within 1 m of one another. While females were on nests, males generally roosted a considerable distance away (mean = 252 m, n = 114), neither at the center nor at the periphery of their home ranges (Gerhardt et al. 1994b).

Incubation apparently began with the laying of the first egg; females remained in the cavity beginning at that time, and young hatched asynchronously. As is typical of owls, females did all of the incubating and brooding, and males did all of the hunting. Even after brooding had ceased, females remained near the nest while males foraged.

BEHAVIOR AND DEVELOPMENT
OF NESTLINGS

At hatching, nestlings had closed and protruding eyes and swollen, yolk-filled abdomens. White natal down originating from the major feather tracts covered most of the body, while areas of bare skin covered the remainder. Cere and feet were flesh colored and talons gray. The beak was gray with a small, white egg tooth that disappeared by day 6. At the age of about 8–10 days, young began to open their eyes and to "beak-click." (This sound, produced by many owl species, is actually made by snapping the tongue against the palate.) At this time the head and, more particularly, the orbital region were the most thickly feathered portions of the young owls, covered in second-generation down.

Prejuvenal feather growth began at 10–12 days. Contour feathers similar to those of adults appeared on the backs and wings, and the tail began slowly growing in. The rest of the owl chick became covered in soft down and semiplumes, the body being of a peach or golden hue and the head creamy white (Plate 21.4). The first prebasic molt was not detected until the age of four months, when adult-plumage contour feathers began appearing on the head and breast. This molt likely began earlier, but we did not capture fledglings to inspect them.

Three young fledged between 27 and 29 days after hatching and another at 32 or 33 days. Ten young weighed within three days of fledging had a mean body mass of 190.6 g, considerably less than that of adults. We did not see young on branches outside the nest prior to fledging. When fledging, the young were incapable of sustained flight and merely glided downward, landing in low underbrush or vines or on the ground. They then climbed up into the understory or leaning trees. Fledglings never returned to the nest. Three months after fledging, young were still in the natal home range, roosting with and being fed by one or both parents.

SPACING MECHANISMS AND POPULATION DENSITY

Territorial Behavior and Displays

We believe that Mexican Wood Owls defended the entire home range, as has been reported for Barred Owls (*Strix varia*: Nicholls and Fuller 1987). Overlap of home range among birds from different pairs amounted to no more than 20% of any pair's home range. Paired owls shared the same home range through three full years, and home ranges remained essentially unchanged even when one of the adults had been replaced by another individual (Bonilla et al. 1992). We observed one instance in which a radio-tagged male, near the outer edge of his home range, flew at another owl, striking it with his talons and then chasing it away. This observation supports the view that the entire range is defended, but such interactions are likely rare—vocalizations are undoubtedly the primary method by which Mexican Wood Owls establish and advertise their territory boundaries (see Vocalizations, earlier in this chapter).

Nest Defense

We climbed to nests during late afternoon. Nest defense varied in intensity among females. Two females struck the climber repeatedly on every occasion, whereas another attacked the climber initially but then apparently became accustomed to the climbing and ceased striking. Four other females never struck but remained nearby, tongue clicking and vocalizing. One female flew a considerable distance from the nest and roosted, whereas another female sat tightly on the nest, and we had to pick her up to examine her eggs. Only on one occasion did a male arrive in apparent response to a female's vocalizing.

Constancy of Territory Occupancy, Use of Alternate Nest Sites

Established pairs were sedentary, remaining within the breeding territory throughout the year and from one

year to the next. Mexican Wood Owls were never documented using the same nest in a subsequent year, even after a successful nesting; this suggested to us that suitable nest cavities were common. Indeed, when searching for nests, after hearing a female give a begging call from a fixed location on a previous night, we would often have to examine several holes that seemed to fit our growing understanding of what constituted an acceptable nest location for these owls before finding the nest. Some other cavity-nesting raptors in Tikal likely do not have suitable nest cavities in such abundance. Barred Forest Falcons (*Micrastur ruficollis*) reused nest cavities only if successful in a previous year (Thorstrom 1993; see Chapter 16), and Collared Forest Falcons (*M. semitorquatus*) and Laughing Falcons (*Herpetotheres cachinnans*) reused nests year after year, even after failures due to predation (Thorstrom 1993; R. Gerhardt, pers. observ.; see Chapters 17, 18).

Home Range Estimates

We estimated home range size for six males during the breeding season (March–August), four in 1990 and two in 1991. Mean home range size was 20.8 ha (85% Harmonic Mean) or 21.7 ha (Minimum Convex Polygon; Gerhardt et al. 1994b). A male Black-and-white Owl (*Strix nigrolineata*: see Chapter 22) had a home range more than twenty times as large. Although this difference is partially explained by the size difference (Mexican Wood Owls weighing half as much as Black-and-white Owls), prey selection undoubtedly plays a role as well. Though both species fed on large insects, there was almost no overlap in the mammalian component of their diets—whereas Mexican Wood Owls ate many rodents, Black-and-white Owls ate numerous bats (Gerhardt et al. 1994a). Of North American owl species, only the much smaller Flammulated Owl (*Psiloscops flammeolus*) has been found to have a smaller (breeding season) home range (estimated at 14.1 ha: Reynolds and Linkhart 1987) than that estimated here for the Mexican Wood Owl.

Inter-nest Spacing and Density of Territorial Pairs

In 1990, breeding density in a 2 km² area was seven pairs, and in 1991, eleven pairs of adults were found in a 2.5 km² area. Hence the density of territorial adults was estimated at 7.0–8.8 adults, or 3.5–4.4 territorial pairs per square kilometer each year. Extrapolating from this, 100 km² of similar habitat (Upland Forest) at Tikal would be predicted to hold roughly 350–440 pairs of these owls. Of course, whether such densities hold true over large areas is unknown. At any rate, the density we observed is far greater than that at which any species of northern temperate owl has been found. Nonetheless, calling surveys conducted at Tikal suggested that the smaller Guatemalan Screech Owl (*Megascops guatemalae*) was perhaps even more common than the Mexican Wood Owl.

DEMOGRAPHY

Age of First Breeding, Attainment of Adult Plumage

Our study was conducted over either too small an area or too short a time frame to document dispersal of young owls from their natal territories to breeding territories. That is, we did not find any breeding adults that had been banded as nestlings. After undergoing the first prebasic molt at approximately four months old, the young owls had attained plumage indistinguishable (to us, at any rate) from that of adults. It is likely that dispersal from the natal territory coincided with attainment of this plumage. We would expect that Mexican Wood Owls are capable of breeding in the year following hatching or, at the latest, by the age of two years. This and other demographic questions remain to be answered.

Frequency of Nesting, Percentage of Pairs Nesting

Not every pair of owls bred each year. In 1991, 5 of 11 monitored pairs did not nest, and by 1992 we were monitoring 14 pairs of owls, of which 6 apparently did not attempt to nest that year. Overall, this gives 14 nesting attempts in 25 pair-years, or 56% of pairs making a nesting effort per pair-year. On an annual basis, 55 and 57% of pairs made nesting attempts in these two years (mean = 55.8%, SD = 1.8%, n = 2). Nonbreeding pairs did, however, engage in courtship feeding, copulation, and territorial advertisement. Some pairs bred in consecutive years (with individuals identified by bands), and young from one breeding season did not remain dependent on their parents by the following season. Nonetheless, that some pairs did not breed each year suggests that for some females the cost of successfully breeding one year may preclude her attaining adequate body condition in time to breed the following year, as appears to be the case in Spotted Owls (R. P. Gerhardt, unpubl. data).

Productivity and Nest Success

Sixteen (76%) of 21 nests fledged a total of 27 young. Successful nests fledged a mean of 1.7 young (range = 1–3), and 1.3 young fledged per nesting attempt. This success rate is fairly high, given the small clutch size of this species. Including territories where pairs apparently did not attempt nesting (some of which may actually have been early failures) yields a fledging rate of 27 young in 32 territory-years, or 0.8 per territory-year. Viewed from this perspective, overall productivity appears rather low.

Of the five nests that failed, one (a late re-nesting) was abandoned during incubation, and the others were depredated, probably by mammals. Two of these depredations occurred during incubation and one when the chicks were about two weeks old. Another nest failed one week after the hatching of the first chick; the second egg had disappeared just before it was due to hatch, and the adult

female, captured at that time, had the quill of an arboreal Mexican Hairy Porcupine *(Sphiggurus mexicanus)* in her feathers. It is possible that this porcupine was involved in the partial or the eventual complete failure of this nest. Other mammals that likely prey from time to time on Mexican Wood Owl eggs and young are the Tayra *(Eira barbara)*, White-nosed Coati *(Nasua narica)*, and Margay *(Leopardus or Felis wiedii)*. One nestling death was attributed to botflies (see Parasites and Pests in this chapter).

Adult Mortality

Five territories were monitored in each of the three years of study. At those sites, one female was replaced between the 1990 and 1991 breeding seasons, and a male was replaced between the 1991 and 1992 breeding seasons. These data suggest a turnover rate, and a probable adult mortality rate, of approximately 10% per year.

Collared Forest Falcons at Tikal captured and ate an adult Mexican Wood Owl (see Chapter 17), and Mexican Wood Owls also appeared in the diet of White Hawks *(Leucopternis albicollis:* see Chapter 10, a juvenile owl), Crested Eagles *(Morphnus guianensis:* see Chapter 13), and Black Hawk-eagles *(Spizaetus tyrannus:* see Chapter 14, an owl of unknown age). Other diurnal raptors such as the Ornate Hawk-eagle *(S. ornatus)* also pose a threat. Indeed, these owls turned up in the diets of other raptors at Tikal more frequently than did any other raptor.

We observed an encounter between an adult Mexican Wood Owl and a Bicolored Hawk *(Accipiter bicolor)* in which the owl, by spreading its wings and making itself appear larger, apparently discouraged the hawk from attacking. This incident occurred near dawn while the owl was going to roost. Mexican Wood Owls were always alert at our approach, and it is probably rare that roosting adults are captured by mammals. Nonetheless, given the relatively small size of Mexican Wood Owls, they are likely more vulnerable to predation (mainly from raptors, boas, and cats) and therefore shorter lived than their larger relative, the Black-and-white Owl.

Parasites and Pests

The death of one young at a 1992 nest was attributed to botfly larvae, which appeared to close off the nasal passages, restricting the chick's respiration. This chick died two weeks after hatching.

CONSERVATION

Mexican Wood Owls can be found in a variety of natural and human-altered landscapes. They are seemingly quite adaptable in terms of habitats and prey exploited, and since suitable nesting cavities are apparently abundant, they are likely to remain common in much of their range. Given this apparent adaptability, Mexican Wood Owls are probably not a sensitive indicator of the health of tropical forest ecosystems. It would be interesting, however, to compare the owls' population density in Tikal with that in other habitats and locations, as it is difficult to imagine a higher density than we observed in the primary forest at Tikal.

HIGHLIGHTS

This research was not only the first such on Mexican Wood Owls but also likely the most in-depth study to date of any tropical forest owl. Many of our findings are, perhaps, not surprising. These owls are abundant and sedentary, and they maintain discrete territories throughout the year. Vocalizations are apparently an important means of communication. Their relatively low fecundity is in keeping with a well-documented trend among birds in general, in that tropical species have smaller clutches than ecologically similar or closely related temperate zone species (Moreau 1944; Lack 1966; Ricklefs 1969b). Even the inclusion in their diet of so many invertebrates is not surprising given the size and abundance of the insects in this ecosystem (Schoener 1971). The high degree of synchronicity of these owls' nesting was rather unexpected, however, and while the ultimate cause probably has to do with seasonal rainfall patterns, the proximate mechanisms involved in initiating breeding remain unclear.

Also unexpected was the degree of size dimorphism exhibited by Mexican Wood Owls. That these owls are highly dimorphic is at odds with important assumptions or predictions of numerous hypotheses regarding the evolution of reversed size dimorphism in owls (Gerhardt and Gerhardt 1997). More than 20 such hypotheses have been advanced (for a summary, see Andersson and Norberg 1981; Mueller and Meyer 1985; Mueller 1986). The data used to support these hypotheses have invariably come from studies of temperate zone, Northern Hemisphere owls, and several ignore the possibility that tropical species might be equally or more dimorphic (Gerhardt and Gerhardt 1997). The starvation hypothesis (Korpimäki 1986; Lundberg 1986) states that larger females are better able to withstand

harsh breeding season conditions, particularly at higher latitudes, and can incubate and brood longer during periods of poor or inconsistent prey deliveries by males. If the starvation hypothesis accurately explains the role of reversed size dimorphism in European owls, it clearly does not do so for these highly dimorphic Mexican Wood Owls. Walter (1979) suggested that nesting in cavities inhibits the degree of dimorphism. Our findings refute this idea, since Mexican Wood Owls always nested in cavities (Gerhardt et al. 1994b). Similarly, a prediction of the nest defense hypothesis (which holds that large females are better able to defend nests: Storer 1966; Reynolds 1972; Snyder and Wiley 1976; Andersson and Norberg 1981) is that since cavity nests experience lower rates of predation than open nests, cavity-nesting species should exhibit little dimorphism. This, too, is inconsistent with our findings. Several hypotheses share the idea that reversed size dimorphism evolved to allow members of a pair to capture different prey types or sizes. These hypotheses predict that highly insectivorous owls should show the least size dimorphism, and the strongest, albeit indirect, argument for these hypotheses is a supposed positive correlation between degree of dimorphism and percentage of vertebrate prey in the diet. This correlation is not supported with the inclusion of the highly insectivorous Mexican Wood Owl (Gerhardt et al. 1994a).

Hypotheses regarding the evolution of reversed size dimorphism would do well to consider tropical owls. Not only did we find both *Strix* species at Tikal to be quite dimorphic, but Wetmore (1968) reports measurements that suggest substantial dimorphism in *Pulsatrix* owls, and substantial dimorphism is also indicated by body masses reported for the African Wood Owl *(C. woodfordii)* and the tropical eagle owl species Spotted Eagle Owl *(Bubo africanus)*, Verreaux's Eagle Owl *(B. lacteus)*, Cape Eagle Owl *(B. capensis)*, and Fraser's Eagle Owl *(B. poensis)* (Kemp 1987; Fry et al. 1988). Many advances in our understanding of ecological and evolutionary phenomena have flowed from the study of tropical birds. With respect to our understanding of the evolution and ecology of owls, a major contribution from the tropics seems likely to continue.

22 BLACK-AND-WHITE OWL

*Richard P. Gerhardt, Dawn M. Gerhardt,
Normandy Bonilla, and Craig J. Flatten*

There is a special place not far from the center of Tikal. Extending east for more than a kilometer from the visitors' center is an old, disused runway, now rapidly being reclaimed by the forest. It leads out into a large bajo or lowland and is immediately bordered by low, woody second growth. Here one can see birdlife that is rare or absent in the tall, primary forest, including Groove-billed Ani *(Crotophaga sulcirostris)*, White-collared Seedeaters *(Sporophila torqueola)*, and Blue Buntings *(Cyanocompsa parellina)*—and perhaps hear the twanging *ching ching ching* of the Mangrove Vireo *(Vireo pallens)*, a true bajo specialist. Near the eastern end of this runway are two small aguadas or ponds, which during the dry season serve as watering holes for a wide variety of animal life. The larger pond is inhabited by a Morelet's Crocodile *(Crocodylus moreletti)*—two meters long the last time we visited—and a small island provides for nests of both Tropical Kingbirds *(Tyrannus melancholicus)* and Social Flycatchers *(Myiozetetes similis)*. An American Pygmy Kingfisher *(Chloroceryle aenea)* is nearly always in attendance at one pond or the other, and a comically colorful Gray-necked Wood-Rail *(Aramides cajanea)* often circumnavigates the shoreline of both ponds.

Dusk is the most exciting time to visit these ponds, and the patient observer may be rewarded with the sight of any of a number of forest dwellers. Some, like Ruddy Quail-Doves *(Geotrygon montana)* and Blue Ground-Doves *(Claravis pretiosa)*, appear nightly as if on cue, arriving in pairs to sneak a last drink before nightfall. Gray-breasted Martins *(Progne chalybea)* descend en masse to circle, taking turns hitting the surface for a drink and alternately snacking on flying insects, and then just as suddenly disappearing, to be replaced by the first of their nocturnal counterparts, the bats. At the same time, a brocket deer *(Mazama* spp.) might creep stealthily from the forest to the margin of the pond. A thirsty Nine-banded Armadillo *(Dasypus novemcinctus)*, by contrast, appears completely unconcerned about stealth or concealment. A small herd of Collared Peccaries *(Tayassu tajacu)* crashing through the forest edge provides an unforgettable experience—but the arrival of a Jaguar *(Panthera onca)* is what one really hopes to see here.

A nighttime visit to these ponds is quite rewarding as well. The openness of the runway provides a view of the starry skies from horizon to horizon, and under a full moon, the Temple of the Grand Jaguar can easily be seen more than a mile to the west. Pauraques *(Nyctidromus albicollis)* are much in evidence, by their red eyeshine in the beam of a flashlight, by their hauntingly beautiful call, and by their mothlike flight as they lift off from the edge of the runway at one's near approach. Early in the dry season, male tarantulas cross the runway in search of a mate, and the soft trill of Guatemalan Screech Owls *(Megascops guatemalae)* suggests that they, too, are about the business of reproducing. The bright glow flying through the bushes is not a lightning bug but rather an elaterid or click beetle; the glow comes from the two false eyespots on the back of this large insect's head. The area surrounding the ponds is frequented by numerous nocturnal insects, by toads larger than a man's fist, and by a fine assortment of bats. Occasionally, a Boat-billed Heron *(Cochlearius cochlearius)* might stop by to try its luck at fishing.

What makes this such a special place to us, however, is another visitor to these ponds—a male Black-and-white Owl *(Strix nigrolineata)*. Indeed, it was at these ponds that we first encountered this beautiful member of a rare and little-known species (Plates 22.1, 22.2). As we began to study this owl, we discovered that he was a regular here; radiotelemetry showed us that he almost never passed a night without visiting one or both ponds while hunting for bats and large beetles. This made sense, as his mate incubated their single egg not far from here on a nest high in the trees—to our knowledge, the first nest of this species ever described. This pair provided us with most of what we learned about this species, since other territories were few and far between and much less accessible. Our understanding of the fascinating ecology of Black-and-white Owls began with our first encounter at the ponds, and we recommend that you visit this special place if ever you find yourself at Tikal.

GEOGRAPHIC DISTRIBUTION AND SYSTEMATICS

Strix nigrolineata is one of five species traditionally placed in the genus *Ciccaba*. Four of these, including the Mexican Wood Owl (*S. squamulata*; see Chapter 21), are Neotropical. All are round-headed, dark-eyed, woodland species now recognized in the genus *Strix*. The Black-and-white Owl inhabits humid lowland and foothill forests from southern Mexico to the northwestern corner of South America, where it occurs in a crescent along the continental margin from Venezuela westward through Colombia and south through Ecuador to northwestern Peru (Grossman and Hamlet 1964; Burton 1973; Enriquez-Rocha et al. 1993; del Hoyo et al. 1999; König et al. 1999). Though itself monotypic, this species forms a superspecies (or may be conspecific) with the smaller Black-banded Owl *(Strix huhula)*, which occupies moist tropical forests throughout South America, except in the northwestern part of the continent, where the Black-and-white Owl occurs (del Hoyo et al. 1999). The Black-and-white Owl occurs from sea level up to 1200 m (Mexico), 2100 m (Panama), or 2400 m (Colombia: del Hoyo et al. 1999).

MORPHOLOGY AND PLUMAGE

A strikingly handsome bird, the Black-and-white Owl is not easily confused with any other. The nape of the rounded head, as well as the back and wings, are black. The facial disc is also black, but this is separated from the crown by eyebrows that are banded white and black. The collar, breast, and underparts have thin horizontal banding of white and black, though close inspection of individual feathers reveals a grayish cast to the whitish portions. The tail is of wide, dark brown bands appearing black, and six or seven narrow gray-white bands. The undersides of the wings are also barred. The beak, cere, and feet are bright yellow-orange and stand out against the plumage in which they are set. The iris is dark brown and the pupil black; when viewed at night by flashlight, the reflection of the eyes is red.

Reported body mass is 440–500 g (Burton 1973) but few actual weights have been published. Appendix 1 reports all published weight records we could find, along with weights of a mated pair at Tikal (Gerhardt and Gerhardt 1997). Combining all weight data yields a mean of 418 ± 16 g for males ($n = 3$), and 487 ± 40 g for females ($n = 4$; Appendix 1). Thus, female Black-and-white Owls weigh about the same amount as a female Common Barn Owl *(Tyto alba)*, but males weigh substantially less than a male Barn Owl. Based on these mean body weights, Black-and-white Owls show a Dimorphism Index (DI) value of 5.1.

The Tikal female and male that composed our focal study pair yielded DI values of 2.4 for wing chord, 12.5 for tail length, and 10.4 for culmen length—for a mean of 8.4 among the three measurements (Appendix 2). Body weight dimorphism for this pair was 7.0. A significant degree of size dimorphism was also visible in the field; the female looked much larger than the male when the two roosted side by side (Gerhardt and Gerhardt 1997). In other pairs that we observed but did not capture, a similar size difference was quite visible.

Appendix 2 gives the largest published series of linear measurements for the Black-and-white Owl—12 specimens reported by Ridgway (1914) and 15 by Wetmore (1968). Ridgway's (1914) data appear inconsistent, with males larger than females for some measurements, and females larger than males for others; these data may be based partly on mis-sexed specimens, or may preclude a true picture of size dimorphism because males and females were in part collected from different localities. Wetmore's data (Appendix 2) appear more trustworthy, and give DI values similar to ours. Averaging Wetmore's DI values with ours, the mean DI values are 3.8 for wing chord, 11.8 for tail length, and 9.1 for culmen length, for a mean of 8.2 among the three measurements.

We measured the external ear openings of both birds captured, since the degree of ear asymmetry has been used in the taxonomy of owls (Peters 1938; Kelso 1940). The male's external ear openings measured 18.9 and 13.6 mm (right and left) in length; the female's were 21.8 and 14.5 mm. No special structures surrounded the ear opening, save perhaps a rudimentary dermal flap.

PREVIOUS RESEARCH

The literature pertaining to this species consists mostly of brief anecdotal accounts (Marshall 1943; Land 1963; Smithe 1966; McCarthy 1976). Food habits have been analyzed based on pellets collected under a single roost in Venezuela (Ibañez et al. 1992) and based on stomach contents of collected individuals (Marshall 1943; Tashian 1952). Systematic calling surveys have been conducted in Costa Rica and Mexico (Enriquez-Rocha and Rangel-Salazar 1996; R. Gerhardt, pers. comm.). Prior to our research, the nests, eggs, and young of this species were undescribed.

RESEARCH IN THE MAYA FOREST

We studied four nestings in two territories within Tikal National Park—one in 1989, two in 1990, and one in 1991. Three nests were in the same ("Airstrip") territory each of the three years, attended by the same banded male (all three years) and female (at least the latter two years; the female wasn't captured until 1990). The 1991 nest was 300 m from the 1989 nest and 450 m from the 1990 nest. A second 1990 nest was in another territory, approximately 3.5 km from the Airstrip nests. There were apparently no active territories between these two. The four nests were 175–200 m above sea level.

After finding the first nest in 1989, we captured the male, using a *bal-chatri* trap with a live mouse as bait. We affixed a tail-mounted, 3.5 g radio transmitter (Holohil Systems, Carp, Ontario, Canada) to the two central rectrices using thread and epoxy. We ascertained this owl's location approximately three days and three nights per week for four months. We collected habitat data at diurnal roosts—discovered with the aid of radiotelemetry—and used all locations to estimate the male's home range area (Gerhardt et al. 1994b). In 1990 we recaptured this male with a good deal more difficulty, finally resorting to placing a noose-covered wire on a favorite diurnal roost limb. We followed him for approximately three months in 1990. We captured the female of this same pair by placing a noose carpet over her nest at a time by which we were certain that the egg was no longer viable.

DIET AND HUNTING BEHAVIOR

Over the course of two years, we collected pellet material—regurgitated, indigestible prey parts—from beneath the roosts of a pair of radio-tagged male (1989 and 1990) and female (1990) Black-and-white Owls (Gerhardt et al. 1994a). We rarely climbed to nests, and only a single prey remain was found in a nest. Nor did we find pellets in or below nests; the female apparently cast her pellets away from the nest and carried off pellets cast by the young. As the male and female often roosted side by side, we could not attribute pellets to one or the other with certainty. Thus, our results are not sex specific, although the male probably cast most of the pellets we collected.

We collected material from 73 Black-and-white Owl pellets (Gerhardt et al. 1994a). Given the intensive search we conducted beneath known roosts, this sample size seems quite small. Researchers in temperate and boreal areas, for example, can reasonably expect to obtain a similar sample size in just a few visits to a nest or roost site. Indeed, the fate of a pellet cast by an owl can be used to highlight the very nature of a tropical forest for those who haven't had the opportunity to visit one.

Tropical forests such as those of Tikal are incredibly diverse in their vegetative composition, both structurally and in species. Areas used for roosting by Black-and-white Owls were typically multilayered, with a canopy of the tallest trees including Black Olive *(Bucida buceras)*, Honduras Mahogany *(Swietenia macrophylla)*, Ramón or Bread-nut Tree *(Brosimum alicastrum)*, and others; the sub-canopy often included palms *(Cryosophila stauracantha* and *Sabal mauritiiformis)* and a variety of other trees. Orchids, bromeliads, and other epiphytes as well as vines abound in the canopy and sub-canopy and in many emergent trees. Beneath the sub-canopy occurs a layer of saplings and low-growing palms, and the ground itself is generally covered by a thick layer of leaf litter. Although Black-and-white Owls often roosted fairly low, cast pellets rarely reached the ground intact. Those that did, and parts of those that did not, were almost immediately set upon by ants, termites, and redu-

viid bugs, which tore apart and carried off the prey parts that were indigestible to the owls. This highlights the dynamic ecology of the tropical forest and the rapid and efficient nutrient cycling and energy flow inherent in this ecosystem.

All 73 pellets contained remains of insects and 19 (26%) consisted only of insect matter (Gerhardt et al. 1994a). Fifty-three (73%) of the pellets contained fur or bones of bats. In two pellets we found rodent parts: the mandibles of a small rat, probably Fulvous Rice Rat *(Oryzomys fulvescens)*, and a single rodent incisor. Five (7%) of the pellets contained feathers of unidentified birds. Also present in 10 pellets (14%) were seeds of *Psidium* species, which we believe to have come from bat stomachs.

Only when a portion of the skull was present were we able to identify bats to species. Of 21 skulls identified, 13 were *Artibeus* species (including several *A. jamaicensis*, a fruit bat common in Tikal), 5 were Black Mastiff Bats *(Molossus ater)*, 2 were Wrinkle-faced Bats *(Centurio senex)*, and 1 was a Peter's Tent-making Bat *(Uroderma bilobatum)* (Gerhardt et al. 1994a). The one prey remain taken from a nest was from a fruit bat.

All insects identified from Black-and-white Owl pellets were either beetles or orthopterans (Gerhardt et al. 1994a). The most commonly identified insects were scarab beetles, which were represented by elytra, legs, pronota, and head capsules. Several scarabaeid subfamilies were found: Scarabaeinae (dung beetles), Geotrupinae (earth-boring dung beetles), Melolonthinae (June beetles), and Dynastinae (unicorn beetles). Snout beetles (Curculionidae) were identified by head capsules and elytra. Long-horned beetles (Cerambycidae) were found less commonly, as elytra parts and heads, and hydrophilid beetles were identified by the sharp spine of the metasternum. Short-horned grasshoppers (Orthoptera; Acrididae) were found in some pellets with femurs and tibia remaining intact, and cockroaches (Orthoptera; Blattidae) were represented by pronota fragments (Gerhardt et al. 1994a).

It is clear that this pair of Black-and-white Owls ate primarily insects and bats during our breeding-season study; we cannot address the owls' relative use of these prey types during the remainder of the year. In numbers, insects were certainly captured much more frequently than bats. Less clear is the relative importance of these two major taxa when considering biomass. We believe that during the breeding season, insects constituted more of the prey biomass than did bats. All pellets contained insect exoskeletal material and 26% contained only insect remains. Moreover, only the hardest, chitinous portions of insects were found, and many soft-bodied insects were likely eaten and digested entirely (i.e., not represented in pellets). Neither katydids (Tettigoniidae) nor cicadas (Cicadae) were found in pellets, although both were abundant in Tikal and had been previously found in stomachs of Black-and-white Owls (Marshall 1943; Tashian 1952)—this absence may have resulted from digestibility biases. In addition, vertebrate bones likely remain below roosts longer than insect parts; in our experience, the latter could not be found after 24 hours. Pellet analysis tends to underestimate insect numbers and overestimate vertebrate numbers (Marti 1987).

That Black-and-white Owls are largely insectivorous is at first surprising, given the size of these owls. It is important to keep in mind, however, that the insects taken were of greater biomass than most of those found in more temperate regions. A large scarab or cockroach is a substantial prey item, and such insects are numerous in Tikal during the Black-and-white Owl nesting season. For example, the larvae of one of the largest beetles at Tikal—*Enema endemion* (Scarabaeidae)—averaged 11.7 ± 1.4 g (n = 13) in weight (D. Whitacre, unpubl. data)—more than many small vertebrates. The greater availability of large insects in tropical than in temperate zone forests has been noted by other researchers (Schoener 1971).

Our findings regarding the diet of the Black-and-white Owl agree with the few previously published reports. In El Salvador, Marshall (1943) collected a male whose stomach contained grasshoppers and two bats, and he collected a female that had been hunting grasshoppers. A pair collected in Chiapas, Mexico, had eaten mainly insects—beetles, tettigoniids (katydids and long-horned grasshoppers), and cicadas—with the remains of a Lesser Naked-backed Bat *(Pteronotus davyi)* in one stomach (Tashian 1952). In Venezuela, Ibañez et al. (1992) collected pellets from beneath the roosts of a pair of Black-and-white Owls during a five-week period. Those pellets contained the same general prey taxa as those we found at Tikal, but bats and birds represented a much larger proportion of prey numbers. Likewise, pellet material from a pair roosting near a forest lake in Quintana Roo, Mexico, was mainly composed of bat skulls and jaws (D. Whitacre and G. Falxa, pers. comm.). Both in Venezuela and Mexico, it is unclear how much time had elapsed between the casting and the finding of the pellets. We believe that only by finding pellets shortly after casting were we able to obtain much of the insect portion of pellets. This difference in sampling may explain the difference in the relative proportions of bats and insects between our study and the observations from Mexico and Venezuela.

Diets of the Black-and-white Owl and the sympatric, congeneric Mexican Wood Owl showed both overlap and divergence (Gerhardt et al. 1994a). Both fed on large insects, including scarabs, short-horned grasshoppers, snout beetles, long-horned beetles, and cockroaches. Black-and-white Owls also captured hydrophilid beetles, and Mexican Wood Owls ate katydids. There was little overlap in the vertebrate component of the diets. Only one Mexican Wood Owl pellet (of 52) contained bat remains; lizard remains were found in two pellets, and rodent parts were found in 50% of pellets (Gerhardt et al. 1994a). Both owl species ate large insects, but Black-and-white Owls ate many bats, whereas Mexican Wood Owls ate numerous small rodents.

Black-and-white Owls hunted exclusively at night. We watched the radio-tagged male hunting on numerous occasions, particularly in the semi-open area near one of three ponds within his home range. He generally perched 2–5 m above ground on a post or branch near the water's edge. From this perch he flew short distances chasing beetles and bats in flight. On a single occasion, this male was observed flying after a bat through the forest. This chase was somewhat longer than other observed prey capture attempts, and the bat eluded its pursuer (Gerhardt et al. 1994a). Observations by other researchers suggest similar behavior, with these owls often hunting from forest edge, flying from a perch to snatch prey from the ground or vegetation and seizing some prey in midair (Stiles et al. 1989). Sometimes, too, these owls are said to be attracted to insects swarming around lights at night (ibid.).

We believe it likely that these owls often caught bats congregated to feed at fruiting trees, although we made no direct observations of this. When a fig tree near our camp was in fruit, the fluttering of bats and the rain of figs through the stiff leaves was sometimes nearly loud enough to preclude sleep—certainly it would have been obvious to any bat-eating owl in the vicinity that a concentration of fruit bats was to be found here.

Just west of Tikal, at a dramatic site known as "Zotz," a pair of Black-and-white Owls and a pair of Common Barn Owls captured bats in flight each dusk as thousands left a communal day roost in crevices of a cliff. During the exodus, both owl species apparently cached bats for later consumption, leaving their hunting site briefly after each capture and returning almost immediately to hunt again (A. Baker, pers. comm.).

HABITAT USE

Published accounts describe Black-and-white Owls occurring in a range of forest types, from wet, evergreen to semi-humid, semi-deciduous tropical forest—as well as along forest edge and in gallery forest and plantations (Howell and Webb 1995). A few accounts stress a proclivity for inhabiting swamp forests, mangroves, and other situations in proximity to water (Dickey and van Rossem 1938; Land 1970; del Hoyo et al. 1999; König et al. 1999), but it is not clear whether such a connection may occur mainly in drier regions. In Oaxaca, Mexico, for example, Binford (1989) describes these owls as occurring in tropical evergreen forest on the wetter, Atlantic slope and in swamp forest on the drier, Pacific slope. Similarly, in drier areas of Costa Rica, these owls are said to prefer evergreen or gallery forest (Stiles et al. 1989).

At Tikal, Black-and-white Owls showed a clear affinity for the low-lying bajos or Scrub Swamp Forest areas of the park. All three nest sites in our main study territory (the Airstrip territory) were near the edge of a large bajo or wooded swamp. Similarly, the single nest in a second territory was in Transitional Forest near a large bajo, about 3.5 km from the Airstrip nests.

Ponds found in these bajos clearly provided important foraging habitat for the owls. Not only were night-time telemetry locations and observations of the focal male frequently close to ponds, but three other pairs of Black-and-white Owls we encountered were all near ponds. Proximity of ponds may be a key factor in nest site selection by these owls, though other aspects of the bajo and the emergent trees therein likely also factor into this choice.

Highlighting this selection of Bajo Forests as nesting and foraging sites is the fact that roost sites were predominantly in Upland Forest (Flatten and Gerhardt 1989). Although several well-used roosts were located in Bajo Forest very near the nest trees, all other roosts were within or at the edge of Upland Forest, averaging 374 m from the nest tree (*n* = 58). Having higher, denser canopies than do the Bajo or Scrub Swamp Forests, Upland Forests afford greater shading and therefore lower daytime temperatures—possibly a factor in selection of roost sites.

We relocated the radio-tagged male 98 times at 37 different diurnal roosts. Twenty roost trees were identified and included 10 species, of which Zapotillo (*Pouteria* spp.), Cedrillo (*Guarea* sp.), and Ramón were most commonly used (Gerhardt et al. 1994b). Many of these perch sites were characterized by overhanging vines. Roost heights averaged 14.0 m (*n* = 98) and ranged from 3.5 to 26.0 m. Mexican Wood Owls radio-tagged during the same period roosted lower than did Black-and-white Owls; roost heights of six male Mexican Wood Owls averaged 5.3 m (*n* = 274 roost locations) and ranged from 0.5 to 18.0 m (Gerhardt et al. 1994b). On extremely hot days, Mexican Wood Owls roosted very low, while Black-and-white Owls, using higher roosts, appeared to seek thermal protection provided by overhanging vines.

BREEDING BIOLOGY AND BEHAVIOR

Nests and Nest Sites

The four nests we found were in three tree species: two were in Black Olive, one in a Honduras Mahogany, and one in a Ramón. These trees were among the largest in the immediate vicinity, with a mean height of 26.3 ± 7.1 m (range = 21–30 m) and diameter at breast height of 87.3 ± 32.6 cm. Mean nest height was 20.5 ± 5.8 m (*n* = 4; Gerhardt et al. 1994b). Nest trees were lone or emergent rather than part of a continuous canopy. This probably afforded owl nests some protection from climbing predators—mammals and reptiles. Indeed, the situation at each of the four nests seemed to make them more vulnerable to flying than to climbing predators. Eggs were laid on bare epiphytes (Gerhardt et al. 1994b); there was no nest construction by the owls (Plate 22.3). Two nests were formed by the orchid *Trigonidium egertonianum*, one by the orchid *Mormolyca ringens*, and one by a bromeliad of the genus *Tillandsia*.

Egg and Clutch Size

Each nest contained a single egg, which was off-white, non-glossy, and elliptical and was laid on the roots or foliage at the center of the epiphyte. About one week after laying, eggs averaged 33.8 ± 2.3 g in mass and had a mean length and width of 46.4 ± 1.1 and 38.4 ± 1.1 mm, respectively (*n* = 4; Gerhardt et al. 1994b). Each egg (and clutch) represented approximately 6.3% of this female's 536 g body mass. Hoyt's (1979) formula, using $K_w = 0.55$ (derived from data for *Ciccaba virgata* [*Strix squamulata*] from Schönwetter 1964), gives a predicted fresh egg mass of 37.6 g, or approximately 7.7% of the mean 487 g female body mass. Several general accounts describe the clutch size as two (König et al. 1999) or one or two eggs (Burton 1973; Hume and Boyer 1991; del Hoyo et al. 1999) and state that nesting takes place in tree cavities or old stick nests—neither of which generalizations accord well with our observations. However, we know of no actual data, other than those published here, on the eggs or nest of this owl, and we surmise that these general accounts are based mainly on extrapolation and supposition.

To date, no other owl species has been reported to have a usual clutch size of one. Among the many species of well-studied temperate zone and subtropical owls of North America and Europe, the smallest average clutch sizes of which we are aware are 2.26 for Great Horned Owl (*Bubo virginianus*: Johnson 1978), 2.41 for Barred Owls (*Strix varia*: Murray 1976), 2.5 for Spotted Owls (*S. occidentalis*: Johnsgard 1988), and 2.6 for Eurasian Eagle Owl (*B. bubo*: Mikkola 1983). The 1-egg clutch size reported here comes from a sample of only four nests, three of them belonging to the same pair. It is known, however, that tropical raptors tend to lay smaller clutches than do closely related temperate zone raptors of similar size and ecology (Moreau 1944), and the same is true also of most other avian groups (Lack 1966; Ricklefs 1969b). It is not altogether surprising, therefore, that a moderately large tropical owl such as the Black-and-white Owl may have a modal clutch of one egg.

Nesting Phenology

Nests were apparently initiated in late March, as all hatching occurred in the last week of April, during the mid to late portion of the dry season. The nests were quite synchronous in this regard (Gerhardt et al. 1994b). It may be important that nesting be completed prior to the arrival of the first hard rains, as all nests were rather exposed above.

Length of Incubation, Nestling, and Dependency Intervals

We estimate that incubation of the single egg required 30–35 days. Since none of the four nests was successful, we can say little about the duration of nestling and dependency periods. We did, however, examine and measure the young in the 1989 nest weekly from about two days after hatching until nest failure in the fourth week. The growth rate of this owlet appeared commensurate with that of other owl species of similar size. We expect, therefore, that fledging would occur in this species at about one month after hatching, give or take a few days.

VOCALIZATIONS

Burton (1973) gave phonetic descriptions for two vocalizations of Black-and-white Owls and described these as (1) a series of 5–11 hoots, each hoot protracted and with an upward inflection; and (2) six hoots rising in pitch with accent on the last two. Besides beak-clacking performed by the female as we neared her nest, all vocalizations we heard from adults were hoots. If these owls are capable of producing other sounds, as are many strigid owls, their use of them appears limited. The primary call we heard was similar to the first vocalization just described. It consisted of 8–12 hoots, of which all but the last had an upward inflection. In addition, the whole series increased in pitch to the penultimate hoot, with the last note returning to a lower pitch. Volume increased through the series, so that the last two notes were the loudest; from a distance, these two hoots were frequently the only ones we heard. Females achieved a higher pitch than did males, and individual pair members could thus be sexed by voice (Flatten and Gerhardt 1989). Both members of the pair were also heard using softer calls of one or two hoots, particularly when close to the nest. The male typically hooted just after nightfall before leaving his diurnal roost and often flew from there to the area of the nest tree (if his roost was nearby). He also called softly upon going to roost just before dawn, seemingly as a way of keeping in contact with his mate. These calling behaviors greatly facilitated our finding nests.

BEHAVIOR OF ADULTS

The female did all incubating and brooding, and the male did most or all of the hunting during our observations (Gerhardt et al. 1994a). When not incubating or brooding, the female often roosted during the day side by side with her mate. During incubation and early brooding, male Black-and-white Owls tended to roost near the nest and to fly to the nest tree as soon as they became active after dusk. Black-and-white Owls did not add any foliage or other nest material to nests.

BEHAVIOR AND DEVELOPMENT OF NESTLINGS

The following descriptions are based on a single young at the 1989 nest (Plate 22.4). We climbed to this nest when the chick was estimated to be 2, 14, and 24 days old.

It appeared to be developing normally but disappeared shortly after the third visit.

On day 2, the chick, weighing 28.0 g, was covered with white down, and its eyes remained closed. Beak and cere, feet and talons were all pink. The youngster gave high-pitched, rather soft peeping sounds. At 14 days the eyes were open, with dark brown iris surrounding a dark blue pupil. The nestling was largely covered in white down, but other body feathers—white with black bars—were also apparent. Flight feathers and rectrices had not yet begun to appear. There was little or no change in the color of the beak and cere, feet and talons. The chick weighed 49.5 g.

Twenty-four days after hatching, the young had attained a weight of 227.0 g. Eye color was unchanged, but the beak and cere were now light orange, the feet light yellow, and the talons dark gray-green. Downy body feathers were light gray, and contour feathers were light gray with black barring. Black wing feathers were protruding, the longest 1.9 cm. Tail feathers were not yet apparent (Flatten and Gerhardt 1989). The young produced "beak-clacking" sounds at both 14 and 24 days.

SPACING MECHANISMS AND POPULATION DENSITY

Apart from the fact that these owls were quite vocal, we have no information on territorial behavior or nest defense.

Constancy of Territory Occupancy, Use of Alternate Nest Sites

Though we studied mainly a single pair, it appears that Black-and-white Owls are both monogamous and sedentary. That our focal pair roosted together after the breeding season suggests that the pair bond is maintained throughout the year. We observed little change in the home range and areas used by the male from one year to the next, and nest sites were within 300 m of one another during all three years of study.

Home Range Estimates

We radio-tagged and followed the Airstrip male from 13 April to 1 August 1989 (118 locations including 56 diurnal roost locations) and from 12 May to 7 July 1990 (90 locations, 42 diurnal roosts: Gerhardt et al. 1994b). Combining these data, we estimated his home range as 437.3 ha using the 85% Harmonic Mean (HM) method (Dixon and Chapman 1980), and as 261.6 ha using the 100% Minimum Convex Polygon (MCP) method (Mohr 1947). A 50% HM estimate yielded an area of high utilization of 78.2 ha. This area included the nest site, several well-used day roosts, and two ponds used as foraging areas (Gerhardt et al. 1994b).

We also estimated this male's home range using only the subset of locations obtained following nest failure in both years (123 locations, 62 of them at diurnal roosts). These data yielded home range estimates of 175.9 ha (85% HM) and 116.4 ha (100% MCP). This male ranged over a much larger area while his mate was on an active nest than he did following nest failure (Gerhardt et al. 1994b).

Inter-nest Spacing and Density of Territorial Pairs

Black-and-white Owls apparently use large home ranges, and three pairs had nearest-neighbor distances of about 3.5 km, implying a density of one pair per 12 km² or so, at least within this localized area. However, other pairs were not encountered in every direction from our focal pair, and such a high local density does not hold true over larger areas of Tikal's forest. This species is apparently absent from vast stretches of Tikal; this likely due to its reliance on specific habitat features—especially bodies of water—that are themselves rare in these forests. The Black-and-white Owl is generally considered rare (at best, uncommon and local) throughout its range, especially northward from Panama (Peterson and Chalif 1973; Hilty and Brown 1986; Ridgely and Gwynne 1989; Stiles et al. 1989; Howell and Webb 1995; Ridgely and Greenfield 2001; Hilty et al. 2003).

DEMOGRAPHY

Frequency of Nesting, Percentage of Pairs Nesting

Our sample does not permit any conclusion as to the percentage of territorial pairs that attempt to nest in a given year. However, it seems likely that a particular pair may often nest in consecutive years. The Airstrip pair laid an egg during each of three consecutive years. Had any of these nestlings survived to fledging, it is possible the adults would have delayed a subsequent nesting effort while caring for the fledgling—however, there is no known precedent of such extended post-fledging parental care in owls.

Productivity and Nest Success

None of our four study nests resulted in fledged young. At the 1989 nest, the young disappeared about 24 days after hatching. In 1990, the egg in this pair's nest failed to hatch, and in 1991, the same pair's young disappeared within four days after hatching. At the second 1990 nest (in a different territory), the nestling disappeared within a week after hatching (Gerhardt et al. 1994b). It is possible that this high failure rate resulted in part from our research activities. In general, however, tropical birds are believed to have lower fecundity and reproductive success than their counterparts in temperate zones (Ricklefs 1969a; Skutch 1985).

Causes of Nesting Failure and Mortality

Three out of four nest failures occurred when nestlings disappeared, apparently as a result of predation. Potential predators observed near Black-and-white Owl nests included Tayras (*Eira barbara*), White-nosed Coatis (*Nasua narica*), Ocelots (*Felis pardalis*), Ornate Hawk-eagles (*Spizaetus ornatus*), Collared Forest Falcons (*Micrastur semitorquatus*), and Great Black Hawks (*Buteogallus urubitinga*; Flatten and Gerhardt 1989). It is unlikely that adult Black-and-white Owls defend their nests at the risk of their own lives. The few times we climbed to nests, females showed very little aggression toward us.

We were not able to gather concrete data on adult survivorship. However, it seems likely that Black-and-white Owls are relatively long-lived. This assumption is consistent with the small clutch size and low reproductive effort and success we observed. The substantial size of this owl presumably must limit the number of potential predators of adults, compared to smaller owl species. Moreover, Black-and-white Owls in day roosts were generally sheltered by overhead vegetation from predation from above—for example, by other raptors—and were always alert and watchful when encountered.

CONSERVATION

By all accounts, Black-and-white Owls have historically been rare throughout their range. Our research at Tikal confirmed that the population supported here is low. Moreover, we suggest that a predilection for water holes and other spots where bats might congregate explains, at least in part, this patchy, low-density distribution.

It is likely that this same predilection may contribute to a decline in this species. Human habitation and attendant deforestation in Central America is often centered on water holes, especially in drier or limestone regions such as the Yucatán Peninsula, where surface water is scarce. Indeed, in the areas immediately outside Tikal National Park, availability of dependable water sources has been, in recent history, a main factor limiting human occupation (see Chapter 2). As more remote water holes are exploited by humans and the surrounding forests cleared for agriculture, Black-and-white Owls stand to lose favored habitat.

Since so little was known of this owl prior to our studies at Tikal, there is no way of knowing historical population numbers or trends. Based on what little we do know, Black-and-white Owls are likely quite sensitive to forest alteration and thus are a species worth monitoring for conservation purposes.

HIGHLIGHTS

Of particular interest is the difference in home range size of the two sympatric, congeneric owls we studied at Tikal. The male Black-and-white Owl had a home range more than 20 times larger than the mean home range size of six male Mexican Wood Owls at Tikal (Gerhardt et al. 1994b). The difference in body size of these two species likely accounts for some of this difference. We suggest, however, that the difference in food habits also plays a role in explaining the great disparity in home range size (Gerhardt et al. 1994b). Whereas both owls ate large insects, particularly scarabs (Gerhardt et al. 1994a), the vertebrate components of their diets were quite different, with Mexican Wood Owls eating many rodents and Black-and-white Owls taking many bats (Ibañez et al. 1992; Gerhardt et al. 1994a). Following nest failure, we observed not only a shrinking of the male Black-and-white Owl's home range size but also a concurrent decrease in the proportion of bats found in pellets (Flatten and Gerhardt 1989). We speculate that a large home range in the Black-and-white Owl may be necessary to incorporate a sufficient number of certain fruiting trees or other areas of bat concentration.

Other surprising findings of our research were the 1-egg clutch and the substrate used for nesting. Although it has long been known that clutch sizes of birds are smaller near the equator (Moreau 1944; Lack 1966; Ricklefs 1969b), and although many diurnal tropical raptors lay 1-egg clutches (see Chapter 23), to date the Black-and-white Owl is the only owl known to exhibit a modal clutch size of one.

Larger temperate zone owls tend to nest on platform-like substrates, but these are most often stick nests constructed by falconiforms, corvids, or squirrels—or else rock ledges or suitable tree forks; only rarely are they epiphytic growths. Similarly, few larger owls nest in cavities, although Spotted Owls and Barred Owls do at times. Given the plethora of orchids and bromeliads relative to other platform types and to large cavities in the forest at Tikal, it makes sense that epiphytes would be utilized for nesting by an owl of this size. We speculate that epiphytes may prove to be of general importance as nesting substrates for this and other large tropical forest owls.

The high degree of sexual size dimorphism we observed in mated pairs of Black-and-white Owls is interesting for two reasons. First, there was no indication from the literature prior to our research that this species exhibited much size dimorphism. Second, many of the hypotheses regarding the evolution of reversed size dimorphism in owls involve assumptions that tropical owls—and highly insectivorous owls—show little dimorphism (for more on this, see Highlights in Chapter 21, and Gerhardt and Gerhardt 1997). Our findings fly in the face of these assumptions.

One last observation is worth noting. We had the opportunity, eight years after completing our studies, to return briefly to Tikal. The first chance we got, we returned after dark to the pond where we had first seen the male who was to become the focus of much of our research. We were blessed by the sight of a Black-and-white Owl, whether the same or another, sitting in the exact same spot known to us from so many years before. The owl in question was not observed closely enough to determine whether it was banded. It is unclear, therefore, whether this observation supports the notions of longevity and site tenacity. If not those, however, it clearly suggests a consistency of the sites used by this species. Indeed, the rare Black-and-white Owl has this in common with other raptors and wildlife of the tropical forest, with the forest itself, and with the Maya civilization that once lived here—it is accessible up to a point. We can see, admire, study, and learn about these things, but we only seem to scratch the surface of the wonders and mysteries they represent.

23 ECOLOGY AND CONSERVATION OF TIKAL'S RAPTOR FAUNA

David F. Whitacre and William A. Burnham

In preceding chapters, we report on the biology of 20 raptor species at a single tropical forest site. Here we search for lessons that can be learned by considering these 20 species together and by making comparisons with raptors in other environments. We divide the chapter into four sections: community ecology, life history patterns, comparative breeding biology, and spatial use and conservation needs of Neotropical lowland forest raptors. Wherever body mass is used in analyses, it was log-transformed (common logarithms).

COMMUNITY ECOLOGY OF TIKAL'S RAPTORS

The raptor species we studied at Tikal occur together over an immense area, from southern Mexico to northern Argentina (see Plate 1.1). While additional species are added closer to the equator, those occurring at Tikal may be considered a "core" set of the most widely coexisting Neotropical forest raptors. Is this a random set of species thrown together by accidents of history, or is there a detectable pattern, perhaps engendered by ecological processes? Is there evidence that competition has filtered this group's membership or shaped the evolution of species's traits? How does Tikal's raptor assemblage compare to those of forests elsewhere? Here we consider such long-standing questions of community ecology.

The degree to which raptors of a given assemblage differ in ecology, and the role of interspecific competition in producing any such differences, remain controversial. Based on a massive data set for raptors in Germany, Lack (1946) concluded that niche separation based on diet and habitat was essentially total—a result, he believed, of past competition. Several subsequent studies of raptor assemblages have also detected evidence of prey partitioning (Herrera and Hiraldo 1976; Bierregaard 1978; Schoener 1984; Bosakowski and Smith 1992; Marti et al. 1993), which has generally been interpreted as evidence of past or present competition among species. In contrast, other studies—especially of raptors eating mainly small mammals at temperate to subarctic latitudes—have found high dietary overlaps and little evidence of assemblage structure attributable to interspecific competition (e.g., Jaksic and Braker 1983). Here we inform this debate by adding the first data for a tropical forest raptor assemblage.

Latitudinal Diversity Gradients in New World Forest Raptors

Among the most striking patterns in ecology is an increase in species richness from high latitudes toward the equator, seen in many groups of organisms (Ricklefs and Schluter 1993). Here we address this trend for New World raptors, focusing on "alpha" diversity—species richness at single points in space.

Most forested localities in temperate North America have no more than a dozen raptor species,[1] of which three to seven typically are owls (Tables 23.1, 23.2; Johnsgard 2002). In contrast, most humid lowland Neotropical forest sites studied to date have 30–34 raptor species: 23–27 (up to 31) falconiforms,[2] plus six to seven owls (Tables 23.2, 23.3, 23.4).

At the most diverse sites, species richness in Neotropical forest falconiforms increases about 50% from the outer tropics to the equator, as noted by Thiollay (1985). However, the number of reasonably common raptor species normally coexisting at any one site varies less markedly from northern Central America to the equator, the most common number being 21–27 falconiforms and six owls. In much of Amazonia, the typical difference between the forest raptor assemblage and that of Tikal is the presence of one or two additional species of *Accipiter*, an additional *Leucopternis* species, one or two additional *Micrastur* species, two caracaras (*Daptrius* and *Ibycter*), sometimes an additional screech owl (*Megascops* sp.), and the Pearl Kite (*Gampsonyx swainsonii*) if it is considered a forest species (Tables 23.3, 23.4).

Neotropical lowland sites studied to date reveal a striking regularity in forest owl species present (Table 23.2). Almost without exception, each site has one screech owl,[3] the Crested Owl (*Lophostrix cristata*), a species of *Pulsatrix*, a member of the pygmy owl (*Glaucidium*

Table 23.1 Number of genera and species in three North American raptor assemblages[a]

Genera	Southeast states[b]		Northeast states		West, local assemblage	
	Forest species	Including open-country species	Forest species	Including open-country species	Forest species	Including open-country species
Accipitridae						
Buteo	3	3	2 or 3	2 or 3	1	3
Accipiter	2	2	3	3	3	3
Elanoides	1	1	—	—	—	—
Ictinia	1	1	—	—	—	—
Aquila	—	—	—	—	—	1
Circus	—	—	—	1	—	1
Falconidae						
Falco	1	1	1–3	1–3	1–3	1–4
Owls						
Tyto	—	1	—	1	—	1
Bubo	1	1	1	1	1	1
Strix	1	1	1	1	1 or 2	1 or 2
Asio	—	—	1	1	1	2
Megascops	1	1	1	1	1 or 2	1 or 2
Aegolius	—	—	1	1	1	1 or 2
Glaucidium	—	—	—	—	1	1
Speotyto	—	—	—	—	—	1
Totals	**11**	**12**	**11–14**	**13–16**	**11–15**	**18–24**

[a] Based on U.S. Geological Survey (USGS) bird lists for national parks, national wildlife refuges, counties, and Bureau of Land Management (BLM) and National Forest districts; the data in table ignore species barely ranging north across the Mexican border; the Bald Eagle and Osprey are omitted throughout; tabulations tend to indicate the maximum number of species occurring sympatrically—it may be uncommon for this many species to actually occur syntopically.

[b] Could include two additional open-country species: Short-tailed Hawk and Crested Caracara, but only in south Florida.

Table 23.2 Species richness of resident, forest-dwelling raptors at five Neotropical sites and in four regions of the United States[a]

Site	Latitude	Total number of forest raptor species	Total minus rare/vagrants	Species in common with Tikal (number)[b]	Species present that are not at Tikal (number)
Western U.S. (local)	40–45 °N	12 (6)	12 (6)	0	All
Western U.S. (regional)	40–45 °N	13–16 (8–9)	13–16 (8–9)	0	All
Northeastern U.S.	40–45 °N	11–14 (5)	11–13 (5)	0	All
Southeastern U.S.	30–35 °N	10–11 (3)	10–11 (3)	0	All
Tikal, Guatemala	17 °N	28 (6)[c]	28 (6)[c]	—	—
La Selva, Costa Rica	10 °N	33 (6)	31 (6)	26 (6)	4–6
Barro Colorado, Panama	9 °N	31 (7)	30 (6)	24 (6)	7
French Guiana	3 °N	37 (6)	37 (6)	29 (6)	7
Manaus, Brazil	3 °S	26 (5)	26 (5)	19 (5)	6
Manu, Peru	12 °S	35 (7)	32 (7)	26 (7)	7

[a] Number of owl species in parentheses (included in totals).

[b] Where a congener replaces another, these are counted as being the same species in both places.

[c] Would be 29 (7) including *Lophostrix*, not yet known from the park but occurring nearby.

complex, and either a Mexican Wood Owl *(Strix squamulata)* or Mottled Owl *(S. virgata)*. Most sites also have a larger *Strix* species—either the Black-and-white Owl *(S. nigrolineata)* or Black-banded Owl *(S. huhula)*. The Ferruginous Pygmy Owl *(Glaucidium brasilianum)* and Ridgway's Pygmy Owl *(G. ridgwayi)*, occurring in early successional habitats, are often excluded from forest avifauna lists. Thus, while forest falconiforms are strikingly more diverse at individual Neotropical sites than at temperate North American sites, forest owl communities show nearly constant species numbers at all latitudes in

the Western Hemisphere (Fig. 23.1). This latitudinal constancy in owl species richness has not, to our knowledge, been previously noted.

Biogeographic Affinities and Phyletic Diversity of Tikal's Raptors

Neotropical forests are rich in accipitrids, with 14 forest-dwelling genera represented by 17 species at Tikal (Table 23.3), compared to two forest-dwelling genera and four species in much of temperate North America

Table 23.3 Falconiform species lists for some Neotropical lowland moist forest sites[a]

Genera	Species[b]	Habitat Distribution Code	Habitat Distribution Code	Tikal[c]	La Selva, Costa Rica[d]	Barro Colorado, Panama[d]	French Guiana[e]	Manaus, Brazil[d,f,g]	Manu, Puru[d,h,l,i]
1	1 Swallow-tailed Kite *Elanoides forficatus*	F	WC	p	p	p	p	p	p
2	2 Gray-headed Kite *Leptodon cayanensis*	F ?	WC	p	p	p	p	v	v
3	3 Hook-billed Kite *Chondrohierax uncinatus*	F ?	WC	p	p	p	p	v	v (r^i)
4	4 Double-toothed Kite *Harpagus bidentatus*	F	WC	p	p	p	p	p	p^k
	5 Rufous-thighed Kite *Harpagus diodon*	F, 2F	WS?	ex	ex	ex	p	a	ex
5	6 Plumbeous Kite *Ictinia plumbea*	F, 2F	WC	p	p	p	p	p	p
6	7 Slender-billed Kite *Helicolestes hamatus*	SF	?	ex	ex	a	r^i	a	v
7	8 Bicolored Hawk *Accipiter bicolor*	F	WC	p	p	a	p	r	p
	9 Tiny Hawk *Accipiter superciliosus*	F	WP	ex	p	p	a	r	v
	10 Gray-bellied Goshawk *Accipiter poliogaster*	F	WP?	ex	ex	ex	p	v	p
8	11 Short-tailed Hawk *Buteo brachyurus*	?	WC	p	p	v	p	p	p
9	12 Roadside Hawk *Rupornis magnirostris*	F-2F	WC	p	p	v	p	p	p
	13 Gray Hawk *Buteo nitidus*	2F	WC	p	p	v	p	p	v
10	14 White Hawk *Leucopternis albicollis*	F	WC	p	p	p	p	p	a (r^i)
	15 Semiplumbeous Hawk *Leucopt. semiplumbea*	?	?	ex	p	p	ex	ex	ex
	16 Plumbeous Hawk *Leucopternis plumbea*	?	?	ex	ex	p	ex	ex	ex
	17 Slate-colored Hawk *Leucopternis schistacea*	O ?	?	ex	ex	ex	r^i	a	p
	18 White-browed Hawk *Leucopternis kuhli/melanops*	?	?	ex	ex	ex	p (melanops)	p (melanops)	p (kuhli)
	19 Barred Hawk *Leucopternis princeps*	?	?	ex	p	a	ex	ex	ex
11	20 Great Black Hawk *Buteogallus urubitinga*	F	WC	p	p	v	p	p	p
12	21 Crested Eagle *Morphnus guianensis*	F	WC	p	p	p	p	p	p
13	22 Harpy Eagle *Harpia harpyja*	F	WC	p^m	a(r?)	p	p	p	p
14	23 Black-and-white Hawk-eagle *Spizaetus melanoleucus*	F	WC	p	p	a	p	v	p
15	24 Ornate Hawk-eagle *Spizaetus ornatus*	F	WC	p	p	p	p	p	p
	25 Black Hawk-eagle *Spizaetus tyrannus*	F	WC	p	p	p	p	v	p
16	26 Crane Hawk *Geranospiza caerulescens*	F	WC	p	p	p	p	a	v
17	27 Collared Forest Falcon *Micrastur semitorquatus*	F	WC	p	p	p	p	p	p
	28 Barred Forest Falcon *Micrastur ruficollis*	F	WC	p	p	p	p	p	p
	29 Lined Forest Falcon *Micrastur gilvicollis*	F	?	ex	ex	ex	p	p	p
	30 Slaty-backed Forest Falcon *Micrastur mirandollei*	F	?	ex	p	p	p	p	a
	31 Buckley's Forest Falcon *Micrastur buckleyi*	F	ES	ex	ex	ex	ex	ex	vi
18	32 Red-throated Caracara *Daptrius americanus*	F	?	ex	p	p	p	p	p
	33 Black Caracara *Daptrius ater*	F	?	ex	ex	ex	p	r	p
19	34 Laughing Falcon *Herpetotheres cachinnans*	F-O	WC	p	p	a ???	p	a	p

	Common name	Scientific name	Habitat	Dist.					hypothetical	
20	35 Bat Falcon	*Falco rufigularis*	F, 2F	WC	p	p	p	p	p	p
	36 Orange-breasted Falcon	*Falco deiroleucus*	F	WC	p	a?	a?	p		a[n],r[i]
Total (including "rare" species):	Species				22	26	19	31	21	23
Total (including "rare" and "vagrant" species):	Species				22[o]	27	24	31	27	31
	Genera				17[d]	18	17	19	17	19

Species not tabulated above because of extreme rarity, mainly non-forest habitat use, absence from lowlands, or presence only as migrant/wintering birds

	Common name	Scientific name	Habitat	Dist.						
21	37 Pearl Kite	*Gampsonyx swainsonii*[p]	O?	?	ex	ex	ex	a	v	v
	38 Broad-winged Hawk	*Buteo platypterus*[q]	F?	BM?	(v)	(p)	(p)	(a)	(p)	(p)
	39 Zone-tailed Hawk	*Buteo albonotatus*[r]	?	?	v	a	v	r[l]	a	p
	40 Common Black Hawk	*Buteogallus anthracinus*[s]	R	?	p[m]	p	p	a	ex	ex
	41 Solitary Eagle	*Harpyhaliaetus solitarius*[t]	F	?	a/v?	p	a	a	a	v
22	42 Yellow-headed Caracara	*Milvago chimachima*[u]	O?	?	ex	a	v	p	p	a
23										

Notes:

Habitat codes: F = largely primary/mature forest, but sometimes successional or forest edge; 2F = mainly successional forest or partly opened forest habitats; SF = swamp forest; R = gallery forest/riparian; O = open country (grassland, savanna, etc.).

Distribution codes: WC = widespread, including much of Central America; WP = widespread, north at least to Panama; WS = widespread but South America only; RS = relatively restricted range, South America; EC = endemic, Central America; ES = endemic, South America; BM = boreal migrant (i.e., wintering here).

Presence codes: ( ) = wintering boreal-breeding migrant; p = present (uncommon or better); a = absent; r = present but rare; v = "vagrant," "accidental," "casual," or "extremely rare"; ex = extralimital, i.e., does not range to this site.

[a] Patently open country and exclusively highland or mid-elevation species are omitted here.

[b] Omitted from list are vultures, *Pandion haliaetus*, *Elanus leucurus*, *Ictinia mississippiensis*, *Rostrhamus sociabilis*, *Buteo swainsoni*, *B. jamaicensis*, *B. albicaudatus*, *Buteogallus (Heterospiza) meridionalis*, *Busarellus nigricollis*, *Circus cyaneus*, *Polyborus plancus*, *Falco femoralis*, *Falco peregrinus*, *Falco sparverius*.

[c] *Source:* this study.

[d] *Source:* Karr et al. 1990.

[e] *Source:* Thiollay 1994.

[f] *Source:* Cohn-Haft et al. 1997; additional source = Karr et al. 1990.

[g] *Source:* Stotz and Bierregaard 1989.

[h] *Source:* Terborgh et al. 1990.

[i] *Source:* Stotz et al. 2001.

[j] *Source:* Robinson 1994.

[k] "rare, in hill forest"; Stotz et al. 2001.

[l] Unrecorded in most of Thiollay's studies; reported in at least one.

[m] = Occurring near Tikal (i.e., in Petén, within 100 km of Tikal, generally in somewhat more mesic forest).

[n] Recorded near Manu National Park but not within.

[o] Excluding *Daptrius americanus*, which occurs/occurred in Guatemala but not at Tikal.

[p] At all sites, detected in open, often marshy habitats; omitted because not truly a forest species.

[q] Wintering, boreal-breeding migrant.

[r] Omitted at all sites because status and distribution very poorly known.

[s] Absent at Tikal for lack of appropriate habitat; ranges much farther north; because omitted at Tikal, omitted at other sites as well.

[t] Omitted at all sites because mainly highland, very rare and local, and status and distribution largely unknown.

[u] Omitted because mostly an open-country bird.

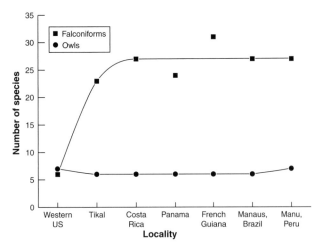

23.1 Latitudinal patterns of alpha diversity (species co-occurring at individual locales) for falconiforms and owls in the Americas.

(Table 23.1). The 29 forest-dwelling raptors breeding at or near Tikal represent 22 genera: 14 of accipitrids, 3 of falconids, and 5 of owls (Tables 23.3, 23.4). Most forest raptor genera and species at Tikal are Neotropical endemics. Fifteen (68%) of Tikal's 22 raptor genera are endemic or near endemic to the Neotropics, and these contribute 16 (55%) of Tikal's 29 forest raptor species. Woodland raptor genera at Tikal that are not of mainly Neotropical affinity include *Accipiter* (one species), *Buteo* (two species), *Falco* (two species), *Strix* (two species), *Megascops* (one species), and *Glaucidium* (two species); these six genera are practically cosmopolitan in distribution.

In temperate North American forests and woodlands, virtually all accipitrid diversity is attributable to two genera—*Accipiter* and *Buteo*—supplemented in open country by additional *Buteo* species, a *Circus* harrier, and the Golden Eagle (*Aquila chrysaetos*: Table 23.1)—all these genera being essentially cosmopolitan. *Ictinia* and

Table 23.4 Owl species lists for some Neotropical lowland moist forest sites

Species[a,b,c]	Habitat code	Distribution code	Tikal	La Selva, Costa Rica[d,e]	Barro Colorado, Panama[e]	French Guiana[f]	Manaus, Brazil[g]	Manu, Peru[e,h,I,j]
1 *Tyto alba*[a]	O	WC	r	—	—	—	r	r
2 *Megascops choliba*	F	WP	ex	a ?	p[k]	?	a	v
3 *Megascops guatemalae*	F	WC	p	p[l]	ex	ex	ex	ex
4 *Megascops watsonii*	F	WS	ex	ex	ex	p	p	p
5 *Lophostrix cristata*	F	WC	n	p	p	p	p	p
6 *Pulsatrix perspicillata*	F	WC	r	p	p	p	p	p
7 *Glaucidium sicki*	F	WC	r	p	p	p	p[m]	p[m]
8 *G. ridgwayi*	2F	WC	p	?[n]	?[n]	?[n]	a	p[o]
9 *Strix squamulata*	F	WP	p	p	p	a	a	a
10 *Strix nigrolineata/huhula*	F	WC	p	p	p	p	p (*huhula*)	v (*huhula*)[p]
Totals (deleting American Barn Owl and vagrant species; larger total includes rare species)[q]								
Species			4–6[r]	6	6	6	5 or 6	6
Genera			3–5	5	5	5	5	5

Notes:
Habitat codes: F = largely primary/mature forest, but sometimes successional or forest edge; 2F = mainly successional forest or partly opened forest habitats; O = open country (grassland, savanna, etc.).
Distribution codes: WC = widespread, including much of Central America; WP = widespread, north at least to Panama; WS = widespread, but South America only.
Presence codes: ex = extralimital, i.e., does not range to this site; p = present (uncommon to common); a = absent; r = present but rare; v = "vagrant," "accidental," or "casual"; n = occurs near site.
Entry 7 *Glaucidium sicki* was formerly placed in *minutissimum* group: includes several forms now recognized as allopatric species (Robbins and Howell 1995; Howell and Robbins 1995).

[a] *Tyto alba* was uncommon near our study site and listed as "rare" at the Brazil site; we assume it may be present at all or many localities; is omitted from all tabulations because not a forest species.
[b] *Rhinoptynx clamator* was reported only for Panama, as a vagrant.
[c] *Speotyto cunicularia*, an open-country species, was listed as "vagrant" only at the Brazil site and is omitted throughout.
[d] *Source:* Enriquez-Rocha and Rangel-Salazar 1996.
[e] *Source:* Karr et al. 1990.
[f] *Source:* Thiollay 1994.
[g] *Source:* Cohn-Haft et al. 1997; additional source = Karr et al. 1990.
[h] *Source:* Terborgh et al. 1990.
[i] *Source:* Stotz et al. 2001.
[j] Stotz et al. (2001) report *Megascops vermiculatus* and *Pulsatrix melanota* as common in hill forest (500–1000 m); because these are not syntopic with other (lowland) species listed here for that locality, we omit them.
[k] Listed as occurring in young second growth.
[l] Listed as *O. guatemalae* in source; now would be regarded *M. vermiculatus*.
[m] *Glaucidium hardyi*.
[n] Not listed in sources used; possibly present in more open or successional habitats.
[o] Reported by Terborgh et al. 1990 as occupying early successional habitats.
[p] One record by Stotz et al. 2001.
[q] If *Glaucidium ridgwayi* is considered present at all sites, some sites rise to 7 species total.
[r] 6 species if 2 rare species are included; 7 if *Lophostrix* is included (known from wetter forests nearby); 3 species were common in the park: *Megascops guatemalae*, *Strix squamulata*, and *Strix nigrolineata*, with *Glaucidium ridgwayi* restricted to low swamp forest.

Elanoides contribute a Neotropical element to woodland raptor diversity in the southeastern states. Thus, most Nearctic accipitrids belong to genera that are widespread globally, several probably having Old World origins.

One striking conclusion is that most falconiform diversity at Tikal occurs at the generic level, with most genera represented by only one species. We suspect this reflects an accumulation of raptor lineages over long stretches of time, with Neotropical forests acting as a phylogenetic "museum." In contrast, falconiforms in North America belong to a smaller set of genera, each with several species, and with *Buteo* contributing one-third of all North American raptor species. We attribute this to fewer ancient lineages being represented here than in the Neotropics.

Our tabulation of falconiform diversity for six Neotropical forest sites—an average of 26.5 species per site—agrees well with Thiollay's (1985) estimate of 25.3. Thus, Neotropical forest falconiform assemblages have about 2.5–2.7 times as many species as do forest assemblages of Africa and southeast Asia (ibid.).

Body size patterns in raptor assemblages: Tikal versus temperate North America. Raptors at Tikal range in size from the 50 g Central American Pygmy-Owl *(G. griseiceps)* to the 4–8 kg Harpy Eagle *(Harpia harpyja)*. The body size distribution shows a strong peak between 100 and 600 g (Fig. 23.2). The smallest falconiform at Tikal is the Bat Falcon *(F. rufigularis;* male weight 130 g). Three owls are yet smaller, including two diurnal pygmy owls (50–75 g) and the 100 g Guatemalan Screech Owl *(M. guatemalae)*. Hence, in Central as in North America, the smallest raptors are owls, and among the smallest owls are the diurnal pygmy-owls.

The most notable differences in body size patterns in Tikal's raptor assemblage compared to those of temperate North American forests are that (1) the smallest size class (< 100 g, all owls) has no more species at Tikal than

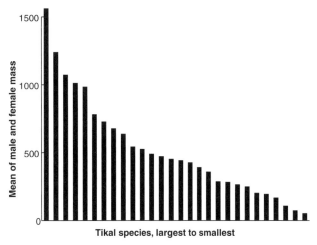
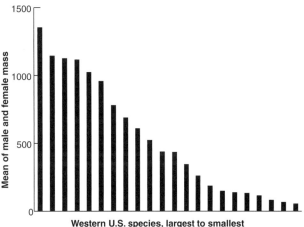

23.3 Comparative body size distributions of 28 forest raptor species at Tikal (minus Harpy Eagle) and 22 raptor species of the western United States, all habitats (minus the Golden Eagle).

in forests of the western United States, where the greatest number of small North American owls occurs; and (2) strikingly more species in the 100–600 g range occur at Tikal than in North American forests (Fig. 23.2). On average, raptor species at Tikal are smaller than in North America (Fig. 23.3).

Diets of Tikal's Raptors

Our dietary data from Tikal are based on prey brought to nests by adults.

Invertebrates. The largest raptor observed bringing invertebrates to the nest was the 1 kg Great Black Hawk *(Buteogallus urubitinga ridgwayi),* in which arthropods made up 2% of items brought to the nest. No raptor over 420 g took more than 3.3% invertebrates—less than 1% on a biomass basis. In the eight smallest species (< 270 g, male mass), the invertebrate contribution to diet biomass ranged from 0 to 90% and was unrelated to body size. Of the 19 species whose diets we studied (only the Gray-headed Kite *[Leptodon cayanensis]* diet was not studied), 10 brought some arthropods to the nest, though

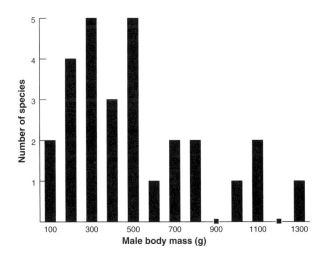

23.2 Frequency histogram of raptor body sizes at Tikal, based on all 28 falconiforms and owls occurring at Tikal, Guatemala, except the Harpy Eagle.

only 6 brought in substantial numbers and biomass: the Plumbeous Kite *(Ictinia plumbea)*, Bat Falcon, Swallow-tailed Kite *(Elanoides forficatus)*, Double-toothed Kite *(Harpagus bidentatus)*, Mexican Wood Owl, and Black-and-white Owl (Tables 23.5, 23.6). These six differed greatly in which insect groups were most heavily used. Cicadas were important prey for both Plumbeous and Double-toothed Kites. Beetles were the most frequent insect prey for both owls and the Swallow-tailed Kite and were also often taken by Plumbeous Kites. Dragonflies were the most frequent prey of Bat Falcons and were taken with some regularity by Plumbeous Kites and rarely by Swallow-tailed Kites. Grasshoppers and katydids were taken most often by the Mexican Wood Owl and Swallow-tailed Kite and a few roaches (Blattidae) by both owls and the Barred Forest Falcon *(Micrastur ruficollis)*. Butterflies were occasionally taken by Bat Falcons and by Plumbeous, Swallow-tailed, and Double-toothed Kites. Many Hymenoptera were taken by Swallow-tailed Kites and smaller numbers by Plumbeous and Double-toothed Kites; Bat Falcons took a number of large bees. The Hook-billed Kite *(Chondroheirax uncinatus)* was the only raptor at Tikal found to prey on snails (Tables 23.5, 23.6).[4]

Fish and amphibians. None of the raptors we studied were strongly associated with water bodies. However, one Laughing Falcon *(Herpetotheres cachinnans)* pair nesting along a lake shore took a few fish (Tables 23.5, 23.6), and frogs and toads were taken by nine species (Tables 23.5, 23.6), especially the Roadside Hawk *(Rupornis magnirostris)* and Crane Hawk *(Geranospiza caerulescens)*.

Reptiles. Fifteen raptor species included reptiles in their diets (dietary percentages on a biomass basis are shown): Laughing Falcon 97% (100% in primary forest), Double-toothed Kite 76%, White Hawk *(Leucopternis albicollis)* 64%, Great Black Hawk 54%, Barred Forest Falcon 38%, Roadside Hawk 32%, and Crested Eagle *(Morphnus guianensis)* 14%, with smaller percentages for eight other species (Tables 23.5, 23.6). Lizards (Plates 23.1, 23.2) featured in the diets of 14 of the 19 raptor species whose diets were studied (Tables 23.5, 23.6). In terms of biomass, use of lizards was heaviest by the Double-toothed Kite, followed by the Barred Forest Falcon, Roadside Hawk, Great Black Hawk, and Plumbeous Kite (Tables 23.5, 23.6). Snakes occurred in the diets of 10 raptor species and were most important to the Laughing Falcon, followed by the White Hawk, Great Black Hawk, and Crested Eagle (Tables 23.5, 23.6).

Birds. Of the 19 raptor species whose diets we studied, all but the Hook-billed Kite included some birds in their diets (Tables 23.5, 23.6). Birds contributed most to the diets (by biomass) of the Bicolored Hawk *(Accipiter bicolor)* 98%, Orange-breasted Falcon *(Falco deiroleucus)* 92%, Swallow-tailed Kite 71%, Ornate Hawk-eagle *(Spizaetus ornatus)* 70%, Bat Falcon 61%, Barred Forest

Falcon 53%, Collared Forest Falcon *(Micrastur semitorquatus)* 52%, and White Hawk 27%; only the Swallow-tailed Kite took many nestling songbirds.

Mammals. All raptors at Tikal except the Hook-billed and Swallow-tailed Kites took some mammals (Tables 23.5, 23.6). Use of mammals was greatest by the Black Hawk-eagle—94% by biomass, followed by the Crane Hawk 84%, Crested Eagle 76%, Black-and-white Owl 75%, Mexican Wood Owl 44%, Collared Forest Falcon 39%, Ornate Hawk-eagle 30%, and Great Black Hawk 21%.

Bats occurred in the diets of 15 raptors, especially the Black Hawk-eagle—51% numeric, 15% by biomass (Tables 23.5, 23.6) and Black-and-white Owl—73% by either measure. While Bat Falcons in some situations take numerous bats, those we studied took very few (see Chapter 19). Opossums (Didelphidae) were the most frequent prey of the Crested Eagle (40% of prey items), but also taken by the Black Hawk-eagle (11%). Members of the raccoon family, mainly the White-nosed Coati *(Nasua narica)*, were taken only by the three largest raptors—the Crested Eagle, Black Hawk-eagle, and Ornate Hawk-eagle. Squirrels (Sciuridae) were important prey for Ornate Hawk-eagles (35% of the numeric diet), Collared Forest Falcon (31% numeric), and Black Hawk-eagle (27%). Arboreal porcupines (Erethizontidae) were important prey of the Crested Eagle (9% of the numeric diet, 28% by biomass), with a few also taken by Black and Ornate Hawk-eagles. The only predation on agoutis (Dasyproctidae) we documented was by Ornate Hawk-eagles. Small rodents of the family Muridae (rats and mice) occurred in the diets of 14 raptor species, contributing most to those of the Crane Hawk (77% biomass) and Mexican Wood Owl (42%).

Raptor Diets at Tikal and in Temperate North America

We compared the diet of Tikal's raptor assemblage to that of temperate North American raptors, using the large data set compiled by Snyder and Wiley (1976) for 17 owl and 27 falconiform species.[5] Prey items in their study were classed as birds, mammals, lower vertebrates (reptiles, amphibians, and fish), and invertebrates.

Diurnal raptor (Falconiform) diets. The most notable finding in this temperate-tropical comparison was that lower vertebrates were the leading prey type for far more diurnal raptors at Tikal than in North America, and this still held true when only forest-dwelling raptors were compared. The greater use of lower vertebrates (amphibians and reptiles) at Tikal than in North America may owe in part to greater diversity and biomass of these prey types in the tropics. Likely more important is that these prey animals are available, at least to some degree, year-round in the tropics, whereas they hibernate during the winter months at north temperate latitudes and become unavailable to raptors.

Table 23.5 Diet of 19 raptors studied at Tikal, based on percentages of numbers of prey items[a]

Species	Snails	Arthropods	Lizards	Snakes	Total number of reptiles	Amphibians	Birds	Bats	Other mammals	Total number of mammals	Total number of vertebrates	Total number of items
Hook-billed Kite	**92.2**	—	7.8	—	7.8	—	—	—	—	—	7.8	64
Swallow-tailed Kite[b]	—	**68.7**	11.0	—	11.0	0.3	**19.8**	—	—	—	31.1	1496
Double-toothed Kite	—	**59.2**	**40.0**	0.2	**40.2**	—	0.2	0.2	0.2	0.4	40.8	463
Plumbeous Kite	—	**93.1**	5.0	0.9	5.9	0.2	0.5	0.3	—	0.3	6.9	655
Bicolored Hawk	—	—	0.9	—	0.9	—	**95.9**	1.8	1.4	3.2	100	218
Crane Hawk	—	—	**19.9**	2.8	22.7	**16.0**	6.1	6.6	**48.6**	**55.2**	100	181
White Hawk	—	3.3	**22.7**	**45.4**	**68.1**	4.3	5.7	5.0	13.6	**18.6**	96.7	210
Great Black Hawk	—	2.3	**30.8**	26.2	**57.0**	6.2	**13.9**	6.9	13.9	**20.8**	97.7	130
Roadside Hawk	—	9.3	**55.7**	1.4	**57.1**	**24.3**	1.4	1.4	6.4	7.9	90.7	140
Crested Eagle	—	—	—	**15.0**	**15.0**	—	**21.0**	—	**64.0**	**64.0**	100	100
Black Hawk-eagle	—	—	—	—	—	—	4.7	**50.6**	**44.7**	**95.3**	100	85
Ornate Hawk-eagle	—	—	—	—	—	—	**56.3**	1.5	**42.2**	**43.7**	100	325
Laughing Falcon[c]	—	—	7.8	**88.6**	**96.4**	—	0.3	—	2.4	2.4	100	747
Barred Forest Falcon[d]	—	7.9	**64.0**	1.1	**65.1**	4.1	**19.5**	0.4	3.0	3.4	**92.9**	267[e]
Collared Forest Falcon	—	—	7.6	11.1	18.7	0.6	**34.5**	9.4	**36.8**	**46.2**	100	170
Bat Falcon	—	**77.7**	0.2	—	0.2	—	**21.7**	0.4	—	0.4	22.3	1422
Orange-breasted Falcon	—	—	—	—	—	—	**85.7**	**14.3**	—	**14.3**	100	105
Mexican Wood Owl	—	**50**	2	—	2	2	2	2	**42**	**44**	50	52
Black-and-white Owl	1	**23**	—	—	—	—	2	**73**	2	**75**	77	73
Number of raptor species heavily using prey category	1	6	6	4	9	2	9	3	7	10	17	Total: 6903

[a] Most heavily used prey types are in bold.
[b] Swallow-tailed Kite diet also included 0.3% fruit; birds in diet were nestlings.
[c] Laughing Falcon diet also included 0.8% fish; 2.6% unidentified vertebrates were proportionally allocated to identified vertebrate categories, except fish.
[d] Barred Forest Falcon diet also included a trace amount of fruits.
[e] If based on n = 405 identified prey, diet = arthropods 8.2%, reptiles 61.5%, amphibians 2.5%, birds 22%, and mammals 5.9%.

Table 23.6 Diet of 19 raptor species studied at Tikal, percentage based on prey biomass

Raptor species	Snails	Arthropods	Lizards	Snakes	Total number of reptiles	Amphibians	Birds	Bats	Other mammals	Total number of mammals	Total number of vertebrates	Total number of prey items
Hook-billed Kite	92.2	—	7.8	—	7.8	—	—	—	—	—	7.8	64
Swallow-tailed Kite[a]	—	18.1	10.4	—	10.4	—	71.2	—	—	—	82.0	1496
Double-toothed Kite	—	22.5	75.9	0.4	76.3	—	0.4	0.4	0.4	0.8	77.5	463
Plumbeous Kite	—	73.1	19.8	3.6	23.4	0.6	1.8	—	—	—	26.1	655
Bicolored Hawk	—	—	0.2	—	0.2	—	97.7	1.0	1.1	2.1	100	218
Crane Hawk	—	—	2.8	3.3	6.1	6.3	4.0	3.5	80.1	83.6	100	181
White Hawk[b]	—	0.1	4.67	59.2	63.9	1.6	26.8	1.75	5.9	7.7	99.0	210
Great Black Hawk	—	2.3	29.0	25.2	54.2	6.1	13.7	6.9	13.7	20.6	97.7	130
Roadside Hawk	—	0.6	29.7	2.6	32.3	21.9	3.4	4.3	37.6	41.9	99.0	140
Crested Eagle	—	—	—	14.2	14.2	—	10.1	—	75.8	75.8	100	100
Black Hawk-eagle	—	—	—	—	—	—	6.1	14.5	79.5	93.9	100	85
Ornate Hawk-eagle	—	—	—	—	—	—	69.8	0.1	30.0	30.2	100	325
Laughing Falcon[c]	—	—	7.8	88.8	96.6	—	0.3	—	2.4	2.4	100	747
Barred Forest Falcon[d]	—	0.5	36.6	1.0	37.6	2.7	52.7	0.8	5.7	6.5	99.5	267[e]
Collared Forest Falcon	—	—	1.4	7.6	9.0	0.04	52.4	2.0	36.5	38.5	100	170
Bat Falcon	—	33.9	0.3	—	0.3	—	63.5	2.3	—	2.3	66.1	1422
Orange-breasted Falcon	—	—	—	—	—	—	92.4	7.6	—	7.6	100	105
Mexican Wood Owl	—	50	2	—	2	2	2	2	42	44	50	52
Black-and-white Owl	—	23	—	—	—	—	2	73	2	75	77	73
Number of raptor species heavily using prey category	1	6	5	4	8	1	9	2	8	9	18	Total: 6903

[a] Swallow-tailed Kite diet also included a trace quantity of fruits; birds in diet were nestlings.

[b] To estimate White Hawk biomass diet from numeric diet, we reduced arthropods from 3.3% to 1.0% and added 0.4% each to lizards and snakes and 1.5% to mammals.

[c] Laughing Falcon diet also included 0.8% fish.

[d] Barred Forest Falcon diet also included a trace quantity of fruits.

[e] These data are based on the most-studied nests; if diet is based on entire data base, n = 405 identified prey; diet = arthropods 8.2%, reptiles 61.5%, amphibians 2.5%, birds 22%, and mammals 5.9%.

Owl diets. All six widely occurring Neotropical lowland forest owls feed partly on insects, and for three or four of them, insects may comprise at least half the diet[6] (see Chapters 21, 22). Owl diets, on the whole, differed dramatically from those of diurnal raptors, both in temperate North America and in the four Neotropical forest owls studied to date. Owls in both regions took very few birds and lower vertebrates (see also Marti et al. 1993). Owls in both regions ate mainly mammals or arthropods to a much greater extent than diurnal raptors (see also Johnsgard 2002). North American forest owls overwhelmingly have mammals as their leading prey type, with only a few southwestern species preying mainly on arthropods. Greater insectivory among tropical than temperate zone owls likely owes to a greater prevalence of large insects in tropical forest (Schoener 1971), as well as to year-round insect activity. At colder latitudes, insects are largely inactive in winter, and the most insectivorous North American owls migrate to warmer climes in winter.

As in North America, neither owl studied at Tikal had lower vertebrates or birds as its principal prey.[7] However, they preyed significantly more on lower vertebrates than did 17 North American owls from forested as well as open habitats.[8] Restricting the North American sample to 12 forest owls, the same tendency held.[9] This result agrees with that among diurnal raptors.

Owl diets and latitudinal patterns of species diversity. The owl species that fed most heavily on lower vertebrates was Ridgway's Pygmy Owl, a mainly tropical species, although dietary data are from Texas. Apparently the diurnal habit of pygmy owls gives them greater access to lower vertebrates, especially lizards, than is the case for most (nocturnal) owls. This might occur for the same reason that diurnal raptors, as a group, take many more lower vertebrates than do the owls. While many frogs, toads, and snakes are nocturnal, perhaps they are more difficult to detect or identify at night than are small mammals, which move about more rapidly and frequently. It is mainly lizards that contribute to the greater use of lower vertebrates by diurnal raptors in the tropics than in the temperate zone. As few lizards are nocturnal, this food source heavily used by tropical falconiforms may be largely inaccessible to tropical owls.

Assemblages of forest owls do not show a pattern of greater species richness in the tropics as diurnal raptors do. Why isn't this pattern observed among the owls? The dietary results given here suggest a possible answer. The greater species richness of diurnal raptor assemblages in the tropics may be facilitated in part by their much greater use of lower vertebrates—especially reptiles—than in the north temperate zone. Because owls tend to prey less often on reptiles, these do not comprise an additional prey source in the tropics as they do for diurnal raptors. At all latitudes, owls, except for diurnal species, prey mainly on small mammals and arthropods, and this constancy of prey types may contribute to the constancy of species richness of owl assemblages across latitudes.

Dietary Overlap

Our most important finding regarding community ecology was a lower degree of dietary overlap among species than has previously been found in any raptor community. This provides a counterpoint to the high overlaps found in many studies at higher latitudes, where several raptor species feed mainly on a few mammal species, especially in winter (Craighead and Craighead 1956; Smith and Murphy 1973; Herrera and Hiraldo 1976; Phelan and Robertson 1978; Yalden 1985; MacLaren et al. 1988; Restani 1991). Moreover, comparing dietary overlap at Tikal with that found in several other raptor communities, we found that dietary overlap decreased as the number of raptor species increased in the assemblage, contrary to earlier findings (Jaksic and Braker 1983). These results suggest that interspecific competition has contributed to dietary separation among raptors in species-rich assemblages such as Tikal's. Interestingly, this greater segregation of diets at Tikal was associated with greater phylogenetic separation among the raptors there (many genera being represented by a single species), compared with temperate North America. The coexistence of many raptor species in Neotropical forests may be facilitated by their greater phylogenetic separation.

Lower dietary overlap among tropical raptors may be facilitated by a greater diversity of prey types available in tropical forests than at higher latitudes—in particular, a greater range of prey types available year-round. In contrast, at northerly latitudes, cold-blooded prey types become largely unavailable to raptors during winter; this likely prevents specialization on such prey types, except among raptors that migrate to warmer climates. In addition, the great dietary overlap often seen among mammal-eating raptors in higher latitudes may result in part simply from the low diversity of such prey types, which limits the possibilities for partitioning this enormous prey biomass by prey taxa. Temperate raptors may often reduce competition by use of different hunting methods and partitioning space by territoriality, mutual avoidance, and habitat, rather than, or in addition to, partitioning prey species (e.g., Schmutz et al. 1980; Carothers and Jaksic 1984; Bosakowski et al. 1992; Gerstell and Bednarz 1999).

Ecological Differences Among Tikal's Raptor Species

Foraging habitat. Among those species restricted to mature forest, some hunted mainly in one or more types of mature forest, while largely neglecting others. Several species hunted largely in Upland and other tall, closed-canopy forests with relatively open understory (Table 23.7). In contrast, Great Black Hawks, Collared Forest Falcons, and Black-and-white Owls were most often found hunting in the lower-canopied, densely understoried Transitional and Scrub Swamp Forests. Black-and-white Owls often hunted along pond margins, and it may have been ponds within these forest types that

Table 23.7 Hunting characteristics of raptors studied at Tikal

Species	Hunting habitat	Hunting strata	Position of quarry	Search style	Prey types
Gray-headed Kite	Forest	Below canopy surface	Often on vegetation	Perch to perch, scans from perch	Poorly known; largely bee/wasp nests
Hook-billed Kite	Forest	Below canopy surface	On vegetation	Perch to perch, scans from perch	Snails, a few small lizards
Swallow-tailed Kite	Mainly forest	Above canopy	Above or in upper canopy	Extensive search flight above forest	Nestling birds, lizards, insects, often in flight
Double-toothed Kite	Mature, mainly Upland Forest	Below canopy surface	Mainly within lower canopy	Perch to perch, scans from perch	Mainly lizards, cicadas; other large insects, small vertebrates
Plumbeous Kite	Forest and non-forest	Above canopy	Above or in upper canopy	Extensive search flight above forest; also attacks from perch	Insects in flight or perched; lizards, other small vertebrates
Bicolored Hawk	Mature, largely Upland Forest	Below canopy surface, mean = 7.5 m above ground	Below canopy	Short-stay perch hunter	Medium-size birds
Crane Hawk	Various forest types	Below canopy surface, all strata, from ground to canopy	In cavities or other retreats, in vegetation or on ground	Active-search cavity prober	Small vertebrates, largely nocturnal, taken from hiding places; invertebrates
White Hawk	Mature, largely Upland Forest	Below canopy surface	On vegetation or ground	Perch hunter	Snakes, lizards, birds, mammals
Great Black Hawk	Forest; some suggestion of heavy use of scrub-swamp, forest edge, pond margins	Often low or on ground	On vegetation or ground	Perch hunter; also search and probe with feet	Great variety of small to medium vertebrates; a few invertebrates
Roadside Hawk	Forest and open habitat	Below canopy surface; in open; on ground	On vegetation or ground	Perch hunter	Great variety of small vertebrates, large invertebrates
Crested Eagle	Mature, largely Upland Forest	Below canopy surface	Largely on vegetation	Perch hunter; possible cavity searcher	Nocturnal, mostly arboreal mammals; large, diurnal snakes
Black Hawk-eagle	Forest	Below canopy surface	Largely in trees, often in diurnal retreats; also on ground	Perch hunter, possible retreat searcher	Squirrels, bats, birds
Ornate Hawk-eagle	Mature, largely Upland Forest	Below canopy surface	On ground or vegetation	Perch hunter; patient, stealthy stalking of detected quarry	Medium to large diurnal birds and mammals
Laughing Falcon	Forest and open habitat	Below canopy surface; open areas	On ground or vegetation	Perch hunter	Snakes; rarely lizards or other vertebrates
Barred Forest Falcon	Mature, mainly Upland Forest	Below canopy surface	On ground or vegetation	Perch hunter	Lizards, small birds; occasional other vertebrates, invertebrates
Collared Forest Falcon	Forest, especially with dense understory	Below canopy surface	On ground or vegetation	Perch hunter; stealth and ambush	Squirrels, medium-size and large birds
Bat Falcon	Forest, forest edge, and mosaic habitats	Above and in canopy surface	In flight over canopy, perched in canopy, or on ground	Perch hunter, aerial prospecting	Small birds, insects in flight; occasionally bats (sometimes many)
Orange-breasted Falcon	Forest, forest edge, and mosaic habitats	Above and in canopy surface	In flight over canopy, perched in canopy	Perch hunter, aerial prospecting	Medium-size birds, bats
Mexican Wood Owl	Forest, forest edge, and mosaic habitats	Below canopy surface	On ground and vegetation	Perch hunter	Small rodents, insects; occasional other vertebrates
Black-and-white Owl	Forest; ponds in forest	Open areas amid forest	In flight, on ground and vegetation	Perch hunter	Bats, large insects

attracted them, rather than forest type per se. Laughing Falcons also hunted frequently along pond margins. In only a few cases did habitat specificity seem to provide a major axis of ecological separation between closely related species (see later). However, raptors at Tikal may have differed subtly in habitat use to a greater degree than we were able to detect.

Aspects of dietary selection. Gross prey taxa, prey size, and prey location in the environment appear to be among the strongest criteria on which raptors select their prey. Selection may also be based on movement patterns and other prey attributes.

Selectivity with regard to prey taxa was clearly apparent in raptors that preyed on birds or other flying prey and in the snake-specialist Laughing Falcon and snail-specialist Hook-billed Kite. Also, certain raptors rarely or never took certain prey types. An example was the total or near-total absence of lizards in the diets of raptors such as the Black Hawk-eagle, Ornate Hawk-eagle, and Crested Eagle. Several species also failed to take any snakes, at least judging from our observations (Tables 23.5, 23.6).

Among the bird-eating raptors, it seemed that the main criteria for prey selection were prey size and availability within a habitat structure—that is, below the forest canopy versus in the sky above. Above the canopy, Bat Falcons and Orange-breasted Falcons partitioned avian prey by size, as did the Barred Forest Falcon, Bicolored Hawk, Collared Forest Falcon, and Ornate Hawk-eagle below the canopy (Table 23.7). Partitioning of prey by size appears especially pronounced in bird-eating raptors, probably because aerodynamic factors demand that these raptors closely match their prey in size, yielding high predatory efficiency over only a narrow range of prey sizes for a given size of predator (Andersson and Norberg 1981). In contrast, predator size appears less critical for capture efficiency in mammal eaters, in which raptors ranging from the 100 g Northern Saw-whet Owl (*Aegolius acadicus*) to the 1700 g Snowy Owl (*Nyctea scandiaca*) may often prey on the same small rodents.

Partitioning of hunting zones by gross habitat structure is especially important in raptors and is generally correlated with marked differences in flight morphology that facilitate use of some foraging zones and prohibit use of others (Jaksic and Carothers 1985). The clearest examples involved raptor species sets in which some foraged above and others below the forest canopy, although for similar prey. Above- and below-canopy foragers at Tikal differed greatly in flight morphology, the former having slender, pointed, falconine wings and the latter shorter, broader, buteonine or accipitrine wings.

Small birds of similar size were taken above and from the canopy surface by Bat Falcons and below the canopy by Barred Forest Falcons (Table 23.7). Likewise, Orange-breasted Falcons and Bicolored Hawks took birds of similar size, the former above the canopy and the latter

below it. To a large extent, raptors taking birds above and below the canopy took different species, with above-canopy bird hunters taking aerial species such as swallows, swifts, and hummingbirds; above-canopy commuters such as parrots and doves; and canopy dwellers such as tanagers, honeycreepers, and some flycatchers. In contrast, below-canopy bird hunters took mainly birds that dwell in or below the canopy or were terrestrial. Dietary overlap occurred when aerial hunters such as Bat Falcons and Orange-breasted Falcons took understory birds as they crossed an opening or seized canopy dwellers from the canopy surface, and when Orange-breasted Falcons and Bicolored Hawks took pigeons and other species in common, possibly during different portions of these prey species's daily activities.

The highly aerial Plumbeous and Swallow-tailed Kites took small lizards from the canopy surface, and Double-toothed Kites and Barred Forest Falcons took small lizards below the canopy (Table 23.7). Plumbeous and Swallow-tailed Kites and Bat Falcons took insects in flight in the open sky and from the canopy surface, whereas the main diurnal insect eaters below the canopy were the Double-toothed Kite and the Crane Hawk. Plumbeous Kites, Swallow-tailed Kites, and Bat Falcons, hunting above the canopy, captured different insects than did raptors hunting below the canopy. For example, dragonflies (*Anisoptera* spp.) were heavily preyed upon by Bat Falcons and rarely taken by sub-canopy hunters. Roaches (Blattidae) were taken more often by sub-canopy than by aerial hunters. Cicadas were important in the diets of Plumbeous and Double-toothed Kites and thus were heavily used by both aerial and sub-canopy hunters.

Below the canopy, some raptor species used certain strata more often than others. Double-toothed Kites hunted mainly in the upper to lower canopy, whereas Barred Forest Falcons, with which they overlapped most in diet, hunted at lower levels, averaging 4 m above ground. Among avian, snake, and rodent prey species, some clearly differed in the heights at which they spent most time, and we expect the same was true of at least some lizards and insects. Thus, differences in sub-canopy foraging height may afford raptors differing prey populations.

In summary, above- and below-canopy foraging provided a dramatic dichotomy in foraging opportunities among raptors at Tikal, and was associated, in several cases, with strong dietary differences. More subtly, some sub-canopy foragers tended to forage at different heights than others.

Time of day. We noted an activity pattern at Tikal that appears essentially absent among north temperate raptors: a distinctly crepuscular pattern in the genus *Micrastur* (see Chapters 16, 17). These raptors began hunting at the earliest hint of dawn and often brought prey to the nest by 0600, in partial daylight (R. Thorstrom, pers. comm.). This falconid genus, with its accipiter-like morphology and secretive, sub-canopy lifestyle, is somewhat owl-like in possessing a slight facial ruff of narrow, stiff, upcurved feathers, and large ear openings (Brown and Amadon 1968). Whether this facial ruff facilitates hunting by ear as in owls and harriers has not, to our knowledge, been investigated. Although some temperate zone falconiforms are also active at dawn and dusk, we are not aware that any are as consistently crepuscular as the *Micrastur* species.

Correlations between activity time of predators and prey. In many cases, raptors' activity times at Tikal correlated with those of their prey, but there were also dramatic departures from this pattern in that some diurnal raptors took nocturnal prey throughout the day. The Crane Hawk, with its specialized, probing style of hunting, often took nocturnal rodents during the day from hidden areas such as tree cavities. The Crested Eagle and Black Hawk-eagle also took many nocturnal, arboreal mammals throughout the day; how they detected them remains a mystery (see Chapter 14).

All lizards at Tikal except the geckos are strictly diurnal. While lizards made up a strikingly large proportion of diurnal raptor diets as a whole, we found few lizards in the owl diets. Likewise, although many snakes are nocturnal, we found no snakes in diet samples of the two owl species studied. Thus, lizards and snakes were less often taken by, and perhaps less available to, nocturnal than diurnal raptors at Tikal. Black-and-white Owls took more bats than any other raptor at Tikal, often sallying from a perch along a water hole to capture them in flight—a tactic available only to nocturnal or crepuscular predators. While several diurnal raptors took a few bats, Black Hawk-eagles took many, throughout the daylight hours. Few birds were taken by Mexican Wood and Black-and-white Owls, whereas several diurnal raptors were dedicated bird hunters. Little emphasis on bird predation appears generally true of owls (see Snyder and Wiley 1976), although exceptions to this pattern exist. Marti et al. (1993) found that falconiforms generally took more prey types at the class level than did owls, and they speculated that birds and perhaps reptiles are generally less available to owls than to diurnal raptors; our results at Tikal agree.

In summary, nocturnal raptors at Tikal showed several dietary differences from the diurnal raptors, which were understandable in terms of the daily activity rhythms of their prey. However, the degree to which several diurnal raptors took nocturnal prey during the day is unparalleled, to our knowledge, among temperate zone raptors. Unlike the north temperate zone, where partly diurnal microtine rodents are often abundant, Neotropical forests are home to few small diurnal mammals other than squirrels (and, closer to the equator, squirrel-sized primates). Thus, diurnal Neotropical forest raptors may be able to exploit most small mammals only by locating nocturnal mammals in their daytime retreats. That some diurnal raptors took many nocturnal mammals may also be related to the relatively low owl diversity in Neotropical forests.

Ecological Separation among Congeneric Species

Because members of a genus share the same evolutionary heritage, they are often similar in ecology and hence of special interest from the standpoint of mechanisms providing ecological separation. Six genera had two or more representatives among Tikal's forest or woodland raptors: *Buteo*, *Spizaetus*, *Micrastur*, *Falco*, *Strix*, and *Glaucidium*. We found clear-cut ecological differences among each set of congeners.

Three buteo-like species occupied forest or somewhat wooded habitats at Tikal: the Roadside Hawk, Gray Hawk (*Buteo nitidus*: Plate 23.3), and Short-tailed Hawk (*Buteo brachyurus*). The Roadside and Gray Hawks, similar in size and morphology, separated clearly on habitat. Whereas Roadside Hawks were widespread residents in certain mature forest types and in open, human-modified habitats, the Gray Hawk was restricted to the latter. Short-tailed Hawks were occasionally seen soaring above Tikal's mature forest. Studies elsewhere indicate that this hawk often uses borders between forest and open habitats and preys much more on birds than do the Gray and Roadside Hawks (Ogden 1974). Thus, Short-tailed Hawks at Tikal may differ in both habitat and diet from the other two species.

The two *Spizaetus* hawk-eagles at Tikal may have differed in habitat use in a subtle fashion that we did not discern, but at a gross scale, they occupied the same forest areas. While both took some squirrels, their diets differed strongly overall. On a biomass basis, Ornate Hawk-eagles' diet comprised 70% birds and 30% mammals, whereas Black Hawk-eagles' diet was 6% birds and 94% mammals.

Tikal's two forest falcons differ greatly in size, the Collared weighing 3.5 times as much as the Barred. They differed as well in the size of nest trees and cavities used and showed virtually no overlap in diet. At a gross level, dietary overlap was substantial for avian prey (53% of the Barred Forest Falcon's diet and 52% for the Collared), but there was little overlap in species taken, since the average mass of Collared Forest Falcon prey (239 g) was 10 times that of Barred Forest Falcon prey (24 g: Thorstrom 2001). The Collared Forest Falcon showed a greater tendency to hunt in low, dense Scrub Swamp Forest than did the Barred Forest Falcon, which nested and hunted in Upland Forest with more open understory. However, these two congeners differed so greatly in prey size and taxa that habitat separation between them was likely of minimal importance to their ecological segregation.

The Bat Falcon and Orange-breasted Falcon are the only *Falco* species of New World tropical moist forests. The Orange-breasted Falcon is 2.8 times heavier than the Bat Falcon, and the diets of these two falcons were separated mainly by prey size. They overlapped little in prey species (mainly birds but some bats, along with many insects by the Bat Falcon), but the difference in bird species taken appeared to result simply from the difference in prey size taken. Neither falcon normally flies beneath the forest canopy, although they may shoot through canyons and gaps in the foliage. For the Orange-breasted Falcon, most diet biomass was contributed by prey items larger than 175 g, and items smaller than 25 g made up only 3.3% of the diet. In contrast, 95% of Bat Falcon prey biomass was provided by items smaller than 40 g, many of these below 25 g. Because of the scant overlap in prey size, the potential for prey competition between these two falcons appears slight.

Among the owls, both congeneric species differed in habitat use and diet. Whereas the Mexican Wood Owl was ubiquitous throughout all forest types except the low Scrub Swamp Forest, the Black-and-white Owl, far more patchy in occurrence, was often found amid Scrub Swamp Forest near forest ponds—possibly because these attracted bats. Their diets differed substantially, with Black-and-white Owls taking many more bats than did Mexican Wood Owls, which took more rodents and insects. The two pygmy owls also differed markedly in habitat use. The Ridgway's Pygmy Owl at Tikal occurred mainly in the largely open farming landscape, with our few mature-forest detections in low Scrub Swamp Forest. In other areas, where the forest was drier and lower, this owl was more common in mature forest, as at Calakmul, 100 km north of Tikal, where much of the forest resembles Tikal's Scrub Swamp Forest. In contrast, the Central American Pygmy Owl was rare and local in tall Upland Forest at Tikal and is usually associated with tall, moist, closed-canopy forest, where we found it common some 100 km west of Tikal. Hence, this genus showed a strong difference in positions along the habitat axis.

In summary, among six congeneric species pairs, two differed mainly in habitat (*Buteo* spp. and *Glaucidium* spp.), two mainly in prey taxa (*Spizaetus* spp. and *Strix* spp.), and two mainly in prey size (*Micrastur* spp. and *Falco* spp.). A third *Buteo* species probably differed from its two congeners in both habitat and diet.

LIFE HISTORY PATTERNS OF TROPICAL AND TEMPERATE ZONE RAPTORS

Birds show a range of life histories aptly described as a "slow-to-fast" continuum (Saether 1987, 1988, 1989; Bennett and Owens 2002). Species near the " fast" end tend to reproduce early, often, and with large clutches and relatively short life expectancies. Birds at the "slow" end of the spectrum show the opposite combination of traits, with later reproduction, smaller clutch sizes, and longer life expectancy. Body size correlates strongly with position along this spectrum, with large birds generally occupying the "slow" end of the range and small birds the "fast" end. Tropical forest birds also occupy the slow end of the spectrum compared to temperate zone birds. It is generally believed that this slow-fast continuum results from a trade-off between current and future reproduction; organisms at the fast end of the spectrum emphasize current reproduction, while those at the slow

end emphasize survival and future reproduction. Such a present/future trade-off arises if there is a cost of reproduction—if current reproduction exacts a toll against future reproduction. A cost of reproduction has been demonstrated in many avian studies (Reznick 1985; Nur 1988; Linden and Moller 1989; Dijkstra et al. 1990; Lessels 1991; Daan et al. 1996).

Compared to temperate zone birds, tropical forest birds are believed to have (1) small clutches; (2) high adult survival rates and physiological longevity (postponed senescence); (3) low nesting success, largely because of heavy nest predation; (4) prolonged incubation and nestling periods (i.e., slow development); (5) many nesting attempts per year; and (6) extended post-fledging parental care.

Most life history comparisons between tropical and temperate zone birds have been based on songbird data. Newton (1979) found that tropical raptors had smaller clutches and slightly longer incubation and nestling periods than their temperate zone counterparts, but the remaining tropical/temperate contrasts listed here have not, to our knowledge, been tested for raptors. Here we test all six generalizations for raptors, combining our data with published data.

1. Small Clutch Size in the Tropics

It is well known that tropical forest birds, on average, lay smaller clutches than do birds of temperate and higher latitudes (Moreau 1944; Lack 1947; Klomp 1970; Yom-Tov et al. 1994). Raptors, too, show an increase in clutch size moving from the equator to higher latitudes, both within and between species. This has been shown within several falconiform (Newton 1979; Jenkins 1991; Van Zyl 1999; Simmons 2000) and owl species (Murray 1976). Among our study species, we found latitudinal clutch size variation in the Plumbeous Kite and Roadside Hawk (see Chapters 7, 12), as well as the Gray Hawk. Among species, Lack (1947) found that 17 falconiforms in mid-Europe showed a mean clutch of 3.3, compared to 1.8 for 11 species in equatorial Africa; for owls, he found a mean clutch of 4.7 for nine species in mid-Europe and 1.8 for three species in equatorial Africa. In Australia, latitudinal clutch size variation in raptors appears much less marked (Olsen and Marples 1993).

To examine possible relationships among body size, latitude, and family membership (accipitrids, falconids), we joined our Tikal data with those of Newton (1979), for a sample of 43 accipitrids and 11 falconids. For the accipitrids, analysis of variance with latitude (temperate vs. tropical), male mass, and an interaction term between the two found all three components significantly related to clutch size. Tropical accipitrids in this sample had smaller clutches than similar-sized temperate zone species (Fig. 23.4).[10] Among 15 temperate zone species, clutch size decreased strongly with increasing body size, according with a well-known pattern among birds.[11] In contrast, among 28 tropical species, the slope of the clutch-size–body mass relationship was indistinguishable

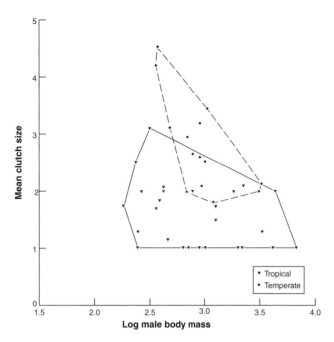

23.4 Species's mean clutch size versus male body mass for tropical and temperate zone accipitrid hawks. Data are from Tikal and from Newton (1979: Table 23).

from zero.[12] In the tropics, raptors over a range of body sizes have a clutch size reduced to a single egg, disrupting the decrease in clutch size with greater body size that is the norm at higher latitudes. Regardless of body size, tropical raptors in this sample laid small clutches.

2. High Adult Survival Rates in the Tropics

The best current evidence indicates that small, tropical land birds have substantially higher survival rates (often 70–80% annual survival) than do comparable birds at temperate and higher latitudes (often around 50–55%) (Faaborg and Arendt 1995; Greenberg and Gradwohl 1997; Johnston et al. 1997; Ricklefs 1997; Jullien and Clobert 2000; Sandercock et al. 2000; Stutchbury and Morton 2001). Is this true also of tropical forest raptors?

The 92–95% annual survival rates estimated for female and male Barred Forest Falcons (see Chapter 16), and 95.6% survival rate estimated for the Collared Forest Falcon (see Chapter 17) are among the highest survival estimates to date for any raptor, particularly for species this small. Similar-sized temperate zone raptors typically have annual survival estimates around 60–70% (Newton 1979).

We collated published adult raptor survival rates for 42 falconiform species, along with each species's body mass and the latitude where survival rates were studied. We found only eight survival estimates for raptors in the tropics, including two from our study. Among these eight species, the Savannah Hawk *(Buteogallus meridionalis)* was an outlier, with an estimated adult annual survival rate of only 71%—all other tropical species had survival rates estimated at 90% or more, averaging 93.9%

(SD = 2.3%, n = 7). The 71% survival estimate for the Savannah Hawk is no higher than rates often estimated for temperate zone species of similar size. However, this is an open country species, and tropical savanna dwellers are believed to show life history traits intermediate between those of tropical forest birds and north temperate zone birds (Moreau 1944; Klomp 1970).

Among these 42 species, body size explained 52% of the variance in survival rates, the larger raptors having higher survival. Controlling for body mass, survival rates were higher toward the equator, with latitude accounting for 12.0% of the remaining variance in survival.

3. Heavy Nest Predation and Low Nesting Success in the Tropics

Nest predation is often the leading cause of avian nest failure (Ricklefs 1969a; Martin 1992), and it has often been generalized that tropical birds suffer greater nest predation than temperate zone birds (Ricklefs 1969b; Skutch 1985). A number of tropical studies show nest predation rates of 80% or more, whereas predation rates of 40–60% are more typical for north temperate birds (Stutchbury and Morton 2001). Although Martin (1996) concluded that evidence for this assertion is inconsistent, Stutchbury and Morton (2001), reviewing largely the same literature, concluded that nest predation is indeed generally heavier for tropical than for north temperate birds. We agree with the latter authors that current data suggest generally higher rates of nest predation in lowland tropical forest than in north temperate localities (Nice 1957; Fogden 1972; Brosset 1990; Martin 1993; Morton and Stutchbury 2000; Robinson et al. 2000).

Among 13 raptor species at Tikal, the leading cause of nest failure was predation on eggs or chicks, with predation on adults also an occasional cause of failure (Table 23.8). Wind was the second most frequent cause of failure, destroying nests, causing nestlings to fall from their nests, and occasionally blowing over nest trees. Ectoparasites (mainly botflies) and harassment by Black-handed Spider Monkeys *(Ateles geoffroyi*: Plate 23.4) were more frequent sources of nest failure than was starvation, which was equaled in frequency by effects of Africanized honeybees. Even if all cases of nest abandonment and chick death are attributed to starvation, nest predation at Tikal was 10 times more frequent than starvation.

With regard to raptors, little attention has been paid to relative nesting success at both tropical and higher latitudes. Here we make an initial comparison, using our data along with those of Newton (1979). Results (Fig. 23.5) suggest that tropical raptors, at least in the smaller size classes, experience higher rates of nest failure than do raptors at higher latitudes. Similar results were obtained when accipitrids alone were examined. Analysis of covariance found nesting success significantly and positively related to male body mass, whereas latitude was near significant, with lower success in the tropics.[13]

Table 23.8 Causes[a] of nesting failure in raptors at Tikal[b]

Species	Nest predation[c]	Wind[d]	Hatching failure	Ectoparasites	Chick fell[e]	Predation on adult	Spider monkeys[f]	Abandonment	Chick died[g]	Starvation	Honeybees	Total
Gray-headed Kite	—	1	—	—	—	1	1	—	1	—	—	4
Swallow-tailed Kite	6	5	—	1	—	—	1	—	—	—	—	13
Double-toothed Kite	2	—	2	—	1	—	—	—	—	—	—	5
Plumbeous Kite	2	2[h]	—	—	—	—	—	—	—	—	—	4
Bicolored Hawk	2	—	2	—	—	—	—	—	—	—	—	4
Crane Hawk	4	3	—	—	—	—	—	—	—	1	—	8
White Hawk	1	1	2	—	—	—	—	—	—	—	—	4
Roadside Hawk	2	2	—	1	—	—	—	—	—	—	—	5
Ornate Hawk-eagle	2	1	—	—	1	—	—	—	—	—	—	4
Barred Forest Falcon	27	—	—	—	—	3	—	—	—	—	—	30
Laughing Falcon	5	—	—	1	—	—	—	1	—	—	1	8
Mexican Wood Owl	4	—	—	1	—	—	—	1	—	—	—	6
Black-and-white Owl	3	—	1	—	—	—	—	—	—	—	—	4
Totals	60	15	7	4	2	4	2	2	1	1	1	99
Mean percentage	50.2	18.5	12.7	3.8	3.5	2.7	2.5	2.2	1.9	1.0	1.0	

a Human disturbance not considered here; 4 Roadside Hawk nests failed due to human disturbance, and 1 Laughing Falcon nest.

b Total and partial nest failures combined; the great majority were total failures. Only species with 4 or more failures are included here; in addition, Hook-billed Kite had 1 failure due to nest predation and 1 due to wind, Great Black Hawk had 1 due to wind, Collared Forest Falcon had 2 due to nest predation, and Bat Falcon had 1 due to nest predation.

c Predation on eggs or nestlings.

d Causing chick, nest, or tree to fall.

e For no apparent reason.

f Harassment of chick, damage to nest.

g Reason unknown.

h At least; probably more.

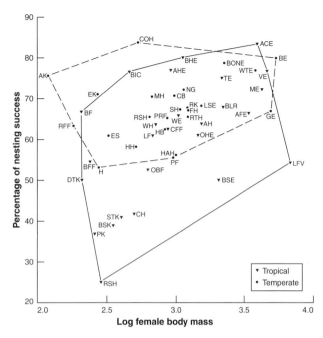

23.5 Mean nesting success of tropical and temperate zone raptors (falconids and accipitrids); percentage of nests fledging one or more young versus common log female body mass. Data are from Tikal and Newton (1979: Table 23). List of abbreviations follows:Tropical: ACE: African Crowned Eagle (now Crowned Hawk-eagle); AFE: African Fish Eagle; AH: African Hawk-eagle); AHE: Ayres's Hawk-eagle; BF: Bat Falcon; BFF: Barred Forest Falcon; BHE: Black Hawk-eagle; BIC: Bicolored Hawk; BLR: Bateleur; BSE: Brown Snake Eagle; BSK: Black-shouldered Kite (now White-tailed Kite); CFF: Collared Forest Falcon; CH: Crane Hawk; DTK: Double-toothed Kite; LF: Laughing Falcon; LFV: Lappet-faced Vulture; ME: Martial Eagle; OBF: Orange-breasted Falcon; OHE: Ornate Hawk-eagle; PK: Plumbeous Kite; RSH: Roadside Hawk; STK: Swallow-tailed Kite; TE: Tawny Eagle; VE: Verreaux's Eagle; WE: Wahlberg's Eagle; WH: White HawkTemperate: AK: American Kestrel; BE: Bald Eagle; BONE: Bonelli's Eagle; CB: Common Buzzard; COH: Cooper's Hawk; EK: European Kestrel (now Common Kestrel); ES: European Sparrowhawk (now Eurasian Sparrowhawk); FH: Ferruginous Hawk; GE: Golden Eagle; H: Hobby (now Eurasian Hobby); HAH: Harris' Hawk; HB: Honey Buzzard; HH: Hen Harrier; MH: Marsh Harrier (now Northern Harrier); NG: Northern Goshawk; PF: Peregrine Falcon; PRF: Prairie Falcon; RFF: Red-footed Falcon; RK: Red Kite; RSH: Red-shouldered Hawk; RTH: Red-tailed Hawk; SH: Swainson's Hawk; WTE: Wedge-tailed Eagle.

4. Prolonged Incubation and Nestling Periods (Slow Growth and Development) in the Tropics

It has often been generalized that tropical birds tend to have longer incubation and nestling periods and slower nestling growth than do birds at higher latitudes (Lack 1968; Ricklefs 1968, 1969b, 1976; Skutch 1976; Woinarski 1985). Is this true for raptors?

Newton (1979) presented evidence that both incubation and nestling periods in tropical raptors are longer than in temperate zone raptors. Using published data along with our data from Tikal, we verified that tropical raptor nestlings grow more slowly than those at higher

latitudes. Starck and Ricklefs (1998) present nestling growth rate constants for 42 falconiform species. To these we added data for eight falconiforms at Tikal, for a total sample of 50 species. As a measure of growth rate, we used the constant (K) used in fitting a logistic growth curve to the nestling growth data, as given by Starck and Ricklefs (1998) and as fitted to our own growth rate data. The logistic growth rate constant is an appropriate measure for comparing growth across species, as it is directly proportional to the rate at which nestling mass approaches its asymptote, is inversely proportional to the duration of the growth period, and is independent of the absolute duration of postnatal development (Starck and Ricklefs 1998). For each species, we calculated a mean value of K, using all replicate studies given by Starck and Ricklefs (1998: their Table 17.2). We calculated a mean body mass for each species from those given by Starck and Ricklefs (1998), using the mean of male and female values where possible. The geographic latitude of each study was estimated and a mean study latitude calculated for each species.

For these 50 species, body mass accounted for 74% of the variation in the logistic growth constant, growth being slower in larger species. To test whether growth rate was related to latitude, we removed the effect of body mass by saving the residuals of the above regression of K values on body mass, then regressed these residuals on the species's mean latitude where studied. The relationship between latitude and growth rate was pronounced, accounting for 38% of the remaining variance in the growth rate/body mass residuals. Nestling growth was faster at high latitudes than in the tropics. These data should be reanalyzed, controlling for phylogeny. However, since families, genera, and species are well distributed throughout the scatter plot, it seems unlikely that this conclusion will be reversed by such an approach. These raptor data support the long-standing generalization of slower nestling growth rates in the tropics than at higher latitudes.

5. Frequency of Nesting Attempts

Although the customary generalization is that tropical birds rear multiple broods per year more often than do birds at higher latitudes, there are two additional aspects of the question of breeding frequency: the proportion of species with nesting cycles longer than a year (and hence breeding every other year, at best), and the percentage of pairs failing to lay eggs in a given year, even among species with short nesting cycles. We address these three factors in turn.

More broods per year in the tropics? In some cases, the number of broods per year is the strongest determinant of annual fecundity, at least among songbirds (Martin 1995; Dhondt 2001). It has often been stated that tropical and south temperate zone birds generally raise more broods per year (and re-nest more often after failure) than do north temperate zone birds, and that this increases tropical fecundity above what might be supposed from clutch size alone (e.g., Murray 1985). If so, this might help compensate for the smaller clutches of tropical than temperate zone birds. For southeast Australia, Woinarski (1985) cited evidence that several species of insectivorous songbirds have more broods per year than do similar north temperate songbirds. Martin (1996) lists several other studies supporting the generalization of more broods per year in the tropics.

The number of successful nestings per year, however, is arguably the least-documented area of possible life history differences between tropical and temperate zone birds (Robinson et al. 2000). Reviewing evidence that re-nesting rates and number of broods per season are higher in tropical birds than in birds elsewhere, Martin (1996) found little evidence for this claim. In equatorial Sarawak, for example, Fogden (1972) found "no indication that any insectivorous forest species ever attempts a second brood after a successful first, the reason being... an extremely prolonged period of parental care," which effectively precludes second broods. For several species of insectivorous songbirds in French Guiana, Jullien and Thiollay (1998) also found that pairs reproduced no more than once yearly, probably because the young were fed for an extended period after fledging. Similar results are given by Willis (1967, 1972, 1973), Robinson et al. (2000), and Russell (2000). In summary, it appears questionable whether tropical songbirds, on average, raise more broods per year than do similar birds in the temperate zone.

Broods per year in tropical raptors. Among raptors, multiple broods per year occur regularly in only a small minority of species, in both the tropics and the temperate zone—necessarily among the smaller species, whose nesting cycles are short enough to permit this. Most tropical raptors, both large and small, raise only one brood per year (see below). Thus, multiple brooding does not help explain the evolution of smaller clutches in tropical raptors as a group.

We observed several raptor species at Tikal re-nest after a failed attempt, but we saw no evidence of multiple broods per year in any species, and we know of no evidence to date of multiple broods per year in any forest-dwelling Neotropical falconiform. Outside the tropics, most raptors also rear no more than one brood per year, and the breeding cycle typically takes up most of the favorable season (Newton 1979). In a few species, however, it is common for some pairs to raise two or more broods per year when prey is abundant. In both temperate and tropical zones, this is seen most regularly in the genera *Tyto* (Barn, Grass, and Bay Owls) and *Elanus* (Black-shouldered Kite and relatives)—both preying on grass-dwelling rodents that are prone to spectacular population irruptions and that often reproduce during all or much of the year.

In some tropical regions there are two annual wet seasons, occurring several months apart (Deshmukh 1986). In equatorial East Africa, under such a bimodal rainfall regime, some small raptors raise more than one brood

per year (Newton 1979). Based on current knowledge (Brown and Britton 1980), however, such species are a distinct minority. Double brooding is so far known only among species weighing less than about 1200 g (which have breeding cycles short enough to permit it), but even among these, it appears common in only a few species. Of 13 East African raptors of 1200 g or less with adequate data, three regularly double brood: the Black-winged Kite *(Elanus caeruleus)*, Common Barn Owl *(Tyto alba)*, and Pale Chanting Goshawk *(Melierax canorus)*. In addition, the African Goshawk *(Accipiter tachiro)* is suspected to double brood, the Black Kite *(Milvus migrans)* is known to do so at least rarely (in Australia), and a few pairs of Augur Buzzards *(Buteo augur)* have been known to produce young three times in 18 months (M. Virani, pers. comm.). For seven of these 13 smaller species, there is no evidence to date of double brooding. For 12 East African raptors larger than 1200 g, there is no indication of double brooding, and the length of their breeding cycles likely prohibits it (Brown and Britton 1980).

The situation is similar in southern Africa. Among 49 falconiform and 11 owl species discussed by Steyn (1982), he mentions only three falconiforms and one owl as sometimes rearing more than one brood per year: again the Black-winged Kite, Common Barn Owl, and Pale Chanting Goshawk (see above), as well as the tiny Pygmy Falcon *(Polihierax semitorquatus)*—all relatively small species with breeding cycles short enough to permit more than one annual brood. Among these 60 species, however, are many small species with short breeding cycles; that only three falconiforms and one owl are known to rear more than one brood per year indicates that such a strategy is uncommon, even among species with short breeding cycles.

In Australia, as well, only a few diurnal raptors are known to produce more than one brood per year. When their rodent prey are abundant, Letter-winged Kites *(Elanus scriptus)* sometimes produce several broods in succession, whereas Australian Kites *(E. axillaris)* may breed once in spring and again in autumn (Olsen 1995). Australian Kestrels *(Falco cenchroides)* sometimes have a spring and a summer brood, and Whistling Kites *(Haliastur sphenurus)* and Black Kites may breed more than once a year, although this is believed rare (ibid.). In North America and Europe a few species often have two or more broods per year, most notably the Harris's Hawk *(Parabuteo unicinctus)*, Southern Caracara *(Caracara plancus)*, and American Kestrel *(Falco sparverius)*. Two annual broods occur rarely in several other north temperate raptors (Cramp 1985).

In summary, rearing more than one brood per year is as unknown for most tropical raptors as it is for most temperate zone raptors, and does not compensate for the small clutches of tropical raptors on the whole. Rearing two or more annual broods is routine among a few hunters of grass-dwelling rodents at both tropical and temperate localities and is apparently common in a few other small, tropical, open country species, but it is not prevalent among tropical raptors in general.

Incidence of non-nesting. The reverse of multiple broods per year is the failure of some pairs to nest. Not all territorial raptor pairs lay eggs in a given year, and the incidence of nonlaying appears to differ between tropical and temperate zone raptors. Based on the data of Newton (1979), among temperate zone accipitrids, 84% of territorial pairs laid eggs in a given year (n = 15 species), while in tropical accipitrids (n = 12 African species), 70% of territorial pairs laid eggs—significantly different rates (Thorstrom et al. 2001; Panasci and Whitacre 2002). At Tikal, among nine species with one-year nesting cycles, $76.0 \pm 12.3\%$ of territorial pairs nested in a given year. Combining data from both studies, tropical raptor pairs nested less frequently than did north temperate raptor pairs. The same was true when body mass was controlled and when species with two-year nesting cycles were excluded. Thus, current evidence suggests that territorial raptor pairs in the tropics more often forego nesting than do similar-sized raptors at higher latitudes.

Length of the nesting cycle. Finally, the prolonged periods of nestling and post-fledging dependency of many tropical raptors—which preclude breeding every year—further diminish their reproductive rates compared to temperate zone raptors. Several raptor species at Tikal require more than a year from egg laying until the juvenile reaches independence. These species did not begin a new nesting attempt until their young reached independence—thus, they bred no more often, when successful, than every other year. Species with such a nesting cycle included the five largest at Tikal: the Crested Eagle, Ornate Hawk-eagle, Black Hawk-eagle, Great Black Hawk, and White Hawk. Among north temperate raptors, the longest nesting cycles are those of large eagles. Bald Eagles (*Haliaeetus leucocephalus*) reached independence 2–11 weeks after fledging (Buehler 2000), and Golden Eagles (*Aquila chrysaetos*) did so 32–85 days after fledging (Kochert et al. 2002). Thus, while we did not statistically investigate this trait in relation to latitude, it is clear that nesting cycles longer than a year are more common in tropical than in temperate zone raptors.

In summary, tropical raptors do not, on the whole, have multiple broods per year more often than temperate zone raptors. On the contrary, a high incidence of two-year breeding cycles in the tropics (resulting from prolonged post-fledging dependency), and a greater tendency toward non-nesting, result in even lower reproductive rates in tropical raptors than those implied by their small clutches.

6. Extended Post-fledging Parental Care in the Tropics

The number of broods raised per season is intimately connected to the duration of post-fledging care. After fledging a brood, a pair of birds can either lay another clutch or provide extended care to fledglings, but not both. Compiling information on 540 passerine species,

Russell (2000) tested and verified the generalization of longer post-fledging care in tropical and Southern Hemisphere birds than in north temperate zone birds, and similar results were obtained by Russell et al. (2004).

Russell (2000) also found evidence suggesting greater post-fledging survival in 34 tropical and Southern Hemisphere species than in 22 north temperate species; presumably, this may be related to the extent of post-fledging parental care. Russell (ibid.) suggests that such prolonged care is a critical aspect of avian life histories in the tropics that enhances juvenile survival and at least partly compensates for small tropical clutch sizes (see also Schaefer et al. 2004). The length of time fledglings remain dependent on adults tends to be quite variable even within a raptor species (Newton 1979). As a rule, however, young of larger species remain dependent longer than those of smaller species (ibid.). Newton (1979) generalized that in north temperate zone raptors, post-fledging dependency typically lasts for 2–3 weeks in small falcons and accipiters, 5–10 weeks in buteos and large kites, and up to several months in large vultures and eagles.

Several raptors at Tikal showed very prolonged post-fledging dependency. Juveniles of the White Hawk remained dependent for 17–19 months after fledging—to our knowledge, a record among raptors, and possibly, among birds. Another lengthy dependency period is that of the Crested Eagle, as our single radio-tagged juvenile showed no sign of nearing independence when we ended our study, exactly one year after it fledged (at age 479 days). Ornate Hawk-eagles reached independence 11–14 months after fledging, whereas the slightly smaller Black Hawk-eagle reached independence at 10 months after fledging. Great Black Hawks appeared to remain dependent for at least 11 months after fledging. These results are not surprising, as these are all large raptors with the exception of the White Hawk. In all the species just mentioned, adults skipped a year of breeding while the previous fledgling remained dependent.

Barred Forest Falcons left their natal area and appeared to reach independence 4–7 weeks after fledging, while the larger Collared Forest Falcon did so at 6–11 weeks or later. Bicolored Hawks and Crane Hawks moved away from natal areas 6–9 weeks after fledging, but in the Crane Hawk, juveniles traveled through the parental home range along with parents and remained dependent for at least four months after fledging. Likewise, Mexican Wood Owls appeared to remain dependent for about four months after fledging.

In contrast, fledglings of the Swallow-tailed Kite, Plumbeous Kite, Bat Falcon, and Orange-breasted Falcon appeared to become independent within a few weeks, although our data were less precise.

While we did not formally compare duration of post-fledging dependency in Tikal's raptors to that of raptors at higher latitudes, a striking proportion of the raptor species at Tikal had notably long post-fledging dependency. Our results support the notion that among raptors, extended parental care may help compensate for low fecundity in tropical species, as suggested for songbirds by earlier researchers (Russell 2000; Schaefer et al. 2004). Regardless of whether the *magnitude* of avian reproductive effort differs between tropical forests and temperate zone, there are latitudinal differences in the *nature* of parental effort, with greater emphasis on post-fledging care in the tropics, at least for some species of raptors.

Discussion

In summary, raptors conform to five of six common generalizations regarding avian life history differences between tropical and higher latitudes. Relative to temperate zone raptors, tropical raptors have smaller clutches, higher adult survival rates, lower nesting success and higher nest predation, longer incubation and nestling periods (slow growth and development), and longer post-fledging care. Tropical raptors do not, however, raise more broods per year than raptors at higher latitudes, as has sometimes been generalized for tropical versus temperate zone songbirds. Rather, among tropical raptors are more species with long post-fledging care and hence very long nesting cycles, resulting in nesting every other year, at best—and even among species with brief nesting cycles, tropical raptors more often fail to nest than do temperate zone raptors. Thus, infrequent nesting by tropical raptors further increases the disparity in fecundity between tropical and temperate zone raptor populations.

Existing data suggest that eggs and chicks of tropical raptors have slower development than those of temperate zone raptors, despite an apparent tendency toward lower nest success and higher nest predation in the tropics. In parallel with data for other kinds of birds, this poses a conundrum: if nest predation is more frequent in the tropics than at higher latitudes, why has it not selected for more rapid development of eggs and young? Ricklefs (1969a) offers a potential answer—that other adaptations for reducing nest predation may be less costly than ones requiring a change in developmental tempo. Alternatively, Skutch (1949) suggests that nest predation in the tropics has selected for slow provisioning rates (by way of small clutches and slow nestling development), which may make it harder for some nest predators to find nests. Or, food during nesting may be more limited in tropical forests than at higher latitudes; some data suggest this to be true (Fogden 1972; Janzen and Pond 1975; Hails 1982; Thiollay 1988; Lovette and Holmes 1995; Jullien and Thiollay 1998; Simmons 2000). Most likely of all, however, is that clutch size, developmental rates, and rates of parental care are all adjusted to both food availability and nest predation pressures (see Martin 1992).

It may also be that prolonged incubation and slow nestling development are intrinsically advantageous, perhaps through increased immunocompetence or enhancement of physiological longevity that arises through mechanisms as yet unknown (Ricklefs 1992, 1993; Tella et al. 2002)—with these hypothetical benefits being

greater in the tropics. Finally, there is the possibility that slow growth and small clutches in the tropics simply reflect greater life expectancy in tropical species, and a resultant, more risk-averse approach to current reproduction, in order to maximize adult survival and lifetime reproduction (Williams 1966). This notion is bolstered by good evidence in a variety of birds, including raptors, of higher survival rates in the tropics (see earlier).

Are the tropics a region most notable for greater avian longevity, greater food scarcity, or greater risk in nesting relative to temperate and boreal zones—or all three? For birds in general, this question is key, but the answer remains elusive (Martin 1996, 2004; Stutchbury and Morton 2001). There is evidence that survival rates (and presumably physiological longevity) in tropical raptors are higher, on average, than in temperate zone raptors of similar size. It is possible that higher tropical survival rates are ultimately manifest in lower food abundance. High survivorship may lead to large numbers of "floaters" (adults without a territory), to compression of territory size, and to depressed food availability. Moreover, the great diversity of potential food competitors in tropical forests—many of them non-raptors and non-avian—may lead to a high degree of diffuse competition, generally depressing food abundance for tropical raptors. The result of this scenario may be that food abundance is generally so low that clutch sizes larger than one egg are not possible or advantageous—and that, fairly often, food abundance is too low to merit the risk of nesting at all.

COMPARATIVE BREEDING BIOLOGY OF TIKAL'S RAPTORS

Raptor species at Tikal differed from one another in both nesting habitat and the type of nest sites used, but many shared a preference for emergent or isolated nest trees (Plate 23.5), and often nested in forest types where such trees were especially common. Two or more species used the same nests or nest sites in turn, and in some cases there was evidence that nest site shortage limited nesting frequency. Nest site reuse (perhaps indicating limited sites) was more frequent in cavity nesters than in stick nesters and more frequent in large versus small raptors (Fig. 23.6). For many species, breeding habitat appeared saturated. Non-territorial "floaters" were evident in a few species, and several incidents suggested intraspecific competition for a place in the breeding population. In short, we believe that competition for limited territories and nest sites is frequent among Tikal's raptors.

Breeding was highly seasonal, mostly beginning in the last weeks of the dry season, with nestlings reared early in the rainy season (Fig. 23.7). As elsewhere, larger raptors began nesting earlier than smaller species, and nesting dates appeared to be associated with diet. The sequence of nesting by raptors with different diets diverged from results found elsewhere and accorded largely with known patterns of timing of peak prey availability at

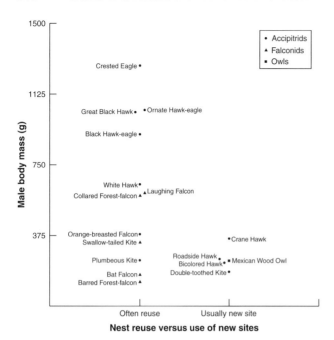

23.6 Use of new nests or cavities each year versus reuse of sites, by body size and family membership, for raptors at Tikal.

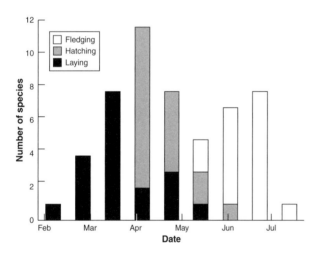

23.7 Frequency histogram for estimated species' peak dates of laying, hatching, and fledging for raptors studied at Tikal.

Tikal. Insect eaters nested early and lizard eaters late (Fig. 23.8). Mammal eaters also nested early, whereas bird eaters showed a wide spread of nesting dates. Most other data on nesting seasons of tropical raptors derive from Africa, where most raptors also fledge young early in the rainy season (Benson et al. 1971; Brown and Britton 1980; Steyn 1982).

Our work added several examples to the group of raptors in which males and females share incubation roles nearly equally, with both sexes meeting their own food demands by hunting throughout the incubation phase (Fig. 23.9). This is in contrast to the more common pattern, in which females perform nearly all incubation and males nearly all hunting during the first half of the

nesting cycle. We hypothesize that certain prey characteristics make self-feeding more efficient than the male delivering prey to the female at the nest. Such prey characteristics include small size, wide dispersal, and unpredictability in time and space. Moreover, raptors at Tikal showing little sexual size dimorphism tended to share nesting roles relatively equally, whereas those showing greater size dimorphism partitioned nesting roles more strongly (Fig. 23.9), as suggested long ago (Hill 1944).

Raptors at Tikal with high degrees of sexual size dimorphism laid larger clutches than did less dimorphic species. This was true when 28 species from three raptor families were analyzed,[14] and also when analysis was restricted to 16 species of the Accipitridae.[15] Clutch mass also appeared to increase, though less markedly, with degree of size dimorphism. Diet and the degree of dimorphism appeared equally associated with clutch size: a diet rich in birds was a significant dietary predictor of a large, heavy clutch. Annual nesting productivity was highest in bird-eating raptors owing to their larger clutches; however, bird eating showed no association with either nesting success or frequency of nesting attempts.

Such a link between a high degree of sexual size dimorphism and larger clutches in raptors has been rarely noted in the past, perhaps only by Mueller and Meyer (1985) for falconiforms of North America and the western Palearctic. To account for these patterns is a challenge to hypotheses seeking to explain the varying degrees of sexual size dimorphism in raptors. We consider it likely that the marked correlation between clutch size, diet, and degree of size dimorphism is causally related to food resources, breeding energetics, and mortality patterns. Conceivably, bird eating both permits and selects for a large clutch relative to other diets. It seems plausible that size dimorphism facilitates harvest and efficient use of maximal avian prey biomass by a pair of raptors, helping to produce the three-way correlation between a high degree of dimorphism, a bird-rich diet, and a large clutch. Bird eating may also entail high mortality, and hence select for high reproductive rates, as suggested by others (Mueller and Meyer 1985).

Across species at Tikal, neither egg success nor nest success was related to clutch size.[16] Of overall egg failures, 81% resulted from complete nest failure and 19% from partial nest failure. This is consistent with other avian studies and suggests that nest predation was the main source of nest failure. The latter was verified by direct tabulation of causes of nest failure: half of all cases were due to predation, followed by wind, failure to hatch (infertile or unviable), ectoparasites, and other causes (Table 23.8). Starvation was rare.

For 15 species, the number of fledglings per nesting attempt averaged 0.82 ± 0.39 (*n* = 15), and the number of fledglings per territorial pair-year averaged 0.61 (*n* = 15). The number of fledglings per territory-year bore a loose, negative relationship with body size, similar to that between clutch size and body size. Not surprisingly, mean clutch size was the variable most strongly correlated with productivity per territory-year.

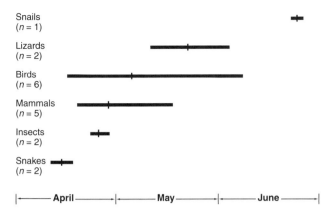

23.8 Median and range of estimated peak hatch dates for 18 Tikal raptor species (16 falconiforms and 2 owls) categorized by main prey type (biomass diet).

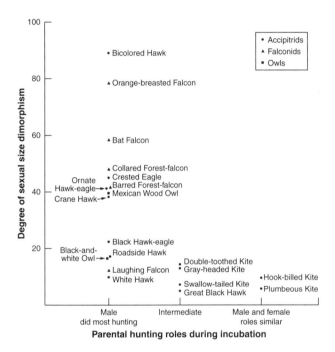

23.9 Parental hunting roles during incubation versus degree of sexual size dimorphism (percentage by which female mass exceeds male mass). In species with similar male and female hunting roles, both played similar incubation roles.

Even among species capable of nesting yearly, not every territorial pair nested each year, which must affect productivity. Among these 15 species, 72% of pairs, on average, nested in a given year (range = 50–93%). The best model describing productivity per territory-year was a regression including clutch size, hatching success, and percentage of pairs nesting[17] (total model variance explained = 91%). Clutch size, hatching success, and percentage of pairs nesting accounted for 70%, 25% and 6% of explained variance, respectively.

Moreover, percentage of pairs nesting appeared to be related to diet—in particular, to dietary niche breadth.

The narrower the diet at the level of prey class, the greater was the percentage of pairs nesting in a given year, with dietary niche breadth accounting for 40% of the variance in percentage of pairs nesting.[18] Thus, dietary niche breadth had a pronounced relationship with frequency of nesting, and a slight relationship with productivity per territory-year, with raptor species having narrower diets nesting more often and producing slightly more young per territory-year, compared with raptors with broader diets. These may be the first data suggesting that among a set of avian predators, more specialized diets are associated with more frequent nesting and higher productivity.

As noted earlier, not all territorial pairs laid eggs every year. If we omit species with two-year nesting cycles, more frequent nesting was associated with large body size, a small, less massive clutch, and low nesting success. Clutch size controlled for body mass did not differ with percentage of vertebrates versus invertebrates in the diet.[19] Female body size, and not male size or degree of size dimorphism, was positively correlated with degree of nesting success.[20] This may suggest that females of larger species were more effective in nest defense against predators.

SPATIAL USE AND CONSERVATION OF NEOTROPICAL FOREST RAPTORS

The largest threat to most Neotropical lowland forest raptors is the ongoing destruction and modification of these forests by humans (Plate 23.6). Evidence suggests that many of these raptors are highly reliant on mature or little-altered forest (Jullien and Thiollay 1996; D. F. Whitacre et al., unpubl. data), and unlikely to persist in the absence of substantial forest stands (Leck 1979; Willis 1979; Willis and Eisenmann 1979; Alvarez-López and Kattan 1995; Robinson 1999). Thus, the spatial needs of these raptors are likely to dictate the area of forest required for viable populations to persist. Here we review existing data and present new information on the spatial needs of Neotropical lowland forest raptors.

Spatial Use and Body Size in Vertebrates

Two aspects of spatial use have been widely studied in vertebrates: size of home range and population density.[21] These two factors need not be tightly coupled because neighboring home ranges may overlap or lie far apart. Virtually all studies to date show larger home ranges with increasing body mass in birds and mammals, with slope (exponent of log-transformed body mass) often between 1.1 and 1.4.[22] Accordingly, evidence suggests that ecological population density (i.e., density in areas actually occupied by a species) in birds and mammals declines with increasing body size, generally scaling with a body mass exponent between –0.5 and –1.0 (Blackburn and Gaston 1997, 1999).

Home Range Size, Nesting Density, and Body Size in Raptors

Previous data, largely from the temperate zone, show that home range size in raptors typically increases, and density of nesting pairs decreases, with larger body mass (Newton 1979; Peery 2000). Home range area and space per pair (the reciprocal of pair density) ranges from 1 km² or less in the smallest hawks to 100 km² and more in some large eagles (Newton 1979). For 32 species of Holarctic falconiforms and owls, Peery (2000) found that home range size increased in direct proportion to body mass (with a mass exponent of 1.01).[23] Newton (1979) noted that raptor breeding densities are generally lower than expected from home range size, apparently because areas of less suitable habitat are not occupied. Larger areas often include a greater range of habitats than do small areas, and thus are likely to contain more habitat that is suboptimal for a given species. Thus, densities may often be lower over large rather than small areas (ibid.).

Spatial Needs of Neotropical Forest Raptors: Prior Research

To date, virtually all knowledge of the spatial needs of Neotropical forest raptors stems from the study of Thiollay (1989a, 1989b) at a French Guiana site, and that of Terborgh et al. (1990) and Robinson and Terborgh (1997) along the Manu River in Amazonian Peru. A few additional data come from Venezuela (Mader 1981) and Brazil (Klein and Bierregaard 1988).

In the French Guiana study, Thiollay (1989a, 1989b) single-handedly assayed populations of diurnal raptors in plots totaling 275 km² of primary forest. Thiollay estimated pair densities within three nested study plots 6, 42, and 100 km² in extent, and assayed presence and absence of 25 raptor species in eleven 25 km² rain forest study plots scattered throughout French Guiana. For most species, Thiollay derived density estimates by plotting on a map over-flights of soaring adults, presumed to reveal approximate territory or home range boundaries. Thiollay's study provided the first estimates of population densities of forest raptors at a Neotropical site. Limitations of sampling intensity and detection methods, however, lend some uncertainty to his findings.

At Manu, in Amazonian Peru, Robinson and Terborgh (1997) estimated raptor populations via spot-mapping on a 1 km² plot in continuous floodplain forest and a 10 km² less intensively studied area. Densities of species with home ranges larger than the 10 km² study plot may have been somewhat overestimated (S. Robinson, pers. comm.).

We based density estimates at Tikal on more concrete data: spacing between neighboring nesting pairs, and home range areas estimated by radiotelemetry—data gained during thousands of hours of fieldwork over an eight-year period. Here we report estimates of pair density for 16 species and home range size for 14 species at Tikal. We then employ these estimates to project the number of raptor pairs that might be initially contained in forest areas of different sizes, in the hope that such projections may prove useful to conservation planners.

Body Size and Space Needs of Neotropical Forest Raptors: Results from Tikal, French Guiana, and Peru

Table 23.9 and Appendix 23.1 give our home range estimates for several raptor species at Tikal, and Table 23.10 our estimates of pair densities. We express pair density as its inverse, the area per pair. For 14 falconiform species at Tikal, space per pair scaled as the 1.216 exponent of log male body mass, with this regression accounting for 61% of the variance in estimated space per pair (Fig. 23.10).[24,25] Based on 11 of the same species,[26] home range size scaled with a male body mass exponent of 1.206,[27] with body mass accounting for 70% of the variance in home range size (Fig. 23.11). For 10 species,[28] home range area accounted for 69% of the variance in space per pair, with an exponent of 0.907.[29] Thus, space per pair scaled 90% as steeply as home range, implying greater overlap among species with larger home ranges, as has been noted in several previous studies of vertebrates. The Y intercept of this regression was 0.21, indicating some unoccupied space between adjacent home ranges. We suggest that our estimates of space per pair, as opposed to home range estimates, are the data most useful for conservation planning.

At Manu, Peru, the 14 species with most reliable density estimates yielded a near-significant relationship of density to body mass,[30] accounting for 14% of the variance in density, with a slope of 0.60. The larger species all occurred at relatively low densities, while small species ranged from low to very high density. The Bat Falcon

Table 23.9 Home range table

Species	Home range estimate (km²)
Gray-headed Kite	(6.2)[a]
Hook-billed Kite	None
Swallow-tailed Kite	None
Double-toothed Kite	(2.0)[a]
Plumbeous Kite	None
Bicolored Hawk	4.8
Crane Hawk	6.93
White Hawk	2.84
Great Black Hawk	None
Roadside Hawk	0.14
Crested Eagle	25[b]
Black Hawk-eagle	12.9[b]
Ornate Hawk-eagle	18.3
Barred Forest Falcon	1.146
Collared Forest Falcon	10.35
Laughing Falcon	15.53
Bat Falcon	None
Orange-breasted Falcon	None
Mexican Wood Owl	0.217
Black-and-white Owl	4.373

[a] Inadequate data for a reliable estimate.
[b] Probably an underestimate.

Table 23.10 Summary table of space per pair (km²) at Tikal

Species	Conservative ("conservation") estimate	Best estimate	High-density (optimistic) estimate	Confidence level
Gray-headed Kite	20	14.5	8	Low
Hook-billed Kite	—	—	—	—
Swallow-tailed Kite	43	—	—	Low
Double-toothed Kite	3.14	3.14	3.14	High
Plumbeous Kite	6.7	6.7	0.9	Low
Bicolored Hawk	4.58	4.15	4.15	High
Crane Hawk	7.82	7.82	7.82	High
White Hawk	7.14	5.2	5.2	High
Great Black Hawk	21.8	20.0	20.0	High
Roadside Hawk, forest	0.872	0.872	0.872	High
Roadside Hawk, farming	0.71	0.71	0.71	High
Crested Eagle	100	60	30	Low
Black Hawk-eagle	25	19	19	Medium
Ornate Hawk-eagle	10.8	10	10	High
Barred Forest Falcon	1.14	1.14	1.14	High
Collared Forest Falcon	9.22	8.5	8.5	High
Laughing Falcon, forest	26	21.6	21.6	High
Laughing Falcon, farming	5.43	4	4	High
Bat Falcon	—	—	—	—
Orange-breasted Falcon	—	—	—	—
Mexican Wood Owl	0.2532	0.2532	—	High
Black-and-white Owl	—	—	—	—
Guatemalan Screech Owl	< 0.2532	< 0.2532	< 0.2532	Low

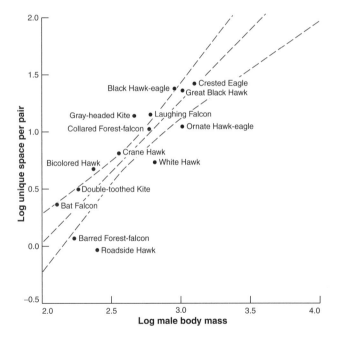

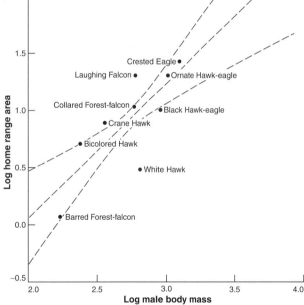

23.10 Space per pair versus male body mass for 14 falconiform species at Tikal, with 95% confidence limits for slope, ordinary least squares regression. Space per pair (km²) is the reciprocal of pair density.

23.11 Home range area versus male body mass for nine falconiform species at Tikal, with 95% confidence limits for slope, ordinary least squares regression.

and Tiny Hawk (*Accipiter superciliosus*) were outliers, with low densities for their body size; with these two omitted, a significant regression accounts for 54% of estimated space per pair, with a male body mass exponent of 1.280.[31]

In French Guiana, for the 24 species occurring in primary forest, Thiollay's data show no relationship between rank density and rank body mass. This may have resulted from limited sampling intensity or methods used in that study, or from inclusion of more species, some having low densities for their body size.

Comparative Raptor Densities at Three Sites

Estimated raptor densities were generally higher at Manu, Peru, than at Tikal, but with broad overlap. Estimated densities in French Guiana were much lower than at Tikal and Manu. While we believe there were some real differences in densities among these sites, divergence in field methods and sampling intensity makes it impossible to be certain how much densities actually differed. As detailed later, however, there is good reason to expect raptor densities to vary among Neotropical forest sites.

Predicting Spatial Needs of Neotropical Forest Raptors

Densities and spatial needs of raptors in disparate localities can be known with certainty only through local research. In the absence of such data for most sites, we can make generalizations about space use that may, with caution, have some predictive value. Below, we use mainly our estimates of pair densities at Tikal to project likely raptor populations in forest areas of different sizes. For the Harpy Eagle, we rely on densities estimated by Alvarez-Cordero (1996) in Venezuela and Panama.

Risks Inherent in Extrapolating Spatial Use Patterns among Sites

There are obvious risks in extrapolating from density or home range information gathered at one or a few sites. Neither home range size nor the density of territorial pairs are fixed species attributes, but rather, tendencies that vary within wide ranges. Home range size and breeding density in raptors have often been found to vary in concert with prey abundance or productivity, with such variations sometimes exceeding an order of magnitude within a single species. In Eurasian Sparrowhawks *(Accipiter nisus)*, mean inter-nest spacing in continuous woodland in different regions of the United Kingdom indicated a 20-fold range in pair density (Newton 1986), with Sparrowhawks nesting farther apart in areas of lower prey abundance. Likewise, pair density in the Common Buzzard *(Buteo buteo)* varied by a factor of 15–22 among local areas in Britain, being highest where prey were most abundant (ibid.). Queen Charlotte Island Goshawks *(Accipiter gentilis laingi)* at times had home ranges 3–70 times larger than those of similarly sized Northern Goshawks *(A. g. atricapillus)* in Oregon and California (Austin 1993; Hargis et al. 1994; Keane and Morrison 1994; Titus et al. 1994; Woodbridge and Detrich 1994), this enormous difference being attributed to differences in habitat and prey abundance in the two regions. In summary, many cases are known among raptors in which pair densities within a species vary by a factor of 20 or more among situations differing in habitat quality or food abundance. Similar patterns are well known in mammalian Carnivora, with 10- to 20-fold density variations common, and up to 100-fold in some cases (Gittleman and Harvey 1982; Lindstedt et al. 1986; Herfindal et al. 2005; Marker and Dickman 2005).

Such potentially large density variation makes it risky to extrapolate conclusions on spatial needs from

one situation to another. Still, there is a clear need to generate rough guidelines regarding spatial needs of Neotropical forest raptors for use in conservation planning, making this risk tolerable.

Population Sizes of Potential Conservation Interest

Before we discuss the ability of different-sized habitat areas to conserve raptor populations, it is necessary to decide what population sizes should be considered. Population sizes of potential conservation interest range from 50 or fewer to many thousand individuals of a given species. Demographic stochasticity is thought to present a threat for populations smaller than roughly 100 individuals (Lande 1993). From a purely genetic perspective, a "50/500" rule of thumb has often been applied: that isolated populations need an effective population size $(N_e)^{32}$ of at least 50 over the short term to avoid inbreeding depression, and at least 500 over the long term to avoid loss of genetic diversity due to genetic drift (Franklin 1980). To achieve an effective population size of 500 may require an actual population much larger—perhaps up to 4500 (Frankham 1995). Lande (1995) suggested that an effective population of 5000 may be needed to avoid loss of potentially adaptive genetic variation and to avoid negative effects of detrimental mutations, implying a possible need for an actual population much larger. Reed et al. (2003) predict that 7000 adult vertebrates are necessary for a 99% probability of persistence over 40 generations (see also Gaillard et al. 2000; Inchausti and Halley 2001).

Lower standards would no doubt be appropriate for smaller protected areas that might contribute to regional persistence, at least over brief time horizons. Using data on 62 bird species, Reed et al. (2003) suggested that about 125 breeding pairs might yield 50% probability of persistence for 40 generations. For birds on California's Channel Islands, species with at least 100 breeding pairs had a 90% probability of persisting for 80 years (Jones and Diamond 1976; Primack 2002). One study suggests that isolated raptor populations as small as 20 pairs may sometimes be capable of long-term viability (Walter 1990).

Maintaining a population of several thousand tropical forest raptors in a single protected area is a demanding goal, given the low densities of many species. Thus, in considering the utility of different sizes of protected areas, we should consider not only optimal, enormous areas, but also areas that may at least avoid rapid local extinction. Truly viable populations of Neotropical forest raptors will likely require reserve networks and human involvement.

Considering the above, we suggest that areas large enough to contain 250 pairs of many raptor species should be of great conservation value, but areas supporting even 50 pairs of many species may be of value, especially if periodic gene flow can be achieved. We focus here on predicting the initial raptor populations of

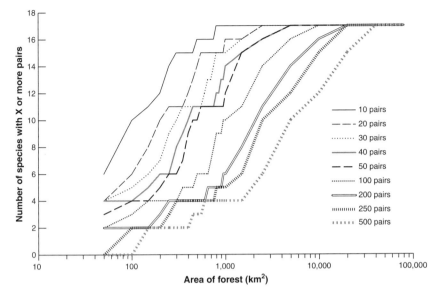

23.12 Number of raptor species estimated to reach specified initial populations in successively larger habitat areas. Based on 16 raptors at Tikal—from small to large, the Mexican Wood Owl, Guatemalan Screech Owl, Roadside Hawk, Barred Forest Falcon, Double-toothed Kite, Bicolored Hawk, Plumbeous Kite, White Hawk, Crane Hawk, Collared Forest Falcon, Ornate Hawk-eagle, Gray-headed Kite, Great Black Hawk, Black Hawk-eagle, Laughing Falcon, and Crested Eagle—plus the Harpy Eagle.

differently sized areas. We ignore the fact that protected areas below a certain size may suffer various ecological changes resulting from small size and isolation, which may reduce their ability to sustain initial raptor populations over time.

Size of Area Required for Conservation of Neotropical Lowland Forest Raptors

Projections of the numbers of various raptor species that might be initially contained in areas of different size can be made using Figure 23.12 and Tables 23.9–23.11. Densities reflected in the figure and tables are best regarded as maximum expected densities, since they were calculated over relatively small areas, often in situations of apparently high density. Larger areas, which are likely to include habitat less suitable for some species, may contain fewer pairs than suggested by these figures.

It seems appropriate to focus on those species occurring at lowest densities, as these represent a limiting factor for conservation of an entire assemblage. Several lines of evidence suggest a density of roughly one pair/100 km² for several of the larger and less common Neotropical lowland forest raptors. The species generally using the largest home range is probably the Harpy Eagle, for which densities have been estimated as one pair/45–79 km² in a Venezuelan study area (Alvarez-Cordero 1996). A conservative approach is to use the upper end of this range, one pair/80 km², or even one pair/100 km², and to assume that densities over large areas might often be half this great, or lower. The Crested Eagle—second-largest widespread Neotropical forest raptor—is likely also to be among the most space-demanding species. A single radio-tagged female at Tikal yielded a minimum home range estimate of 50 km², and in French Guiana, Thiollay (1989a) estimated one or two pairs/100 km².

Some other species appear often to occur at densities as low as, or lower than, those of the Harpy and Crested

Eagles. In French Guiana, these included the Gray-bellied Hawk *(Accipiter poliogaster)*, Black Caracara *(Daptrius ater)*, Bat Falcon, Tiny Hawk, and Hook-billed Kite, all detected at densities of one pair/90–100 km² or lower (Thiollay 1989a, 1989b). We suggest it is more conservative to assume half these densities in conservation planning, and some truly rare species typically occur at far lower densities. For example, Orange-breasted Falcons occurred at an estimated one pair/3000 km² in French Guiana and at roughly one pair/1770 km² in our study area (see Chapter 20).

If one uses 500 individuals (roughly 250 pairs) as a potentially viable population, then 25,000–50,000 km² or more may be required in order to have a high probability of initially containing this many pairs of the less common species. In areas of highest Harpy Eagle density, such as Darien, Panama, local densities of one pair/10–20 km² (Alvarez-Cordero 1996) or 18–20 km² (J. Vargas, pers. comm.) may occur. In areas of such high density, it is more conservative to assume a density of one pair/40 km² or more, since these density estimates excluded areas where no pairs were known. In areas with Harpy Eagle densities this high, this eagle may not be the lowest-density raptor, and other uncommon raptors may be more appropriate targets for conservation planning. In sum, we suggest that an area of 100 km²/pair for the less common raptors is an appropriate figure for use in planning, or 200 km²/pair for a maximally cautious approach.

While an area of 25,000–50,000 km² is the smallest predicted to potentially contain 250 pairs of the less common raptor species, much smaller areas may have conservation value for many other raptors. Areas of 5000 km² may contain 50 pairs or more of all 17 species evaluated here, and 250 pairs of 12 species (Table 23.11). Thus, an area of 5000–10,000 km² might contain 500 or more adults of all but the several lowest-density raptor species (Fig. 23.12, Table 23.11). The potential for conservation of the majority of the raptor assemblage drops rapidly

Table 23.11 Minimum habitat area needed to initially contain different numbers of pairs for 17 raptor species, based on data from Tikal

Species	Space/pair estimated at Tikal (km²/pair)	Minimum area predicted to initially contain the following numbers of raptor pairs (km²)													
		10	20	30	40	50	75	100	200	250	500	750	1000
Mexican Wood Owl	0.2532	2.5	5.1	7.6	10.1	12.7	19.0	25.3	50.6	63	127	190	253
Guatemalan Screech Owl	0.2532	2.5	5.1	7.6	10.1	12.7	19.0	25.3	50.6	63	127	190	253
Roadside Hawk, forest	0.87	8.7	17.4	26.1	34.8	43.5	65	87	174	218	435	653	870
Barred Forest Falcon	1.14	11.4	22.8	34.2	45.6	57	86	114	228	285	570	855	1140
Double-toothed Kite	3.14	31.4	62.8	94.2	125.6	157	236	314	628	785	1570	2355	3140
Bicolored Hawk	4.58	45.8	91.6	137.4	183	229	344	458	916	1145	2290	3435	4580
Plumbeous Kite	6.7	67	134	201	268	335	503	670	1340	1675	3350	5025	6700
White Hawk	7.14	71	143	214	286	357	536	714	1428	1785	3570	5355	7140
Crane Hawk	7.82	78	156	235	313	391	587	782	1564	1955	3910	5865	7820
Collared Forest Falcon	9.22	92	184	277	369	461	692	922	1844	2305	4610	6915	9220
Ornate Hawk-eagle	10.8	108	216	324	432	540	810	1080	2160	2700	5400	8100	10,800
Gray-headed Kite	20	200	400	600	800	1000	1500	2000	4000	5000	10,000	15,000	20,000
Great Black Hawk	21.8	218	436	654	872	1090	1635	2180	4360	5450	10,900	16,350	21,800
Black Hawk-eagle	25	250	500	750	1000	1250	1875	2500	5000	6250	12,500	18,750	25,000
Laughing Falcon, forest	26	260	520	780	1040	1300	1950	2600	5200	6500	13,000	19,500	26,000
Crested Eagle	50	500	1000	1500	2000	2500	3750	5000	10,000	12,500	25,000	37,500	50,000
Harpy Eagle	80	800	1600	2400	3200	4000	6000	8000	16,000	20,000	40,000	60,000	80,000

below about 3000 km² (or 6000 km², assuming densities half of those at Tikal: Fig. 23.12, Table 23.11). If we take 50 pairs as the lowest objective of conservation interest, then areas of 400–1000 km² might initially contain this many pairs of at least half the 17 species evaluated here. To contain 20 pairs of all but the least common species, an area of 500 km² may suffice (or 1000 km², assuming half the densities we estimated). This is roughly the size of our study area, Tikal National Park (576 km²).

While there is no single minimum area predicted to yield conservation benefits for Neotropical forest raptors, one might roughly state that areas of 400–500 km² are of conservation interest, areas of 1000 km² are of far greater interest, and areas of 2500 km² and larger should be of substantial value. While the greatest raptor conservation value accrues in areas of 25,000–50,000 km² or larger, even areas of 100 km² should have value for the smallest, highest-density raptor species. Because densities at any given locale could be lower than what is suggested here, local density data would provide a much stronger basis for planning the size of a particular protected area.

Geographic Variation in Raptor Abundance

Particular raptor species appear to be more abundant in some areas than in others and may thus be more easily conserved in areas where they achieve higher density. The Great Black Hawk was much less abundant at Tikal than at both South American sites, and the Black-and-white Hawk-eagle (Spizaetus melanoleucus) was much more common in French Guiana than at Tikal or Manu. Crane Hawks were rare or absent in mature forest at the two South American sites, but common at Tikal. The density of Orange-breasted Falcons we found in Belize is so far unparalleled elsewhere (see Chapter 20).

Species Occurring at Unexpectedly Low Densities

Some species occurred at low densities for their body size at two or at all three sites, possibly because they require specific conditions found only at a few locations within the forest. Such eclectic density patterns limit our ability to predict spatial needs from body size. Orange-breasted Falcons were rare at all sites, and Bat Falcons uncommon. Short-tailed Hawks were among the least common species at all sites, and Hook-billed Kites rather uncommon. At both South American sites, Black Caracaras and Gray-bellied Hawks were rare, and Tiny Hawks uncommon for their body size.

For several species in French Guiana, Thiollay (1989b) attributed sparse populations to habitat specificity, while several other species he found inexplicably rare. At Tikal, we found several of the same species to be locally patchy in occurrence as did Thiollay. Species whose patchiness at Tikal appeared to be due to habitat specificity or nest site rarity included the Swallow-tailed Kite,[33] Plumbeous Kite, Bat Falcon, Orange-breasted

Falcon, and Black-and-white Owl. Species that appeared uncommon at Tikal for reasons unknown to us were the Gray-headed Kite, Hook-billed Kite, and Black Hawk-eagle. Either pairs often went undetected, used larger home ranges than we estimated, or required special habitat situations, or their populations were limited by something other than habitat availability.

Variation in Soil Fertility, Net Primary Production, and Animal Abundance among Neotropical Forest Sites

There is good reason to expect raptor densities to vary among Neotropical lowland forest sites, because these sites vary tremendously in the density of many faunal elements comprising raptor prey. Many studies have found animal densities higher at sites with richer soil (Freese et al. 1982; Emmons 1984; Gentry and Emmons 1987; Bierregaard 1990; Eisenberg 1990; Lovejoy and Bierregaard 1990; Malcolm 1990; Peres 1997, 1999; Spironello 2001; Haugaasen and Peres 2005; Palacios and Peres 2005). Correlations between soil fertility and net primary production (NPP: plant productivity) provide a basis for such variations in faunal density. Above-ground coarse wood productivity of 104 Neotropical forest plots varied by a multiple of 3.5 among sites (Malhi et al. 2004). Although this measure comprises only a fraction of total production, current data indicate it is proportional to total above-ground NPP. There was no apparent relationship between productivity and rainfall, length of dry season, or amount of sunshine, but a strong positive relationship between wood production and soil fertility, which showed strong regional patterns correlated with soil fertility (ibid.). These results suggest that differences in soil fertility may be responsible for a more than threefold difference in the energy flow driving different Neotropical lowland forest ecosystems. One might expect this to affect densities of raptors, which are at the apex of ecosystem energy flow pyramids.

Conservation Implications of Regional Differences in Vertebrate Densities

About 95% of Amazonian forests occupy nutrient-poor soils (Palacios and Peres 2005). Since these forests typically support low biomass of mammals and some other fauna, they may support relatively low densities of raptors as well. If so, this implies that larger protected areas will be required for raptor conservation in such forests than in more productive alluvial or flooded forests along rivers or those benefited by volcanic deposits. It is possible that raptor densities over much of Amazonia will yet prove more similar to the very low densities Thiollay reported in French Guiana than to those we found at Tikal or those at Manu, Peru.

As the Harpy Eagle is among the most space-demanding Neotropical forest raptors, it is noteworthy that the large mammals providing much of its diet vary greatly in density among neighboring forest types. Several studies show that obligate folivores including sloths and howler monkeys (frequent Harpy Eagle prey; Plate 23.7) are far more abundant in frequently flooded riparian forests than in unflooded forests nearby (Peres 1997, 1999; Haugaasen and Peres 2005). Peres (1999) notes that the abundance of these folivores "becomes drastically reduced beyond a critical distance from rivers throughout western Amazonia." This pattern suggests that Harpy Eagle carrying capacity may be substantially greater in riparian than in nearby terra firme forests.

Conservation Strategies for Patchily Distributed Species

Certain raptors that show great regional differences in abundance, or that are highly patchy in occurrence at local scales, may not be best served simply by large protected areas. For such species, protection of areas of known occurrence or high density may be most important. One example is the Orange-breasted Falcon. This rare falcon nests largely on cliffs, which are absent in most areas. An appropriate conservation strategy for this species may be to identify the densest populations and ensure protection of nest sites and surrounding habitat. A similar strategy may be useful for species associated with wetlands and other discrete, recognizable habitat features.

Human-Caused Mortality

For the larger and more conspicuous Neotropical raptors, shooting of raptors and their prey species appears to be an important conservation issue. In his Venezuelan study area, Alvarez-Cordero (1996) documented 43 Harpy Eagles shot or eaten and 20 captured alive over a several-year period. He concluded that shooting is a serious threat to Harpy Eagle populations. Likewise,

Thiollay (1984) estimated that the 50 inhabitants of a village in French Guiana shot 50 raptors per year of a range of species. He also speculated that human hunters reduced prey populations for some of the larger raptors. Shooting of raptors is probably the single raptor conservation issue that is easiest to address, for example, through public education.

Spatial Needs and Conservation: Summary

There is no single minimum size of protected area that offers utility in conserving Neotropical lowland forest raptors; rather, a range of sizes may play different conservation roles. An area of 25,000–50,000 km^2 or much larger may be required to initially support 250 pairs of Harpy Eagles and some other uncommon raptors. Currently conceived minimum viable population sizes of several thousand adults are unlikely to be achieved for most tropical forest raptors within a single protected area. Retaining these species within many protected areas may require corridors permitting population interchange, direct human assistance in gene flow or dispersal, and public education to reduce persecution.

On the other hand, areas as small as 400–500 km^2 should have considerable conservation value for many raptor species, and even smaller areas should house many pairs of the less space-demanding raptors. Areas of 2500 km^2 should be of great value. In the interest of caution, the areas just stated should be considered minima, and might be doubled.

Because NPP varies by a factor of three or more among Neotropical lowland forest sites, we should perhaps expect substantial geographic variation in raptor densities, and current, imperfect data suggest this to be the case. Thus, spatial needs for raptor conservation may vary greatly across the lowland Neotropics, even between neighboring sites.

Home range estimates

Species	Best home range estimate (km²)	Home range method	Number of locations	Adequate estimate?	Notes
Gray-headed Kite	6.2	MCP	12	No	12 locations during a 2-week period
Hook-billed Kite	None	—	—	—	—
Swallow-tailed Kite	None	—	—	—	—
Double-toothed Kite	2.0	MCP	15	No	15 points over 2 months, post-fledging
Plumbeous Kite	None	—	—	—	—
Bicolored Hawk	5.0	85% HM	68	Yes	Adult male during breeding, 1991
	5.9	100% MCP	68	Yes	Same as above
	4.0	85% HM	50	Yes	1992 male during breeding season; some overlap among neighbors
	3.8	100% MCP	50	Yes	
	4.5	—	—	—	= Mean of two 85% HM estimates
	4.8	—	—	—	= **Mean of two MCP estimates**
	4.68	—	—	—	= Mean of all four estimates
Crane Hawk	7.6	100% MCP	27	Yes	Arroyo Negro male, 1994, over 9 months, nonbreeding, after losing chick
	3.9	85% HM	27	Yes	Same data as above
	10.45	100% MCP	71	Yes	Chikintikal female, 1994 over 15 months (most of two breeding seasons, plus intervening nonbreeding period; using the entire 15 months may overestimate her annual home range; thus, we prefer the male's estimate)
	7.75	85% HM	71	Yes	
	9.03	—	—	—	Mean of male's and female's MCP estimates
	6.93	—	—	—	**Mean of all four estimates**
White Hawk	2.48	MCP	55	Yes	1991 Baren male, breeding season
	2.32	MCP	27	Yes	Brecha Sur male, 10 months, nonbreeding, no dependent juvenile
	3.35	95% HM	27	Yes	Same as above
	2.9	MCP	42	Not quite	1993 Brecha 4 male; breeding season, still slowly increasing w/more points
	2.81	95% HM	42	Not quite	Same as above
	2.57	—	—	—	Mean of all three MCP estimates
	2.72	—	—	—	Mean of three estimates from the two best data sets
	2.64	—	—	—	Mean of three males above (using MCP and HM for each bird)
	2.84	—	—	—	**Median of three males above (using MCP and HM for each bird);** we suspect actual home ranges may be a bit larger than these estimates; space/pair estimates (5.13 km²/pair) may reflect real home range size
Great Black Hawk	none	—	—	—	—
Roadside Hawk	0.09	MCP	31	Probably	**Non-nesting male, paired, during breeding season, over 2.5-month period**
	0.12	85% HM	31	Probably	Same male; he remained within 350 m of center of his usage area
	0.14	95% HM	31	Probably	Same male
Crested Eagle	24.2	MCP	28	Underestimate	Adult female during eight months, mostly post-fledging, with dependent fledgling
	23.5	85% HM	28	Underestimate	Same as above
	25.0	rounded up	28	Underestimate	Estimate is very possibly too small
Black Hawk-eagle	**9.65**	MCP	19	Underestimate	**Adult male arroyo Negro, 7 months, with dependent fledgling**
	12.9	95% HM	19	Underestimate	Same data as above
Ornate Hawk-eagle	9.6	MCP	42	Underestimate	Arroyo Negro male, over 11.5 months with dependent fledgling (1993/94)
Male, poor estimate	9.5	95% HM	42	Underestimate	Same as above
	5.7	85% HM	42	Underestimate	Same as above
	8.27	—	59	Underestimate	Mean of these three for this male
Good male 1	14.2	MCP	59	Yes	Naranjal male, during one nesting effort, over 6 months; no other estimates for this male this year (1992)
Good male 2	18.8	MCP	108	Yes, maybe Overestimate	Naranjal male, spanning 16 months including some of fledgling dependency period and throughout next nesting effort in 1991/92
	14.8	85% HM	108	Yes, maybe Overestimate	Same as above

APPENDIX 23.1—*cont.*

Species	Best home range estimate (km²)	Home range method	Number of locations	Adequate estimate?	Notes
Good female 1	8.3	95% HM	108	Overestimate	Same as above
	16.5	—	—	—	Mean of MCP estimates for Naranjal males in 2 years
	13.2	MCP	64	Yes	Naranjal female, 8 months, with dependent fledgling
	11.3	95% HM	64	Yes	Same
	7.7	85% HM	64	Yes	Same
	10.7	—	—	—	Mean of these three estimates for this female
Good female 2	20.3	MCP	87	Yes	Brecha Norte female, 7 months with a dependent juvenile, and 1 month after it dispersed; plus the following pre-nesting period
	18.1	95% HM	87	Yes	Same
	11.6	85% HM	87	Yes	Same
	16.7	—	—	—	Mean of these three estimates for this female
Good female 3	25	MCP	48	Yes	Temple 4/5 female, 1990/91, same time span as above
	26.1	95% HM	48	Yes	Same
	14.2	85% HM	48	Yes	Same
	21.8	—	—	—	Mean of these three estimates for this female
	19.5	—	—	Yes	Mean of MCP estimates for these three females
	16.4	—	—	—	Mean of the means of three methods for these three females
	15.35	—	—	—	Mean of mean estimates for two males
	15.9	—	—	—	Mean of male and female means
	18.3	—	—	—	**Mean of the MCP estimate for each of these five birds**
	16.0	—	—	—	Mean of the mean estimate for each of these five birds
Barred Forest Falcon	1.228	MCP	—	Yes	Mean of 11 breeding males, with mean of 5.5% home range overlap
	0.98	95% HM	—	Yes	Same as above
	1.146	85% HM	—	Yes	**Same as above with mean of 2.8% home range overlap**
Collared Forest Falcon	7.1	MCP	90	—	Breeding male
	11.8	MCP	160	—	Breeding male
	9.45	Mean	—	—	Mean of these two males
	11	MCP	103	—	Breeding female
	13.5	85% HM	103	—	Same
	10	MCP	—	Yes	Mean of two breeding males and one breeding female
	10.7	85% HM	—	Yes	Mean of two breeding males and one breeding female
	10.35	Mean	—	Yes	**Mean of these MCP and 85% HM means for these three birds**
Laughing Falcon, primary forest	25.3	MCP	74	Yes	Male, breeding season, Casita 1990; some clearings around ruins included
	14.6	MCP	91	Yes	Female, Casita 1990, breeding season; some clearings around ruins included
	20.0	Mean	—	Yes	Mean of male and female 1990 Casita estimates
	6.7	MCP	61	?	Casita male 1992, late nesting and post-fledging period
	13.3	Mean	—	Yes	Mean of the 1990 two-sex mean and the 1992 male mean above
	15.5	Mean	—	Yes	**Mean of all three Casita estimates, equally weighted**
	6.5	—	41	Probably an underestimate	San Antonio male, 1990; 40% of times we got no signal
	13.3	Mean	—	—	Mean of all four estimates above
Bat Falcon	none	—	—	—	—
Orange-breasted Falcon	none	—	—	—	—
Mexican Wood Owl	**0.217**	MCP	—	Yes	**Mean of six males in breeding season**
	0.208	85% HM	—	Yes	Same
Black-and-White Owl	2.616	MCP	208	Yes	One male during breeding seasons of 2 years
	4.373	85% HM	208	Yes	One male during breeding seasons of 2 years

Chapter 23 Notes

1. Including low-density species such as the Peregrine Falcon *(Falco peregrinus)* and Merlin *(F. columbarius)*.

2. Including low-density species such as the Harpy Eagle *(Harpia harpyja)* and Orange-breasted Falcon *(Falco deiroleucus)*.

3. And sometimes a second *Megascops* species, often less common and/or in habitat different from the first.

4. The Gray-headed Kite *(Leptodon cayanensis)*, however, is known to take some snails elsewhere; we gathered very few dietary data on this kite.

5. Snyder and Wiley (1976) compiled dietary data for 17 owls and 27 falconiforms in North America, mostly from files of the U.S. Fish and Wildlife Service, and mainly from stomach contents. Numbers of prey items, rather than biomass, were given. To better approximate biomass, we weighted vertebrate prey by a factor of 5 relative to invertebrate prey. Sample sizes were generally large, with 17 of the 44 raptor species treated having at least 1000 prey items and all but 12 species having at least 100. The Osprey *(Pandion haliaetus)* and Bald Eagle *(Haliaeetus leucocephalus)* were omitted from comparisons, as no water-associated raptors were studied at Tikal. The snail-eating Everglades Kite *(Rostrhamus sociabilis)* was included in those comparisons not restricted to forest-dwelling raptors, as was the snail-eating Hook-billed Kite *(Chondroheirax uncinatus)* at Tikal.

6. The Common Barn Owl *(Tyto alba)* has been ignored here, as it is mainly an open country species.

7. Although Ridgway's *(Glaucidium ridgwayi)* and Northern Pygmy Owls *(G. californicum)* likely have lower vertebrates or birds as their leading prey type at times, and other owl species as well, under special conditions.

8. $P = 0.026$.

9. $P = 0.070$.

10. For 43 accipitrids, tropical and temperate, both latitude code $(P = 0.002)$ and male body mass $(P = 0.0002)$ were significant predictors, along with the interaction between the two $(P = 0.011)$.

11. For 15 temperate zone accipitrid species, male mass accounted for 49% of the variance in clutch size (smaller clutches with increasing body size), with $F_{1,13} = 14.323$, $P = 0.002$, adjusted $r^2 = 0.488$).

12. For tropical accipitrids, regression of clutch size on body mass gave a nonsignificant relationship $(P = 0.096)$.

13. For 50 raptor species (11 falconids and 39 accipitrids— our data, data from Newton's [1979] Table 23, and data from Vargas [1995]), analysis of covariance found male body mass a significant predictor of nesting success $(P = 0.002)$, with latitude code (temperate versus tropical) near significant $(P = 0.11)$, success tending to increase with latitude; overall $F_{3,46} = 5.893$, $P = 0.002$, adjusted $r^2 = 0.231$.

14. For 28 raptor species occurring at Tikal (16 accipitrids, 5 falconids, and 7 owls; Harpy Eagle excluded), regression of mean clutch size on male body mass and degree of sexual mass dimorphism (percentage by which female mass exceeds male mass) gave a significant regression with both predictor variables highly significant. Clutch size decreased with increasing male size $(P = 8.3 \times 10^{-7})$ and increased slightly with degree of size dimorphism $(P = 5.9 \times 10^{-4})$. Overall $P = 4.7 \times 10^{-7}$, adjusted $r^2 = 0.663$, SE = 0.379, $F_{2,25} = 27.56$.

15. For 16 accipitrid species occurring at Tikal, regression of mean clutch size on female body mass and degree of sexual mass dimorphism gave a significant regression with both predictor variables highly significant. Clutch size decreased with increasing female size $(P = 0.031)$ and increased slightly with degree of size dimorphism $(P = 0.004)$. Overall $P = 0.008$, adjusted $r^2 = 0.449$, SE = 0.367, $F_{2,13} = 7.10$. Results were similar using male rather than female mass.

16. For 16 Tikal species with adequate data, regressing egg success on mean clutch size gave a nonsignificant regression, with $P = 0.841$ and adjusted $r^2 = 0.000$.

17. Young produced per territory-year, regressed on mean clutch size, egg success, and percentage of pairs nesting, gave a significant regression $(P = 1.18 \times 10^{-6})$ with all three predictor variables significant, with adjusted $r^2 = 0.911$, SE = 0.09, and $F_{3,11} = 49.0$.

18. If we omit species with two-year nesting cycles and the Roadside Hawk (for which our results on percentage of pairs nesting seemed unlikely to be typical), for 12 Tikal species, percentage of pairs nesting, when regressed on dietary niche breadth, gave a significant negative relationship, with $P = 0.015$, adjusted $r^2 = 0.408$, SE = 11.83, $F_{1,10} = 8.569$.

19. Clutch size/body mass residuals regressed on biomass percentage of vertebrates in diet gave a nonsignificant regression, with $P = 0.435$ and adjusted $r^2 = 0.000$.

20. For 15 Tikal species with adequate data, egg success regressed on female body mass gave a near-significant regression, with $P = 0.056$, adjusted $r^2 = 0.195$, $F_{1,13} = 4.384$, with larger species having higher success.

21. Home range is defined as the area traversed by an animal in its normal activities of food gathering, mating, and caring for young (Burt 1943).

22. McNab 1963; Armstrong 1965; Schoener 1968; Harestand and Bunnell 1979; Gittleman and Harvey 1982; Mace and Harvey 1983; Owen-Smith 1988; Swihart et al. 1988; du Toit 1990; Mysterud et al. 2000; Nunn and Barton 2000; Kelt and Van Vuren 2001.

23. Using ordinary least squares regression.

24. Regression of log space per pair (km²/pair) on log male body mass gave the following: regression coefficient (male body mass exponent) = 1.253, adjusted $r^2 = 0.611$, $F_{1,12} = 21.457$, $P = 0.0006$, $n = 14$.

25. Male body mass was a better predictor of space per pair than was female body mass. Regression of log space per pair on log female body mass gave the following: regression coefficient = 1.216, adjusted $r^2 = 0.552$, $F_{1,12}$, $P = 0.001$, $n = 14$.

26. Here we omitted 3 of the 14 species used above: the Roadside Hawk, for which only one home range was estimated, and which was a serious outlier, and the Plumbeous Kite and Great Black Hawk, for which we had no home range estimates.

27. Regression of log home range size (km²) on log male body mass gave the following: regression coefficient (male mass exponent) = 1.206, adjusted $r^2 = 0.700$, $F_{1,9} = 24.323$, $P = 0.0008$, $n = 11$.

28. Here the Crested Eagle is omitted because this species's pair density estimate is based on its estimated home range size; that is, the two are not independent.

29. Regression of log space per pair on log home range size $(n = 10$ species, Crested Eagle omitted), gave the following: regression coefficient (home range exponent) = 0.907, adjusted $r^2 = 0.724$, $F_{1,8} = 21.014$, $P = 0.002$, $n = 10$.

30. $P = 0.103$.

31. Regression of log space per pair on log male body mass for raptors at Manu, Peru, data from S. Robinson (1994 and pers. comm.) gave the following: regression coefficient (male mass exponent) = 1.280, adjusted $r^2 = 0.543$, $F_{1,10} = 14.067$; SE = 0.337, $P = 0.004$, $n = 12$.

32. Effective population size is the size of an "ideal" population that would be affected by genetic drift at the same rate as the actual population in question (Kalinowski and Waples 2002). The effective population is smaller than the actual population to the extent that the latter departs from non-overlapping generations, an even sex ratio, constant population size, random survival of progeny, random mating, and equal reproductive success of individuals (ibid.).

33. Colonial nesting also contributed to patchy distributions of this kite.

APPENDIX 1

Body mass and sexual size dimorphism data for Maya Forest raptor species

Sex	Mass (g)[a]	Mean mass (g) ± SD	Locality	Source	Percentage by which females exceed males (%)	(Cube root) value of dimorphism index (%)	Mean of dimorphism indexes for wing, culmen, body mass (%)
Gray-headed Kite (*Leptodon cayanensis cayanensis*)							
F	416, 530	473 + 80.6 (n = 2)	Surinam	3	—	—	—
M	range = 414–465	444 (n = 7)	Surinam	3	—	—	—
F	540[b]	540 (n = 1)	Panama	13	—	—	—
M	435, 445[b]	440 + 7.1 (n = 2)	Panama	13	—	—	—
M	524[b]	524 (n = 1)	Trinidad	5	—	—	—
F	475	475 (n = 1)	Belize	6	—	—	—
F	643[c]	643 (n = 1)	Tikal	1	—	—	—
M	553[c]	553 (n = 1)	Tikal	1	—	—	—
F	range = 416–643	*521 + 84.4 (n = 5)	All data combined	—	13.0	4.1	2.24
M	range = 414–553	*461 (n = 11)	All data combined	—			
Hook-billed Kite (*Chondrohierax uncinatus uncinatus*)							
F	235–300 (n = 4)	256 (n = 4)	Surinam	3, 31, 34	—	—	—
M	251–275 (n = 6)	255 (n = 6)	Surinam	3, 31	—	—	—
F	—	296 + 22.2 (n = 3)	Panama	4	—	—	—
M	—	265 + 11.6 (n = 4)	Panama	4	—	—	—
F	350 (n = 1)	—	Venez/Brazil border	22	—	—	—
M	310 (n = 1)	—	Venez/Brazil border	22	—	—	—
F	282 (n = 1)	—	Belize	6	—	—	—
F	258 (n = 1)	—	Near Tikal	30	—	—	—
M	252 (n = 1)	—	Near Tikal	30	—	—	—
F	—	296.3 (n = 6)	All data combined except Haverschmidt and Mees (1994)	—	—	—	—
M	—	270.3 (n = 6)	All data combined except Haverschmidt and Mees (1994)	—	—	—	—
F	—	*280.2 (n = 10)	All data combined	—	—	—	—
M	—	*263.3 (n = 11)	All data combined	—	6.4	2.1	1.44
Swallow-tailed Kite (*Elanoides forficatus yetapa*)							
F	range = 372–435 (n = 4)	midpoint = 404	—	2	—	—	—
M	390, 407	399 + 23 (n = 2)	—	2	—	—	—
F	range = 372–435 (n = 6)	*401 (n = 6)	Surinam	3	—	—	—
M	range = 354–407 (n = 5)	*382 (n = 5)	Surinam	3	5.0	1.6	0.94
F	448 (n = 1)	448 (n = 1)	Tikal	11	—	—	—
M	410 (n = 1)	410 (n = 1)	Tikal	11	—	—	—
F	500 (n = 1)	500 (n = 1)	Uaxactún (near Tikal)	30	—	—	—
Both	—	"450"	Tikal	32	—	—	—
Swallow-tailed Kite (*Elanoides forficatus*; both subspecies combined)							
F	—	434 (n = 6)	Florida, Guatemala, Surinam	19	—	—	—
M	—	441 (n = 6)	Florida, Panama, Surinam	19	—	—	—
Double-toothed Kite (*Harpagus bidentatus fasciatus*)							
F	190–229 (n = ?)	midpoint = 210 (n = ?)	—	2	—	—	—
M	175–198 (n = ?)	midpoint = 187 (n = ?)	—	2	—	—	—

APPENDIX 1—*cont.*

Sex	Mass (g)[a]	Mean mass (g) + SD	Locality	Source	Percentage by which females exceed males (%)	(Cube root) value of dimorphism index (%)	Mean of dimorphism indexes for wing, culmen, body mass (%)
F	228.8 (n = 1)	—	Oaxaca	7	—	—	—
M	197.7 (n = 1)	—	Oaxaca	7	—	—	—
M	190 (n = 1)	—	Belize	6	—	—	—
F	206 (n = 1)	—	Panama	4	—	—	—
M	165 (n = 1)	—	Panama	4	—	—	—
F	189.5 (n = 1)	—	Tikal	16	—	—	—
M	174.7 (n = 1; immature)	—	Tikal	16	—	—	—
F	—	*208.1 ± 20 (n = 3)	Data above	—	14.4	4.5	4.63
M	—	*181.9 ± 15 (n = 4)	Data above	—	—	—	—
Double-toothed Kite (*Harpagus bidentatus bidentatus*: South American subspecies)							
F	196–212 (n = 5)	201.5 (n = 5)	Surinam	3	20.3	6.2	5.19
M	152–187 (n = 11)	167.5 (n = 11)	Surinam	3	—	—	—
Plumbeous Kite (*Ictinia plumbea*)							
F	232–280 (n = 6)	midpoint = 256	—	2	—	—	—
M	190–267 (n = 5)	midpoint = 228.5	—	2	—	—	—
F	232–294 (n = 7)	*257 (n = 7)	Surinam	3	5.8	1.9	1.99
M	190–297 (n = 16)	*243 (n = 16)	Surinam	23, 34	—	—	—
F	374.1 (n = 1) very fat	—	Oaxaca	7	—	—	—
M	257 (n = 1) stomach full	—	NE Nicaragua	33	—	—	—
F	302 (n = 1)	—	Belize	6	—	—	—
M	282, 251, 261, 255 (n = 4)	262 ± 14 (n = 4)	Belize	6	—	—	—
Bicolored Hawk (*Accipiter bicolor bicolor*)[d]							
F	584.8,[e] 446.1[f] (n = 2)	515.5	Oaxaca, Mexico	7	—	—	—
F	453.6 (n = 1)	—	Campeche, Mexico	8	—	—	—
F	430 (n = 1)	—	Panama	9	—	—	—
F	425 (n = 1)	—	Colombia	10	—	—	—
F	342 (n = 1)	—	Surinam	3	—	—	—
F	467, 450, 474,[f] 475, 480	469 ± 11.7 (n = 5)	Tikal	1	—	—	—
M	249.6 (n = 1)	—	Campeche, Mexico	8	—	—	—
M	235 (n = 1)	—	Panama	4	—	—	—
M	190, 250 (n = 2)	220 (n = 2)	Colombia	10	—	—	—
M	204, 206 (n = 2)	205 (n = 2)	Surinam	3	—	—	—
M	233,[f] 250, 235, 250, 273[g]	248.2 ± 16.0 (n = 5)	Tikal	1	—	—	—
F	Overall, version 1	447 ± 78.5 (n = 6)	All data except Tikal	—	—	—	—
M	Overall, version 1	222.4 ± 25.8 (n = 6)	All data except Tikal	—	—	—	—
F	Overall, version 2	*457 ± 57.2 (n = 11)	All data combined	—	—	—	—
M	Overall, version 2	*234.1 ± 24.8 (n = 11)	All data combined	—	95	22.2	18.67
Black Crane Hawk (*Geranospiza caerulescens nigra*)							
F	519.5 (n = 1)	519.5 (n = 1)	Oaxaca	7	—	—	—
M	340, 358 (n = 2)	349 ± 12.7 (n = 2)	—	2	—	—	—
F	434, 532 (n = 2)	483 ± 69 (n = 2)	Tikal	1	—	—	—
M	375 (n = 1)	375 (n = 1)	Tikal	1	—	—	—

Sex	Mass	Mean mass	Locality	Source(s)			
F	Overall	*495 ± 53 (n = 3)	Above data combined	1, 2, 7	38.3	10.8	8.23
M	Overall	*358 ± 17.5 (n = 3)	Above data combined	1, 2, 7	—	—	—
Crane Hawk (*Geranospiza caerulescens caerulescens*)							
F	Range = 273–353 (n = 6)	311 (n = 6)	Surinam	3	18.3	5.6	
M	Range = 230–295 (n = 3)	263 (n = 3)	Surinam	3, 34	—	—	
White Hawk (*Leucopternis albicollis ghiesbreghti*)							
F	695, 715, 720[b] (n = 3)	710[b] ± 13 (n = 3)	Tikal	1	9.9[i]	3.1	4.5
M	592, 655, 683 (n = 3)	643 ± 47 (n = 3)	Tikal	1	—	—	—
M	652 (n = 1)	—		2			
F	Overall	*710 ± 13 (n = 3)	All of above data	1			
M	Overall	*645.5 ± 38.3 (n = 4)	All of above data	2			
White Hawk (*Leucopternis albicollis costaricensis*)							
F	650 (n = 1)	—	E. Panama	9			
White Hawk (*Leucopternis albicollis albicollis*)							
F	780–908 (n = 4)	837 (n = 4)	Surinam	3			
M	605, 653 (n = 2)	629 (n = 2)	Surinam	3			
M	600, 670 (n = 2)	—	1 probably Surinam; other ?	2, 34			
Great Black Hawk (*Buteogallus urubitinga ridgwayi*)							
F	995 g	995 g (n = 1)	Tikal	1			
M	1013 g	1013 g (n = 1)	Tikal	1			
F	625–1400 (n = 7)	1042 ± 254 g (n = 7)	1 each from Tikal, Belize, Yucatán Peninsula, and an unspecified site; 3 from Panama	1, 2, 4, 6, 8, 13			
F	900–1400 (n = 6)	*1111 ± 192 g (n = 6)	Same as above, but deleting 625 g female from Panama	1, 2, 4, 6, 8, 13	7.2	2.3	4.7
M	854–1306 (n = 5)	*1036 ± 165 g (n = 5)	1 each from Tikal, Belize, Yucatán Peninsula, Tamaulipas (Mexico), and an unspecified site	1, 2, 6, 8, 12			
M	652	6461 ± 38 (n = 4)	site	2			
Gray Hawk (*Buteo (Asturina) nitidus*: presumably these are masses for *B. n. plagiatus*)							
F	*637 (n = 4)		—	18	53.1	14.2	11.1
M	*416 (n = 5)		—	18	—	—	—
F	636 (n = 1)		Arizona	19			
M	434 (n = 1) (these two a mated pair)		Arizona	19			
F	572, 605 (n = 2)		—	2			
M	364, 404 (n = 2)		—	2			
Gray Hawk (*Buteo (Asturina) nitidus constaricensis*)							
F	420 (n = 1)		Panama	4			
Roadside Hawk (*Rupornis magnirostris griseocauda*)							
F	263–315 (n = 4)	*286 ± 22.3 (n = 4)	Tikal	1, 42	17.2	5.3	5.0
M	239–246 (n = 4)	*244 ± 3.32 (n = 4)	Tikal	1, 42	—	—	—

Sex	Mass (g)[a]	Mean mass (g) + SD	Locality	Source	Percentage by which females exceed males (%)	(Cube root) value of dimorphism index (%)	Mean of dimorphism indexes for wing, culmen, body mass (%)
F	291, 314	—	—	2, 42	—	—	—
F	303.0 (juvenile); 307.3 (adult)	—	Tikal	16, 42	—	—	—
Roadside Hawk (Rupornis magnirostris probably griseocauda)							
F	—	272 ± 11.7 (n = 7)	Panama	4, 42	—	—	—
M	—	266 ± 8.6 (n = 9)	Panama	4, 42	—	—	—
F	—	284.2 (n = 15)	All griseocauda data combined				
M	—	259.2 (n = 13)			9.7	3.1	4.2
Roadside Hawk (Rupornis magnirostris magnirostris)							
F	257–350 (n = 8)	*289 (n = 8)	Surinam	3, 42	—	—	—
M	206–290 (n = 9)	*247 (n = 9)	Surinam	3, 34, 42	17.0	5.2	—
Short-tailed Hawk (Buteo brachyurus)[j]							
F	520, 425, 457	*467 ± 48 (n = 3)	Florida, Mexico, Colombia	3, 19	—	—	—
M	417, 363, 470	*417 ± 54 (n = 3)	Florida, Mexico, Suriname	3, 19	12.0	3.8	7.8
Crested Eagle (Morphnus guianensis)							
F	1750 (n = 1)	—	—	2	—	—	—
F	1948 (n = 1)	—	Tikal	1	—	—	—
F	range = 1750–1948	*1849 ± 140 (n = 2)	Mean of 2 above	above	—	—	—
F	1697 (n = 1)[k]	—	Tikal	13	—	—	—
F	2226 (n = 1)[l]	—	[i]	3	—	—	—
M	*1275 (n = 1)	—	Surinam	3	—	—	—
M	≈ 1200 (n = 1)[m]	—	[m]	14	45	12.4	7.8
Harpy Eagle (Harpia harpyja)							
F	7600–9000 g	(midpoint = 8300 g*)	—	20	—	—	—
M	4000–4800 g	(midpoint = 4400 g)	—	20	—	—	—
M	4000–4600 (n = 4)	mean ≈ 4400*	—	2 (probably includes Surinam data)			
M	4000, 4494 g	4250 (n = 2)	Surinam	3	ca. 90	21.1	12.6
F	7598 (n = 1, adult)	—	Guyana	35	—	—	—
F	7541 (n = 1, juvenile)	—	Guyana	35	—	—	—
M	4819 (n = 1, juvenile)	—	Guyana	35	—	—	—
Black-and-white Hawk-eagle (Spizaetus melanoleucus)							
F	*1191 (n = 1)	—	Belize	6	—	—	—
M	*780 (n = 1)	—	Surinam	3	53	14.1	11.3
Black Hawk-eagle (Spizaetus tyrannus serus)							
F	1092, 1123, 1150 (n = 3)	—	Surinam	3, 2	—	—	—
F	1090, 1122	—	Tikal	1	—	—	—
F	—	*1115 ± 25 (n = 5)	The above, combined	3, 1	—	—	—
F	1141 (juvenile)	[n]	Tikal	1	—	—	—
M	904 (n = 1)	—	Surinam	3, 2			

Sex			Locality	n			
M	955, 875 (juveniles)		Tikal	1	—	—	—
M	—	*911 + 41 (n = 3)°	Above combined	3, 1	22.4	6.7	8.7

Ornate Hawk-eagle (*Spizaetus ornatus vicarius*)

Sex			Locality	n			
F	1389, 1400, 1607	1465 ± 123 (n = 3)	—	2	—	—	—
M	964, 1004	984 ± 28 (n = 2)	—	2	—	—	—
F	1432 (n = 1; adult)	—	Yucatán Peninsula	8	—	—	—
M	1035 (n = 1; juvenile)	—	Yucatán Peninsula	8	—	—	—
F	adults, 1321–1630 (n = 7)	1449 ± 108 (n = 7)	Tikal	1	—	—	—
F	juveniles, 1200–1582 (n = 6)	1353 ± 130.6 (n = 6)	Tikal	1	—	—	—
M	1012, 1132	1072 ± 85 (n = 2)	Tikal	1	—	—	—
M	906, 997, 1000 (fledglings)	968 ± 53 (n = 3)	Tikal	1	—	—	—
F	All adults given above	*1452 ± 101 (n = 11)	See above	2, 8, 1	41.2	11.5	11.7
M	All adults given above	*1028 ± 72 (n = 4)	See above	2, 8, 1	—	—	—
F	All juveniles given above	1353 ± 130.6 (n = 6)	See above	2, 8, 1	—	—	—
M	All juveniles given above	984.5 ± 55.1 (n = 4)	See above	2, 8, 1	—	—	—

Barred Forest Falcon (*Micrastur ruficollis guerilla*)

Sex			Locality	n			
F	range = 207–307 g (n = 17)	*238 ± 23 g (n = 17)	Tikal	1	41.7	11.4	9.1
M	range = 156–176 g (n = 13)	*168 ± 6 g (n = 13)	Tikal	1	—	—	—

Collared Forest Falcon (*Micrastur semitorquatus naso*)[p]

Sex			Locality	n			
F	792–940 (n = 6)	*869 ± 63 g (n = 6)	Tikal	1	48.0	13.1	9.5
M	563–605 (n = 4)	*587 ± 18 g (n = 4)	Tikal	1	—	—	—
F	900 (n = 1)	—	Panama	4	—	—	—
M	646, 479 (n = 2)	563 (n = 2)	Belize	6	—	—	—
F	660, 750 (n = 2)	705 (n = 2)	Panama	2	—	—	—
M	511–642 (n = 4)	564 (n = 4)	—	2	—	—	—
M	641.7 (immature), 547.7 (adult)	603 (n = 2)	Tikal	16	—	—	—

Laughing Falcon (*Herpetotheres cachinnans cachinnans*) (includes *chapmani* of some authors)

Sex			Locality	n			
F	668	668 (n = 1)	Tikal	1	—	—	—
M	643, 645	644 (n = 2)	Tikal	1	—	—	—
F	590–655 (n = 3)	623 (n = 3)	Surinam	3	—	—	—
M	544–597 (n = 7)	570 (n = 7)	Surinam	3, 34	—	—	—
F	626 (n = 1)	—	Nicaragua	33	—	—	—
F	800 (n = 1)	—	Panama	4	—	—	—
M	675 (n = 1)	—	Panama	4	—	—	—
F	732, 703 (n = 2)	—	Belize	6	—	—	—
M	589, 583, 686 (n = 3)	—	Belize	6	—	—	—
M	408 (n = 1|an outlier, best deleted?)	—	Tikal	16	—	—	—
F	—	*675 (n = 8)	All above data, except 671.5 g Louisiana State Univ. Mus. of Natural Science–Ornithol. (LSU) female		—	—	—
F	—	671.5 (n = 9)	All above data, including 671.5 g LSU female		—	—	—
M	—	*601 (n = 13)	All above data, except 408 g male		12.3	3.9	0.35

Bat Falcon (*Falco rufigularis rufigularis*)

Sex			Locality	n			
F	177–242 g (n = 9)	*206 g (n = 9)	Surinam	3	58.5	15.3	18.0
M	108–148 g (n = 14)	*130 g (n = 14)	Surinam	3	—	—	—

365

APPENDIX 1—cont.

Sex	Mass (g)[a]	Mean mass (g) + SD	Locality	Source	Percentage by which females exceed males (%)	(Cube root) value of dimorphism index (%)	Mean of dimorphism indexes for wing, culmen, body mass (%)
Bat Falcon (*Falco rufigularis petoensis*)							
F	208.8 (*n* = 1)	—	Yucatán	8	—	—	—
M	127.8 (*n* = 1‖a mated pair)	—	Yucatán	8	—	—	—
F	195, 190 (*n* = 2)	192.5 (*n* = 2)	Belize	6	—	—	—
M	129, 141 (*n* = 2)	135 (*n* = 2)	Belize	6	—	—	—
F	195 (*n* = 1)	—	Near Tikal	30	—	—	—
M	140 (*n* = 1)	—	Near Tikal	30	—	—	—
M	149.9 (*n* = 1)	—	Panama	4	—	—	—
F	—	197.2 (*n* = 4)	Mean of all *petoensis* data		—	—	—
M	—	137.5 (*n* = 5)	Mean of all *petoensis* data		—	—	—
F	—	203 (*n* = 13)	All *petoensis* and *rufigularis* data combined		—	—	—
M	—	132 (*n* = 19)	All *petoensis* and *rufigularis* data combined		—	—	—
Orange-breasted Falcon (*Falco deiroleucus*)							
F	550–650	midpoint = 600	—	15	—	—	—
M	330–360	midpoint = 345	—	15	—	—	—
F	653.7		Tikal	16	—	—	—
F	595, 620		Surinam	3, 36	—	—	—
M	338 (adult), 340 (juvenile)		Surinam	3, 36	—	—	—
F	550–654 (*n* = 4)	*605 (*n* = 4)	Includes Surinam specimen above	2	78.5	20.2	18.1
M	338, 340	*339 (*n* = 2)		2			
F	480 (*n* = 1, subadult)		Bolivia	Field Museum 295,961	—	—	—
F	393 (*n* = 1, late subadult)		Peru	Field Museum 153,621	—	—	—
Guatemalan Screech Owl (*Megascops guatemalae*—sensu del Hoyo et al. 1999)							
F	105.2 (*n* = 1)		Tikal	16, 41	—	—	—
F	139.0 (*n* = 1)		Baja Verapaz, Guatemala	37 (MVZ # 142,430)	—	—	—
B	"100–150 g"		—	38, 41	—	—	—
Guatemalan Screech Owl (*Megascops guatemalae hastatus*—sensu del Hoyo et al. 1999)							
F	108 g (*n* = 1)		Michoacan, Mexico	Univ. Michigan Mus. Zool., 130,523			
M	105 g (*n* = 1)		Michoacan, Mexico	Univ. Michigan Mus. Zool., 130,524			
Guatemalan Screech Owl (*Megascops guatemalae thompsoni*—sensu del Hoyo et al. 1999)							
F	95 g (*n* = 1)		Chichen Itza, Yucatan, Mexico	Univ. Michigan Mus. Zool. 95,090			
F	116.0 g (*n* = 1)		Chichen Itza, Yucatan, Mexico	Univ. Michigan Mus. Zool. 95,091			
Guatemalan Screech Owl (*Megascops guatemalae*, all subspecies—sensu del Hoyo et al. 1999)							
F	95–139 (*n* = 5)	*112.6 ± 16.5 (*n* = 5)	All above data, combined	41	—	2.3	4.1
M	105 (*n* = 1)	*105 (*n* = 1)	All above data, combined		7.2		
Both sexes		*111.4 (*n* = 6)	Above data combined, both sexes pooled		—	—	—

Vermiculated Screech Owl (considered as *Otus guatemalae* by Burton [1988]—now considered specifically distinct, *Megascops vermiculatus*) (referred to as *Otus guatemalae* in the work cited; now known as *Megascops vermiculatus napensis*)

Sex	Weight	Mean weight	Locality	Source			
Sex?	94–128 (n = 4)	112.8 (n = 4)	Andean foothills, Peru	21, 41			—

Vermiculated Screech Owl (referred to as *Otus guatemalae roraimae* in the work cited; now known as *Megascops vermiculatus roraimae*)

Sex	Weight	Mean weight	Locality	Source			
F	100 g (n = 1)	—	Venezuela/Brazil	22, 41		—	—
M	102 g (n = 1)	—	Venezuela/Brazil	22, 41		—	—

Vermiculated Screech Owl (referred to as *Otus guatemalae* in the work cited; now known as *Megascops vermiculatus vermiculatus*)

Sex	Weight	Mean weight	Locality	Source			
F	125 (n = 1)	—	E. Panama	9, 41			—

Vermiculated Screech Owl (referred to as *Otus guatemalae=Otus vermiculatus* in the work cited; now known as *Megascops vermiculatus napensis*)

Sex	Weight	Mean weight	Locality	Source			
Sex?	91–123 g (n = 9)	107 g (n = 9)	All or mostly Peru	39, 41			—

Crested Owl (*Lophostrix cristata*) (from specimens in LSU collection)

Sex	Weight	Mean weight	Locality	Source			
F	*620 (n = 1)	—	Mexico or Peru	23			—
M	*425, 510 (n = 2)	468 (n = 2)	Mexico or Peru	23	32.5	9.4	4.2
?	"400–600 g"	—	—	38			—

Spectacled Owl (*Pulsatrix perspicillata*)

Sex	Weight	Mean weight	Locality	Source			
F	765–980	—	—	24			—
M	590–760	—	—	24			—
Both	591–1250 (n = 13)	873 (n = 13)	Many localities; includes very large Argentine subspecies	23			
Both	590–980	—	—	38			—

Spectacled Owl (*Pulsatrix perspicillata saturata*)

Sex	Weight	Mean weight	Locality	Source			
F	816, 982 (n = 2)	899 (n = 2)	Belize	6			—
M	750 (n = 1)	—	E. Panama	9			—
F	900, 970 (n = 2)	—	Panama	4			—
F	710, 800 (n = 2)	—	Panama	4			—
Sex?	885 (n = 1)	—	Panama	24			—
F	500 g (n = 1)	—	NW Costa Rica	Berkeley Mus. Vert. Zoo. (MVZ); 166,391; 24			—
M	655 g (n = 1)	—	Guatemala	Field Museum, 212,758			—
F	*833.6 ± 198 (n = 5)	500–982 (n = 5)	All the above data for this subspecies, combined		14.3		
M	*729 ± 61.4 (n = 4)	655–800 (n = 4)	All the above data for this subspecies, combined		4.5		

Spectacled Owl (*Pulsatrix perspicillata perspicillata*)

Sex	Weight	Mean weight	Locality	Source			
F	850 (n = 1; found dead)	—	French Guiana	26 (Dick et al. 1984; birds and weights from Saul, French Guiana)			—
F	597–980 (n = 4)	824 (n = 4)	Surinam	3	30.0		—
M	550–761 (n = 4)	634 (n = 4)	Surinam	3	8.7		—

Spectacled Owl (*Pulsatrix perspicillata pulsatrix*)

Sex	Weight	Mean weight	Locality	Source			
F	1250 (n = 1)	—	SE Brazil	25			—
M	1075 (n = 1)	—	SE Brazil	25	16.3	5.0	—

Central American Pygmy Owl (*Glaucidium griseiceps*: sensu del Hoyo et al. 1999)[q]

Sex	Weight	Mean weight	Locality	Source			
F	*≈ 56 g			38	≈ 10		—
M	*≈ 51 g			38	16.3	5.0	—
M	50.6 ± 1.1 g (n = 3)		(includes *occultum* and *rarum*)	29			—

APPENDIX 1—cont.

Sex	Mass (g)[a]	Mean mass (g) + SD	Locality	Source	Percentage by which females exceed males (%)	(Cube root) value of dimorphism index (%)	Mean of dimorphism indexes for wing, culmen, body mass (%)
M	58.8 (n = 1)	—	Limon, Costa Rica	Univ. Michigan Mus. Zool. 132,251 (G. g. rarum)	—	—	—
M	—	*52.65 ± 4.2 g (n = 4)	Mean of 4 males listed above		—	—	—
Ridgway's Pygmy Owl (Glaucidium ridgwayi)							
F	62–95 (n = 16)	75.1 (n = 16)	—	28		—	—
M	46–74 (n = 29)	61.4 (n = 29)	—	28	22.3	6.7	
F	64.0–105.8 (n = 22)	*80.4 ± 9.0 (n = 22)	Mexico, Berkeley MVZ specimens		17.2	5.3	5.6
M	56.5–86.0 (n = 73)	*68.6 ± 5.8 (n = 73)	Mexico, Berkeley MVZ specimens		—	—	
F	66–102 (n = 38)	77.2 ± 8.1 g (n = 38)	Texas	40		—	—
M	53–79 (n = 68)	64.3 ± 5.4 g (n = 68)	Texas	40	20.1	6.1	
F	74.6, 77.7, 94.8, 64.4 (n = 4)	77.9 ± 12.6 (n = 4)	Belize	6	26.5	7.8	
M	60.5, 62.6 (n = 2)	61.6 ± 1.5 (n = 2)	Belize	6	—	—	
F	—	67.2 (n = 23)[r]	Mexico	27		—	—
M	—	63.4 (n = 36)	Mexico	27	—	—	—
Mexican Wood Owl (Strix squamulata centralis)							
F	308–385 g (n = 11)	*335 g (n = 11)	Tikal	1, 41		—	—
M	220–256 g (n = 7)	*240 g (n = 7)	Tikal	1, 41	39.6	12.1	8.5
F	316.8 (n = 1)		Tikal	16, 41	—	—	
M	234.5, 236.6, 253.0 (n = 3)		Tikal	16, 41	—	—	
M	236 (n = 1)		Near Tikal	30, 41		—	—
F	251, 345		Belize	4, 41		—	—
M	242, 230		Belize	4, 41	—	—	
F	405.1		Veracruz	Berkeley MVZ 155,225		—	—
M	298.2		Veracruz	Berkeley MVZ 155,224 (these 2 a mated pair)		—	—
F	365.0		Guatemala	Berkeley MVZ 142,432		—	—
F	304.8		Costa Rica	Berkeley MVZ 155,593		—	—
M	298.3		Nicaragua	Berkeley MVZ 161,923		—	—
M	291.4		Nicaragua	Berkeley MVZ 161,924		—	—
F	356.2		Veracruz	Field Museum 208,708		—	—
F	251–405.1	334.8 ± 49.4 (n = 7)	All above data combined, except our own data from Tikal			—	—
M	230–298.2	257.8 ± 29.4 (n = 9)	All above data combined, except our own data from Tikal		—	—	—
Black-and-white Owl (Strix nigrolineata)							
F	443, 468	455.5 ± 17.7 (n = 2)	Southern Mexico	17		—	—
F	500		Panama	4		—	—
F	536[s]		Tikal	1		—	—
M	403		Belize	6		—	—
M	417		Panama	4		—	—
M	435[s]		Tikal	1		—	—
F	—	*487 ± 40 (n = 4)	All of above	17, 4, 6, 1		—	—
M	—	*418 ± 16 (n = 3)	All of above	17, 4, 6, 1	16.5	5.1	5.1

Sources: 1 = present study; 2 = Brown and Amadon 1968; 3 = Haverschmidt and Mees 1994; 4 = Hartman 1961; 5 = ffrench 1980; 6 = Russell 1964; 7 = Binford 1989; 8 = Paynter 1955; 9 = Robbins et al. 1985; 10 = Miller 1952; 11 = Washburn, unpubl. data; 12 = Martin et al. 1954; 13 = Hartman 1955; 14 = unpubl. data from Oklahoma City Zoological Park, courtesy Barbara Howard; 15 = Cade 1982; 16 = Smithe and Paynter 1963; 17 = Tashian 1952; 18 = Snyder and Wiley 1976; 19 = Palmer 1988; 20 = del Hoyo et al. 1994; 21 = Davis 1986; 22 = Dickerman and Phelps 1982; 23 = Dunning 1993; 24 = Burton 1984; 25 = Belton 1984; 26 = Dick et al. 1984; 27 = Voous 1988; 28 = Earhart and Johnson 1970; 29 = Howell and Robbins 1995; 30 = Van Tyne 1935; 31 = Haverschmidt 1964a; 32 = Smithe 1966; 33 = Howell 1972; 34 = Haverschmidt 1948; 35 = Fowler and Cope 1964; 36 = Haverschmidt 1963; 37 = collection of Museum of Vertebrate Zoology, University of California, Berkeley; 38 = del Hoyo et al. 1999; 39 = Weske and Terborgh 1981; 40 = Proudfoot and Johnson 2000; 41 = König and Weick 2008; 42 = Riesing et al. 2003.

Notes: Where multiple data sets are given for a species, these are not always independent. For example, specimen series used by Brown and Amadon (1968) and Wetmore (1965, 1968) may overlap and may include some individuals listed in other series. Whether data sets overlap in this fashion will in some cases be apparent from the collection localities listed.

Asterisks indicate those data used in characterizing the Maya Forest species population.

[a]Adult mass unless otherwise noted.

[b]Likely after being frozen for some time.

[c]These were a nesting pair; sex assigned by relative mass and dimensions.

[d]All Bicolored Hawk weights from Tikal, except as noted, are from incubation and nestling period.

[e]Enlarged follicles.

[f]Immature

[g]Retrapped early in the nestling period, this female weighed 483 g.

[h]A female weight of 821 g from Tikal was excluded on the off chance it was in error; if included, females were 14.2 % heavier than males.

[i]Females significantly heavier than males: t = 2.74, df = 5, P = 0.04.

[j]Haverschmidt and Mees (1994) state that additional specimens attributed to *Buteo brachyurus* in Haverschmidt (1968) were misidentified.

[k]Tikal nestling weighed 1630 g 72 days after hatching and 1697 g 92 days after hatching.

[l]Wild-caught female from NE Peru weighed 2226 g on arrival in excellent condition and juvenal plumage at Oklahoma City Zoological Park; weighed 2950 g after 6 years in captivity, clear evidence of captive weight gain. Another wild-caught South American female at the same zoo, after at least 1 year in captivity, weighed 3 kg.

[m]Hatchling male at Oklahoma City Zoological Park (progeny of a Peruvian female and a South American male) reached 1139 g at 42 days of age and stabilized at 1117 g at 42–53 days of age; when surgically sexed at 27 months of age, weighed 1.8 kg, no doubt reflecting obesity induced by the captive environment. Large raptors typically gain an additional 5–10% body mass after fledging; allowing this bird 7.5% additional mass after fledging yields an estimated 1200 g adult body mass.

[n]Juvenile female weighed 1101 g at 42 days of age and 1141 g at 61 days.

[o]At Tikal, one adult male was recorded as 1178 g, possibly an error or with full crop. Two nestling males weighed 955 and 875 g at approximately 6 and 11 weeks old, respectively. Omitting the heavy Tikal and assuming these nestlings were at roughly adult weight, mean male mass = 911 + 41 g (*n* = 3).

[p]Wetmore (1965) states that many museum specimens have the sex wrongly marked, which recommends caution in using published weights.

[q]*Glaucidium griseiceps* mass data from Dunning (1993) and Voous (1988) were deleted, as they probably contained many data on *G. g. palmarum,* which are about 5 g lighter than *G. g. griseiceps.*

[r]Voous (1988) data on mass of *Glaucidium brasilianum ridgwayi* give far different results from other data sets, hence should probably be ignored; likely there was a typographical error in his text, and female mass should have been 77 rather than 67 g.

[s]This male and female were a mated pair.

369

Linear measurements and sexual size dimorphism for Maya Forest raptor species

	Sex	Wing chord[a]	DI	Tail length[a]	DI	Culmen from cere[a]	DI	Tarsus length[a]	DI	Middle toe length[a,b]	DI
Gray-headed Kite (*Leptodon cayanensis*, mostly *cayanensis*)											
Source: 1; Locale: Costa Rica, Panama, Guyana, Brazil, Bolivia	F	314.1 (303–338) n = 10	—	236 (220–263) n = 10	—	24.3 (22.5–26.0) n = 10	—	46–54 (49.8) n = 10	—	37.6 (34–42) n = 10	—
Source: 1; Locale: El Salvador, Panama, Colomb., Guyana, Brazil	M	309.8 (290–338) n = 9	1.38	227.6 (208–250) n = 9	3.62	24 (22.5–25.0) n = 9	1.24	47–51 (49) n = 9	1.62	40.6 (35.0–42.5) n = 9	-7.67
Source: 2; Locale: Tikal	F	332 n = 1	—	261 n = 1	—	48 n = 1	—	48 n = 1	—	33 n = 1	—
Source: 2; Locale: Tikal	M	321 n = 1	—	248 n = 1	—	—	—	48 n = 1	—	—	—
Hook-billed Kite (*Chondrohierax uncinatus uncinatus*)											
Source: 1; Locale: Guatemala to Brazil	F	289.4 (268–321) n = 31	—	202.8 (191–228) n = 31	—	31.6 (28.0–37.0) n = 31	—	33.8 (31–37) n = 31	—	30.9 (28–34) n = 31	—
Source: 1; Locale: Guatemala to Argentina	M	285.8 (265–301) n = 26	1.3	191.1 (173–210) n = 26	5.9	31.3 (27.0–35.5) n = 26	0.95	35.1 (32–37) n = 26	-3.8	31.1 (28–35) n = 26	-0.64
Swallow-tailed Kite (*Elanoides forficatus yetapa*)											
Source: 1; Locale: ?	F	410.9 (390–427) n = 14	—	304 (275–326) n = 14	—	20.2 (19.5–21.0) n = 14	—	32.3 (32.0–33.5) n = 14	—	28.9 (23.0–29.5) n = 14	—
Source: 1; Locale: ?	M	418.2 (405–447) n = 26	-1.8	318.4 (298–330) n = 26	-4.6	19.6 (19–20) n = 26	3.0	32.4 (31.5–33.0) n = 26	-0.3	28.8 (28.0–29.5) n = 26	0.4
Double-toothed Kite (*Harpagus bidentatus fasciatus*)											
Source: 7; Locale: Panama	F	218.2 (207–227) n = 8	—	148.3 (139.1–158.0) n = 8	—	16.3 (15.4–17.5) n = 8	—	40.5–43.8 (42.6) n = 4	—	—	6
Source: 7; Locale: Panama	M	202.4 (198–210) n = 9	7.5	139 (132.8–146.0) n = 9	6.5	16 (15.1–17.3) n = 9	1.9	42.8 (41.5–44.3) n = 5	-4.7	—	—
Source: 4; Locale: ?	F	224 n = 6	—	156 n = 6	—	16.1 n = 6	—	44.5 n = 6	—	28.2 n = 6	—
Source: 4; Locale: ?	M	203 n = 7	9.8	143 n = 7	8.7	15.7 n = 7	2.5	41.7 n = 7	6.5	26.7 n = 7	5.5
Plumbeous Kite (*Ictinia plumbea* monotypic)											
Source: 1; Locale: ?	F	301.6 (274.0–320.5) n = 11	—	145.3 (139–161) n = 11	—	17.1 (16.0–19.5) n = 11	—	37.7 (34–42) n = 11	—	26.3 (24–29) n = 11	—
Source: 1; Locale: ?	M	298.2 (270–313) n = 13	1.1	148.5 (123–167) n = 13	-2.2	16.6 (15.5–18.0) n = 13	3.0	38.5 (37.0–42.4) n = 13	-2.1	26.8 (26.0–27.5) n = 13	-1.9
Source: 4; Locale: ?	F	302 n = 16	—	145 n = 16	—	17.6 n = 16	—	37.7 n = 16	—	26.3 n = 16	—
Source: 4; Locale: ?	M	298 n = 18	1.3	149 n = 18	-2.7	16.6 n = 18	5.8	38.5 n = 18	-2.1	26.8 n = 18	-1.9
Bicolored Hawk (*Accipiter bicolor bicolor*)											
Source: 7; Locale: Panama	F	238 (234–243) n = 7	—	188.7 (177.0–204.5) n = 7	—	18.2 (16.7–19.5) n = 7	—	68.4 (67.3–69.3) n = 5	—	44.35 n = 4	—
Source: 7; Locale: Panama	M	205.9 (200–215) n = 10	14.5	172.8 (166–180) n = 10	8.8	15 (14.5–15.8) n = 10	19.3	61.7 (59.9–64.1) n = 5	10.3	37.55 n = 9	—
Source: 2; Locale: Tikal	F	252 (247–256) n = 2	—	210 (209–210) n = 2	—	19.8 (19.4–20.2) n = 2	—	57 (55–59) n = 2	—	—	—
Source: 2; Locale: Tikal	M	218 (212–220) n = 4	—	163 (160–170) n = 4	—	15.9 (15.5–17.0) n = 4	—	48.5 (47–50) n = 2	—	—	—

	Sex	Wing chord[a]	DI	Tail length[a]	DI	Culmen from cere[a]	DI	Tarsus length[a]	DI	Middle toe length[a,b]	DI
Black Crane Hawk (*Geranospiza caerulescens nigra*)											
Source: 1; Locale: Costa Rica, El Salvador	F	326.1 (315–340) n = 6	—	244.3 (233.5–254.0) n = 6	—	21.4 (20.1–23.0) n = 6	—	91.4 (85–98); n = 6	—	38.7 (35–42); n = 6	—
Source: 1; Locale: Mexico, Guatemala, Costa Rica, El Salvador	M	302.2 (282–318) n = 8	7.6	238.7 (224–247) n = 8	2.3	20.1 (19.5–21.0) n = 8	6.3	84.2 (78–88); n = 8	8.2	34.6 (32–37) n = 8	11.2
Source: 2; Locale: Tikal	F	310 n = 2	—	250 n = 2	—	20.2 n = 2	—	102 n = 2	—	33.7 n = 2	—
Source: 2; Locale: Tikal	M	320 sic; n = 1	—	240 n = 1	—	20.7 n = 1	—	92.7 n = 1	—	36.6 n = 1	—
Source: above data combined	F	322 n = 8	—	245.7 n = 8	—	21.1 n = 8	—	94.1 n = 8	—	37.5 n = 8	—
Source: above data combined	M	304.2 n = 9	5.7	238.8 n = 9	5.7	20.2 n = 9	2.8	85.1 n = 9	10.0	34.8 n = 9	7.5
White Hawk (*Leucopternis albicollis ghiesbreghti*)											
Source: 1; Locale: Guatemala, Belize	F	375 (362–388) n = 5	—	226.8 (222–234) n = 5	—	27.5 (26.5–29.0) n = 5	—	84.6 (84–87) n = 5	—	41.6 (38–44) n = 5	—
Source: 1; Locale: Mexico, Guatemala, Belize	M	353.3 (336–366) n = 6	—	225.3 (221–235) n = 6	—	25.7 (24–27) n = 6	—	82.7 (80–85) n = 6	—	39.2 (35–45) n = 6	—
Source: 2; Locale: Tikal	F	376 (366–382) n = 3	—	222 (205–230) n = 3	—	26.8 (25.2–30.0) n = 4	—		—	38.4 n = 1	—
Source: 2; Locale: Tikal	M	354 (345–367) n = 3	—	212 (205–225) n = 3	—	26.7 (24.9–30.1) n = 3	—	95.3 n = 1	—	38.3 n = 1	—
Source: above data combined	F	375.4 (362–388) n = 8	—	225 (205–234) n = 8	—	27.2 (25.2–30.0) n = 9	—	84.6 (84–87) n = 5	—	41.1 (38–44) n = 6	—
Source: above data combined	M	354 (336–367) n = 9	5.9	221 (205–235) n = 9	1.8	26.0 (24.0–30.1) n = 9	4.5	84.5 (80.0–95.3) n = 7	0.1	39.1 (35–45) n = 7	5.0
Great Black Hawk (*Buteogallus urubitinga ridgwayi*)											
Source: 1; Locale: Mexico, Honduras, Costa Rica	F	388.8 (363–417) n = 7	—	248.5 (237–270) n = 7	—	30.3 (29–32) n = 7	—	116.4 (108–127) n = 7	—	48.3 (45–50) n = 7	—
Source: 1; Locale: Mexico, Belize, Costa Rica, Panama	M	376.4 (367–403) n = 11	2.6	243.3 (226–274) n = 11	1.2	28.8 (26–31) n = 11	1.2	116.4 (112–126) n = 11	0.2	47.2 (44.5–49.0) n = 11	0.4
Source: 2; Locale: Tikal	F	390	—	—	—	—	—	119	—	—	—
Source: 2; Locale: Tikal	M	340	—	—	—	—	—	108	—	—	—
Source: above data combined	F	389.1 n = 9	—	—	—	—	—	117.0 n = 9	—	—	—
Source: above data combined	M	370.8 n = 13	—	—	—	—	—	115.1 n = 13	—	—	—
Gray Hawk (*Buteo (Asturina) nitidus plagiatus*)											
Source: 6 (source = 1 for middle toe)	F	273.2 (254–289) n = 17	—	178.9 (161–195) n = 17	—	24.1 (22.5–25.0) n = 17	—	75 (71.0–79.5) n = 17	—	42.0 n = 6	—
Source: 6 (source = 1 for middle toe)	M	252.8 (232.5–272.0) n = 28	7.8	165.9 (146–187) n = 28	7.5	21.5 (20.0–24.50) n = 28	11.4	71 (67.5–75.0) n = 28	5.5	36.7 n = 12	13.5
Roadside Hawk (*Rupornis magnirostris griseocauda*) [Blake [1977] refers to these as *B. m. direptor*]											
Source: 6 (5 for toe length)?	F	236.8 (222–250) n = 12	—	167.2 (157–175) n = 12	—	20.7 (16–23) n = 12	—	64.8 (63–66) n = 3	—	32.8 n = 6	—
Source: 6 (5 for toe length)?	M	228.4 (217–239) n = 13	3.6	160.7 (154–169) n = 13	4.0	19.5 (18–21) n = 13	6.0	66.5 (66.0–68.9) n = 3	—	31.7 n = 13	—
Source: 2; Locale: Tikal	F	—	—	—	—	—	—	65.0 n = 3	—	—	—
Source: 2; Locale: Tikal	M	—	—	—	—	—	—	61.2 n = 3	6.0	—	—

Source / Locale	Sex										
Source: the above combined	F	—	—	—	—	—	—	64.9 n = 6	—	—	—
Source: the above combined	M	—	—	—	—	—	—	63.9 n = 6	—	—	—
Short-tailed Hawk (*Buteo brachyurus fuliginosus*)											
Source: 6 (toe length from 5); Locale: ?	F	322.2 (295-335) n = 7	—	166.8 (144-190) n = 7	—	20.7 (19-22) n = 7	—	60.6 (59.0-61.9) n = 5	—	38.7 n = 7	—
Source: 6 (toe length from 5); Locale: ?	M	284.7 (265-310) n = 8	12.4	145.1 (133-161) n = 8	13.9	18.7 (18-20) n = 8	10.2	58.5 (58.5-58.5) n = 2	3.5	34.6 n = 4	10.6
Crested Eagle (*Morphnus guianensis*) monotypic											
Source: 6 (toe length from 5); Locale: ?	F	464.7 (425-484) n = 10	—	392 (373-430) n = 10	—	36.4 (34-39) n = 10	—	115 (108.1-118.0) n = 10	—	50.4 n = 6	—
Source: 6 (toe length from 5); Locale: ?	M	436.2 (425-449) n = 10	6.3	369.6 (340-385) n = 10	5.9	31.9 (29-34) n = 10	13.2	109.6 (103.0-117.7) n = 10	4.8	47.2 n = 9	6.6
Harpy Eagle (*Harpia harpyja*) monotypic											
Source: 1; Locale: Costa Rica, Panama, Surinam, Bolivia	F	578.6 (583-610) n = 4	—	418 (417-420) n = 4	—	53 (46-63) n = 4	—	123 (118-130) n = 4	—	86.0 (83-90) n = 4	—
Source: 1; Locale: Mexico, Nicaragua, Costa Rica, Panama, Brazil	M	556.5 (543-580) n = 6	3.9	392 (372-412) n = 6	6.4	48.3 (41.5-54.0) n = 6	9.3	115.8 (114-120) n = 6	6.0	77.0 (73-80) n = 6	11.0
Source: 4; Locale: mostly same as above	F	579 n = 6	—	418 n = 6	—	53.6 n = 6	—	123.0 n = 6	—	86.0 n = 6	—
Source: 4; Locale: mostly same as above	M	557 n = 9	3.9	392 n = 9	6.4	47.1 n = 9	12.9	115.8 n = 9	6.0	77.0 n = 9	11.0
Black-and-white Hawk-eagle (*Spizaetus melanoleucus*) monotypic											
Source: 1; Locale: Mexico, Nicaragua, Costa Rica, Colombia, Brazil	F	411.7 (393.7-423.0) n = 5	—	242 (230-253) n = 5	—	28 (26-30) n = 5	—	93.4 (88-99) n = 5	—	57.1 (56-58.5) n = 5	—
Source: 1; Locale: Mexico, Costa Rica, Panama, Guyana	M	364.6 (340.0-386.2) n = 5	12.1	238.5 (230-245) n = 5	1.5	25.9 (24.5-28.0) n = 5	7.8	77 (72-84) n = 5	19.2	50.5 (49.0-52.5) n = 5	12.3
Black Hawk-eagle (*Spizaetus tyrannus serus*)											
Source: 5; Locale: ?	F	400.4 (353-445) n = 8	—	320.1 (289-386) n = 8	—	30 (29-31) n = 8	—	87.9 (84.6-92.0) n = 3	—	—	—
Source: 5; Locale: ?	M	380.2 (354-401) n = 11	5.2	309.2 (291-325) n = 11	3.5	28.2 (26-30) n = 11	6.2	82 (78.1-86.0) n = 5	6.9	—	—
Source: 1; Locale: Mexico, Honduras, Costa Rica, Brazil	F	430.9 (406.4-458.2) n = 6	—	379.4 (342.9-398.7) n = 6	—	29.7 (29.0-31.5) n = 6	—	92 (90-95) n = 6	—	49.5 (46-53) n = 6	—
Source: 1; Locale: Mexico, Panama, Surinam	M	374.6 (368.3-381.0) n = 5	14.0	309.4 (291.1-325.0) n = 5	20.3	28.1 (27.5-29.0) n = 5	5.5	84 (81-88) n = 5	9.1	45.6 (44-47) n = 5	8.2
Source: 4; Locale: mostly same as above	F	431 n = 8	—	379 n = 8	—	30.5 n = 8	—	92.0 n = 8	—	49.5 n = 8	—
Source: 4; Locale: mostly same as above	M	375 n = 7	13.9	309 n = 7	20.3	28.4 n = 7	7.1	84.0 n = 7	9.1	45.6 n = 7	8.2
Ornate Hawk-eagle (*Spizaetus ornatus vicarius*)											
Source: 1; Locale: Mexico to Panama	F	377.8 (353.3-388.0) n = 8	—	281.6 (266-290) n = 8	—	30 (27.0-31.5) n = 8	—	94.1 (89.5-100.0) n = 8	—	53 (51-55) n = 8	—
Source: 1; Locale: Mexico to Ecuador	M	339.8 (337.8-349.3) n = 7	—	255.6 (244-268) n = 7	—	27.1 (25.5-29.0) n = 7	—	89 (87-92) n = 7	—	49.7 (47-53) n = 7	—

APPENDIX 2—*cont.*

	Sex	Wing chord[a]	DI	Tail length[a]	DI	Culmen from cere[a]	DI	Tarsus length[a]	DI	Middle toe length[a,b]	DI
Source: 4; Locale: mostly same as above	F	377 n = 13	—	276 n = 13	—	31.6	—	94.9	—	53.4	—
Source: 4; Locale: mostly same as above	M	338 n = 11	10.9	249 n = 11	10.3	27.8	12.8	89.8	5.5	49.1	8.4
Barred Forest Falcon (*Micrastur ruficollis guerilla*)											
Source: 2 (toe length from 5); Locale: Tikal	F	178 n = 20	—	171 n = 28	—	15.3 n = 24	—	48.8 n = 17	—	30.0 n = 6	—
Source: 2 (toe length from 5); Locale: Tikal	M	170 n = 21	6.6	164 n = 23	6.1	14.5 n = 20	9.3	47.3 n = 12	3.1	28.7 n = 8	4.4
Collared Forest Falcon (*Micrastur semitorquatus naso*)											
Source: 2; Locale: Tikal	F	283 (275–319) n = 6	—	298 (281–315) n = 6	—	23.4 (22.7–23.9) n = 6	—	—	—	—	—
Source: 2; Locale: Tikal	M	258 (255–261) n = 4	9.2	265 (261–268) n = 4	11.7	22.0 (21.6–22.4) n = 4	6.2	—	—	—	—
Source: 1; Locale: Mexico, Honduras, Panama, Colombia	F	273.1 (265–285) n = 7	—	280.9 (270–300) n = 7	—	23.1 (22–24) n = 7	—	82.0 (78–87) n = 7	—	47.1 (43.0–50.5) n = 7	—
Source: 1; Locale: Mexico, Belize, Honduras, Costa Rica, Colombia	M	263.7 (256–275) n = 11	3.5	269.9 (260–300) n = 11	4.0	22.1 (20.0–23.5) n = 11	4.4	84.2 (82–87) n = 11	-2.6	45.4 (43–47) n = 11	3.7
Laughing Falcon (*Herpetotheres cachinnans chapmani*: regarded as *H. c. chapmani* by Friedmann [1950], but as *H. c. cachinnans* by many authors)											
Source: 1; Locale: Mexico, Guatemala	F	279.6 (255–305) n = 13	—	216.9 (188–242) n = 13	—	24.2 (22.0–26.4) n = 1	—	62.0 (59.0–65.8) n = 13	—	41.7 (39-45) n = 13	—
Source: 1; Locale: Mexico, Guatemala, Honduras	M	278.3 (262–307) n = 23	—	212.7 (187–232) n = 23	—	24.9 (22.8–27.0) n = 23	—	65.4 (58–77) n = 23	—	41.2 (36–45) n = 23	—
Laughing Falcon (*Herpetotheres cachinnans cachinnans*)											
Source: 1; Locale: Nicaragua to Ecuador	F	267.3 n = 27	—	201.7 n = 27	—	23.8 sic; n = 27	—	60.2 sic; n = 27	—	41.4 sic; n = 27	—
Source: 1; Locale: Nicaragua to Peru	M	267.1 n = 29	0.07	201.6 n = 29	0.05	24.5 sic; n = 29	-2.9	60.3 sic; n = 29	-0.17	42.7 sic; n = 29	-3.1
Bat Falcon (*Falco rufigularis*, mostly *petoensis*)											
Source: 1; Locale: Mexico, Honduras, Belize, Guatemala, Costa Rica, Venezuela, Colombia	F	220 (209–229) n = 18	—	111.7 (103–118) n = 18	—	13.9 (13–15) n = 18	—	36.5 (35–39) n = 18	—	33.8 (32–35) n = 18	—
Source: 1; Locale: Mexico, Guatemala, Belize, Honduras, Panama, Venezuela, Guyana	M	189.1 (173–197) n = 11	—	95 (88–102) n = 11	—	12.1 (11.5–12.5) n = 11	—	33.6 (32–36) n = 11	—	29.7 (28–31) n = 11	—
Source: 4; Locale: mostly same as above	F	220 n = 22	—	112 n = 22	—	14.3 n = 22	—	36.5 n = 22	—	33.8 n = 22	—
Source: 4; Locale: mostly same as above	M	189 n = 15	15.2	95 n = 15	16.4	11.3 n = 15	23.4	33.6 n = 15	8.3	29.7 n = 15	12.9

Orange-breasted Falcon (*Falco deiroleucus*) monotypic

Source; Locale	Sex										
Source: 5; Locale: ?	F	—	279.2 (265–289) n = 10	—	137.9 (133–146) n = 10	—	22.6 (21–24) n = 10	45.5 (47–56) n = 4	—	—	—
Source: 5; Locale: ?	M	13.5	243.8 (234–249) n = 6	19.4	119.6 (113–130) n = 6	14.2	18.6 (18–19) n = 6	40.2 (39.6–41.2) n = 3	12.4	—	—
Source: 4; Locale: ?	F	—	281 n = 8	—	140 n = 8	—	22.5 n = 8	49.5 n = 8	—	—	49.0 n = 8
Source: 4; Locale: ?	M	—	240 n = 3	—	115 n = 3	—	18.5 n = 6	43.5 n = 3	—	—	42.3 n = 3

Guatemalan Screech Owl (*Megascops guatemalae*) (including only individuals from geographic range of the subspecies *M. g. guatemalae*, as described by del Hoyo et al. [1999])

Source; Locale	Sex										
Source: 6; Locale: see above note	F	—	170.1 n = 4	—	87.4 n = 4	—	14.8 n = 4	—	—	—	—
Source: 6; Locale: see above note	M	7.2	158.3 n = 5	6.0	82.3 n = 5	2.7	14.4 n = 5	—	—	—	—

Guatemalan Screech Owl (*Megascops guatemalae*) (including all individuals in Ridgway 1914 from geographic range of the species *M. guatemalae*, as described by del Hoyo et al. [1999])

Source; Locale	Sex										
Source: 6; Locale: see above note	F	—	162.3 n = 11	—	83.0 n = 11	—	13.8 n = 11	—	—	—	—
Source: 6; Locale: see above note	M	4.5	155.2 n = 12	3.4	80.2 n = 12	2.9	13.4 n = 12	—	—	—	—

Vermiculated Screech Owl (*Megascops vermiculatus*)

Source; Locale	Sex										
Source: 8; Locale: Panama	F	—	156.5 (153.3–162.2) n = 8	—	76.8 (73.1–79.4) n = 8	—	15.1 (14.0–16.7) n = 8	28.9 (27.9–30.6) n = 8	—	—	—
Source: 8; Locale: Panama	M	1.0	155.0 (148.0–163.0) n = 6	3.6	74.1 (71.2–76.4) n = 6	4.8	14.4 (14.0–14.9) n = 6	28.3 (26.4–30.4) n = 6	2.1	—	—

Crested Owl (*Lophostrix cristata stricklandi*)

Source; Locale	Sex										
Source: 8; Locale: Chiapas, Costa Rica, Panama	F	—	287.6 (280–292) n = 6	—	174.1 (167–180) n = 6	—	18.9 (18–20) n = 6	48.9 (47.9–50.6) n = 6	?	—	—
Source: 8; Locale: Nicaragua, Costa Rica, Panama	M	1.1	284.5 (278–292) n = 6	4.1	167.1 (159–179) n = 6	2.1	18.5 (18.0–19.5) n = 6	45.3 (43.0–48.4) n = 6	7.6	—	—

Spectacled Owl (*Pulsatrix perspicillata saturata*)

Source; Locale	Sex										
Source: 6; Locale: Mexico, Belize, El Salvador, Costa Rica, Nicaragua, Panama	F	—	339 (317–360) n = 10	—	192.9 (164–204) n = 10	—	30 (27.0–32.5) n = 10	—	—	—	—
Source: 6; Locale: Nicaragua, Costa Rica, Panama	M	1.7	333.4 (314–347) n = 11	–1.2	195.2 (177.5–215.0) n = 11	—	28.7 (27–30) n = 11	4.4	—	—	—
Source: 8; Locale: Mexico to W Panama	F	—	337.0 (320–347) n = 6	—	193.8 (181–211) n = 6	—	28.9 (26.7–31.5) n = 5	—	—	—	—
Source: 8; Locale: Honduras	M	4.2	323.2 (320–329) n = 4	7.9	179.0 (177–182) n = 4	—	30.3 (28.9–31.1) n = 4	–4.7	—	—	—

Spectacled Owl (*Pulsatrix perspicillata perspicillata*)

Source; Locale	Sex										
Source: 3; Locale: Surinam	F	—	331 (325–340) n = 3	—	—	—	—	—	—	—	—
Source: 3; Locale: Surinam	M	2.1	324 (317–340) n = 6	—	—	—	—	—	—	—	—

Spectacled Owl (*Pulsatrix perspicillata chapmani*)

Source; Locale	Sex										
Source: 8; Locale: Panama	F	—	338.8 (332–344) n = 7	—	186.5 (179–197) n = 7	—	29.4 (28.0–31.4) n = 7	—	—	—	—
Source: 8; Locale: Panama	M	4.3	324.6 (316–333) n = 10	2.6	181.7 (175–193) n = 10	—	27.9 (26.2–29.2) n = 10	5.2	—	—	—

APPENDIX 2—cont.

	Sex	Wing chord[a]	DI	Tail length[a]	DI	Culmen from cere[a]	DI	Tarsus length[a]	DI	Middle toe length[a,b]	DI
Central American Pygmy Owl (*Glaucidium griseiceps*: sensu del Hoyo et al. 1999: previously *G. minutissimum occultum, griseiceps,* and *rarum*)											
(G. g. rarum)											
Source: 8; Locale: Colombia	F	93.4 n = 1	—	43.5 n = 1	—	11.2 n = 1	—	18.0 n = 1	—	—	—
Source: 8; Locale: Panama	M	89.6 n = 3	—	47.5 n = 3	—	11.0 n = 3	—	18.5 n = 3	—	—	—
(includes both G. g. occultum and rarum)											
Source: 10; Locale: range of the subspecies	F	—	—	—	—	—	—	—	—	—	—
Source: 10; Locale: range of the subspecies	M	87.2 n = 10	—	47.8 n = 10	—	—	—	—	—	—	—
Ridgway's Pygmy Owl (*Glaucidium ridgwayi*)											
Source: 9; Locale: range of the subspecies	F	95.1 n = 8	—	—	—	—	—	—	—	—	—
Source: 9; Locale: range of the subspecies	M	93.5 n = 11	1.7	—	—	—	—	—	—	—	—
Source: 6; Locale: range of the subspecies	F	97.1 (93.0–102.5) n = 21	—	62.5 (55.5–69.5) n = 21	—	11.6 (10.5–13.0) n = 21	—	—	—	—	—
Source: 6; Locale: range of the subspecies	M	92.1 (86.5–97.0) n = 32	5.3	59.7 (52.5–65.5) n = 32	4.6	11 (9.5–12.0) n = 32	5.3	—	—	—	—
Source: 8; Locale: Panama	F	93.6 (92.9–95.0); n = 4	—	57.2 (55.2–59.0); n = 4	—	11.3 (10.9–11.7); n = 4	—	23.3 (23.0–23.8) n = 4	—	—	—
Source: 8; Locale: Panama	M	88.5 (86.0–93.4) n = 8	5.6	54.2 (52.9–55.5) n = 8	5.4	10.5 (10.0–11.0) n = 8	7.3	21.2 (20.4–22.4) n = 8	9.4	—	—
Source: 11; Locale: range of subspecies	F	98.3 ± 2.7 n = 50	—	62.3 ± 2.5 n = 50	—	11.2 ± 0.7 n = 50	—	—	—	—	—
Source: 11; Locale: range of subspecies	M	94.0 ± 3.9 n = 93	7.5	60.2 ± 4.3 n = 93	3.4	10.9 ± 0.7 n = 93	2.7	—	—	—	—
Source: 12; Locale: south Texas	F	96.5 ± 2.4 n = 38	—	67.9 ± 2.4 n = 34	—	11.5 ± 0.4 n = 16	—	—	—	—	—
Source: 12; Locale: south Texas	M	92.9 ± 2.8 n = 68	3.8	64.6 ± 2.0 n = 73	5.0	10.9 ± 0.5 n = 23	5.4	—	—	—	—
Mean of 3 largest data sets above:			5.5		4.3		4.5				
Mexican Wood Owl (*Strix squamulata centralis*) specimens from the geographic range of this subspecies											
Source: 6; Locale: Veracruz & Campeche to W Panama	F	239.9 (236–249) n = 19	—	137.9 (132.2–147.2) n = 19	—	19.8 (19.0–21.5) n = 19	—	—	—	—	—
Source: 6; Locale: Veracruz & Campeche to W Panama	M	225.9 (221–233) n = 17	6.0	133.0 (124.5–140.0) n = 17	3.6	18.4 (17.7–20.5) n = 17	7.3	—	—	—	—
Source: 2; Locale: Tikal	F	246 n = 12	—	150 n = 11	—	—	—	—	—	—	—
Source: 2; Locale: Tikal	M	233 n = 8	—	141 n = 7	—	—	—	—	—	—	—
Black-and-white Owl (*Strix nigrolineata*) monotypic											
Source: 2; Locale: Tikal	F	293 n = 1	—	187 n = 1	—	23.2 n = 1	—	—	—	—	—
Source: 2; Locale: Tikal	M	286 n = 1	2.42	165 n = 1	12.5	20.9 n = 1	10.4	—	—	—	—
Source: 6; Locale: Chiapas to N and W Colombia	F	274.4 (255–293) n = 8	—	164.5 (154.0–179.5) n = 8	—	21.6 (19.5–23.0) n = 4	—	—	—	—	—
Source: 6; Locale: Nicaragua to W Colombia	M	277.2 (272–285) n = 4	—	164.6 (161.0–171.5) n = 4	—	21 (18.5–22.0) n = 4	—	—	—	—	—

Source: 8; Locale: Mexico, Honduras, Panama, Colombia	F	275.2 (270–279) $n = 9$	—	170.2 (160–178) $n = 9$	—	21.6 (19.5–22.8) $n = 9$	—	—	—
Source: 8; Locale: Panama, Colombia	M	261.6 (254–268) $n = 6$	5.1	152.3 (144–165) $n = 6$	11.1	20.0 (19.0–21.1) $n = 5$	7.7	—	—
Mean of ours and Wetmore:		3.8	—	11.8	—	9.1	—	—	—

Sources: 1= Friedmann 1950; 2 = present study; 3 = Haverschmidt and Mees 1994; 4 = Bierregaard 1978; 5 = Blake 1977; 6 = Ridgway 1914; 7 = Wetmore 1965; 8 = Wetmore 1968; 9 = Earhart and Johnson 1970; 10 = Howell and Robbins 1995; 11 = Proudfoot and Johnson 2000, study skins; 12 = Proudfoot and Johnson 2000, birds captured in Texas.
[a]In mm; mean; range in parentheses; sample size.
[b]Without claw.

LITERATURE CITED

Albuquerque, J. L. B. 1986. Conservation and status of raptors in southern Brazil. Birds Prey Bull. No. 3:88–94.

Albuquerque, J. L. B. 1995. Observations of rare raptors in southern Atlantic rainforest of Brazil. J. Field Ornithol. 66:363–369.

Aldrich, J. W., and B. P. Bole Jr. 1937. The birds and mammals of the western slope of the Azuero Peninsula (Republic of Panama). Sci. Publ. Cleveland Mus. Nat. Hist. 7.

Allen, J. A. 1905. Supplementary notes on birds collected in the Santa Marta District, Colombia, by Herbert H. Smith, with descriptions of nests and eggs. Bull. Am. Mus. Nat. Hist. 21:275–295.

Altmann, J. 1974. Observational study of behavior: sampling methods. Behavior 49:227–267.

Altringham, J. D. 1996. Bats: biology and behavior. Oxford: Oxford Univ. Press.

Alvarez-Cordero, E. 1996. Biology and conservation of the Harpy Eagle in Venezuela and Panama. Ph.D. dissertation, Univ. of Florida, Gainesville.

Alvarez del Toro, M. A. 1971. Las aves de Chiapas. Tuxtla Guttierrez, Mexico: Gobierno del estado de Chiapas.

Alvarez del Toro, M. A. 1980. Las aves de Chiapas, 2nd ed. Tuxtla Guttierrez, Mexico: Univ. Autonoma Chiapas.

Alvarez-López, H., and G. H. Kattan. 1995. Notes on the conservation status of resident diurnal raptors of the middle Cauca Valley, Colombia. Bird Conserv. Int. 5:341–348.

Amadon, D. 1961. Relationships of the falconiform genus *Harpagus*. Condor 63:178–179.

Amadon, D. 1964. Taxonomic notes on birds of prey. Am. Mus. Novit. 2166:1–24.

Amadon, D. 1982. A revision of the sub-buteonine hawks *(Accipitridae, Aves)*. Am. Mus. Novit. 2741:1–20.

Amadon, D., and J. Bull. 1988. Hawks and owls of the world. A distributional and taxonomic list. Proc. West. Found. Vertebr. Zool. 3:295–357.

Anderson, D. J. 1990. Evolution of obligate siblicide in boobies: 1. A test of the insurance egg hypothesis. Am. Nat. 135:334–350.

Anderson, D. L. 1999. Tawahka Project, Honduras: 1999 field season report. Boise, ID: The Peregrine Fund.

Andersson, M., and R. A. Norberg. 1981. Evolution of reversed sexual size dimorphism and role partitioning among predatory birds, with a size scaling of flight performance. Biol. J. Linn. Soc. 15:105–130.

Andrews, R. M. 1979. Evolution of life histories: a comparison of *Anolis* lizards from matched island and mainland habitats. Breviora 454:1–51.

Andrews, R. M. 1983. *Norops polylepis. In* D. H. Janzen, ed., Costa Rican natural history, pp. 409–110. Chicago: Univ. of Chicago Press.

Andrews, R. M., and A. S. Rand. 1982. Seasonal breeding and long-term population fluctuations in the lizard *Anolis limifrons. In* E. G. Leigh Jr., A. S. Rand, and D. M. Windsor, eds., The ecology of a tropical forest: seasonal rhythms and long-term changes, pp. 405–412. Washington, DC: Smithsonian Institution Press.

Andrle, R. F. 1967. Birds of the Sierra de Tuxtla in Veracruz, Mexico. Wilson Bull. 79:163–187.

AOU. 1983. Check-list of North American birds, 6th ed. Washington, DC: American Ornithologists' Union.

AOU. 1998. Check-list of North American birds, 7th ed. Washington, DC: American Ornithologists' Union.

Armstrong, J. T. 1965. Breeding home range in the nighthawk and other birds; its evolutionary and ecological significance. Ecology 46:619–629.

Austin, G. T., N. M. Haddad, C. Méndez, T. D. Sisk, D. D. Murphy, A. E. Launer, and P. R. Ehrlich. 1996. Annotated checklist of the butterflies of the Tikal National Park area of Guatemala. Trop. Lepid. 7:21–37.

Austin, K. K. 1993. Habitat use and home range size of breeding Northern Goshawks in the southern Cascades. M.Sc. thesis, Oregon State Univ., Corvallis, OR.

Azevedo, M. A. G., and M. Di-Bernardo. 2005. Natural history and conservation of the Swallow-tailed Kite, *Elanoides forficatus*, on Santa Catarina Island, southern Brazil. Ararajuba 13:81–88. (in Portuguese with English summary)

Azevedo, M. A. G., A. L. Roos, J. L. B. Albuquerque, and V. d. Q. Piacentini. 2000. Breeding and food habits of the Swallow-tailed Kite, *Elanoides forficatus* (Falconiformes: Accipitridae), on Santa Catarina Island, SC—Brazil. Melopsittacus 3:122–127. (in Portuguese)

Baker, A. 1998. Status and breeding biology, ecology, and behavior of the Orange-breasted Falcon *(Falco deiroleucus)* in Guatemala and Belize. M.S. thesis, Brigham Young Univ., Provo, UT.

Baker, A. J., J. P. Jenny, and D. F. Whitacre. 1992. Orange-breasted Falcon reproduction, density, and behavior in Guatemala and Belize. *In* D. F. Whitacre and R. K. Thorstrom, eds., Progress report V, 1992, Maya Project, pp. 217–224. Boise, ID: The Peregrine Fund.

Baker, A. J., D. F. Whitacre, O. A. Aguirre-Barrera, J. López A., and C. M. White. 1999. Observations of a Double-toothed Kite *Harpagus bidentatus* hawking bats. J. Raptor Res. 33:343–344.

Baker, A. J., O. A. Aguirre-Barrera, D. F. Whitacre, and C. M. White. 2000a. First record of a Barred Forest-Falcon *(Micrastur ruficollis)* nesting in a cliff pothole. Ornitol. Neotrop. 11:81–82.

Baker, A. J., D. F. Whitacre, O. A. Aguirre-Barrera, and C. M. White. 2000b. The Orange-breasted Falcon *(Falco deiroleucus)* in Mesoamerica: a vulnerable, disjunct population? Bird Cons. Int. 10:29–40.

Balgooyen, T. G. 1976. Behavior and ecology of the American Kestrel *(Falco sparverius)* in the Sierra Nevada of California. Univ. Calif. Publ. (Zool.) 103:1–83.

Barlow, J. C., J. A. Dick, D. H. Baldwin, and A. Davis. 1969. New records of birds from British Honduras. Ibis 111:399–402.

Baroni-Urbani, C., G. Josen, and G. J. Peakin. 1978. Empirical data and demographic parameters. *In* M. V. Brian, ed., Production ecology of ants and termites, pp. 5–44. Cambridge: Cambridge Univ. Press.

Barreto, A. T. 1968. Observations on the behavior and changes in captivity of *Falco rufigularis rufigularis*. Lozania (Acta Zoologica Colombiana).

Beaumont, J. 2005. An instance of cooperative hunting in Great Black Hawks in Brazil. Int. Hawkwatcher 10:16.

Beavers, R. A. 1992. The birds of Tikal: an annotated checklist for Tikal National Park and Petén, Guatemala. College Station: Texas A&M Univ. Press.

Beavers, R. A., D. J. Delaney, C. W. Leahy, and F. Oatman. 1991. New and noteworthy bird records from Petén, Guatemala, including Tikal National Park. Bull. Br. Ornithol. Club 111:77–90.

Becker, J. 1987. Revision of "*Falco*" ramenta Wetmore and the Neogene evolution of the Falconidae. Auk 104:270–276.

Beckoff, M., and L. D. Mech. 1984. Computer simulation, simulation analyses of space use: home range estimates, variability, and sample size. Behav. Res. Methods Instrum. Comput. 16:32–37.

Bednarz, J. C. 1995. Harris' hawk. Birds N. Am. 146:1–23.

Bednarz, J. C. 1987. Pair and group reproductive success, polyandry, and cooperative breeding in Harris' Hawks. Auk 104:393–404.

Beebe, W. 1950. Home life of the bat falcon, *Falco albigularis albigularis* Daudin. Zoologica 35:69–86.

Beeman, K. 1993. Signal user's guide, version 2.2. Belmont, MA: Engineering Design.

Behrensmeyer, A. K., J. D. Damuth, W. A. DiMichele, R. Potts, H.-D. Sues, and S. L. Wing. 1992. Terrestrial ecosystems through time: evolutionary paleoecology of terrestrial plants and animals. Chicago: Univ. of Chicago Press.

Belcher, C., and G. D. Smooker. 1934. Birds of the colony of Trinidad and Tobago. Part 1. Ibis 4:572–595.

Belcher, C., and G. D. Smooker. 1936. Birds of the colony of Trinidad and Tobago. III. Ibis (Series 13) 6:1–35.

Belton, W. 1984. Birds of Rio Grande do Sul, Brazil, part 1. *Rheidae* through *Furnariidae*. Bull. Am. Mus. Nat. Hist. 178:369–636.

Beltzer, A. H. 1990. Biologia alimentaria del gavilan comun *Buteo magnirostris saturatus* (Aves: Accipitridae) en el valle aluvial del Rio Parana Medio, Argentina. Ornitol. Neotrop. 1:3–8.

Bennett, P. M., and I. P. F. Owens. 2002. Evolutionary ecology of birds: life histories, mating systems and extinction. Oxford: Oxford Univ. Press.

Benson, C. W., R. K. Brooke, R. J. Dowsett, and M. P. S. Irwin. 1971. The birds of Zambia. London: Collins.

Bent, A. C. 1938. Life histories of North American birds of prey. Washington, DC: U.S. Govt. Printing Office.

Berger, D. D., and H. C. Mueller. 1959. The *bal-chatri*: a trap for birds of prey. Bird-Banding 30:18–26.

Berry, K. J., and P. W. Mielke Jr. 1986. R by chi-square analyses with small expected cell frequencies. Educ. Psychol. Meas. 46:169–173.

Berry, R. B., C. W. Benkham, A. Muela, Y. Seminario, and M. Curti. 2010. Isolation and decline of a population of the Orange-breasted Falcon. Condor 112:479–489.

Bierregaard, R. O. 1978. Morphological analyses of community structure in birds of prey. Ph.D. dissertation, Univ. of Pennsylvania.

Bierregaard, R. O., Jr. 1984. Observations of the nesting biology of the Guiana Crested Eagle *(Morphnus guianensis)*. Wilson Bull. 96:1–5.

Bierregaard, R. O., Jr. 1990. Species composition and trophic organization of the understory bird community in a Central Amazonian terra firme forest. *In* A. H. Gentry, ed., Four neotropical rainforests, pp. 217–236. New Haven: Yale Univ. Press.

Bierregaard, R. O., Jr. 1994. Neotropical falconiform accounts. *In* J. del Hoyo, A. Elliot, and J. Saragatal, eds., Handbook of the birds of the world, vol. 2. Barcelona: Lynx Edicions.

Bildstein, K. L., and K. Meyer. 2000. Sharp-shinned Hawk *(Accipiter striatus)*. *In* A. Poole and F. Gill, eds., The birds of North America, No. 482. Philadelphia: Academy of Natural Sciences; Washington, DC: American Ornithologists' Union.

Binford, 1989. A distributional survey of the birds of the Mexican state. Ornithol. Monogr. 43.

Bird, D. M., and Y. Aubrey. 1982. Reproductive and hunting behavior in Peregrine Falcon, *Falco peregrinus*, in southern Quebec. Can. Field-Nat. 96:167–171.

BirdLife International. 2000. Threatened birds of the world. Barcelona: Lynx Edicions; Cambridge, UK: BirdLife International.

Blackburn, T. M., and K. J. Gaston. 1997. A critical assessment of the form of the interspecific relationship between abundance and body size in animals. J. Anim. Ecol. 66:233–249.

Blackburn, T. M., and K. J. Gaston. 1999. The relationship between animal abundance and body size: a review of the mechanisms. Adv. Ecol. Res. 28:182–210.

Blake, E. R. 1953. Birds of Mexico. Chicago: Univ. of Chicago Press.

Blake, E. R. 1958. Birds of Volcán de Chiriquí, Panama. Fieldiana Zool. 36.

Blake, E. R. 1970. Birds of Mexico. Chicago: Univ. of Chicago Press.

Blake, E. R. 1977. Manual of Neotropical birds, vol. 1. Spheniscidae (penguins) to Laridae (gulls and allies). Chicago: Univ. of Chicago Press.

Blockstein, D. E. 1988. Two endangered birds of Grenada, West Indies: Grenada Dove and Grenada Hook-billed Kite. Carib. J. Sci. 24:127–136.

Blockstein, D. E. 1991. Population declines of the endangered endemic birds on Grenada, West Indies. Bird Conserv. Int. 1:83–91.

Boinski, S., and N. L. Fowler. 1989. Seasonal patterns in a tropical lowland forest. Biotropica 21:223–233.

Boinski, S., and P. E. Scott. 1988. Association of birds with monkeys in Costa Rica. Biotropica 20:136–143.

Boinski, S., and R. M. Timm. 1985. Predation by squirrel monkeys and Double-toothed Kites on tent-making bats. Am. J. Primatol. 9:121–127.

Boinski, S., L. Kauffman, A. Westoll, C. M. Stickler, S. Cropp, and E. Ehmke. 2003. Are vigilance, risk from avian predators and group size consequences of habitat structure? A comparison of three species of squirrel monkey *(Saimiri oerstedii, S. boliviensis,* and *S. sciureus).* Behaviour 140:1421–1467.

Bond, J. 1936. Birds of the West Indies. Philadelphia: Academy of Natural Sciences.

Bond, J. 1961. Extinct and near extinct birds of the West Indies. Pan-American Section, ICBP, Res. Rep. No. 4.

Bond, J. 1979. Birds of the West Indies, 4th ed. London: Collins.

Bond, J. 1984. Twenty-fifth supplement to the check-list of birds of the West Indies. (1956), pp. 1–22. Philadelphia: Academy of Natural Sciences.

Bonilla G., N., L. A. Oliveros F., and R. C. Hernández. 1992. Reproductive biology and home range of the Mottled Owl. *In* D. F. Whitacre and R. K. Thorstrom, eds., Maya Project progress report V, pp. 225–228. Boise, ID: The Peregrine Fund.

Booth-Binczik, S. D., G. A. Binczik, and R. F. Labisky. 2004. A possible foraging association between White Hawks and White-nosed Coatis. Wilson Bull. 116:101–103.

Bosakowski, T., and D. G. Smith. 1992. Comparative diets of sympatric nesting raptors in the eastern deciduous forest biome. Can. J. Zool. 70:984–992.

Bosakowski, T., D. G. Smith, and R. Speiser. 1992. Niche overlap of two sympatric-nesting hawks *Accipiter* spp. in the New Jersey-New York highlands. Ecography 15:358–372.

Boyce, D. A., Jr. 1980. Hunting and prenesting behavior of the Orange-breasted Falcon. J. Raptor Res. 14:35–39.

Boyce, D. A., and L. F. Kiff. 1981. Have the eggs of the Orange-breasted Falcon *(Falco deiroleucus)* been described? J. Raptor Res. 15:89–93.

Braun, M. J., and W. E. Holznagel. 2002. (Abstract) Molecular phylogeny of falconiform birds. *In* American Ornithologists' Union, ed., Third North American ornithological conference, p. 9. North American Ornithological Conference, New Orleans, LA.

Brodkorb, P. 1948. Taxonomic notes on the Laughing Falcon. Auk 65:406–410.

Brodkorb, P. 1964. Catalogue of fossil birds: part 2 (Anseriformes through Galliformes). Bull. Fla. State Mus. Biol. Sci. 8:195–335.

Brosset, A. 1990. A long term study of the rain forest birds in M'Passa (Gabon). *In* A. Keast, ed., Biogeography and ecology of forest bird communities, pp. 259–274. The Hague: SPB Academic Publishing.

Brown, G. P., R. Shine, and T. Madsen. 2002. Responses of three sympatric snake species to tropical seasonality in northern Australia. J. Trop. Ecol. 18:549–568.

Brown, L. 1976a. Birds of prey: their biology and ecology. New York: A. W. Publishers.

Brown, L. 1976b. Eagles of the world. Newton Abbott, UK: David and Charles.

Brown, L., and D. Amadon. 1968. Eagles, hawks, and falcons of the world. Feltham, Middlesex, UK: Country Life Books.

Brown, L., and D. Amadon. 1989. Eagles, hawks, and falcons of the world. Secaucus, NJ: Wellfleet Press.

Brown, L. H., and P. L. Britton. 1980. The breeding seasons of East African birds. Nairobi, Kenya: East African Natural History Society.

Brush, T. 1999. The Hook-billed Kite: a reclusive, snail-eating raptor of the lower Rio Grande valley. Texas Birds 1 (Fall–Winter 1999):26–32.

Brush, T. 2005. Nesting birds of the tropical frontier: the Lower Rio Grande Valley of Texas. College Station: Texas A&M Univ. Press.

Buckley, N. J. 1999. Black Vulture *(Coragyps atratus). In* A. Poole and F. Gill, eds., The birds of North America, No. 411. Philadelphia: Birds of North America.

Buehler, D. A. 2000. Bald Eagle *(Haliaeetus leucocephalus). In* A. Poole and F. Gill, eds., The birds of North America, No. 506. Philadelphia: Birds of North America.

Bull, J., and J. T. Marshall. 1988. Otus. *In* D. Amadon and B. F. King, eds., Hawks and owls of the world: a distributional and taxonomic list. Proc. West. Found. Vertebr. Zool. 3:331–336.

Bullock, S. H., H. A. Mooney, and E. Medina. 1995. Seasonally dry tropical forests. Cambridge: Cambridge Univ. Press.

Burnham, W. A., J. P. Jenny, and C. W. Turley. 1988. Progress report, Maya Project: investigation of raptors and their habitats as environmental indicators for preserving biodiversity and tropical forests of Latin America. Boise, ID: The Peregrine Fund.

Burnham, W. A., D. F. Whitacre, and J. P. Jenny. 1994. The Maya Project: use of raptors as tools for conservation and ecological monitoring of biological diversity. In B.-U. Meyburg and R. D. Chancellor, eds., Raptor conservation today, pp. 257–264. Berlin: WWGBP and Pica Press.

Burt, W. H. 1943. Territoriality and home range concepts as applied to mammals. J. Mammal. 24:346–352.

Burton, A. M. 2006. Recent breeding records and sightings of Neotropical forest eagles in southeastern Mexico. *In* R. Rodríguez-Estrella, ed., Current raptor studies in México, pp. 59–70. La Paz, Baja California Sur, Mexico: Centro de Investigaciones Biológicas del Noreste, S. C. and Comisión Nacional para el Conocimiento y Uso de la Biodiversidad.

Burton, J. A., ed. 1973. Owls of the world. New York: E. P. Dutton.

Burton, J. A., ed. 1984. Owls of the world. Glasgow: Peter Lowe.

Burton, P. J. K. 1978. The intertarsal joint of the harrier-hawks *Polyboroides* spp. and the crane hawk *Geranospiza caerulescens.* Ibis 120:171–177.

Buskirk, W. H. 1976. Social systems in a tropical forest avifauna. Am. Nat. 110:293–310.

Buskirk, W. H., and M. Lechner. 1978. Frugivory by Swallow-tailed Kites in Costa Rica. Auk 95:767–768.

Buskirk, W. H., G. V. N. Powell, J. F. Wittenberger, R. E. Buskirk, and T. U. Powell. 1972. Interspecific bird flocks in tropical highland Panama. Auk 89:612–624.

Cabanne, G. S. 2005. Notes on advertisement flights of three kites of the Atlantic forest: the Gray-headed (*Leptodon cayanensis*), the Plumbeous (*Ictinia plumbea*), and the Rufous-thighed (*Harpagus diodon*) Kites. Ornitol. Neotrop. 16:197–204.

Cabot, J., and P. Serrano. 1986. Data on the distribution of some species of raptors in Bolivia. Bull. Br. Ornithol. Club 106:170–173.

Cade, T. J. 1960. Ecology of the peregrine and gyrfalcon populations in Alaska. Univ. Calif. Publ. Zool. 63:151–290.

Cade, T. J. 1982. The falcons of the world. Ithaca, NY: Comstock/Cornell Univ. Press.

Campbell, J. A. 1998. Amphibians and reptiles of Northern Guatemala, the Yucatán, and Belize. Norman: Univ. of Oklahoma Press.

Campbell, J. A., and W. W. Lamar. 1989. The venomous reptiles of Latin America. Ithaca, NY: Comstock/Cornell Univ. Press.

Campbell, J. A., and J. P. Vannini. 1989. Distribution of amphibians and reptiles in Guatemala and Belize. Proc. West. Found. Vertebr. Zool. 4:1–21.

Campbell, J. A., D. R. Formanowicz, and P. B. Medley. 1989. The reproductive cycle of *Norops uniformis* (Sauria: Iguanidae) in Veracruz, Mexico. Biotropica 21:237–243.

Canuto, M. 2008. Observations of two hawk-eagle species in a humid lowland tropical forest reserve in central Panama. J. Raptor Res. 42:287–292.

Carlier, P. 1993. Sex differences in nesting site attendance by Peregrine Falcons. J. Raptor Res. 27:31–34.

Carlos, C.J., and W. Girão. 2006. The history of the Ornate Hawk-eagle *Spizaetus ornatus* in the Atlantic forest of northeastern Brazil. Rev. Bras. Ornitol. 14:405–409.

Carothers, J.H., and F.M. Jaksic. 1984. Time as a niche difference: the role of interference competition. Oikos 42:403–406.

Carrara, L.A., P. d. T. Zuquim Antas, and R. de Souza Yabe. 2007. Nesting of the Collared Forest-falcon *Micrastur semitorquatus* (Aves: Falconidae) in the Pantanal, Brazil: biometry, nestling diet and competition with macaws. Rev. Bras. Ornitol. 15:25–33. (in Portuguese with English summary)

Carriker, M.A., Jr. 1910. An annotated list of the birds of Costa Rica, including Cocos Island. Ann. Carnegie Mus. 6:314–915.

Carvalho, C.E.A., and G.A. Bohórquez. 2007. (Abstract) Breeding biology of Plumbeous Kite (*Ictinia plumbea*) in southeast Brazil. *In* K.L. Bildstein, D.R. Barber, and A. Zimmerman, eds., Neotropical raptors, pp. 272. Orwigsburg, PA: Hawk Mountain Sanctuary.

Carvalho, C.E.A., G. Zorzin, M. Canuto, and E.P.M. Carvalho Filho. 2007. (Abstract) A preliminary study of the diet of the Bat Falcon (*Falco rufigularis*) in southwestern Brazil. *In* K.L. Bildstein, D.R. Barber, and A. Zimmerman, eds., Neotropical raptors, p. 275. Orwigsburg, PA: Hawk Mountain Sanctuary.

Carvalho Filho, E.P.M., G.D. Mendes de Carvalho, and C.E.A. Carvalho. 2005. Observations on nesting Gray-headed Kites (*Leptodon cayanensis*) in southeastern Brazil. J. Raptor Res. 39:89–92.

Carvalho Filho, E.P.M., M. Canuto, and G. Zorzin. 2006. Breeding biology and diet of the Great Black Hawk (*Buteogallus u. urubitinga*) in southeast Brazil. Rev. Bras. Ornitol. 14:445–448. (in Portuguese with English summary)

Cash, K.J., and R.M. Evans. 1986. Brood reduction in the American White Pelican. Behav. Ecol. Sociobiol. 18:413–418.

Ceballos, G. 1995. Vertebrate diversity, ecology and conservation in Neotropical dry forests. *In* S.H. Bullock, H.A. Mooney, and E. Medina, eds., Seasonally dry tropical forests, pp. 195–220. Cambridge: Cambridge Univ. Press.

Cely, J.E. 1973. A nest of the Swallow-tailed Kite at Wambaw Creek, Charleston County, South Carolina. Chat 37:23.

Cely, J.E., and J.A. Sorrow. 1990. The American Swallow-tailed Kite in South Carolina. Nongame and Heritage Trust Fund publ. No. 1. South Carolina Wildlife and Marine Resources Department, Columbia, SC.

Chapman, F.M. 1929. My tropical air castle. New York: D. Appleton.

Chase, R., and D.G. Robinson. 2001. The uncertain history of land snails on Barbados: implications for conservation. Malacologia 43:33–57.

Chavez-Ramirez, F., and E.C. Enkerlin. 1991. Notes on the food habits of the Bat Falcon (*Falco rufigularis*) in Tamaulipas, Mexico. J. Raptor Res. 25:142–143.

Cherrie, G.K. 1916. A contribution to the ornithology of the Orinoco region. Brooklyn Inst. Arts Sci. Mus. Sci. Bull. 2:144a–374.

Chubb, C. 1910. On the birds of Paraguay. Ibis (Series 9) 4:53–78.

Clark, L. 1991. The nest protection hypothesis: adaptive use of plant secondary compounds by European starlings.

In J.E. Loye and M. Zuk, eds., Bird-parasite interactions: ecology, evolution and behaviour, pp. 205–221. Oxford: Oxford Univ. Press.

Clinton-Eitniear, J., M.R. Gartside, and M.A. Kainer. 1991. Ornate Hawk-eagle feeding on Green Iguana. J. Raptor Res. 25:18–19.

Clum, N.J., and T.J. Cade. 1994. Gyrfalcon (*Falco rusticolus*). *In* A. Poole and F. Gill, eds., The birds of North America, No. 114. Philadelphia: Academy of Natural Sciences; Washington, DC: American Ornithologists' Union.

Coates, A.G. 1997. The forging of Central America. *In* A.G. Coates, ed., Central America: a natural and cultural history, pp. 1–37. New Haven, CT: Yale Univ. Press.

Cohn-Haft, M., A. Whittaker, and P.C. Stauffer. 1997. A new look at the "species-poor" Central Amazon: the avifauna north of Manaus, Brazil. *In* J.V. Remsen Jr., ed., Studies in Neotropical ornithology honoring Ted Parker. Ornithological Monographs No. 48. Washington, DC: American Ornithologists' Union.

Coley, P.D. 1983. Herbivory and defensive characteristics of tree species in a lowland tropical forest. Ecol. Monogr. 53:209–233.

Coley, P.D., and J.A. Barone. 1996. Herbivory and plant defenses in tropical forests. Annu. Rev. Ecol. Syst. 27:305–335.

Colinvaux, P.A. 1997. The ice-age Amazon and the problem of diversity. NWOHuygenslezing 1997, pp. 7–30. The Hague: Netherlands Organisation for Scientific Research.

Collar, N.J. 1986. Threatened raptors in the Americas: work in progress from the ICBP/IUCN red data book. Birds of Prey Bulletin 3:13–20.

Collar, N.J., L.P. Gonzaga, N. Krabbe, A. Madroño Nieto, L.G. Naranjo, T.A. Parker III, and D.C. Wege. 1992. Threatened birds of the Americas; the ICBP/IUCN red data book, 3rd ed., part 2. Washington, DC: Smithsonian Institution Press.

Coltart, N.B. 1951. Nests of the Orange-breasted Falcon. Ool. Rec. 26:43.

Congreve, W.M. 1954. The work of some oologists in the field in 1954. Ool. Rec. 28:37–42.

Coulson, J.A. 2006. Intraguild predation, low reproductive potential, and social behaviors that may be slowing the recovery of a Northern Swallow-tailed Kite population. Ph.D. dissertation. Tulane University, New Orleans, LA.

Cowie, R.H. 2001. Decline and homogenization of Pacific faunas: the land snails of American Samoa. Biol. Conserv. 99:207–222.

Craighead, J.J., and F.C. Craighead Jr. 1956. Hawks, owls and wildlife. New York: Dover.

Cramp, S., ed. 1985. Handbook of the birds of Europe, the Middle East and North Africa. The birds of the western Palearctic. Vol. 4—Terns to woodpeckers. Oxford: Oxford Univ. Press.

Cramp, S.K., and K.E.L. Simmons, eds. 1980. Handbook of the birds of Europe, the Middle East and North Africa. The birds of the western Palearctic. Vol. 2. Oxford: Oxford Univ. Press.

Cresswell, W. 1996. Surprise as a winter hunting strategy in sparrowhawks *Accipiter nisus*, peregrines *Falco peregrinus* and merlins *F. columbarius*. Ibis 138:684–692.

Crocoll, S.T. 1994. Red-shouldered Hawk (*Buteo lineatus*). *In* A. Poole and F. Gill, eds., The birds of North America, No. 107. Philadelphia: Academy of Natural Sciences; Washington, DC: American Ornithologists' Union.

Crooks, K.R., and M.E. Soule. 1999. Mesopredator release and avifaunal extinctions in a fragmented system. Nature 400:563–566.

Crowell, K.L. 1962. Reduced interspecific competition among the birds of Bermuda. Ecology 43:75–88.

Curtis, J.H., D.A. Hodell, and M. Brenner. 1996. Climate variability on the Yucatán Peninsula (Mexico) during the last 3500 years, and implications for Maya cultural evolution. Q. Res. 46:37–47.

Curtis, J.H., M. Brenner, D.A. Hodell, R.A. Balser, G.A. Islebe, and H. Hooghiemstra. 1998. A multi-proxy study of Holocene environmental change in the Maya Lowlands of Peten, Guatemala. J. Paleolimnol. 19:139–159.

Cuvier, G. 1817. Le Regne animal Distribué d/aprés son Organisation, 1st ed., vol. 1. Paris: Fortin.

Daan, S., C. Deerenberg, and C. Dijkstra. 1996. Increased daily work precipitates natural death in the kestrel. J. Anim. Ecol. 65:539–544.

Daltry, J.C., T. Ross, R.S. Thorpe, and W. Wüster. 1998. Evidence that humidity influences snake patterns: a field study of the Malayan pit viper *Calleselasma rhodostoma*. Ecography 21:25–34.

Da Silva, N.J., Jr., and J.W. Sites Jr. 1995. Patterns of diversity of Neotropical squamate reptile species with emphasis on the Brazilian Amazon and the conservation potential of indigenous reserves. Conserv. Biol. 9:873–901.

Daudin, F.M. 1800. Traité élémentair d'entomologie 2:78.

Davis, D.D. 1953. Behavior of the lizard *Corytophanes cristatus*. Fieldiana Zool. 35:1–8.

Davis, T.J. 1986. Distribution and natural history of some birds from the departments of San Martin and Amazonas, northern Peru. Condor 88:50–56.

Dekker, D. 1980. Hunting success rates, foraging habits, and prey selection of Peregrine Falcons migrating through central Alberta. Can. Field-Nat. 94:371–382.

Dekker, D. 1988. Peregrine Falcon and Merlin predation on small shorebirds and passerines in Alberta. Can. J. Zool. 66:925–928.

Delannoy, C.A., and A. Cruz. 1988. Breeding biology of the Puerto Rican Sharp-shinned Hawk *(Accipiter striatus venator)*. Auk 105:649–662.

Delannoy, C.A., and A. Cruz. 1991. Philornid parasitism and nestling survival of the Puerto Rican Sharp-shinned Hawk. *In* J.E. Loye and M. Zuk, eds., Bird-parasite interactions: ecology, evolution, and behaviour, pp. 93–103. Oxford: Oxford Univ. Press.

de la Peña, M.R. 1983. Notas nidologicas sobre aves Argentinas. El Hornero, no. extraordinario 170–173.

de la Peña, M.R. 2004. Nests of the Jabiru *(Jabiru mycteria)* and Plumbeous Kite *(Ictinia plumbea)* in northeastern Argentina. Nuestras Aves 20:15–16. (in Spanish)

de la Peña, M.R. 2006. Photo guide to the nests, eggs, and young of Argentina birds. Buenos Aires: Editorial LOLA. (in Spanish)

de la Peña, M.R. undated. Guia de aves Argentinas, 2nd ed., Tomo II. Buenos Aires: Literature of Latin America.

del Hoyo, J., A. Elliott, and J. Sargatal. 1994. Handbook of the birds of the world, vol. 2: New World vultures to guineafowl. Barcelona: Lynx Edicions.

del Hoyo, J., A. Elliott, and J. Sargatal. 1999. Handbook of the birds of the world, vol. 5: Barn-owls to hummingbirds. Barcelona: Lynx Edicions.

Delnicki, D. 1978. Second occurrence and first successful nesting record of the Hook-billed Kite in the United States. Auk 95:427.

Deshmukh, I. 1986. Ecology and tropical biology. Boston: Blackwell.

Dhondt, A.A. 2001. Life history trade-offs in tits. Ardea 89:153–166.

Dick, J.A., W.B. McGillivray, and D.J. Brooks. 1984. A list of birds and their weights from Saul. French Guiana. Wilson Bull. 96:347–514.

Dickerson, R.W., and W.H. Phelps Jr. 1982. An annotated list of the birds of Cerro Urutani on the border of Estado Bolivar, Venezuela, and Territorio Roraima, Brazil. Am. Mus. Novit. No. 2732, pp. 1–20.

Dickey, D.R., and A.J. van Rossem. 1938. The birds of El Salvador. Field Mus. Nat. Hist. Zool. Ser. 23, Publ. 406:1–609.

Di Giacomo, A.G. 2000. Nesting of some little known raptors in the eastern Chaco of Argentina. Hornero 15:135–139. (in Spanish with English summary)

Di Giacomo, A.G. 2005. Birds of El Bagual Reserve. *In* A.G. Di Giacomo and S.F. Krapovickas, eds., Natural history and landscape of the El Bagual Reserve, Formosa Province, Argentina: inventory of the vertebrate fauna and vascular flora of a protected area of the humid chaco, pp. 202–465. Buenos Aires: Aves Argentinas/Asociación del Plata. (in Spanish)

Dijkstra, C., A. Bult, S. Bijlsma, S. Daan, T. Meijer, and M. Zijlstra. 1990. Brood size manipulations in the kestrel *(Falco tinnunculus)*: effects on offspring and parent survival. J. Anim. Ecol. 59:269–285.

Dinerstein, E. 1986. Reproductive ecology of fruit bats and the seasonality of fruit production in a Costa Rican cloud forest. Biotropica 18:307–318.

Dixon, K.R., and J.A. Chapman. 1980. Harmonic mean measure of animal activity areas. Ecology 61:1040–1044.

Dorward, D.F. 1962. Comparative biology of the white booby and the brown booby, *Sula* spp. at Ascension. Ibis 103b:174–220.

Drent, R. 1973. The natural history of incubation. *In* D.S. Farner, ed., Breeding biology of birds, pp. 262–311. Washington, DC: National Academy of Sciences.

Duellman, W.E. 1965. Amphibians and reptiles from the Yucatán Peninsula, México. Univ. Kansas Publications of the Museum of Nat. Hist. 15 (12):577–614.

Dugelby, B.L. 1995. Chicle latex extraction in the Maya Biosphere Reserve: behavioral, institutional, and ecological factors affecting sustainability. Ph.D. dissertation, Duke Univ., Durham, NC.

Dugelby, B.L. 1998. Governmental and customary arrangements guiding chicle latex extraction in the Petén, Guatemala. *In* R.B. Primack, D.B. Bray, H.A. Galletti, and I. Ponciano, eds., Timber, tourists, and temples: conservation and development in the Maya Forest of Belize, Guatemala, and Mexico, pp. 155–178. Washington, DC: Island Press.

Dunham, A.E. 1978. Food availability as a proximate factor influencing individual growth rates in the iguanid lizard *Sceloporus merriami*. Ecology 59:770–778.

Dunning, J.B., Jr. 1993. CRC handbook of avian body masses. Boca Raton, FL: CRC Press.

Dunstan, T.C. 1972. Radio-tagging Falconiform and Strigiform birds. Raptor Res. 6:93–102.

du Toit, J.T. 1990. Home range-body mass relations: a field study on African browsing ruminants. Oecologia 85:301–303.

DuVal, E.H., H.W. Greene, and K.L. Manno. 2006. Laughing Falcon *(Herpetotheres cachinnans)* predation on coral snakes *(Micrurus nigrocinctus)*. Biotropica 38:566–568.

Earhart, C.M., and N.K. Johnson. 1970. Size dimorphism and food habits of North American owls. Condor 72:251–264.

Echternacht, A.C. 1983. *Ameiva* and *Cnemidophorus*. *In* D.H. Janzen, ed., Costa Rican natural history, pp. 375–379. Chicago: Univ. of Chicago Press.

Edwards, T.C., and M.W. Collopy. 1983. Obligate and facultative brood reduction in eagles: an examination of factors that influence fratricide. Auk 100:630–635.

Egler, S.G. 1991. Double-toothed Kites following tamarins. Wilson Bull. 103:510–512.

Eisenberg, J.F. 1981. The mammalian radiations. Chicago: Univ. Chicago Press.

Eisenberg, J.F. 1990. Neotropical mammal communities. In A.H. Gentry, ed., Four neotropical rainforests, pp. 358–368. New Haven, CT: Yale Univ. Press.

Eisenmann, E. 1963. Mississippi Kite in Argentina; with comments on migration and plumages in the genus Ictinia. Auk 80:74–77.

Eisenmann, E. 1971. Range expansion and population increases in North and Middle America of the White-tailed Kite (Elanus leucurus). Am. Birds 25:529–536.

Ellis, D.H., and D.G. Smith. 1986. An overview of raptor conservation in Latin America. Birds Prey Bull. 3:21–25.

Ellis, D.H., and W.H. Whaley. 1981. Three Crested Eagle records for Guatemala. Wilson Bull. 93:284–285.

Ellis, D.H., R.L. Glinski, and D.G. Smith. 1990. Raptor road surveys in South America. J. Raptor Res. 24:98–106.

Ellis, D.H., C. James, D.G. Smith, and S.P. Flemming. 1993. Social foraging classes in raptorial birds. Bioscience 43:14–20.

Emmons, L.H. 1984. Geographic variation in densities and diversities of non-flying mammals in Amazonia. Biotropica 16:210–222.

Emmons, L.H. 1987. Comparative feeding ecology of felids in a neotropical rainforest. Behav. Ecol. Sociobiol. 20:271–283.

Emmons, L.H. 1990. Neotropical mammals. Chicago: Univ. of Chicago Press.

Emmons, L.H., and F. Feer. 1990. Neotropical rainforest mammals. Chicago: Univ. of Chicago Press.

Emmons, L.H., and F. Feer. 1997. Neotropical rainforest mammals. 2nd ed. Chicago: Univ. of Chicago Press.

Enderson, J.H. 1964. A study of the Prairie Falcon in the central Rocky Mountain region. Auk 81:332–352.

Enderson, J.H., S.A. Temple, and L.G. Swartz. 1972. Time-lapse photographic records of nesting Peregrine Falcons. Living Bird 11:113–128.

Enriquez-Rocha, P.L., and J.L. Rangel-Salazar. 1996. Owl abundance in a tropical humid forest in Costa Rica. In Abstracts, Second international conference on raptors, Urbino, Italy, 2–5 October 1996, pp. 34–35. Urbino: Raptor Research Foundation, Univ. of Urbino.

Enriquez-Rocha, P.L., J.L. Rangel-Salazar, and D.W. Holt. 1993. Presence and distribution of Mexican owls: a review. J. Raptor Res. 27:154–160.

Euler, C. 1900. Descripcao de ninho e ovos das aves do Brazil. Rev. Mus. Paul. 4:9–164.

Evans, S.A. 1981. Ecology and behavior of the Mississippi Kite (Ictinia mississippiensis) in southern Illinois. M.Sc. thesis, Southern Illinois Univ., Carbondale, IL.

Faaborg, J., and W.J. Arendt. 1995. Survival rates of Puerto Rican birds: are islands really that different? Auk 112:503–507.

Feduccia, A. 1996. The origin and evolution of birds. New Haven, CT: Yale Univ. Press.

Feeny, P. 1970. Seasonal changes in oak leaf tannins and nutrients as a cause of spring feeding by winter moth caterpillars. Ecology 51:565–581.

Feldsine, J.W., and L.W. Oliphant. 1985. Breeding behavior of the Merlin: the courtship period. J. Raptor Res. 19:60–67.

Felten, H., and J. Steinbacher. 1955. Contributions to the knowledge of the avifauna of El Salvador. Comunicaciones de la Instituto de Tropicales Investigaciones Cientificas 4(1/2):1–36.

Ferguson-Lees, J., and D.A. Christie. 2001. Raptors of the world. Princeton, NJ: Princeton Univ. Press.

Ferrari, S.F. 1990. A foraging association between two kite species (Ictinia plumbea and Leptodon cayanensis) and buffy-headed marmosets (Callithrix flaviceps) in southeastern Brazil. Condor 92:781–783.

ffrench, R. 1973. A guide to the birds of Trinidad and Tobago. Wynnewood, PA: Livingston.

ffrench, R. 1976. A guide to the birds of Trinidad and Tobago. Revised ed. Wynnewood, PA: Livingston.

ffrench, R. 1980. A guide to the birds of Trinidad and Tobago. Revised ed. Newton Square, PA: Harrowood Books.

ffrench, R. 2004. Post-fledging dependence of a young Ornate Hawk-eagle. Living World, Journal of the Trinidad and Tobago Field Naturalists' Club 2004:39.

Fitch, H.S. 1970. Reproductive cycles in lizards and snakes. Univ. Kans. Mus. Nat. Hist. Misc. Publ. 52:1–247.

Fitch, H.S. 1982. Resources of a snake community in prairie-woodland habitat of northeastern Kansas. In N.J. Scott, ed., Herpetological communities. Fish and Wildlife Service, Wildlife Research Report 13:83–97.

Fitch, H.S. 1983. Sphenomorphus cherrei. In D.H. Janzen, ed., Costa Rican natural history, pp. 422–425. Chicago: Univ. of Chicago Press.

Fjeldsa, J., and N. Krabbe. 1990. Birds of the high Andes. Svendborg, Denmark: Apollo Books.

Flatten, C.J., and R.P. Gerhardt. 1989. Nest description, home range, and food habits of the Black-and-white Owl (Ciccaba nigrolineata). In W.A. Burnham, J.P. Jenny, and C.W. Turley, eds., Maya Project, progress report II, 1989, pp. 61–68. Boise, ID: The Peregrine Fund.

Flatten, C.J., J.A. Madrid M., H.D. Madrid M., S.H. Funes A., A.E. Hernandez C., and R.R. Botzoc G. 1990. Biology of the Ornate Hawk-eagle (Spizaetus ornatus). In W.A. Burnham, D.F. Whitacre, and J.P. Jenny, eds., Progress report III, 1990: Maya Project: use of raptors as environmental indices for design and management of protected areas and for building local capacity for conservation in Latin America, pp. 129–144. Boise, ID: The Peregrine Fund.

Fleetwood, R.J., and J.L. Hamilton. 1967. Occurrence and nesting of the Hook-billed Kite (Chondrohierax uncinatus) in Texas. Auk 84:598–601.

Fleishman, L.J. 1985. Cryptic movement in the vine snake Oxybelis aeneus. Copeia 1985:242–245.

Fleming, T.H. 1988. The short-tailed fruit bat. Chicago: Univ. of Chicago Press.

Fogden, M.P.L. 1972. The seasonality and population dynamics of equatorial forest birds in Sarawak. Ibis 114:307–343.

Fontaine, R. 1980. Observations of the foraging association of Double-toothed Kites and White-faced Capuchin monkeys. Auk 97:94–98.

Ford, R.G. 1995. CAMRIS: computer aided mapping and resource inventory system. Portland, OR: Ecological Consulting.

Forsman, E.D., E.C. Meslow, and H.M. Wight. 1984. Distribution and biology of the Spotted Owl in Oregon. Wildl. Monogr. 87.

Foster, M.S. 1971. Rain, feeding behavior, and clutch size of tropical birds. Auk 91:722–726.

Foster, M.S., and R.M. Timm. 1976. Tent-making by Artibeus jamaicensis (Chiroptera: Phyllostomatidae) with comments on plants used by bats for tents. Biotropica 8:254–269.

Foster, M.S., and R.W. McDiarmid. 1983. Rhinophrynus dorsalis. In D.H. Janzen, ed., Costa Rican natural history, pp. 419–421. Chicago: Univ. of Chicago Press.

Fowler, J. M., and J. B. Cope. 1964. Notes on the Harpy Eagle of British Guiana. Auk 81:257–273.

Fox, N. 1977. The biology of the New Zealand Falcon (*Falco novaeseelandiae* Gmelin 1788). Ph.D. dissertation, Univ. of Canterbury. Christchurch, New Zealand.

Frankham, R. 1995. Effective population size/adult population size ratios in wildlife: a review. Genet. Res. Cambr. 66:95–107.

Franklin, I. R. 1980. Evolutionary change in small populations. *In* M. E. Soule and B. A. Wilcox, eds., Conservation biology: an evolutionary-ecological perspective, pp. 135–149. Sunderland, MA: Sinauer Associates.

Freese, C. H., P. G. Heltne, N. Castro R., and G. Whitesides. 1982. Patterns and determinants of monkey densities in Peru and Bolivia, with notes on distributions. Int. J. Primatol. 3:53–89.

Friedmann, H. 1950. The birds of North and Middle America. Part 11. U.S. Natl. Mus. Bull. 50.

Friedmann, H., and F. D. Smith Jr. 1950. A contribution to the ornithology of northeastern Venezuela. Proc. U.S. Natl. Mus. 100:411–538.

Friedmann, H., and F. D. Smith Jr. 1955. A further contribution to the ornithology of northeastern Venezuela. Proc. U.S. Natl. Mus. 104:463–524.

Friedmann, H., L. Griscom, and R. T. Moore. 1950. Distributional check-list of the birds of Mexico. Pacific Coast Avifauna No. 29, Cooper Ornithological Club.

Frumkin, R. 1994. Intraspecific brood-parasitism and dispersal in fledgling Sparrowhawks *Accipiter nisus.* Ibis 136:426–433.

Fry, C. H., S. Keith, and E. K. Urban. 1988. The birds of Africa, vol. 3. London: Academic Press.

Gaillard, J.-M., M. Festa-Bianchet, N. G. Yoccoz, A. Loison, and C. Tiogo. 2000. Temporal variation in fitness components and population dynamics of large herbivores. Annu. Rev. Ecol. Syst. 31:367–393.

Gainzarain, J. A., R. Arambarri, and A. F. Rodriguez. 2000. Breeding density, habitat selection and reproductive rates of the Peregrine Falcon *Falco peregrinus* in Álava (northern Spain). Bird Study 47:225–231.

Gargett, V. 1970. Black Eagle experiment no. 2. Bokmakierie 22:32–35.

Gargett, V. 1993. The Black Eagle. London: Academic Press.

Garrido, O. H. 1985. Cuban endangered birds. Ornithol. Monogr. 36:992–999.

Garrido, O. H., and A. Kirkconnell. 2000. Field guide to the birds of Cuba. Ithaca, NY: Comstock/Cornell Univ. Press.

Gaunt, S. L. L., L. F. Baptista, J. E. Sanchez, and D. Hernandez. 1994. Song learning as evidenced from song sharing in two hummingbird species *(Colibri coruscans* and *C. thalassinus).* Auk 111:87–103.

Gehlbach, F. R. 1994. The Eastern Screech Owl: life history, ecology, and behavior in the suburbs and countryside. College Station: Texas A&M Univ. Press.

Gentry, A. H., and L. H. Emmons. 1987. Geographical variation in fertility, phenology, and composition of the understory of Neotropical forests. Biotropica 19:216–227.

Gerhardt, R. P. 1991a. Mottled Owls *(Ciccaba virgata):* response to calls, breeding biology, home range, and food habits. M.S. thesis, Boise State Univ., Boise, ID.

Gerhardt, R. P. 1991b. Response of Mottled Owls to broadcast of conspecific call. J. Field Ornithol. 62:239–244.

Gerhardt, R. P. 2004. Cavity nesting in raptors of Tikal National Park and vicinity, Petén, Guatemala. Ornitol. Neotrop. 15 (Suppl):477–483.

Gerhardt, R. P., and D. M. Gerhardt. 1997. Size, dimorphism, and related characteristics of *Ciccaba* owls from Guatemala. *In* J. R. Duncan, D. H. Johnson, and T. H. Nicholls, eds., Biology and conservation of owls of the Northern Hemisphere, pp. 190–196. U.S. For. Serv. Gen. Tech. Rep. NC-190.

Gerhardt, R. P., M. A. Vásquez, and D. M. Gerhardt. 1991. Breeding biology, food habits, and siblicide of Swallow-tailed Kites *(Elanoides forficatus). In* D. F. Whitacre, W. A. Burnham, and J. P. Jenny, eds., Maya Project, progress report IV, pp. 65–71. Boise, ID: The Peregrine Fund.

Gerhardt, R. P., P. M. Harris, and M. A. Vásquez. 1993. Food habits of nesting Great Black Hawks in Tikal National Park, Guatemala. Biotropica 25:349–352.

Gerhardt, R. P., N. Bonilla G., D. M. Gerhardt, and C. J. Flatten. 1994a. Breeding biology and home range of two *Ciccaba* owls. Wilson Bull. 106:629–639.

Gerhardt, R. P., D. M. Gerhardt, C. J. Flatten, and N. Bonilla G. 1994b. The food habits of sympatric *Ciccaba* owls in northern Guatemala. J. Field Ornithol. 65:258–264.

Gerhardt, R. P., D. M. Gerhardt, and M. A. Vásquez. 1997. Siblicide in Swallow-tailed Kites. Wilson Bull. 109:112–120.

Gerhardt, R. P., D. M. Gerhardt, and M. A. Vásquez. 2004. Food delivered to nests of Swallow-tailed Kites in Tikal National Park, Guatemala. Condor 106:177–181.

German, R. Z., and L. L. Meyers. 1989. The role of time and size in ontogenetic allometry: II. An empirical study of human growth. Growth Dev. Aging 53:107–115.

Gerstell, A. T., and J. C. Bednarz. 1999. Competition and patterns of resource use by two sympatric raptors. Condor 101:557–565.

Gibbons, J. W., and R. D. Semlitsch. 1987. Activity patterns. *In* R. A. Seigel, J. T. Collins, and S. S. Novak, eds., Snakes: ecology and evolutionary biology, pp. 396–421. New York: McGraw-Hill.

Gill, F. B., and L. L. Wolf. 1975. Economics of feeding territoriality in the golden-winged sunbird. Ecology 56:333–345.

Girard, P. 1933. Notas sobre algunas aves de Tucuman. El Hornero 5:223–225.

Gittleman, J. L., and P. H. Harvey. 1982. Carnivore home-range size, metabolic needs and ecology. Behav. Ecol. Sociobiol. 10:57–63.

Giudice, R. 2007. First nesting report of the Ornate Hawk-Eagle *(Spizaetus ornatus)* in Peru. *In* K. L. Bildstein, D. R. Barber, and A. Zimmerman, eds., Neotropical raptors, pp. 9–13. Orwigsburg, PA: Hawk Mountain Sanctuary.

Glinski, R. L., and R. D. Ohmart. 1983. Breeding ecology of the Mississippi kite in Arizona. Condor 85:200–207.

González-García, F. 1993. Avifauna de la Reserva de la Biosfera "Montes Azules," Selva Lacandon, Chiapas, Mexico. Acta Zool. Mex., Nueva Ser., No. 55. Xalapa, Veracruz: Instituto de Ecologia, A.D.

Goodrich, L. J., E. R. Inzunza, and S. W. Hoffman. 1993. Raptor migration through Veracruz, Mexico. Final report, Project No. 92–033. Washington, DC: National Fish and Wildlife Foundation.

Gower, J. C., and G. J. S. Ross. 1969. Minimum spanning trees and single linkage cluster analysis. Appl. Stat. 18:54–64.

Gradwohl, J., and R. Greenberg. 1982. The breeding season of antwrens on Barro Colorado Island. *In* E. G. Leigh Jr., A. S. Rand, and D. M. Windsor, eds., The ecology of a tropical forest: seasonal rhythms and long-term changes, pp. 245–351. Washington, DC: Smithsonian Institution Press.

Graham, R. R. 1930. Safety devices in wings of birds. Br. Birds 24:2–65.

Greenberg, R., and J. Gradwohl. 1997. Territoriality, adult survival, and dispersal in the Checker-throated Antwren in Panama. J. Avian Biol. 28:103–110.

Greene, H. W. 1988. Species richness in tropical predators. *In* F. Almeda and C. M. Pringle, eds., Tropical rainforests: diversity and conservation, pp. 259–280. San Francisco: California Academy of Sciences.

Greene, H. W. 1997. Snakes: the evolution of mystery in nature. Berkeley: Univ. of California Press.

Greene, H. W., and M. Santana. 1983. Field studies of hunting behavior by bushmaster. Am. Zool. 23:897.

Greeney, H. F., R. A. Gelis, and R. White. 2004. Notes on breeding birds from an Ecuadorian lowland forest. Bull. Br. Ornithol. Club 124:28–37.

Greeney, H. F., and R. A. Gelis. 2008. Further breeding records from the Ecuadorian Amazonian lowlands. Cotinga 29:62–68.

Greenlaw, J. S. 1967. Foraging behavior of the Double-toothed Kite in association with White-faced Monkeys. Auk 84:596–597.

Gregory, P. T., J. M. Macartney, and K. W. Larsen. 1987. Spatial patterns and movements. *In* R. A. Siegel, J. T. Collins, and S. S. Novak, eds., Snakes: ecology and evolutionary biology, pp. 366–395. New York: Macmillan.

Griffiths, C. S. 1994a. Monophyly of the Falconiformes based on syringeal morphology. Auk 111:787–805.

Griffiths, C. S. 1994b. Syringeal morphology and the phylogeny of the Falconidae. Condor 96:127–140.

Griffiths, C. S. 1999. Phylogeny of the Falconidae inferred from molecular and morphological data. Auk 116:116–130.

Griffiths, C. S., G. F. Barrowclough, J. G. Groth, and L. Mertz. 2004. Phylogeny of the Falconidae (Aves): a comparison of the efficacy of morphological, mitochondrial, and nuclear data. Mol. Phylogenet. Evol. 32:101–109.

Griffiths, H., H. Lutter, and S. J. S. Debus. 2002. Breeding and diet of a pair of Square-tailed Kites *Lophoictinia isura* on the mid-north coast of New South Wales. Aust. Bird Watcher 19:184–193.

Griscom, L. 1932. The distribution of bird-life in Guatemala. Bull. Am. Mus. Nat. Hist. 64.

Groom, M. J. 1992. Sand-colored nighthawks parasitize the antipredator behavior of three nesting bird species. Ecology 73:785–793.

Grosselet, M., and D. Gutierrez Carbonet. 2007. First confirmed observation of the Crested Eagle *Morphnus guianensis* for Mexico. Cotinga 28:74–75. (in Spanish with English summary)

Grossman, M. L., and J. Hamlet. 1964. Birds of prey of the world. New York: Clarkson N. Potter.

Guedes, N. M. R. 1993. Nesting of the Collared Forest Falcon *(Micrastur semitorquatus)* in Pantanal. *In* Anais do III Congreso Brasileiro de Ornitologia, pp. 57. Sociedade Brasileira de Ornitologia, Pelotas, Brazil. (in Portuguese)

Gundlach, J. 1874. Neue Beiträge ur Ornithologie Cubas, nach eigenen 30jährigen Beobachtungen zusammelgestellt. J. Ornithologie 22:113–166.

Gundlach, J. 1893. Ornitologia Cubana. La Habana, Cuba: Imprenta La Moderna.

Gurney, J. H. 1879. Notes on *Falco atriceps* and *Falco peregrinator.* Stray Feathers 8:423–437.

Gwinner, H. 1997. The function of green plants in nests of European Starlings *(Sturnus vulgaris).* Behaviour 134:337–351.

Hails, C. J. 1982. A comparison of tropical and temperate aerial insect abundance. Biotropica 12:310–313.

Hall, E. R., and K. R. Kelson. 1959. The mammals of North America, vol. 2. New York: Ronald Press.

Hall, J. L. 1995. Lucky shot. Living Bird 14:8–9.

Haller, H. 1996. The Golden Eagle in the Grisons. Long-term studies on the population ecology of *Aquila chrysaetos* in the centre of the Alps (Switzerland). Ornithol. Beob. Suppl. 9.

Haney, J. C. 1983. First sight record of Orange-breasted Falcon for Belize. Wilson Bull. 95:314–315.

Hardy, J. W., R. J. Raitt, J. Orejuela, T. Webber, and B. Edinger. 1975. First observation of the Orange-breasted Falcon in the Yucatán Peninsula of Mexico. Condor 77:512.

Harestand, A. S., and F. L. Bunnell. 1979. Home range and body weight—a reevaluation. Ecology 60:389–402.

Hargis, C. D., C. McCarthy, and R. D. Perloff. 1994. Home ranges and habitats of Northern Goshawks in eastern California. Stud. Avian Biol. 16:66–74.

Hartley, R. R. 2000. Falconry as a conservation tool in Africa. *In* R. D. Chancellor and B.-U. Meyburg, eds., Raptors at risk, pp. 373–378. Berlin: World Working Group on Birds of Prey; Blaine, WA: Hancock House.

Hartman, F. A. 1955. Heart weight in birds. Condor 57:221–238.

Hartman, F. A. 1961. Locomotor mechanisms of birds. Smithsonian Misc. Coll. 143, No. 1.

Hartman, F. A. 1966. Egg of the Great Black Hawk, *Buteogallus urubitinga ridgwayi.* Condor 68:515.

Hatshorn, G. S. 1980. Neotropical forest dynamics. Biotrop. Suppl. Trop. Succession 12 (Suppl):23–30.

Haugaasen, T., and C. A. Peres. 2005. Primate assemblage structure in Amazonian flooded and unflooded forests. Am. J. Primatol. 67:243–258.

Haverschmidt, F. 1948. Bird weights from Surinam. Wilson Bull. 60:230–239.

Haverschmidt, F. 1962. Notes on the feeding habits and food of some hawks in Surinam. Condor 64:154–158.

Haverschmidt, F. 1963. *Falco deiroleucus* Temminck in Surinam. J. Ornithologie 104:443–445.

Haverschmidt, F. 1964a. Investigations on *Chondrohierax uncinatus* (Temminck) in Surinam. J. Ornithologie 105:64–66.

Haverschmidt, F. 1964b. Nesting of the crane hawk in Surinam. Condor 66:303–305.

Haverschmidt, F. 1965. Eggs of *Chondrohierax uncinatus* (Temminck). J. Ornithologie 106:223.

Haverschmidt, F. 1968. Birds of Surinam. Edinburgh: Oliver and Boyd.

Haverschmidt, F. 1977. Roosting habits of the Swallow-tailed Kite. Auk 94:392.

Haverschmidt, F., and G. F. Mees. 1994. Birds of Suriname. Uitgeversmaatschappij, Paramaribo, Suriname: VACO.

Hayashida, F. M. 2005. Archaeology, ecological history, and conservation. Annu. Rev. Anthropol. 34:43–65.

Hector, D. P. 1986. Cooperative hunting and its relationship to foraging success and prey size in an avian predator. Ethology 73:247–257.

Heideman, P. D. 1995. Synchrony and seasonality of reproduction in tropical bats. Sympos. Zool. Soc. Lond. 67:151–165.

Heinzman, R., and C. Reining. 1990. Sustained rural development: extractive forest reserves in the northern Petén of Guatemala. TRI Working Paper No. 37, Tropical Resources Institute, Yale School of Forestry and Environmental Studies, New Haven, CT.

Helbig, A. J., A. Kocum, I. Seibold, and M. J. Braun. 2005. A multi-gene phylogeny of aquiline eagles (Aves: Accipitriformes) reveals extensive paraphyly at the genus level. Mol. Phylogenet. Evol. 35:147–164.

Hellebrekers, W. 1942. Revision of the Penard oological collection from Surinam. ZoologischeMededelingen 24:240–275.

Hellmayr, C. E., and B. Conover. 1949. Catalogue of birds of the Americas and the adjacent islands. Pt. 1. Field Mus. Nat. Hist. Zool. Ser. 13 (pt. 1, no. 4):1–358.

Henderson, R.W., and L.G. Hoevers. 1977. The seasonal incidence of snakes at a locality in northern Belize. Copeia 1977:349–355.

Henny, C.J., R.A. Olson, and T.L. Fleming. 1985. Breeding chronology, molt, and measurements of *Accipiter* hawks in northeastern Oregon. J. Field Ornithol. 56:97–112.

Henwood, K., and A. Fabrick. 1979. A quantitative analysis of the dawn chorus: temporal selection for communicatory optimization. Am. Nat. 114:260–274.

Herbert, R.A., and K.G.S. Herbert. 1965. Behavior of Peregrine Falcons in the New York City region. Auk 82:62–94.

Herfindal, I., J.D.C. Linnell, J. Odden, E.B. Nilsen, and R. Andersen. 2005. Prey density, environmental productivity and home-range size in the Eurasian lynx *(Lynx lynx).* J. Zool. Lond. 265:63–71.

Herklots, G.A.C. 1961. The birds of Trinidad and Tobago. London: Collins.

Herman, T.B., and I. Hedstron. 1990. The Orange-breasted Falcon *(Falco deiroleucus)* in Costa Rica: gone for thirty years? Brenesia 34:153–154.

Herrera, C.M., and F. Hiraldo. 1976. Food-niche and trophic relationships among European owls. Ornis Scand. 5:181–191.

Hewitt, V. 1937. A description of the eggs of Grey Crane-Hawk, Guiana White-tailed Hawk, Eyebrowed Hawk and Four-banded Sparrow Hawk. Ool. Rec. 17:12–14.

Heymann, E.W. 1992. Associations of tamarins *(Saguinas mystax* and *Saguinas fuscicollis)* and Double-toothed Kites *(Harpagus bidentatus)* in Peruvian Amazon. Folia Primatol. 59:51–55.

Hill, N.P. 1944. Sexual dimorphism in the Falconiformes. Auk 61:228–234.

Hilty, S.L. 2003. Birds of Venezuela. 2nd ed. Princeton, NJ: Princeton Univ. Press.

Hilty, S.L., and W.L. Brown. 1986. A guide to the birds of Colombia. Princeton, NJ: Princeton Univ. Press.

Holdaway, R.N. 1994. An exploratory phylogenetic analysis of the genera of the Accipitridae, with notes on the biogeography of the family. *In* B.U. Meyburg and R.D. Chancellor, eds., Raptor conservation today, pp. 601–637. London: World Working Group on Birds of Prey and Owls/Pica Press.

Holdridge, L. 1967. Life zone ecology. San José, Costa Rica: Tropical Science Center.

Hooge, P.N., and B. Eichenlaub. 1997. Animal movement extension to ArcView. Version 1.1 and later. Anchorage: Alaska Biological Science Center, U.S. Geological Survey.

Houston, D.C. 1976. Breeding of the White-backed and Ruppell's Griffon vultures, *Gyps africanus* and *G. rueppellii.* Ibis 118:14–40.

Hovis, J., T.D. Snowman, V.L. Cox, R. Fay, and K.L. Bildstein. 1985. Nesting behavior of Peregrine Falcons in west Greenland during the nestling period. J. Raptor Res. 19:15–19.

Howell, S. 1995. More on black-hawks. Birding 27:434–436.

Howell, S.N.G., and M.B. Robbins. 1995. Species limits of the Least Pygmy-Owl *(Glacidium minutisium)* complex. Wilsons Bull. 107:7–25.

Howell, S.N.G., and S. Webb. 1995. A guide to the birds of Mexico and northern Central America. Oxford: Oxford Univ. Press.

Howell, S.N.G., and A. Whittaker. 1995. Field identification of Orange-breasted and Bat Falcons. Cotinga 4:36–43.

Howell, T.R. 1972. Birds of the lowland pine savanna of northeastern Nicaragua. Condor 74:316–340.

Hoyt, D.F. 1979. Practical methods of estimating volume and fresh weight of bird eggs. Auk 96:73–77.

Hubert, C., A. Gallo, and G. Le Pape. 1995. Modification of parental behavior during the nesting period in the Common Buzzard *(Buteo buteo).* J. Raptor Res. 29:103–109.

Hume, R., and T. Boyer. 1991. Owls of the world. Philadelphia: Running Press.

Hustler, K. 1983. Breeding biology of the Peregrine Falcon in Zimbabwe. Ostrich 54:161–177.

Ibañez, C., C. Ramo, and B. Busto. 1992. Notes on food habits of the Black-and-white Owl. Condor 94:529–531.

Inchausti, P., and J. Halley. 2001. Investigating long-term ecological variability using the global population dynamics database. Science 293:655–657.

Iñigo, E., M. Ramos, and F. Gonzalez. 1989. Some ecological aspects of two primary evergreen forest raptor communities compared with cultivated tropical areas in southern Mexico. *In* B.-U. Meyburg and R.D. Chancellor, eds., Raptors in the modern world, pp. 529–543. Berlin: World Working Group on Birds of Prey.

Iñigo-Elias, E., M. Ramos, and F. Gonzalez. 1987. Two recent records of Neotropical eagles in southern Veracruz, Mexico. Condor 89:671–672.

Jackson, J.A. 1988a. American Black Vulture. *In* R.S. Palmer, ed., Handbook of North American birds, vol. 4, pp. 11–24. New Haven, CT: Yale Univ. Press.

Jackson, J.A. 1988b. Turkey Vulture. *In* R.S. Palmer, ed., Handbook of North American birds, vol. 4, pp. 25–42. New Haven, CT: Yale Univ. Press.

Jackson, J.B.C., and L. D'Croz. 1997. The ocean divided. *In* A.G. Coates, ed., Central America: a natural and cultural history, pp. 38–71. New Haven, CT: Yale Univ. Press.

Jaksic, F.M., and H.E. Braker. 1983. Food-niche relationships and guild structure of diurnal birds of prey: competition versus opportunism. Can. J. Zool. 61:2230–2241.

Jaksic, F.M., and J.H. Carothers. 1985. Ecological, morphological, and bioenergetic correlates of hunting mode in hawks and owls. Ornis Scand. 16:165–172.

James, C., and R. Shine. 1985. The seasonal timing of reproduction: a tropical-temperate comparison in Australian lizards. Oecologia 67:464–474.

James, P.C., I.G. Wartkentin, and L.W. Oliphant. 1989. Turnover and dispersal in urban Merlins *(Falco columbarius).* Ibis 131:426–429.

Janzen, D.H. 1973a. Rate of regeneration after a tropical high elevation fire. Biotropica 5:117–122.

Janzen, D.H. 1973b. Sweep samples of tropical foliage insects: description of study sites, with data on species abundances and size distributions. Ecology 54:659–686.

Janzen, D.H. 1973c. Sweep samples of tropical foliage insects: effects of seasons, vegetation types, elevation, time of day, and insularity. Ecology 54:687–708.

Janzen, D.H., and C.M. Pond. 1975. A comparison by sweep sampling of the arthropod fauna of secondary vegetation in Michigan, England, and Costa Rica. Trans. Royal Entomol. Soc. Lond. 127:33–50.

Janzen, D.H., and T.W. Schoener. 1968. Differences in insect abundance and diversity between wetter and drier sites during a tropical dry season. Ecology 49:96–110.

Janzen, D.H., and D.E. Wilson. 1983. Mammals. *In* D.H. Janzen, ed., Costa Rican natural history, pp. 426–442. Chicago: Univ. of Chicago Press.

Jehl, J.R., Jr. 1968. Foraging behavior of *Geranospiza nigra,* the Blackish Crane-Hawk. Auk 85:493–494.

Jenkins, A. 1991. Latitudinal prey productivity and potential density in the Peregrine Falcon. Gabar 6:20–24.

Jenkins, A.R. 1995. Morphometrics and flight performance of southern African Peregrine and Lanner Falcons. J. Avian Biol. 26:49–58.

Jenkins, A. R. 2000. Characteristics of Peregrine and Lanner Falcon nesting habitats in South Africa. Ostrich 71:416–424.

Jenkins, A. R., and P. A. R. Hockey. 2001. Prey availability influences habitat tolerance: an explanation for the rarity of Peregrine Falcons in the tropics. Ecography 24:359–367.

Jenkins, M. A. 1978. Gyrfalcon nesting behavior from hatching to fledging. Auk 95:122–127.

Jenny, J. P., and T. J. Cade. 1986. Observations of the biology of the Orange-breasted Falcon *(Falco deiroleucus)*. Birds Prey Bull. No. 3:119–124.

Johnsgard, P. A. 1988. North American owls: biology and natural history. Washington, DC: Smithsonian Institution Press.

Johnsgard, P. A. 1990. Hawks, eagles, and falcons of North America: biology and natural history, 1st ed. Washington, DC: Smithsonian Institution Press.

Johnsgard, P. A. 2002. North American owls: biology and natural history, 2nd ed. Washington, DC: Smithsonian Institution Press.

Johnson, D. R. 1978. The study of raptor populations. Moscow: Univ. Press of Idaho.

Johnson, N. K., and H. J. Peeters. 1963. The systematic position of certain hawks in the genus *Buteo*. Auk 80:417–446.

Johnston, J. P., W. J. Peach, R. D. Gregory, and S. A. White. 1997. Survival rates of temperate and tropical passerines: a Trinidadian perspective. Am. Nat. 150:771–789.

Jollie, M. 1977. A contribution to the morphology and phylogeny of the Falconiformes, part 3. Evol. Theory 3:285–298.

Jolón Morales, M. R. 1996. Patrones de estratificación vertical de mamidreos menores en el Parque Nacional Tikal, Petén, Guatemala. Documento Técnico. Wildlife Conservation Society Guatemala Program: Guatemala.

Jolón Morales, M. R. 1997. Fauna invisible de la Reserva de la Biosfera Maya. Guía de campo para algunos roedores menores. Wildlife Conservation Society, Mesoamerican and Caribbean Program: Guatemala.

Jones, H. L. 2002. Central America. N. Am. Birds 56:115–117.

Jones, H. L., and J. M. Diamond. 1976. Short-time-base studies of turnover in breeding bird populations on the California Channel Islands. Condor 78:526–549.

Jones, H. L., E. McRae, M. Meadows, and S. N. G. Howell. 2000. Status updates for selected bird species in Belize, including several species previously undocumented from the country. Cotinga 13:17–31.

Jones, L. E., and J. Sutter. 1992. Results and comparisons of two years of census efforts at three units of the Maya Biosphere Reserve/Calakmul Biosphere Reserve Complex. *In* D. Whitacre and R. K. Thorstrom, eds., Progress report V, Maya Project: use of raptors and other fauna as environmental indicators for design, management, and monitoring of protected areas and for building local capacity for conservation in Latin America, pp. 63–80. Boise, ID: The Peregrine Fund.

Jullien, M., and J. Clobert. 2000. The survival value of flocking in Neotropical birds: reality or fiction? Ecology 81:3416–3430.

Jullien, M., and J.-M. Thiollay. 1996. Effects of rain forest disturbance and fragmentation: comparative changes of the raptor community along natural and human-made gradients in French Guiana. J. Biogeogr. 23:7–25.

Jullien, M., and J-M. Thiollay. 1998. Multi-species territoriality and dynamic of Neotropical forest understory bird flocks. J. Anim. Ecol. 67:227–252.

Jullien, M., and J.-M. Thiollay. 2001. The adaptive significance of flocking in tropical understorey forest birds: the

field evidence. *In* F. Bongers, P. Charles-Dominique, P.-M. Forget, and M. Thery, eds., Dynamics and plant-animal interactions in a Neotropical rainforest, pp. 143–159. Dordrecht, Netherlands: Kluwer Academic.

Julliot, C. 1994. Predation of a young spider monkey *(Ateles paniscus)* by a Crested Eagle *(Morphnus guianensis)*. Folia Primatol. 63:75–77.

Kalinowski, S. T., and R. S. Waples. 2002. Relationship effective to census size in fluctuating populations. Conserv. Biol. 16:129–136.

Karr, J. R., S. K. Robinson, J. G. Blake, and R. O. Bierregaard Jr. 1990. Birds of four Neotropical forests. *In* A. H. Gentry, ed., Four Neotropical rainforests, pp. 237–269. New Haven, CT: Yale Univ. Press.

Keane, J. J., and M. L. Morrison. 1994. Northern Goshawk ecology: effects of scale and levels of biological organization. Stud. Avian Biol. 16:3–11.

Kelso, L. H. 1940. Variation of the external ear-opening in the Strigidae. Wilson Bull. 52:24–29.

Kelt, D. A., and D. H. Van Vuren. 2001. The ecology and macroecology of mammalian home range area. Am. Nat. 157:637–645.

Kemp, A. 1987. The owls of southern Africa. London: Struik Winchester.

Kemp, A. C., and T. M. Crowe. 1990. A preliminary phylogenetic and biogeographic analysis of the genera of diurnal raptors. *In* G. Peters and R. Hutterer, eds., Vertebrates in the tropics, pp. 161–175. Bonn: Museum Alexander Koenig.

Kemp, A. C., and T. M. Crowe. 1994. Morphometrics of falconets and hunting behaviour of the Black-thighed Falconet *Microhierax fringillarius*. Ibis 136:44–49.

Kenward, R. E., and K. H. Hodder. 1996. RANGES V. An analysis system for biological location data. Wareham, UK: Institute of Terrestrial Ecology.

Kiff, L. F. 1979a. Bird egg collections in North America. Auk 96:746–755.

Kiff, L. F. 1979b. The eggs of the Black Hawk-Eagle. Raptor Res. 13:15.

Kiff, L. F. 1981. Notes on eggs of the Hook-billed Kite, *Chondrohierax uncinatus*, including two overlooked nesting records. Bull. Br. Ornithol. Club 101:318–323.

Kiff, L. F. 1988. Eggs of the Orange-breasted Falcon *(Falco deiroleucus)*. J. Raptor Res. 22:117–118.

Kiff, L. F., and M. Cunningham. 1980. The egg of the Ornate Hawk-Eagle *(Spizaetus ornatus)*. J. Raptor Res. 14:51.

Kiff, L. F., D. B. Peakall, and D. P. Hector. 1980. Eggshell thinning and organochlorine residues in the Bat and Aplomado Falcons in Mexico. *In* Acta 17th Congressus Internationalis Ornithologici, pp. 949–952. Berlin: Deutsche Ornithologen-Gesellschaft.

Kiff, L. F., M. P. Wallace, and N. B. Gale. 1989. Eggs of captive Crested Eagles *(Morphnus guianensis)*. J. Raptor Res. 23:107–108.

Kilham, L. 1978. Alarm call of Crested Guan when attacked by Ornate Hawk-Eagle. Condor 80:347–348.

King, W. B. 1979. Red data book, volume 2: Aves. 2nd ed. Morges, Switzerland: International Union for Conservation of Nature and Natural Resources.

King, W. B. 1981. Endangered birds of the world: the ICBP bird red data book. Washington, DC: Smithsonian Institution Press/International Council for Bird Preservation (ICBP).

Kirven, M. L. 1976. The ecology and behavior of the Bat Falcon *(Falco rufigularis)*. Ph.D. dissertation, Univ. of Colorado, Boulder.

Klein, B. C., and R. O. Bierregaard. 1988. Movement and calling behavior of the Lined Forest-Falcon *(Micrastur gilvicollis)* in the Brazilian Amazon. Condor 90:497–499.

Klein, B. C., L. H. Harper, R. O. Bierregaard, and G. V. N. Powell. 1988. The nesting and feeding behavior of the Ornate Hawk-Eagle near Manaus, Brazil. Condor 90:239–241.

Klomp, H. 1970. The determination of clutch-size in birds: a review. Ardea 58:1–121.

Kochert, M. N., K. Steenhof, C. L. McIntyre, and E. H. Craig. 2002. Golden Eagle *(Aquila chrysaetos). In* A. Poole and F. Gill, eds., The birds of North America, No. 684. Philadelphia: Birds of North America.

König, C., and F. Weick. 2008. Owls of the world, 2nd ed. New Haven, CT: Yale Univ. Press.

König, C., F. Weick, and J.-H. Becking. 1999. Owls: a guide to the owls of the world. New Haven, CT: Yale Univ. Press.

Komar, O. 1998. Avian diversity in El Salvador. Wilson Bull. 110:511–533.

Komar, O. 2003. Predation on birds by the White Hawk *(Leucopternis albicollis).* Ornitol. Neotrop. 14:541–543.

Korpimäki, E. 1986. Reversed size dimorphism in birds of prey, especially in Tengmalm's Owl *Aegolius funereus:* a test of the "starvation hypothesis." Ornis Scand. 17:326–332.

Korpimäki, E., and K. Norrdahl. 1991. Numerical and functional responses of kestrels, Short-eared Owls, and Long-eared Owls to vole densities. Ecology 72:814–826.

Kreuger, R. 1963. Details of three previously undescribed South American raptor eggs. Ool. Rec. 43:5–6.

Krügel, M. M. 2003. Records documenting *Chondrohierax uncinatus* (Temminck, 1822) (Falconiformes: Accipitriformes) for Rio Grande do Sul. Ararajuba 11:83–84. (in Portuguese with English summary)

Lack, D. 1946. Competition for food by birds of prey. J. Anim. Ecol. 15:123–129.

Lack, D. 1947. The significance of clutch-size, parts 1 and 2. Ibis 89:302–352.

Lack, D. 1966. Population studies of birds. Oxford: Clarendon Press.

Lack, D. 1968. Ecological adaptations for breeding in birds. London: Chapman and Hall.

Lamm, D. W. 1974. White Hawk preying on the Great Tinamou. Auk 91:845–846.

Land, H. C. 1963. A collection of birds from the Caribbean lowlands of Guatemala. Condor 65:49–65.

Land, H. C. 1970. Birds of Guatemala. Wynnewood, PA: Livingston.

Lande, R. 1993. Risks of population extinction from demographic and environmental stochasticity and random catastrophes. Am. Nat. 142:911–927.

Lande, R. 1995. Mutation and conservation. Conserv. Biol. 9:782–791.

Lasley, G. W., C. Sexton, and D. Hillsman. 1988. First record of Mottled Owl *(Ciccaba virgata)* in the United States. Am. Birds 42:23–24.

Latham, R. M. 1952. The fox as a factor in the control of weasel populations. J. Wildl. Manage. 16:516–517.

Lawrence, G. N. 1876. Catalogue of birds collected by Prof. Francis Sumicrasti in southwestern Mexico, and now in the National Museum at Washington, DC. Bull. U.S. Natl. Mus. 4:5–56.

Lebeau, B., P. Jolicover, and G. Pageau. 1968. Asymptotic growth egg production and trivariate allometry in *Esox masquinongy mitehill.* Growth 50:185–200.

Leck, C. F. 1979. Avian extinction in an isolated tropical wet-forest preserve, Ecuador. Auk 96:343–352.

Lee, J. C. 1980. An ecogeographic analysis of the herpetofauna of the Yucatán Peninsula. Univ. Kansas Mus. Nat. Hist. Misc. Publ. 67:1–75.

Lee, J. C. 1996. The amphibians and reptiles of the Yucatán Peninsula. Ithaca, NY: Comstock/Cornell Univ. Press.

Lee, J. C. 2000. A field guide to the amphibians and reptiles of the Maya world. Ithaca, NY: Cornell Univ. Press.

Lehmann, F. C. 1943. El genero *Morphnus.* Caldasia 2:165–179.

Lehmann, F. C. 1970. Avifauna in Colombia. *In* H. K. Buechner and J. H. Buechner, eds., The avifauna of northern Latin America, pp. 88–92. Smith. Cont. Zool., No. 26. Washington, DC: Smithsonian Institution Press.

Lemke, T. O. 1979. Fruit-eating behavior of Swallow-tailed Kites *(Elanoides forficatus)* in Colombia. Condor 81:207–208.

Leopold, A. S. 1950. Vegetation zones of Mexico. Ecology 31:507–518.

Lerner, H. R. L., and D. P. Mindell. 2005. Phylogeny of eagles, Old World vultures, and other Accipitridae based on nuclear and mitochondrial DNA. Mol. Phylogenet. Evol. 37:327–346.

Lessells, C. M. 1991. The evolution of life histories. *In* J. R. Krebs and N. B. Davies, eds., Behavioural ecology: an evolutionary approach, 3rd ed., pp. 32–68. Oxford: Blackwell Scientific.

Levings, S. C., and D. M. Windsor. 1985. Litter arthropod populations in a tropical deciduous forest: relationships between years and arthropod groups. J. Anim. Ecol. 54:61–69.

Levins, R. 1968. Evolution in changing environments. Princeton, NJ: Princeton Univ. Press.

Lewis, S. E., and R. M. Timm. 1991. Predation on nestling Bare-throated Tiger-Herons by a Great Black Hawk. Ornitol. Neotrop. 2:37.

Leyden, B. W. 1984. Guatemala forest synthesis after Pleistocen aridity. Proc. Natl. Acad. Sci. USA 81:4856–4859.

Lillywhite, H. B., and R. W. Henderson. 1993. Behavioral and functional ecology of arboreal snakes. *In* R. A. Seigel and J. T. Collins, eds., Snakes: ecology and behavior, pp. 1–48. New York: McGraw-Hill.

Linden, M., and A. P. Moller. 1989. Cost of reproduction and covariation of life history traits in birds. Trends Ecol. Evol. 4:367–371.

Lindstedt, S. L., B. J. Miller, and S. W. Buskirk. 1986. Home range, time, and body size in mammals. Ecology 67:413–418.

Lister, B. C., and A. G. Aguayo. 1992. Seasonality, predation, and the behavior of a tropical mainland anole. J. Anim. Ecol. 61:717–733.

Loiselle, B. A., and W. G. Hoppes. 1983. Nest predation in insular and mainland lowland rainforest in Panama. Condor 85:93–95.

López-Lanús, B. 2000. Collared Forest-Falcon *Micrastur semitorquatus* courtship and mating, with take-over of a macaw nest. Cotinga 14:9–11.

Lopez-Ornat, A., J. F. Lynch, and B. MacKinnon de Montes. 1989. New and noteworthy records of birds from the eastern Yucatán Peninsula. Wilson Bull. 101:390–409.

Lourdes-Ribeiro, A., M. R. Gimenes, and L. don Anjos. 2004. Observations on the reproductive behavior of *Ictinia plumbea* (Falconiformes: Accipitridae) on the campus of the Maringa State University, Paraná, Brazil. Ararajuba 11. (in Portuguese with English summary)

Lovejoy, T. E., and R. O. Bierregaard Jr. 1990. Central Amazonian forests and the Minimum Critical Size of Ecosystems Project. *In* A. H. Gentry, ed., Four Neotropical rainforests, pp. 60–71. New Haven, CT: Yale Univ. Press.

Lovette, I. J., and R. T. Holmes. 1995. Foraging behavior of American Redstarts in breeding and wintering habitats: implications for relative food availability. Condor 97:782–791.

Lowery, G.H., Jr., and W.W. Dalquest. 1951. Birds from the state of Veracruz, Mexico, vol. 3, no. 4, pp. 531–649. Lawrence: Univ. of Kansas Publications, Museum of Natural History.

Lundberg, A. 1986. Adaptive advantages of reversed sexual size dimorphism in European owls. Ornis Scand. 17:133–140.

Lundell, C.L. 1937. The vegetation of Peten. Carnegie Institute of Washington, Publication No. 478. Washington, DC.

Lyon, B., and A. Kuhnigk. 1985. Observations on nesting Ornate Hawk-Eagles in Guatemala. Wilson Bull. 97:141–264.

MacArthur, R.H., J. Diamond, and J.R. Karr. 1972. Density compensation in island faunas. Ecology 53:330–342.

Mace, G.M., and P.H. Harvey. 1983. Energetic constraints on home-range size. Am. Nat. 121:120–132.

MacLaren, P.A., S.H. Anderson, and D.E. Runde. 1988. Food habits and nest characteristics of breeding raptors in southwestern Wyoming. Great Basin Nat. 48:548–553.

Mader, W.J. 1979. First nest description for the genus *Micrastur* (Forest-Falcons). Condor 81:320.

Mader, W.J. 1981. Notes on nesting raptors in the llanos of Venezuela. Condor 83:48–51.

Mader, W.J. 1982. Ecology and breeding habits of the Savanna Hawk in the llanos of Venezuela. Condor 84:261–271.

Madrid M., J.A., H.D. Madrid M., S.H. Funes A., J. López, R. Botzoc G., and A. Ramos. 1991. Reproductive biology and behavior of the Ornate Hawk-Eagle *(Spizaetus ornatus)* in Tikal National Park. *In* D. Whitacre and R.K. Thorstrom, eds., Progress report IV, Maya Project: use of raptors and other fauna as environmental indicators for design, management, and monitoring of protected areas and for building local capacity for conservation in Latin America, pp. 93–114. Boise, ID: The Peregrine Fund.

Madrid M., H.D., R.A. Madrid M., J.R. Cruz E., J.L. Córdova A., W.E. Martínez A., and A. Ramos C. 1992. Behavior and breeding biology of the Ornate Hawk-Eagle. *In* D. Whitacre and R.K. Thorstrom, eds., Progress report V, Maya Project: use of raptors and other fauna as environmental indicators for design, management, and monitoring of protected areas and for building local capacity for conservation in Latin America, pp. 179–192. Boise, ID: The Peregrine Fund.

Madrid M., J.A., C. Marroquín V., T. Dubon O., M.D. Schulze, J. Hunt, and D.F. Whitacre. 1995. Monitoring population parameters of a wintering migrant songbird, the Kentucky Warbler: persistence pays. *In* J.A. Bissonette and P.R. Krausman, eds., Integrating people and wildlife for a sustainable future, pp. 479–483. Proceedings of the First International Management Congress. Bethesda, MD: The Wildlife Society.

Madsen, T. 1984. Movements, home range size and habitat use of radio-tracked grass snakes *(Natrix natrix)* in southern Sweden. Copeia 1984:707–713.

Mager, W. 1967. Nestling Swallow-tailed Kite banded on Key Largo, recovered in Brazil. Florida Field Naturalist 40:13–14.

Malcolm, J.R. 1990. Estimation of mammalian densities in continuous forest north of Manaus. *In* A.H. Gentry, ed., Four Neotropical rainforests, pp. 339–357. New Haven, CT: Yale Univ. Press.

Malhi, Y., and 27 coauthors. 2004. The above-ground coarse wood productivity of 104 Neotropical forest plots. Global Change Biol. 10:563–591.

Marchant, S. 1960. The breeding of some S.W. Ecuadorian birds. Ibis 102:349–382.

Marini, M. Â., T.M. Aguilar, R.D. Andrade, L.O. Leite, M. Ancães, C.E.A. Carvalho, C. Duca, M. Maldonado-Coelho, F. Sebaio, and J. Gonçalves. 2007. Nesting biology of birds from southeastern Minas Gerais, Brazil. Rev. Bras. Ornitol. 15:367–376.

Marker, L.L., and A.J. Dickman. 2005. Factors affecting leopard *(Panthera pardus)* spatial ecology, with particular reference to Namibian farmlands. S. Afr. J. Wildl. Res. 35:105–115.

Marks, J.S., D.L. Evans, and D.W. Holt. 1994. Long-eared Owl *(Asio otus)*. *In* A. Poole and F. Gill, eds., The birds of North America, No. 133. Philadelphia: Academy of Natural Sciences; Washington, DC: American Ornithologists' Union.

Marler, P., and W.J. Hamilton III. 1966. Mechanisms of animal behavior. New York: John Wiley and Sons.

Marshall, J.T. 1943. Additional information concerning the birds of El Salvador. Condor 45:21–33.

Marti, C. 1987. Raptor food habits studies. *In* B.G. Pendelton, B.A. Milsap, K.W. Kline, and D.A. Bird, eds., Raptor management techniques manual, pp. 67–69. Washington, DC: Natl. Wildl. Fed. Tech. Ser. No. 10.

Marti, C.D., K. Steenhof, and J.S. Marks. 1993. Community trophic structure; the roles of diet, body size, and activity time in vertebrate predators. Oikos 67:6–18.

Martin, P.S., C.R. Robbins, and W.B. Heed. 1954. Birds and biogeography of the Sierra de Tamaulipas, an isolated pine-oak habitat. Wilson Bull. 66:38–57.

Martin, T.E. 1992. Interaction of nest predation and food limitation in reproductive strategies. Curr. Ornithol. 9:163–197.

Martin, T.E. 1993. Nest predation and nest sites: new perspectives on old patterns. Bioscience 43:523–532.

Martin, T.E. 1995. Avian life history evolution in relation to nest sites, nest predation, and food. Ecol. Monogr. 65:101–127.

Martin, T.E. 1996. Life history evolution in tropical and south temperate birds: what do we really know? J. Avian Biol. 27:263–272.

Martin, T.E. 2004. Avian life-history evolution has an eminent past: does it have a bright future? Auk 121:289–301.

Massoia, E. 1988. Presas de *Buteo magnirostris* en el partido de General Rodriguez, provincia de Buenos Aires. Aprona Bol. Cientifica 10:8–11.

Matthiessen, P. 1972. The tree where man was born. New York: Dutton.

Mayr, E., and G.W. Cottrell. 1979. Checklist of birds of the world, vol. 1, 2nd ed. Cambridge, MA: Museum of Comparative Zoology.

Mays, N.M. 1985. Ants and foraging behavior of the Collared Forest-Falcon. Wilson Bull. 97:231–232.

McCarthy, T. 1976. Black-and-white Owl in Belize (British Honduras). Southwest. Nat. 20:585–586.

McCarthy, T.J. 1982. Bat records from the Caribbean lowlands of El Petén, Guatemala. J. Mamm. 63:683–685.

McNab, B.K. 1963. Bioenergetics and the determination of home range size. Am. Nat. 97:133–139.

Meyer, J.R., and C.F. Foster. 1996. A guide to the frogs and toads of Belize. Malabar, FL: Krieger.

Meyer, K.D. 1995a. Communal roosts of the American Swallow-tailed Kite in Florida: habitat associations, critical sites, and a technique for monitoring population status. Final Rep. Fla. Game and Fresh Water Fish Comm., Tallahassee, FL.

Meyer, K.D. 1995b. Swallow-tailed Kite *(Elanoides forficatus)*. *In* A. Poole and F. Gill, eds., The birds of North America, No. 138. Philadelphia: Academy of Natural

Sciences; Washington, DC: American Ornithologists' Union.

Meyer, K.D., and M.W. Collopy. 1995. Status, distribution, and habitat requirements of the American Swallow-tailed Kite (Elanoides forficatus) in Florida. Fla. Game and Fresh Water Fish Comm. Nongame Wild. Program Project Rep., Tallahassee, FL.

Meyer de Schauensee, R. 1964. The birds of Colombia and adjacent areas of South and Central America. Philadelphia: Academy of Natural Sciences.

Meyer de Schauensee, R., and W.H. Phelps Jr. 1978. A guide to the birds of Venezuela. Princeton, NJ: Princeton Univ. Press.

Mikkola, H. 1983. Owls of Europe. Vermillion, SD: Buteo Books.

Miller, A.H. 1952. Supplementary data on the tropical avifauna of the arid upper Magdalena Valley of Colombia. Auk 69:450–457.

Miller, A.H. 1963. The vocal apparatus of two South American owls. Condor 65:440–441.

Miller, L.H. 1937. Skeletal studies of the tropical hawk Harpagus. Condor 39:219–221.

Millsap, B. 1981. Distributional status of falconiformes in west central Arizona with notes on ecology, reproductive success, and management. U.S. Bureau of Land Management Technical Note 355.

Minitab. 1996. Minitab user's guide. Release 11 for Windows. State College, PA: Minitab.

Mock, D.W. 1985. Siblicidal brood reduction: the prey-size hypothesis. Am. Nat. 125:327–343.

Mohr, C.O. 1947. Table of equivalent populations of North America small mammals. Am. Midl. Nat. 37:233–249.

Monroe, B.L., Jr. 1968. A distributional survey of the birds of Honduras. Ornithological Monograph No. 7, American Ornithologists' Union. Lawrence, KS: Allen Press.

Montiel de la Garza, F.G., and A.J. Contreras-Balderas. 1990. First Hook-billed Kite specimen from Nuevo Leon, Mexico. Southwestern Naturalist 35:370.

Moreau, R.E. 1944. Clutch size: a comparative study, with special reference to African birds. Ibis 86:286–347.

Morrison, D.W. 1980. Foraging and day-roosting dynamics of canopy fruit bats in Panama. J. Mammal. 61:10–29.

Morrison, D.W. 1983. Artibeus jamaicensis. In D.H. Janzen, ed., Costa Rican natural history, pp. 449–451. Chicago: Univ. of Chicago Press.

Morton, E.S. 1975. Ecological sources of selection on avian sounds. Am. Nat. 109:17–34.

Morton, E.S., and B.J.M. Stutchbury. 2000. Demography and reproductive success in the dusky antbird, a sedentary tropical passerine. J. Field Ornithol. 71:493–500.

Moynihan, M. 1976. The New World primates. Princeton, NJ: Princeton Univ. Press.

Mueller, H.C. 1986. The evolution of reversed sexual dimorphism in owls: an empirical analysis of possible selective factors. Wilson Bull. 98:387–406.

Mueller, H.C., and K. Meyer. 1985. The evolution of reversed sexual dimorphism in size: a comparative analysis of the Falconiformes of the western Palearctic. Curr. Ornithol. 2:65–101.

Munn, C.A., and J.W. Terborgh. 1979. Multi-species territoriality in Neotropical foraging flocks. Condor 81:338–347.

Murray, B.G., Jr. 1985. Evolution of clutch size in tropical species of birds. Ornithol. Monogr. 36:505–519.

Murray, D.L., 1994. Cultivating crisis: the human cost of pesticides in Latin America. Austin: Univ. of Texas Press.

Murray, G.A. 1976. Geographic variation in the clutch sizes of seven owl species. Auk 93:602–613.

Mysterud, A., F.J. Perez-Barbería, and I.J. Gordon. 2000. The effect of season, sex, and feeding style on home range area versus body mass scaling in temperate ruminants. Oecologia 127:30–39.

Navarro R., R., G. Marin E., and J. Muñoz G. 2007. Notes on the breeding ecology of three Venezuelan accipitrid species. Ornitol. Neotrop. 18:453–457.

Naveda-Rodríguez, A. 2007. Distribution and habitat of hawk-eagles in Venezuela. In K.L. Bildstein, D.R. Barber, and A. Zimmerman, eds., Neotropical raptors, pp. 99–105. Orwigsburg, PA: Hawk Mountain Sanctuary.

Nelson, R.W. 1970. Some aspects of the breeding behavior of Peregrine Falcons on Langara Island, B.C. M.S. thesis, Univ. of Calgary.

Newman, K. 1980. Birds of southern Africa and Kruger National Park., Johannesburg, South Africa: MacMillan South Africa Publishers.

Newton, I. 1978. Feeding and development of sparrowhawk Accipiter nisus nestlings. J. Zool. 184:465–487.

Newton, I. 1979. Population ecology of raptors. Berkhamsted, UK: T. & A.D. Poyser.

Newton, I. 1986. The Sparrowhawk. Calton, UK: T. & A.D. Poyser.

Newton, I. 1989. The control of sparrowhawk Accipiter nisus nesting densities. In B.-U. Meyburg and R.D. Chancellor, eds., Raptors in the modern world, pp. 169–180. Berlin: World Working Group on Birds of Prey.

Nice, M.M. 1957. Nesting success in altricial birds. Auk 74:305–321.

Nicholls, T.H., and M.R. Fuller. 1987. Territorial aspects of Barred Owl home range and behavior in Minnesota. In R.W. Nero, R.J. Clark, R.J. Knapton, and R.H. Hamre, eds., Biology and conservation of northern forest owls: symposium proceedings, pp. 121–128. U.S. For. Serv. Gen. Tech. Rep. RM-142.

Nijman, V., B.S. van Balen, and R. Sözer. 2000. Breeding biology of the Javan Hawk-Eagle Spizaetus bartelsi in West Java, Indonesia. Emu 100:125–132.

Norberg, U.M. 1985. Flying, gliding, and soaring. In M. Hildebrand, D.M. Bramble, K.F. Liem, and D.B. Wake, eds., Functional vertebrate morphology, pp. 129–158. Cambridge, MA: Belknap Press, Harvard Univ. Press.

Norris, J.P., Jr. 1926. A catalogue of sets of Accipitres' eggs in the collection of Joseph Parker Norris, Jr., Philadelphia, PA U.S.A. Oolog. Rec. 6:25–41.

Nowak, R., and J. Paradiso. 1983. Walker's mammals of the world. Baltimore: John Hopkins Univ. Press.

Nunn, C.L., and R.A. Barton. 2000. Allometric slopes and independent contrasts: a comparative test of Kleiber's Law in primate ranging patterns. Am. Nat. 156:519–533.

Nunnery, T., and M.R. Welford. 2002. Barred Forest-Falcon (Micrastur ruficollis) predation on a hummingbird. J. Raptor Res. 36:239–240.

Nur, N., 1988. The cost of reproduction in birds: an examination of the evidence. Ardea 76:155–168.

Oberholser, H.C., and E.B. Kincaid Jr. 1974. The bird life of Texas. Austin: Univ. of Texas Press.

Ogden, J.C. 1974. The Short-tailed Hawk in Florida, I. Migration, habitat, hunting techniques, and food habits. Auk 91:95–110.

Olivares, A. 1962. Aves de la región sur de la Sierra de la Macarena, Meta, Colombia. Rev. Acad. Colombiana Ciencias Exactas, Fisicas, Naturales 11:305–344.

Olmos, F. 1990a. Harrier-like hunting behavior by a crane hawk Geranospiza caerulescens. Bull. Br. Ornithol. Club 110:225–226.

Olmos, F. 1990b. Nest predation of Plumbeous Ibis by capuchin monkeys and Greater Black Hawk. Wilson Bull. 102:169–170.

Olrog, C.C. 1985. Status of wet forest raptors in northern Argentina. *In* I. Newton and R.D. Chancellor, eds., Conservation studies on raptors, pp. 191–197. International Council for Bird Preservation (ICBP) Tech. Publ. No. 5.

Olsen, P. 1995. Australian birds of prey. Sydney, Australia: Univ. of New South Wales Press; Baltimore, MD: Johns Hopkins Univ. Press.

Olsen, P., and T.G. Marples. 1993. Geographic variation in egg size, clutch size, and date of laying of Australian raptors (Falconiformes and Strigiformes). Emu 93:167–179.

Olson, S.L. 1985. The fossil record of birds. Avian Biol. 8:79–238.

Olwagen, C.D. 1984. Breeding behaviour of the rednecked falcon in captivity. *In* J.M. Mendelsohn and C.W. Sapsford, eds., Proceedings of the 2nd symposium on African predatory birds, pp. 23–30. Durban, South Africa: Natal Bird Club.

Orians, G.H., and D.R. Paulson. 1969. Notes on Costa Rican birds. Condor 71:426–431.

Ortiz-Crespo, F.I. 1986. Notes on the status of diurnal raptor populations in Ecuador. Birds Prey Bull. No. 3:71–79.

Ouellet, H. 1991. Description of the courtship and copulation behavior of the crane hawk. J. Field Ornithol. 62:403–406.

Oversluijs Vasquez, M.R., and E.W. Heymann. 2001. Crested Eagle *(Morphnus guianensis)* predation on infant tamarins *(Saguinus mystax)* and *(Saguinus fuscicollis,* Callitrichinae). Folia Primatol. 72:301–303.

Owen-Smith, R.N. 1988. Megaherbivores: the influence of very large body size on ecology. Cambridge: Cambridge Univ. Press.

Palacios, E., and C.A. Peres. 2005. Primate population densities in three nutrient-poor Amazonian terra firme forests of south-eastern Colombia. Folia Primatol. 76:135–145.

Palmer, R.S., ed. 1988. Handbook of North American birds, vol. 4, 5: diurnal raptors, parts 1 and 2. New Haven, CT: Yale Univ. Press.

Pampel, F.C. 2000. Logistic regression: a primer. Thousand Oaks, CA: Sage Univ. Papers Series on Quantitative Applications in the Social Sciences, Series No. 07–132.

Panasci, T.A., and D.F. Whitacre. 2000. Diet and foraging behavior of nesting Roadside Hawks in Petén, Guatemala. Wilson Bull. 112:555–558.

Panasci, T.A., and D.F. Whitacre. 2002. Roadside Hawk breeding ecology in forest and farming landscapes. Wilson Bull. 114:114–121.

Parker, A. 1979. Peregrines at a Welsh coastal eyrie. Br. Birds 72:104–114.

Parker, J.W. 1988. Mississippi Kite *Ictinia mississippiensis.* *In* Palmer, R.S., ed., Handbook of North American birds, vol. 4, pp. 166–186. New Haven, CT: Yale Univ. Press.

Parker, J.W., and M. Ports. 1982. Helping at the nest by yearling Mississippi Kites. J. Raptor Res. 16:14–17.

Parry, V. 1968. Seasonality, territoriality and breeding biology of the Kookaburra, *Dacelo gigas* (Boddaert). M.Sc. thesis, Monash University, Clayton, Victoria, Australia.

Paulson, D.R. 1983. Flocking in the Hook-billed Kite. Auk 100:749–750.

Paynter, R. 1955. The ornithogeography of the Yucatán Peninsula. Peabody Museum of Natural History, Yale Univ., Bulletin No. 9, New Haven, CT.

Pearson, D.L., and J.A. Derr. 1986. Seasonal patterns of lowland forest floor arthropod abundance in southeastern Peru. Biotropica 18:244–256.

Peery, M.Z. 2000. Factors affecting interspecies variation in home-range size of raptors. Auk 117:511–517.

Peeters, H.-J. 1963. Einiges uber den Waldfalken *Micrastur semitorquatus.* J. Ornithologie 104:357–364.

Pennington, T.D., and J. Sarukhan. 1968. Arboles tropicales de Mexico. Instituto Nacional de Investigaciones Forestales, Mexico, D.F.

Pennycuick, C.J. 1975. Mechanics of flight. *In* D.S. Farner and J.R. King, eds., Avian biology, vol. 5, pp. 1–75. New York: Academic Press.

Pennycuick, C.J. 1989. Bird flight performance: a practical calculation manual. New York: Oxford Univ. Press.

Peres, C.A. 1997. Primate community structure at twenty western Amazonian flooded and unflooded forests. J. Trop. Ecol. 13:381–405.

Peres, C.A. 1999. The structure of nonvolant mammal communities in different Amazonian forest types. *In* J.F. Eisenberg and K.H. Redford, eds., Mammals of the central Neotropics, vol. 3: Ecuador, Peru, Bolivia, Brazil, pp. 564–581. Chicago: Univ. of Chicago Press.

Peters, J.L. 1931. Check-list of birds of the world, vol. 1. Cambridge, MA: Harvard Univ. Press.

Peters, J.L. 1938. Systematic position of the genus *Ciccaba* Wagler. Auk 55:179–186.

Peters, J.L., and L. Griscom. 1929. The Central American races of *Rupornis magnirostris.* Proc. New Engl. Zool. Club 11:43–48.

Peterson, R.T., and E.L. Chalif. 1973. A field guide to Mexican birds. Boston: Houghton Mifflin.

Phelan, F.J.S., and R.J. Robertson. 1978. Predatory responses of a raptor guild to changes in prey density. Can. J. Zool. 56:2565–2572.

Pianka, E.R. 1970. On *r-* and *K*-selection. Am. Nat. 104:592–597.

Pianka, E.R. 1974. Niche overlap and diffuse competition. Proc. Natl. Acad Sci. USA 71:2141–2145.

Poole, K.G., and R.G. Bromely. 1988. Natural history of the Gyrfalcon in the central Canadian Arctic. Arctic 41:31–38.

Porter, T.A., and G.S. Wilkinson. 2001. Birth synchrony in greater spear-nosed bats *(Phyllostomus hastatus).* J. Zool. Lond. 253:383–390.

Porter, W.F. 1994. Meleagrididae (Turkeys). *In* J. del Hoyo, A. Elliot, and J. Saragatal, eds., Handbook of the birds of the world, vol. 2. Barcelona: Lynx Edicions.

Poulin, B., G. Lefebvre, and R. McNeil. 1992. Tropical avian phenology in relation to abundance and exploitation of food resources. Ecology 73:2295–2309.

Poulin, B., G. Lefebvre, R. Ibáñez, C. Jaramillo, C. Hernández, and A.S. Rand. 2001. Avian predation upon lizards and frogs in a Neotropical forest understorey. J. Trop. Ecol. 17:21–40.

Primack, R.B. 2002. Essentials of conservation biology. Third edition. Sunderland, MA: Sinauer Associates.

Primack, R.B., D.B. Bray, H.A. Galletti, and I. Ponciano, eds. 1998. Timber, tourists, and temples: conservation and development in the Maya Forest of Belize, Guatemala, and Mexico. Washington, DC: Island Press.

Proudfoot, G.A., and R.R. Johnson. 2000. Ferruginous Pygymy-Owl *(Glaucidium brasiliamum).* *In* A. Poole and F. Gill, eds., The Birds of North America, No. 498. Philadelphia, Academy of Natural Sciences; Washington, DC: American Ornithologists' Union.

Pulliam, H.R. 1988. Sources, sinks, and population regulation. Am. Nat. 132:652–661.

Pulliam, H.R., and G.C. Millikan. 1982. Social organization in the non-breeding season. *In* D.S. Farner, J.R. King, and

K. C. Parkes, eds., Avian biology, vol. 6., pp. 169–197. New York: Academic Press.

Raffaele, H., J. Wiley, O. Garrido, A. Keith, and J. Raffaele. 1998. A guide to the birds of the West Indies. Princeton, NJ: Princeton Univ. Press.

Ramos, M. A. 1986. Birds in peril in Mexico: the diurnal raptors. Birds Prey Bull. 3:26–42.

Rangel-Salazar, J. L., and P. L. Enriquez-Rocha. 1993. Nest record and dietary items for the Black Hawk-Eagle *(Spizaetus tyrannus)* from the Yucatán Peninsula. J. Raptor Res. 27:121–122.

Raposo do Amaral, F. S., M. J. Miller, L. F. Silveira, E. Bermingham, and A. Wajntal. 2006. Polyphyly of the hawk genera *Leucopternis* and *Buteogallus* (Aves, Accipitridae): multiple habitat shifts during the Neotropical buteonine diversification. BMC Evol. Biol. 6:10.

Rappole, J. H., M. A. Ramos, and K. Winker. 1989. Wintering Wood Thrush movements and mortality in southern Veracruz. Auk 106:402–410.

Ratcliffe, D. A. 1993. The Peregrine Falcon. London: T. & A. D. Poyser.

Reagan, D. P., and R. B. Waide. 1996. The food web of a tropical rain forest. Chicago: Univ. of Chicago Press.

Reed, D. H., J. J. O'Grady, B. W. Brook, J. D. Ballou, and R. Frankham. 2003. Estimates of minimum viable population sizes for vertebrates and factors influencing those estimates. Biol. Conserv. 113:23–34.

Regalado, P. 1981. Nuestras aves de presa. Magazine Juventud Técnica, 28–35.

Reining, C. S., R. M. Heinzman, M. Cabrera M., S. López, and A. Solórzano. 1992. Non timber forest products of the Maya Biosphere Reserve, Petén, Guatemala. Washington, DC: Conservation International Foundation.

Restani, M. 1991. Resource partitioning among three *Buteo* species in the Centennial Valley, Montana. Condor 93:1007–1010.

Rettig, N. L. 1978. Breeding behavior of the Harpy Eagle *(Harpia harpyja)*. Auk 95:629–643.

Reyer, H.-U., and D. Schmidl. 1988. Helpers have little to laugh about: group structure and vocalization in the Laughing Kookaburra *Dacelo novaeguineae*. Emu 88:150–160.

Reynolds, R. T. 1972. Sexual dimorphism in accipiter hawks: a new hypothesis. Condor 74:191–197.

Reynolds, R. T., and B. D. Linkhart. 1987. Fidelity to territory and mate in flammulated owls. *In* R. W. Nero, R. J. Clark, R. J. Knapton, and R. H. Hamre, eds., Biology and conservation of northern forest owls: symposium proceedings, pp. 234–238. U.S. For. Serv. Gen. Tech. Rep. RM-142.

Reynolds, R. T., and H. M. Wight. 1978. Distribution, density, and productivity of accipiter hawks breeding in Oregon. Wilson Bull. 90:182–196.

Reznick, D. 1985. Cost of reproduction: an evaluation of the empirical evidence. Oikos 44:267.

Rice, D. S., and Rice, P. M. 1990. Population size and population change in the central Petén lakes region, Guatemala. *In* T. P. Culbert and D. S. Rice, eds., Precolumbian population history in the Maya lowlands, pp. 123–148. Albuquerque: Univ. of New Mexico Press.

Rice, W. R. 1982. Acoustical location of prey by the Marsh Hawk: adaptation to concealed prey. Auk 99:403–413.

Ricklefs, R. E. 1967. A graphical method of fitting equations to growth curves. Ecology 48:978–983.

Ricklefs, R. E. 1968. Patterns of growth in birds. Ibis 110:419–451.

Ricklefs, R. E. 1969a. An analysis of nesting mortality in birds. Smithson. Contrib. Zool. 9:1–28.

Ricklefs, R. E. 1969b. The nesting cycle of songbirds in tropical and temperate regions. Living Bird 8:165–175.

Ricklefs, R. E. 1976. Growth rates of birds in the humid New World Tropics. Ibis 118:179–207.

Ricklefs, R. E. 1992. Embryonic development period and the prevalence of avian brood parasites. Proc. Natl. Acad. Sci. USA 89:4722–4725.

Ricklefs, R. E. 1993. Sibling competition, hatching asynchrony, incubation period, and lifespan in altricial birds. Curr. Ornithol. 11:199–276.

Ricklefs, R. E. 1997. Comparative demography of New World populations of thrushes *(Turdus* spp.). Ecol. Monogr. 67:23–43.

Ricklefs, R. E., and D. Schluter, eds. 1993. Species diversity in ecological communities: historical and geographical perspectives. Chicago: Univ. of Chicago Press.

Ridgely, R. S. 1976. A guide to the birds of Panama. Princeton, NJ: Princeton Univ. Press.

Ridgely, R. S., and P. J. Greenfield. 2001. The birds of Ecuador: status, distribution, and taxonomy. Ithaca, NY: Comstock/Cornell Univ. Press.

Ridgely, R. S., and J. A. Gwynne, Jr. 1976. A guide to the birds of Panama. Princeton, NJ: Princeton Univ. Press.

Ridgely, R. S., and J. A. Gwynne, Jr. 1989. A guide to the birds of Panama with Costa Rica, Nicaragua, and Honduras, 2nd ed. Princeton, NJ: Princeton Univ. Press.

Ridgway, R. 1873. Revision of the falconine genera, *Micrastur, Geranospiza* and *Rupornis,* and the strigine genus, *Glaucidium.* Proc. Boston Soc. Nat. Hist. 16:73–106.

Ridgway, R. 1875. Studies of the American Falconidae: monograph of the genus *Micrastur.* Proc. Acad. Nat. Sci. Phila. 3:470–502.

Ridgway, R. 1914. The birds of North and Middle America. Part I of U.S. Natl. Mus. Bull. 50:598–617.

Riesing, M. J., L. Kruckenhauser, A. Gamauf, and E. Haring. 2003. Molecular phylogeny of the genus *Buteo* (Aves: Acciptridae) based on mitochondrial marker sequences. Mol. Phylogenet. Evol. 27:328–342.

Roalkvam, R. 1985. How effective are hunting Peregrines? Raptor Res. 19:27–29.

Robbins, M. B., and S. N. G. Howell. 1995. A new species of Pygmy-Owl (Strigidae: *Glaucidium*) from the eastern Andes. Wilson Bull. 107:1–6.

Robbins, M. B., and D. A. Wiedenfeld. 1982. Observations at a Laughing Falcon nest. Wilson Bull. 94:83–84.

Robbins, M. B., T. A. Parker III, and S. E. Allen. 1985. The avifauna of Cerro Pirre, Darién, eastern Panama. Ornithol. Monogr. 36:198–232.

Robertson, W. B., Jr. 1988. American Swallow-tailed Kite. *In* R. S. Palmer, ed., Handbook of North American birds, vol. 4, pp. 109–131. New Haven, CT: Yale Univ. Press.

Robinson, S. K. 1994. Habitat selection and foraging ecology of raptors in Amazonian Peru. Biotropica 26:443–458.

Robinson, S. K., and J. Terborgh. 1997. Bird community dynamics along primary successional gradients of an Amazonian whitewater river. *In* J. V. Remsen Jr., ed., Studies in Neotropical ornithology honoring Ted Parker, pp. 641–672. Washington, DC: American Ornithologists' Union.

Robinson, W. D. 1999. Long-term changes in the avifauna of Barro Colorado Island, Panama, a tropical forest isolate. Conserv. Biol. 13:85–97.

Robinson, W. D., T. R. Robinson, S. K. Robinson, and J. D. Brawn. 2000. Nesting success of understory forest birds in central Panama. J. Avian Biol. 31:151–164.

Rockwood, L. L. 1974. Seasonal changes in the susceptibility of *Cresentia alata* leaves to the flea beetle, *Oedionychus,* sp. Ecology 55:142–148.

Röhe, F., and A. Pinassi Antunes. 2008. Barred Forest Falcon *(Micrastur ruficollis)* predation on relatively large prey. Wilson J. Ornithol. 120:228–230.

Roling, G. 1992. Some notes on small mammals of the Maya Biosphere Reserve. *In* D. Whitacre and R. K. Thorstrom, eds., Progress report IV, Maya Project: use of raptors and other fauna as environmental indicators for design, management, and monitoring of protected areas and for building local capacity for conservation in Latin America, pp. 93–114. Boise, ID: The Peregrine Fund.

Roling G. 1995. Programa piloto de manejo de vida silvestre en Uaxactún, Petén, Guatemala. ARCAS/UICN/CONAP/USAC. Escuela Biol, Guatemala.

Rosenfield, R. N., and J. Bielefeldt. 1993. Cooper's Hawk *(Accipiter cooperii). In* A. Poole and F. Gill, eds., The birds of North America, No. 75. Philadelphia: Academy of Natural Sciences; Washington, DC: American Ornithologists' Union.

Rosenfield, R. N., and J. Bielefeldt. 1997. Reanalysis of relationships among eye color, age and sex in the Cooper's hawk. J. Raptor Res. 31:313–316.

Rowley, J. S. 1984. Breeding records of land birds in Oaxaca, Mexico. Proc. West. Found. Vertebr. Zool. 2:73–224.

Rudebeck, G. 1950. The choice of prey and modes of hunting of predatory birds with special reference to their selective effect. Oikos 2:65–88; 3:201–231.

Rudebeck, G. 1951. The choice of prey and modes of hunting of predatory birds with special reference to their selective effect. Oikos 3:201–231.

Rudolph, S. G. 1982. Foraging strategies of American Kestrels during breeding. Ecology 63:1268–1276.

Ruelas, E. I., L. J. Goodrich, S. W. Hoffman, and J. E. Montejo-Diaz. 2002. (Abstract #26) The migration of the Hook-billed Kite *(Chondrohierax uncinatus).* Third North American Ornithological Conference.

Russell, E. M. 2000. Avian life histories: is extended parental care the southern secret? Emu 100:377–399.

Russell, E. M., Y. Yom-Tov, and E. Geffen. 2004. Extended parental care and delayed dispersal: northern, tropical, and southern passerines compared. Behav. Ecol. 15:831–838.

Russell, S. M. 1964. A distributional study of the birds of British Honduras. Ornithol. Monogr. 1, American Ornithologists' Union.

Saether, B.-E. 1987. Pattern of covariation between life-history traits of European birds. Oikos 48:79–88.

Saether, B.-E. 1988. Patterns of covariation between life-history traits of European birds. Nature 331:616–618.

Saether, B.-E. 1989. Survival rates in relation to body weight in European birds. Ornis Scand. 20:13–21.

Saint Girons, H. 1982. Reproductive cycles of male snakes and their relationships with climate and female reproductive cycles. Herpetologica 38:5–16.

Salvador, S. A. 1990. Nidificacion de rapaces argentinos (Falconiformes y Strigiformes). Nuestras Aves 23:28–29.

Salvin, O., and F. D. Godman. 1897. Biologia Centrali-Americana. Aves. London: Taylor & Francis.

Sandercock, B. K., S. R. Beissinger, S. H. Stoleson, R. R. Melland, and C. R. Hughes. 2000. Survival rates of a Neotropical parrot: implications for latitudinal comparisons of avian demography. Ecology 81:1351–1370.

Schaefer, H.-C., G. W. Eshiamwata, F. B. Munyekenye, and K. Böhning-Gaese. 2004. Life-history of two African Sylvia warblers: low annual fecundity and long post-fledging care. Ibis 146:427–437.

Schaldach, W. J., Jr. 1963. The avifauna of Colima and adjacent Jalisco, Mexico. Proc. West. Found. Vertebr. Zool. 2:73–224.

Schmutz, J. K., S. M. Schmutz, and D. A. Boag. 1980. Coexistence of three species of hawks *(Buteo* spp.) in the prairie-parkland ecotone. Can. J. Zool. 58:1075–1089.

Schoener, T. W. 1968. Sizes of feeding territories among birds. Ecology 49:123–141.

Schoener, T. W. 1970. Nonsynchronous spatial overlap of lizards in patchy habitats. Ecology 51:408–418.

Schoener, T. W. 1971. Large-billed insectivorous birds: a precipitous diversity gradient. Condor 73:154–161.

Schoener, T. W. 1974. The compression hypothesis and temporal resource partitioning. Proc. Natl. Acad. Sci. USA 71:4169–4172.

Schoener, T. W. 1981. An empirically based estimate of home range. Theor. Popul. Biol. 20:281–325.

Schoener, T. W. 1984. Size differences among sympatric, bird-eating hawks: a worldwide survey. *In* D. R. Strong, D. Simberloff, L. G. Abele, and A. B. Thistle, eds., Ecological communities: conceptual issues and the evidence, pp. 254–281. Princeton, NJ: Princeton Univ. Press.

Schönwetter, M. 1961. Handbuch der Oologie, Lieferung 3. Berlin: Akademie Verlag.

Schönwetter, M. 1964. Handbuch der Oologie 9–10:530–598.

Schulenberg, T. S., and T. A. Parker III. 1981. Status and distribution of some northwest Peruvian birds. Condor 83:209–216.

Schulze, M. D. 1992. A preliminary description of woody plant communities of Tikal National Park. *In* D. F. Whitacre and R. K. Thorstrom, eds., Progress report V, 1992, Maya Project, pp. 53–61. Boise, ID: The Peregrine Fund.

Schulze, M., and D. F. Whitacre. 1999. Classification and ordination of the tree community in Tikal National Park, Petén, Guatemala. Bull. Fla. Mus. Nat. Hist. 41:169–297.

Schulze, M. D., J. L. Córdova, N. E. Seavy, and D. F. Whitacre. 2000a. Behavior, diet, and breeding biology of Double-toothed Kites at a Guatemalan lowland site. Condor 102:113–126.

Schulze, M. D., N. E. Seavy, and D. F. Whitacre. 2000b. A comparison of the Phyllostomid Bat assemblages in undisturbed Neotropical forest and in forest fragments of a slash-and-burn farming mosaic in Petén, Guatemala. Biotropica 32:174–184.

Schwartz, N. B. 1990. Forest society: a social history of Petén, Guatemala. Philadelphia: Univ. of Pennsylvania Press.

Schwartz, P. 1972. *Micrastur gilvicollis*, a valid species sympatric with *M. ruficollis* in Amazonia. Condor 74:399–415.

Scott, N. J. 1983. *Oxybelis aeneus. In* D. H. Janzen, ed., Costa Rican natural history, pp. 410–411. Chicago: Univ. of Chicago Press.

Scott, N. J., and S. Limerick. 1983. Reptiles and amphibians. *In* D. H. Janzen, ed., Costa Rican natural history, pp. 351–367. Chicago: Univ. of Chicago Press.

Seavy, N. E., and R. P. Gerhardt. 1998. Breeding biology and nestling diet of the Great Black-Hawk. J. Raptor Res. 32:175–177.

Seavy, N. E., M. D. Schulze, and D. F. Whitacre. 1997a. Two Plumbeous Kites *(Ictinia plumbea)* capture swallow. J. Raptor Res. 31:289.

Seavy, N. E., M. D. Schulze, D. F. Whitacre, and M. A. Vasquez. 1997b. Diet and hunting behavior of the Plumbeous Kite. Wilson Bull. 109:526–532.

Seavy, N. E., M. D. Schulze, D. F. Whitacre, and M. A. Vásquez. 1998. Breeding biology and behavior of the Plumbeous Kite. Wilson Bull. 110:77–85.

Seigel, R. A., and N. B. Ford. 1987. Reproductive ecology. *In* R. A. Seigel, J. T. Collins, and S. S. Novak, eds., Snakes: ecology and evolutionary biology, pp. 210–252. New York: McGraw-Hill.

Selas, V. 1997. Influence of prey availability on re-establishment of Goshawk *Accipiter gentilis* nesting territories. Ornis Fennica 74:113–120.

Sexton, O.J., E.P. Ortleb, L.M. Hathaway, R.E. Ballinger, and P. Licht. 1971. Reproductive cycles of three species of anoline lizards from the isthmus of Panama. Ecology 52:201–215.

Shackleford, C.E., and G.G. Simons. 2000. A two-year report of the Swallow-tailed Kite in Texas: a survey and monitoring project for 1998 and 1999. Texas Parks and Wildlife PWD BK W7000–496 (6/00).

Sheffler, W.J., and A.J. van Rossem. 1944. Nesting of the Laughing Falcon. Auk 61:141–142.

Sherrod, S.K. 1983. Behavior of fledgling Peregrines. Ft. Collins, CO: The Peregrine Fund.

Short, L.L. 1975. A zoogeographic analysis of the South American Chaco avifauna. Bull. Am. Mus. Nat. Hist. 154:165–352.

Shufeldt, R.W. 1891. Some comparative osteological notes on the North-American kites. Ibis 6 series 3:228–232.

Shy, M.M. 1982. Interspecific feeding among birds: a review. J. Field Ornithol. 53:370–393.

Sibley, C.G., and J.E. Ahlquist. 1990. Phylogeny and classification of birds. New Haven, CT: Yale Univ. Press.

Sick, H. 1993. Birds of Brazil. Princeton, NJ: Princeton Univ. Press.

Simmons, R. 1988. Offspring quality and the evolution of cainism. Ibis 130:339–357.

Simmons, R.E. 2000. Harriers of the world: their behaviour and ecology. Oxford: Oxford Univ. Press.

Simmons, R.E., and J.M. Mendelsohn. 1993. A critical review of cartwheeling flights of raptors. Ostrich 64:13–24.

Skutch, A.F. 1947. A nesting of the Plumbeous Kite in Ecuador. Condor 49:25–31.

Skutch, A.F. 1949. Do tropical birds rear as many young as they can nourish? Ibis 92:430–455.

Skutch, A.F. 1950. The nesting seasons of Central American birds in relation to climate and food supply. Ibis 92:185–222.

Skutch, A.F. 1954. Life histories of Central American birds. Pacific Coast Avifauna 31:1–448.

Skutch, A.F. 1960a. The laughing reptile hunter of tropical America. Animal Kingdom 63:115–119.

Skutch, A.F. 1960b. Life histories of Central American birds. II. Pacific Coast Avifuana No. 34. Cooper Ornithological Society, Berkeley, CA.

Skutch, A.F. 1965. Life history notes on two tropical American kites. Condor 67:235–246.

Skutch, A.F. 1969. Life histories of Central American birds. III. Pacific Coast Avifauna 35:1–580.

Skutch, A.F. 1971. A naturalist in Costa Rica. Gainesville: Univ. of Florida Press.

Skutch, A.F. 1976. Parent birds and their young. Austin: Univ. of Texas Press.

Skutch, A.F. 1983. *Herpetotheres cachinnans.* In Janzen, D.H., ed., Costa Rican natural history. Chicago: Univ. of Chicago Press.

Skutch, A.F. 1985. Clutch size, nesting success, and predation on nests of Neotropical birds, reviewed. In P.A. Buckley, M.S. Foster, E.S. Morton, R.S. Ridgely, and F.G. Buckley, eds., Neotropical ornithology, pp. 575–594. Ornithol. Monogr. 36. Washington, DC: American Ornithologists' Union.

Slauson, W.L., B.S. Cade, and J.D. Richards. 1991. User manual for BLOSSOM statistical software. Ft. Collins, CO: National Ecology Research Center, U.S. Fish and Wildlife Service.

Slud, P. 1960. The birds of Finca 'La Selva,' Costa Rica. Bull. Am. Mus. Nat. Hist. 121:49–148.

Slud, P. 1964. The birds of Costa Rica: distribution and ecology. Bull. Am. Mus. Nat. Hist. 128.

Slud, P. 1980. The birds of Hacienda Palo Verde, Guanacaste, Costa Rica. Smithson. Contrib. Zool. 292.

Smith, D.G., and J.R. Murphy. 1973. Breeding ecology of raptors in the eastern Great Basin of Utah. Brigham Young Univ. Sci. Bull. Biol. Ser. 18(3):1–76.

Smith, J.P., J. Simon, S.W. Hoffman, and C. Riley. 2001. New full-season autumn hawkwatches in coastal Texas. In K.L. Bildstein and D. Klem, eds., Hawkwatching in the Americas, pp. 67–91. North Wales, PA: Hawk Migration Association of North America.

Smith, N.G. 1968. The advantage of being parasitized. Nature 219:690–694.

Smith, N.G. 1969. Provoked release of mobbing—a hunting technique of *Micrastur* Falcons. Ibis 111:241–243.

Smith, N.G. 1970. Nesting of King Vulture and Black Hawk-Eagle in Panama. Condor 72:247–248.

Smith, T.B. 1982. Nests and young of two rare raptors in Mexico. Biotropica 14:79–80.

Smith, T.B. 1988. Hook-billed Kite. In R.S. Palmer, ed., Handbook of North American birds, vol. 4, pp. 102–108. New Haven, CT: Yale Univ. Press.

Smith, T.B., and S.A. Temple. 1982a. Feeding habits and bill polymorphism in Hook-billed Kites. Auk 99:197–207.

Smith, T.B., and S.A. Temple. 1982b. Grenada Hook-billed Kites: recent status and life history notes. Condor 84:131.

Smithe, F.B. 1966. The birds of Tikal. Garden City, NY: Natural History Press.

Smithe, F.B., and R.A. Paynter Jr. 1963. Birds of Tikal, Guatemala. Bull. Mus.. Comp. Zool. Harv. Coll. 128:245–324.

Smythe, N. 1982. The seasonal abundance of night-flying insects in a Neotropical forest. In E.G. Leigh Jr., A.S. Rand, and D.M. Windsor, eds., The ecology of a tropical forest: seasonal rhythms and long-term changes, pp. 309–318. Washington, DC: Smithsonian Institution Press.

Snyder, N.F.R. 1974. Breeding biology of Swallow-tailed Kites in Florida. Living Bird 13:73–97.

Snyder, N.F.R., and H.W. Kale, II. 1983. Mollusk predation by Snail Kites in Colombia. Auk 100:93–97.

Snyder, N., and H. Snyder. 1991. Raptors: North American birds of prey. Stillwater, MN: Voyageur Press.

Snyder, N.F.R., and H.A. Snyder. 1969. A comparative study of mollusc predation by Limpkins, Everglade Kites, and Boat-tailed Grackles. Living Bird 8:177–222.

Snyder, N.F.R., and J.W. Wiley. 1976. Sexual size dimorphism in hawks and owls of North America. Ornithol. Monogr. 20:1–96.

Sodhi, N.S., I.G. Warkentin, and L.W. Oliphant. 1991. Hunting techniques and success rates of urban Merlins *(Falco columbarius).* J. Raptor Res. 25:127–131.

Soule, M.E., D.T. Bolger, A.C. Alberts, J. Wright, M. Sorice, and S. Hill. 1988. Reconstructed dynamics of rapid extinctions of chaparral-requiring birds in urban habitat islands. Conserv. Biol. 2:75–92.

Spironello, W.R. 2001. The Brown Capuchin Monkey *(Cebus apella):* ecology and home range requirements in Central Amazonia. In R.O. Bierregaard Jr., C. Gascon, T.E. Lovejoy, and R.C.G. Mesquita, eds., Lessons from Amazonia: the ecology and conservation of a fragmented forest, pp. 271–283. New Haven, CT: Yale Univ. Press.

Stafford, P.J., and J.R. Meyer. 1999. A guide to the reptiles of Belize. New York: Academic Press.

Stamps, J.A. 1976. Rainfall, activity and social-behavior in lizard, *Anolis aeneus.* Anim. Behav. 24:603–608.

Starck, J.M., and R.E Ricklefs. 1998. Avian growth and development. Evolution within the altricial-precocial spectrum. New York: Oxford Univ. Press.

Stattersfield, A.J., M.J. Crosby, A.J. Long, and D.C. Wege. 1998. Endemic bird areas of the world: priorities for biodiversity conservation. Cambridge: BirdLife International.

Stehli, F.G., and S.D. Webb. 1985. The great American biotic interchange. New York: Plenum Press.

Steyn, P. 1982. Birds of prey of southern Africa. Dover, NH: Tanager Books.

Stiles, F.G. 1985. Conservation of forest birds in Costa Rica: problems and perspectives. *In* A.W. Diamond and T.E. Lovejoy, eds., Conservation of tropical forest birds, pp. 141–168. International Council for Bird Preservation (ICBP), Cambridge, UK.

Stiles, F.G., and D.H. Janzen. 1983. *Buteo magnirostris. In* D.H. Janzen, ed., Costa Rican natural history, pp. 551–552. Chicago: Univ. of Chicago Press.

Stiles, F.G., and Levey, D.J. 1994. Birds of La Selva and vicinity. La Selva: ecology and natural history of a Neotropical rain forest. *In* L.A. McDade, K.S. Bawa, H.A. Hespenheide, and G.S. Hartshorn, eds., pp. 384–393. Chicago: Univ. of Chicago Press.

Stiles, F.G., A.F. Skutch, and D. Gardner. 1989. A guide to the birds of Costa Rica. Ithaca, NY: Comstock/Cornell Univ. Press.

Storer, R.W. 1961. Two collections of birds from Campeche, Mexico. Occasional Papers of the Museum of Zoology, Univ. of Michigan 621:1–20.

Storer, R.W. 1966. Sexual dimorphism and food habits in three North American accipiters. Auk 83:423–436.

Stotz, D.F., and R.O. Bierregaard Jr. 1989. The birds of the Fazendas Porto Alegre, Esteio and Dimona north of Manaus, Amazonas, Brazil. Rev. Bras. Biol. 49:861–872.

Stotz, D.F., J.W. Fitzpatrick, and D.E. Willard. 2001 (Feb. 26). Birds of Amazonia Lodge and vicinity (Manu, Madre de Dios, Peru). http://www.amazonialodge.com/birds.html.

Stutchbury, B.J.M., and E.S. Morton. 2001. Behavioral ecology of tropical birds. San Diego: Academic Press.

Sutter, J. 1999. Ecology of the Crane Hawk *(Geranospiza caerulescens)* in Tikal National Park, Guatemala. M.Sc. thesis, Boise State Univ., Boise, ID.

Sutter, J., W.E. Martínez A., F. Oliva T., N. Oswaldo J., and D.F. Whitacre. 2001. Diet and hunting behavior of the Crane Hawk in Tikal National Park, Guatemala. Condor 103:70–77.

Sutton, G.M. 1944. The kites of the genus *Ictinia.* Wilson Bull. 56:3–8.

Sutton, G.M. 1951. Mexican birds: first impressions. Norman: Univ. of Oklahoma Press.

Sutton, G.M. 1954. Blackish crane hawk. Wilson Bull. 66:237–242.

Sutton, G.M., and O.S. Pettingill. 1942. Birds of the Gomez Farias region, southwestern Tamaulipas. Auk 59:1–34.

Sutton, G.M., R.B. Lea, and E.P. Edwards. 1950. Notes on the ranges and breeding habits of certain Mexican birds. Bird-Banding 21:45–59.

Sutton, I.D. 1955. Nesting of the Swallow-tailed Kite. Everglades Nat. Hist. 3:72–84.

Swann, H.K. 1923. Notes on the Gordon collection of eggs of the Accipiters. Ool. Rec. 3:25–30.

Tanaka, L.K., and S.K. Tanaka. 1982. Rainfall and seasonal changes in arthropod abundance on a tropical oceanic island. Biotropica 14:114–123.

Tashian, R.E. 1952. Some birds from the Pelenque region of northeastern Chiapas, Mexico. Auk 69:60–66.

Tella, J.L., A. Scheuerleim, and R.E. Ricklefs. 2002. Is cell-mediated immunity related to the evolution of life-history strategies in birds? Proc. Royal Soc. Lond. B 269:1059–1066.

Temeles, E.J. 1985. Sexual size dimorphism of bird-eating hawks: the effect of prey vulnerability. Am. Nat. 125:485–499.

Terborgh, J. 1983. Five New World primates: a study in comparative ecology. Princeton, NJ: Princeton Univ. Press.

Terborgh, J., S. Robinson, T.A. Parker III, C. Munn, and N. Pierpont. 1990. Structure and organization of an Amazonian forest bird community. Ecol. Monogr. 60:213–238.

Thiollay, J.-M. 1970. L'exploitation par les oiseaux des essaimages de fourmis et termites dans une zone de contact savane-foret en Cote-d'Ivoire. Alauda 38:255–273.

Thiollay, J.-M. 1984. Raptor community structure of a primary rain forest in French Guiana and effect of human hunting pressure. J. Raptor Res. 18:117–122.

Thiollay, J.-M. 1985. Composition of Falconiform communities along successional gradients from primary rainforest to secondary habitats. *In* I. Newton and R.D. Chancellor, eds., Conservation studies on raptors, pp. 181–191. International Council for Bird Preservation (ICBP) Tech. Pub. No. 5.

Thiollay, J.-M. 1988. Comparative foraging success of insectivorous birds in tropical and temperate forests: ecological implications. Oikos 53:17–30.

Thiollay, J.-M. 1989a. Area requirements for the conservation of rain forest raptors and game birds in French Guiana. Conserv. Biol. 3:128–137.

Thiollay, J.-M. 1989b. Censusing of diurnal raptors in a primary rain forest: comparative methods and species detectability. J. Raptor Res. 23:72–84.

Thiollay, J.-M. 1991a. Altitudinal distribution and conservation of raptors in southeastern Colombia. J. Raptor Res. 25:1–8.

Thiollay, J.-M. 1991b. Foraging, home range use, and social behaviour of a group-living rainforest raptor, the Red-throated Caracara *Daptrius americanus.* Ibis 133:386–393.

Thiollay, J.-M. 1994. Structure, density and rarity in an Amazonian rain forest bird community. Journal of Tropical Ecology 10:449–481.

Thomas, B.T. 1979. The birds of a ranch in the Venezuelan llanos. *In* J.F. Eisenberg, ed., Vertebrate ecology of the northern Neotropics, pp. 213–232. Washington, DC: Smithsonian Institution Press.

Thorpe, W.H. 1956. Learning and instinct in animals. Cambridge, MA: Harvard Univ. Press.

Thorstrom, R. 1993. The breeding ecology of two species of forest-falcons *(Micrastur)* in northeastern Guatemala. M.S. thesis, Boise State Univ., Boise, ID.

Thorstrom, R. 1997. A description of nests and behavior of the Gray-headed Kite. Wilson Bull. 109:173–177.

Thorstrom, R. 2000. The food habits of sympatric forest-falcons during the breeding season in northeastern Guatemala. J. Raptor Res. 34:196–202.

Thorstrom, R. 2001. Nest-site characteristics and breeding density of two sympatric forest-falcons in Guatemala. Ornitol. Neotrop. 12:337–343.

Thorstrom, R., and L.F. Kiff. 1999. Notes on eggs of the Bicolored Hawk *Accipiter bicolor.* J. Raptor Res. 33:244–247.

Thorstrom, R., and D. McQueen. 2008. Breeding and status of the Grenada Hook-billed Kite *(Chondriohierax uncinatus mirus).* Ornitol. Neotrop. 19:221–228.

Thorstrom, R., and A. Quixchan. 2000. Breeding biology and nest site characteristics of the Bicolored Hawk in Guatemala. Wilson Bull. 112:195–202.

Thorstrom, R., E. Massiah, and C. Hall. 2001. Nesting biology, distribution, and population estimate of the Grenada Hook-billed Kite *Chondrohierax uncinatus mirus.* Carib. J. Sci. 37:278–281.

Thorstrom, R., R. Watson, A. Baker, S. Ayers, and D. Anderson. 2002. Preliminary ground and aerial surveys for Orange-breasted Falcons in Central America. J. Raptor Res. 36:39–44.

Thorstrom, R.K. 1996a. Fruit-eating behavior by a Barred Forest-Falcon. J. Raptor Res. 30:44.

Thorstrom, R.K. 1996b. Methods for capturing tropical forest birds of prey. Wildl. Soc. Bull. 24:516–520.

Thorstrom, R.K., C.W. Turley, F.G. Ramirez, and B.A. Gilroy. 1990. Description of nests, eggs and young of the Barred Forest-Falcon *(Micrastur ruficollis)* and of the Collared Forest-Falcon *(M. semitorquatus).* Condor 92:237–239.

Thurber, W.A., J.F. Serrano, A. Sermeno, and M. Benitez. 1987. Status of uncommon and previously unreported birds of El Salvador. Proc. West. Found. Vertebr. Zool. 3:109–293.

Titus, K., C.J. Flatten, and R.E. Lowell. 1994. Northern Goshawk ecology and habitat relationships on the Tongass National Forest. Report prepared for the Forest Service, Alaska Department of Fish and Game, Division of Wildlife Conservation, Juneau.

Todd, W.E.C., and M.A. Carriker Jr. 1922. The birds of the Santa Marta region of Colombia: a study in altitudinal distribution. Ann. Carnegie Mus. 14.

Tostain, O., J.-L. Dujardin, C. Erard, and J.-M. Thiollay. 1992. Oiseaux de Guyane: the birds of French Guiana. Brunoy, France: Societé d'Etudes Ornithologiques.

Trail, P.W. 1987. Predation and antipredator behavior of Guianan Cock-of-the-Rock leks. Auk 104:496–507.

Traylor, M.A., Jr. 1941. Birds from the Yucatán Peninsula. Field Mus. Nat. Hist. Zool. Ser. 24:195.

Treleaven, R.B. 1961. Notes on the Peregrine in Cornwall. Br. Birds 54:136–142.

Treleaven, R.B. 1980. High and low intensity hunting in raptors. Zeitschrift für Tierpsychologie 54:339–345.

Tucker, V.A. 1998. Gliding flight: speed and acceleration of ideal falcons during diving and pull out. J. Exp. Biol. 201:403–414.

Tucker, V.A., T.J. Cade, and A.E. Tucker. 1998. Diving speeds and angles of a Gyrfalcon *(Falco rusticolus).* J. Exp. Biol. 201:2061–2070.

Turley, C.W. 1989. Evaluation of raptor survey techniques. *In* W.A. Burnham, J.P. Jenny, and C.W. Turley, eds., Progress report II, 1989; Maya Project: use of raptors as environmental indices for design and management of protected areas and for building local capacity for conservation in Latin America, pp. 21–32. Boise, ID: The Peregrine Fund.

Valiela, I., J.L. Bowen, and J.K. York. 2001. Mangrove forests: one of the world's threatened major tropical environments. BioScience 51:807–815.

Vallejos, M.A.V., M. Lanzer, M. Aurélio-Silva, and L.F. Silva-da-Rocha. 2008. Collared Forest-Falcon *Micrastur semitorquatus* (Vieillot, 1817) nesting in a cave in southern Brazil. Rev. Bras. Ornitol. 16:268–270. (in Portuguese with English summary)

Vaninni, J. 1989a. Neotropical raptors and deforestation: notes on diurnal raptors at Finca El Faro, Quetzaltenango, Guatemala. J. Raptor Res. 23:27–38.

Vannini, J. 1989b. Reintroduction of Bat Falcons *(Falco rufigularis)* on Guatemala's Pacific versant. *In* W.A. Burnham, J.P. Jenny, and C.W. Turley, eds., Maya Project:

progress report II, 1989; use of raptors as environmental indices for design and management of protected areas and for building local capacity for conservation in Latin America, pp. 99–103. Boise, ID: The Peregrine Fund.

Van Tyne, J. 1935. The birds of northern Petén, Guatemala. Univ. of Michigan Museum of Zoology, Misc. Publ. 27. Ann Arbor: Univ. of Michigan Press.

Van Tyne, J. 1950. Bird notes from Barro Colorado Island, Canal Zone. Occasional Papers of the Museum of Zoology, Univ. of Michigan, No. 525.

Van Zyl, A.J. 1999. Breeding biology of the Common Kestrel in southern Africa (32S) compared to studies in Europe (53N). Ostrich 70:127–132.

Vargas-González, J. de J., R. Mosquera, and M. Watson. 2006. Crested Eagle *(Morphnus guianensis)* feeding a post-fledged young Harpy Eagle *(Harpia harpyja)* in Panama. Ornitol. Neotrop. 17:581–584.

Vargas, H. 1995. Food habits, breeding biology, and status of the Gray-backed Hawk *(Leucopternis occidentalis)* in western Ecuador. M.S. thesis, Boise State Univ., Boise, ID.

Vásquez, M.A., and E.R. Moreno. 1992. Nesting biology of the Roadside Hawk in Tikal National Park. *In* D.F. Whitacre and R.K. Thorstrom, eds., Progress report V, 1992 Maya Project: use of raptors and other fauna as environmental indicators for design and management of protected areas and for building local capacity for conservation in Latin America, pp. 169–172. Boise, ID: The Peregrine Fund.

Vásquez, M.A., E.R. Moreno, and T.D. Ortíz. 1992. Nesting biology of three species of kites. *In* D.F. Whitacre and R.K. Thorstrom, eds., Progress report V, 1992 Maya Project: use of raptors and other fauna as environmental indicators for design and management of protected areas and for building local capacity for conservation in Latin America, pp. 145–151. Boise, ID: The Peregrine Fund.

Verea, C., A. Solórzano, M. Díaz, L. Parra, M.A. Araujo, F. Antón, O. Navas, J.L. Ruiz, and A. Fernández-Badillo. 2009. Breeding records and molt activities in some birds of northern Venezuela. Ornitol. Neotrop. 20:181–201.

Village, A. 1990. The kestrel. London: T. & A.D. Poyser.

Vitt, L.J. 1983. Ecology of an anuran-eating guild of terrestrian tropical snakes. Herpetologica 39:52–66.

Vitt, L.J., and L.D. Vangilder. 1983. Ecology of a snake community in northeastern Brazil. Amphibia-Reptilia 4:273–296.

Voous, K.H. 1969. Predation potential in birds of prey from Surinam. Ardea 57:117–148.

Voous, K.H. 1988. Owls of the Northern Hemisphere. Cambridge, MA: MIT Press.

Walter, H. 1979. Eleanora's Falcon: adaptations to prey and habitat in a social raptor. Chicago: Univ. of Chicago Press.

Walter, H.S. 1990. Small viable population: the Red-tailed Hawk of Socorro Island. Conserv. Biol. 4:441–443.

Ward, P., and A. Zahavi. 1973. The importance of certain assemblages of birds as "information-centres" for food finding. Ibis 119:517–534.

Warham, J. 1975. The crested penguins. *In* B. Stonehouse, ed., The biology of penguins, pp. 189–269. Baltimore, MD: Univ. Park Press.

Warkentin, I.G., P.C. James, and L.W. Oliphant. 1991. Influence of site fidelity on mate switching in urban-breeding Merlins. Auk 108:294–302.

Wattel, J. 1973. Geographical differentiation in the genus *Accipiter.* Publications of the Nuttall Ornithological Club, No. 13, Cambridge, MA.

Wauer, R.H. 1980. Naturalist's Big Bend: An introduction to the trees and shrubs, wildflowers, cacti, mammals, birds, reptiles and amphibians, fish, and insects. College Station, TX: Texas A&M University Press.

Webster, F.S., Jr. 1978. South Texas Region. American Birds 32:1183.

Weske, J.S., and J.W. Terbough. 1981. *Otus marshalli,* a new species of screech-owl from Peru. Auk 98:1–7.

Wetmore, A. 1926. Observations on the birds of Argentina, Paraguay, Uruguay, and Chile. U.S. Nat. Mus. Bull. 133.

Wetmore, A. 1943. The birds of southern Veracruz, Mexico. Proc. U.S. Natl. Mus. 93:215–340.

Wetmore, A. 1944. A collection of birds from northern Guanacaste, Costa Rica. Proc. U.S. Natl. Mus. 95:25–80.

Wetmore, A. 1957. The birds of Isla Coiba, Panama. Smithsonian Misc. Coll., vol. 134. Smithsonian Institution, Pub. 4295.

Wetmore, A. 1965. The birds of the Republic of Panama. Part 1. Tinamidae (tinamous) to Rhynchophidae (skimmers). Smithsonian Misc. Coll., vol. 150. Smithsonian Institution, Pub. 4617.

Wetmore, A. 1968. The birds of the Republic of Panama. Part 2. Columbidae (pigeons) to Picidae (woodpeckers). Smithsonian Misc. Coll., vol. 150, part 2. Smithsonian Institution, Pub. 4732.

Wetmore, A. 1974. The egg of a Collared Forest-Falcon. Condor 76:103.

Weyer, D. 1984. Diurnal birds of prey of Belize. Hawk Trust Annu. Rep. 14:22–40.

Wheeler, B., and W. Clark. 1995. Photographic field guide to North American raptors. San Diego: Academic Press.

Whitacre, D.F. 1998. The Peregrine Fund's Maya Project: ecological research, habitat conservation efforts, and development of human resources in the Maya Forest. *In* R.B. Primack, D.B. Bray, H.A. Galletti, and I. Ponciano, eds., Timber, tourists, and temples: conservation and development in the Maya Forest of Belize, Guatemala, and Mexico, pp. 241–266. Washington, DC: Island Press.

Whitacre, D.F., and W.A. Burnham. 2001. Building human resources for conservation on the tropical forest frontier: an example from Guatemala's Maya Project. *In* J.L.B. Albuquerque, J.F. Cândido Jr., F.C. Straube, and A.L. Roos, eds., Ornitologia e conservação da ciência às estratégias, pp. 207–222. Tubarão, Brazil: Editora Unisul.

Whitacre, D.F., C.W. Turley, and E.C. Cleaveland. 1990. Correlations of diurnal raptor abundance with habitat features in Tikal National Park. *In* W. Burnham, D. Whitacre, and J.P. Jenny, eds., Progress report III, 1990, Maya Project: use of raptors as environmental indices for design and management of protected areas and for building local capacity for conservation in Latin America, pp. 53–65. Boise, ID: The Peregrine Fund.

Whitacre, D.F., P.M. Harris, and I. Córdova M. 1992a. Relative abundance of raptors and selected other bird species in two natural forest types and the slash-and-burn agricultural mosaic in and near Tikal National Park, Petén, Guatemala. *In* D.F. Whitacre and R.K. Thorstrom, eds., Progress report V, 1992: Maya Project, pp. 93–106. Boise, ID: The Peregrine Fund.

Whitacre, D.F., C. Turley, A.E. Hernández, and F. Osorio. 1992b. A comparison of raptor communities of primary tropical forest and the slash-and-burn/cattle ranching agricultural mosaic analysis of 1989 data. *In* D.F. Whitacre and R.K. Thorstrom, eds., Progress report V, 1992: Maya Project, pp. 81–92. Boise, ID: The Peregrine Fund.

Whitacre, D.F., J. Madrid, C. Marroquín, M. Schulze, L. Jones, J. Sutter, and A. Baker. 1993. Migrant songbirds, habitat change, and conservation prospects in northern Petén, Guatemala: some initial results. *In* D.M. Finch and P.W. Strangel, eds., Status and management of Neotropical migratory birds, pp. 339–345. USDA Forest Serv. Gen. Tech. Rep. RM-229.

Whitacre, D.F., J. Madrid M., C. Marroquín V., T. Dubón O., N.O. Jurado, W.R. Tobar, B. Gonzáles C., A. Arévalo O., G. Garcia C., M. Schulze, L. Jones, J. Sutter, and A. Baker. 1995a. Slash-and-burn farming and bird conservation in northern Petén, Guatemala. *In* M. Wilson and S. Sader, eds., Conservation of Neotropical migratory birds in Mexico, pp. 215–226. Maine Agricultural and Forest Experiment Station, Misc. Pub. 27, Orono, ME.

Whitacre, D.F., M.D. Schulze, and N.E. Seavy. 1995b. Habitat affinities of a Central American forest avifauna: implications for conservation in Neotropical slash-and-burn farming landscapes. Technical report to the U.S. Man and the Biosphere Program. Boise, ID: The Peregrine Fund.

Whitacre, D.F., J.L. Avila, and G.L. Avila. 2002. Behavioral and physical development of a nestling Crested Eagle *(Morphnus guianensis).* J. Raptor Res. 36:77–81.

Whittaker, A. 1996. First records of the Orange-breasted Falcon *Falco deiroleucus* in central Amazonian Brazil, with short behavioural notes. Cotinga 6:65–68.

Whittaker, A. 2002. A new species of forest-falcon (Falconidae: *Micrastur*) from southeastern Amazonia and the Atlantic rainforests of Brazil. Wilson Bull. 114:421–445.

Wiegert, R.G., and D.C. Coleman. 1970. Ecological significance of low oxygen consumption and high fat accumulation by *Nasutitermes costalis* (Isoptera: Termitidae). BioScience 20:663–665.

Wiley, J.W. 1985. Status and conservation of forest raptors in the West Indies. *In* I. Newton and R.D. Chancellor, eds., Conservation studies on raptors, pp. 199–203. International Council for Bird Preservation (ICBP) Tech. Pub. No. 5.

Wiley, J.W. 1986. Status and conservation of raptors in the West Indies. Birds Prey Bull. 3:57–70.

Wiley, J.W., and B.N. Wiley. 1981. Breeding season ecology and behavior of Ridgway's Hawk *(Buteo ridgwayi).* Condor 83:132–151.

Wiley, R.H., and D.G. Richards. 1982. Adaptations for acoustic communications in birds: sound transmission and signal detection. *In* D.E. Kroodsma and E.H. Miller, eds., Acoustic communication in birds: production, perception, and design features of sounds, pp. 131–181. New York: Academic Press.

Wilkerson, L. 1990 and 1991. SYSTAT: the system for statistics. SYSTAT, Evanston, Illinois.

Williams, G.C. 1966. Adaptation and natural selection. Princeton, NJ: Princeton Univ. Press.

Willis, E.O. 1967. The behavior of bicolored antbirds. Univ. Calif. Pub. Zool. 79:1–125.

Willis, E.O. 1972. The behavior of spotted antbirds. Ornithol. Monogr. 10:1–162.

Willis, E.O. 1973. The behavior of ocellated antbirds. Smithsonian Contrib. Zool. 144:1–57.

Willis, E.O. 1976. A possible reason for mimicry of a bird-eating hawk by an insect-eating kite. Auk 93:841–842.

Willis, E.O. 1979. The composition of avian communities in remanescent woodlots in southern Brasil. Papeis Avulsos de Zoologia (Sao Paulo) 33:1–25.

Willis, E.O., and E. Eisenmann. 1979. A revised list of birds of Barro Colorado Island, Panama. Smithsonian Contrib. Zool. 291.

Willis, E.O., D. Wechsler, and F.G. Stiles. 1983. Forest-falcons, hawks, and a pygmy-owl as ant followers. Rev. Bras. Biol. 43:23–28.

Wimberger, P.H. 1984. The use of green plant material in bird nests to avoid ectoparasites. Auk 101:615–618.

Woinarski, J.C.Z. 1985. Breeding biology and life history of small insectivorous birds in Australian forests: response to a stable environment. Proc. Ecol. Soc. Aust. 14:159–168.

Wolda, H. 1978. Seasonal fluctuations in rainfall, food and abundance of tropical insects. J. Anim. Ecol. 47:369–381.

Wolda, H. 1982. Seasonality of Homoptera on Barro Colorado Island. *In* E.G. Leigh Jr., A.S. Rand, and D.M. Windsor, eds., The ecology of a tropical forest: seasonal rhythms and long-term changes, pp. 319–330. Washington, DC: Smithsonian Institution Press.

Wolfe, L.R. 1938. Eggs of the Falconiformes. Ool. Rec. 18:74–87.

Wolfe, L.R. 1954. Nesting of the Laughing Falcon. Condor 56:161–162.

Wolfe, L.R. 1957. Nesting of the Laughing Falcon. Ool. Rec. 31:52–53.

Wolfe, L.R. 1959. Nesting of the Laughing Falcon. Ool. Rec. 33:6–9.

Wolfe, L.R. 1962. Food of the Mexican White Hawk. Auk 79:488.

Wolfe, L.R. 1964. Eggs of the Falconiformes, further part. Ool. Rec. 38:50–57.

Woodbridge, B., and P.J. Detrich. 1994. Territory occupancy and habitat patch size of Northern Goshawks in the Southern Cascades of California. Stud. Avian Biol. 16:83–87.

Wotzkow, C. 1985. Status and distribution of Falconiformes in Cuba. Bull. World Working Group on Birds of Prey 3:1–10.

Wotzkow, C. 1994. Status, distribution, current research and conservation of forest birds of prey in Cuba. *In* B.-U. Meyburg and R.D. Chancellor, eds., Raptor conservation today, pp. 291–299. Berlin: World Working Group on Birds of Prey; and London: Pica Press.

Wrege, P.H., and T.J. Cade. 1977. Courtship behavior of large falcons in captivity. J. Raptor Res. 11:1–46.

Wright, M., R.O. Green Jr., and N.D. Reed. 1970. The nesting activities of Swallow–tailed Kites in Everglades National Park. Unpublished manuscript.

Wright, S.J. 1980. Density compensation in island avifaunas. Oecologia 45:385–389.

Yalden, D.W. 1985. Dietary separation of owls in the Peak District. Bird Study 32:122–131.

Yamazaki, T. 1990. Distribution and breeding ecology of the Mountain Hawk-Eagle in the Suzuka Mountains of Japan. Talk given at 1990 Raptor Research Foundation annual meeting, Allentown, PA.

Yom-Tov, Y., M.I. Christie, and G.J. Iglesias. 1994. Clutch size in passerines of southern South America. Condor 96:170–177.

Zhang, S., and L. Wang. 2000. Following of brown capuchin monkeys by White Hawks in French Guiana. Condor 102:198–201.

Zug, G. 1983. *Bufo marinus*. *In* D.H. Janzen, ed., Costa Rican natural history, pp. 386–387. Chicago: Univ. of Chicago Press.

Zug, G. 1993. Herpetology: an introductory biology of amphibians and reptiles. New York: Academic Press.

Zug, G.R., L.J. Vitt, and J.P. Caldwell. 2001. Herpetology: an introductory biology of amphibians and reptiles. New York: Academic Press.

INDEX

Page numbers followed by *f* refer to figures, by *t* to tables, and by *n* to notes.